AF572727

Inelastic Deformation of Composite Materials

International Union of Theoretical and Applied Mechanics

George J. Dvorak (Ed.)

Inelastic Deformation of Composite Materials

IUTAM Symposium, Troy, New York
May 29–June 1, 1990

With 101 Illustrations

Springer-Verlag
New York Berlin Heidelberg London
Paris Tokyo Hong Kong Barcelona

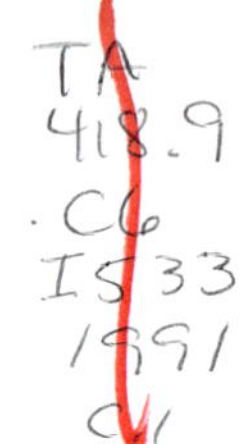

George J. Dvorak
Department of Civil Engineering
Center for Composite Materials and Structures
Rensselaer Polytechnic Institute
Troy, New York 12180-3590
USA

Printed on acid-free paper.

ISBN 0-387-52011-2/1991 $0.00 + 0.20

Camera-ready copy provided by the editor.
Printed and bound by Edwards Brothers, Inc., Ann Arbor, Michigan.
Printed in the United States of America.

9 8 7 6 5 4 3 2 1

ISBN 0-387-97458-X Springer-Verlag New York Berlin Heidelberg
ISBN 3-540-97458-X Springer-Verlag Berlin Heidelberg New York

Preface

In the last 25 years, the science and technology of composite materials have experienced a period of substantial development. The initial goal was to provide light, strong, and stiff materials for the aerospace industry. That was met by the introduction of polymer matrix composites with continuous fiber reinforcement, and with certain discontinuous reinforcements. Such materials are now routinley used not only in aerospace, but also in numerous other applications, e.g., in automobile and construction industries.

Meanwhile, composite materials have been introduced, or are expected to serve, in many other functions which cannot be fulfilled by conventional materials, particularly in extreme environments. Accordingly, the research focus has been broadened to include not only new polymer systems, but also metal, intermetallic, and ceramic matrix materials. This has brought forth a number of new problems in fabrication and processing, and in analysis of composite material behavior and properties.

The latter set of problems is usually approached by various micromechanical techniques. In recent years, their scope has been expanded from prediction of overall properties of elastic, perfectly bonded systems, to include problems associated with inelastic deformation of the phases, debonding at interfaces, and growth of distributed damage. Many familiar aspects of mechanical behavior, such as fracture, fatigue, compressive strength and buckling have been reexamined and adapted for application to the new material systems.

This volume contains a selection of recent work by leading researchers in micromechanics that was presented at the IUTAM Symposium on Inelastic Deformation of Composite Materials at Rensselaer. The Symposium was made possible by the generous support of AFOSR, ARO, NSF, IUTAM and RPI. Thanks are due to the sponsors, and to the local organizing committee for their support and work on behalf of the Symposium. Special thanks are due to Christine Stephenson for her coordination of the local arrangements, and for her contribution to the preparation of this volume.

George J. Dvorak
Troy, New York

June 1990

Sponsors of the IUTAM Symposium on Inelastic Deformation of Composite Materials

International Union of Theoretical and Applied Mechanics (IUTAM)

National Science Foundation
Solid Mechanics and Geomechanics Program
Dr. Oscar W. Dillon, Program Director

Department of the Air Force
Air Force Office of Scientific Research
Division of Aerospace Sciences
Lt. Colonel George K. Haritos, Acting Director

Department of the Army
Army Research Office
Materials Science Division
Dr. Edward S. Chen

The Civil Engineering Department and
The Institute Center for Composite Materials and Structures
Rensselaer Polytechnic Institute

International Scientific Committee of the Symposium

Professor George J. Dvorak (USA), Chairman
Professor Z. Hashin (USA-Israel)
Professor J. Backlund (Sweden)
Professor T. Hayashi (Japan)
Professor J.P. Boehler (France)
Professor T. Lehmann (FRG)
Professor R.M. Christensen (USA)
Professor A.J.M. Spencer (UK)
Professor D.C. Drucker (USA)
Professor V.V. Vasiliev (USSR)
Professor S.S. Wang (USA)

Local Arrangements

Christine Stephenson
Judy Brownell
Janet Pertierra
Jo Ann Grega
Michelle Peattie

List of Participants

Ahzi, S., Massachusetts Institute of Technology, USA
Allen, David H., Texas A&M University, USA
Alpa, Giovanni, University of Genova, Italy
Argon, Ali S., Massachusetts Institute of Technology, USA
Bassani, John L., University of Pennsylvania, USA
Bahei-El-Din, Yehia, Rensselaer Polytechnic Institute, USA
Bathias, Claude, C.N.A.M., Paris, France
Benveniste, Yakov, Tel-Aviv University, Israel
Berveiller, Marcel, Laboratoire P.M.M., Metz, France
Budiansky, Bernard, Harvard University, USA
Cardon, Albert H., Vrije Universiteit Brussel, Belgium
Chen, Peter, Department of the Army, USA
Chen, Tungyang, Rensselaer Polytechnic Institute, USA
Christensen, Richard M., Lawrence Livermore Labs, USA
de Buhan, Patrick, Ecole Polytechnique Palaiseau, France
Dvorak, George J., Rensselaer Polytechnic Institute, USA
Eggleston, Michael, General Electric Corporation, USA
Fares, Nabil F., Rensselaer Polytechnic Institute, USA
Fleck, Norman A., Cambridge University, England
Gambarotta, Luigi, University of Genova, Italy
Gosz, Michael R., Northwestern University, USA
Gupta, Vijay, Dartmouth College, USA
Hall, Richard, Rensselaer Polytechnic Institute, USA
Haritos, George K., Department of the Air Force, USA
Hashin, Zvi, University of Pennsylvania, USA; Tel-Aviv Univeristy, Israel
Herakovich, Carl T., University of Virginia, USA
Herrmann, Klaus, Gesamthochschule Paderborn, FRG
Huang, Chien-Ming, Rensselaer Polytechnic Institute, USA
Hui, C.Y. Herbert, Cornell University, USA
Ikegami, Kozo, Tokyo Institute of Technology, Japan
Krempl, Erhard, Rensselaer Polytechnic Institute, USA
Kuo, Wen-Shyong, University of Delaware, USA
Ladeveze, Pierre, E.N.S. de Cachan, France
Lagoudas, Dimitris C., Rensselaer Polytechnic Institute, USA
Laws, Norman, University of Pittsburgh, USA
Llorca, Javier, Brown University, USA
Mallon, Patrick J., University College Galway, Ireland
Moran, Brian, Northwestern University, USA
Mura, Toshio, Northwestern University, USA

Murakami, Sumio, Nagoya University, Japan
Nakajo, Yuichi, Ashikaga Institute of Technology, Japan
Needleman, Alan, Brown University, USA
Nemat-Nasser, Sia, University of California, San Diego, USA
Nigam, Himanshu, Rensselaer Polytechnic Institute, USA
Ortiz, Michael, Brown University, USA
Parks, David M., Massachusetts Institute of Technology, USA
Pindera, Marek-Jerzy, University of Virginia, USA
Pipes, R. Byron, University of Delaware, USA
Ponte Castaneda, Pedro, Johns Hopkins University, USA
Pyrz, Ryszard, University of Aalborg, Denmark
Rajapakse, Yapa D.S., Office of Naval Research, USA
Reddy, J.N., Virginia Polytechnic Institute, USA
Rogers, Tryfan G., University of Nottingham, England
Saleh, Ahmed, Rensselaer Polytechnic Institute, USA
Santare, Michael H., University of Delaware, USA
Schapery, Richard A., Texas A&M University, USA
Selvadurai, A.P.S. Patrick, Carlton University, USA
Shah, Rahul S., Rensselaer Polytechnic Institute, USA
Sham, T.L. Sam, Rensselaer Polytechnic Insitute, USA
Shephard, Mark S., Rensselaer Polytechnic Institute, USA
Spencer, Anthony J.M., University of Nottingham, England
Steif, Paul S., Carnegie-Mellon University, USA
Sternstein, Sanford S., Rensselaer Polytechnic Institute, USA
Storakers, Bertil, The Royal Institute of Technology, Sweden
Suresh, Sabra, Brown University, USA
Taggart, David, University of Rhode Island, USA
Talbot, David R.S., Coventry Polytechnic, England
ten Busschen, Albert, Delft University of Technology, Holland
Teply, Jan L., Aluminum Company of America, USA
Tvergaard, Viggo, Technical University of Denmark, Denmark
Wafa, Amr, Rensselaer Polytechnic Institute, USA
Walker, Kevin, Engineering Science Software, Inc., USA
Weitsman, Y. Jack, University of Tennessee, USA
Weng, George J., Rutgers University, USA
Willis, John R., Bath University, England
Wisnom, Michael, University of Bristol, England
Wu, Jer-Fang, Rensselaer Polytechnic Institute, USA
Xu, Simon W., University of Cincinnati, USA
Zarzour, Joseph, Rensselaer Polytechnic Institute, USA
Zbib, Hussein M., Washington State University, USA

Participants of the 1990 IUTAM Symposium

Contents

Interfaces I

Composite Materials with Interphase: Thermoelastic and Inelastic Effects

Zvi Hashin[1]
Department of Mechanical Engineering
and Applied Mechanics
University of Pennsylvania
Philadelphia, PA

ABSTRACT

The effect of thin interphase between constituents of a composite material is described in terms of imperfect interface conditions which involve interface parameters. Elastic, viscoelastic and elastoplastic interphases are considered and their effect on the mechanical properties of composites is evaluated on the basis of the composite cylinder/spheres assemblage models and the generalized self consistent scheme approximation.

INTRODUCTION

The effective properties of a composite material depend on two kinds of information: the properties of the constituents and the

---

[1] On leave of absence from Tel Aviv University, Tel Aviv, Israel

interface geometry. The interface imposes certain restrictions on the deformations and stresses in the constituents which are called interface conditions. The concept of interface and interface conditions is of course a simplified description of a complex microscopic state in the region where the constituents come in contact, but introduction of such details into analysis of a composite material would lead to prohibitively difficult problems. Therefore investigation of the effect of the nature of the interface on composite material properties should be divided into two stages. In the first stage the nature of the interface is translated into interface conditions, possibly on the basis of micromechanics of the interface region, and in the second stage the composite is analyzed subject to the derived interface conditions.

The classical interface conditions assume that displacements and tractions are continuous at the interface. This will be referred to as **perfect interface conditions** and every other kind of interface condition will be called imperfect. Unless we have reason to be concerned with surface tension interface traction continuity must be retained for reasons of equilibrium. It is the displacement continuity requirement which is abandoned when the interface is imperfect and to understand the underlying physics let us imagine that there is a thin region (compared to typical constituent dimensions such as fiber diameters) between the constituents, called the **interphase**, which has properties which are very different from those of the constituents. The interphase may be a thin layer of another material introduced by design or by chemical interaction of the constituents, or it may be a region containing many small defects such as pores or cracks. The case of interest is an interphase with stiffness much smaller than that of an adjoining constituent. This will be the case when the interphase material stiffness is very small or when micro-defects reduce stiffness considerably. The deformations of such an interphase, though thin, may be of the order of the deformations of stiffer constituents. The interphase deformation is the difference between displacements of adjoining constituents. If the interphase, by virtue of its small thickness, is idealized to become a surface - i.e. the interface, then this displacement difference becomes an interface displacement discontinuity.

If the interphase is elastic the simplest assumption is that normal and tangential displacement discontinuities are proportional to associated traction components. Thus with respect to a local orthogonal system of axes n,s,t originating at some point on the interface, Fig. 1, where n is normal direction and s,t are tangential directions we describe elastic imperfect interface conditions by the relations

$$T_n^{(1)} = T_n^{(2)} = D_n[u_n] \qquad [u_n] = u_n^{(1)} - u_n^{(2)}$$

$$T_s^{(1)} = T_s^{(2)} = D_s[u_s] \qquad [u_s] = u_s^{(1)} - u_s^{(2)} \qquad (1)$$

$$T_t^{(1)} = T_t^{(2)} = D_t[u_t] \qquad [u_t] = u_t^{(1)} - u_t^{(2)}$$

on S_{12}, where D_n, D_s and D_t are spring constant type parameters which will be called **interface parameters**. It is seen that infinite values of these parameters imply vanishing of the interface displacement jumps and therefore perfect interface conditions. At the other extremity, zero values imply vanishing of the interface tractions and therefore disbond of the adjoining media. Any finite positive values of the interface parameters define an imperfect interface.

A special case of (1) is $D_n \to \infty$ which implies normal bond and imperfect shear bond. If furthermore $D_s = D_t = 0$ there is no shear bond and the case is referred to as free sliding. The case of imperfect shear bond only was considered by Mal and Bose (1975) and Benveniste (1985) for the case of spherical particle elastic composites. Mura et al. (1985) have analyzed the axisymmetric problem of a spheroidal inclusion with interface

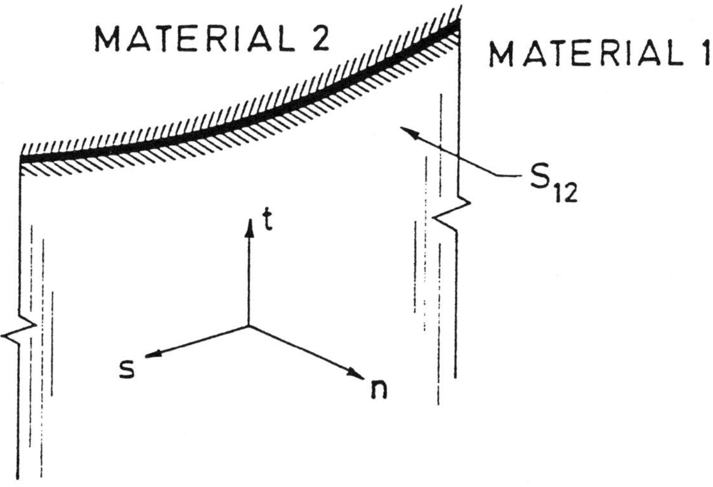

Fig. 1 - Interface

conditions of free sliding.

The more general interface conditions of type (1) with imperfect normal and shear bond were used by Achenbach and Zhu (1989, 1990) for numerical analysis of periodic fiber arrays and by Hashin (1990a,b,c) for analysis of thermoelastic properties of fiber and spherical particle composites and the elastic problem of the spherical inclusion.

INTERFACE CONDITIONS AND INTERPHASE

Elastic Interface

It should be noted that the interface conditions (1) have been introduced into the literature as a convenient assumption without examination of their general validity or origin. It has been shown in Hashin (1990a,b,c) that such interface conditions are indeed valid for fibers of circular cross section coated with transversely isotropic interphase and for spherical inclusions which are coated with isotropic interphase, provided that interphase thickness and stiffness are very much smaller than diameters and stiffness, respectively, of fibers and inclusions. It has been shown on the basis of detailed model analysis that for fibers the interface parameters are given by

$$D_n = (k_i + G_{Ti})/\delta$$
$$D_s = G_{Ti}/\delta \tag{2}$$
$$D_t = G_{Ai}/\delta$$

where n is radial outward direction to fiber surface and s,t are transverse and axial tangential directions, respectively, t is axis of transverse isotropy, δ is interphase thickness, k_i, G_{Ti}, G_{Ai} are interphase material transverse bulk modulus, transverse shear modulus and axial shear modulus, respectively, which are supposed to be much smaller than fiber elastic moduli

Note that D_t in Hashin (1990a) was given with a factor 2. This is a matter of interpretation of the displacement continuity due to axial shear. The present interpretation seems to be preferable.

For spherical inclusions with isotropic interphase it was shown in Hashin (1990b,c) that

$$D_n = (K_i + \frac{4}{3} G_i)/\delta$$
$$D_s = D_t = G_i/\delta \tag{3}$$

where K_i is isotropic bulk modulus. Note that for isotropic interphase (2) become the same as (3).

We can derive (2,3) in very simple

fashion without recourse to model analysis. Furthermore, the derivation is not restricted to fibers of circular cross sections nor to spherical inclusions but is valid for generally cylindrical fibers and inclusions of any (smooth) shape. Recall that the interface is a very thin layer which is loaded by assummedly smooth distribution of normal and tangential tractions, equal in magnitude on both sides. The variation of interphase stresses with r, distance normal to interphase bounding surfaces, can therefore be neglected. It follows that if the interphase is homogeneous the strains are also independent of r. Therefore the displacement jumps produced by the interphase are given by

$$
\begin{aligned}
[u_n] &= \epsilon_{nn}^{(i)}\,\delta \\
[u_s] &= 2\epsilon_{ns}^{(i)}\delta \\
[u_t] &= 2\epsilon_{nt}^{(i)}\delta
\end{aligned}
\qquad (4)
$$

where index i indicates interphase. Since ϵ_{nn} is the normal derivative of u_n the first of (4) is obvious. Establishment of the remaining two, for curved interphase, is based on neglect of the small number δ/a with respect to 1, where a is radius of curvature of fiber or inclusion.

At constituent/interphase interfaces each

of the tangential interphase strains $\epsilon_{ss}^{(i)}$, $\epsilon_{tt}^{(i)}$ must be equal to its constituent counterpart. Assuming that particles and fibers are very much stiffer than the interphase material the tangential interphase strains can be neglected relative to the normal interphase strain. Thus

$$\epsilon_{ss} = \epsilon_{ss}^{(i)} << \epsilon_{nn}^{(i)} \quad \epsilon_{tt} = \epsilon_{tt}^{(i)} << \epsilon_{nn}^{(i)} \qquad (5)$$

where strains with no index are in fibers or particles. Assuming isotropic interphase it follows from the elastic stress-strain law of the material that

$$\begin{aligned} \sigma_{nn}^{(i)} &= (K_i + 4G_i/3)\,\epsilon_{nn}^{(i)} \\ \sigma_{ns}^{(i)} &= 2G_i\,\epsilon_{ns}^{(i)} \\ \sigma_{nt}^{(i)} &= 2G_i\,\epsilon_{nt}^{(i)} \end{aligned} \qquad (6)$$

These stresses are the normal and tangential tractions. Expressing them in terms of (4) and using (1), the result (3) follows at once. Derivation of (2) is entirely analogous. There is in fact no difficulty to find the interfac constants for interphase which is orthotropic with respect to the n,s,t axes. For more complicated interphase anisotropy, when normal and shear strain/stress are no longer decoupled, the interface conditions will no longer have the form (1) but each traction component will be linearly related to all displacement jumps in the n,s,t coordinate system.

Work on the effect of interphase on the properties of composite materials has often been concerned with variability of interphase properties in the normal direction; see e.g. Sideridis (1988). The present approach can be readily adapted to this case. The interphase stresses remain constant through the thickness for the same reasons given above but since interphase stiffness now varies through the thickness the strains will behave similarly. The relations (4) are now replaced by

$$
\begin{aligned}
[u_n] &= \int_0^{\delta} \epsilon_{nn}(r)\, dr \\
[u_s] &= 2\int_0^{\delta} \epsilon_{ns}(r)\, dr \\
[u_t] &= 2\int_0^{\delta} \epsilon_{nt}(r)\, dr
\end{aligned}
\tag{7}
$$

Assuming transversely isotropic interphase coating of a fiber we then have from the elastic stress-strain relations

$$
\begin{aligned}
\frac{1}{D_n} &= \int_0^{\delta} \frac{dr}{k_i(r)+G_{Ti}(r)} \\
\frac{1}{D_s} &= \int_0^{\delta} \frac{dr}{G_{Ti}(r)}
\end{aligned}
\tag{8}
$$

$$\frac{1}{D_t} = \int_0^{\delta} \frac{dr}{G_{Ai}(r)}$$

with obvious specialization to isotropic interphase.

Finally, we consider the case of thin interphase with no stiffness restrictions in which case the tangential strains (5) are no longer negligible with respect to normal strain. To be specific we consider the case of isotropic interphase adjoining an isotropic constituent. Expressing interphase normal stress in terms of interphase strains and using (4) and the left part of (5) the normal interface condition now becomes

$$[u_n]/\delta = \left[\frac{1}{K_i + 4G_i/3} + \frac{2\nu_i}{E_2(1 - \nu_i)} \right] T_n -$$

$$\frac{(1 - \nu_2)\nu_i}{E_2(1 - \nu_i)} (\sigma_{ss}^{(2)} + \sigma_{tt}^{(2)}) \tag{9}$$

There is of course no difficulty to derive similar relations for anisotropic interphase and/or anisotropic constituents. It is seen that (9) is no longer of the form (1) while the shear interface conditions remain unchanged.

Examination of (9) reveals that the normal displacement jump is of the order of interphase thickness multiplied by the ratio of stress to interphase stiffness. For interphase stiffness which is not small (compared to constituent stiffness) this ratio will be a very small number which implies very small normal displacement jump, a situation which could be adequately represented by normal displacement continuity - thus perfect interface condition. Similar considerations apply to shear interface conditions. For isotropic interphase normal and shear stiffness will be of the same order, thus normal and shear interface conditicns will both be either perfect or imperfect. However, in the case of an anisotropic interface it is possible to have shear stiffness which is very much smaller than normal stiffness. This may lead to perfect normal interface condition and imperfect shear interface conditions.

If the interphase is such that perfect interface conditions are appropriate, (9) or its anisotropic counterparts are still useful for purpose of determination of the stresses in the interphase.

Viscoelastic Interphase

We consider the case of linear viscoelastic isotropic interphase and make the usual assumption that the bulk modulus K_i is time independent and thus the viscoelastic effect is confined to shear and is characterized by the relaxation modulus $G_i(t)$

and the associated creep compliance $g_i(t)$. Then the viscoelastic counterpart of (1,3) is given by

$$T_n(t) = D_n(t)[u_n(0)] + \int_0^t D_n(t - t') \frac{\partial}{\partial t'} [u_n(t')]dt' \quad .$$

$$T_s(t) = D_s(t)[u_s(0)] + \int_0^t D_s(t - t') \frac{\partial}{\partial t'} [u_s(t')]dt' \quad . \tag{10}$$

$$D_n(t) = [K_i H(t) + \frac{4}{3} G_i(t)]/\delta$$

$$D_s(t) = G_i(t)/\delta$$

where H(t) is the Heaviside function. For applications it is useful to introduce the Laplace Transform (LT) of (10) which is

$$\hat{T}_n = p\hat{D}_n[\hat{u}_n] \qquad \hat{T}_s = p\hat{D}_s[\hat{u}_s]$$

$$p\hat{D}_n = \left[K_i + \frac{4}{3} p \hat{G}_i\right]/\delta \qquad p\hat{D}_s = p\hat{G}_i/\delta \tag{11}$$

where $\hat{}$ indicates LT and p is the transform variable. It follows from Tauberian theorems for LT that the initial and final values of the interface functions $D_n(t)$ and $D_s(t)$ are

$$D_n(0/\infty) = \left[K_i + \frac{4}{3} G_i(0/\infty)\right]/\delta \tag{12}$$

$$D_s(0/\infty) = G_i(0/\infty)/\delta$$

Often $G_i(\infty)$ is so small that it may be assumed to vanish. In this event the final shear interface conditions reduce to free sliding.

The above interface functions are appropriate for quasi-state effects of relaxation and creep. In the case of steady state oscillations it is convenient to use complex moduli. Retaining the assumption of elastic K_i the complex interface parameters are

$$\tilde{D}_n(\iota\omega) = \left[K_i + \frac{4}{3}\tilde{G}_i(\iota\omega)\right]/\delta$$

$$\tilde{D}_s(\iota\omega) = \tilde{G}_i(\iota\omega)/\delta \tag{13}$$

where ω is frequency, $\iota = \sqrt{-1}$ and $\tilde{G}_i(\iota\omega)$ is the complex shear modulus of the interphase.

Elastoplastic Interphase

In this case the interphase may develop plastic yielding. The situation is more complex than the previous ones considered because of the nonlinear stress interaction and it is therefore not clear if it is possible to develop interface conditions of as general a nature as before. Here the development of the interface conditions for elastoplastic interphase will be confined to a specific geometrical model in the context of unidirectional fiber composites to be considered below.

THERMOELASTIC EFFECTS

Thermoelastic properties of fiber composites and of particulate composites have been analyzed in Hashin (1990a,b) on the basis of the composite cylinder assemblage (CCA) model for fiber composites, the composite spheres assemblage (CSA) model for particulate composites and the generalized self consistent scheme (GSCS) for both kinds of composites. It will be recalled that the CCA and CSA are described by filling out space with geometrically similar composite cylinders and composite spheres, respectively. The GSCS is based on the assumption that the state of strain/stress in any fiber or particle can be approximated by embedding a composite cylinder or composite sphere in an infinite medium which has the properties of the composite material. The GSCS, CCA and CSA give the same results in all cases when an exact solution of the two latter models can be obtained. However, the GSCS approximation can also be applied when exact solutions for the CCA and CSA are not available, thus for transverse shear of fiber composites and shear of particular composites.

The analysis can be much facilitated by introduction of the concept of equivalent fiber or particle. To explain this consider fiber/particle which is coated with a thin interphase layer, or more abstractly-exhibits the displacement jumps (1) when subjected to surface tractions. It has been shown that for the purpose of model analysis such a fiber can be replaced by a homogeneous fiber with the following equivalent thermoelastic properties

$$k_e = \frac{k_2}{1 + \frac{2\,k_2}{k_i + G_{Ti}}\frac{\delta}{a}}$$

$$E_{Ae} = E_{A2} \qquad \nu_{Ae} = \nu_{A2} \tag{14}$$

$$G_{Ae} = \frac{G_{A2}}{1 + \frac{G_{A2}}{G_{Ai}}\frac{\delta}{a}}$$

$$\alpha_{Ae} = \alpha_{A2} \qquad \alpha_{Te} = \alpha_{T2}$$

Here k is transverse bulk modulus, E, ν, G and α are Young's modulus, Poisson's ratio, shear modulus and thermal expansion coefficient, respectively, A denotes axial - in fiber-direction and T denotes transverse direction and 2 and e indicate fiber and equivalent property, respectively.

Similarly, for a spherical particle the equivalent bulk modulus is given by

$$K_e = \frac{K_2}{1 + \frac{9K_2}{3K_i + 4G_i} \frac{\delta}{a}} \tag{15}$$

where 2 now indicates particle. In case of defined interphase the interface constants appearing in (14, 15) are given by (2,3).

The implication of these results is that in all cases where CCA and CSA results are available for perfect interface conditions, conversion to imperfect interface conditions is simply effected by replacement of fiber or particle properties by (14) or (15). In other cases such as transverse shear for a unidirectional fiber composite and shear for a particulate composite the equivalent fiber/particle concept is not rigorously valid and the GSCS analysis has to be generalized to accommodate imperfect interface conditions.

It turns out that in the case of a fiber composite with transversely isotropic fibers the effect of imperfect interface enters through the nondimensional parameters

$$e = k_2/D_n a \qquad q_A = G_{A2}/D_t a$$
$$m = G_{T2}/D_n a \qquad q_T = G_{T2}/D_s a \tag{16}$$

with similar parameters in the case of a particulate composite. Analysis has shown that for a unidirectional fiber composite effective axial Young's modulus, axial

Poisson's ratio and axial thermal expansion coefficient are insignificantly affected by interface imperfection, transverse bulk modulus and transverse thermal expansion coefficient are significantly affected throough the parameter e only, axial shear modulus and transverse shear modulus are significantly affected - the former through q_A and the latter through m and q_A. These effects are illustrated in Figs. 2-4. The large effect of imperfect normal bond on transverse thermal expansion coefficient is of particular interest since it raises the possibility of obtaining e, experimentally, in simple fashion. Similarly, measurement of effective axial shear modulus can be used to determine q_A.

For transverse shear modulus the relation between m and q gives rise to various possibilities. In the case of isotropic interphase it follows from (2,3) that

$$m/q_T = \frac{1-2\nu_i}{2(1-\nu_i)} \tag{17}$$

where ν_i is the Poisson's ratio of the interphase. Figure 4 shows the case of perfect normal bond-imperfect shear bond, in which case m = 0, and the case of imperfect normal and shear bonds for $m = q_T/5$. If there is an isotropic interphase then the first case corresponds to $\nu_i = .5$, thus an incompressible

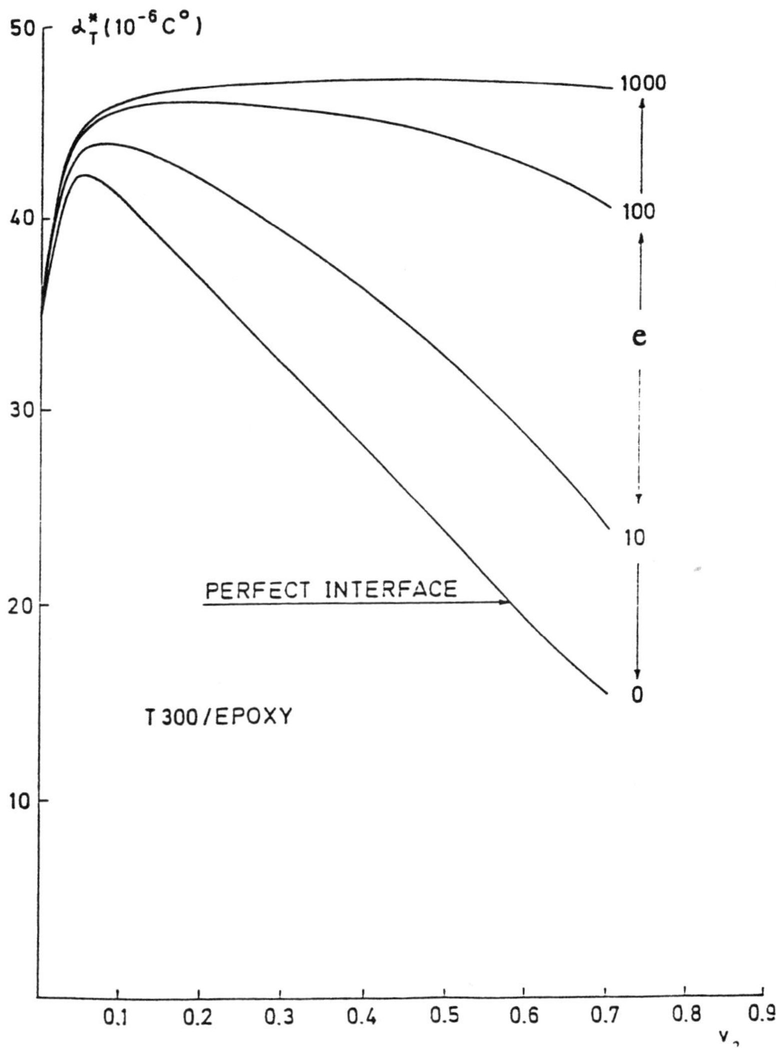

Fig.2 - Transverse Thermal Expansion Coefficient :

Elastic Interphase

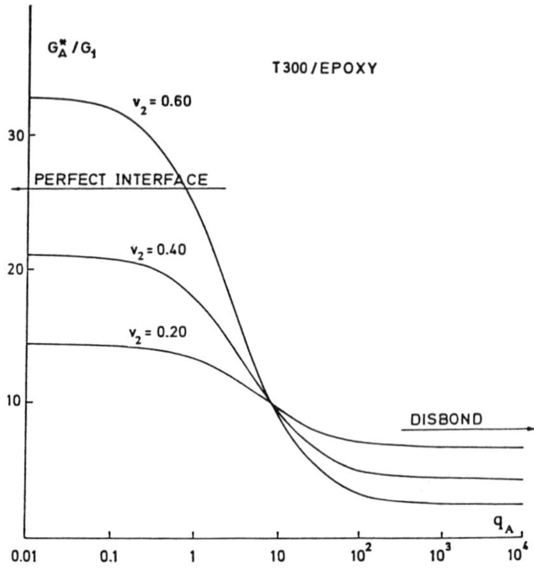

Fig.3 - Axial Shear Modulus : Elastic Interphase

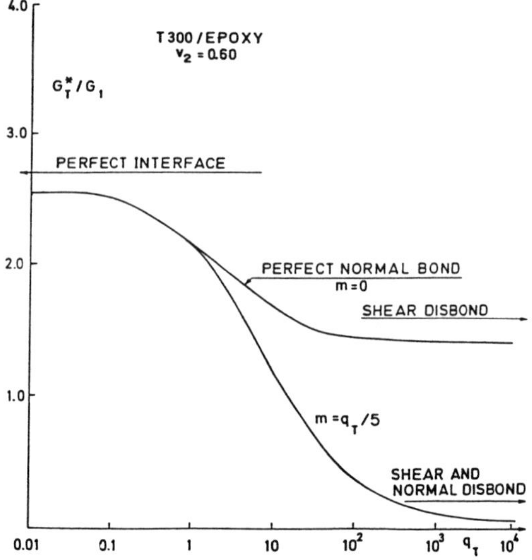

Fig.4 - Transverse Shear Modulus : Elastic Interphase

interphase, while the second case corresponds to ν_i = .375.

In the case of particulate composites the effective elastic moduli and the thermal expansion coefficient are significantly affected by interfac bond while the effect on the Poisson's ratio is moderate. See Hashin (1990b).

VISCOELASTIC EFFECTS

Time dependent interphases are of interest for modeling elevated temperature behavior. They are also a convenient device to produce relaxation, creep an damping effects into an elastic brittle material such as a ceramic composite. The interface conditions resulting from a linear viscoelastic interphase have been discussed above. We outline here a method of evaluation of the time dependence of composite properties due to such interphase and we discuss some of the results. For further details see Hashin (1990d).

Analysis is much facilitated if we recall the correspondence principle for viscoelastic omposites, Hashin (1965), according to which the LT of effective viscoelastic relaxation moduli (creep compliances) by replacement of phase elastic moduli (compliances) by p multiplied LT of phase relaxation moduli (creep compliances). The latter have been named for convenience TD (transform domain) moduli (compliances). In the present case the only source of viscoelasticity is the interphase and inspection of (11) reveals that these equations define TD interface parameters $p\hat{D}_n$ and $p\hat{D}_s$. It follows that to obtain LT of effective viscoelastic properties the elastic interface parameters appearing in expressions for effective elastic properties are replaced by these TD interface parameters.

Furthermore, it has been shown by application of Tauberian theorems, Hashin (1966), that the initial/final values of effective relaxation moduli (creep compliances) can be simply obtained by replacement of phase moduli (compliances) by initial final values of phase relaxation moduli (creep compliances). Thus in the present case we can use initial/final values of the interphase functions (12) to obtain initial/final values of effective viscoelastic properties, assuming that the interphase relaxation shear modulus becomes vanishingly small after infinite time.

Performing an elastic analysis with these interphase parameters it is found that the initial and final values of some properties are numerically very close implying that the viscoelastic effect cn be neglected. Properties which are of this nature are:

axial Young's relaxation modulus and creep compliance, transverse bulk relaxation modulus and creep compliance, axial Poisson's effect and the axial and transverse thermal expansion coefficients.

There is significant viscoelastic effect for axial and transverse shear. Representing interphase viscoelasticity by a differential time operator the effective axial shear relaxation modulus and creep compliance can be evaluated analytically on the basis of the correspondence principle and LT inversion. Figures 5-6 show such results for Nikalon fibers embedded in a Boron Silicate matrix. The shear viscoelasticity of the interphase is represented by the Maxwell model with shear modulus G_i and viscosity coefficient η_i. Time has been normalized with respect to the characteristic time of the Maxwell model which is given by

$$\tau_i = \eta_i/G_i \tag{18}$$

The different plots of relaxation moduli and creep compliances are characterized by the nondimensional interface parameter

$$q_A = \frac{\delta}{a}\frac{G_{A2}}{G_i} \tag{19}$$

It is seen that the smaller q_A the more time is required to develop the viscoelastic effect. It is also seen that whatever the value of q_A, ultimately the fibers behave as

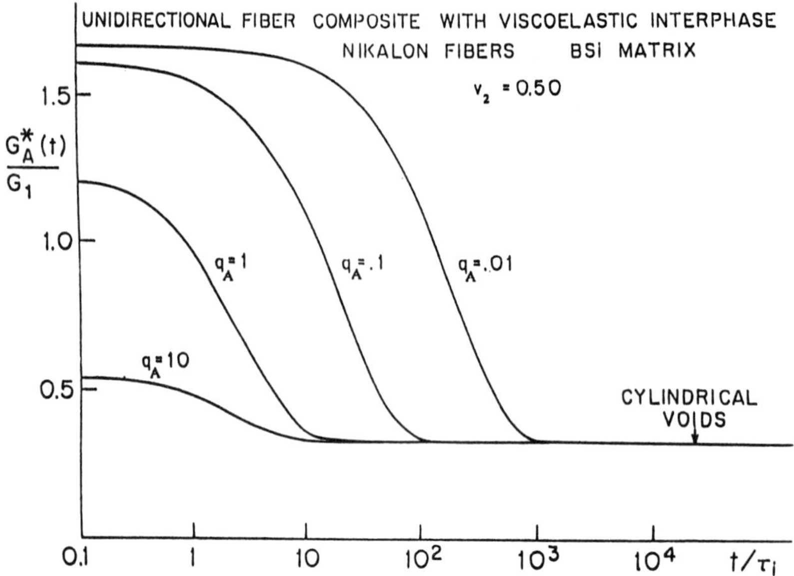

Fig.5 - Axial Shear Relaxation Moduli : Viscoelastic Interphase

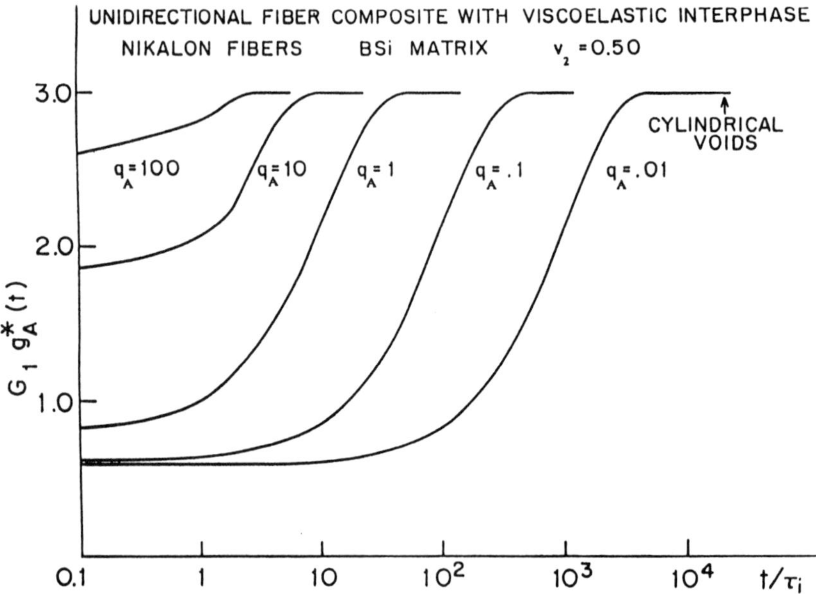

Fig.6 - Axial Shear Creep Compliances : Viscoelastic Interphase

cylindrical voids because of the free sliding interface condition which develops as a final state.

For transverse shear and transverse uniaxial stressing/straining analytical LT inversion is not possible but can in principle be performed numerically. The Tauberian theorems are readily applicable to this case. The initial/final values are obtained by using the initial/final values (11) with zero value of final shear modulus in the elastic analysis. Such results are shown in Fig. 7 for the effective transverse Young's compliance $E_T^*(t)$.

Similar results for particulate composites have been given in Hashin (1990b).

PLASTICITY EFFECTS

If the interphase material is subject to plastic yielding the composite will behave as an anisotropic elasto-plastic material. This is again of particular interest for ceramic composites as such an interphase introducs ductility into an otherwise brittle material.

As in the case of viscoelastic interphase it is to be expected that platicity effects for a fiber composite with elasto-plastic interphase will be signficant only for axial and transverse shear and other loadings which incorporate any of then, e.g. uniaxial stress. In particular all axisymmetric states such as axial stress long fibers, isotropic transverse stress and thermal expansion will be adequately described by elastic analysis.

We focus attention on the important case of axial shear in terms of the CCA or GSCS models, which in the elastic case give the same results for axial shear. The state of stress for axial shear is antiplane and therefore the only surviving stresses in cylindrical coordinates are σ_{rz} and $\sigma_{\theta z}$. The form of the stresses in CCA analysis is

$$\begin{aligned} \sigma_{rz} &= s_{rz} \cos \theta \\ \sigma_{\theta z} &= s_{\theta z} \sin \theta \end{aligned} \tag{20}$$

Since the interphase is very thin r dependence of interphase stresses can be neglected. Introducing (20) into the surviving equilibrium equation we find that the stresses assume the simple form

$$\begin{aligned} \sigma_{rz} &= s \cos \theta \\ \sigma_{\theta z} &= -s \sin \theta \end{aligned} \tag{21}$$

If initial yielding is governed by the Mises criterion then the whole interphase will yield when

$$s = \tau_y = \sigma_y/\sqrt{3} \tag{22}$$

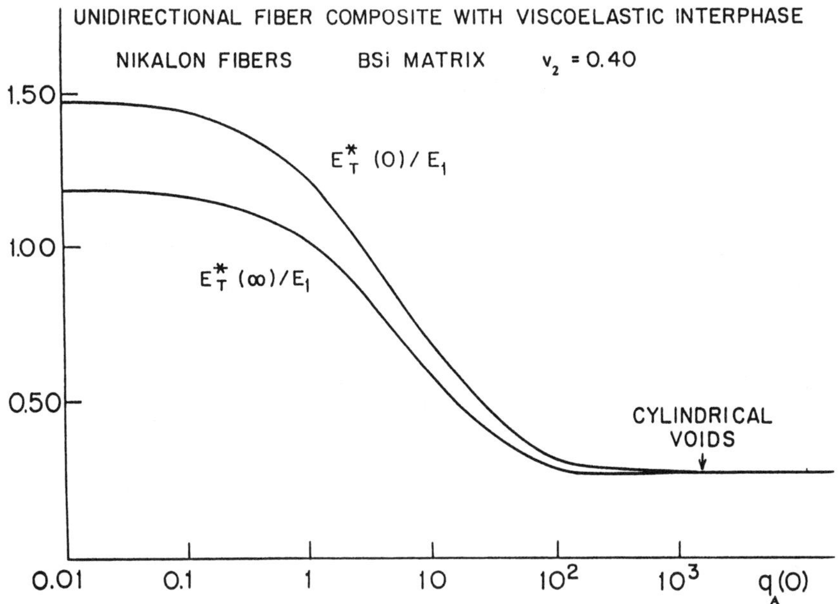

Fig.7 - Initial and Final Transverse Young's Relaxation Modulus

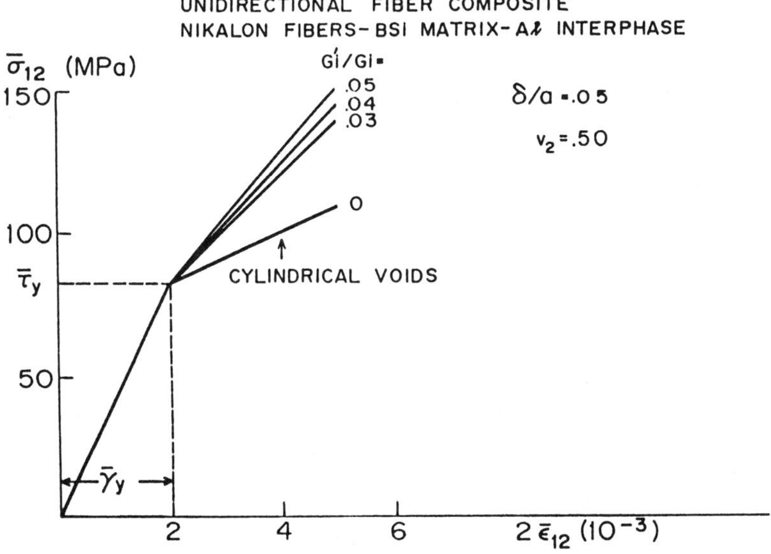

Fig.8 - Axial Shear Stress-Strain Relation : Elastoplastic Interphase

where σ_y is the uniaxial yield stress. Also note that (21) is equivalent to a principal pure shear of magnitude τ_y. Therefore continued yielding is governed by the one dimensional shear stress-strain relation of the interphase. This is taken in the simple bilinear form

$$
\begin{aligned}
s &= G_i \gamma & s &< \tau_y \\
s - \tau_y &= G_i'(\gamma - \gamma_y) & s &\geq \tau_y
\end{aligned}
\qquad (23)
$$

It follows that the effective axial shear stress-strain relation for the CCA or the GSCS models is also bilinear and can be simply obtained from the results for elastic interphase with appropriate interphase shear modulus in the two different linear ranges. The macroscopic stress-strain relation relating average axial shear stress strain $\bar{\sigma}_{12}$ and $\bar{\epsilon}_{12}$ is then also bilinear and is as follows:

$$
\bar{\sigma}_{12} = 2G^* \bar{\epsilon}_{12} \qquad 2\bar{\epsilon}_{12} < \bar{\gamma}_y
$$

$$
\bar{\sigma}_{12} = 2G^{*'} \bar{\epsilon}_{12} + (G^* - G^{*'})\, \bar{\gamma}_y \qquad 2\bar{\epsilon}_{12} \geq \bar{\gamma}_y
$$

$$
G^* = G_1 \frac{g(1+v_2) + (1+q)\, v_1}{g\, v_1 + (1+q)(1+v_2)}
$$

$$
G^{*'} = G_1 \frac{g(1+v_2) + (1-q')\, v_1}{g\, v_1 + (1+q')(1+v_2)}
$$

$$\bar{\gamma}_y = \frac{\tau_y}{G_2}\left[g\, v_1 + (1+q))(1+v_2)\right] v_2$$

$$q = \frac{G_{A2}}{G_i}\frac{\delta}{a} \qquad q' = \frac{G_{A2}}{G_i'}\frac{\delta}{a}$$

$$g = G_{A2}/G_1$$

and v_1, v_2 are the matrix and fiber volume fractions. Note that G^* is the elastic axial shear modulus for the case of interphase as given in Hashin (1990a).

A plot of (24) for a composite consisting of 50% Nikalon fibers and 50%BS matrix and elasto-plastic interphase is shown in Fig. 8 for various values of the ratio $G1/G_i$. The value .05 is appropriate for Aluminum interphase while the value 0 implies ideal plasticity.

The case of transverse shear can be analyzed in similar simple fashion but evaluation of stress-strain relations for loading programs requires incremental analysis.

CONCLUSION

It has been shown that the presence of thin interphase between the constituents of a composite material can be described in terms of imperfect interface conditions which are formulated in terms of interface displacement jumps and interface parameters. This has been applied to elastic, vicoelastic and elasto-plastic interphase. The appropriate interface parameters can in each case be evaluated in terms of interphase material properties and thickness. Thermoelastic properties of composite materials for these various kinds of interphases have been evaluated with special emphasis of unidirectional fiber composites. It has been pointed out in which cases the interphse has significant effect on properties.

ACKNOWLEDGMENT

This work has been supported by the Air Force Office of Scientific Research (AFOSR) Lt. Col. Haritos, and by the Office of Naval Research (ONR), Dr. Rembert Jones, through contract between the National Center for Composite Materials Research at the University of Illinois, Urbana, Dr. Su Su Wang, Director, and the University of Pennsylvania. I am indebted to the aforementioned for support and encouragement.

REFERENCES

Achenbach, J. D. and Zhu, H., 1989, "Effect of Interfacial Zone on Mechanical Behavior and Failure of Fiber-Reinforced Composites," Journal of the Mechanics and Physics of

Solids, Vol. 37, pp. 381-393.

Achenbach, J. D. and Zhu, H., 1990, "Effect of Interphases on Micro- and Macro-Mechanical Behavior of Hexagonal Array Fiber Composites", to be published.

Benveniste, Y., 1985, "The Effective Mechanical Behavior of Composite Materials with Imperfect Contact Between the Constituents", Mechanics of Materials, Vol. 4, pp. 197-208.

Gosz, M., Moran, B. and Achenbach, J. D., 1990, "Effect of Viscoelastic Interface on the Transverse Behavior of Fiber Reinforced Composites", to be published.

Hashin, Z., 1965, "Viscoelastic Behavior of Heterogeneous Media", ASME Journal of Applied Mechanics, Vol. 32, pp. 630-636.

Hashin, Z., 1966, "Viscoelastic Fiber Reinforced Materials", AIAA Journal, Vol. 4, pp. 1411-1417.

Hashin, Z., 1990a, "Thermoelastic Properties of Fiber Composites with Imperfect Interface", Mechanics of Materials, Vol. 8, pp. 333-348.

Hashin, Z., 1990b, "Thermoelastic Properties

of Particulate Composites with Imperfect Interface", Journal of the Mechanics and Physics of Solids, in press.

Hashin, Z., 1990c, "The Spherical Inclusion with Imperfect Interface Conditions", ASME Journal of Applied Mechanics, in press.

Mal, A. K. and Bose, S. K., 1975 "Dynamic Elastic Moduli of a Suspension of Imperfectly Bonded Spheres", Proceedings of the Cambridge Philosophical Society, Vol. 76, pp. 587-600.

Mura, T., Jasiuk, I. and Tsuchida, B., 1985, "The Stress Field of a Sliding Inclusion", International Journal of Solids and Structures, Vol. 21, pp. 1165-1179.

Sideridis, E., 1988 "The In-plane Shear Modulus of Fiber Reinforced Composite as Defined by the Concept of Interphase", Composites Science and Technology, Vol. 31, pp. 35-53.

Effect of a Viscoelastic Interfacial Zone on the Mechanical Behavior and Failure of Fiber-Reinforced Composites

B. Moran, M. Gosz and J. D. Achenbach
Center for Quality Engineering and Failure Prevention
Northwestern University, Evanston, IL 60208, USA

Abstract

The overall behavior and strength of fiber–reinforced composites is significantly affected by the interfacial bonding between the fiber and matrix material. The assumption of perfect bonding between constituents (i.e., continuity of interfacial tractions and displacements) may not be suitable in the presence of a thin interfacial zone (fiber coating) or cohesive type bonding (intermolecular bonding). In this paper a simple viscoelastic model is used to characterize the stiffness and viscosity of the interphase, and the finite element method is used to obtain the mechanical response of the composite. Calculations are carried out for a unit cell in a hexagonal array of fibers. The influence of loading rate on interfacial crack initiation and growth is investigated for both displacement and traction controlled failure processes. Stress distributions in the cracked interphase and in the matrix material contiguous to the interphase are also obtained.

1. Introduction

Unlike the axial strength and stiffness properties which are primarily governed by the axial properties of the fiber, the behavior of the fiber-reinforced composite in the transverse direction is dominated by a relatively low stiffness matrix material and the nature of the bond between the fiber and matrix phases. This may place severe limitations on the overall performance of the composite and thus it is desirable to accurately characterize the transverse properties. In most analytical and numerical work, investigators have assumed a perfect bond between the fibers and the matrix material which is modeled by continuity of interfacial tractions and displacements. In reality, however, the assumption of perfect bonding may not be suitable in the presence of a thin interfacial zone which connects the two phases (e.g. fiber coating or intermolecular bonding). In this analysis it is assumed that the bond between the fibers and the matrix is effected across an infinitesimally thin interfacial zone which supports a traction field with both normal and tangential components. Continuity of tractions is assumed across the interphase, however, displacements may be discontinuous from fiber to matrix due to the presence of the interphase in-between (see also Aboudi, 1987; Needleman, 1987; Steif and Hoysan, 1987; Achenbach and Zhu, 1989a,89b; and Hashin, 1989).

2. Problem Formulation

In this paper, the finite element method is used to investigate interfacial crack initiation and growth in a transversely loaded composite. We consider a unit cell in a uniform hexagonal array of fibers as illustrated in figure 1 and, noting the hexagonal symmetry, analyze the trapezoidal region *ABEF*. See Gosz et al. (1990) for details of loading and boundary conditions. It is assumed that the matrix is isotropic and linearly elastic. The fiber is taken to be linearly elastic and transversely isotropic. The elastic constants employed in this analysis were obtained by Kriz and Stinchcomb (1979) and are given in table 1. For the case of plane strain, the stress-strain relations for the matrix phase can be written as

$$\sigma_{\alpha\beta} = 2\mu\varepsilon_{\alpha\beta} + \lambda\varepsilon_{\gamma\gamma}\delta_{\alpha\beta} \tag{1}$$

where the parameters λ and μ are the Lame' constants and the Greek indices α, β, and γ range over 1 and 2. For the transversely isotropic fiber phase, the elastic stress-strain relations for the case of plane strain are given by Hashin (1979). The in-plane components are written as

$$\sigma_{\alpha\beta} = (K_T - G_T)\varepsilon_{\gamma\gamma}\delta_{\alpha\beta} + 2G_T\varepsilon_{\alpha\beta} \tag{2}$$

where K_T and G_T are the transverse bulk and shear moduli respectively.

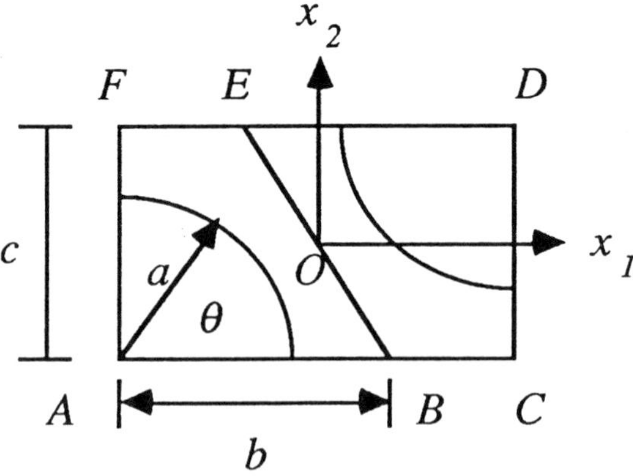

Fig. 1. Schematic of the unit cell of the hexagonal array composite.

Table 1. Elastic constants of the graphite/epoxy material system

	$E_A(GPa)$	ν_A	$E_T(GPa)$	ν_T	$G_A(GPa)$	$G_T(GPa)$	$K_T(GPa)$
Graphite fiber	232	0.279	15.0	0.490	24.0	5.03	15.0
Epoxy matrix	5.35	0.354	5. 35	0.354	1.976	1.976	6.76

2.1 Interphase Model.

Both a linearly elastic and a linearly viscoelastic constitutive relation are considered for the interfacial zone. For the linearly elastic interphase, it is assumed that the normal traction between the fiber and matrix phases is proportional to the jump in the normal displacement across the interphase. Similarly, the tangential traction is taken to be proportional to the jump in the tangential displacement across the interphase. Thus,

$$\boldsymbol{T}_n = k_n[\boldsymbol{u}_n]_I$$

$$\boldsymbol{T}_t = k_t[\boldsymbol{u}_t]_I \tag{3}$$

where

$$\boldsymbol{u}_n = u_i n_i \boldsymbol{n} \text{ and } \boldsymbol{T}_n = (\sigma_{ij} n_i n_j)\boldsymbol{n}$$

$$\boldsymbol{u}_t = \boldsymbol{u} - \boldsymbol{u}_n \text{ and } \boldsymbol{T}_t = \boldsymbol{T} - \boldsymbol{T}_n \tag{4}$$

are the normal and tangential displacement and traction vectors respectively, $[\,.\,]_I$ denotes the jump in the quantity across the interphase, k_n and k_t are normal and tangential stiffness parameters and $\boldsymbol{T}$ is the traction vector ($T_i = \sigma_{ij}n_j$). Here $\sigma_{ij} = \sigma_{ji}$ is the Cauchy stress tensor. A positive jump in normal displacement, $[\boldsymbol{u}_n]_I$, denotes normal separation between the fiber and matrix phases. However, we assume that a negative jump in normal displacement would correspond to a physically unrealistic interpenetration of the matrix phase into the fiber phase and, thus, we enforce the impenetrability constraint

$$[u_n]_I \geq 0. \tag{5}$$

For the linearly viscoelastic interphase, the time dependent response of the interfacial zone is taken into account. The material response in both the normal and tangential directions is considered to be that of a standard linear solid (SLS). The SLS qualitatively represents the behavior of an idealized cross-linked polymer and can be viewed as a spring in parallel with a spring and a dashpot. The normal and tangential tractions $\boldsymbol{T}_n$ and $\boldsymbol{T}_t$ are then related to the displacement jumps, $[\boldsymbol{u}_n]_I$ and $[\boldsymbol{u}_t]_I$, across the interphase by

$$\dot{T}_n + \frac{T_n}{\tau} = k_{gn}[\dot{u}_n]_I + \frac{k_{\infty n}}{\tau}[u_n]_I$$

$$\dot{T}_t + \frac{T_t}{\tau} = k_{gt}[\dot{u}_t]_I + \frac{k_{\infty t}}{\tau}[u_t]_I \tag{6}$$

where τ is the relaxation time, k_{gn} and k_{gt} are the instantaneous (glassy) stiffness components and $k_{\infty n}$ and $k_{\infty t}$ are the long term (rubbery) stiffness components.

2.2 Finite Element Implementation

We assume small displacements and the strain displacement relation is written as

$$\varepsilon_{ij} = (u_{i,j} + u_{j,i})/2 \tag{7}$$

while the equilibrium equation (assuming no body forces) is given by

$$\sigma_{ij,j} = 0. \tag{8}$$

The Principle of Virtual Work is written as

$$\int_{\Omega} \sigma_{ij}\delta\varepsilon_{ij} d\Omega + \int_{S} \delta\phi dS = \int_{\Gamma} T_i \delta u_i d\Gamma \tag{9}$$

where Ω denotes the interior of the trapezoidal region shown in figure 1, Γ is the external traction boundary, and S is the interfacial traction boundary. The δu_i are the kinematically admissible displacements (satisfying the periodic displacement boundary conditions and vanishing on the prescribed displacement boundary). The second term in the above equation is the virtual work of separation of the matrix and fiber phases, i.e.

$$\delta\phi = T_n \delta[u_n]_I + T_t \delta[u_t]_I . \tag{10}$$

Due to the time dependence of the linearly viscoelastic interphase, the virtual work expression, (9), must be discretized in both space and time. The approach we use is based on the method presented by Taylor et al. (1970) and is discussed more fully by Gosz et al. (1990). In order to satisfy the impenetrability constraint (5) full Newton Raphson equilibrium iteration is employed with a penalty-like stress update scheme in which the interphase is taken to have a suitably high normal stiffness parameter in compression.

3. Macroscopic Response

The macroscopic response of the composite is influenced by the impenetrability condition (5) which acts as a unilateral constraint on the deformation of the interphase. When the constraint is not imposed the response is transversely isotropic as expected for a configuration with linearly elastic constituents and hexagonal symmetry (see Love, 1927; Lekhnitskii, 1963). When the constraint is imposed the response is not transversely isotropic but depends upon the direction and character of the loading. For example, the constraint gives rise to significant differences in the tensile and compressive response. These differences are most pronounced when the interfacial stiffness parameters are low. In the present investigation we consider uniaxial tensile loading only, and in this case the deviations from transverse isotropy are slight.

The effective transverse properties of the fiber-reinforced composite may be calculated using the numerical procedure briefly outlined above. For the composite with a linearly elastic interphase, the numerical results for the effective transverse bulk and shear moduli are compared with the analytical results of Hashin (1989) where the composite cylinder assemblage (CCA) model is employed to obtain the effective transverse bulk modulus, and the generalized self consistent scheme (GSCS) model is used to determine the effective transverse shear modulus.

The numerical results for the effective transverse bulk and shear moduli and the CCA and GSCS results of Hashin (1989) are shown in figures 2 and 3. In figure 2, the effective transverse bulk modulus is plotted as a function of normalized interfacial stiffness. The normal and tangential stiffness components are taken to be equal ($k_n=k_t$). The normalization is chosen such that $k=k_n a/G_{Tm}$ where, a, is the fiber radius and G_{Tm} is the transverse shear modulus of the matrix. In figure 3, the effective transverse shear modulus normalized with respect to the shear modulus of the matrix is plotted versus normalized interfacial stiffness. Again, the normal and tangential stiffness components are taken to be equal, and the normalization is chosen as before. As shown in figure 2, when the impenetrability constraint (5) is not enforced, the numerical and CCA results virtually coincide over the entire range of interphase stiffness parameters. When this constraint is enforced, the numerical results only deviate from the CCA results at relatively low interfacial stiffness parameters. For the effective transverse shear modulus, when the impenetrability constraint (5) is enforced, the numerical results deviate significantly from the GSCS results at relatively low interfacial stiffness parameters, but the deviation at low stiffnesses is less significant when this constraint is not enforced. It should be noted that the magnitude of the deviation which results when the constraint (5) is enforced depends on the loading condition considered in the numerical procedure. In the

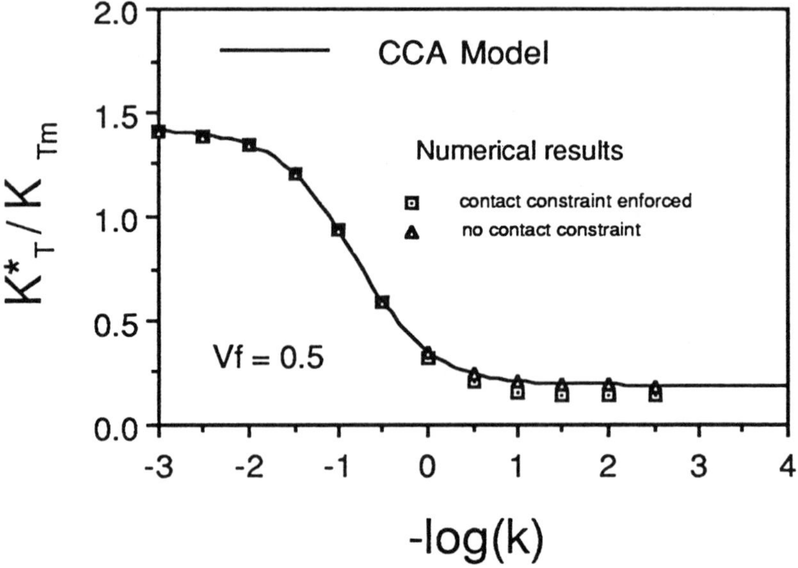

Fig. 2. Normalized effective transverse bulk modulus versus normalized interphase stiffness.

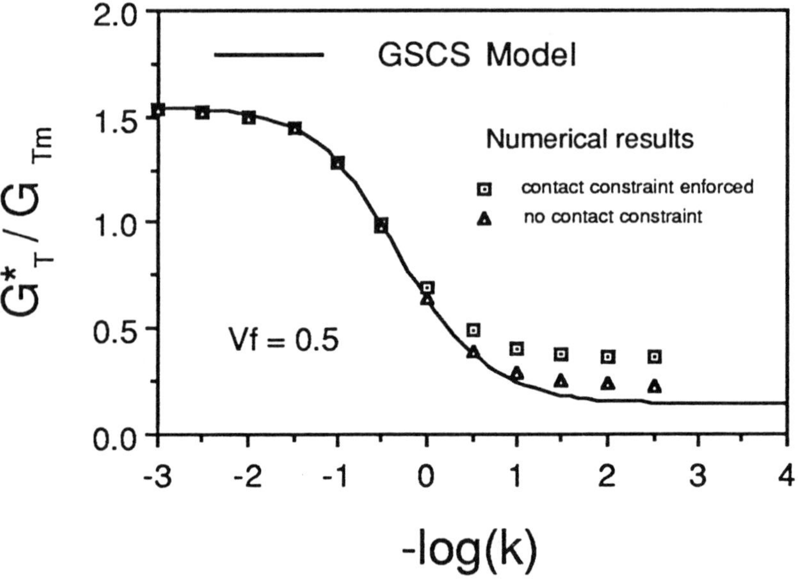

Fig. 3. Normalized effective transverse shear modulus versus normalized interphase stiffness.

present analysis K_T^* and G_T^* are determined by subjecting the hexagonal array model to tensile uniaxial loading. When the condition (5) is not imposed, the numerical values obtained for both moduli are independent of the loading condition considered.

Similar results for the time-dependent behavior of the composite may be obtained. In this case suitable application of the correspondence principle allows approximate analytical representations of the relaxation behavior to be derived and compared with the numerical results (see Gosz et al. , 1990 for details). For example, as shown in figure 4, the analytic results are compared with those obtained numerically for the effective relaxation modulus in bulk. In this case, the effective time dependent bulk modulus, $K_T^*(t)$, normalized with respect to the transverse bulk modulus of the matrix, is plotted versus time. The normal and tangential stiffness components of the SLS are assumed to be synchronous. The normalized glassy stiffness is chosen such that $k_g = k_{gn}a/G_{Tm}$ and the ratio of glassy to long term stiffness is taken to be ($k_g/k_\infty = 10$).

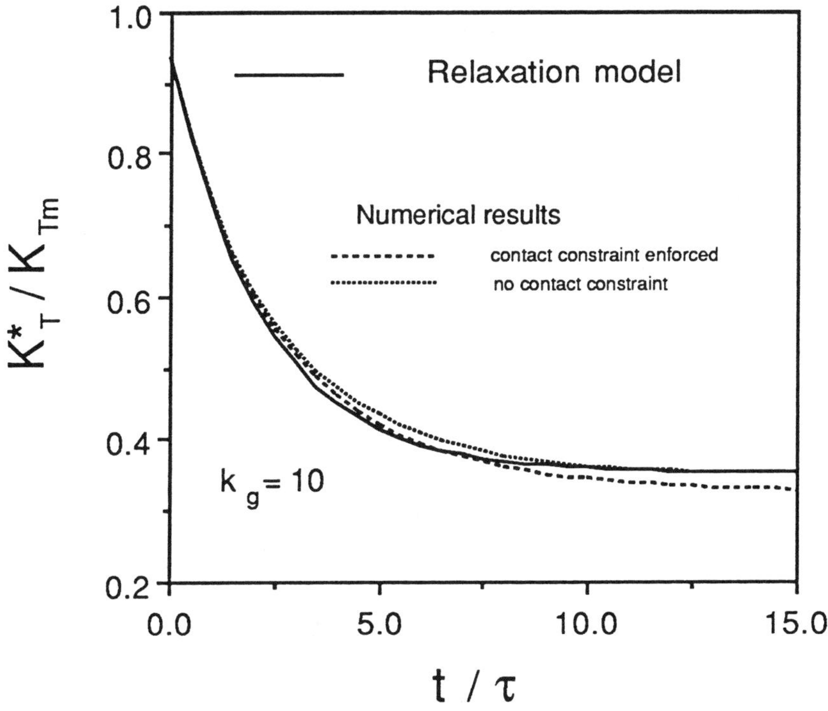

Fig. 4. Comparison of relaxation model in transverse bulk with the numerical results.

4. Failure Simulation

We now focus on the microscopic behavior of a transversely loaded fiber-reinforced composite. In particular, for the graphite/epoxy composite with a linearly viscoelastic interfacial zone, the effect of loading rate on interfacial crack initiation and growth is examined. As in the previous section, the numerical calculations are carried out for a unit cell in a periodic array of fibers (the trapezoidal region of figure 1). The cell is loaded in the mid–closest packing direction (see figure 5) and strained at a constant rate. Interfacial crack initiation, i.e. debonding of the fiber and matrix, is assumed to occur in either a displacement controlled or a traction controlled manner as discussed below. The consideration of a periodic unit cell yields qualitative information about micromechanical failure processes and the associated macroscopic response.

4.1 Failure Criteria

It is assumed that interphase failure can occur in either a displacement controlled or a traction controlled manner. In the traction controlled process it is assumed that failure occurs when the normal traction component, T_n, between the fiber and matrix phases reaches a critical value, $T_n{}^*$. The failure criterion is then written simply as

$$T_n - T_n{}^* = 0. \tag{11}$$

In the displacement controlled failure process, it is assumed that bond failure initiates when the normal jump in displacement across the interphase reaches a critical value, δ. Thus, interfacial bond failure occurs when

$$[u_n]_I - \delta = 0. \tag{12}$$

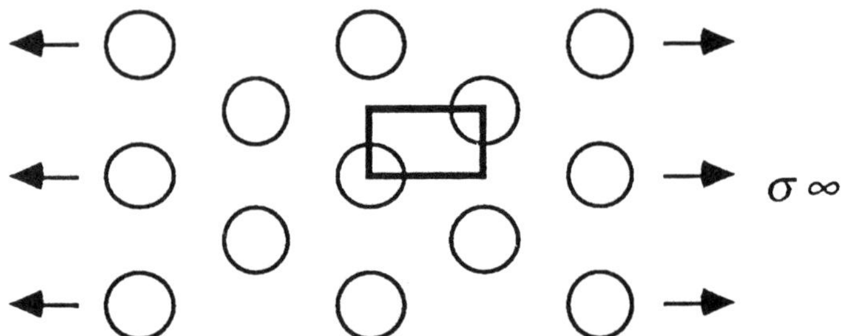

Fig. 5. Hexagonal array composite under Mid–CPD loading.

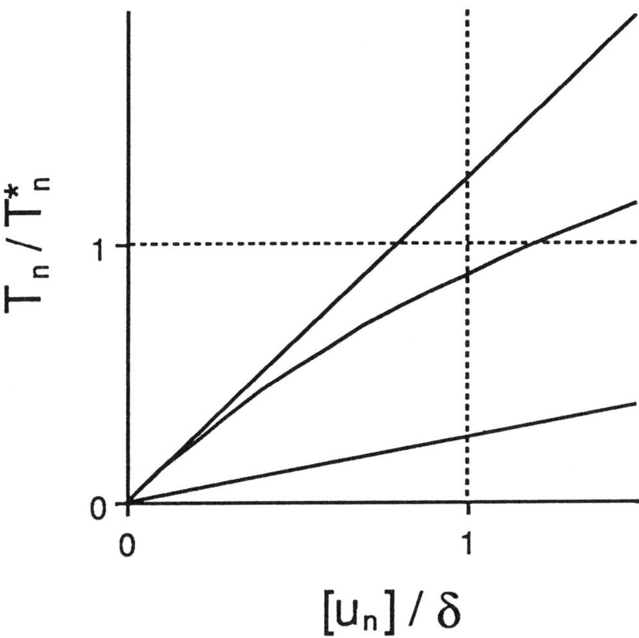

Fig. 6. Displacement and traction controlled failure of the standard linear solid. The lower solid line represents the constitutive response of the SLS as the time rate of change of the jump in normal displacement, $[\dot{u}_n]$, approaches zero. The upper solid line represents the response as $[\dot{u}_n] \rightarrow \infty$. The solid line in-between represents the response at an intermediate rate.

The isolated response of the SLS subject to both failure criteria is shown in figure 6. As shown in the figure, for the traction controlled failure process, failure occurs at points along the horizontal dashed line, and an increasingly brittle response of the SLS is observed as the loading rate is increased. For the displacement controlled process, failure occurs at points along the vertical dashed line and a tougher response is observed with increasing loading rate. When the two failure processes are considered to be competing, failure will occur in either a traction controlled mode or a displacement controlled mode depending on the loading rate and the critical values of T_n^* and δ.

4.2 Macroscopic Constitutive Response

The macroscopic constitutive response of the fiber–reinforced composite is illustrated in figure 7. The fiber volume fraction is taken to be $V_f = 0.5$. For the displacement controlled process, the critical normal separation, δ, is arbitrarily chosen such that $\delta/a = 0.001$. For a fiber diameter of $20\mu m$, for

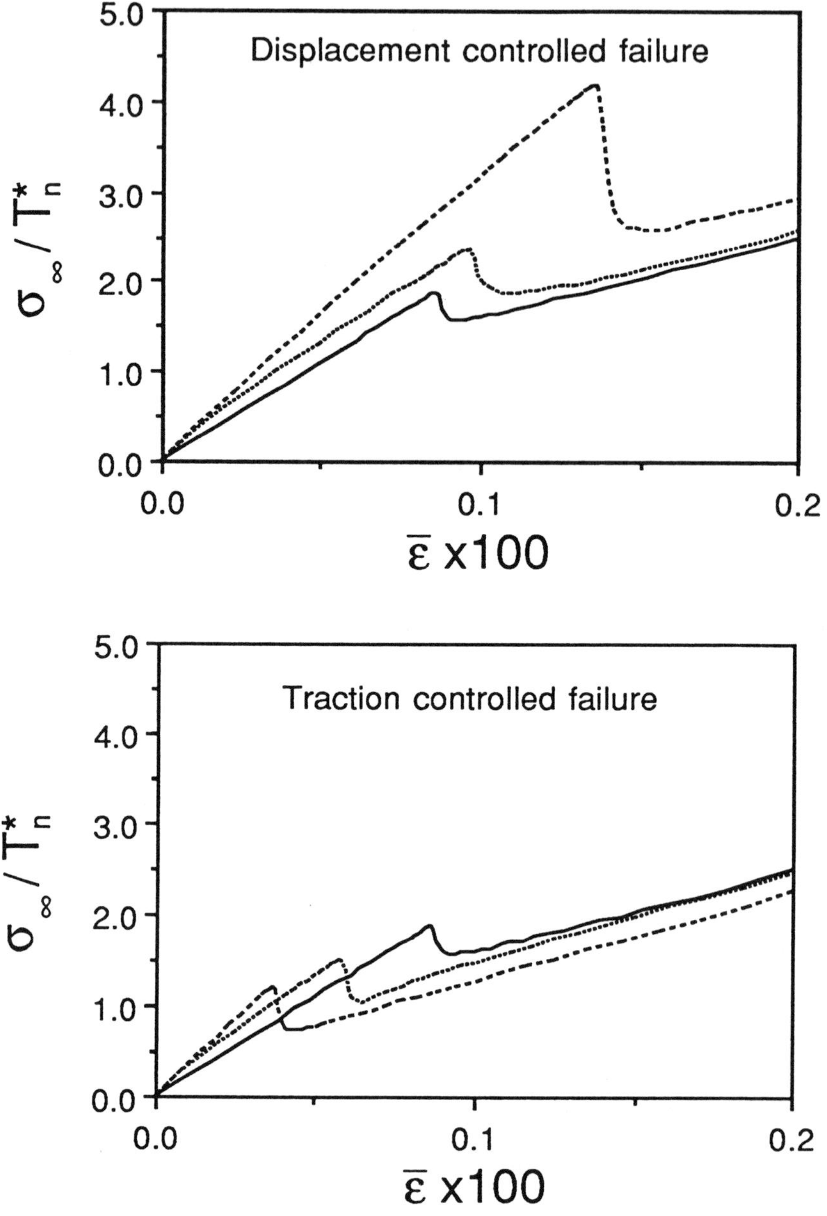

Fig. 7. Macroscopic constitutive response for displacement and traction controlled failure processes. In both figures, the solid line represents the response of the composite as the macroscopic strain rate $\dot{\varepsilon} \rightarrow 0$. The dotted line represents the response for $\dot{\varepsilon} = 1.0E-04$, while $\dot{\varepsilon} = 1.0E-03$ for the dashed line.

example, δ can be considered of the order *10nm*. For the traction controlled process, the critical normal traction is chosen to be $T_n^* = k_{\infty n}\delta$. The normal interfacial stiffness parameters, normalized with respect to the fiber radius and the transverse shear modulus of the matrix, are taken to be $k_{gn}a/G_{Tm} = 5$ and $k_{\infty n}a/G_{Tm} = 1$. It is assumed that the interphase exhibits a stiffer response in shear than for normal separation. We thus choose the tangential stiffness parameters such that $(k_{gn}/k_{gt} = k_{\infty n}/k_{\infty t} = 0.1)$. The time constant associated with each component is chosen to be unity for convenience.

As shown in the figure, the remote applied stress, σ_∞, normalized with respect to the critical normal traction, T_n^*, is plotted versus percent macroscopic strain for both failure criteria. In the displacement controlled process, the point of failure initiation occurs at a progressively higher remote applied stress and at a greater macroscopic strain as the macroscopic strain rate is increased. Qualitatively, under these circumstances, both the toughness and the ductility of the composite increase with loading rate. In the traction controlled process, the opposite trend is observed. The effect of increasing the strain rate tends to embrittle the composite.

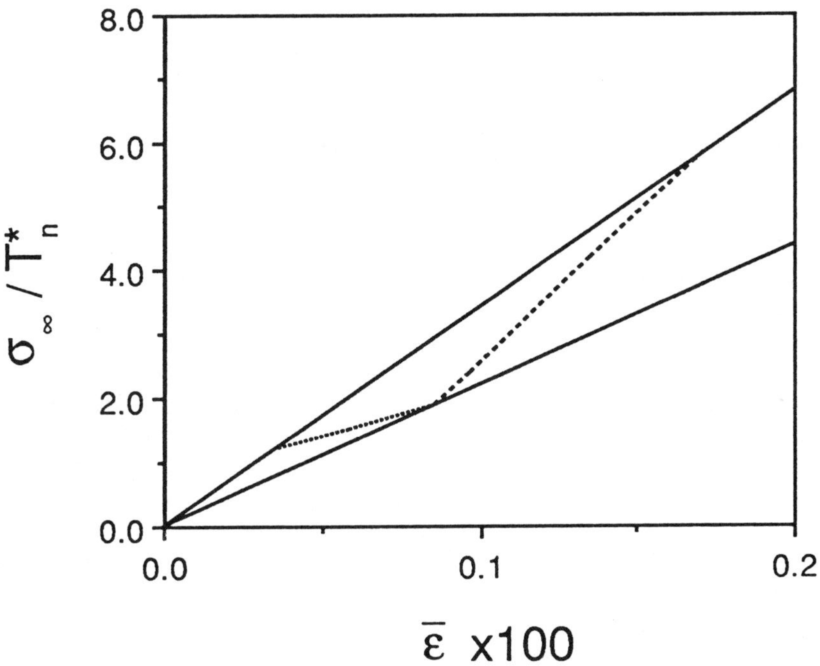

Fig. 8. Illustration of the linear relationship between points of failure initiation for displacement and traction controlled failure criteria.

It is interesting to note that a linear relationship exists between points of failure initiation at a given strain rate for both failure criteria as shown in figure 8. The lower solid line represents the response of the composite as the macroscopic strain rate $\dot{\varepsilon} \rightarrow 0$. The upper solid line represents the response as $\dot{\varepsilon} \rightarrow \infty$. When the composite is strained at a given rate in-between these two extremes, interfacial bond failure will initiate at a point on the dotted line for the traction controlled process and at a point on the dashed line for the displacement controlled process. Due to our choice of critical values δ and T_n^*, the dotted and dashed lines intersect at a point on the stress-strain curve corresponding to $\dot{\varepsilon} \rightarrow 0$. Thus, the composite would always fail in a traction controlled manner if the two processes were competing.

4.3 Stress Distributions

Significant changes in the stress distributions in the interphase and in the matrix material just outside of the interphase are observed during the failure process. The stress distributions shown in figure 9 are obtained during the traction controlled simulation at a macroscopic strain rate of $\dot{\varepsilon} = 0.001$. In the upper diagram, the normal traction in the interphase normalized with respect to the critical value, T_n^*, is plotted versus angle, θ. In the lower diagram, the circumferential stress, $\sigma_{\theta\theta}$, normalized with respect to the critical normal traction is plotted versus angle. The distributions represented by the solid lines are obtained at the point on the stress-strain curve (the dashed line of figure 7) just prior to failure initiation. The distributions represented by the dotted lines are obtained after the onset of failure at a macroscopic strain of approximately *0.04* percent – the point where the remote applied stress begins to increase monotonically. The distributions represented by the dashed lines are obtained at a macroscopic strain of *0.2* percent.

The distributions plotted in the upper diagram indicate that interphase failure initiates at an angle of zero degrees (see also figure 1), and an interphase crack develops and propagates relatively quickly to an angle of approximately *35* degrees. As the composite is macroscopically strained further, the interphase crack grows at a much slower rate to an angle of approximately *54* degrees when a macroscopic strain of *0.2* percent is reached. It is expected that when the macroscopic strain of the composite is increased beyond *0.2* percent, the crack will continue to propagate until it reaches the compressive region which develops in the interphase at approximately *82* degrees.

The corresponding circumferential stress distributions in the matrix material just outside the interphase are illustrated in the lower diagram. These distributions indicate that a large stress concentration develops at the crack tip as the crack propagates around the fiber. This large circumferential stress concentration may give rise to radial matrix cracking leading to severe degradation of the mechanical properties of the composite.

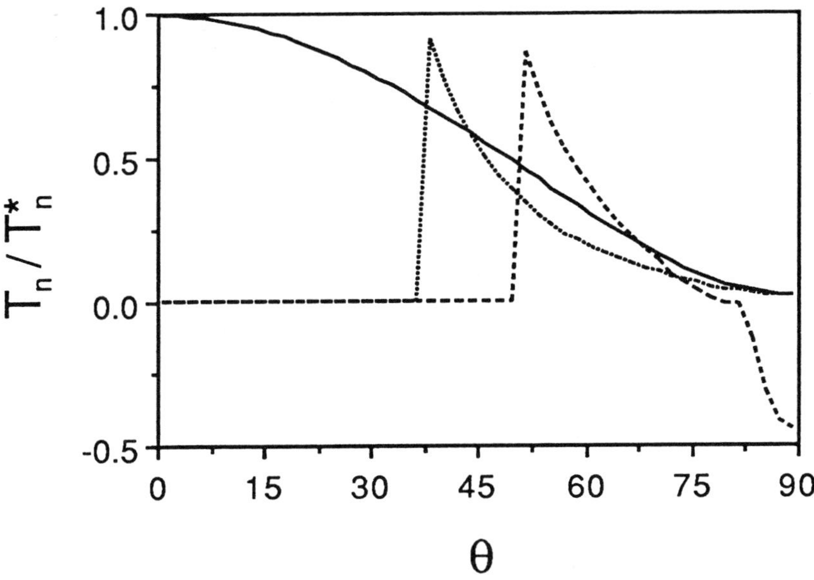

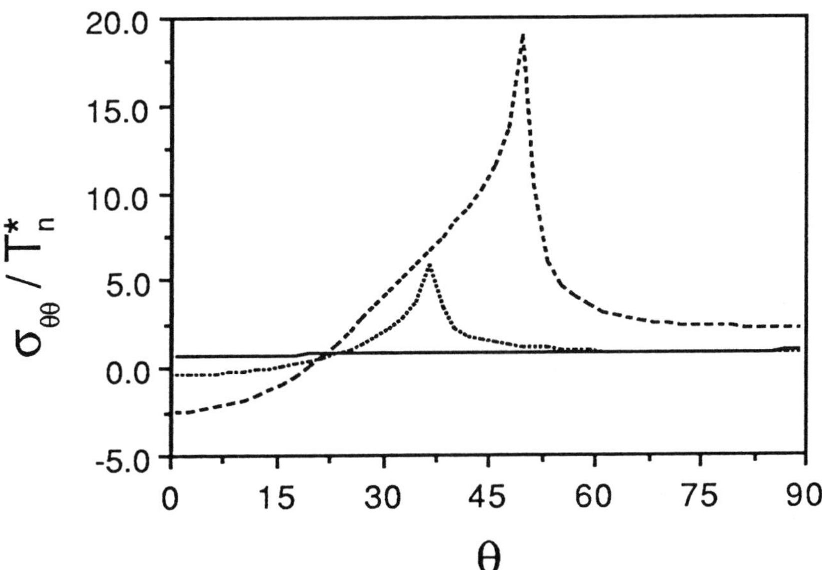

Fig. 9. Normal traction distributions in the interphase and circumferential stress distributions in the matrix material just outside of the interphase during the traction controlled failure process.

5. Concluding Remarks

In the present work, the transverse loading of the hexagonal array composite is examined from both macroscopic and microscopic points of view. Interphase failure initiation and growth are examined for displacement and traction controlled failure processes. The macroscopic constitutive response of the composite is obtained for both failure criteria. As the macroscopic strain rate of the composite is increased, the toughness and ductility increase in the displacement controlled process, while the composite becomes more brittle in the traction controlled process. Stress distributions in the interphase and in the matrix material contiguous to the interphase are obtained during the failure simulations. Large circumferential stress concentrations in the matrix form at the onset of interphase failure and redistribute and increase in magnitude as failure progresses. This phenomenon may give rise to radial matrix cracking leading to severe degradation of the macroscopic properties of the composite.

Acknowledgements – This work was carried out in the course of research sponsored by the U.S. Navy (Contract No. N00014-86-K-0799) under Subcontract No. 89-122 to Northwestern University.

References

Aboudi, J., 1987, "Damage in Composites-- Modeling of Imperfect Bonding," *Composites and Science Technology*, Vol. 28, pp. 102–128.

Achenbach, J. D., and Zhu, H., 1989a, "Effect of Interfacial Zone on Mechanical Behavior and Failure of Fiber-Reinforced Composites," *J. Mech. Phys. Solids*, Vol. 37, pp. 381–393.

Achenbach, J. D., and Zhu, H., 1989b, "Effect of Interphases on Micro- and Macro-Mechanical Behavior of Hexagonal Array Fiber Composites," To be published in *J. Appl. Mech.*

Gosz, M., Moran, B., and Achenbach, J. D., 1990, "Effect of a Viscoelastic Interface on the Transverse Behavior of Fiber-Reinforced Composites," Submitted for publication.

Hashin, Z., 1979, "Analysis of Properties of Fiber Composites With Anisotropic Constituents," *J. Appl. Mech.*, Vol. 46, pp. 543–550.

Hashin, Z., 1989, "Thermoelastic Properties of Fiber Composites with Imperfect Interface," To be published in *Mechanics of Materials*.

Kriz, R. D., and Stinchcomb, N. W., 1979, "Elastic Moduli of Transversely Isotropic Graphite Fibers and Their Composites," *Experimental Mechanics*, Vol. 19, pp. 41–49.

Lekhnitskii, S. G., 1963, *Theory of Elasticity of an Anisotropic Elastic Body*, Holden-Day, Inc., San Francisco.

Love, A. E. H., 1927, *A Treatise on the Mathematical Theory of Elasticity*, Cambridge.

Needleman, A., 1987, "A Continuum Model for Void Nucleation by Inclusion Debonding," *J. Appl. Mech.*, Vol. 54, pp. 525–531.

Steif, P., and Hoysan, S. F., 1987, "An Energy Method For Calculating the Stiffness of Aligned Short-Fiber Composites," *Mechanics of Materials*, Vol. 6, pp. 197–210.

Taylor, R. L., Pister, K. S., and Goudreau, G. L., 1970, "Thermomechanical Analysis of Viscoelastic Solids," *Int. J. Numer. Meth. Engng.*, Vol. 2, pp. 45–59.

Measurement of Strength of Thin Film Interfaces by Laser Spallation Experiment

V. Gupta* and A. S. Argon

(Massachusetts Institute of Technology, Cambridge, MA 02139.)

Abstract

Laser Spallation Experiment is developed to measure the strength of planar interfaces between a substrate and a thin coating(.3 to 3μm). In this technique a laser pulse of a high enough energy and a pre-determined length is converted into a pressure pulse of a critical amplitude and width that is sent through the substrate toward the free surface with the coating. The reflected tensile wave from the free surface of the coating pries off the coating. The critical stress amplitude that accomplishes the removal of the coating is determined from a computer simulation process. The simulation itself is verified by means of a piezoelectric crystal probe that is capable of mapping out the profile of the stress pulse generated by the laser pulse. Interface strength values ranging from 3.7 to 10.53 GPa are determined for the Si/SiC system, whereas for the carbon/SiC system, an average value of 6.91 GPa is obtained. Furthermore, sufficient experimental evidences are provided to show the potential of the laser technique to determine the interface toughness, provided well characterizable flaws can be planted on the interface.

I Introduction

It is now well recognized that considerable toughness in composites with brittle, but strong reinforcing fibers, can be achieved by controlled debonding of the fibers from the matrix to prevent premature fiber fracture. In composites with brittle matrices, such as ceramics and glass, this is often the only means of obtaining toughness (Evans 1989). In metal matrix composites, an additional problem of interface reaction between the matrix and the fiber is present, which

*Presently at the Thayer School of Engineering, Dartmouth College, Hanover, NH 3755.

sooner or later, leads to the failure of fiber. To neutralize this problem, it is now a standard practice to protect the fibers with a less reactive sacrificial coating (Gupta *et al.* 1989). This also permits controlled delamination of the coating from the fiber along the fiber-coating interface to prevent cracks external to the fiber or residual reaction products on the fiber from damaging the latter. Thus, it has been suggested (Gupta *et al.* 1989, Argon *et al.* 1989 *a*) that such fiber-coating interfaces with tailored strength and toughness properties can be used as mechanical fuses to decouple the undamaged fibers from their damaged surroundings to prevent catastrophic failure of the part. An important dimension of utilizing tailored properties of interfaces is the measurement of the tensile strength and intrinsic delamination toughness of the fiber-coating interface. How the intrinsic toughness can be determined from the spontaneous delamination from the substrates of stressed coatings, where the coating is under positive or negative residual stress, has been discussed earlier by us (Argon *et al.* 1989 *a* &*b*). Here, we discuss how the *tensile strength* of such fiber-coating interfaces can be measured by a laser spallation technique.

Although the development of the laser spallation experiment was motivated by our specific interest in engineering the interfaces in composites, the technique can be used to determine the strength of planar interfaces between coatings (> .3 μm in thickness) and substrates of any material. Hence, it should be of considerable interest to researchers in the device and thermal spray industries, where the mechanical problems stemming from the interface failure are of critical importance.

Measurement of interface properties is not a new problem in the thin film and coating technology, and the literature is replete with practical techniques (Jacobsson 1976; Jacobsson and Kruse 1973; Chapman 1974; Chiang *et al.* 1981; Davutoglu and Aksay 1981; Chow *et al* 1976) for the measurement of some average properties of thin films or coating interfaces, recently reviewed by Mittal (1978). But, none of the tests mentioned above measure the strength of interfaces. In order to measure the strength of interfaces between very thin coatings and substrates, a new laser spallation technique has been developed that had been initially introduced by Vossen (1978), utilizing shock waves produced by short laser pulses (Lang 1974; Fox 1974; Ready 1965; Anderholm 1970; Peercy *et al.* 1970). The technique involves impinging a high energy laser pulse (pulse width of nanosecond duration), from a Nd-YAG laser (1.06 μm wavelength) onto a thin absorption layer on the back surface of the substrate to which the coating of interest is attached. The sudden expansion of the absorbing film generates a compressive shock wave directed towards the test coating/substrate interface. It is the reflection of the compressive wave packet from the surface of the test coating that gives rise to a tensile pulse and leads to the removal of the coating, if the amplitude is high enough. The technique was used earlier by Park (1986) to detach films from the substrates. In his experiments, the absorbing film was sandwiched by a transparent fused quartz disc in order to increase the amplitude of the generated stress pulse. In most of the previous investigations of this phenomenon, ample evidence for the feasibility of the technique was provided, but insufficient details were given for the quantitative use of the approach as a research tool.

Hence, the technique in its previous form, prior to the present development, could be used at best only to provide a figure of merit of the interface. The thrust of the present research has been to develop the laser spallation experiment in a more definitive way capable of quantifying the interface strength. Furthermore, the goal here is to detach thin films (0.3-2 μm thick) from various substrates in contrast to the experiments done by Vossen (1978) and Park (1976), wherein test coatings of 20-30 μm thickness were considered. In our investigation to be presented here, we demonstrate that by careful use of the technique accurate determinations of tensile strength of interfaces are possible.

II Strategy of Interface Strength Measurement

2.1 Experimental Arrangement

Since the actual interfaces of interest are between a cylindrical fiber and its coating, which are not readily accessible to measurement by the spallation technique, the experimental approach to be presented here utilizes a planar arrangement of a model substrate and coating combination which is shown in Fig. 1. The collimated laser pulse is made to impinge on a thin energy-absorbing film sandwiched between the back surface of the substrate of interest and a fused quartz confining plate, transparent to the laser wavelength (1.06 μm in this experiment). The characteristics of the ideal energy absorbing film are: high absorptivity; a very small critical absorption depth (much smaller than the thickness of the film); a high melting temperature, high coefficient of thermal expansion, high elastic modulus, low thermal diffusivity, and finally a

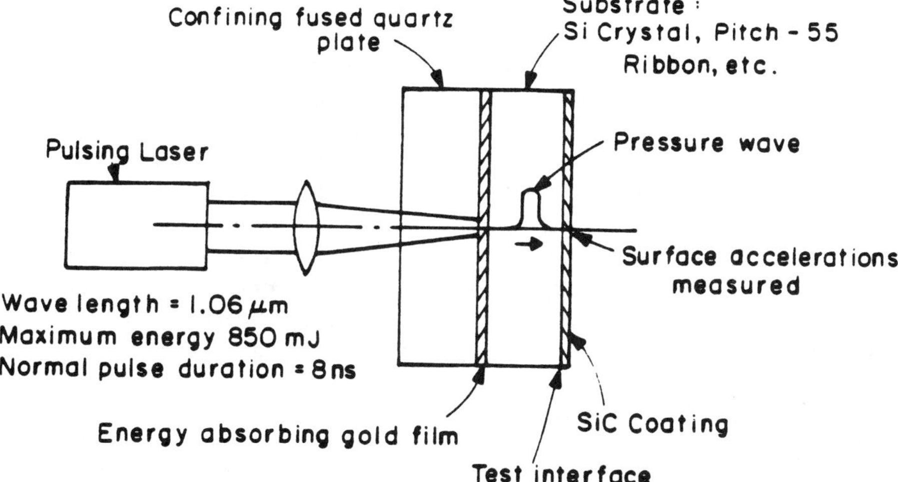

Figure 1. Schematic of the Laser Spallation Experiment.

thickness, Λ, roughly equal to the test coating of interest applied to the front surface of the substrate. Based on a figure of merit analysis that captures the essence of the thermoelastic stress generation for various alternative energy absorbing materials, it was concluded that gold is the most effective material that maximizes the generated stress pulse amplitudes.

A key element of the experiment is to determine the amplitude of the tension wave that is formed by reflection of the pressure wave from the free surface of the coating attached to the rear surface of the substrate. The preferred method of accomplishing this is to measure the time rate of change of displacement of the free surface of the coating as the main compression pulse is reflected. This is usually done by the laser Doppler interferometry used widely in the plate impact research (Clifton, 1978). In the experiments to be reported here, however, only pulsing lasers of limited power were available, requiring to focus the beam over relatively small areas of roughly 1 mm^2. This limits the planar portion of the pressure pulse to be reflected from the free surface of the test coating to a similarly small area, which makes the accurate measurement of the accelerations and decelerations of these small planar portions of the free surface rather difficult by the laser Doppler interferometry. In view of this, quantification of the measurements of the interface strength used here was based on a three part strategy.

The first part of the strategy was the development of a finite element computer simulation of the conversion of the laser light pulse into a pressure pulse. Next, the traverse of this pulse through any desired substrate of interest is monitored and finally the resulting amplitude and history of the tensile stress at the interface is determined as the stress wave is reflected from the free surface of the coating of given properties and thickness.

In the second part of our strategy, the pressure pulses were measured in a micro-electronic device in which the conditions of the computer simulation study were experimentally achieved. In this device, to be described below, the substrate and its test coating were replaced by a X-cut piezoelectric (PE) crystal equipped on its back face with an energy-absorbing gold film and on the front surface with a very thin gold electrode for signal pickup. The PE crystal with its attached fused quartz confining plate then became a pressure transducer operating in the short circuit mode, capable of pressure wave determination with a time resolution of 0.7 *ns*. The measured pressure signal profiles were then compared with those obtained with the computer simulation in which the substrate is appropriately given the elastic and thermal properties of the X-cut PE quartz crystal. This permits verifying and fine tuning of the computer simulation.

In the third part of the strategy, actual spallation experiments were carried out. The laser fluence necessary for the removal of the probed portion of the coating at the interface was recorded, and the tensile stress across the interface that accomplishes this was determined from the computer program.

In what follows, we discuss first in Section III the computer simulation of the laser-pulse-generated pressure pulses. In Section IV, we present the experimental details of the micro-electronic device for the measurement of pressure pulses with the PE crystal probe. We then present the results of the pressure wave profiles that have been obtained. Results for the actual spallation experiment and the specific interface strength values are given for the SiC/Si and SiC/Carbon interface systems in Section V. In Section VI, the basic results of the computer simulation are given as generalized interface stress charts for various coating/substrate systems. Finally, Section VII provides a discussion of the approach and its potential.

III Simulation of the Stress Pulses (Part I of Strategy)

3.1 Statement of the Problem

Figure 2(a) shows schematically the configuration of the laser spallation probe consisting of, from left to right: the fused quartz confining plate (Q), the thin

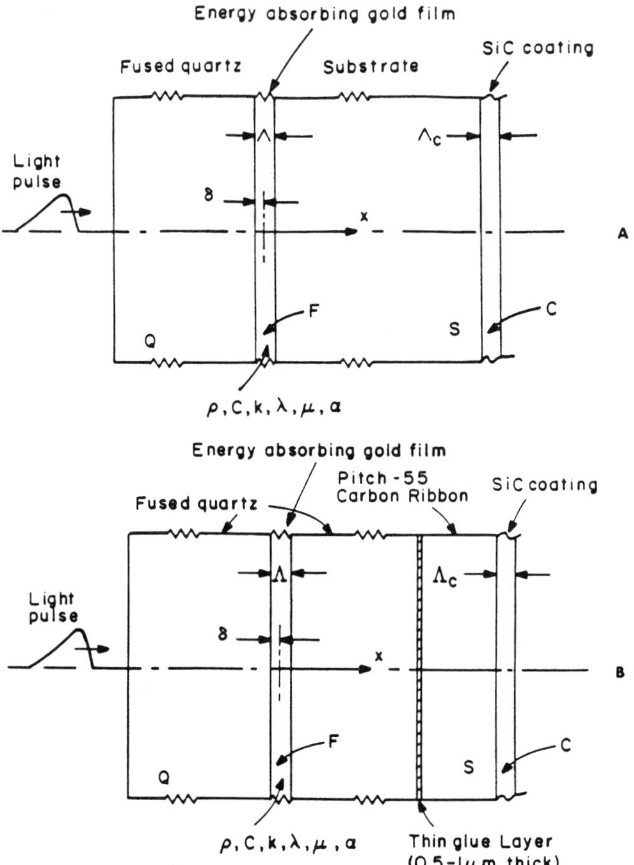

Figure 2. Configuration for testing (a) Si/SiC and (b) Pitch-55 Ribbon/SiC systems.

energy absorbing film (F) of thickness Λ, the substrate (S) and the coating (C) of interest. The laser pulse impinging on the energy absorbing film (gold) is generated by a Nd-YAG laser emitting at a wavelength of 1.06 μm. The laser used was in the MIT George Harrison Spectroscopy Laboratory. The peak power of the laser is 3 x 10^8 watts. The spatial profile of the laser pulse is Gaussian in nature. The temporal profile, however, is a distorted Gaussian (Ready 1971). The laser can be operated at 2.5 ns or 8 ns nominal pulse widths with a maximum achievable energy of 800 mJ in a nominal laser beam diameter of 7 mm.

The critical penetration depth δ of laser light at this wavelength into the gold is only 20 nm (Ready 1965). Since this is very much smaller than the thickness $\Lambda(\sim 1 \mu m)$ of the energy absorbing film, the simplifying assumption is made that the energy is deposited on the interface between the film and the fused quartz confining plate. The cross-sectional area of the laser beam incident on this interface is of the order of 1-2 mm^2.

Additional specifications in the simulation include: ignoring temperature dependence in all the relevant physical properties, such as thermal conductivity, coefficient of thermal expansion, specific heat, density, and elastic moduli. Such variation can, in principle, be taken into account in the finite element solution but were ignored in the initial simulation. Actual measurements and the simulation indicated that melting of the energy-absorbing film is undesirable since it broadens and diffuses the sharpness of the pressure wave, but is necessary to endure in order to obtain the required level of interface stress. Therefore, the simulation considers melting in the energy-absorbing film but only solid-like behavior in all the other media. The effect of volume changes associated with the phase change in the gold film is considered. This is quite significant for the gold film which undergoes a volume expansion of 5.1% (Smithells Metals Handbook 1983) on melting. The time constant for viscous flow relaxations in the narrow molten region of the energy absorbing film is much longer than the duration of the laser pulse. Therefore, even if melting of portions of the film occurs, it is adequate to account in the simulation for only the changes in the physical properties due to the melting transition, but to ignore all viscous flow relaxations. Furthermore, the thermal diffusion distance in most absorbing films during the laser heating time of 2.5 ns varies between 1 to 10 μm. This distance is negligible as compared to the radius of the laser heating spot of 1 mm. The absence of fluid-like flow of the molten material and small thermal diffusion distance suggests that the individual material points are blind to their neighbors in the radial direction and their deformation is strictly governed by the amplitude of the laser energy impinging directly on them. Hence, the generation of the stress pulse can be modelled as a one-dimensional phenomenon.

The associated phenomenon of the pulse propagation in the substrate, and the eventual spalling of the test coating are also modelled using the one-dimensional stress wave equations. The validity of this assumption will be verified later by the actual measured pressure pulse shapes to be discussed in

Section IV below.

Ignoring the thermoelastic effect as being quite small, the simulation considers two related, but uncoupled phenomena of transient heat transfer, giving temperature distributions in and around the energy-absorbing gold film as a function of time, and the related transient elastic wave propagation in all four media of Q, F, S and C, that results from the time rate of change of the thermal misfit, induced by transient heating. Since the Pitch-55 carbon ribbon and the pyrolytic graphite substrate used to model the Pitch-55 fiber are orthotropic in their elastic properties, the anisotropic effects are also considered in the present formulation. The details of the governing equations and the solution technique can be found in Gupta (1990). Here, we discuss briefly the key features of the solution.

3.2 Solution of the Governing Equations

Figures 3 a & b show the traverse of pressure fronts in both the Q and S spaces for two different normalized times of 70 and 109 for the constants H_q, K_q and H_s and K_s(to be defined below) given in the figure. In Figure 3, the time t; distance x, temperature T, displacement u, and the stress σ are normalized by introducing the variables:

$$\zeta = x/\Lambda;\quad \tau = t/t_o;\quad \theta = T/T_o;\quad \varphi = u/u_o;\quad \Sigma = \sigma/\sigma_o \tag{1}$$

where Λ, t_o, T_o, u_o and σ_o are the fundamental units of length, time, temperature, displacement and stress respectively, and are defined as

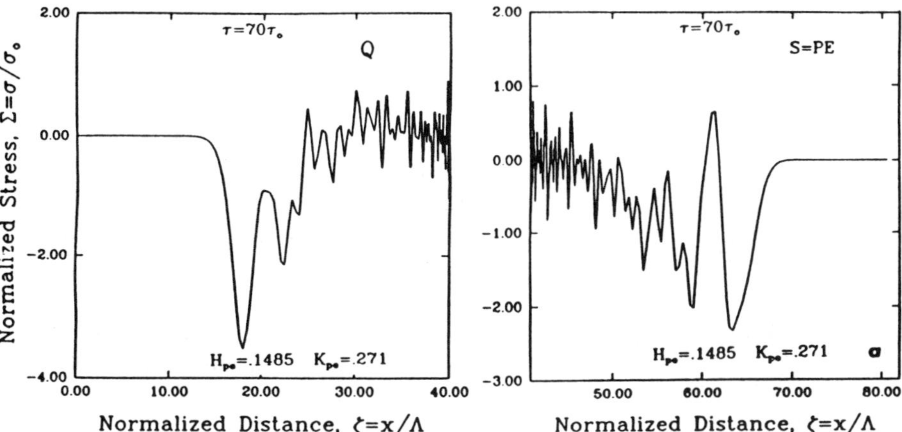

Figure 3. Generation and propagation of stress pulses in Q and S spaces.

$$t_o = (\Lambda/c)_f;\ T_o = \frac{q\Delta t_l}{(C\rho\Lambda)_f};\ u_o = \left(\frac{\Lambda\gamma}{\rho\delta c^2}\right)_f q\Delta t_l$$

$$\sigma_o = \left(\frac{(2\mu + \lambda)\gamma}{\rho\delta c^2}\right)_f q\Delta t_l \quad \text{and} \quad \gamma = (3\lambda + 2\mu)(\alpha/\rho C) \tag{2}$$

where in turn, Λ, c, C, ρ, δ, γ are the thickness, the one-dimensional pressure wave velocity, the specific heat, the density, the characteristic penetration depth of light energy and the Grueneisen constant respectively, all for the energy-absorbing film, which is taken as the medium for the basis of normalization. Δt_l is duration of the laser pulse and q is the absorbed laser fluence.The parameters H and K enter the formulation when the different physical properties of the substrate and confining quartz media are represented in terms of the properties of the energy absorbing film as

$$H_i = \left(\frac{c_f}{c_i}\right)^2 \ ; \ K_i = \frac{(\gamma C/c)_i}{(\gamma C/c)_f} \tag{3}$$

In equations 2&3 above, the subscripts i and f refer to the substrate and the energy-absorbing film respectively.

A more expanded figure of the pressure wave profile in the piezoelectric substrate is shown in Fig. 4, which reveals, in addition to the expected main pressure wave *abc*, an initially unexpected tension wave *cde* immediately following the compression wave. This tension wave has been found in all simulations with metallic energy absorbing films and should have useful properties in providing additional tension across the test interface between S and C. It appears to result from a space heating effect. As the thermal front penetrates into the energy absorbing film away from the Q-F interface, the interior of the film undergoes a "flash" expansion during the time increment Δt_l of the short laser pulse, while the material in the forward direction still experiences no effect of the pulse. The pressure wave released from such interior slabs in the film and travelling back towards the Q-F interface will be partially reflected from the interface back into the film and in the forward direction as a tension wave, because of the lower modulus and much lower density of the fused quartz in comparison to those properties of the gold film. This was also confirmed by a test simulation where the elastic properties of the two media were matched at the Q-F interface, resulting in the absence of the tension peak. Furthermore, such a "rebounding" condition should be favored in a system in which the "flash" expansion of the energy-absorbing packet cannot be significantly counteracted by the elastic wave tending to disperse the misfit during the time, the laser pulse is on; i.e., systems in which

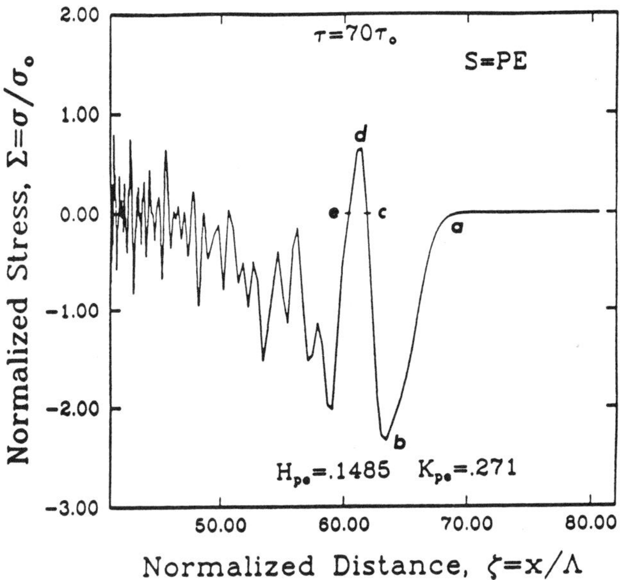

Figure 4. Detailed structure of the stress pulse in the piezoelectric substrate.

$$\left(\frac{k}{\Delta t_l C\rho c^2}\right)^{1/2} = \frac{k}{\Lambda_c C\rho c}, \tag{4}$$

the ratio of the average thermal velocity to the sound velocity is high. In eqn. (4) above, Λ_c is the thickness of the coating to be blown off and k is the thermal conductivity of the energy absorbing film. In systems with metallic energy-absorbing films for which the simulations have been performed, where the tension peak was found, the ratio

$$k/\Lambda_c C\rho c$$

in Eqn. (4) was found to be around 0.05-0.1, while for a carbon film which does not show the peak, the ratio is as low as $3\text{x}10^{-3}$.

In the simulation, the propagation of the stress pulse was monitored in the substrate until it completed the tensile loading of the coating/substrate interface. The stress history of the Si/SiC interface due to the pulse of Fig. 4 and for a SiC coating of 1.5 μm thickness is shown in Fig. 5. Zero values prior to the main signal correspond to the time lag in the arrival of the stress pulse at the interface. Thus, for a given threshold laser fluence, the interface strength can be calculated from the amplitude of such plots.

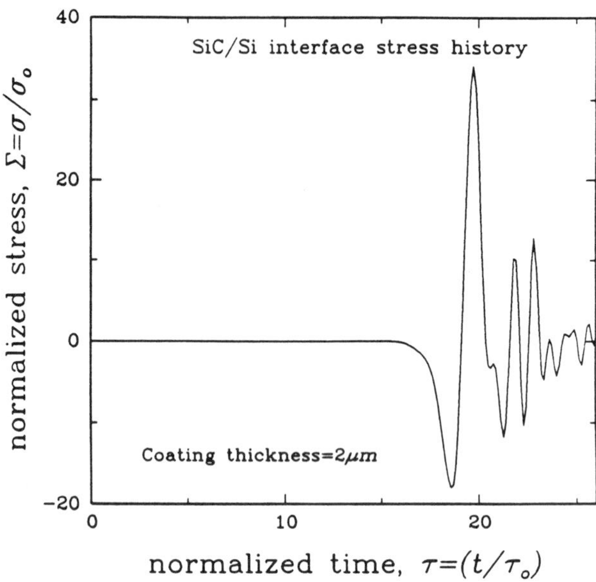

Figure 5. Stress history at the interface.

IV Experimental Stress Pulse Measurements (Part II of the Strategy)

4.1 The Micro-electronic Device

A schematic view of the micro-electronic device with the X-cut piezoelectric crystal is shown in Fig. 6. The assembly of fused quartz plate (*Q*), energy-absorbing(gold) film (*F*), and the PE crystal, taking the place of a substrate, is shown encased in a Bakelite housing. 1 μm thick gold films were chosen (Ditchburn 1963) to completely absorb the laser fluence in order to avoid the well-known radial cracking phenomenon (Volkova 1967) in the substrate disc due to its heating by the transmitted laser fluence. The laser-absorbing gold film also acts as a ground electrode for the piezoelectric crystal. A rubber "O" ring between the housing and the PE crystal radially compresses the latter when the copper electrode housing (*C*) is attached to the bakelite housing. The radial pre-compression of the substrate disc counteracts cracking of the substrate disc subjected to the sudden impact of the generated stress pulse. Figure 7 shows a perspective view of a typical sandwich element in the PE device. The PE device has a gold energy-absorbing film sandwiched between the PE crystal and the *Q* plate, acting also as the ground electrode of the PE device. As shown, this extends all the way to the front surface of the PE crystal, where it terminates as an outer ring, to which the copper electrode housing makes contact. Six equally spaced round gold electrode tabs of 2 *mm* diameter communicating with a central round tab through thin conducting spokes (25 μm wide) constitute the secondary electrode. These electrodes were deposited by photolithographic techniques. This geometry of the secondary electrode was designed to pick up the one-dimensional structure of the current signal. This was ensured by focussing the laser beam to a 2-3 *mm* spot on the laser-absorbing film in line with one of the 2 *mm* diameter secondary electrodes as shown in Fig. 7. This assembly permits as many as six separate measurements of the shapes of arriving pressure pulses by indexing a new tab into the line of sight of the collimated beam.

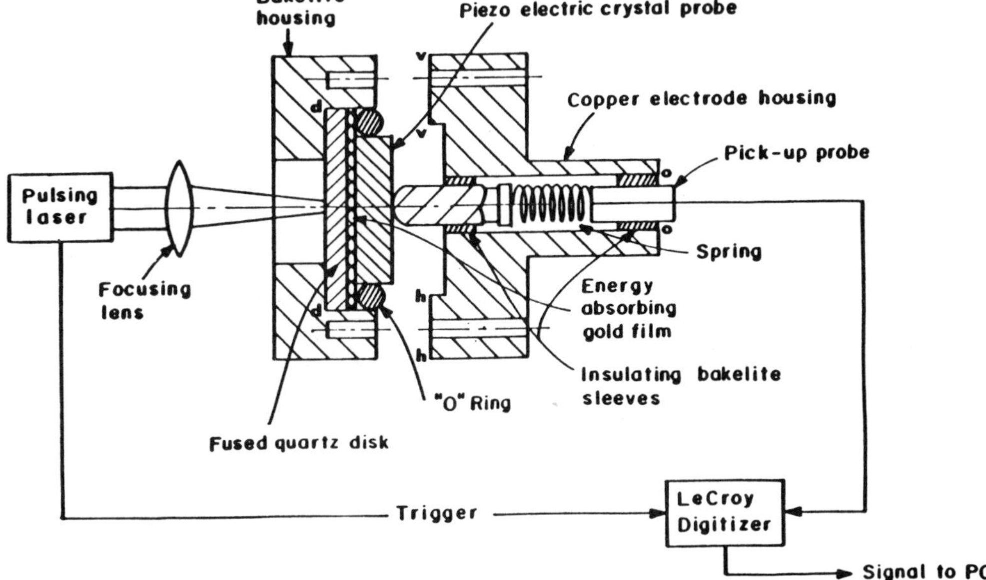

Figure 6. A schematic view of the microelectronic device.

This assembly of fused and piezo-electric quartz is then encapsuled by a copper electrode housing by lapping it onto the edges of the bakelite housing so that the legs *vv* and *hh* of the copper piece (see Fig. 6) touch the outer ground electrode ring on the rear face of the crystal. Thus, the whole copper enclosure acts as a ground electrode. A specially constructed pickup probe with a copper leg, insulated from the copper piece, is then inserted via the opening *oo*, so that the copper leg touches the central secondary electrode. The short circuit current derived between the two electrodes is picked up via a BNC plug fastened on top of the pick-up probe and fed into a Lecroy high speed digitizer by a BNC cable. A high speed Lecroy digitizer (1.3 GHz sampling rate) is needed to record the stress pulses with sub-nanosecond rise times. Since the experiment is performed in the single shot mode, the digitizer is triggered by an electronic pulse that is fired just before the lasing of the optical pulse. The details of the micro-circuitry and various considerations in designing the micro-electronic device can be found in Gupta (1990).

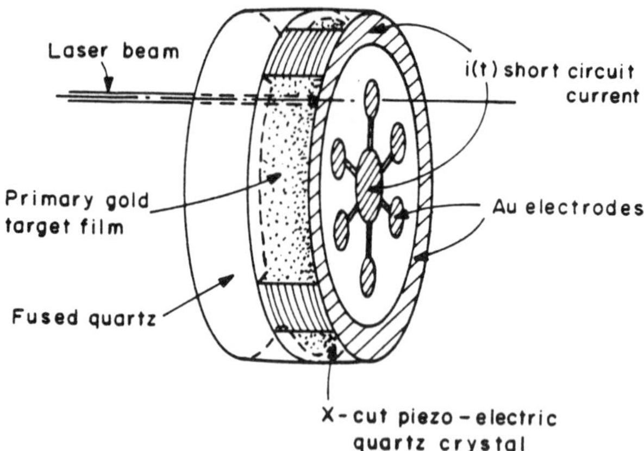

Figure 7. Perspective view of the micro-electronics device.

4.2 Pressure Pulses in the Piezo-electric Crystals

The procedure of using the X-cut PE quartz crystal to record ultra-short duration high amplitude pulses follows the technique described by Graham *et al.* 1965. The measurement is made in the short circuit mode, between the ground electrode that is struck by the laser beam and the front electrode from which the pressure wave is reflected, through an external circuit incorporating the digitizer with a 50Ω impedance. The short circuit current $i(t)$ obtained from the crystal as the stress wave propagates in the crystal is given by (Graham *et al.* 1965)

$$i(t) = \frac{fAc}{l}\left[\, \sigma_o(t) - \sigma_l(t)\,\right] \tag{5}$$

where A is the effective cross-sectional area of the laser beam, l is the thickness of the crystal, c is the velocity of the planar pressure wave, f is the polarization coefficient of the crystal and, $\sigma_o(t)$ and $\sigma(t)$ are the amplitudes of the stress pulses on the ground (impacted energy-absorbing film) and front surface electrodes of the crystal. The term $\sigma_l(t)$ is to be interpreted as the amplitude that hits the free surface, but is annulled instantly by an equal and opposite virtual amplitude in order to satisfy the free surface boundary condition. In the absence of such interpretations, which were apparently not explained by (Graham *et al.* 1965), equation (5) could be misleading as the term $\sigma_l(t)$ is always zero. Nevertheless, if the wavelength of the stress packet is smaller than the crystal thickness, equation (5) reduces to

$$i(t) = \frac{fAc}{l}\, \sigma_o(t) \quad for\ t < l/c \tag{6}$$

which gives the time dependent profile of the stress wave. Equation (6) predicts that the stress at the input electrode is directly proportional to the instantaneous current for times less than the wave transit time.

Most of the researchers (Lang 1974; Fox 1974; Ready 1965; Anderholm 1970; Park 1986) in the past have used the above equation to measure the stresses at the input electrode of the crystal. However, if the thickness of the crystal is greater than the width of the stress pulse, then two short-circuit current signals corresponding to each electrode are obtained. The stress is related to the current signal at the ground electrode via equation (6), but interestingly, the current signal output when the pulse hits the rear electrode is exactly twice that obtained from the ground electrode. This is explained in details by Gupta and Epstein (1990), where the physical phenomenon described by the above equations is elucidated. Because the laser fluences were sufficient to melt a part of the input electrode, it became necessary to record the current signal emitted from the rear electrode for more reliable measurement of the generated stress pulse.

4.4 Comparison of Measurements with Predictions

The simulation presented in Section III considered only a one-dimensional case of propagation of elastic waves due to the laser-pulse-induced, rate-of-change-of-thermal-expansion misfit. That this is close to reality is demonstrated in Fig. 8, where the input current profile resulting from the impingement of the laser beam on the ground electrode is plotted above the current profile emitted from the secondary electrode due to the first reflection of the compressive wave. Apart from the expected amplification of the reflected current by a factor of 2 above the

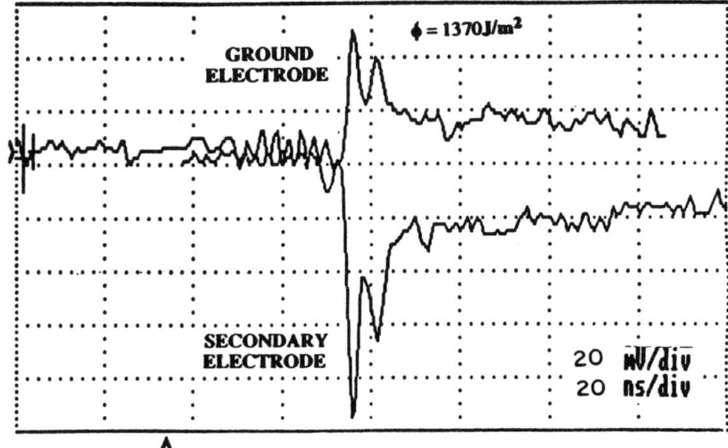

Figure 8. Evidence of the planarity of the travelling stress wave.

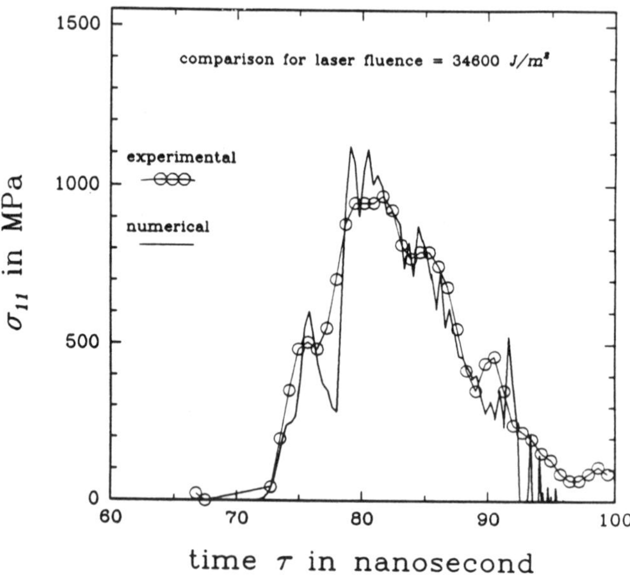

Figure 9. Comparison of the predicted and measured profiles of the stress pulse.

input signal, the two profiles are quite similar in the main portion, but differ in the structure of their tails. This similarity is taken as evidence of planarity of the main pulse.

Figure 9 shows a direct comparison of the predicted and measured profiles of the input stress pulse for a fluence of 34,600 J/m^2 (with distinct melting behavior of the gold film). Figure 9 shows that the predicted profile of the stress pulse for an emissivity of 0.18 for the gold film is clearly similar in broad outline to the profile measured by the *PE* device. The amplitudes of the experimentally recorded pulse and the predicted ones are remarkably similar. This is very encouraging for the simulation, as the ultimate aim in this exercise is to predict the stress amplitudes at the substrate/coating interface. Figure 9, which shows the structure of the recorded pressure pulses for an absorbing gold film, demonstrates the presence of two compression peaks with an intervening deep depression which in the simulation of Fig. 4 appeared as a tension peak. The depression in Fig. 9 has remained in the compression region and has not gone into tension as the simulation predicts. This currently remains a discrepancy that must be resolved. This, however, is of little interest in the present strategy as the laser fluence required to spall the test coatings is sufficiently high to cause the melting of the gold film. Therefore, the computer model can be used with sufficient confidence to predict the interface stress histories in other substrate/coating pairs of interest to researchers in the thin film area. The interface stress is determined from the amplitude of the interface stress history plots of the type shown in Fig. 5.

V Spallation Experiments

The spallation technique was applied to three substrates consisting of Si single-crystal wafers with (100) plane surfaces, Pyrolytic graphite (PG) platelets with the principal axis of layer normals lying in the plane of the platelet, and Pitch-55 (P-55) type carbon ribbons of 600μm width and 35μm thickness having a meso-phase morphology very similar to the Pitch-55 carbon fibers[1]. Detailed description of the morphology of the Pitch-55 ribbon can be found in Gupta and Argon (1990) and for PG in Gupta *et al.* (1988). In all these cases, the coatings to be removed were amorphous SiC coatings deposited by the plasma-assisted chemical vapor deposition technique. The coatings had a thickness of about 1-3 μm. They were deposited in a nearly stress free manner by maintaining the substrates at a certain temperature to prevent the entrapment of a significant concentration of hydrogen, which otherwise would have resulted in high residual compressive stresses and premature delamination in a manner described in detail by us earlier (Argon *et al.*1988).

The Si/SiC, PG/SiC interface systems were tested in an experimental configuration of the type shown in Fig. 2 (a), whereas for the P-55/SiC system, the setup of Fig. 2 (b) was employed. The coating could be successfully spalled off in all cases with appropriate levels of laser pulse energy. The threshold laser energy at which spallation occurs was recorded using a light power meter. Since

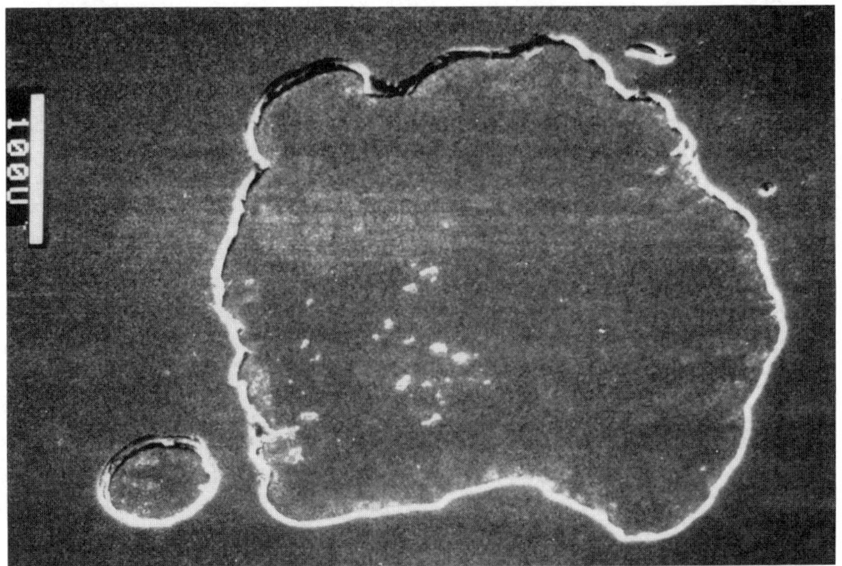

Figure 10. Spalled spot of SiC coating from Si single crystal.

---

[1]These exploratory Pitch-55 ribbons were specially prepared and furnished to us by the textile Fibers Department of the DuPont Company for which we are grateful to Dr. E. M. Schulz.

the amplitude of the stress pulse should depend upon the absorbed laser fluence, the diameter of the laser beam was recorded on photographic film for the tabulation of the threshold laser fluence, as mentioned above.

(a) Si/SiC interface: Figure 10 shows a micrograph of the spot from which the SiC coating has spalled off. The laser fluence necessary to achieve this spallation is 7,388 J/m^2. The interface strength corresponding to this laser fluence as calculated from the computer simulation is 5.29 GPa. The thickness of the amorphous SiC coating of low modulus, deposited at low ion beam energy, was 1.5 μm. The dark portions of the micrographs are the regions where the SiC coating still adheres to the Si substrate. The islands of SiC within the spalled spots are due to the statistical variability of the interface strength over the delaminated spot. The interface strength varies from 10.19 GPa to 10.88 GPa, which is surprisingly low (only 7%) for a brittle interface.
The clean flat structure of the interface shows that the SiC coating has come off uniformly from the substrate at the interface. In order to verify this, Auger electron spectroscopy (AES) was performed on the spalled spot to determine the location of the failure. The irregular pattern at the rim of the spalled spot is due to the partial delamination and breaking of the coating at the circumference of the spalled coating. It was possible to successfully delaminate coatings of thicknesses ranging from 0.3 to 3 μm.

Finally, the actual interface strength values for the various SiC/Si systems produced by using different deposition parameters (of the SiC coating) vary from 3.7 GPa to 10.88 GPa. While such a variation may appear disappointing, this degree of freedom is exactly what is required to carry out the interface delamination scheme in composites as outlined in Section I. The interface strength between the carbon and the SiC interface is of interest in such an endeavor. These measurements are discussed next.

(b) PG/SiC coating interface: Figure 11 shows a spot from which the SiC coating of 2.1 μm thickness is spalled off from the PG surface. The substrate disc was 1.69 mm thick. Interestingly, the spalled pattern is elongated along the edges of the graphitic planes that terminate perpendicular to the surface. This is also the stiffest direction in that plane. For this case the interface strength is determined to be 3.68 GPa. These calculations also include the anisotropic character of the PG substrate. Since the surface techniques available were not able to distinguish the carbon of the SiC from that of the PG substrate, the depth of the crater as determined by a mechanical profilometer (to an accuracy of 2.5 to 10 nm) compared remarkably well with the thickness of the deposited SiC coatings, thereby confirming the failure at the interface. Interface strength values ranging from 3.4 to 7.48 GPa were obtained for different SiC coatings.

(c) Ribbon/SiC Coating Interface: Due to the presence of inhomogeneities and the weak (transverse) strength of these ribbons across lamellae, failure was predominantly observed to be within the ribbon. However, in some cases, failure at the interface was also observed. Figure 11 shows a high magnification view of an edge of a spalled coating. The coating is intact on the left of the micrograph and is removed from the right side. For this sample, an average interface strength value of .240 GPa was obtained. These calculations also

include the anisotropic character of the ribbon material. A part of the same ribbon was tested without the coating in order to determine the transverse strength of the ribbon. A value of 0.26 GPa was obtained. As expected, this value is higher than the interface strength observed on the same ribbon/coating system. Interface strength values ranging from 0.22 to 0.24 GPa were obtained

Figure 11. Spalled spot of SiC coating from the Pitch-55 ribbon.

Figure 12. Spalled spot of SiC coating from PG substrate.

for the Pitch-55 ribbon/SiC system. In all the above cases, the location of the failure was confirmed by matching the coating thickness to the depth of the spalled spot determined from a high-precision profilometer. The interface strength determined from this test is significantly lower than that obtained from the PG/SiC system. This is probably due to the inhomogeneities on the ribbon surface (see Gupta 1990 for the structure of the ribbon), which act as sizable interface flaws and lead to lower strength values. In view of the poor structural integrity of the ribbon material, the PG/SiC system appears to be a better candidate for determining the properties of the Pitch-55 fiber/SiC interface. Nevertheless, the values obtained from both the systems are greater than the desired level of interface strength of 200 MPa for interface delamination in the SiC coated Pitch-55 fiber/aluminum matrix composites (Gupta 1990). Thus, it is necessary to impair the interface strength by implanting embrittling agents at the interface during the coating deposition process. This could be achieved by planting a few atomic layers of either Na, Sb or As. Recently, Rice and Wang (1989) have theoretically explored the effect of such foreign agents on the interface toughness. Similar effects need to be experimentally achieved in the present work in an attempt to bring down the interface strength levels.

Since the interface is loaded by a stress pulse that is external to the material system, the laser spallation experiment is capable of determining the interface strength for any thin film interface. To encourage the wide applicability of the laser spallation experiment, the results of the computer code, which has been verified by the piezo-electric probe, are furnished as generalized interface stress charts so that researchers interested in using the technique do not have to reproduce the computer code. These results should make such strength measurements possible in most systems of interest to workers on composite materials.

VI Interface Stress Charts

The intent of this section is to present generalized interface stress charts for various substrate/coating pairs so that researchers interested in using this technique do not have to reproduce the computer model. It is not possible to give results for exhaustive sets of substrates. Here, we give results for four substrates only, spanning wide ranges of H_S and K_S values. An empirical scheme for obtaining the stresses at other substrates (with arbitrary values of H_S and K_S values)/coating interfaces can be found in Gupta (1990). Since the stress pulse generated in the substrate is independent of the test coating on its front surface, the interface stress can be further normalized as

$$\tilde{\sigma}_i = \sigma_o/T \tag{7}$$

where σ_o is the normalized stress, defined before in Section 3.2 and T is the acoustic transmission coefficient at the substrate/test coating interface, defined as

$$T = \frac{2\sqrt{E_c \rho_c}}{\sqrt{E_s \rho_s} + \sqrt{E_c \rho_c}} \tag{8}$$

where ρ_C, E_C and ρ_S, E_S are respectively the density and Young's modulus of the test coating and the substrate medium. With the above normalization, it is possible to determine the stress at the interface between a substrate and coating of *any* material provided the elastic properties of the two media are known *apriori* in order to calculate the acoustic transmission coefficient T. A typical interface stress chart is shown in Fig. 13. This chart can be used for any substrate with the same H_S and K_S values. A catalogue of interface stress charts valid for other systems can be found in Gupta (1990). Thus, the threshold laser fluence determined from the experiment can now be converted into the actual strength values via the interface stress charts over a wide range of substrate/coating systems of interest to researchers in the device and thermal spray industries.

Due to the absence of any gripping effects, the spallation experiment also provides a good measure of the intrinsic strength of thin coatings and single crystals. The results of such measurements can be found in Gupta *et al.* 1990.

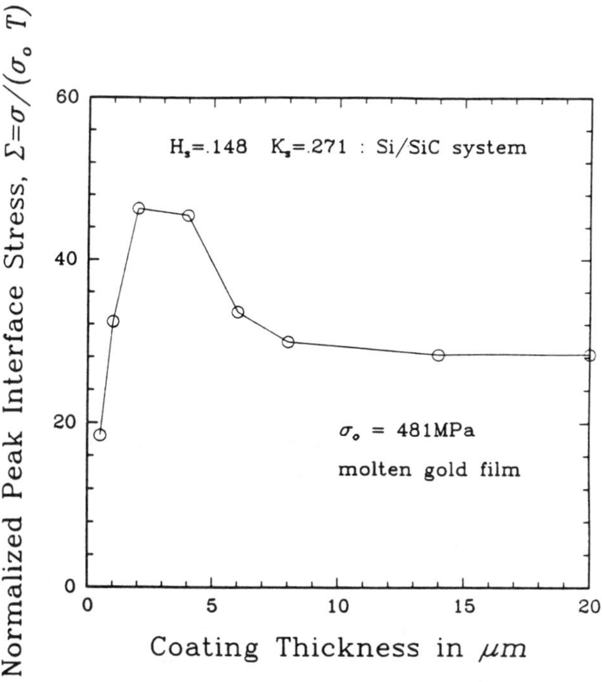

Figure 13. A typical interface stress chart.

VII Discussion

The tensile strength of the interfaces between two media in many instances determines the overall strength of a heterogeneous solid. For example, considerations of initiation of cavities on interfaces of particles is of interest in understanding the initial phases of the ductile fracture in metals and alloys. Similar considerations abound in composite materials where the tensile strength of interfaces between reinforcing fibers and the surrounding matrix, or between the fiber and a protective coating is often of major importance. The measurements of such strengths in macroscopic experiments has been attempted by many investigators but is not free of problems. In most of the reported cases, part of the interface is subjected to high stresses in an inhomogeneous deformation field. The magnitudes of these stresses need to be determined through the solutions of complex boundary value problems of elastic and plastic deformation. While these attempts have often given operationally useful answers, the factor of uncertainty in them has been considerable because of the inadequacy of the solution of the local deformation problems. The measurement of strength between a substrate and a thin coating presents particular difficulties in the application of the stress by an inhomogeneous local deformation field where premature failure elsewhere is likely, before the desired interface can be probed.

It is in these latter cases that the laser spallation technique is an attractive alternative, provided that the interface to be probed can be obtained in planar form. If the coating to be pried off is of thickness Λ_c , the required stress pulses with widths of roughly the same magnitude as the coating thickness can be achieved by a laser pulse of duration Δt_l ,

$$\Delta t_l \cong \Lambda_c^2 \rho C/k \tag{9}$$

where ρ, C, and k are the density, specific heat and thermal conductivity of the energy-absorbing film in which the primary thermal misfit is being generated. Thus, for an energy-absorbing film of Au, the required laser pulse duration to pry off a coating of 1 μm thickness is about 8 ns. This is readily achievable with many pulse laser systems. The three-part strategy which we have outlined here furnishes an operationally attractive means of measurement of the strength of planar interfaces that can be made part of a laser spallation probe. In this, the fundamental tool is the computer code which permits a reliable means of simulating the generation and propagation of elastic waves in the substrate. The basic results of this code, which have been broadly verified by the piezo-electric probe for various material pairs, are provided in the form of interface stress charts presented in Section VI. These results should make such strength measurements possible in most systems of interest to workers on composite materials by using the interpolation scheme furnished in Section 6.4.

Tensile strength of interfaces between amorphous SiC coatings and substrates of Pyrolytic graphite, Pitch-55 carbon ribbon and Silicon single crystal discs were explored in the present investigation. The average interface strength for the SiC/Si pair determined from the experiment ranges from 3.7 GPa to 10.88 GPa, depending upon the deposition conditions for the SiC coating. Average interface strength values ranging from 3.68 to 7.48 GPa for the PG/SiC system and 0.22 to 0.26 GPa for the Pitch-55 fiber/SiC system were obtained.

Since the tensile stress causing the interface delamination is built up as a result of the reflection of the main compressive pulse from the free surface of the coating, it is necessary that the length of the initial compressive pulse is of the same order as the thickness of the SiC coating. This will ensure building of sufficient tensile stress at the interface to cause interface failure. It was found that SiC coatings of thicknesses 0.3 to 3 μm could be readily removed from substrates. In order to spall even thinner coatings from the substrate, it will become necessary to use picosecond laser pulses in place of nanosecond ones as employed in the present investigation. Results of the numerical exercise in Section III indicate other interesting possibilities to delaminate very thin coatings (<1 μm thick) from substrates, that are of considerable interest to researchers in the semiconductor and device industry.

While the interface tensile strength is an important quantity required to determine the critical initiation conditions in interface separation, in flawed interfaces or in cases when the propagation of an existing crack along the interface is the determining factor, the most relevant quantity required is the fracture toughness K_C or the critical energy release rate G_C for the propagation of a crack along the interface. A measure of this can also be obtained with the laser spallation method, if well characterizable and reproducible flaws can be

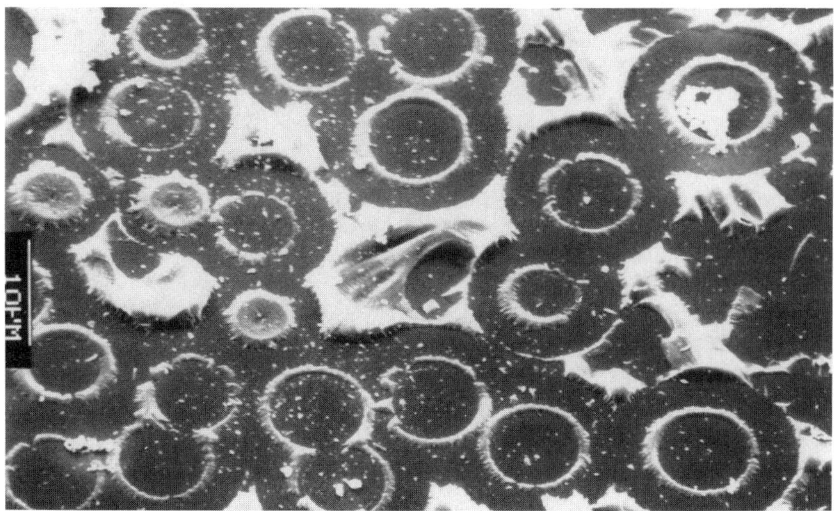

Figure 14. Spalled pattern on a contaminated interface.

placed on the interface, so that the cracks emanating from these can be propagated under a critical reflected tensile stress pulse that can provide the required G_c or produce the K_c. In a preliminary experiment where some local interface contamination of an undetermined nature had apparently occurred during the deposition of the SiC coating, flaws of a potentially interesting type were implanted on the interface. When this specimen was subjected to a typical laser pulse, the coating spalled off independently from individual spots as shown in Fig. 14. Clearly, under the condition of the reflected tensile stress pulse, which probes the interface, a central crack is formed in the area of the interface flaw where a much reduced level of interface strength must have been present. After the crack had propagated radially outward a certain distance, the effective ratio of the crack tip stress intensity factors K_I and K_{II} had apparently undergone almost two full reversals of sign before the delamination is complete. This has resulted first in an excursion of the crack away from the interface into the coating, to be followed by a sharp reversal returning the crack back to the interface, only to be followed by another reversal putting the crack into the coating again where it continued to propagate to final fracture. Evidently, two separate effects: the bi-material nature of the field in which the crack propagates (Rice, 1988), and the dynamic nature of the crack propagation process (Freund, 1976) produce important changes in the driving forces that propagate the crack must be better understood, before an interpretation of the phenomenon can be attempted to extract a precise value of the interface fracture toughness out of the experiment. Nevertheless, it is clear that the potential for the measurement of the interface toughness is present in the laser spallation experiment.

References

Ahn J., Mittal, K.L. and Macqeen, R.H., 1978, "Hardness and Adhesion of Filmed Structures as Determined by the Scratch Technique," ASTM STP **640**, pp. 134-157.

Anderholm, N.C., 1970, "Laser Generated Stress Waves," *Appl. Phys. Lett*, **16**, 3, pp. 113-115.

Aravas, N., Kim, K. S.and Loukis, M. J., 1989, "On the Mechanics of Adhesion Testing of Thin Films," *Matl. Sci & Engng*, **A107**, pp. 159-168.

Argon, A. S., Gupta, V. Landis, H. S. and Cornie, J. A., 1989*a*, "Intrinsic Toughness of Interfaces Between SiC Coatings and Substrates of Si or C Fiber," *J. Mater. Sci.*,**24**, pp. 1207-1218.

Argon, A. S., Gupta, V., Landis, H. S. and Cornie, J. A., 1989*b*, "Intrinsic Toughness of Interfaces," *Mater. Sci.&Engg.*, A 107, pp. 41-47.

Chambers, B., 1988, PHD Thesis in progress, Department of Material Science and Engineering, MIT, Cambridge, Massachusetts.

Chambers, B. and Rothery,W. H.,1969, Liquid Metals and Vapor Pressure, in *progress in Materials Science*, Eds. B.Chambers and W. H. Rothery, **14**, no. 4.

Chapman,B.N., 1974, "Thin Film Adhesion," *J. Vac. Sci. Tech*, **11**, pp. 106-113.

Charalambides, P.G., J. Lund, Evans, A.G. and McMeeking, R.M., 1989, "A Test Specimen for Determining the Fracture Resistance of Bi-material Interfaces,"

J. Appl. Mech., **56**, pp. 77-82.

Chiang, S. S., Marshal, D. B. and Evans. A.G., 1981, "Simple Method of Adhesion Measurement," *Surfaces and Interfaces in Ceramic and Ceramic-Metal Systems*, pp. 603-617, Eds. J. Pask and A. G. Evans, Materials Science Research, Vol. **14**, Plenum Press, New York.

Chow, T. S., Liu, C.A. and Pennell, R.C., 1976, "Direct Determination of Interfacial Energy Between Brittle and Polymeric Film," *J . Polm. Sci.*,**14**, pp. 1305-1310.

Clifton, R. J.., 1978, "Analysis of the Laser Velocity Interferometer," *J . Appl. Phys.*, **41**, pp. 5335-5337.

Davutuglu, A. and Aksay, A. I., 1981, "Work of Adhesion by a Periodic Cracking Technique," *Surfaces and Interfaces in Ceramic and Ceramic-Metal Systems*, pp. 641-649, Eds. J. Pask and A. G. Evans, Materials Science Research, Vol. **14** , Plenum Press, New York.

Ditchburn, R. W., 1963, *Light*, 2nd Ed., Interscience Publishers, John Wiley & Sons, Inc. New York.

Evans, A. G., 1989, "The Mechanical Performance of Fiber-Reinforced Ceramic Matrix Composites," *Matl. Sci and Engng*, **A107**, pp. 227-239.

Fox, J.A., 1974, "Effect of Pulse Shaping on Laser-Induced Spallation," *Appl. Phys. Lett.*,**24**, pp. 340-343.

Freund, L. B., 1976, "The Analysis of Elastodynamic Crack Tip Stress Fields," *Mechanics Today*, pp. 55-91, Ed. S. Nemat-Nasser, Pergamon Press: Oxford, **3**, Cambridge.

Graham, R.A., Neilson, F.W. and Benedick,W.B., 1965, "Piezoelectric Current From Shock Loaded Crystal," *J. Appl. Phys.*,**36**, pp. 1775-1783.

Gupta, V., Argon, A.S.and Cornie, J.A., 1989, "Interfaces with Controlled Toughness as Mechanical Fuses to Isolate Fiber from Damage," *J. Mater. Sci*, **24**, pp. 2031-2040..

Gupta, V. and Argon, A. S., 1990 "Measurement of Strength and Toughness of Pitch-55 Carbon Ribbons," manuscript intended for publication.

Gupta, V., 1990, *Ph.D. Thesis*, Department of Mechanical Engg., MIT, Cambridge, Massachusetts, USA.

Gupta, V. and Epstein, D. J., 1990, "Current from Shock Loaded Piezo-electric Crystals," *J. Appl. Physics*, **67**,4, pp. 2185-2188.

Jacobbson, R., 1976, "Measurement of Adhesion of Thin Films," *Thin Solid Films*, **34**, pp. 191-199.

Jacobsson, R. and Kruse, B., 1973 "Measurement of Adhesion of Thin Evaporated Films on Glass Substrates by Means of a Direct Pull Method," *Thin Solid Films*, **15**, pp. 71-77.

Landolt, H- Bornstein, R, 1971, *Landolt- Bornstein Handbook*, Springer-Verlag Publisher, New York, **II**, pp. 431.

Mittal, K.L., 1978, "Adhesion Measurement: Recent Progress, Unsolved Problems, and Prospects," ASTM STP **640**, pp. 5-17.

Park, H.S., 1986, *MS Thesis* , Graduate School of Engineering, The University of Texas at Austin, Texas.

Peercy, P.S., Jones, E.D., Bushnell, J.C. and Gobeli, G.W, 1970 , "Ultrafast Rise Time Laser Induced Stress Waves," *Appl. Phys. Lett.*, **16**, pp. 120-122.

Ready, J.F., 1965, "Effects Due to Absorption of Laser Radiation," *J. Appl. Phys*, **36**, pp. 462-468.

Ready, J.F., 1971, *Effects of High -power Laser radiation,* Academic Press. New York.

Rice, J. R., 1988, "Elastic Fracture Mechanics Concepts for Interfacial Cracks," *J. Appl. Mech.*, **110**, pp. 98-103..

Salvadori, M. G. and Baron, M. L., 1961, *Numerical Methods in Engineering*, pp. 261, Englewood Cliff, N. J. : Prentice Hall Inc.

Smithells, C. J.,1983, *Smithells Metals Reference Book*, 6th edition, Ed. E. A. Brandes, Robert Hartnoll Ltd. Bodmin, Cornwall.

Volkova, N. V., 1967, "Fracture of LiF Single Crystals Under the Action of Laser Radiation," *Sov. Phys.-Solid State*, **8**, pp. 2133-2135.

Vossen, J.L., 1978, ASTM STP **640**, pp.122-133.

Yang, L.C., 1974, "Stress Waves Generated in Thin Metallic Films by a Q Switched Ruby Laser," *J. Appl. Phys.*,**45**, pp. 2601-2698.

Weibull, W.,1939, "A Statistical Theory of the Strength of Materials," *Ingeniors Vetenskaps Akademien, Handlingar*, no. 151.

Acknowledgments

This research has been supported initially by IST/SDIO through the ONR under Contract No. N00014-85-K-0645 and more recently by ONR under Grant No. N00014-89-J-1609. For this support and his continued keen interest in this research we are grateful to Dr. S. Fishman of that agency. The simulation made use of the DARPA/ONR computational facility supported under ONR Grant No. N00014-86-K-0768. The ABAQUS finite element code that was utilized was made available under the academic license by HKS Inc. of Providence, R. I. We are also grateful to Professor D. Epstein for advise in the development of the piezo-electric probe; to the M.I.T George Harrison Spectroscopy Laboratory for the use of their facilities; to Professor D. M. Parks for discussions on numerical simulation exercise; to Professor J. W. Hutchinson for some useful discussions, to Mr. B. Chambers for preparing the coatings and finally to Dr. James Cornie for partial support of the work.

Interfaces II

On a Correspondence Between Mechanical and Thermal Fields in Composites with Slipping Interfaces

Y. Benveniste and G. J. Dvorak

Department of Solid Mechanics, Material and Structures
Faculty of Engineering, Tel–Aviv University
Ramat Aviv, 69978
Israel

Department of Civil Engineering
Rensselaer Polytechnic Institute
Troy, New York 12180–3590
USA

Abstract

The present paper is concerned with composites in which the constituent interfaces are weak in shear and therefore exhibit shear deformation associated with sliding. Thermomechanical loadings of such systems are considered which consist of homogeneous traction or displacement boundary conditions and a uniform temperature change on the outside surface of the composite. For binary systems with isotropic constituents, it is shown that the actual fields in the purely thermal problem can be uniquely determined from the solution of the purely mechanical problem. This correspondence relation is used to determine the effective thermal strain and stress tensors on the basis of the effective mechanical properties. For multi–phase systems with anisotropic constituents undergoing interface slip and separation, the theorem of virtual work is used to establish a similar relation between the effective thermal tensors and the mechanical concentration factors and constituent properties of the composite.

Introduction

Thermal problems in heterogeneous media have drawn much interest in the last years due to the increasing importance of high temperature composites. Several fundamental aspects in the micromechanics of composites in the context of thermomechanical problems have recently been investigated by the authors, Dvorak (1986), Dvorak and Chen (1989), Benveniste and Dvorak (1989) where the reader can find a list of references in the field.

Most of the work dealing with composites assumes perfect bonding between the constituents. However, due to poor bonding between the phases, a jump in the displacement field may occur at internal boundaries, and it is of interest to study thermomechanical problems in composites under such circumstances. Determination of the effective properties requires special attention in the presence of imperfect bonding, and a proper framework for the investigation of such problems has been laid down by Benveniste (1985). Interfaces which are weak in shear may be modeled by demanding that the normal displacements are continuous, but the tangential displacements exhibit a jump which is proportional to the shear tractions. For limiting values of the constant of proportionality, the special cases of perfect bonding and lubricated contact are obtained. Such models of a flexible interface which may also include imperfect bonding in the normal direction have been previously used in the literature, see for example Lené and Leguillon (1982), Benveniste and Aboudi (1984), Aboudi (1987), Benveniste and Miloh (1986), Jasiuk and Tong (1989), Achenbach and Zhu (1989), and Hashin (1990). The reader is referred to these works for a further list of references on imperfect interfaces. Recently, several problems of inclusions which undergo pure slip at interfaces have been considered by Mura et al. (1985), Tsuchida et al. (1986), and Jasiuk et al. (1988).

The present paper is concerned with binary systems with flexible interfaces in shear, and isotropic constituents. It starts by establishing a correspondence relation between local fields induced in such two–phase composites by purely mechanical and purely thermal problems. These relations are obtained by using a decomposition scheme originally proposed by Dvorak (1983, 1986), and further employed by Benveniste and Dvorak (1989) in binary composites with anisotropic constituents, arbitrary phase geometry, but perfect bonding between the phases. Recently, Dvorak (1990) has thoroughly explored the implications of this concept in regard to the existence of uniform fields in heterogeneous media. We show here that this decomposition scheme can be generalized to the case of two–phase media undergoing slip at interphase boundaries, but with isotropic constituents. The implementation of the scheme shows that local fields in such composites which are induced by a uniform temperature change at external boundaries can be uniquely determined from the solution of the same system subjected to uniform overall mechanical loading. In the second part of the first section of the paper, the established correspondence principle is used to derive the effective thermal strain and stress tensors on the basis of the effective mechanical properties of the composite. The second section of the paper is concerned with multiphase composites with anisotropic constituents undergoing slip of the above described nature at interphase boundaries. Only effective properties are considered in this section, and a generalization of Levin's (1967) and Rosen and Hashin's (1970) result is derived using the theorem of virtual work. The obtained results reduce correctly to those obtained in the previous section for the case of binary composites with isotropic constituents.

1. Correspondence Between Purely Mechanical and Purely Thermal Problems in Binary Composites with Interfaces Weak in Shear

1a. General Theory

Consider a two–phase composite with isotropic constituents, but arbitrary phase geometry. Let the thermoelastic constitutive relations of the homogeneous phases r = 1,2 be given by:

$$
\begin{aligned}
\sigma_r &= L_r\, \epsilon_r + \ell_r \theta \quad , \qquad r = 1, 2 \\
\epsilon_r &= M_r\, \sigma_r + m_r \theta \quad ,
\end{aligned}
\tag{1}
$$

where σ_r, ϵ_r and θ denote respectively the stress, strain tensors and temperature field, L_r and $M_r = L_r^{-1}$ are the phase stiffness and compliance tensors, m_r is the thermal strain tensor (of expansion coefficients), and $\underset{\sim}{\ell}_r$ is the thermal stress tensor such that $\underset{\sim}{\ell}_r = -\underset{\sim}{L}_r \underset{\sim}{m}_r$. In this paper we will denote the matrix phase by the index r = 1 and the inclusion phase by index r = 2.

The two–phase composite is assumed to have constituents interfaces which are weak in shear and are modeled by a jump in the tangential displacement which is prescribed as proportional to the shear traction there. Perfect bonding in the normal direction is assumed in this part of the work; however, in the second part open cracks at interfaces are allowed. Let $\underset{\sim}{p}$ denote the unit normal vector at S_{12} pointing from phase r = 2 to phase r = 1, and let $\underset{\sim}{u}$ and $\underset{\sim}{t}$ denote respectively the displacement and traction vectors. The interface conditions at S_{12} may be expressed in the following manner. Let $(\underset{\sim}{p}, \underset{\sim}{q}, \underset{\sim}{s})$ be an orthogonal set of unit vectors at S_{12} where $\underset{\sim}{p}$ denotes the unit normal vector. The components of the traction and displacements vectors in this coordinate system are respectively expressed as $\underset{\sim}{t} = \underset{\sim}{t}_p + \underset{\sim}{t}_q + \underset{\sim}{t}_s$, $\underset{\sim}{u} = \underset{\sim}{u}_p + \underset{\sim}{u}_q + \underset{\sim}{u}_s$. The interface is then modeled by the following set of equations:

$$[\underset{\sim}{u}_p]_{S_{12}} = 0 \quad , \quad [\underset{\sim}{t}]_{S_{12}} = 0$$
$$[\underset{\sim}{u}_q]_{S_{12}} = R\, \underset{\sim}{t}_q \qquad [\underset{\sim}{u}_s]_{S_{12}} = Q\, \underset{\sim}{t}_s \tag{2}$$

where R and Q are constants of proportionality for the interface which is flexible in shear and a square bracket [ ] on a quantity $\underset{\sim}{\psi}$ denotes the jump in that quantity across S_{12}, that is

$$\left[\underset{\sim}{\psi}\right] = \underset{\sim}{\psi}^{(2)}\Big|_{S_{12}} - \underset{\sim}{\psi}^{(1)}\Big|_{S_{12}} \quad , \tag{3}$$

It is noted that for $R \rightarrow 0$, $Q \rightarrow 0$, perfect bonding in shear is obtained, and that $R \rightarrow \infty$, $Q \rightarrow \infty$ yield the case of lubricated contact. The analysis which follows in this section is also valid if at part of the interfaces there exists imperfect bonding ($R \neq 0$, $Q \neq 0$), and at other parts perfect bonding prevails ($R = 0$, $Q = 0$); in fact different values of R and Q may exist at different points in the interface.

Consider now purely mechanical problems in which the outside surface of the composite is subjected to homogeneous displacement or traction boundary conditions described by:

$$\begin{aligned} \underset{\sim}{u}(S) &= \underset{\sim}{\epsilon}_0 \underset{\sim}{x} \quad , \qquad \theta(S) = 0 \quad , \\ \underset{\sim}{t}(S) &= \underset{\sim}{\sigma}_0 \underset{\sim}{n} \quad , \qquad \theta(S) = 0 \quad , \end{aligned} \tag{4}$$

where $\underset{\sim}{u}(S)$ and $\underset{\sim}{t}(S)$ denote the displacement and traction vector at S, $\underset{\sim}{n}$ is the outside normal to S, $\underset{\sim}{\epsilon}_0$ and $\underset{\sim}{\sigma}_0$ are constant strain and stress tensors, and finally $\underset{\sim}{x}$ denotes the components of a Cartesian system.

Let the local strain and stress fields induced in the phases by these boundary conditions be denoted by

$$\underset{\sim}{\epsilon}_r(\underset{\sim}{x}) = \underset{\sim}{A}_r(\underset{\sim}{x})\underset{\sim}{\epsilon}_0 \, , \qquad \underset{\sim}{\sigma}_r(\underset{\sim}{x}) = \underset{\sim}{L}_r \, \underset{\sim}{A}_r(\underset{\sim}{x})\underset{\sim}{\epsilon}_0 \, , \tag{5}$$

$$\underset{\sim}{\sigma}_r(\underset{\sim}{x}) = \underset{\sim}{B}_r(\underset{\sim}{x})\underset{\sim}{\sigma}_0 \, , \qquad \underset{\sim}{\epsilon}_r(\underset{\sim}{x}) = \underset{\sim}{M}_r \, \underset{\sim}{B}_r(\underset{\sim}{x})\underset{\sim}{\sigma}_0 \, , \tag{6}$$

with (5) and (6) corresponding to $(4)_1$ and $(4)_2$ respectively. Furthermore, let us denote the jump in the displacement vector at S_{12} by

$$\begin{aligned} \left[\underset{\sim}{u}(\underset{\sim}{x})\right]_{S_{12}} &= \underset{\sim}{D}(\underset{\sim}{x}) \, \underset{\sim}{\epsilon}_0 \quad , \\ \left[\underset{\sim}{u}(\underset{\sim}{x})\right]_{S_{12}} &= \underset{\sim}{F}(\underset{\sim}{x}) \, \underset{\sim}{\sigma}_0 \quad , \end{aligned} \tag{7}$$

again, under $(4)_1$ and $(4)_2$ respectively. Of course, the fields (5), (6), and (7) satisfy the interface conditions in (2). Local fields are denoted in this paper by the argument $(\underset{\sim}{x})$, whereas expressions without such an argument will refer to average quantities.

Next, consider thermal loading problems in which the surface of the composite is subjected to a uniform temperature rise and to zero displacement or traction boundary conditions.

$$\theta(S) = \theta_0 \qquad \underset{\sim}{u}(S) = \underset{\sim}{0} \quad , \tag{8}$$

$$\theta(S) = \theta_0 \qquad \underset{\sim}{t}(S) = \underset{\sim}{0} \quad . \tag{9}$$

The local fields under (8) and (9) will respectively be denoted by:

$$\begin{aligned} \underset{\sim}{\epsilon}_r(\underset{\sim}{x}) &= \underset{\sim}{a}_r(\underset{\sim}{x})\theta_0 \quad , \\ \underset{\sim}{\sigma}_r(\underset{\sim}{x}) &= (\underset{\sim}{L}_r\underset{\sim}{a}_r(\underset{\sim}{x}) + \underset{\sim}{\ell}_r)\theta_0 \quad , \end{aligned} \tag{10}$$

$$\left[\underset{\sim}{u}(\underset{\sim}{x})\right]_{S_{12}} = \underset{\sim}{d}(\underset{\sim}{x})\theta_0$$

$$\begin{aligned} \underset{\sim}{\sigma}_r(\underset{\sim}{x}) &= \underset{\sim}{b}_r(\underset{\sim}{x})\,\theta_0 \quad , \\ \underset{\sim}{\epsilon}_r(\underset{\sim}{x}) &= (\underset{\sim}{M}_r\,\underset{\sim}{b}_r(\underset{\sim}{x}) + \underset{\sim}{m}_r)\theta_0 \quad , \end{aligned} \tag{11}$$

$$\left[\underset{\sim}{u}(\underset{\sim}{x})\right]_{S_{12}} = \underset{\sim}{f}(\underset{\sim}{x})\theta_0 \quad ,$$

where the vectors $\underset{\sim}{d}(\underset{\sim}{x})$ and $\underset{\sim}{f}(\underset{\sim}{x})$ satisfy the interface conditions in (2). We also note that a uniform temperature field will prevail in the composite under (8) and (9).

It will be shown now that in the two–phase composite with isotropic constituents characterized by the constitutive relations (1) and the interface conditions (2), knowledge of the tensors $\underset{\sim}{A}_r(\underset{\sim}{x})$, $\underset{\sim}{D}(\underset{\sim}{x})$ uniquely determines $\underset{\sim}{a}_r(\underset{\sim}{x})$, $\underset{\sim}{d}(\underset{\sim}{x})$, and $\underset{\sim}{B}_r(\underset{\sim}{x})$, $\underset{\sim}{F}(\underset{\sim}{x})$ determine $\underset{\sim}{b}_r(\underset{\sim}{x})$, $\underset{\sim}{f}(\underset{\sim}{x})$.

Let us first establish the correspondence between the fields induced by $(4)_1$ and (8). This is achieved by using the decomposition scheme described by Dvorak (1986), and Benveniste and Dvorak (1989) for the case of perfectly bonded composites. We will see here that this procedure can be used to establish the desired correspondence relations in the case of interface conditions (2) for two–phase composites with isotropic phases. In the first stage of this decomposition scheme we seek a strain field $\hat{\epsilon}$ which is uniform in V under a uniform temperature change θ_0. This can be achieved by demanding that $\hat{\epsilon}$ and θ_0 result in a uniform stress field:

$$L_1 \hat{\epsilon} + \ell_1 \theta_0 = L_2 \hat{\epsilon} + \ell_2 \theta_0 \quad , \tag{12}$$

so that the tractions at S_{12} are continuous. Equation (12) yields for $\hat{\epsilon}$:

$$\hat{\epsilon} = (L_1 - L_2)^{-1} (\ell_2 - \ell_1)\theta_0 \quad . \tag{13}$$

At this stage of the procedure, uniform strain and stress fields prevail in the composite, and both the displacements and tractions are continuous at S_{12}. Also, it turns out that for the isotropic constituents, the created uniform stresses are hydrostatic, and shear tractions at S_{12} vanish. Therefore, the interface conditions at S_{12} described in (2) are automatically satisfied. At the outside boundary S, displacements arising from (13) have been now induced and, as demanded by (8) they need to be reduced to zero. To accomplish this, we apply the following displacements on S:

$$u\,(S) = -\hat{\epsilon}\, x \quad , \tag{14}$$

and obtain

$$\epsilon_r\,(x) = -A_r(x)\hat{\epsilon} \quad ,$$
$$\left[u\,(x)\right]_{S_{12}} = -D(x)\hat{\epsilon} \quad . \tag{15}$$

By superposition with the uniform fields, the resulting fields at the end of the decomposition scheme are therefore:

$$\underset{\sim}{\epsilon}_r(\underset{\sim}{x}) = (\underset{\sim}{I} - \underset{\sim}{A}_r(\underset{\sim}{x}))(\underset{\sim}{L}_1 - \underset{\sim}{L}_2)^{-1}(\underset{\sim}{\ell}_2 - \underset{\sim}{\ell}_1)\theta_0 \quad , \tag{16}$$

$$\left[\underset{\sim}{u}(\underset{\sim}{x})\right]_{S_{12}} = -\underset{\sim}{D}(\underset{\sim}{x})(\underset{\sim}{L}_1 - \underset{\sim}{L}_2)^{-1}(\underset{\sim}{\ell}_2 - \underset{\sim}{\ell}_1)\theta_0 \quad . \tag{17}$$

with $\underset{\sim}{I}$ being the fourth order identity tensor. The concentration factors $\underset{\sim}{a}_r(\underset{\sim}{x})$ and $\underset{\sim}{d}(\underset{\sim}{x})$ can be therefore read out as:

$$\underset{\sim}{a}_r(\underset{\sim}{x}) = (\underset{\sim}{I} - \underset{\sim}{A}_r(\underset{\sim}{x}))\,(\underset{\sim}{L}_1 - \underset{\sim}{L}_2)^{-1}(\underset{\sim}{\ell}_2 - \underset{\sim}{\ell}_1) \quad , \tag{18}$$

$$\underset{\sim}{d}(\underset{\sim}{x}) = -\,\underset{\sim}{D}(\underset{\sim}{x})\,(\underset{\sim}{L}_1 - \underset{\sim}{L}_2)^{-1}(\underset{\sim}{\ell}_2 - \underset{\sim}{\ell}_1) \quad . \tag{19}$$

The difficulty of extending the above procedure to anisotropic constituents becomes now apparent. For such constituents, shear tractions at S_{12} would exist after the reassembly of the aggregate. To remove these shear tractions, one would have to solve a boundary value problem in which the S_{12} interfaces are loaded by the negative of the shear tractions induced therein. Even though the solution of such a boundary value problem can be formulated in principle, it is not clear at this time that such a solution can be related to a purely mechanical problem with prescribed overall strain.

The correspondence between the fields resulting from $(4)_2$ and (9) can be similarly established. In the first step, a uniform stress field $\hat{\underset{\sim}{\sigma}}$ is sought which together with a temperature change θ_0 causes a in uniform strain field, and therefore continuous displacements throughout. The condition is

$$\underset{\sim}{M}_1\,\hat{\underset{\sim}{\sigma}} + \underset{\sim}{m}_1\,\theta_0 = \underset{\sim}{M}_2\,\hat{\underset{\sim}{\sigma}} + \underset{\sim}{m}_2\,\theta_0 \quad ; \tag{20}$$

it yields

$$\hat{\underset{\sim}{\sigma}} = (\underset{\sim}{M}_1 - \underset{\sim}{M}_2)^{-1}(\underset{\sim}{m}_2 - \underset{\sim}{m}_1)\theta_0 \quad . \tag{21}$$

Since some the constituents are isotropic, these stresses do not result in shear tractions or displacement jumps at S_{12}. To comply with (9), we now remove the tractions induced on the outside surface by (21), by application of:

$$\underset{\sim}{t}\,(S) = -\hat{\underset{\sim}{\sigma}}\,\underset{\sim}{n} \quad , \tag{22}$$

which by themselves cause the local effects

$$\begin{aligned} \underset{\sim}{\sigma}_r(\underset{\sim}{x}) &= -\underset{\sim}{B}_r(\underset{\sim}{x})\hat{\underset{\sim}{\sigma}} \quad , \\ [\underset{\sim}{u}(\underset{\sim}{x})]_{S_{12}} &= -\underset{\sim}{F}(\underset{\sim}{x})\hat{\underset{\sim}{\sigma}} \quad . \end{aligned} \tag{23}$$

This is superimposed with the uniform field $\hat{\sigma}$ to yield:

$$\underset{\sim}{\sigma}_r(\underset{\sim}{x}) = (\underset{\sim}{I} - \underset{\sim}{B}_r(\underset{\sim}{x}))\,(\underset{\sim}{M}_1 - \underset{\sim}{M}_2)^{-1}(\underset{\sim}{m}_2 - \underset{\sim}{m}_1)\;\theta_0 \quad , \tag{24}$$

$$\Big[\underset{\sim}{u}(\underset{\sim}{x})\Big]_{S_{12}} = -\underset{\sim}{F}(\underset{\sim}{x})\,(\underset{\sim}{M}_1 - \underset{\sim}{M}_2)^{-1}(\underset{\sim}{m}_2 - \underset{\sim}{m}_1)\;\theta_0 \quad , \tag{25}$$

The concentration factors thus are

$$\underset{\sim}{b}_r(\underset{\sim}{x}) = (\underset{\sim}{I} - \underset{\sim}{B}_r(\underset{\sim}{x}))\,(\underset{\sim}{M}_1 - \underset{\sim}{M}_2)^{-1}(\underset{\sim}{m}_2 - \underset{\sim}{m}_1) \tag{26}$$

$$\underset{\sim}{f}(\underset{\sim}{x}) = -\underset{\sim}{F}(\underset{\sim}{x})\,(\underset{\sim}{M}_1 - \underset{\sim}{M}_2)^{-1}(\underset{\sim}{m}_2 - \underset{\sim}{m}_1)\,. \tag{27}$$

We have therefore established the desired correspondence relations. It is of interest to note here that the structure of (18) and (26), is similar to that given in Benveniste and Dvorak (1989) for perfect bonding between the constituents.

1b. Application: Effective Thermal Stress and Strain Tensors

One of the applications of the correspondence principle described in the previous section is the determination of the effective thermal tensors of the composite based solely on the information obtained from the mechanical problem. Suppose therefore that the effective constitutive law of the composite is described by

$$\underset{\sim}{\sigma} = \underset{\sim}{L}\,\underset{\sim}{\epsilon} + \underset{\sim}{\ell}\,\theta \quad ,$$
$$\underset{\sim}{\epsilon} = \underset{\sim}{M}\,\underset{\sim}{\sigma} + \underset{\sim}{m}\,\theta \quad , \tag{28}$$

where $\underset{\sim}{L}$ and $\underset{\sim}{M}$ with $\underset{\sim}{M} = \underset{\sim}{L}^{-1}$ denote respectively the effective stiffness and compliance tensors, $\underset{\sim}{\ell}$ and $\underset{\sim}{m}$ with $\underset{\sim}{\ell} = -\underset{\sim}{L}\,\underset{\sim}{m}$ are the effective thermal strain and stress tensors. The tensors $\underset{\sim}{\sigma}$ and $\underset{\sim}{\epsilon}$, and the temperature θ refer to average quantities.

The tensors $\underset{\sim}{\ell}$ and $\underset{\sim}{m}$ are determined in principle by subjecting the composite to boundary conditions (8) and (9) respectively. Let us first consider the determination of $\underset{\sim}{\ell}$. It is important to note here that since displacement jumps occur at constituent interfaces, special care should be taken in defining average quantities in the composite, and the reader is referred to Benveniste (1985) for a proper framework for the computation of effective properties in these situations. Under (8), the average strain in the composite vanishes, therefore in accordance with the quoted paper

$$\underset{\sim}{\epsilon} = c_1\,\underset{\sim}{\epsilon}_1 + c_2\,\underset{\sim}{\epsilon}_2 - c_2\,\underset{\sim}{J} = 0 \quad , \tag{29}$$

where $\underset{\sim}{J}$ is a second order tensor representing the deformation at internal boundaries, and is given by:

$$J_{ij} = \frac{1}{2V}\int_{S_{12}} ([u_i]\,p_j + [u_j]\,p_i)\,dS_{12} \quad , \tag{30}$$

where p_j was defined in (2); c_r and $\underset{\sim}{\epsilon}_r$ with $r = 1, 2$ denote the phase volume fractions and phase strain averages respectively, and V is the volume of the composite. The average stress in the composite, in view of $(1)_1$, $(28)_1$, and (29) is given by

$$\underset{\sim}{\sigma} = c_1 \underset{\sim}{\sigma}_1 + c_2 \underset{\sim}{\sigma}_2 = \\ = c_1(\underset{\sim}{L}_1 \underset{\sim}{\epsilon}_1 + \underset{\sim}{\ell}_1 \theta_0) + c_2(\underset{\sim}{L}_2 \underset{\sim}{\epsilon}_2 + \underset{\sim}{\ell}_2 \theta_0) = \underset{\sim}{\ell}\, \theta_0 \quad , \tag{31}$$

where we have used the fact that a uniform temperature prevails throughout. Elimination of $\underset{\sim}{\epsilon}_2$ from (29) and substitution into (31) provides:

$$\underset{\sim}{\ell} = c_1 \underset{\sim}{\ell}_1 + c_2 \underset{\sim}{\ell}_2 + c_2\, (\underset{\sim}{L}_2 - \underset{\sim}{L}_1)\, \underset{\sim}{a}_2 + c_2\, \underset{\sim}{L}_1\, \underset{\sim}{a} \quad , \tag{32}$$

where the concentration factors are defined as in (10):

$$\underset{\sim}{\epsilon}_2 = \underset{\sim}{a}_2\, \theta_0 \quad , \quad \underset{\sim}{J} = \underset{\sim}{a}\, \theta_0 \quad . \tag{33}$$

The tensor $\underset{\sim}{a}_2$ is simply the average of $\underset{\sim}{a}_2(\underset{\sim}{x})$ in (18), and is given by

$$\underset{\sim}{a}_2 = (\underset{\sim}{I} - \underset{\sim}{A}_2)(\underset{\sim}{L}_1 - \underset{\sim}{L}_2)^{-1}(\underset{\sim}{\ell}_2 - \underset{\sim}{\ell}_1) \quad , \tag{34}$$

where $\underset{\sim}{A}_2$ is again the phase volume average of $\underset{\sim}{A}_2\, (\underset{\sim}{x})$. The tensor $\underset{\sim}{a}$ is obtained by substituting $(10)_3$ and (19) into (30):

$$\underset{\sim}{a} = -\underset{\sim}{A}\, (\underset{\sim}{L}_1 - \underset{\sim}{L}_2)^{-1}\, (\underset{\sim}{\ell}_2 - \underset{\sim}{\ell}_1) \quad , \tag{35}$$

with the concentration factor $\underset{\sim}{J} = \underset{\sim}{A}\, \underset{\sim}{\epsilon}^{0}$ defined as:

$$A_{ijkl} = \frac{1}{2V} \int_{S_{12}} (D_{ikl}\, (\underset{\sim}{x})\, p_j + D_{jkl}\, (\underset{\sim}{x})\, p_i)\, dS_{12} \quad . \tag{36}$$

Substitution of (34) and (35) into (32) provides

$$\underset{\sim}{\ell} = c_1\underset{\sim}{\ell}_1 + c_2\underset{\sim}{\ell}_2 + \\ + c_2(\underset{\sim}{L}_2 - \underset{\sim}{L}_1)(\underset{\sim}{I} - \underset{\sim}{A}_2)(\underset{\sim}{L}_1 - \underset{\sim}{L}_2)^{-1}(\underset{\sim}{\ell}_2 - \underset{\sim}{\ell}_1) \\ - c_2\,\underset{\sim}{L}_1\,\underset{\sim}{A}\,(\underset{\sim}{L}_1 - \underset{\sim}{L}_2)^{-1}\,(\underset{\sim}{\ell}_2 - \underset{\sim}{\ell}_1)\ , \qquad (37)$$

hence $\underset{\sim}{\ell}$ has been determined in terms of the constituent phase properties and the mechanical concentration factors $\underset{\sim}{A}_2$ and $\underset{\sim}{A}$.

Equation (37) can be further simplified. To this end, recall that the effective stiffness $\underset{\sim}{L}$ of the composite is obtained by subjecting the external surface S to $(4)_1$ and using the fact that

$$\underset{\sim}{\epsilon} = c_1\underset{\sim}{\epsilon}_1 + c_2\underset{\sim}{\epsilon}_2 - c_2\underset{\sim}{J} = \underset{\sim}{\epsilon}_0\ . \qquad (38)$$

After some manipulations this leads to (Benveniste (1985)):

$$\underset{\sim}{L} = \underset{\sim}{L}_1 + c_2\,(\underset{\sim}{L}_2 - \underset{\sim}{L}_1)\,\underset{\sim}{A}_2 + c_2\underset{\sim}{L}_1\underset{\sim}{A}\ . \qquad (39)$$

Solving for $\underset{\sim}{A}_2$ in (39), and substituting into (37), we obtain:

$$\underset{\sim}{\ell} = \underset{\sim}{\ell}_1 + (\underset{\sim}{L} - \underset{\sim}{L}_1)(\underset{\sim}{L}_2 - \underset{\sim}{L}_1)^{-1}(\underset{\sim}{\ell}_2 - \underset{\sim}{\ell}_1) \qquad (40)$$

Equation (40), interestingly enough, is the same as equation (3.11) in Benveniste and Dvorak (1989). Note however that imperfect bonding at S_{12} as described in equation (2) still affects the effective thermal tensor $\underset{\sim}{\ell}$, since $\underset{\sim}{L}$ itself is affected, as in (39).

The determination of $\underset{\sim}{m}$ follows similar steps, this time under the stress boundary conditions (9). It leads to a set of equations which are counterparts to (29), (31) and (32):

$$\underset{\sim}{\sigma} = c_1\underset{\sim}{\sigma}_1 + c_2\underset{\sim}{\sigma}_2 = 0\ , \qquad (41)$$

$$\underset{\sim}{\epsilon} = c_1\underset{\sim}{\epsilon}_1 + c_2\underset{\sim}{\epsilon}_2 - c_2\underset{\sim}{J} = c_1(\underset{\sim}{M}_1\underset{\sim}{\sigma}_1 + \underset{\sim}{m}_1\theta_0) + \\ + c_2(\underset{\sim}{M}_2\underset{\sim}{\sigma}_2 + \underset{\sim}{m}_2\theta_0) = \underset{\sim}{m}\theta_0\ , \qquad (42)$$

$$\underset{\sim}{m} = c_1\underset{\sim}{m}_1 + c_2\underset{\sim}{m}_2 + c_2(\underset{\sim}{M}_2 - \underset{\sim}{M}_1)\underset{\sim}{b}_2 - c_2\underset{\sim}{b} \quad , \tag{43}$$

where we have defined the concentration factors $\underset{\sim}{b}_2$ and $\underset{\sim}{b}$ as follows:

$$\underset{\sim}{\sigma}_2 = \underset{\sim}{b}_2\theta_0. \qquad \underset{\sim}{J} = \underset{\sim}{b}\theta_0 \quad . \tag{44}$$

In analogy to (34) and (35), these tensors can be written in the form:

$$\underset{\sim}{b}_2 = (\underset{\sim}{I} - \underset{\sim}{B}_2)\,(\underset{\sim}{M}_1 - \underset{\sim}{M}_2)^{-1}\,(\underset{\sim}{m}_2 - \underset{\sim}{m}_1) \quad , \tag{45}$$

$$\underset{\sim}{b} = -\,\underset{\sim}{B}\,(\underset{\sim}{M}_1 - \underset{\sim}{M}_2)^{-1}\,(\underset{\sim}{m}_2 - \underset{\sim}{m}_1) \quad , \tag{46}$$

with

$$B_{ijkl} = \frac{1}{2V}\int_{S_{12}} (F_{ikl}(x)p_j + F_{jkl}(x)p_i)\,dS_{12} \quad . \tag{47}$$

The equation for $\underset{\sim}{m}$, finally becomes:

$$\begin{aligned}\underset{\sim}{m} = {} & c_1\underset{\sim}{m}_1 + c_2\underset{\sim}{m}_2 + \\ & + c_2(\underset{\sim}{M}_2 - \underset{\sim}{M}_1)(\underset{\sim}{I} - \underset{\sim}{B}_2)(\underset{\sim}{M}_1 - \underset{\sim}{M}_2)^{-1}(\underset{\sim}{m}_2 - \underset{\sim}{m}_1) \\ & + c_2\underset{\sim}{B}(\underset{\sim}{M}_1 - \underset{\sim}{M}_2)^{-1}(\underset{\sim}{m}_2 - \underset{\sim}{m}_1) \quad .\end{aligned} \tag{48}$$

The expression for the effective compliance tensor (Benveniste (1985)),

$$\underset{\sim}{M} = \underset{\sim}{M}_1 + c_2\,(\underset{\sim}{M}_2 - \underset{\sim}{M}_1)\underset{\sim}{B}_2 - c_2\,\underset{\sim}{B} \quad , \tag{49}$$

helps to reduce equation (48) to the form

$$\underset{\sim}{m} = \underset{\sim}{m}_1 + (\underset{\sim}{M} - \underset{\sim}{M}_1)\,(\underset{\sim}{M}_2 - \underset{\sim}{M}_1)^{-1}\,(\underset{\sim}{m}_2 - \underset{\sim}{m}_1) \quad , \tag{50}$$

which is the counterpart of (40). Using $\underset{\sim}{\ell}_r = -\,\underset{\sim}{L}_r\,\underset{\sim}{m}_r$ and the fact that $\underset{\sim}{L} = \underset{\sim}{M}^{-1}$, one can verify that $\underset{\sim}{\ell}$ and $\underset{\sim}{m}$ as given by (40) and (50) fulfill the relation $\underset{\sim}{\ell} = -\underset{\sim}{L}\underset{\sim}{m}$.

We have utilized here the correspondence relations established in the previous section to derive expressions for the effective thermal tensors $\underset{\sim}{\ell}$ and $\underset{\sim}{m}$ in terms of effective mechanical properties of the composite. The correspondence relations are limited to isotropic constituents, and therefore, the derived equations (40) and (50) apply also to such systems only.

Expressions for the effective thermal tensors in terms of effective mechanical properties have been given before in the literature for the case of composites with perfectly bonded anisotropic phases. The basic idea was due to Levin (1967) which used the principle of virtual work to this end. Levin's paper was extended to anisotropic constituents by Rosen and Hashin (1970), see also Laws (1973) and Schulgasser (1989) for an alternative derivation of these relations. We will show in the next section that the virtual work theorem can again lead to equations similar to (40) and (50) for the case of multiphase materials with anisotropic phases and imperfect interfaces of the type described in (2). It should be of course made clear that in spite of its limitation to isotropic constituents in the present case, the decomposition scheme is in a sense more general than the results provided by the virtual work theorem since it provides results *on fields* and not only on average properties.

2. Effective Thermal and Stress Tensors in Multiphase Composites with Anisotropic Constituents and Interfaces Weak In Shear

We consider now multiphase composites described by (1) and (2), but allow this time for general anisotropic behavior in for the phases. As in Section 1, different parts of the interfaces may possess different values of $0 \leq R \leq \infty$ and $0 \leq Q \leq \infty$. An expression for the effective thermal stress tensor $\underset{\sim}{\ell}$ in terms of purely mechanical properties will be first derived by considering the boundary conditions $(4)_1$ and (8). For convenience, we let the fields induced by $(4)_1$ be denoted by primed quantities and those resulting from (8) by unprimed quantities. The principle of virtual work for composites with imperfect interfaces can be found in Benveniste (1985). When applied to the boundary value problems $(4)_1$ and (8), it can be written as:

$$\int_V \sigma_{ij}(\underset{\sim}{x})\epsilon'_{ij}(\underset{\sim}{x})dV$$

$$= \int_S t_i(\underset{\sim}{x})u'_i(\underset{\sim}{x})dS + \sum_{r=2}^{N}\int_{S_{1r}} t_i(\underset{\sim}{x})[u'_i(\underset{\sim}{x})]dS_{1r} \ , \tag{51}$$

where t_i and u_i denote the traction and displacement vector, $r = 1$ stands for the matrix, and S_{1r} denotes the boundaries of the inclusion phases with the matrix.

Substitution of σ_{ij} from (1) into (51) yields:

$$\int_V L_{ijkl}\ \epsilon_{kl}(\underset{\sim}{x})\ \epsilon'_{ij}(\underset{\sim}{x})\ dV + \int_V \ell_{ij}\ \epsilon'_{ij}(\underset{\sim}{x})\ \theta_0 dV$$

$$= \int_S t_i(\underset{\sim}{x})u'_i(\underset{\sim}{x})dS + \sum_{r=2}^{N}\int_{S_{1r}} t_i(\underset{\sim}{x})[u'_i(\underset{\sim}{x})]dS_{1r} \ , \tag{52}$$

with the material properties assuming the index $r = 1, 2, \ldots N$ depending on the position of the point $\underset{\sim}{x}$ in the composite. For the boundary condition $(4)_1$, the first integral on the right hand side of (52) can be simplified as:

$$\int_S t_i(\underset{\sim}{x})u'_i(\underset{\sim}{x})dS = \int_S t_i(\underset{\sim}{x})\overset{0}{\epsilon}_{ij}x_j dS$$

$$= \overset{0}{\epsilon}_{ij}\int_S \sigma_{ik}\ (\underset{\sim}{x})n_k x_j dS$$

$$= \overset{0}{\epsilon}_{ij} \, \sigma_{ij} \, V = \overset{0}{\epsilon}_{ij} \, \ell_{ij} \, \theta_0 V \quad , \tag{53}$$

where, σ_{ij} denotes the overall average stress, and the fact that the average strain vanishes under (8) has been used. Substitution of (53) into (52) gives:

$$\int_V L_{ijkl} \, \epsilon_{kl} \, (\underset{\sim}{x}) \epsilon'_{ij} \, (\underset{\sim}{x}) dV + \int_V \ell_{ij} \, \epsilon'_{ij}(\underset{\sim}{x}) \theta_0 \, dV$$

$$= \sum_{r=2}^{N} \int_{S_{1r}} t_i(\underset{\sim}{x})[u'_i(\underset{\sim}{x})] \, dS_{1r} + \overset{0}{\epsilon}_{ij} \, \ell_{ij} \, \theta_0 \, V \, . \tag{54}$$

The virtual work theorem is now applied to the boundary value problems $(4)_1$ and (8) with the alternative choice of admissible displacement and stress fields; the fields in $(4)_1$ are denoted by primed quantities and those in (8) by unprimed ones. The result is:

$$\int_V \sigma'_{ij} \, (\underset{\sim}{x}) \, \epsilon_{ij}(\underset{\sim}{x}) dV$$

$$= \int_S t'_i(\underset{\sim}{x}) u_i(\underset{\sim}{x}) dS + \sum_{r=2}^{N} \int_{S_{1r}} t'_i(\underset{\sim}{x})[u_i(\underset{\sim}{x})] dS_{1r}, \tag{55}$$

$$\int_V L_{ijkl} \, \epsilon'_{kl}(\underset{\sim}{x}) \, \epsilon_{ij} \, (\underset{\sim}{x}) dV$$

$$= \sum_{r=2}^{N} \int_{S_{1r}} t'_i(\underset{\sim}{x})[u_i(\underset{\sim}{x})] dS_{1r} \quad , \tag{56}$$

where we used the condition $u_i(S) = 0$. Subtraction of (56) from (54) yields

$$\sum_{r=1}^{N} c_r \ell_{ij}^{(r)} \epsilon_{ij}^{(r)} \theta_0$$

$$= \frac{1}{V} \sum_{r=2}^{N} \int_{S_{1r}} \left\{ t_i(\tilde{x})[u_i'(\tilde{x})] - t_i'(\tilde{x})[u_i(\tilde{x})] \right\} dS_{1r}$$

$$+ \ell_{ij} \overset{0}{\epsilon}_{ij} \theta_0 \quad . \tag{57}$$

The integral on the right hand side involves the scalar product between the traction vector in one loading system and the displacement jump vector in the second system. The interface conditions described in (2) make this term vanish. Equation (57) therefore yields:

$$\tilde{\ell} = \sum_{r=1}^{N} c_r \tilde{A}_r^T \tilde{\ell}_r \quad , \tag{58}$$

where we have invoked the definition of the concentration factors $\tilde{A}_r$ and reverted again to the bold face tensorial notation. The transpose sign in (58) denotes:

$$(\tilde{A}_r^T)_{ijkl} = (\tilde{A}_r)_{klij} \quad . \tag{59}$$

It is somewhat surprising to see that equation (58) is the same as Rosen and Hashin's (1970) result for perfect bonding between the phases. Note however that due to interface slip, the tensors $\tilde{A}_r^T$ are not equal to those which would be obtained under perfect bonding conditions. We finally mention that if part of the interfaces at S_{12} contain open cracks,(58) remains valid since the tractions at these boundaries vanish identically if all crack closure effects are neglected.

For the case of binary composites equation (58) can also be written in other equivalent forms with one among them making contact with the results obtained in the previous section. Under $(4)_1$, note that

$$\underset{\sim}{c_1 A_1} + c_2 \underset{\sim}{A}_2 - c_2 \underset{\sim}{A} = \underset{\sim}{I} \quad , \tag{60}$$

where $\underset{\sim}{A}$ was defined in (36). Solving for $c_1 \underset{\sim}{A}_1^T$ in (60) and substituting in (58) provides:

$$\underset{\sim}{\ell} = \underset{\sim}{\ell}_1 + c_2 \underset{\sim}{A}_2^T (\underset{\sim}{\ell}_2 - \underset{\sim}{\ell}_1) + c_2 \underset{\sim}{A}^T \underset{\sim}{\ell}_1 \quad . \tag{61}$$

Another form can be obtained by writing first (39) as:

$$\underset{\sim}{L} = \underset{\sim}{L}_1 + c_2 \underset{\sim}{A}_2^T (\underset{\sim}{L}_2 - \underset{\sim}{L}_1) + c_2 \underset{\sim}{A}^T \underset{\sim}{L}_1 \quad , \tag{62}$$

where the diagonal symmetry of the stiffness tensors has been invoked. Solving for $\underset{\sim}{A}_2^T$ in (62) and substituting into (61) provides:

$$\begin{aligned} \underset{\sim}{\ell} = \underset{\sim}{\ell}_1 &+ (\underset{\sim}{L} - \underset{\sim}{L}_1)(\underset{\sim}{L}_2 - \underset{\sim}{L}_1)^{-1} (\underset{\sim}{\ell}_2 - \underset{\sim}{\ell}_1) + \\ &+ c_2 \underset{\sim}{A}^T \left\{ \underset{\sim}{\ell}_1 - \underset{\sim}{L}_1 (\underset{\sim}{L}_2 - \underset{\sim}{L}_1)^{-1} (\underset{\sim}{\ell}_2 - \underset{\sim}{\ell}_1) \right\} \quad . \end{aligned} \tag{63}$$

Equation (63) is the counterpart of (40) of the previous section for the case of anisotropic constituents. Let us next prove that in the special of isotropic phases the last term in (63) vanishes.

For isotropic phases let,

$$\begin{aligned} (\underset{\sim}{\ell}_1)_{ij} &= \alpha \, \delta_{ij} \quad , \\ (\underset{\sim}{L}_1)_{ijrs} &= \beta \, \delta_{ij} \, \delta_{rs} + \gamma \left(\delta_{ir} \, \delta_{js} + \delta_{is} \, \delta_{jr} - \frac{2}{3} \, \delta_{ij} \, \delta_{rs} \right) \quad , \\ (\underset{\sim}{L}_2 - \underset{\sim}{L}_1)^{-1}_{rsmn} &= \xi \delta_{rs} \delta_{mn} + \zeta \left(\delta_{rm} \delta_{sn} + \delta_{rn} \delta_{sm} - \tfrac{2}{3} \delta_{rs} \delta_{mn} \right) \quad , \\ (\underset{\sim}{\ell}_2 - \underset{\sim}{\ell}_1)_{mn} &= \lambda \, \delta_{mn} \quad , \end{aligned} \tag{64}$$

where α, β, γ, ξ, ζ, λ are constants. Writing $\underset{\sim}{A}^T$ in indicial notation and carrying out the summation in (63) according to

(64) shows that the tensor $\underset{\sim}{A}^T$ enters only as $(\underset{\sim}{A}^T)_{pqii}$. However, since according to the interface conditions (2) the normal displacements are continuous at S_{12} it follows from (36) that $A_{iikl} \equiv 0$ or, in fact, $(\underset{\sim}{A}^T)_{pqii} \equiv 0$. We have therefore shown that for the case of isotropic constituents (63) reduce to (40).

A similar implementation of the virtual work theorem (51) to the boundary value problems $(4)_2$ and (9) yields equations for the thermal strain tensor $\underset{\sim}{m}$. For the sake of brevity we will give only the final results, counterparts to equations (58), (61) and (63). These are:

$$\underset{\sim}{m} = \sum_{r=1}^{N} c_r \underset{\sim}{B}_r^T \underset{\sim}{m}_r \tag{65}$$

$$\underset{\sim}{m} = \underset{\sim}{m}_1 + c_2 \underset{\sim}{B}_2^T(\underset{\sim}{m}_2 - \underset{\sim}{m}_1) \tag{66}$$

$$\underset{\sim}{m} = \underset{\sim}{m}_1 + (\underset{\sim}{M} - \underset{\sim}{M}_1)(\underset{\sim}{M}_2 - \underset{\sim}{M}_1)^{-1}(\underset{\sim}{m}_2 - \underset{\sim}{m}_1) + \\ + c_2 \underset{\sim}{B}^T(\underset{\sim}{M}_2 - \underset{\sim}{M}_1)^{-1}(\underset{\sim}{m}_2 - \underset{\sim}{m}_1) \tag{67}$$

where the last two equations refer to binary systems only. It is noted that the structure of (66) and (67) are not exactly similar to (61) and (63) respectively. This is due to the fact that $\underset{\sim}{\epsilon}$ and $\underset{\sim}{\sigma}$ in (29) and (31) and also (39) and (49) have a different structure. For the same reasons mentioned above equation (67) reduces to (50) for the case of isotropic constituents.

Finally, it is easy to show that $\underset{\sim}{\ell}$ and $\underset{\sim}{m}$, as given by (58) and (65) for example, satisfy $\underset{\sim}{\ell} = -\underset{\sim}{L}\underset{\sim}{m}$. From the definitions of the $\underset{\sim}{A}_r$ and $\underset{\sim}{B}_r$ tensors and also due to $\underset{\sim}{L} = \underset{\sim}{M}^{-1}$, it results that

$$\underset{\sim}{B}_r = \underset{\sim}{L}_r \underset{\sim}{A}_r \underset{\sim}{L}^{-1} \qquad r = 1, \ldots N \tag{68}$$

which provides

$$\underset{\sim}{L}\, \underset{\sim}{B}_r^T = \underset{\sim}{A}_r^T\, \underset{\sim}{L}_r \qquad r = 1, \ldots N \tag{69}$$

Multiplying (65) by $(-\underset{\sim}{L})$ from the left, using (69) and $\underset{\sim}{\ell}_r = -\underset{\sim}{L}_r \underset{\sim}{m}_r$ shows readily that (58) is in fact recovered.

Acknowledgement

Funding for this work was provided, in part, by the ONR/DARPA–HiTASC composites program at Rensselaer, and by the Air Force Office of Scientific Research.

References

Aboudi, J., 1987, "Damage in Composites — Modeling of Imperfect Bonding," *Comp. Sci. and Tech.* *28*, pp. 103–128.

Achenbach, J.D. and Zhu, H., 1989, "Effect of Interfacial Zone on Mechanical Behavior and Failure of Fiber–Reinforced Composites," *J. Mech. Phys. Solids*, 37, pp. 381–393.

Benveniste, Y., 1985, "The Effective Mechanical Behaviour of Composite Materials with Imperfect Contact Between the Constituents," *Mechanics of Materials*, 4, pp. 197–208.

Benveniste, Y. and Aboudi, J., 1984 "A Continuum Model for Fiber Reinforced Materials with Debonding," *Int. J. Solids Structures*, 20, 11/12, pp. 935–951.

Benveniste, Y. and Dvorak, G. J., 1990, "On a Correspondence Between Mechanical and Thermal Effects in Two–Phase Composites," in Toshio Mura Anniversary Volume *Micromechanics and Inhomogeneity*, Editors: Weng, G.J., Taya, M., and Abe, H, Springer Verlag, pp. 65–81.

Benveniste, Y. and Miloh, T., 1986, "The Effective Conductivity of Composites with Imperfect Thermal Contact at Constituent Interfaces," *Int. J. Eng. Sci.* 24, pp. 1537–1552.

Dvorak, G.J., 1983, "Metal Matrix Composites: Plasticity and Fatigue," in *Mechanics of Composite Materials -- Recent Advances*, edited by Z. Hashin and C.T. Herakovich, Pergamon Press, New York, pp. 73—92.

Dvorak, G.J., 1986, "Thermal Expansion of Elastic—Plastic Composite Materials," *J. Appl. Mech.* 53, pp. 737—743.

Dvorak, G.J. and Chen, T., 1989, "Thermal Expansion of Three—Phase Composite Materials" *J. Appl. Mech.* 56, pp. 418—422.

Dvorak, G.J., 1990, "On Uniform Fields in Heterogeneous Media," to appear in *Proceedings* A *of the Royal Society.*

Hashin, Z., 1990, "Thermoelastic Properties of Fiber Composites with Imperfect Interface," *Mechanics of Materials* 8, pp. 333—348.

Jasiuk, I., Mura, T., and Tsuchida, E., 1988, "Thermal Stresses and Thermal Expansion Coefficients of Short Fiber Composites with Sliding Interfaces," *J. of Engng. Mat. and Tech.*, 110, pp. 96—100.

Jasiuk, I. and Tong, Y., 1989, "The Effect of Interface on the Elastic Stiffness of Composites," in *Mechanics of Composite Materials and Structures*, edited by J.N. Reddy and J.L. Teply, AMD—Vol. 100, pp. 49—54.

Laws, N., 1973, "On the Thermostatics of Composite Materials," *J. Mech. Phys. Solids* 21, pp. 9—17.

Lené, F. and Leguillon, D., 1982, "Homogeneized Constitutive Law for a Partially Cohesive Composite Material," *Int. J. Solids Structures* 18, pp. 443—458.

Levin, V.M., 1967, "Thermal Expansion Coefficients of Heterogeneous Materials", *Mekhanika Tverdago Tela*, 2, pp. 88 — 94. English Translation *Mechanics of Solids*, 11, pp. 58 — 61.

Mura, T., Jasiuk, I. and Tsuchida, E. 1985, "The Stress Field of a Sliding Inclusion," *Int. J. Solids and Struct.*, 12, pp. 1165—1179.

Rosen, B.W. and Hashin Z., 1970, "Effective Thermal Expansion Coefficients and Specific Heats of Composite Materials", *Int. J. Engng. Sci.* 8, pp. 157—173.

Schulgasser, K., 1989, "Environmentally—Induced Expansion of Heterogeneous Media," *J. Appl. Mech.* 56, pp. 546—549.

Tsuchida, E., Mura, T. and Dundurs, J., 1986, "The Elastic Field of an Elliptic Inclusion with a Slipping Interface," *J. Appl. Mech.* 53, pp. 103—107.

Micromechanical Modelling of Fibre Debonding in a Metal Reinforced by Short Fibres

Viggo Tvergaard

Department of Solid Mechanics

The Technical University of Denmark, Lyngby, Denmark

ABSTRACT — Failure of a whisker–reinforced metal–matrix composite by decohesion of the fibre–matrix interface is analysed numerically. A cohesive zone model that accounts for decohesion by normal separation as well as by tangential separation is used to model failure at the interface. The composite material containing a periodic array of aligned fibres is represented in terms of a unit cell model analysis. The effect of varying fibre aspect ratio and varying fibre volume fraction is investigated, and the sensitivity of the predictions to mesh refinements is studied.

1. INTRODUCTION

When aluminium alloys are reinforced by SiC–whiskers, in order to improve the tensile properties, the ductility and the fracture toughness are simultaneously reduced, due to early void formation by debonding at the matrix–fibre interface (Divecha *et al.*, 1981; McDanels, 1985). A number of investigations of this debonding behaviour have been carried out to improve the understanding of the influence of different material parameters.

Needleman (1987) has modeled the debonding of an inclusion from a metal matrix in terms of a potential that specifies the dependence of the interface tractions on the interfacial separation. The formulation was used by Nutt and Needleman (1987) to study the onset of failure by decohesion at the fibre ends, and good qualitative agreement was found between the theoretical predictions of initial void shapes and experimental observations. Further analyses of the effect of fibre volume fraction and fibre spacing on debonding predictions were carried out by Needleman and Nutt (1989). In a more recent paper Tvergaard (1989) has proposed an alternative cohesive zone model that describes decohesion by purely tangential separation as well as the decohesion by normal separation considered previously (Needleman, 1987). In addition to the failure initiation by void formation at the flat fibre ends this alternative cohesive zone model makes it possible to also study the continued failure process by fibre pull–out.

The assumption will be made here that all fibres are parallel to the principal tensile direction. This is realistic, since processing based on extrusion can lead to almost perfectly aligned fibres (Nieh, 1984; German and Bose, 1989). The same assumption has been used in a number of analyses of the tensile properties of perfectly bonded composites. Teply and Dvorak (1988) have applied minimum principles of plasticity to derive upper and lower bounds on instantaneous stiffnesses for various periodic models of fibrous and particulate composites. Christman *et al.* (1989a) have used an axisymmetric unit cell model, representing a regular array of end–to–end fibres, to account for the cylindrical whisker shape with sharp 90–degree corners at the ends that is usually observed experimentally. An alternative model developed by Tvergaard (1990) represents fibres that are somewhat shifted relative to one another, both in the axial and transverse directions. The shifted fibres result in significant plastic shearing of the matrix material between adjacent fibre ends, and the predictions of this alternative model are in reasonable agreement with experimental uniaxial stress–strain curves found by Christman *et al.* (1989a) for a 2124 Al–SiC whisker reinforced composite. The effect of shifted fibres and of fibre clustering have also been studied in terms of a plane strain model by Christman *et al.* (1989b).

The present paper gives an extension of the previous fibre–matrix debonding study (Tvergaard, 1989) to consider the effect of varying fibre volume fraction and varying fibre aspect ratio. Furthermore, the sensitivity of the debonding predictions to refinements of the mesh used for the numerical solution is studied in some detail.

2. DEBONDING MODEL

Debonding of an inclusion from a metal matrix has been modeled by Needleman (1987) in terms of an interface potential that specifies the dependence of the tractions T_n and T_t on the normal and tangential components, u_n and u_t, of the displacement difference across the interface. A positive u_n corresponds to increasing interfacial separation. These interface constitutive relations specify the nonlinear variation of the normal traction T_n from the value 0 at $u_n = 0$ through a maximum value σ_{max} and again down to the value 0 at $u_n = \delta_n$, where final separation is assumed to occur. This debonding model describes only debonding by normal separation; but during fiber pull–out under significant normal compression the positive normal separation u_n required for debonding in Needleman's model will not occur. Therefore, an alternative debonding model was proposed by Tvergaard (1989), which coincides with that of Needleman (1987) for an interface undergoing purely normal separation $(u_t \equiv 0)$. No potential exists in general for this alternative debonding model.

A nondimensional parameter λ is defined as

$$\lambda = \sqrt{\left[\frac{u_n}{\delta_n}\right]^2 + \left[\frac{u_t}{\delta_t}\right]^2} \tag{2.1}$$

and a function $F(\lambda)$ is chosen as

$$F(\lambda) = \frac{27}{4}\,\sigma_{max}(1-2\lambda+\lambda^2)\ ,\ \text{for}\ \ 0 \le \lambda \le 1 \tag{2.2}$$

Then, as long as λ is monotonically increasing, the interface tractions are taken to be given by the expressions

$$T_n = \frac{u_n}{\delta_n}\,F(\lambda)\ \ ,\ \ T_t = \alpha\,\frac{u_t}{\delta_t}\,F(\lambda) \tag{2.3}$$

In purely normal separation $(u_t \equiv 0)$ the maximum traction is σ_{max}, total separation occurs at $u_n = \delta_n$, and the work of separation per unit interface area is $9\sigma_{max}\delta_n/16$. In purely tangential separation $(u_n \equiv 0)$ the maximum traction is $\alpha\sigma_{max}$, total separation occurs at $u_t = \delta_t$, and the work of separation per unit interface area is $9\alpha\sigma_{max}\delta_t/16$. For a given interface the values of the four parameters δ_n, δ_t, σ_{max} and α will have to be chosen such that the maximum traction and the work of separation in different situations are reasonably well approximated.

The expression (2.3a) for T_n resembles the dependence of interatomic forces on interatomic separation; but in the present paper the cohesive zone formulation is viewed as a phenomenological model, which represents the average effect of debonding mechanisms on a somewhat larger length–scale than atomic. These mechanisms include the effect of small flaws or patches of poor bonding, or void formation in the matrix material near the interface as observed experimentally by Christman *et al.* (1989a). As has been discussed by Tvergaard (1989), it is expected that the behaviour due to plastic failure mechanisms at a metal/ceramic interface is analogous to the behaviour for ceramic/ceramic interface cracks, where the critical energy release rate has been found to be larger in mode II conditions than in mode I conditions. The present debonding model represents this in terms of the factor α for the strength increase in purely tangential separation and the factor $\alpha\delta_t/\delta_n$ for the increase of the work of separation.

The incremental expressions for the cohesive zone tractions are obtained from (2.3) as

$$\left.\begin{aligned} \dot{T}_n &= \frac{\dot{u}_n}{\delta_n} F(\lambda) + \frac{u_n}{\delta_n}\frac{\partial F}{\partial \lambda}\dot{\lambda} \\ \dot{T}_t &= \alpha\frac{\dot{u}_t}{\delta_t} F(\lambda) + \alpha\frac{u_t}{\delta_t}\frac{\partial F}{\partial \lambda}\dot{\lambda} \end{aligned}\right\} \text{ for } \lambda = \lambda_{max} < 1 \text{ and } \dot{\lambda} \geq 0 \tag{2.4}$$

where

$$\frac{\partial F}{\partial \lambda} = \frac{27}{2}\sigma_{max}(-1+\lambda) \ , \ \dot{\lambda} = \frac{1}{\lambda}\left[\frac{u_n}{\delta_n}\frac{\dot{u}_n}{\delta_n} + \frac{u_t}{\delta_t}\frac{\dot{u}_t}{\delta_t}\right] \tag{2.5}$$

For decreasing λ a type of elastic unloading is used to represent the partly damaged interface

$$T_n = \frac{u_n}{\delta_n} F(\lambda_{max}) \ , \ T_t = \alpha\frac{u_t}{\delta_t} F(\lambda_{max}) \ , \text{ for } \lambda < \lambda_{max} \text{ or } \dot{\lambda} < 0 \tag{2.6}$$

Finally, under normal compression elastic springs with high stiffness are used to approximately represent contact (instead of (2.4a) or (2.6a)), thus taking

$$T_n = \frac{27}{4}\sigma_{max}\frac{u_n}{\delta_n} \ , \ \lambda = \left|\frac{u_t}{\delta_t}\right| \ , \text{ for } u_n < 0 \tag{2.7}$$

Friction between fibre and matrix after the occurrence of debonding is often an important effect in fibre pull–out problems. Such friction is readily incorporated in the present formulation, as has been shown by Tvergaard (1989) for the case of Coulomb friction. However, since the influence of friction found by Tvergaard (1989) was rather small, this effect is not included in the present study.

3. CELL MODEL ANALYSIS

A convected coordinate, Lagrangian formulation of the field equations is used for the analysis, in which g_{ij} and G_{ij} are metric tensors in the reference configuration and the current configuration, respectively, with determinants g and G, and $\eta_{ij} = \frac{1}{2}(G_{ij} - g_{ij})$ is the Lagrangian strain tensor. The contravariant components τ^{ij} of the Kirchhoff stress tensor on the current base vectors are related to the components of the Cauchy stress tensor σ^{ij} by

$$\tau^{ij} = \sqrt{G/g}\,\sigma^{ij} \tag{3.1}$$

The matrix material deformations are taken to be described by a finite strain generalization of J_2 flow theory, in which the incremental stress–strain relationship is of the form $\dot{\tau}^{ij} = L^{ijkl}\,\dot{\eta}_{kl}$, with the tensor of instantaneous moduli given by

$$L^{ijkl} = \frac{E}{1+\nu}\left\{\frac{1}{2}\left(G^{ik}G^{jl} + G^{il}G^{jk}\right) + \frac{\nu}{1-2\nu}G^{ij}G^{kl} - \beta\frac{3/2(E/E_t-1)}{E/E_t - (1-2\nu)/3}\frac{s^{ij}s^{kl}}{\sigma_e^2}\right\} - \frac{1}{2}\left(G^{ik}\tau^{jl} + G^{jk}\tau^{il} + G^{il}\tau^{jk} + G^{jl}\tau^{ik}\right) \tag{3.2}$$

Here, the effective Mises stress is $\sigma_e = (3s_{ij}s^{ij}/2)^{\frac{1}{2}}$, $s^{ij} = \tau^{ij} - G^{ij}\tau_k^k/3$ is the stress deviator, and the value of β is 1 or 0 for plastic yielding or elastic unloading, respectively. Furthermore, E is Young's modulus, ν is Poisson's ratio, and E_t is the slope of the true stress vs. natural strain curve at the stress level σ_e. The uniaxial stress–strain behaviour is represented by

$$\epsilon = \begin{cases} \dfrac{\sigma}{E} & , \text{ for } \sigma \le \sigma_y \\[2ex] \dfrac{\sigma_y}{E}\left[\dfrac{\sigma}{\sigma_y}\right]^n & , \text{ for } \sigma > \sigma_y \end{cases} \tag{3.3}$$

where σ_y is the uniaxial yield stress, and n is the strain hardening exponent.

An axisymmetric cell model analysis is used to approximately represent a material with a periodic array of aligned fibres as that shown in Fig. 1 (Tvergaard, 1989). A cross–section perpendicular to the fibres (Fig. 1b) shows a square array of fibres with spacing $2a_c$, and the initial radius $r_c = (2/\sqrt{\pi})a_c$ of the axisymmetric model

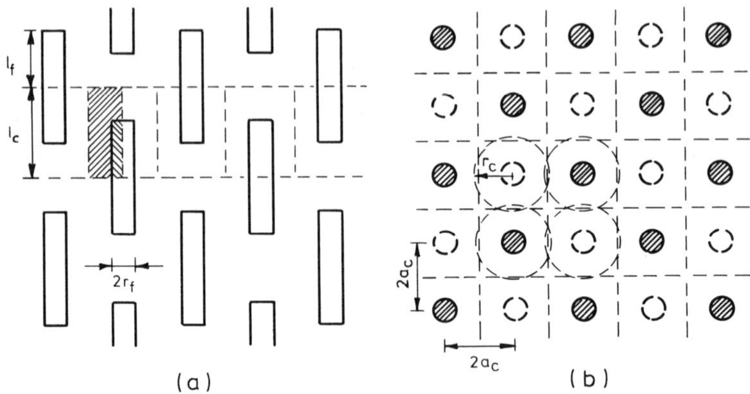

Fig. 1. Periodic array of aligned fibres. (a) Cross–section along fibres. (b) Cross-section normal to fibres.

problem is chosen such that the fibre volume fraction of the cell is equal to that of the material illustrated in Fig. 1. With the initial cell length ℓ_c, and the fibre geometry specified by the initial half length ℓ_f and radius r_f, the fibre volume fraction f is

$$f = \frac{r_f^2}{r_c^2}\frac{\ell_f}{\ell_c} \tag{3.4}$$

The initial fibre aspect ratio α_f and cell aspect ratio α_c, respectively, are specified by

$$\alpha_f = \ell_f/r_f \quad , \quad \alpha_c = \ell_c/r_c \tag{3.5}$$

On the curved side of the circular cylindrical cell equilibrium and compatibility with the neighbouring cells has to be represented. A neighbouring cell is identical to that analysed, but is rotated 180 degrees so that it points in the opposite direction (see Fig. 1a). Compatibility and equilibrium in the axial direction are directly specified in terms of the axial edge displacements and nominal tractions. In the radial direction compatibility is represented by the requirement that the total cross–sectional area (consisting of an equal number of cross–sections of the two types of neighbouring cells considered) is independent of the axial coordinate. The detailed formulation of the boundary conditions is not repeated here (see Tvergaard, 1989, 1990).

The average straining of the material is determined from the average displacement gradients F_{ij} , referring to a Cartesian frame, which are calculated as

$$F_{ij} = \delta_{ij} + \frac{1}{V}\int_S u_i n_j dS \tag{3.6}$$

Here, δ_{ij} is the Kronecker delta, V and S are the volume and surface, respectively, in the reference frame, and u_i and n_j are the Cartesian components of the displacements and the outward unit normal. The average nominal stresses Σ_{ij} are computed as the appropriate area averages of the microscopic nominal stress components on the surface (considering both the cell analysed and one of the neighbouring cells of opposite kind). The axial stress and gradient components are Σ_{11} and F_{11} , the transverse Cartesian components are $\Sigma_{22} = \Sigma_{33}$ and $F_{22} = F_{33}$ for the axisymmetric problem, and all shear components vanish. The average true stresses σ_1 and σ_2 and logarithmic strains ϵ_1 and ϵ_2 in axial and transverse direction are calculated from these values.

The fibres are approximated as rigid, to simplify the debonding analysis. It has been found for perfectly bonded whisker reinforced composites (Tvergaard, 1990) that predictions for elastic or rigid fibres differ rather little when plastic yielding has occurred.

Numerical solutions are obtained by a finite element approximation, using a linear incremental method based on the incremental principle of virtual work (described in more detail by Tvergaard, 1989). The elements employed are quadrilaterals each built up of four triangular, axisymmetric, linear displacement elements. The meshes used for the computations are somewhat finer than those used by Tvergaard (1989), and an example is shown in Fig. 2. A special Rayleigh Ritz – finite element method is applied to implement the boundary conditions and to enforce a fixed ratio $\rho = \sigma_2/\sigma_1$ of the average true stresses in the transverse and axial directions, respectively.

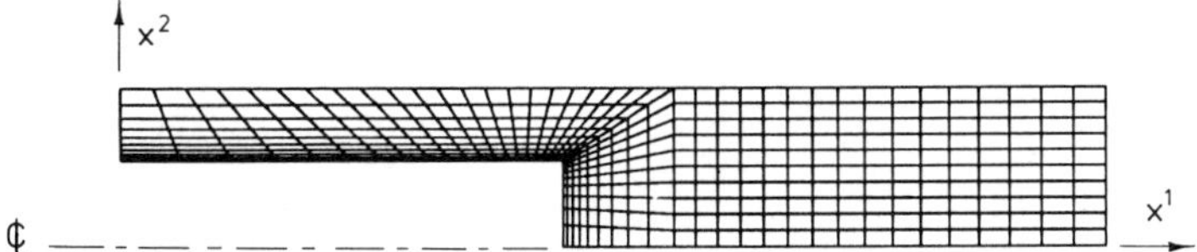

Fig. 2. Mesh with 1792 elements used for some of the numerical analyses.

4. RESULTS

The 2124 Al–SiC whisker reinforced composite investigated experimentally by Christman *et al.* (1989a) had the fibre volume fraction $f = 0.13$ and the average fibre aspect ratio $\alpha_f \simeq 5$. Furthermore, the cell aspect ratio $\alpha_c \simeq 6$ gives a reasonable representation of the observed average fibre spacings in the axial and transverse directions. The uniaxial stress–strain curve for this matrix material can be approximated by the power law (3.3) with $\sigma_y/E = 0.005$ and $n = 7.66$. In the analyses to be presented in the following these material parameters will be used, and the influence of varying the fibre volume fraction or the fibre aspect ratio will be investigated. In all cases uniaxial tension in the fibre direction is considered $(\sigma_2 = 0)$.

In the first cases analysed the fibre geometry and spacing is taken to be specified by $f = 0.13$, $\alpha_f = 5$ and $\alpha_c = 6$. Different sets of parameters in the debonding model (2.1)–(2.7) are investigated in Fig. 3. For $\sigma_{max} = 5\,\sigma_y$, $\delta_n = \delta_t = 0.02\,r_f$ and $\alpha = 1$ Fig. 3 shows rather good agreement between predictions obtained previously (Tvergaard, 1989) and those obtained here by the finer mesh of Fig. 2 (1792 triangular elements instead of 768 elements). Such good agreement has also been found in other cases; but not for $\alpha = 4$, where the finer mesh gives significantly earlier debonding (see Fig. 3). No tangential debonding at all is predicted for $\alpha =$

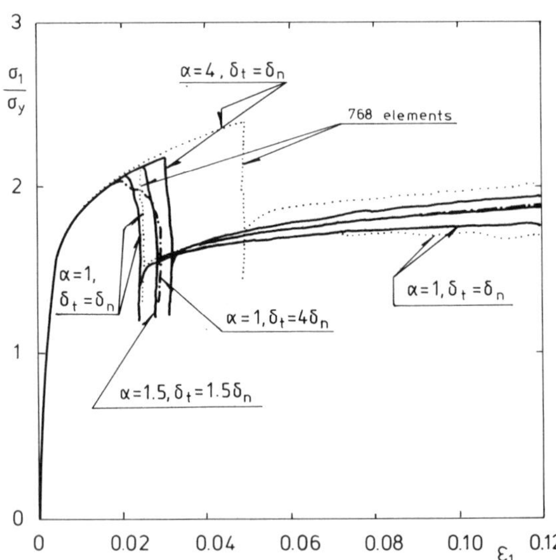

Fig. 3. Stress–strain curves predicted for $\sigma_{max} = 5\,\sigma_y$, $\delta_n = 0.02\,r_f$, $\alpha_f = 5$ and $f = 0.13$.

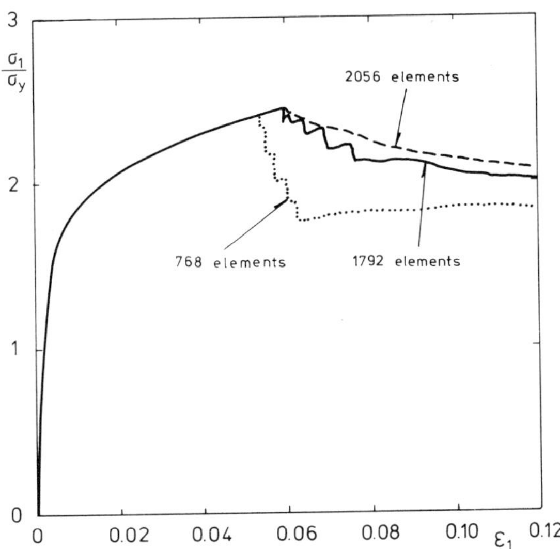

Fig. 4. Stress–strain curves for $\sigma_{max} = 7\ \sigma_y$, $\delta_n = \delta_t = 0.02\ r_r$, $\alpha = 1$, $\alpha_f = 5$ and $f = 0.13$.

4 , but the mesh in Fig. 2 is significantly refined near the sharp fibre edge, and the corresponding better resolution of the stress peak near the edge results in earlier debonding. Fig. 3 also shows that taking either $\alpha = 1$ and $\delta_t = 4\ \delta_n$ or $\alpha = 1.5$ and $\delta_t = 1.5$ gives debonding at a value of the overall strain ϵ_1 between those found for the first two cases considered. For $\alpha = 1$ and $\delta_t = 4\ \delta_n$ the work of purely tangential separation is identical to that for $\alpha = 4$ and $\delta_t = \delta_n$; but the simultaneous increase of the peak stress in the latter case has the effect of eliminating fibre pull–out.

A significant effect of the mesh refinement has also been found for $\sigma_{max} = 7\ \sigma_y$, $\delta_n = \delta_t = 0.02\ r_f$ and $\alpha = 1$, as shown in Fig. 4. Initial debonding occurs on the flat fibre end near the sharp edge, and subsequently this toroidal void grows large as illustrated in Fig. 5, without leading to complete debonding at the fibre end. By contrast, the cruder mesh computation (768 elements) predicts complete debonding at the fibre end with a corresponding steeper load drop and a lower load level during the subsequent fibre pull–out. To check this further, the computation has been repeated with a mesh specially designed to give a fine resolution at the fibre end (2056 elements). This special mesh has 28 elements of equal length along the flat fibre end. Due to the further mesh refinement, the decaying part of the stress-strain curve (during debonding) is less jumpy; but otherwise the predictions obtain-

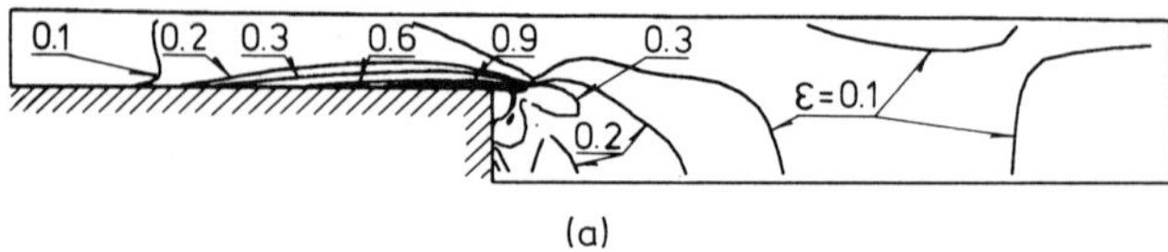

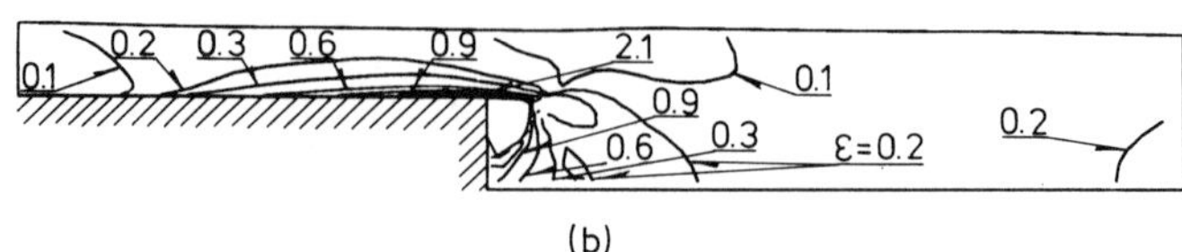

Fig. 5. Debonding behaviour and contours of maximum principal logarithmic strain for $\sigma_{max} = 7\ \sigma_y$, $\delta_n = \delta_t = 0.02\ r_f$, $\alpha = 1$, $\alpha_f = 5$ and $f = 0.13$; for 1792 elements. (a) $\epsilon_1 = 0.078$. (b) $\epsilon_1 = 0.112$.

ed by the two finer meshes are in good agreement. It is noted that the development of a small void near the sharp fibre edge, corresponding to the initial stages of that shown in Fig. 5, has also been found by Nutt and Needleman (1987).

The effect of varying the fibre aspect ratio α_f is investigated Fig. 6. Here, $\sigma_{max} = 6\ \sigma_y$ and $\delta_n = 0.02\ r_f$ are assumed, while a larger resistance to tangential separation than normal separation at the interface is modelled by taking $\alpha = 1.5$ and $\delta_t = 1.5\ \delta_n$. For $f = 0.13$ the three values 2.5 , 5 and 10 of the aspect ratio α_f

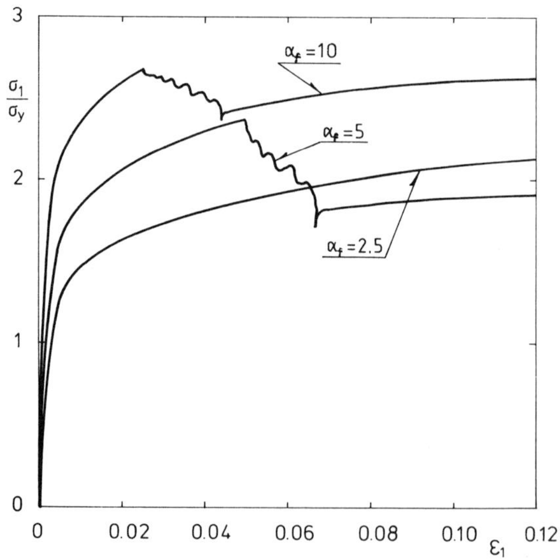

Fig. 6. Stress–strain curves for $\sigma_{max} = 6\ \sigma_y$, $\delta_n = 0.02\ r_f$, $\delta_t = 1.5\ \delta_n$, $\alpha = 1.5$ and $f = 0.13$.

have been analysed that were also considered for perfectly bonded composites by Tvergaard (1990). Since the stress level in the material increases with increasing α_f, it was expected (Tvergaard, 1990) that the ductility would decrease with increasing value of α_f. In agreement with this Fig. 6 shows that the strain at which debonding starts for $\alpha_f = 10$ is less than half that for $\alpha_f = 5$, and for $\alpha_f = 2.5$ no debonding is found at all in the range considered. Both for $\alpha_f = 5$ and $\alpha_f = 10$ debonding starts at the sharp fibre edge and gradually spreads all over the flat end. At $\epsilon_1 = 0.12$ the length of the tangential debonding region on the cylindrical fibre surfaces, measured from the sharp edge, is $0.67\ r_f$ for $\alpha_f = 10$ and $0.13\ r_f$ for $\alpha_f = 5$.

The effect of varying the fibre volume fraction, for fixed fibre aspect ratio $\alpha_f = 5$, is studied in Fig. 7. The parameter values used for the debonding model are identical to those considered in Fig. 6, and the fibre volume fractions analysed are 0.25, 0.13 and 0.08. For comparison, the uniaxial stress–strain curve of the matrix material $(f = 0)$ is included in the figure. Here, the relatively late onset of debonding for $f = 0.25$ is in contradiction to the expected reduction of ductility

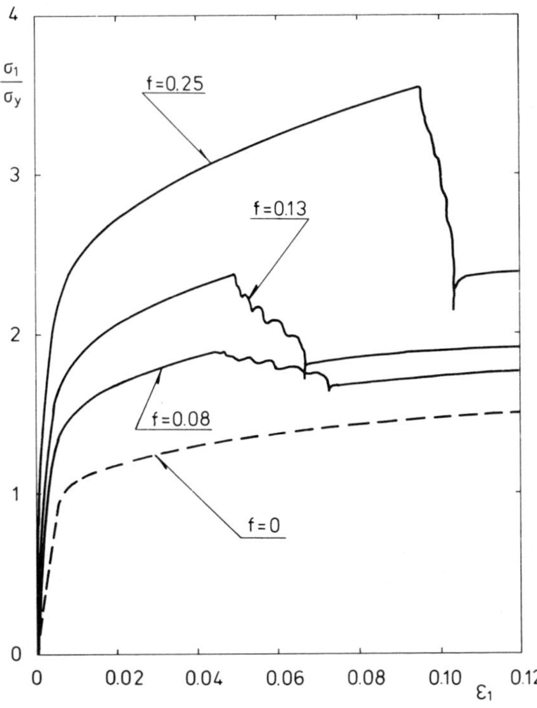

Fig. 7. Stress–strain curves for $\sigma_{max} = 6\ \sigma_y$, $\delta_n = 0.02\ r_f$, $\delta_t = 1.5\ \delta_n$, $\alpha = 1.5$ and $\alpha_f = 5$.

associated with the increased stress level in the material. This is due to interaction between neighbouring fibres in the particular periodic arrangement of whiskers analysed here (Fig. 1). For $f = 0.25$ the neighbouring fibres overlap, and the distance between them is quite small, so that high shear stresses in a small overlap region of the cylindrical fibre surfaces transmit a significant part of the axial load. In fact, for $f = 0.25$ the first debonding is tangential, in this highly stressed overlap–region, even though a higher resistance to tangential debonding has been assumed by taking $\alpha = 1.5$ and $\delta_t = 1.5\ \delta_n$. The redistribution of stresses following this tangential debonding leads immediately after to normal debonding at the fibre end and the associated abrupt reduction of the average tensile stress shown in Fig. 7 (at $\epsilon_1 = 0.095$) .

For a regular array of end to end fibres Needleman and Nutt (1989) do find earlier debonding when f is increased, in agreement with the experimentally observed reduction of ductility for increasing fibre volume fraction (e.g. see McDanels, 1985). This regular array does not allow for shielding of fibre ends due to whisker overlap. In a real material the whiskers are more or less randomly distributed so that some fibre ends are shielded by overlap, while others are not, and clearly void formation by debonding will tend to start at unshielded fibre ends. It has been found (Tvergaard, 1990) that the periodic fibre arrangement in Fig. 1 gives a good representation of the overall stress strain behaviour. However, from the point of view of debonding the present results show the possibility of a strong effect of local interactions. This emphasizes the need for more detailed studies of the effect of different fibre distributions.

REFERENCES

Christman, T., Needleman, A., Nutt, S., and Suresh, S., 1989a, "On Microstructural evolution and Micromechanical Modelling of Deformation of a Whisker-Reinforced Metal–Matrix Composite," *Materials Science and Engineering*, Vol. A107, pp. 49–61.

Christman, T., Needleman, A., and Suresh, S., 1989b, "An Experimental and Numerical Study of Deformation in Metal–Ceramic Composites," Division of Engineering, Brown University.

Divecha, A.P., Fishman, S.G., and Karmarkar, S.D., 1981, "Silicon Carbide Reinforced Aluminum – A Formable Composite," *J. of Metals*, Vol. 33, pp. 12–17.

German, R.M., and Bose, A., 1989, "Fabrication of Intermetallic Matrix Composites," *Materials Science and Engineering*, Vol. A107, pp. 107–116.

McDanels, D.L., 1985, "Analysis of Stress–Strain, Fracture, and Ductility Behaviour of Aluminum Matrix Composites Containing Discontinuous Silicon Carbide Reinforcement," *Metallurgical Transactions*, Vol. 16A, pp. 1105–1115.

Needleman, A., 1987, "A Continuum Model for Void Nucleation by Inclusion Debonding," *J. Appl. Mech.*, Vol. 54, pp. 525–531.

Needleman, A., and Nutt, S.R., 1989, "Void Formation in Short–Fibre Composites," *Advances in Fracture Research*, K. Salama et al., eds., Pergamon Press, pp. 2211–2220.
Nieh, T.G., 1984, "Creep Rupture of a Silicon Carbide Reinforced Aluminum Composite," *Metallurgical Transactions*, Vol. 15A, pp. 139–146.
Nutt, S.R., and Needleman, A., 1987, "Void Nucleation at Fiber Ends in Al–SiC Composites," *Scripta Metallurgica*, Vol. 21, pp. 705–710.
Teply, J.L., and Dvorak, G.J., 1988, "Bounds on Overall Instantaneous Properties of Elastic–Plastic Composites," *J. Mech. Phys. Solids*, Vol. 36, pp. 29–58.
Tvergaard, V., 1989, "Effect of Fibre Debonding in a Whisker–Reinforced Metal," Danish Center for Appl. Math. and Mech., Report No. 400.
Tvergaard, V., 1990, "Analysis of Tensile Properties for a Whisker–Reinforced Metal–Matrix Composite," *Acta Metallurgica* (to appear).

Damage

Fiber Stress Enhancement Due to Initial Matrix Cracking

A. Dollar and P.S. Steif
Mechanical Engineering Department
Carnegie Mellon University
Pittsburgh, PA 15213

ABSTRACT

The tendency for a bridged matrix crack to induce fiber breakage is studied theoretically. We contemplate a composite which has a single matrix crack bridged by all fibers, and which is subjected to tension parallel to the reinforcement. Of interest in judging whether the fibers will fail is the degree to which the fiber stress deviates from its mean value, and the dependence of this deviation on interface parameters. This issue is pursued here for an idealized two-dimensional composite with widely spaced fibers and with a fiber-matrix interface which is governed by Coulomb friction. It is shown that a stronger interface causes a higher stress concentration at the fiber surface thereby raising the likelihood of premature fiber failure.

INTRODUCTION

In ceramic-matrix composites subjected to tensile stresses parallel to the fibers, large matrix cracks are sometimes observed. In fact, a single matrix crack can traverse the entire specimen leaving the fibers intact (Aveston, Cooper and Kelly, 1970). Depending on the composite, the applied load can often be raised to the point that matrix cracks parallel to the first one can appear. The composite can then go on to sustain stresses that are typical of the fiber strengths. On the other hand, if the first matrix crack imposes undue stresses on the fibers, then the fibers may break, causing premature composite failure. Hence, further matrix cracking and, consequently, the ultimate strength are dependent on the state of stress prevailing once a single bridged matrix crack has traversed the entire specimen. From a microstructural point of view, the tendency for fiber breakage to interrupt multiple matrix cracking is widely believed to be dependent upon the characteristics of the fiber-matrix interface. In particular, the higher the degree of bonding, or the greater the resistance to slippage, the more likely it is for the fibers to break prematurely.

The source of the variety of behaviors can be readily appreciated. Regardless of the interface properties, the *average* tensile stress in the fibers at the matrix crack plane is the remote stress divided by the fiber volume fraction. Hence, assuming the fiber properties are held constant, the difference between a composite that sustains multiple matrix cracking and one which suffers premature fiber breakage can be associated with the degree to which the tensile stress in the fiber near the matrix crack plane is *non-uniform*. For example, a composite with a strong interface (in which the fibers break prematurely) will have high stress concentrations where the matrix crack impinges upon the fibers.

A preliminary, highly idealized treatment of this problem is presented here in which the composite is viewed as two-dimensional, the moduli of the fiber and matrix are assumed to be equal, the influence of neighboring fibers is neglected, and the fiber-matrix interface is characterized by a Coulomb friction interface law. When interactions between fibers are neglected, the stresses in the vicinity of the matrix-crack-impinged fiber can be deduced by studying the problem shown in Figure 1. This problem is in some respects similar to one recently treated by the authors (Dollar and Steif, 1989) in which a finite crack impinges on Coulomb friction interfaces. There it was shown that the stress at the crack tip is finite, and that the stress concentration is greater for interfaces that are more resistant to slip. The model problem depicted in Figure 1 will give some insight into the tendency for the interface to control the breakage of fibers after matrix cracking has begun.

ANALYSIS

The two-dimensional problem considered here can be equivalently recast as a half-plane problem, as shown schematically in Figure 2. The fiber, occupying the region $-a < x < a$, is subjected to a uniform normal displacement and zero shear stress; the tensile load transmitted by the fiber is P, which should be viewed as equal to the tension applied to the composite divided by the fiber volume fraction. This problem is similar to that of a half plane upon which is pressed a flat, rigid, frictionless punch, except that the force applied to the punch is tensile instead of compressive. The second, now crucial, difference is that the fiber is *not* perfectly bonded to the matrix material; the particular interface law employed here is the following. Relative motion at the interface is modeled with Coulomb friction, an approach which has been used by the authors in two recent papers (Dollar and Steif, 1988;1989). According to this interface law, each point along the interface is either sticking, slipping, or opening. Specifically, these three states along the interface $x = a$ are described as follows:

$$\text{stick condition} \quad \sigma < 0, |\tau| < \mu\,|\sigma|, \ \frac{dg}{dt} = 0\ ,\ h = \frac{dh}{dt} = 0 \tag{1a}$$

$$\text{slip condition} \quad \sigma < 0, |\tau| = \mu\,|\sigma|, \ \mathrm{sgn}\left(\frac{dg}{dt}\right) = \mathrm{sgn}(\tau)\ ,\ h = \frac{dh}{dt} = 0 \tag{1b}$$

$$\text{open condition} \quad \sigma = \tau = 0\ ,\ h > 0 \tag{1c}$$

with

$$\sigma = \sigma_{xx} \quad \tau = \sigma_{xy}$$

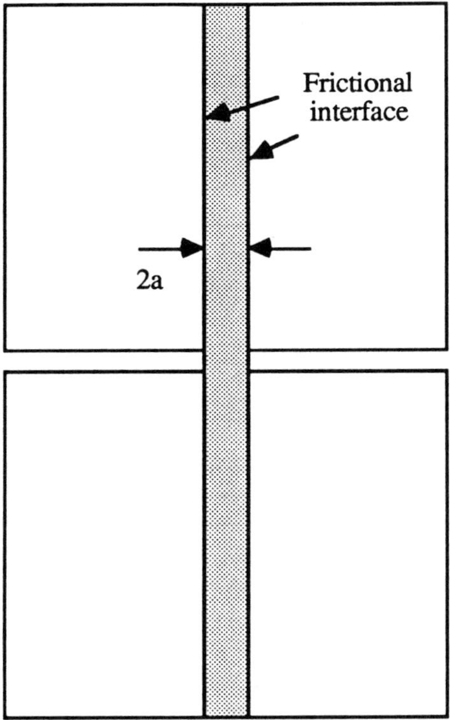

Figure 1. Schematic of a fiber bridging a matrix crack.

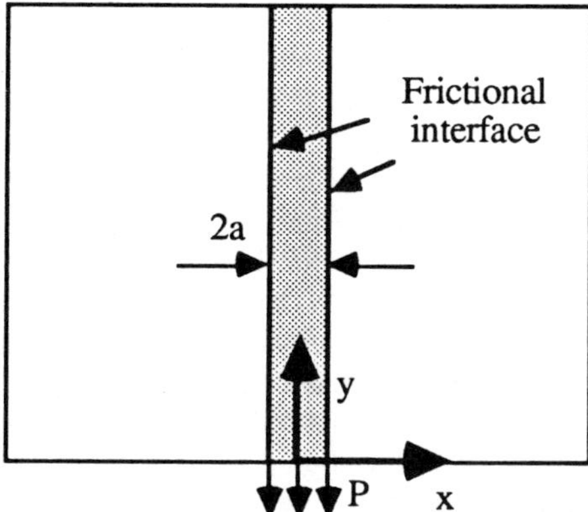

Figure 2. Bridged matrix crack recast as a half-plane problem.

$$g = \lim_{\varepsilon \to 0^+} \left[v(a - \varepsilon, y) - v(a - \varepsilon, y) \right]$$

$$h = \lim_{\varepsilon \to 0^+} \left[v(a - \varepsilon, y) - v(a - \varepsilon, y) \right]$$

In these equations σ_{xx} and σ_{xy} denote the usual Cartesian components of stress, u and v denote the x- and y-components of displacement, respectively, and μ is the friction coefficient. An analogous interface law would hold for $x = -a$. In applying equations (1), one must be careful to use the total stresses, including any residual stresses. In real composites, residual stresses inevitably arise at the interface, often due to differences in thermal expansion coefficient. Here we simulate a residual compression at the interface by first subjecting the body to uniform compression $\sigma_{xx} = - \sigma_o$. Thus, two parameters characterize each point on the interface: the residual stress σ_o and the coefficient of friction μ. In the problem posed below, both quantities will be assumed to be constant along the interfaces. Symmetry of the problem shown in Figure 2 dictates that the extent of slip is the same for both interfaces; however, it is dependent on the load in a manner which comes out of the analysis.

Before we outline the solution method for the case of slippage at the interface, we consider briefly the nature of the fields in the vicinity of the crack tips. The near-tip behavior when a crack impinges upon a slipping interface was discussed in some detail in a previous paper by the authors (Dollar and Steif, 1989); we found that the stress at the crack tip was finite. That is, there are no admissible crack-tip eigenfunctions, satisfying the conditions of traction-free crack faces and frictional slippage at the interface, which exhibit singular stresses. The dominant eigenfunction is one involving piece-wise constant stresses, the only non-zero stress component being the tensile stress σ_{yy} ahead of the crack tip. Our solution method is such that this near-tip behavior can be simulated. (Actually, the additional condition imposed in Dollar and Steif (1989) was that slippage was to occur such that the crack opened; for a closing crack, a singular compressive stress ahead of the crack tip is possible.)

The method of solution follows closely that used in other papers focusing the effects of frictional slippage (Dollar and Steif, 1988, 1989). Slippage at the interface is represented by a continuous distribution of dislocations. The total stresses are the sums of the stresses associated with the perfectly bonded solution (the negative of the flat, rigid, frictionless punch on the half plane) and the stresses associated with the distributed dislocations. One obtains a singular integral equation for the dislocation density by enforcing the friction condition (1b) along the slip zone. The length of the slip zone is unknown and is found as part of the solution.

The terms in the governing equations are conveniently expressed in terms of the Muskhelishvili (1953) complex analytic functions $\phi(z)$ and $\psi(z)$, which are related to the stress and displacements according to

$$\sigma_{xx} + \sigma_{yy} = 2(\phi' + \overline{\phi'}) \tag{2a}$$

$$\sigma_{yy} - \sigma_{xx} + 2i\,\sigma_{xy} = 2(\bar{z}\,\phi'' + \psi') \tag{2b}$$

$$2G(u + iv) = \kappa\phi - z\overline{\phi'} - \overline{\psi} \tag{2c}$$

where $z = x + iy$, $(\ )'$ denotes complex differentiation with respect to z, an overbar denotes complex conjugation, G is the elastic shear modulus, and $\kappa = 3 - 4\nu$ in plane strain.

To use superposition as indicated above, one must have, as the kernel solution, the solution to the problem of a dislocation in an infinite medium which has two semi-infinite, traction-free cracks. The technique to find such a solution is given by Lo (1978) (among others), who used distributed dislocations to represent the kink of a kinked crack. For a dislocation in the y-direction with Burger's vector b_y, one finds this kernel solution to be given by

$$\phi' = \phi'_\infty + \phi'_R \tag{3a}$$

$$\psi' = \psi'_\infty + \psi'_R \tag{3b}$$

where

$$\phi'_\infty(z,z_0) = \frac{\alpha}{z-z_0} \qquad \psi'_\infty = \frac{\alpha}{z-z_0} - \frac{\alpha\,\bar{z}_0}{(z-z_0)^2} \tag{4a,b}$$

$$\phi'_R(z,z_0) = -\alpha\left[F(z,z_0) + F(z,\bar{z}_0) + (z_0-\bar{z}_0)\,H(z,\bar{z}_0)\right] + \frac{c}{X(z)} \tag{5}$$

$$\psi'_R(z,z_0) = \phi'_R(z,z_0) - \phi'_R(z,z_0) - z\phi''_R(z,z_0) \tag{6}$$

$$F(z,z_0) = \frac{1 - \dfrac{X(z_0)}{X(z)}}{2(z-z_0)} \tag{7}$$

$$H(z,z_0) = \frac{\partial F(z,z_0)}{\partial z_0} \tag{8}$$

$$X(z) = \sqrt{z^2-a^2} \tag{9}$$

$$\alpha = \frac{G\,b_y}{\pi\,(k+1)} \tag{10}$$

This solution appears quite similar to the kernel solution for a finite crack (see Lo (1978)). One important difference here is that the branch of the square root $\sqrt{z^2-a^2}$ is the one which has discontinuities along the branch cuts $-\infty < x < -a$ and $a < x < \infty$. Here, the constant c can be evaluated by noting that the function $c/\sqrt{z^2-a^2}$ is the solution to the problem of a flat, rigid, frictionless punch applied to the lower (or upper) half-plane. It is obvious that a solution to the punch problem (with an arbitrary load) may be superposed on the dislocation solution. Here, since we are superposing the distributed dislocations with the punch loading of the perfectly bonded half-plane, the kernel dislocation solution must involve zero net force transmitted across $-a < x < a$. Hence, c must be zero.

The perfectly bonded solution, ϕ'_p, is given by (Mushkelishvili, 1953)

$$\phi'_p = \frac{i P}{2\pi X(z)} \tag{11}$$

where X(z) is defined by (9). As expected, the stresses are square-root singular as the crack tip is approached. Once we allow frictional slip to occur, however, the interface serves to "blunt" the impinging matrix crack.

To formulate an integral equation, we now assume (as was done by Dollar and Steif, 1989) that slip occurs along a single portion of each interface ($x = \pm a$, $-L < y < L$), with the remainder of the interface being in a stick condition. Then, it is necessary to distribute dislocations (with Burger's vector in the y-direction) only along the slipped portions of the interface. The distribution of dislocations is chosen to satisfy the following integral equation which enforces the friction condition $\sigma_{xy} = \pm\mu|\sigma_{xx}|$ along the slip zones:

$$\int_0^L b(y_0)\left[R_0(y,y_0) + R_1(y,y_0) + \mu R_2(y,y_0)\right] dy_0 + f(y) = 0 \tag{12}$$

where b(y) is the dislocation density, and the functions R_0 (the singular part), R_1, R_2 and f are given in the Appendix. To write the equation along only $0 < y < L$, we have taken advantage of the symmetries of the problem. As in Dollar and Steif (1989), the slip is adjusted (for a given P and σ_0) so that the net stress intensity factor at the crack tips ($x = \pm a$, $y = 0$) is zero, and so that the dislocation density vanishes at the ends of the slip zones.

Of the various quantities which may be computed from the solution, the tensile stress immediately ahead of the crack tips is the most important one considered here. This tensile stress, $(\sigma_{yy})_{tip}$, can be obtained in two ways. First, it is readily shown that this stress is related to the dislocation density as one approaches the crack tip from the slip zone according to

$$(\sigma_{yy})_{tip} = \frac{E}{1-\nu^2} b(0) \tag{13}$$

Alternatively, the stress at various points ahead of the crack tip can be computed from the entire distribution of dislocations, followed by an extrapolation to the crack tip. The degree to which these two methods yield the same number is a measure of the accuracy of the numerical solution. Generally, agreement to within a few percent was found.

RESULTS

The principal result here is the tendency for the frictional interface to effectively "blunt" the matrix crack (see Figure 3). Therefore, the stress at the tip, σ_{tip}, normalized by the average fiber stress P/2a, is plotted as a function of the average fiber stress relative to the nominal friction stress $\mu\sigma_0$. The dependence on the fiber load, insofar as it is normalized by $\mu\sigma_0$, is typical of problems involving frictional interfaces. As the applied stress increases relative to $\mu\sigma_0$, the stress concentration diminishes; this is the blunting effect of the frictional interface. For loads that are small compared with the $\mu\sigma_0$, the slip zone becomes vanishingly small compared with the fiber diameter; in this limit one recovers the infinitely large stress concentration of the perfectly bonded case. The limit of small scale slipping was studied explicitly in Dollar and Steif (1989), where it was found that the stress at the crack tip is proportional to σ_0, and independent of the applied load which is small compared with σ_0.

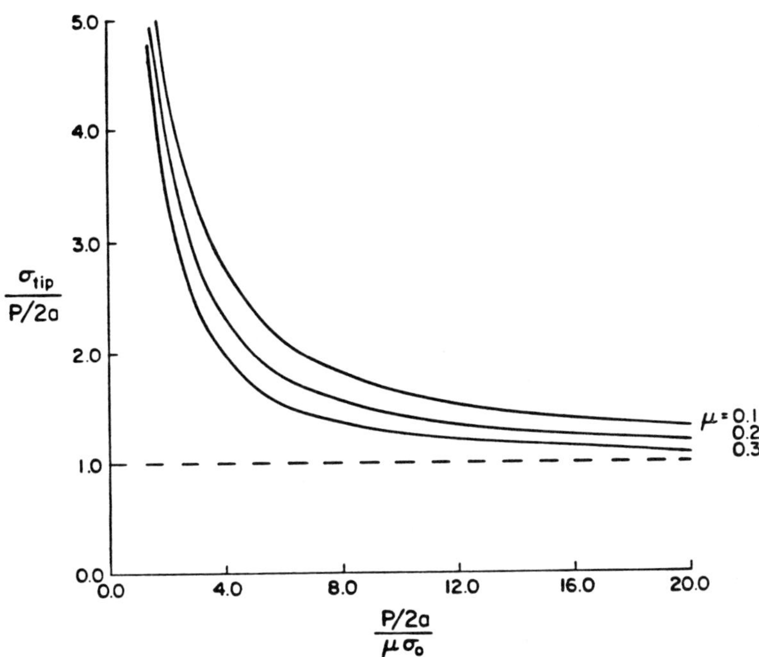

Figure 3. Stress concentration as a function of fiber stress.

We consider now the results for quantities which are commonly of interest, and which can be deduced by approximate calculation. In Figure 4, the slip length is plotted (in solid lines) as a function of the normalized fiber stress. For comparison, we show the same quantity calculated by a highly approximate method in which the shear stress at the interface is taken to be exactly equal to $\mu\sigma_0$. The approximate slip length, then, is that distance over which the fiber load P is completely transfered to the matrix. This slip length is plotted as the dotted line. Note the relatively modest discrepancy between the numerical results and the highly approximate calculation. Note also that the curves for different friction coefficients μ are distinct. This is characteristic of similar problems involving Coulomb friction (see Dollar and Steif, 1988, 1989); namely, that the results depend to a greater or lesser degree on the parameters μ and σ_0 individually, and not just on their product.

Consider also the crack-tip opening displacement which is shown in Figure 5. This opening corresponds to the maximum amount of slip at the interface (as $y \to 0$). As in Figure 4, the solid lines are the results of the calculations carried out here, and the dotted line is an approximate calculation of the maximum slip based on the assumption of a constant interfacial shear stress (equal to $\mu\sigma_0$). There is reasonable agreement between the numerical and approximate results though the latter does not, of course, predict any dependence on μ alone.

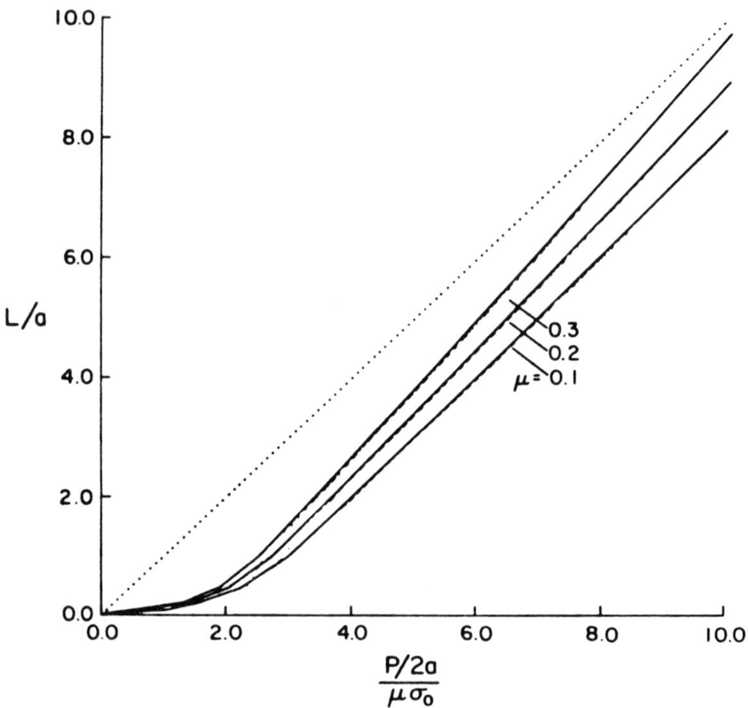

Figure 4. Slip length as a function of fiber stress.

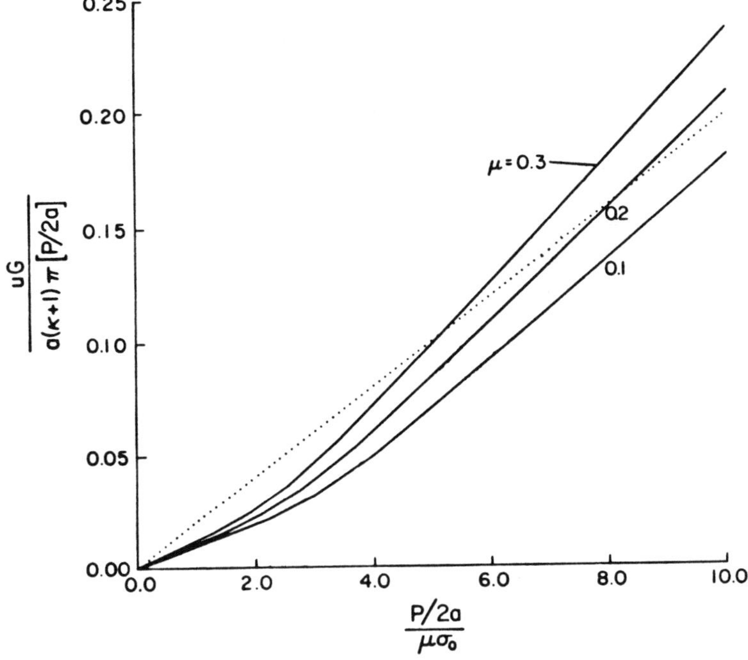

Figure 5. Crack-tip opening displacement as a function of fiber stress.

CONCLUSIONS

An idealized two-dimensional fiber composite with a single bridged matrix crack has been studied, particularly for the degree to which the tensile stress in the fiber deviates from its mean value. For loads which are small compared with the friction stress $\mu\sigma_0$, the maximum stress is proportional to σ_0 (as was found previously) and, thus, highly concentrated. Even when the applied stress is several times the friction stress, the stress at the fiber surface can be twice the mean fiber stress. However, for reasonable material parameters, the blunting that continues with increasing load appears to be sufficient to preclude premature fiber failure, at least for an interface that is characterized by Coulomb friction. It is likely that actual bonding (chemical) at the interface would raise the stress concentration, though in that case one wonders how the first matrix crack managed to get across at all. These questions are being pursued.

ACKNOWLEDGMENTS

This research has been supported by the Department of Energy under Grant DE-FG02-89ER45404, by the General Electric Engine Business Group, and by the Department of Mechanical Engineering, Carnegie Mellon University.

REFERENCES

Aveston, J., Cooper, G.A., and Kelly, A., 1971,."Single and Multiple Fracture," in Conference Proceedings, National Physical Laboratory, IPC Science and Technology Press Ltd., pp. 15-26.

Dollar, A. and Steif, P.S., 1988,."A Tension Crack Impinging Upon Frictional Interfaces," ASME J. Appl. Mech., Vol. 56, pp. 291-298.

Dollar, A. and Steif, P.S., 1988,."Load Transfer in Composites with a Coulomb Friction Interface," Int. J. Sol. Structures, Vol. 24, pp. 789-803.

Lo, K.K., 1978,."Analysis of Branched Crack," ASME J. Appl. Mech., Vol. 45, pp. 797-802.

Muskhelishvili, N.I., 1963,."Some Basic Problems of the Mathematical Theory of Elasticity," Noordhoff, the Netherlands.

APPENDIX

In this Appendix we give expressions for the terms in the dimensionless version of equation (12). Stresses have been nondimensionalized by P/2a, spatial variables by a, and the dislocation intensity $b(y_o)$ by $P\pi(\kappa+1)/(2Ga)$.

$$R_0(y,y_0) = \frac{-2}{y-y_0}$$

$$R_1(y,y_0) = 2\,\mathrm{Im}\,[h_2] + \mathrm{Im}\left[\frac{\bar{h}_3^2 + h_3^2}{\bar{h}_3} + \frac{\bar{h}_4^2 + h_4^2}{\bar{h}_4}\right] + 2y\,\mathrm{Re}\left[H'(z,z_0)\right]$$

$$R_2(y,y_0) = -\mathrm{Re}\left[\frac{\bar{h}_3^2 + h_3^2}{\bar{h}_3} + \frac{\bar{h}_4^2 + h_4^2}{\bar{h}_4}\right] + 2y\,\mathrm{Im}\left[H'(z,z_0)\right] - 2\,\mathrm{Re}\left[H(z,z_0)\right]$$

where

$$h_2 = \frac{1}{z-\bar{z}_0} \qquad h_3 = \frac{1}{z+z_0} \qquad h_4 = \frac{1}{z+\bar{z}_0}$$

$$H(z,z_0) = \frac{2}{[X(z)+X(z_0)]^2}\left[1 + \frac{z_0^3-1}{X(z)\,X(z_0)}\right] + \frac{2}{[X(z)-\overline{X(z_0)}]^2}\left[1 - \frac{\bar{z}_0^3-1}{X(z)\,\overline{X(z_0)}}\right]$$

$$H'(z,z_0) = \frac{\partial H(z,z_0)}{\partial z}$$

The function f(y) is written as

$$f(y) = f_1(y) + \mu f_2(y)$$

where

$$f_1(y) = \frac{-2}{\pi}\, y\, \mathrm{Im}\left[\frac{z}{[X(z)]^3}\right]$$

$$f_2(y) = \frac{2}{\pi}\left\{-\mathrm{Im}\left[\frac{1}{X(z)}\right] + y\,\mathrm{Re}\left[\frac{z}{[X(z)]^3}\right]\right\} - \frac{\sigma_o}{P/2a}$$

Fracture Toughness Enhancement Due to Particle Transformation

Zhanjun Gao and Toshio Mura
Department of Civil Engineering
Northwestern University
Evanston, IL 60208, U.S.A.

Abstract

In certain ceramics, the high stress in the tip of a macroscopic crack induces martinsitic type transformation of second phase particles (e.g., Zirconia, Z_rO_2). The transformation changes the near-tip stress by decreasing the net stress intensity factor and the toughness of the material is thereby enhanced. To evaluate the decrease of the stress intensity factor, one needs to know the distribution of transformation strain around the crack tip. Therefore, an incremental analysis of plasticity, which takes into account the microstructural properties and mechanism of the particle transformation, is generally required.

In this paper, we present a method of calculating the decrease of stress intensity factor without considering the microstructural details or performing an incremental analysis. Regardless of the history of loading and unloading, the current crack opening displacement is all the information needed for the calculation. It is proven that under a given distribution of crack opening displacement, there are infinite numbers of possible configurations of transformation zones with transformation strains inside. Identifying the actual transformation strain and transformation zone is impossible unless additional information is provided. However, all these transformation strains inside the corresponding transformation zones induce the same decrease of stress intensity factor. In this sense, they are all equivalent. If we can obtain any of these transformation strains with the corresponding transformation zone, the decrease of stress intensity factor is then determined.

The problem is formulated as a system of integral equations of the first kind with transformation strains as unknowns and crack opening displacement as input data. The regularization method is employed to obtain a stable solution of these ill-posed integral equations. The stress intensity factor is then computed by using Bueckner's weight function.

1. Introduction

The high stress in the vicinity of a macroscopic crack induces local plastic deformation of particles, such as martensitic type transformation of second phase particles. The transformation decreases the net crack tip intensity factor and thereby the toughness of the material is enhanced. This fracture toughness enhancement has been observed in a number of ceramic materials (e.g., Claussen, 1976; Evans and Heuer, 1980).

McMeeking and Evans (1982) and Budiansky et. al. (1983) have studied the problem by taking into account only the dilatant part of the transformation. As they have pointed out, due to the neglect of shear transformation, the increases in toughness predicted are less than experimental values. The effects of shear components of transformation with different shapes and orientations are further accounted for by Lambropoulos (1986). In Lambropoulos' analysis, a constitutive model is proposed. The constitutive model is similar to the incremental theories of metal plasticity in which the behavior of material is characterized by a yield function, a loading criterion and a set of flow equations.

In this paper, measured crack opening displacement is used to evaluate the increase of fracture toughness due to particle transformation (dilation as well as shear). No particular constitutive law and loading history are involved explicitly in our analysis. When crack opening displacement is provided, no matter what loading (cyclic or monotonic) was the material loaded, the changes of stress intensity factor can be obtained by solving a set of integral equations.

2. Basic Equations

For simplicity of notations, we consider only Mode I problem. A material D with a crack is loaded at its surface ∂D (Fig. 1(a)). Transformation zone Ω is developed near the crack tips. It follows from the principle of superposition (Fig. 1) that the net stress intensity factor K_I^{net} (SIF for problem shown in Fig. 1(a)) is given by

$$K_I^{net} = K_I^{(0)} + K_I$$

where $K_I^{(0)}$ is the stress intensity factor that would be induced at the tips by the applied loading in the absence of transformation zone (i.e., SIF for problem shown in Fig. 1(b)) and K_I is the stress intensity factor for problem 1(c).

There is a well developed body of knowledge for computing $K_I^{(o)}$. The effort of this paper is to evaluate K_I, the change of stress intensity factor due to the existence of particle transformation (SIF for problem shown in Fig. 1(c). The net stress intensity factor is then obtained as the sum of $K_I^{(0)}$ and K_I.

Let us first derive the integral equation for problem shown in Fig. 1(c). It has been shown in our previous papers (Gao and Mura, 1989a and 1989b) that the integral equation for the problem with transformation strain ϵ^{p}_{ij} in a subdomain of D (Fig. 2) is

$$\int_{\Omega} C_{ijk\ell}\, G_{km,\ell}\,(\underset{\sim}{x} - \underset{\sim}{x}')\, \epsilon^{p}_{ij}\,(\underset{\sim}{x})\, d\underset{\sim}{x}$$

$$= \int_{\partial D} C_{ijk\ell}\, G_{km,\ell}\,(\underset{\sim}{x} - \underset{\sim}{x}')\, u_i(\underset{\sim}{x})\, n_j\, ds(\underset{\sim}{x}) \qquad (1)$$

$$+ \beta\, u_m(\underset{\sim}{x}')$$

where $\beta = 1$ for $\underset{\sim}{x}' \epsilon D$ and $\beta = 1/2$ for $\underset{\sim}{x}' \epsilon \partial D$ and $u_i(\underset{\sim}{x})$ is displacement field. $C_{ijk\ell}$ is elastic moduli of the material. $G_{km}(\underset{\sim}{x} - \underset{\sim}{x}')$ is Green's function for an infinite elastic medium and satisfies the equation of equilibrium for a unit point force

$$C_{ijk\ell}\, G_{km,\ell j}\,(\underset{\sim}{x} - \underset{\sim}{x}') = -\,\delta_{im}\,\delta(\underset{\sim}{x} - \underset{\sim}{x}').$$

δ_{im} is the Kronecker data and $\delta(\underset{\sim}{x} - \underset{\sim}{x}')$ is the Dirac's delta function.

If a crack is developed together with transformation strain ϵ^{p}_{ij} (Fig. 1(c)), the entire boundary consists of ∂D plus upper and lower crack faces. Therefore, the integral equation is

$$\int_{\Omega} C_{ijk\ell}\, G_{km,\ell}\,(\underset{\sim}{x} - \underset{\sim}{x}')\, \epsilon^{p}_{ij}\,(\underset{\sim}{x})\, d\underset{\sim}{x}$$

$$= \int_{\partial D} C_{ijk\ell}\, G_{km,\ell}\,(\underset{\sim}{x} - \underset{\sim}{x}')\, u_i(\underset{\sim}{x})\, n_j\, ds(\underset{\sim}{x})$$

$$+ \beta\, u_m(\underset{\sim}{x}') + \int_{\Gamma} C_{ijk\ell}\, G_{km,\ell}\,(\underset{\sim}{x} - \underset{\sim}{x}')\, u^{+}_i(\underset{\sim}{x})\, n_j\, ds(\underset{\sim}{x})$$

$$+ \int_{\Gamma} C_{ijk\ell}\, G_{km,\ell}\,(\underset{\sim}{x} - \underset{\sim}{x}')\, u^{-}_i\,(\underset{\sim}{x})\, n_j\, ds(\underset{\sim}{x}) + \alpha[u^{+}_m(\underset{\sim}{x}')+u^{-}_m(\underset{\sim}{x}')] \qquad (2)$$

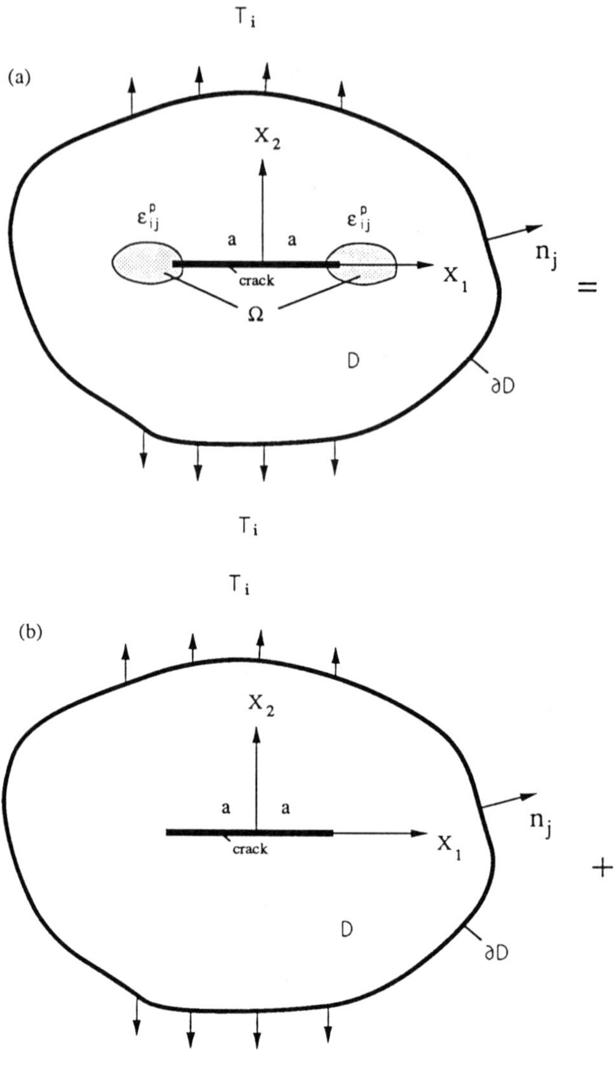

(a)
T_i
X_2
ε^p_{ij}
ε^p_{ij}
a
a
crack
X_1
n_j
Ω
D
∂D
T_i
=
(b)
T_i
X_2
a
a
crack
X_1
n_j
D
∂D
T_i
+

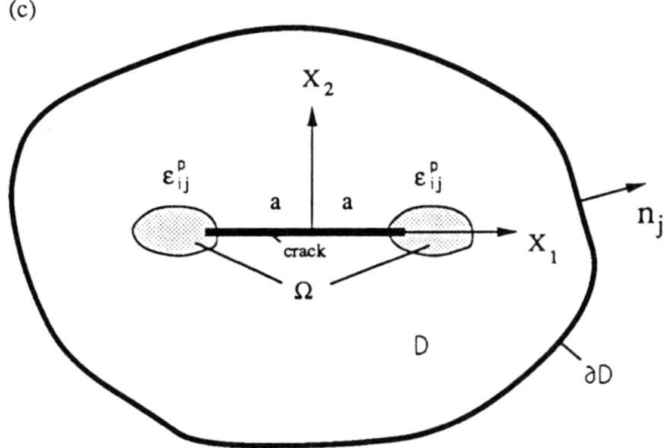

Fig. 1 Transformation zone Ω with transformation strains $\epsilon^p_{k\ell}$ developed near the crack tips in a body D. ∂D is the boundary of D. Problem (a) is the superposition of problems (b) and (c).

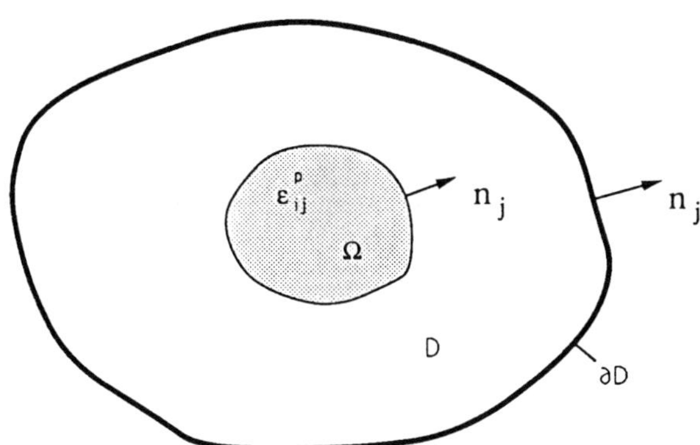

Fig. 2 Transformation strains are accumulated in Ω, a subdomain of body D.

where $\Gamma = \{\underset{\sim}{x} \mid x_1 \epsilon [-a, a], x_2 = 0\}$ is the crack face, $u_m^+(\underset{\sim}{x}')$ and $u_m^-(\underset{\sim}{x}')$ are displacements on the upper and lower crack faces, respectively. The parameters α and β in equation (2) are defined as

$$\beta = \begin{cases} 1 & \text{for } \underset{\sim}{x}' \in D \\ 0.5 & \text{for } \underset{\sim}{x}' \in \partial D \\ 0 & \text{for } \underset{\sim}{x}' \in \Gamma \end{cases}$$

$$\alpha = \begin{cases} 0.5 & \text{for } \underset{\sim}{x}' \in \Gamma \\ 0 & \text{for } \underset{\sim}{x}' \notin \Gamma. \end{cases}$$

Note that

$$(n_1, n_2) = (0, 1) \qquad \text{on upper crack face}$$

and

$$(n_1, n_2) = (0, -1) \qquad \text{on lower crack face.}$$

We further simplify equation (2) to

$$\int_\Omega C_{ijk\ell} G_{km,\ell} (\underset{\sim}{x} - \underset{\sim}{x}') \epsilon_{ij}^p (\underset{\sim}{x}) d\underset{\sim}{x}$$

$$= \int_{\partial D} C_{ijk\ell} G_{km,\ell} (\underset{\sim}{x} - \underset{\sim}{x}') u_i(\underset{\sim}{x}) n_j \, ds(\underset{\sim}{x})$$

$$+ \beta u_m(\underset{\sim}{x}') + \int_{-a}^{a} C_{i2k\ell} G_{km,\ell} (\underset{\sim}{x} - \underset{\sim}{x}') \Big|_{x_2=0} [u_i^+(x_1) - u_i^-(x_1)] dx_1$$

$$+ \alpha [u_m^+(\underset{\sim}{x}') + u_m^-(\underset{\sim}{x}')]. \tag{3}$$

In the problem of transformation toughness, the size of transformation zone is quite small compared with any characteristic length of the material. The material is modeled as an infinite medium, which means the integral along ∂D in equation (3) disappears.

For model I problem

$$u_m^+(\underset{\sim}{x}') + u_m^-(\underset{\sim}{x}') = 0.$$

Therefore, when $\underset{\sim}{x}' \in \Gamma$, we rewrite equation (3) as

$$\int_{\Omega} C_{ijk\ell}\, G_{km,\ell}\, (\underset{\sim}{x} - \underset{\sim}{x}')\, \epsilon^{p}_{ij}\, (\underset{\sim}{x})\, d\underset{\sim}{x}$$

$$= \int_{-a}^{a} C_{i2k\ell}\, G_{km,\ell}\, (\underset{\sim}{x} - \underset{\sim}{x}')\, \Big|_{x_2=0}\, b_i(x_1)\, dx_1 \tag{4}$$

where $b_i(x_1) = u^{+}_i(x_1) - u^{-}_i(x_1)$ is the crack opening displacement.

Recalling the superposition shown in Fig. 1, we have

$$b_i = b^{(a)}_i - b^{(b)}_i \tag{5}$$

where $b^{(a)}_i$ and $b^{(b)}_i$ are the crack opening displacements for problems shown in Fig. 1(a), and 1(b), respectively. $b^{(a)}_i$ are measured experimentally. $b^{(b)}_i$ can be computed either analytically or numerically for a given load. Therefore, b_i, as well as the right hand side term of equation (4) is known after (5) is applied.

3. Fracture Toughness as a Global Property of Transformation Strains

Our goal is to compute transformation strain ϵ^{p}_{ij} by using given values of $b_i(x_1)$, and derive stress intensity factor K_I by the computed ϵ^{p}_{ij}. However, equation (4) for unknown ϵ^{p}_{ij} cannot be solved uniquely from given crack opening displacement. In fact, given crack opening displacement $b_i(x_1)$, equation (4) has infinite numbers of solutions for any chosen Ω^*, which covers the transformation zone Ω.

The nonuniqueness of solutions for (4) is consistent with the fact that particle transformation is irreversible plastic deformation, which is loading history dependent. The computation of transformation strain ϵ^{p}_{ij} requires models accounting for microstructural details and for constitutive equations characterizing the development of transformation strain ϵ^{p}_{ij} and transformation zone. Incremental analysis is generally needed because local unloading may occur even for monotonic loads.

It is important to notice that catastropic failure of material results from propagation of the main crack. Our ultimate goal, therefore, is to evaluate the effects of the transformation strain on the toughness of the main crack, which is a global property of the transformation strain ϵ^p_{ij} and can be related to the crack opening displacement.

It can be shown that all the solutions of equation (4) have some common characteristic properties. One of these properties is that they induce the same stress intensity factor at the crack tip (see Appendix for details). This leads to a new idea to find any one of the solutions of equation (4) and compute the stress intensity factor from the solution.

The question is, therefore, how to obtain a solution of equation (4). It is impossible to solve equation (4) without specifying the transformation zone Ω. Since Ω is unknown, we choose another domain Ω^* such that Ω^* contains Ω (Fig. 3), which guarantees the existence of solution. By replacing Ω with Ω^*, we write equation (4) as

$$\begin{aligned} &\int_{\Omega^*} C_{ijl\ell}\, G_{km,\ell}\, (\underset{\sim}{x} - \underset{\sim}{x}')\, \epsilon^p_{ij}\, (\underset{\sim}{x})\, d\underset{\sim}{x} \\ &= \int_{-a}^{a} C_{i2k\ell}\, G_{km,\ell}\, (\underset{\sim}{x} - \underset{\sim}{x}')\, \Big|_{x_2=0}\, b_i\, (x_1)\, dx_1. \end{aligned} \tag{6}$$

The solutions of equations (4) and (6) are denoted by $\epsilon^p_{ij}(\underset{\sim}{x})$ and $\tilde{\epsilon}^p_{ij}$, respectively. Although $\epsilon^p_{ij}(\underset{\sim}{x})$ and $\tilde{\epsilon}^p_{ij}(\underset{\sim}{x})$ may be substantially different, they induce the same stress intensity factor (Appendix). This enables us to compute the crack intensity factor by using $\tilde{\epsilon}^p_{ij}(\underset{\sim}{x})$, without knowing the actual transformation strain $\epsilon^p_{ij}(\underset{\sim}{x})$.

The stress intensity factor for transformation strain $\tilde{\epsilon}^p_{ij}$ distributed in Ω^* is equal to that of the problem shown in Fig. 4, where $\tilde{\epsilon}^p_{ij}$ is replaced by body force X_i and traction force t_i

$$\begin{aligned} X_i &= - C_{ijk\ell}\, \tilde{\epsilon}^p_{k\ell,j}(\underset{\sim}{x}) && \text{in } \Omega^* \\ t_i &= C_{ijk\ell}\, \tilde{\epsilon}^p_{k\ell}(\underset{\sim}{x})\, n_j && \text{in } \partial\Omega^*. \end{aligned} \tag{7}$$

$\partial\Omega^*$ in the above equation is the boundary of Ω^*.

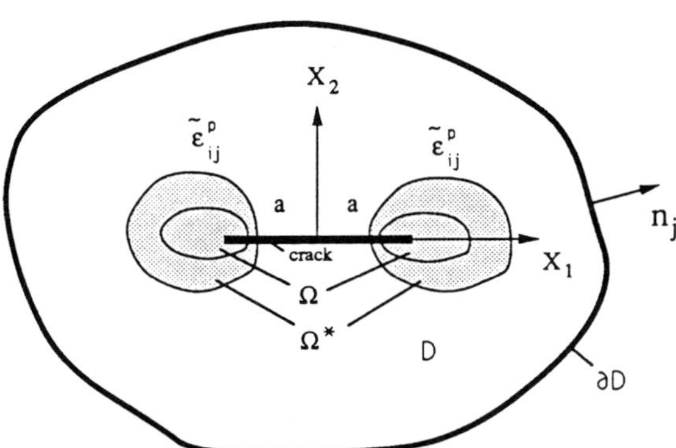

Fig. 3. Domain Ω^* is introduced to contain the transformation zone Ω. The properly chosen transformation strains $\tilde{\epsilon}_{k\ell}^{p}$ in Ω^* induce the actual decrease of stress intensity factor K_I.

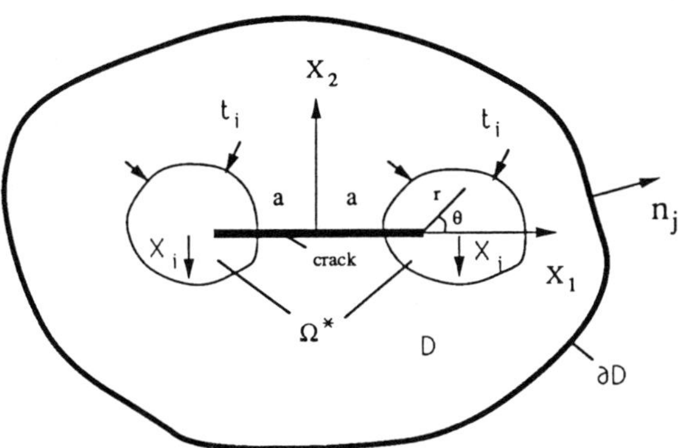

Fig. 4. The transformation strains $\tilde{\epsilon}_{k\ell}^{p}$ are replaced by body forces $X_i = C_{ijk\ell}\,\tilde{\epsilon}_{k\ell}^{p}$ in Ω^* and traction forces $t_i = C_{ijk\ell}\,\tilde{\epsilon}_{k\ell}^{p}\,n_j$ on $\partial\Omega^*$, the boundary of Ω^*.

A useful device for calculating the stress intensity factor due to body force and traction (Fig. 4) is the weight function of Bueckner (1970). The function h_i, a weight function, is defined in such a way that

$$K_I = \int_{\partial\Omega^*} h_i(\underset{\sim}{x},a)t_i(\underset{\sim}{x})ds + \int_{\Omega^*} h_i(\underset{\sim}{x},a)X_i(\underset{\sim}{x})d\underset{\sim}{x} \tag{8}$$

where $t_i(\underset{\sim}{x})$ is the traction applied through the line $\partial\Omega^*$ and $X_i(\underset{\sim}{x})$ is the body force applied inside domain Ω^*. K_I is the crack intensity factor due to $t_i(\underset{\sim}{x})$ and $X_i(\underset{\sim}{x})$. The agrument "a" of $h_i(\underset{\sim}{x},a)$ is the half of the crack length. The weight function is defined by

$$h_i(\underset{\sim}{x},a) = \frac{H}{4K_I^w(a)} \frac{\partial u_i^w(\underset{\sim}{x},a)}{\partial a} \tag{9}$$

where $u_i^w(\underset{\sim}{x},a)$ is the displacement field under a given load, $K_I^w(a)$ is the stress intensity factor under the load.

$$H = \begin{cases} \dfrac{E}{1-\nu^2} & \text{for plane strain} \\ E & \text{for plane stress.} \end{cases}$$

E, ν are Young's modulus and Poisson's ratio, respectively.

It should be pointed out that $h_i(\underset{\sim}{x},a)$ is a universal function for a given geometry and composition and bears no particular load system to which the body may be subjected.

By plugging equation (7) into equation (8), we have

$$\begin{aligned} K_I = &\int_{\partial\Omega^*} h_i(\underset{\sim}{x},a)\ C_{ijk\ell}\ \tilde{\epsilon}^p_{k\ell}(\underset{\sim}{x})\ n_j\ ds \\ &- \int_{\Omega^*} h_i(\underset{\sim}{x},a)\ C_{ijk\ell}\ \tilde{\epsilon}^p_{k\ell,j}(\underset{\sim}{x})\ d\underset{\sim}{x} \end{aligned} \tag{10}$$

After applying Gauss' theorem, equation (10) is rewritten as

$$K_I = \int_{\Omega^*} U_{k\ell}(\underset{\sim}{x},a)\ \tilde{\epsilon}^p_{k\ell}(\underset{\sim}{x})\ d\underset{\sim}{x} \tag{11}$$

where $U_{k\ell} = C_{ijk\ell}\, h_{i,j}(\underset{\sim}{x},a)$. For a two dimensional half plane crack problem, the expressions of $U_{k\ell}$ were given by Hutchinson (1974) and Gao (1989) as

$$
\begin{aligned}
U_{11} &= \frac{1}{16(1-\nu)\sqrt{2\pi}}\, r^{-3/2}\left[\cos\frac{3\theta}{2} + 3\cos\frac{7\theta}{2}\right] \\
U_{21} = U_{12} &= \frac{3}{16(1-\nu)\sqrt{2\pi}}\, r^{-3/2}\left[\sin\frac{7\theta}{2} - \sin\frac{3\theta}{2}\right] \qquad (12) \\
U_{22} &= \frac{1}{16(1-\nu)\sqrt{2\pi}}\, r^{-3/2}\left[7\cos\frac{3\theta}{2} - 3\cos\frac{7\theta}{2}\right]
\end{aligned}
$$

where r and θ are shown in Fig 4.

4. Computation of ϵ_{ij}^{p}

There are several experimental techniques for measuring crack opening displacement at any points on the crack face except those near the crack tip. For the points near the crack tip, we use the asymptotic expansion of the displacement field. In a Mode I problem,

$$u_1 = \frac{2(1+\nu)}{E} K_I \sqrt{\frac{r}{2\pi}}\,(1 - 2\nu + \sin^2\frac{\theta}{2})\cos\frac{\theta}{2}$$

$$u_2 = \frac{2(1+\nu)}{E} K_I \sqrt{\frac{r}{2\pi}}\,(2 - 2\nu - \cos^2\frac{\theta}{2})\sin\frac{\theta}{2}$$

$$0 \le r \le \delta << 1$$

where δ is a chosen small parameter. For $-a \le x_1 \le -a + \delta$ and $a - \delta \le x_1 \le a$, we derive $b_i(x_1)$ from the above expression. Therefore, equation (6) is written as

$$\int_{\Omega^*} C_{ijk\ell}\, G_{km,\ell}(\underset{\sim}{x} - \underset{\sim}{x}')\, \epsilon_{ij}^{p}(\underset{\sim}{x})\, d\underset{\sim}{x}$$

$$= K_I \int_{-a}^{-a+\delta} C_{22k\ell}\, G_{km,\ell}(\underset{\sim}{x} - \underset{\sim}{x}')\,\Big|_{x_2=0} \frac{8(1-\nu^2)}{E}\sqrt{\frac{a+x_1}{2\pi}}\, dx_1$$

$$+ K_I \int_{a-\delta}^{a} C_{22k\ell} G_{km,\ell} (\underset{\sim}{x} - \underset{\sim}{x}') |_{x_2=0} \frac{8(1-\nu^2)}{E} \sqrt{\frac{a-x_1}{2\pi}} dx_1$$

$$+ \int_{-a+\delta}^{a-\delta} C_{i2k\ell} G_{km,\ell} (\underset{\sim}{x} - \underset{\sim}{x}') |_{x_2=0} b_i(x_1) dx_1. \qquad (13)$$

K_I in the right hand side of (13) is the unknown stress intensity factor we are looking for. Hence, equation (13) must be solved by trial and error method or other iterative procedures so that K_I is consistent with the unknown $\epsilon^p_{ij}(\underset{\sim}{x})$ of the left hand side in equation (13).

The solution of equation (13) is nonunique and unstable. There are several ways to convert equation (13) into a new well-posed problem. Let us write equation (13) in the following compact form

$$\int_{\Omega^*} \underset{\sim}{C}(\underset{\sim}{x} - \underset{\sim}{x}') \underset{\sim}{V} (\underset{\sim}{x}) d\underset{\sim}{x} = \underset{\sim}{U}(\underset{\sim}{x}' K_I) \qquad (14)$$

$$\underset{\sim}{x}' \in \Gamma$$

where

$$C(\underset{\sim}{x}-\underset{\sim}{x}') = \begin{bmatrix} C_{11k\ell} G_{k1,\ell}(\underset{\sim}{x}-\underset{\sim}{x}') , & C_{22k\ell} G_{k1,\ell}(\underset{\sim}{x}-\underset{\sim}{x}') , & C_{12k\ell} G_{k1,\ell}(\underset{\sim}{x}-\underset{\sim}{x}') \\ C_{11k\ell} G_{k2,\ell}(\underset{\sim}{x}-\underset{\sim}{x}') , & C_{22k\ell} G_{k2,\ell}(\underset{\sim}{x}-\underset{\sim}{x}') , & C_{12k\ell} G_{k2,\ell}(\underset{\sim}{x}-\underset{\sim}{x}') \end{bmatrix}$$

$$\underset{\sim}{V}(\underset{\sim}{x}) = [\epsilon^p_{11}, \epsilon^p_{22}, 2\epsilon^p_{12}]^T$$

$$\underset{\sim}{U}(\underset{\sim}{x}', K_I) = [g_1 , g_2]^T$$

$$g_m = K_I \int_{-a}^{-a+\delta} C_{22k\ell} G_{km,\ell} (\underset{\sim}{x}-\underset{\sim}{x}') |_{x_2=0} \frac{8(1-\nu^2)}{E} \sqrt{\frac{a+x_1}{2\pi}} dx_1$$

$$+ K_I \int_{a-\delta}^{a} C_{22k\ell} G_{km,\ell} (\underset{\sim}{x}-\underset{\sim}{x}') |_{x_2=0} \frac{8(1-\nu^2)}{E} \sqrt{\frac{a-x_1}{2\pi}} dx_1$$

$$+ \int_{-a+\delta}^{a-\delta} C_{i2k\ell} G_{km,\ell} (\underset{\sim}{x}-\underset{\sim}{x}') |_{x_2=0} b_i (x_1) dx_1.$$

One way to obtain a well-posed problem is to change problem (14) into

$$\begin{cases} \text{Min } \| V(\underset{\sim}{x}) \|^{2} \\ \text{Subject to } \| \int_{\Omega^*} \underset{\sim}{C}(\underset{\sim}{x}-\underset{\sim}{x}')\underset{\sim}{V}(\underset{\sim}{x})d\underset{\sim}{x} - \underset{\sim}{U}(\underset{\sim}{x}', K_I) \|^2 = \epsilon \end{cases} \tag{15}$$

where $\| \; \|^2$ is the square of L_2 norm, i.e., the inner product of a function with itself over the domain it is defined. For example,

$$\| \underset{\sim}{V} (\underset{\sim}{x}) \|^2 = \int_{\Omega^*} \underset{\sim}{V}^T (\underset{\sim}{x}) \underset{\sim}{V} (\underset{\sim}{x}) d\underset{\sim}{x}.$$

The positive parameter ϵ measures the discrepancy of our crack opening displacement data from the exact values. The detail discussions can be found in Gao and Mura (1989a, 1989b).

An alternative way is to solve problem

$$\begin{cases} \text{Min } \| \int_{\Omega^*} \underset{\sim}{C}(\underset{\sim}{x}-\underset{\sim}{x}') \underset{\sim}{V} (\underset{\sim}{x}) d\underset{\sim}{x} = \underset{\sim}{U} (\underset{\sim}{x}', K_I) \|^2 \\ \text{Subject to } \| V(\underset{\sim}{x}) \|^2 = \gamma \end{cases} \tag{16}$$

where γ is a constant used to specify the norm of the solution. When we know the order of norm of exact solution, γ is a given value.

The Euler equations for problems (15) and (16) are

$$\begin{aligned} &\int_{\Omega^*} \underset{\sim}{C}^*(\underset{\sim}{x}, \underset{\sim}{\eta}) \underset{\sim}{V} (\underset{\sim}{x}) d\underset{\sim}{x} + \alpha \underset{\sim}{V} (\underset{\sim}{\eta}) = \underset{\sim}{U}^* (\underset{\sim}{\eta}, K_I) \\ &\| \int_{\Omega^*} \underset{\sim}{C}(\underset{\sim}{x} - \underset{\sim}{x}') \underset{\sim}{V} (\underset{\sim}{x}) d\underset{\sim}{x} - \underset{\sim}{U} (\underset{\sim}{x}', K_I) \|^2 = \epsilon \end{aligned} \qquad \underset{\sim}{\eta} \in \Omega^* \tag{17}$$

and

$$\begin{aligned} &\int_{\Omega^*} \underset{\sim}{C}^*(\underset{\sim}{x}, \underset{\sim}{\eta}) \underset{\sim}{V} (\underset{\sim}{x}) d\underset{\sim}{x} + \alpha \underset{\sim}{V} (\underset{\sim}{\eta}) = \underset{\sim}{U}^* (\underset{\sim}{\eta}, K_I) \\ &\| \underset{\sim}{V} (\underset{\sim}{x}) \|^2 = \gamma \end{aligned} \qquad \underset{\sim}{\eta} \in \Omega^* \tag{18}$$

respectively, where

$$\underset{\sim}{C}^*(\underset{\sim}{x}, \underset{\sim}{\eta}) = \int_{\partial D} \underset{\sim}{C}^* (\underset{\sim}{\eta}, \underset{\sim}{x}') \underset{\sim}{C} (\underset{\sim}{x}, \underset{\sim}{x}') d\underset{\sim}{x}' \qquad x, \underset{\sim}{\eta} \in \Omega^*,$$

$$\underset{\sim}{U}^*(\underset{\sim}{\eta}, K_I) = \int_{\partial D} \underset{\sim}{C}^* (\underset{\sim}{\eta}, \underset{\sim}{x}') \underset{\sim}{U} (\underset{\sim}{x}', K_I) d\underset{\sim}{x}' \qquad \underset{\sim}{\eta} \in \Omega^*.$$

When the order of measurement error ϵ is known, problem (17) can be employed. In some cases, we roughly know the order of norm of the transformation strain, i.e., γ is known, problem (18) is suitable.

5. Numerical Examples

The method of imposing (15) or (16) to obtain a stable solution of an ill-posed system is called regularization method, suggested first by Tikhonov (1963). Tikhonov also showed that $||V^{(\alpha)}||^2$ is a monotonically decreasing function of $\alpha \in [0^+, \infty)$, where $V^{(\alpha)}$ satisfies the first equation of problem (18). Therefore, the solution of problem (18) can be obtained by a simple iteration procedure.

An example is presented as shown in Fig. 5. The half crack length "a" is taken as the unit length. The material is undergoing transformation strain,

$$\epsilon_{11}^P(\underset{\sim}{x}) = \epsilon_{22}^P(\underset{\sim}{x}) = 1 \qquad \underset{\sim}{x} \in \Omega \tag{19a}$$

$$\epsilon_{12}^P(x) = \begin{cases} -1 & x_2 \geq 0 \text{ and } \underset{\sim}{x} \in \Omega \\ 1 & x_2 \leq 0 \text{ and } \underset{\sim}{x} \in \Omega \end{cases} \tag{19b}$$

where Ω is the domain,

$$r \leq \frac{8}{3\sqrt{3}} w \cos^2 \frac{\theta}{2}, \quad w = 0.1, \quad -\pi \leq \theta \leq \pi. \tag{19c}$$

In this test example, the crack opening displacement $b_i(x_1)$ is obtained by solving equation (4) for $b_i(x_1)$ with $\epsilon_{ij}^P(\underset{\sim}{x})$ and Ω given in (19a) $\approx$ (19c). For an engineering problem, however, $b_i(x_1)$ should be obtained by using equation (5).

In solving for transformation strain $\epsilon_{ij}^P(\underset{\sim}{x})$, (18) is employed. The scheme is as follows.

1. γ and Ω^* are chosen from our knowledge about the order of the transformation strain and the location of transformation zone Ω; choose K_I^* as an initial value of K_I.

2. Solve the first equation in (18) numerically with the right hand side replaced by $U(\underset{\sim}{x}, K_I) = U(\underset{\sim}{x}, K_I^*)$. The parameter α is selected such that the second equation of (18) is also satisfied. This can be done because $||V^{(\alpha)}||^2$ is a monotonically decreasing function of α, where $V^{(\alpha)}$ is the solution of the first equation in (18) for the given value of α.

3. Calculate the stress intensity factor K_I deduced by the computed transformation strain ϵ_{ij}^p (i.e., $\underset{\sim}{V}$). In this calculation, equation (11) is applied. The method of trial and error is used to select K_I^* such that $K_I \approx K_I^*$.

In the example shown in Fig. 5, Ω^* is chosen as Ω. The comparison of computer results with actual values of normalized stress intensity factor is shown in Fig. 6. P is a normalized factor given by

$$P = \sqrt{w}\, E\, \{ \int_\Omega \epsilon_{ij}^p \epsilon_{ij}^p \, d\underset{\sim}{x}/A\}^{1/2}$$

where A is the area of the transformation zone Ω, E is the Young's modulus of the material. The horizontal axis of Fig. 6 is the distance that the crack tip has advanced into the prior transformation zone, normalized by w (see Eq. (19c)).

A similar problem is also solved (Fig. 7). Ω is the actual transformation zone

$$\Omega: \quad r \le \frac{8}{3\sqrt{3}} w \cos^2 \frac{\theta}{2} \quad , \quad w = 0.06.$$

The distribution of transformation strain is

$$\epsilon_{11}^p = \epsilon_{22}^p = \begin{cases} 1 & \text{in } \Omega \\ 0 & \text{outside } \Omega, \end{cases}$$

$$\epsilon_{12}^p = 0$$

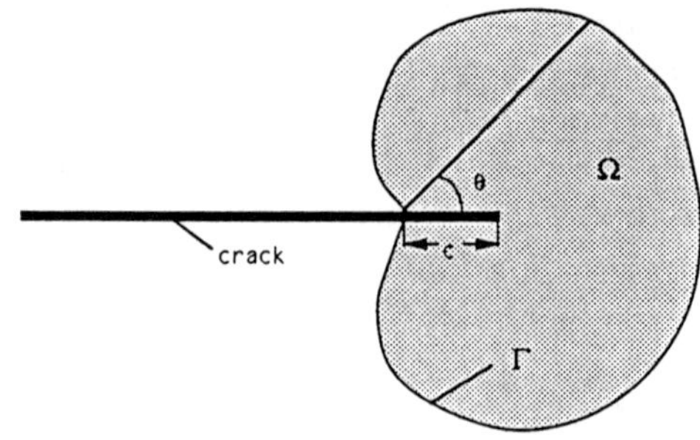

Fig. 5. A local configuration of crack tip. The equation of the boundary $\partial\Omega$ of transformation zone is $r = \frac{8}{3\sqrt{3}} w \cos^2 \frac{\theta}{2}$ $-\pi \le \theta \le \pi$ with $w = 0.1$.

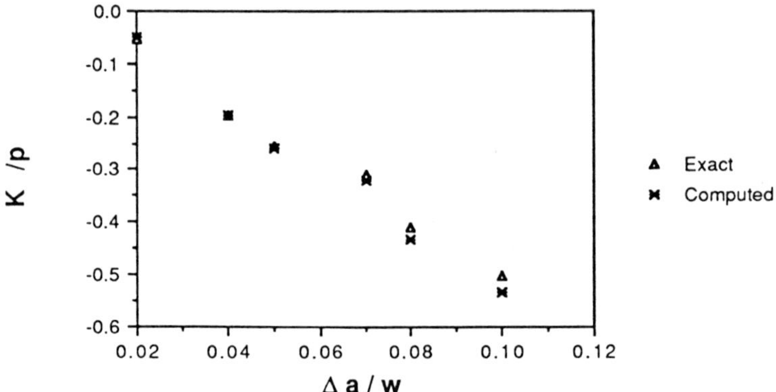

Fig. 6. K_I, the decrease of stress intensity factor, versus Δa, the distance the crack tip has advanced into the prior transformation zone. $w = 0.1$ is a characteristic length of the transformation zone shown in Fig. 5. $P = \sqrt{w}\, E \left(\int_\Omega \epsilon_{ij}^p \epsilon_{ij}^p \, d\underset{\sim}{x} / A \right)^{1/2}$, where A is the area of the transformation zone Ω, E is the Young's modulus of the material.

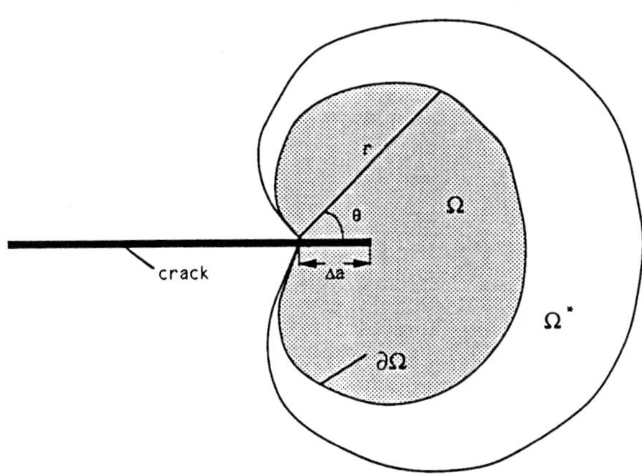

Fig. 7 Ω^* is chosen to cover the transformation zone Ω.

$$\Omega : r \le \frac{8}{3\sqrt{3}} w \cos^2 \frac{\theta}{2} \qquad \text{with } w = 0.06$$

$$\Omega^* : r \le \frac{8}{3\sqrt{3}} w \cos^2 \frac{\theta}{2} \qquad \text{with } w = 0.09$$

Domain Ω^* is chosen to cover the transformation zone Ω. Using the scheme described above, a distribution of transformation strain $\tilde{\epsilon}_{k\ell}^{p}$ inside Ω^* is obtained. The stress intensity factor K_I is then calculated by applying Eq. (11). With $\Delta a/w = 0.5$, the actual and computed values of K_I/P are 0.39 and 0.40, respectively. P is defined as

$$P = \sqrt{w}\, E \left\{ \int_{\Omega^*} \tilde{\epsilon}_{ij}^{p}\, \tilde{\epsilon}_{ij}^{p}\, d\underset{\sim}{x}/A^* \right\}^{1/2}$$

where A^* is the area of Ω^*.

6. Conclusion

The decrease of stress intensity factor due to particle transformation has been analyzed by using measured crack opening displacement. We have shown that crack opening displacement is not sufficient to determine the details of particle transformation (the shape of transformation zone, the distribution of transformation strain). This is easy to understand because the process of the transformation is plastic deformation. Nevertheless, the stress intensity factor induced by the transformation strain is a global property of the transformation and uniquely determined by the crack opening displacement.

According to our previous results (Gao and Mura, 1989b), crack opening displacement can only determine the stress field outside the transformation zone Ω. Information about constitutive laws and loading history is required to calculate the exact distribution of stress inside Ω. In this paper, we have extended these results by calculating the stress intensity factor K_I, which is an important quantity associated with the crack tip stress field inside Ω. The unique determination of K_I simply means that the leading term of the stress field near the crack tip is uniquely determined by the crack opening displacement.

The advantage of this method is that no constitutive law and loading history are explicitly involved in the calculation. The loadings can be monotonic as well as cyclic. The stress intensity factor can be computed as long as current crack opening displacement is provided.

The existance of cracks can be simulated by a proper distribution of transformation strain (Mura, 1982). Therefore, the present analysis can be applied to calculate the decrease of stress intensity factor due to the existance of micro-cracks near a main crack.

Acknowledgment

This research was supported under U. S. Army Research Office Contract No. DAAL03-88-C-0027 through a subcontract with Rockwell International Science Center. The authors are grateful for valuable discussions with Dr. B. Cox.

References

1. Budiansky, B., Hutchinson, J. W. and Lambropoulos, J. C., 1983, "Continuum Theory of Dilatant Transformation Toughening in Ceramics," Int. J. Solids Structures, Vol. 19, pp. 337-355.

2. Bueckner, H. F., 1970, "A Novel Principle for the Computation of Stress Intensity Factors," Z. argew. Math. Mech. Vol. 50, pp. 529-546.

3. Claussen, N., 1976, "Fracture Toughness of $A\ell_2 0_3$ with an Unstabilized $Z_r 0_2$ Dispessed Phase", J. Am. Ceram. Soc. Vol. 59, pp. 49-51.

4. Evans, A. G. and Heuer, A. H., 1980, "REVIEW - Transformation - Toughening in Cermics: Martensitic Transformations in Crack-Tip Stress Fields", J. Am. Ceram. Soc., Vol. 63, pp. 241-248.

5. Gao. H., 1989, "Application of 3-D Weight Functions - I. Formulations of Crack Interactions with Transformation Strains and Dislocations", J. Mech. Phys. Solids, Vol. 37, pp. 133-153.

6. Gao, Z. and Mura, T., 1989a, "Nondestructive Evaluation of Interfacial Damages in Composite Materials", Int. J. Solids Structures (in press).

7. Gao, Z. and Mura, T., 1989b, "On the Inversion of Residual Stresses from Surface Displacements", J. Appl. Mech. (in press).

8. Hutchinson, J. W., 1974, "On Steady Quasi-Static Crack Growth", Harvard University Report, Division of Applied Science, DEAP S-8.

9. Lambropoulos, J. C., 1986, "Shear Shape and Orientation Effects in Transformation Toughening, Int. J. Solids Structures Vol. 22, pp. 1083-1106.

10. McMeeking, R. M. and Evans, A. G., 1982, "Mechanics of Transformation Toughening in Brittle Materials", J. Am. Ceram. Soc. Vol. 65, pp. 242-246.

11. Tikhonov, A. N., 1963, "Regularization of Incorrectly Posed Problems", Doklady Akad. Nauk SSSR, Vol. 153, pp. 1624-1627.

Appendix

Let $\epsilon_{ij}^{p}(\underset{\sim}{x})$ and $\tilde{\epsilon}_{ij}^{p}(\underset{\sim}{x})$ be defined in Ω and Ω^* and satisfy equations (4) and (6), respectively. We want to prove that the stress intensity factors induced by ϵ_{ij}^{p} and $\tilde{\epsilon}_{ij}^{p}$ are the same.

Define

$$\Delta\epsilon_{ij}^{p}(\underset{\sim}{x}) = \tilde{\epsilon}_{ij}^{p}(\underset{\sim}{x}) - \epsilon_{ij}^{p}(\underset{\sim}{x})$$

and denote the corresponding displacement and stress as Δu_i, $\Delta\sigma_{ij}$, respectively. We have

$$\int_{\Omega^*} C_{ijk\ell}\, G_{km,\ell}(\underset{\sim}{x} - \underset{\sim}{x}')\, \Delta\epsilon_{ij}^{p}(\underset{\sim}{x})\, d\underset{\sim}{x} = 0$$

which implies that $\Delta\epsilon_{ij}^{p}(\underset{\sim}{x})$ causes zero displacement on the crack face.

It has been pointed out (Gao and Mura, 1989a) that if both displacement and traction are zero on a part of the boundary, then the displacement and stress are zero in the elastic domain which the boundary belongs to. Here, in the domain $D - \Omega^*$, the material is elastic and

$$\Delta u_i = 0$$

$$\Delta\sigma_{ij}\, n_j = 0 \quad \text{for} \quad x_1' \in [-a, a] \ , \ x_2' = 0.$$

Therefore,

$$\Delta u_i = 0 \qquad \text{in } D - \Omega^*$$

$$\Delta\sigma_{ij} = 0 \qquad \text{in } D - \Omega^*.$$

Now we prove that the stress intensity factor ΔK_I induced by $\Delta\epsilon_{ij}^{p}(\underset{\sim}{x})$ is zero.

As we have mentioned (see equation (9)), the weight function h_i is a universal function which can be obtained from displacement field and the corresponding nonzero stress intensity factor under any loading. We choose this loading as the residual stress field caused by $\Delta\,\epsilon_{ij}^{p}$. Therefore, if $\Delta\,K_I$ (due to ϵ_{ij}^{p}) is not zero, we have

$$h_i(\underset{\sim}{x}, a) = \frac{H}{4\Delta K_I(a)} \frac{\partial}{\partial a} \Delta u_i\,(\underset{\sim}{x}\ , a). \tag{A1}$$

We prove $\Delta\,K_I = 0$ by deriving a contradiction from (A1).

Consider the solid the same as the one shown in Fig. 3, but with no transformation strain in Ω or Ω^*. The loading is a traction $t_i^{(o)}$ applied through a contour Σ outside Ω^*. The corresponding crack intensity factor $K_I(t_i^{(o)})$ is obtained from equations (8) and (A1) as

$$K_I(t_i^{(o)}) = \frac{H}{4\Delta K_I(a)} \int_\Sigma t_i^{(o)} \frac{\partial}{\partial a} \Delta u_i(\underset{\sim}{x}, a)\, d\Sigma$$

$$= \frac{H}{4\Delta K_I(a)} \frac{\partial}{\partial a} \int_\Sigma t_i^{(o)} \Delta u_i (\underset{\sim}{x}, a)\, d\Sigma$$

$$= 0.$$

The last equality holds since Δu_i is identically zero in domain $D - \Omega^*$.

Note that the concentrated load $t_i^{(o)}$ is arbitrary and thereby the stress intensity factor $K_I(t_i^{(o)})$, induced by $t_i^{(o)}$ cannot be identically zero. The contradiction is caused by equation (A1). Therefore, we conclude that

$$\Delta K_I(a) = 0,$$

which means $\epsilon_{ij}^p(\underset{\sim}{x})$ and $\tilde{\epsilon}_{ij}^p(\underset{\sim}{x})$ cause the same stress intensity factor.

Cracks at the Extremities of Cylindrical Fibre Inclusions

A.P.S. Selvadurai
Department of Civil Engineering
Carleton University
Ottawa, Ontario, Canada K1S 5B6

ABSTRACT

Studies on short fibre reinforced brittle composite materials indicate that the fibre ends act as crack initiators. An assessment of the behaviour of such cracks is of considerable interest to the adequate design of short fibre reinforced composites. This paper examines the boundary element modelling of the behaviour of penny-shaped cracks that are developed at the extremities of cylindrical elastic fibre inclusions.

INTRODUCTION

The integrity of the bond between a fibre and the surrounding matrix is of fundamental importance to the development of adequate reinforcing action in fibre reinforced composites. Debonding and cracking at a fibre-matrix interface can be initiated by a variety of factors including stress concentrations at sharp edges, inhomogeneities, thermal mismatch between the matrix and the reinforcement and other environmentally induced loading effects (Figure 1). The evaluation of the influences of such defects on fracture propagation, stiffness degradation, etc., can contribute to the accurate modelling and efficient design of fibre reinforced composites (Sih and Tamuzs, 1979; Selvadurai, 1981; Hashin and Herakovich, 1983; Kelly

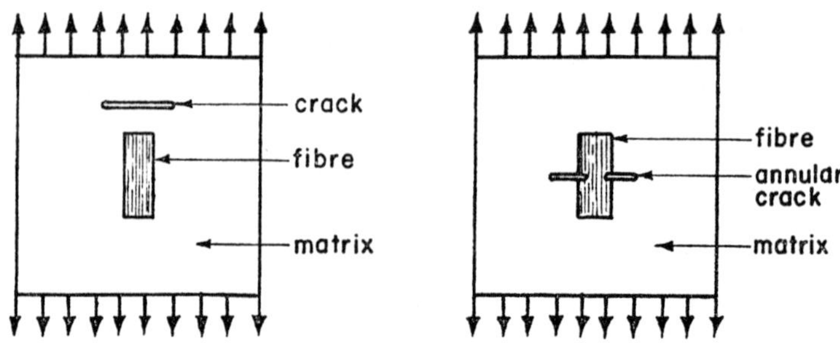

(a) Interaction of fibre inclusion with crack.

(b) Fibre and matrix fracture.

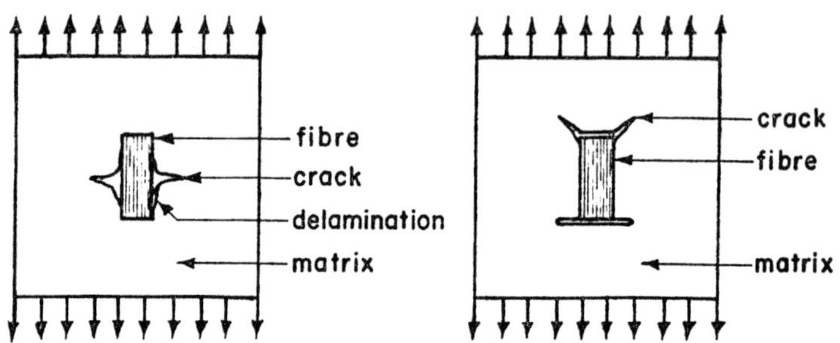

(c) Matrix fracture and interface delamination.

(d) Cracking at fibre extremities.

Figure 1. Interaction of fibre reinforcement with defects

and Rabotnov, 1983). This paper examines the elastostatic crack-fibre interaction problem for a single cylindrical elastic fibre which is embedded in an isotropic elastic matrix of infinite extent. The composite is weakened by penny-shaped cracks which are located at the plane end of the cylindrical fibre. In this instance the sharp boundaries at the plane end of the fibre act as crack initiators (see e.g., Taya and Mura (1981) and Mura (1982)). The composite region containing the fibre with plane end cracks is subjected to a uniaxial stress field (Figure 2) which induces stress concentrations at the penny-shaped crack boundaries. The analysis of the problem focusses on the evaluation of the stress intensity factors at the tip of the penny-shaped crack. A boundary element technique is used to determine the flaw opening mode and flaw shearing mode stress intensity factors at the crack tip. The numerical scheme is used to assess the manner in which these stress intensity factors are influenced by the elasticity mismatch between the fibre and the surrounding matrix and other geometrical parameters such as the length to diameter ratio of the fibre and the radius of the crack in relation to the radius of the fibre. The accuracy of the numerical scheme is verified by comparison with analytical solutions developed for classical penny-shaped crack problems related to an isotropic elastic medium (Sneddon, 1946; Kassir and Sih, 1975).

BOUNDARY ELEMENT METHODS

The formulation of the boundary element method for elastostatic problems is given by Brebbia (1978) and Banerjee and Butterfield (1981). In this section we shall present a brief account of the boundary element equations applicable to a bi-material elastic region. The generalization to a bi-material region will enable the examination of the crack-fibre interaction problem shown in Figure 2. Attention is restricted to isotropic elastic materials which satisfy the linear elastic stress-strain relationships

$$\sigma_{ij}^{(\alpha)} = \lambda_\alpha \delta_{ij} u_{k,k}^{(\alpha)} + G_\alpha \left\{ u_{i,j}^{(\alpha)} + u_{j,i}^{(\alpha)} \right\} \tag{1}$$

and the Navier equations

$$G_\alpha \nabla^2 u_i^{(\alpha)} + (\lambda_\alpha + G_\alpha)\, u_{k,ki}^{(\alpha)} = 0 \tag{2}$$

where G_α and λ_α are Lame's constants; the subscript or superscript 'α' refer to the matrix (m) or fibre (f) regions; u_i and σ_{ij} are respectively the displacement components and stress tensor referred to the rectangular Cartesian coordinate system x, y, z; $i, j = x, y, z$; $\lambda_\alpha = 2G_\alpha \nu_\alpha / (1 - 2\nu_\alpha)$; ν_α are Poisson's ratios; ∇^2 is Laplace's operator referred to the rectangular Cartesian coordinate system and δ_{ij} is Kronecker's delta function. Here and in the sequel the Greek indices and superscripts will refer to quantities pertaining to the matrix and

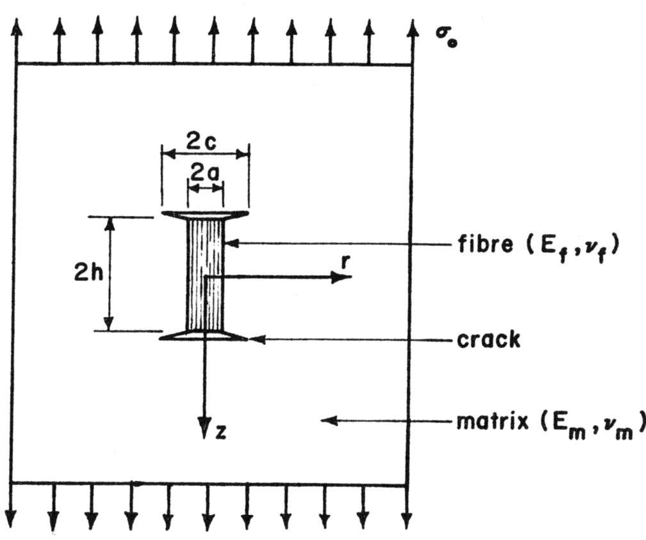

Figure 2. Delamination and fracture at the plane ends of an elastic fibre

fibre regions.

The boundary integral equation for the axisymmetric problem pertaining to the fibre-matrix composite region can be written in the form (see e.g. Kermanidis (1975); Cruse and Wilson (1977))

$$c_{\ell k} u_k^{(\alpha)} + \int_{\Gamma_\alpha} \left\{ P_{\ell k}^{*(\alpha)} u_k^{(\alpha)} - u_{\ell k}^{*(\alpha)} P_k^{(\alpha)} \right\} \frac{r}{r_i} d\Gamma = 0 \tag{3}$$

where Γ_α is the boundary of the region α; $u_k^{(\alpha)}$ and $P_k^{(\alpha)}$ are respectively the displacements and tractions on the boundary Γ_α and $u_{ik}^{*(\alpha)}$ and $P_{ik}^{*(\alpha)}$ are fundamental solutions. Also in (3) c_{ij} is a constant ($= 0$, if the point is outside the body; $= \delta_{ij}$ if the point is inside the body; $= \delta_{ij}/2$ if the point is located at a smooth boundary and $=$ a function of discontinuity at a corner and of Poisson's ratio (Banerjee and Butterfield, 1981)).

For axial symmetry

$$\begin{aligned} u_{rr}^{*(\alpha)} &= C_1 \left\{ \frac{4(1-\nu_\alpha)(\rho^2+\overline{z}^2)-\rho^2}{2r\overline{R}} \right\} K(\overline{m}) \\ &- \left\{ \frac{(7-8\nu_\alpha)}{4r}\overline{R} - \frac{(e^4-\overline{z}^4)}{4r\overline{R}^3 m_1} \right\} E(\overline{m}) \end{aligned} \tag{4}$$

$$u_{rz}^{*(\alpha)} = C_1 \overline{z} \left[\frac{(e^2+\overline{z}^2)}{2\overline{R}^3 m_1} E(\overline{m}) - \frac{1}{2\overline{R}} K(\overline{m}) \right] \tag{5}$$

$$u_{zr}^{*(\alpha)} = C_1 r_i \overline{z} \left[\frac{(e^2 - \overline{z}^2)}{2r\overline{R}^3 m_1} E(\overline{m}) + \frac{1}{2r\overline{R}} K(\overline{m}) \right] \tag{6}$$

$$u_{zz}^{*(\alpha)} = C_1 r_i \left[\frac{(3 - 4\nu_\alpha)}{\overline{R}} K(\overline{m}) + \frac{\overline{z}^2}{\overline{R}^3 m_1} E(\overline{m}) \right] \tag{7}$$

where

$$\overline{z} = (z - z_i) \quad ; \quad \overline{r} = (r + r_i) \quad ; \quad \rho^2 = \left(r^2 + r_i^2\right)$$

$$e^2 = \left(r^2 - r_i^2\right) \quad ; \quad \overline{R}^2 = \overline{r}^2 + \overline{z}^2 \quad ; \quad C_1 = \frac{1}{4\pi G_\alpha (1 - \nu_\alpha)} \tag{8}$$

$$\overline{m} = \frac{2rr_i}{\overline{R}^2} \quad ; \quad m_1 = 1 - \overline{m}$$

and $K(\overline{m})$ and $E(\overline{m})$ represent, respectively the complete elliptic integrals of the first and second kind. The corresponding terms for the traction fundamental solution $P_{\ell k}^{*(\alpha)}$ can be obtained by the manipulation of the results (4) to (7).

Upon discretization of the boundaries Γ_α into boundary elements (Figure 3), the integral equation (3) can be represented in the form of a boundary element matrix equation as follows:

$$\begin{bmatrix} \mathbf{H}^{(\alpha)} & \mathbf{H}_I^{(\alpha)} \end{bmatrix} \begin{bmatrix} \mathbf{u}^{(\alpha)} \\ \mathbf{u}_I^{(\alpha)} \end{bmatrix} = \begin{bmatrix} \mathbf{M}^{(\alpha)} & \mathbf{M}_I^{(\alpha)} \end{bmatrix} \begin{bmatrix} \mathbf{P}^{(\alpha)} \\ \mathbf{P}_I^{(\alpha)} \end{bmatrix} \tag{9}$$

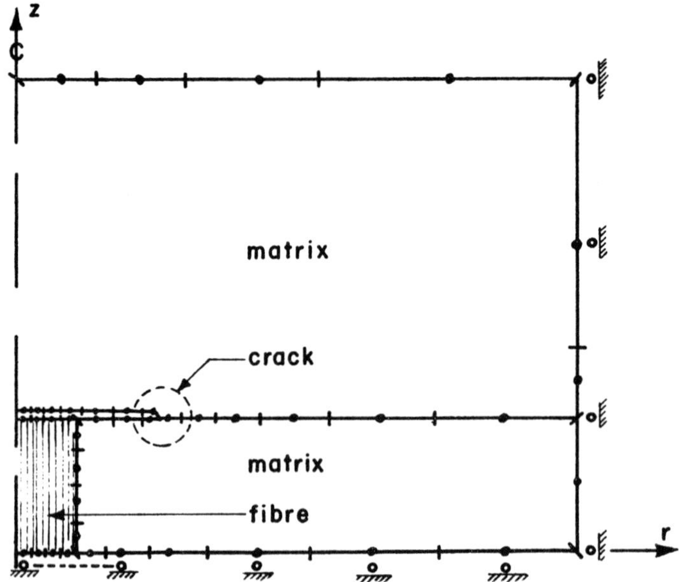

Figure 3. The boundary element discretization

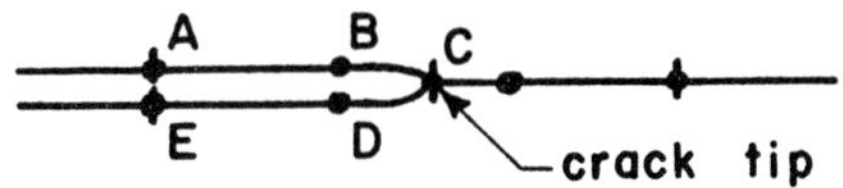

Figure 4. Detail at the crack tip

where **H**'s and **M**'s are the influence coefficient matrices derived from the integration of the fundamental solutions $P_{\ell k}^{*(\alpha)}$ and $u_{\ell k}^{*(\alpha)}$ respectively. In the instance where there is complete bonding between the fibre-matrix interface we have

$$\begin{aligned} \mathbf{u}_I^{(f)} &= \mathbf{u}_I^{(m)} = \mathbf{u}_I \\ \mathbf{P}_I^{(f)} &= -\mathbf{P}_I^{(m)} = P_I \end{aligned} \tag{10}$$

Using the above result, the complete matrix equation governing the fibre composite-crack interaction problem can be expressed in the form

$$\begin{bmatrix} \mathbf{H}^{(f)} & \mathbf{H}_I^{(f)} & 0 \\ 0 & \mathbf{H}_I^{(m)} & \mathbf{H}^{(m)} \end{bmatrix} \begin{bmatrix} \mathbf{u}^{(f)} \\ \mathbf{u}_I \\ \mathbf{u}^{(m)} \end{bmatrix} =$$

$$\begin{bmatrix} \mathbf{M}^{(f)} & \mathbf{M}_I^{(f)} & 0 \\ 0 & \mathbf{M}_I^{(m)} & \mathbf{M}^{(m)} \end{bmatrix} \begin{bmatrix} \mathbf{P}^{(f)} \\ \mathbf{P}_I \\ \mathbf{P}^{(m)} \end{bmatrix} \tag{11}$$

MODELLING OF CRACK TIP BEHAVIOUR

In the boundary element discretizations discussed in the previous section, quadratic elements will be employed to model the boundaries of the matrix and fibre regions. That is, the variation of the displacements and tractions within an element can be described by

$$\left.\begin{array}{l} u_i^{(\alpha)} \\ \\ P_i^{(\alpha)} \end{array}\right\} = a_0 + a_1\zeta + a_2\zeta^2 \tag{12}$$

where ζ is the local coordinate of the element and a_r ($r = 0, 1, 2$) are constants of interpolation. However, in the context of linear elastic fracture mechanics the stress field at the crack tip should contribute to a $1/\sqrt{r}$ type singularity. In the finite element technique, the quarter point element of the type proposed by Henshell and Shaw (1975) and Barsoum (1976) can be used to model the required $\sqrt{r}$ type variation of the displacements. That is

$$\left.\begin{array}{l} u_i^{(\alpha)} \\ \\ P_i^{(\alpha)} \end{array}\right\} = b_0 + b_1\sqrt{r} + b_2 r \tag{13}$$

if the same type of element is implemented in a boundary element method. Since the $P_i^{(\alpha)}$ in (13) does not produce a $1/\sqrt{r}$ type singularity, Cruse and Wilson (1977) developed the so-called "singular traction quarter point boundary element", where the traction variations in (13) are multiplied by a non-dimensional $\sqrt{\ell/r}$ where ℓ is the length of the crack tip element. The variations of tractions can be expressed in the form

$$P_i = \frac{c_0}{\sqrt{r}} + c_1 + c_2\sqrt{r} \tag{14}$$

where b_i and c_i ($i = 0, 1, 2$) are constants. The performance of both types of quarter point elements have been studied by Blandford et al. (1981), Smith and Mason (1982) and Selvadurai and Au (1987,1989) and their accuracy established by comparison with known exact solutions.

In the crack-fibre interaction problem examined in this paper the axial straining induces a state axial symmetry in the fibre-matrix composite region. Consequently only the Mode I and Mode II stress intensity factors are present at the tips of the penny-shaped crack region (Figure 2). The flaw opening mode stress intensity factor can be evaluated by applying the displacement correlation method which utilizes the nodal displacements at four locations A, B, E, D and the crack tip (Figure 4) i.e.

$$K_I^{(\alpha)} = \frac{G_\alpha}{(k_\alpha + 1)}\sqrt{\frac{2\pi}{\ell_0}}\,\{4\,[u_z(B) - u_z(D)] + u_z(E) - u_z(A)\} \tag{15}$$

where $k_\alpha = (3 - 4\nu_\alpha)$ and ℓ_0 is the length of the crack tip element. Similarly the flaw shearing mode stress intensity factor can be written in the form

$$K_{II}^{(\alpha)} = \frac{G_\alpha}{(k_\alpha + 1)}\sqrt{\frac{2\pi}{\ell_0}}\,\{4\,[u_r(B) - u_r(D)] + u_r(E) - u_r(A)\} \tag{16}$$

NUMERICAL RESULTS

Owing to the symmetry of the crack-fibre interaction problem about the plane $z = 0$, it is sufficient to consider a boundary element discretization which is restricted to the region $z \geq 0$. The Figure 3 shows the associated boundary element discretization. The matrix region is modelled as a region of infinite extent. The accuracy of the boundary element modelling has been verified by comparison with exact analytical solution to the problem of a penny-shaped crack located in an isotropic elastic medium which is subjected to a uniaxial state of stress σ_0 (Sneddon, 1946). In this case, the flaw opening mode stress intensity factor is given by

$$K_I = \frac{2\sigma_0\sqrt{c}}{\pi} \tag{17}$$

where c is the radius of the penny-shaped crack. The boundary element scheme provides numerical estimates for this stress intensity factor to within an accuracy of 5 percent.

In the numerical evaluation of the stress intensity factors at the extremities of the end cracks a number of factors need to be taken into consideration. These include (i) the length/diameter ratio of the elastic fibre (h/a), (ii) the radius of the crack in relation to the radius of the fibre (c/a), (iii) Poisson's ratios of the matrix and fibre materials (ν_m, ν_f) and (iv) the modular ratio of the fibre in relation to the matrix (E_f/E_m). For purposes of illustration, these non-dimensional parameters are assigned the following explicit or range of values; $(h/a)\epsilon(3,10)$; $(c/a)\epsilon(1.5,3.0)$; $\nu_f = \nu_m = 0.2$; $(E_f/E_m)\,\epsilon\,(1,10^3)$.

The Figures 5, 6 and 7 illustrate the variations in which the flaw opening mode stress intensity factor at the crack tip. From the results given in Figure 7, it becomes evident that as the modular ratio $E_f/E_m \to 1$, $c/a > 1$ and when $h/a > 10$, the interaction between the cracks at the extremities of the fibre inclusion is less significant and we obtain from the numerical results, the relevant stress intensity factor for the classical problem of a penny-shaped crack in an elastic solid of infinite extent. As h/a becomes small (in the range 3 to 5) the interaction between the cracks influences the result for K_I even for the case when $(E_f/E_m) \to 1$ and $(c/a) > 1$. The results of the numerical investigations also indicate that an increase in the modular ratio E_f/E_m has the effects of amplifying the flaw opening mode stress intensity factor at the crack tip. This amplification becomes particularly significant as (h/a) increases and as $(c/a) \to 1$.

The Figures 8, 9 and 10 illustrate the manner in which the flaw shearing mode stress intensity factor K_{II} is influenced by the geometric and material parameters indicated previously. It is evident that the flaw shearing or mode II stress intensity factor is considerably smaller than the corresponding mode I stress intensity factor. The effects of mode II stress intensity factor become appreciable only as $(E_f/E_m) \to 1$. For values of $(E_f/E_m) > 10^2$, the stiffness of the cylindrical inclusion is

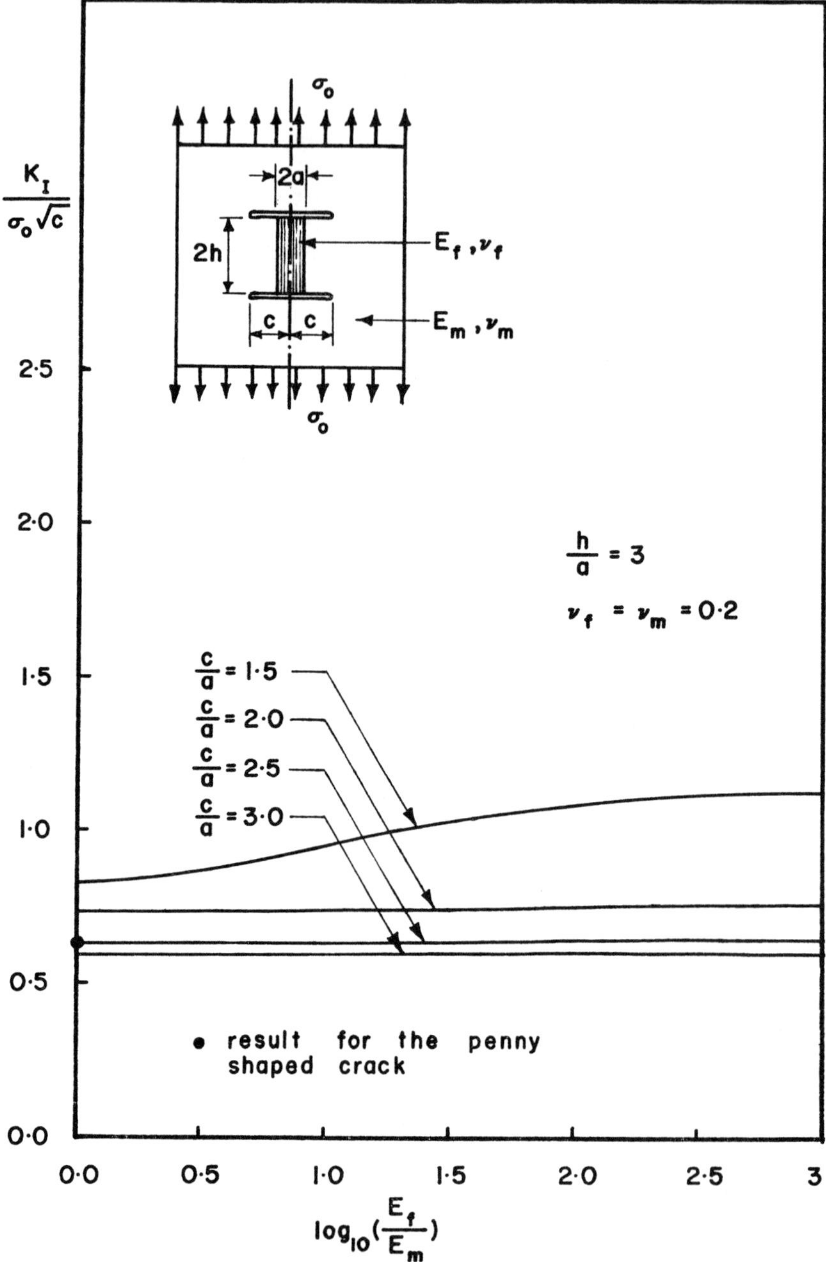

Figure 5. Mode I stress intensity factor at the crack tip

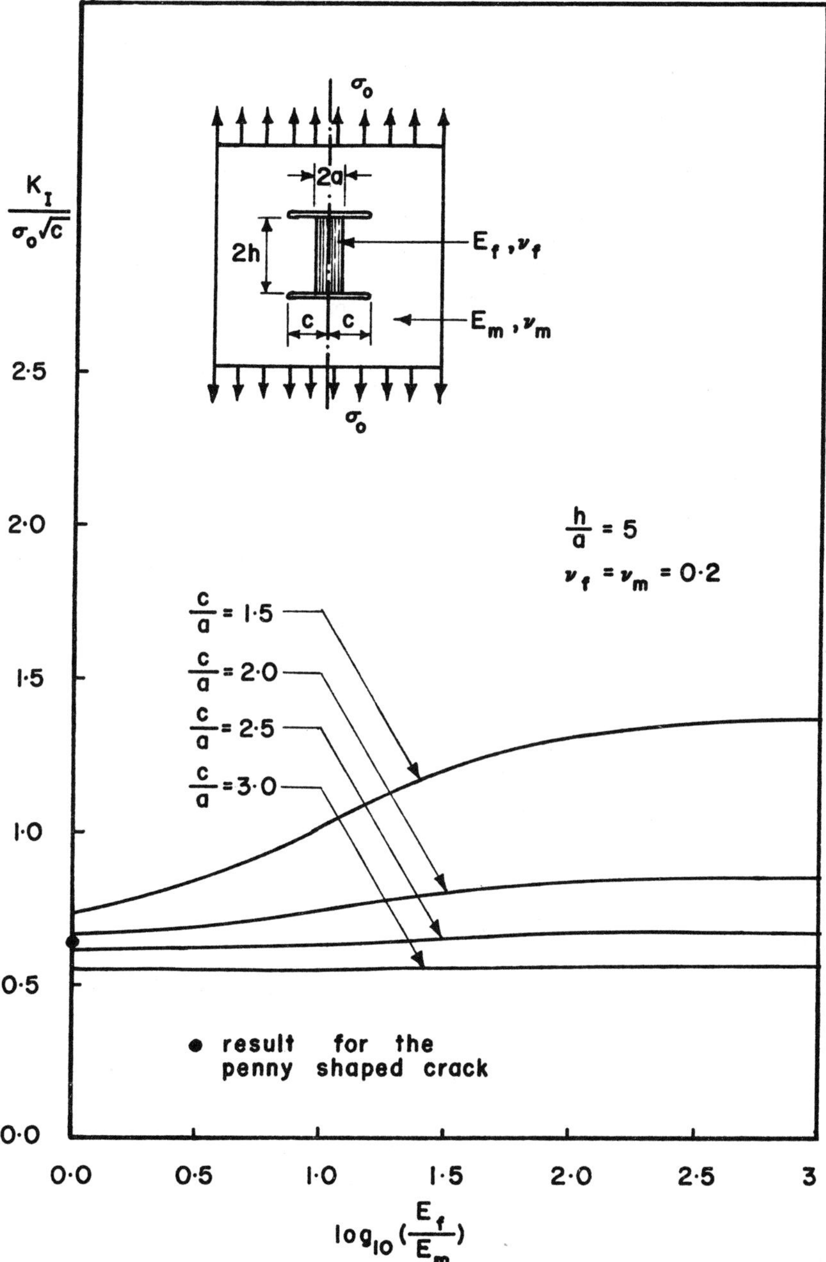

Figure 6. Mode I stress intensity factor at the crack tip

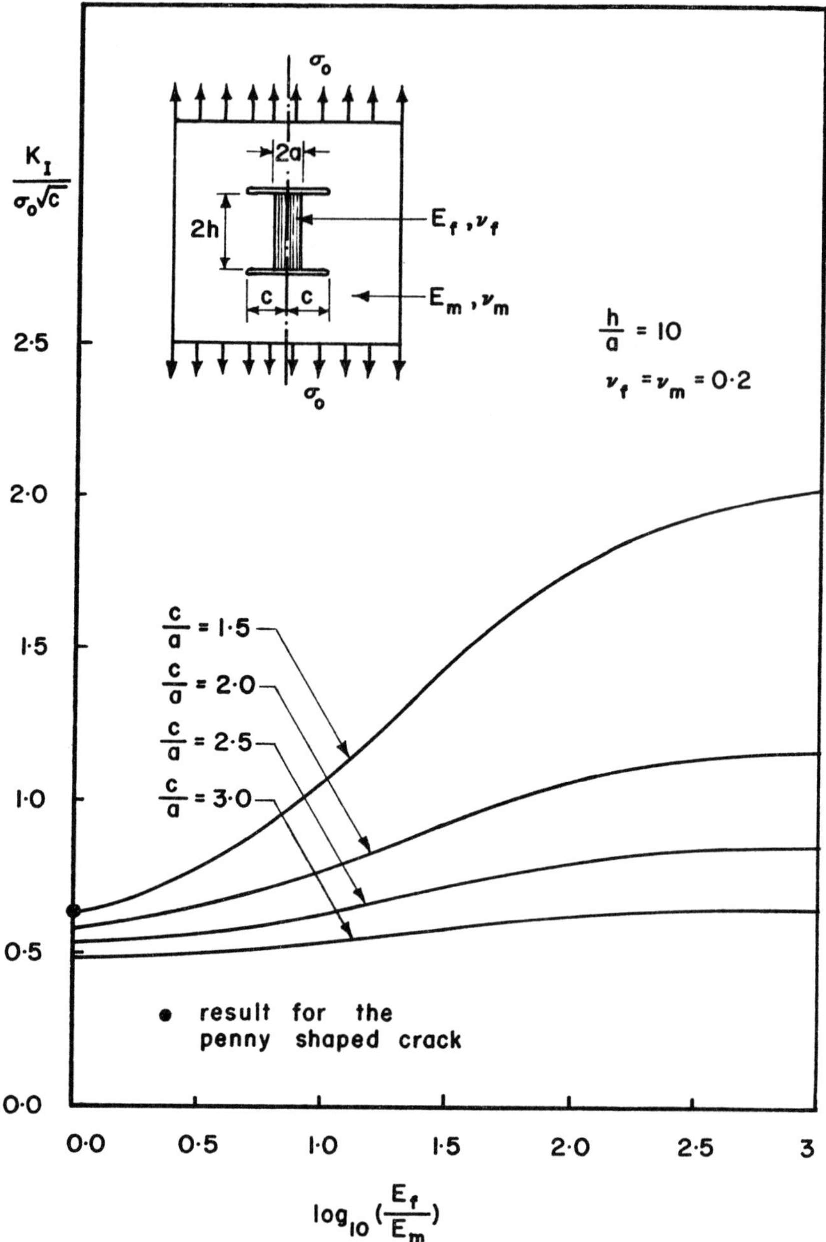

Figure 7. Mode I stress intensity factor at the crack tip

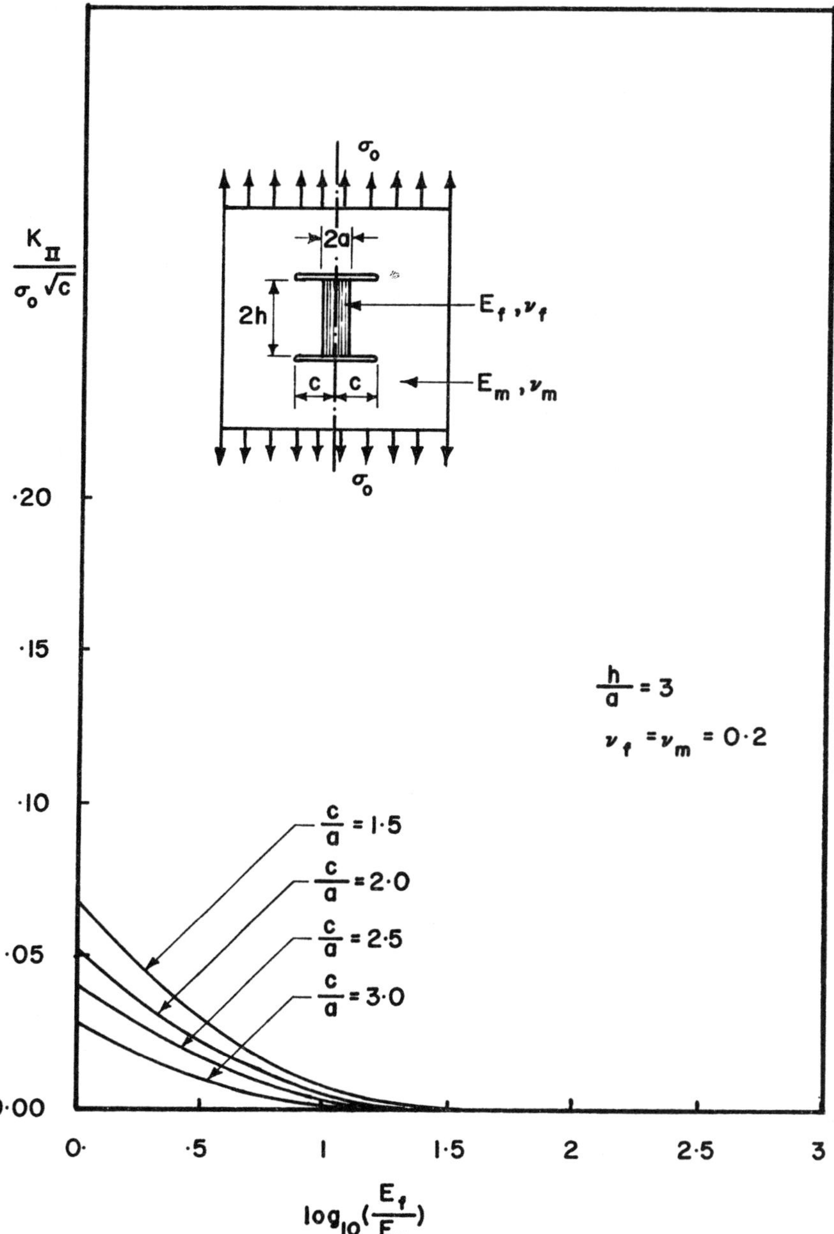

Figure (8. Mode II stress intensity factor at the crack tip

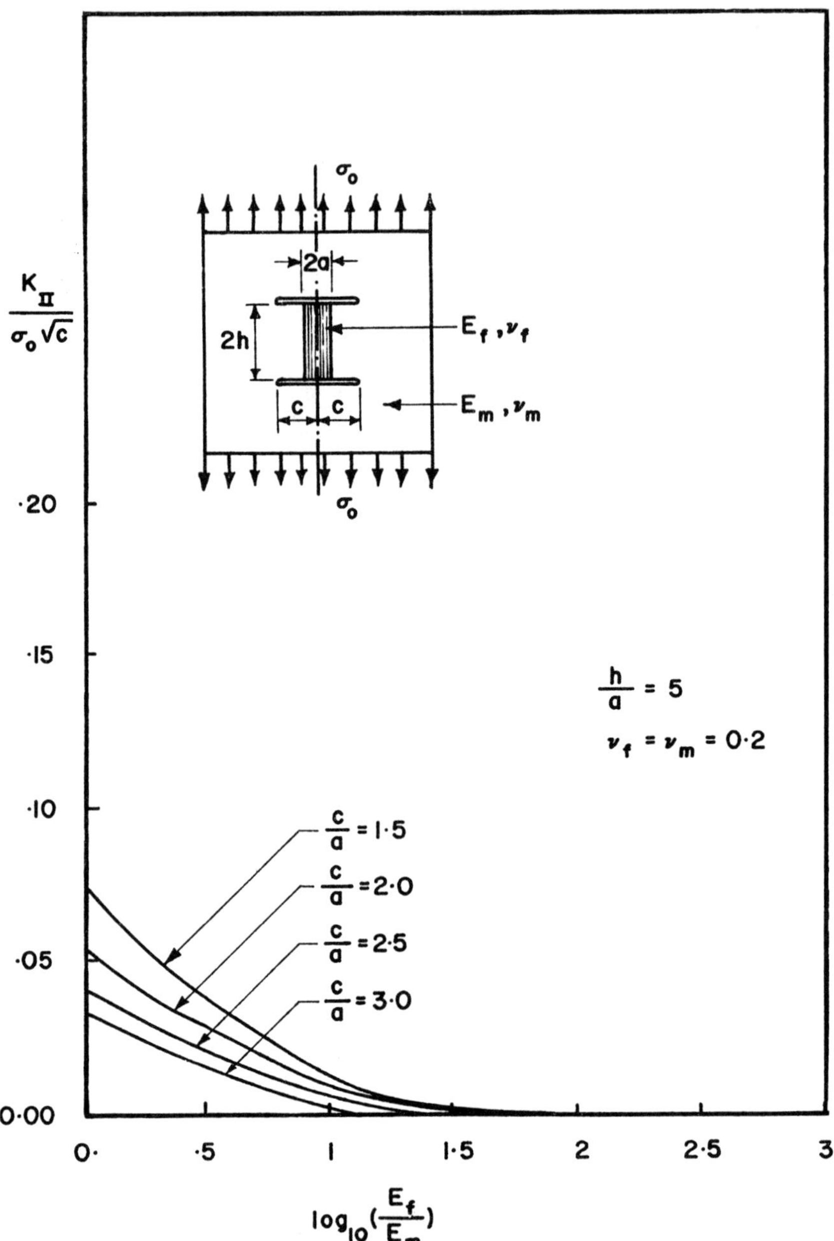

Figure 9. Mode II stress intensity factor at the crack tip

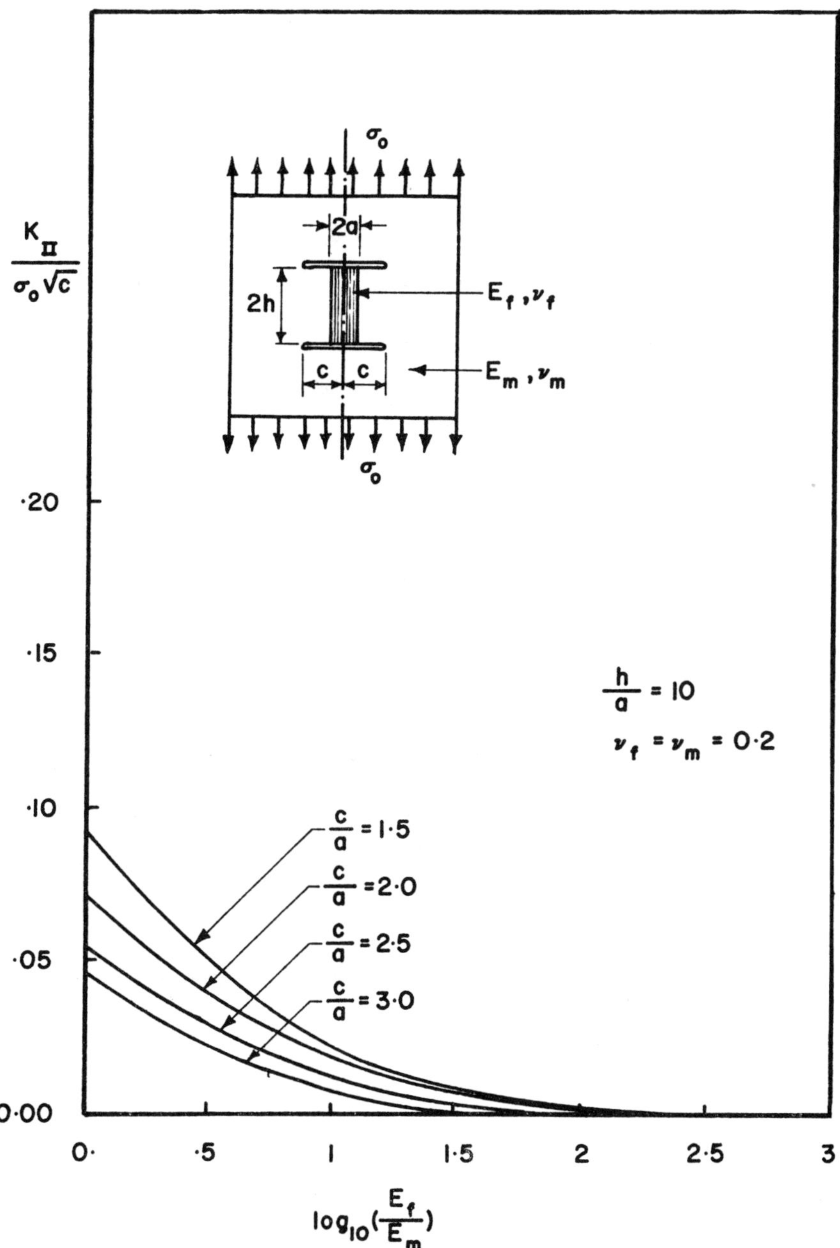

Figure 10. Mode II stress intensity factor at the crack tip

sufficient to suppress the flaw shearing mode for all choices of (h/a) and (c/a).

DISCUSSION

This paper presents a boundary element approach to the study of elastostatic interaction between a reinforcing fibre inclusion and penny-shaped cracks which are developed at its plane ends. The analysis focusses on the idealized situation where two penny-shaped cracks are located symmetrically at the ends of the cylindrical fibre inclusion. The boundary element formulation accounts for elasticity mismatch between the fibre and the matrix and other parameters which relate to the geometric aspect ratio of the inclusion and that of the crack. The boundary element scheme accounts for the stress singularity that is present in a crack located in a homogeneous solid. It is shown that both the flaw opening and flaw shearing mode stress intensity factors are considerably influenced by the elasticity mismatch between the fibre and the matrix and the geometry parameters. In the special case where the crack reduces to a delamination region at the plane ends of the cylindrical inclusion, the crack tip is located at the junction of a bi-material elastic region. Any stress singularities that are computed via the current scheme will give only approximate estimates for the oscillatory stress singularities that can occur at a crack tip located at the junction of a bi-material region (Atkinson, 1979). In such situations the boundary element procedure needs to be modified to account for the oscillatory phenomena (see e.g., Lee and Choi, 1988; Yuuki and Cho, 1988,1989). Similar considerations apply for situations where the debonding at the plane ends is restricted to a limited region within the cross sectional area of the fibre.

REFERENCES

Atkinson, C., 1979, Stress singularities and fracture mechanics, *Applied Mechanics Reviews*, Vol.32, pp.123-135.

Banerjee, P.K., and Butterfield, R., 1981, *Boundary Element Methods for Engineering Science*, McGraw-Hill, New York.

Barsoum, R.S., 1976, "On the use of isoparametric finite elements in linear fracture mechanics", *Int. J. Num. Meth. Engng.*, Vol.10, pp.25-37.

Blandford, G.E., Ingraffea, A.R., and Liggett, J.A., 1981., "Two-dimensional stress intensity factor computations using the boundary element method", *Int. J. Num. Meth. Engng.*, Vol.17, pp.387-404.

Brebbia, C.A., 1978, *Boundary Element Methods for Engineers*, Pentech Press, London.

Cruse, T., and Wilson, R.B., 1977, *Boundary Integral Equation Method for Elastic Fracture Mechanics*, AFSOR-TR-78-0355, 10-11.

Hashin, Z., and Herakovich, C.T., Eds., 1983, *Mechanics of Composite Materials: Recent Advances*, Proceedings of the IUTAM Symposium on Mechanics of Composite Materials, Blacksburg, Va., Pergamon Press, New York.

Henshell, R.D., and Shaw, K.G., 1975, "Crack tip finite elements are unnecessary", *Int. J. Num. Math. Eng.*, Vol.9, pp.495-507.

Kassir, M.K., and Sih, G.C., 1975, *Three-Dimensional Crack Problems*, Vol.2, *Mechanics of Fracture* (Sih, G.C., Ed.), Noordhoft Int. Publ. Leyden, The Netherlands.

Kelly, A., and Rabotnov, Yu, N., Eds., 1983, *Handbook of Composites*, Vol.1-4, North Holland, Amsterdam, The Netherlands.

Kermanidis, Th., 1975, "Numerical solution for axially symmetrical elasticity problems", *Int. J. Solids Struct.*, Vol.11, pp.495-500.

Lee, K.Y., and Choi, H.J., 1988, "Boundary element analysis of stress intensity factors for bi-material interface cracks", *Engng. Fract. Mech.*, Vol.29, pp.461-472.

Mura, T., 1982, *Micromechanics of Defects in Solids*, Martinus Nijhoff Publ., The Hague, Netherlands.

Selvadurai, A.P.S., (Ed.), 1981, *Mechanics of Structural Media, Proceedings of the Int. Symp. on the Mechanical Behaviour of Structured Media*, Ottawa, Studies in Applied Mechanics, Vols. 5A and 5B, Elsevier Scientific Publ. Co., Amsterdam, The Netherlands.

Selvadurai, A.P.S., and Au, M.C., 1988, "Cracks with frictional surfaces: A boundary element approach", *Proceedings of the 9th Boundary Element Conference*, (Ed. Brebbia, C.A., Southampton, UK, Springer-Verlag, Berlin, pp.211-230.

Selvadurai, A.P.S., and Au, M.C., 1989, "Crack behaviour associated with contact problems with non-linear interface constraints", *Boundary Element Techniques: Applications in Engineering*, (Brebbia, C.A. and Zamani, N., Eds.), Proceedings of the Boundary Element Technology Conference, Windsor, Ontario, Computational Mechanics Publ., pp.1-17.

Sih, G.C., and Tamuzs, V.P., (Eds.), 1979, *Fracture of Composite Materials, Proceedings of the 1st USA-USSR Symposium*, Sijthoff and Noordhoff, The Netherlands.

Smith, R.N.L., and Mason, J.C., 1982, "A boundary element method for curved crack problems in two dimensions", in *Boundary Element Methods in Engineering*, (Brebbia, C.A., Ed.), Springer-Verlag, Berlin, pp.472-484.

Sneddon, I.N., 1946, "The stress distribution in the neighbourhood of a crack in an elastic solid", *Proc. Roy. Soc., Ser.A*, Vol.187, pp.229-260.

Taya, M., and Mura, T., 1981, "On the stiffness and strength of an aligned short-fibre reinforced composite containing fibre-end cracks under uniaxial applied stress", *J. Appl. Mech.*, Vol.48, pp.361-367.

Yuuki, R., and Cho, S.B., 1988, "Boundary element analysis of the stress intensity factors for an interface crack in dissimilar materials", in *Boundary Element Methods in Applied Mechanics*, (Tanaka, M., and Cruse, T.A., Eds.), Pergamon Press,

New York, pp.179-190.

Yuuki, R., and Cho, S.B., 1989, "Efficient boundary element analysis of stress intensity factors for interface cracks in dissimilar materials", *Engng. Fracture Mech.*, Vol.34, pp.179-188.

Inelastic Behavior I

Bounds for the Creep Behaviour of Polycrystalline Materials

R. Dendievel,[1] G. Bonnet[1] and J. R. Willis[2]

[1]Laboratoire Génie Physique et Mécanique des Matériaux,
INPG ENSPG,
Domaine Universitaire,
BP 46,
38402 Saint-Martin d'Heres cedex,
FRANCE.

[2]School of Mathematical Sciences,
University of Bath,
Bath, BA2 7AY,
U.K.

Abstract

A polycrystalline material is considered, in which any individual crystal deforms by slip on a set of crystallographic slip systems. The rate of shear on any one system is assumed to be a function of the resolved shear stress on that system. A power law is used for illustration and this allows the overall stress-strain-rate behaviour to be defined in terms of a characterizing stress $\bar{\tau}_0$. Elementary bounds, corresponding to the Voigt and Reuss bounds of linear elasticity, are easy to generate for $\bar{\tau}_0$, and are displayed. The main result, however, is the development of a new upper bound for $\bar{\tau}_0$, which corresponds to the Hashin-Shtrikman upper bound in the linear case. The derivation makes use of elementary results from convex analysis.

1. Introduction

During the course of work directed mainly towards finding self-consistent estimates for the overall creep behaviour of cubic polycrystals, Hutchinson (1976) also presented some elementary bounds, analogous to the Voigt and Reuss bounds of linear elasticity. The self-consistent method involves an approximation and so, in a sense, the elementary bounds represent the only mathematically rigorous information that is available for the polycrystals considered by Hutchinson. The present work is devoted to the development of new bounds, of Hashin-Shtrikman type, for the overall strain-rate potential of polycrystals. The formalism allows for single-crystal behaviour more general than that assumed by Hutchinson (1976) and the polycrystal itself could display texture. Detailed results, however, are presented here for precisely the type of polycrystal considered by Hutchinson. For these, single crystal behaviour is modelled as pure power-law creep on crystallographic slip systems, the polycrystal is taken as isotropic (with all crystallite orientations equally likely) and the relation between rate of shear and resolved shear stress is taken as the same for each slip system of the same crystallographic type. Implementation of the self-consistent method is easier for uniaxial loading than for general loading and Hutchinson's self-consistent results were restricted to this case. In contrast, the present bound can be calculated for any loading and a complete (upper bound) characterization of the polycrystal's response is obtained. The elementary (Voigt-type) upper bound follows directly from the present formalism as a limiting case and is presented, again for any loading, for comparison. Also, for completeness, the elementary lower bound (of Reuss type) is calculated for any loading.

2. General framework

The polycrystals to be considered fall within a general class of composite materials, with n distinct constituents firmly bonded across interfaces (the possibility $n \to \infty$ being included). Material of type r has constitutive relation

$$\varepsilon_{ij} = \frac{\partial F_r}{\partial \sigma_{ij}} \tag{2.1a}$$

in suffix notation or, symbolically,

$$\varepsilon = F_r'(\sigma) \quad . \tag{2.1b}$$

In equations (2.1), ε is the Eulerian strain-rate tensor and σ is Cauchy stress. The function F_r is the strain-rate potential for material of type r; it is a function of the current stress σ (with respect to which it is differentiated in equations (2.1)) but it may also (in general) depend on the history of stress and deformation. In any event, F_r is taken to be a convex function of σ (Rice 1971, Mandel 1972).

The composite material is heterogeneous, with creep potential $F(\sigma, x)$, given by

$$F(\sigma, x) = \sum_{r=1}^{n} F_r(\sigma)\, f_r(x) \quad , \tag{2.2}$$

where f_r is the characteristic function of the region occupied by material r and is zero otherwise.

The overall creep potential $\tilde{F}(\bar{\sigma})$ of the composite is defined as the mean value of $F(\sigma, x)$, when the composite is subjected to boundary velocities

$$v_i = \bar{\varepsilon}_{ij} x_j \quad , \quad x \in \partial\Omega \quad . \tag{2.3}$$

Here, the composite is taken to occupy a domain Ω and $\bar{\varepsilon}$ is a constant tensor. The average value of the stress σ over Ω is denoted by $\bar{\sigma}$ and $\bar{\varepsilon}$ is chosen to generate a prescribed value for $\bar{\sigma}$. Then, as discussed by Willis (1989), $\tilde{F}(\bar{\sigma})$ is given *precisely* by

$$\tilde{F}(\bar{\sigma}) = \inf \int_\Omega F(\sigma, x)\, dx \quad , \tag{2.4}$$

the infimum being taken over stress fields σ which have zero divergence and whose mean value over Ω is $\bar{\sigma}$. In (2.4), units are chosen, for convenience, so that Ω has unit volume. Of course, for a composite, Ω should be chosen to be at least as large as a 'representative volume element', for which response to any 'macroscopically uniform' boundary condition should be the same: thus, in particular, the alternative boundary condition

$$\sigma_{ij}\, n_j = \bar{\sigma}_{ij}\, n_j \quad , \quad x \in \partial\Omega \tag{2.5}$$

should lead to the same overall response, to whatever accuracy is specified, though (2.4) in fact always yields the *smallest* $\tilde{F}(\bar{\sigma})$.

An elementary upper bound for $\tilde{F}(\bar{\sigma})$ follows directly by substituting the admissible stress $\sigma = \bar{\sigma}$ into the integral in (2.4):

$$\tilde{F}(\bar{\sigma}) \leq \int_\Omega F(\bar{\sigma}, x)\, dx \quad . \tag{2.6}$$

This is the analogue of the Reuss bound of linear elasticity; it is discussed further in Section 6.

A family of lower bounds is now developed, following Willis (1983, 1986), Talbot and Willis (1985, 1987) and Ponte Castañeda and Willis (1988), by introducing a function $F_0(\sigma)$ and defining

$$V(\eta, x) = \sup_\sigma \{\sigma \cdot \eta - F(\sigma, x) + F_0(\sigma)\} \quad . \tag{2.7}$$

It follows that

$$F(\sigma, x) \geq \sigma \cdot \eta + F_0(\sigma) - V(\eta, x) \tag{2.8}$$

and hence that

$$\tilde{F}(\bar{\sigma}) \geq \inf \int_\Omega [\sigma \cdot \eta + F_0(\sigma) - V(\eta, x)]\, dx \quad , \tag{2.9}$$

the infimum being taken over stress fields σ that are admissible for (2.4), for any η.

The infimum in (2.9) is easily calculated (at least formally) if F_0 is chosen to be quadratic:

$$F_0(\sigma) = \tfrac{1}{2}\, \sigma \cdot M_0 \sigma \quad , \tag{2.10}$$

where M_0 is a fourth-order 'tensor of compliances'. The infimum is then attained when

$$\sigma = \bar{\sigma} - \Delta\eta \quad , \tag{2.11}$$

where Δ is a linear operator, related to the Green's function for a linear 'comparison medium' with tensor of compliances M_0 (Willis, 1981). Employing (2.11) in (2.9) then gives

$$\tilde{F}(\bar{\sigma}) \geq \int_\Omega (\bar{\sigma}\cdot\eta - \eta\cdot\Delta\eta + \tfrac{1}{2}\eta\cdot\Delta M_0 \Delta\eta - \bar{\sigma}\cdot\Delta\eta + F_0(\bar{\sigma}) - V(\eta, x))\, dx$$

or, upon using the identities

$$\int_\Omega \Delta\eta \, dx = 0 \quad \text{and} \quad \Delta M_0 \Delta = \Delta \quad ,$$

$$\tilde{F}(\bar{\sigma}) \geq \int_\Omega (\bar{\sigma}\cdot\eta - \tfrac{1}{2}\, \eta\cdot\Delta\eta - V(\eta, x))\, dx + \tfrac{1}{2}\, \bar{\sigma}\cdot M_0 \bar{\sigma} \,. \tag{2.12}$$

In the case that each F_r is a quadratic function of σ, (2.12) is one of the bounds, for linearly elastic behaviour, of Hashin and Shtrikman (1962); the general case of convex functions F_r was developed by Willis (1983, 1986) and Talbot and Willis (1985).

Reverting explicitly to the form (2.2) for $F(\sigma, x)$,

$$V(\eta, x) = \sum_{r=1}^{n} V_r(\eta)\, f_r(x) \quad , \tag{2.13}$$

with V_r related to F_r in the obvious way, and $\eta(x)$ is now chosen to have the special form

$$\eta(x) = \sum_{r=1}^{n} \eta_r \, f_r(x) \quad , \tag{2.14}$$

with η_r constants. The bound (2.12) then takes the form

$$\tilde{F}(\bar{\sigma}) \geq \sum_r c_r\{\bar{\sigma}\cdot\eta_r - \tfrac{1}{2}\sum_s c_s\eta_r\cdot B_{rs}\eta_s - V_r(\eta_r)\} + \tfrac{1}{2}\,\bar{\sigma}\cdot M_0\bar{\sigma}\ , \tag{2.15}$$

where

$$c_r = \int_\Omega f_r(x)\ dx \tag{2.16}$$

is the volume fraction of material r and B_{rs} is defined so that

$$c_r c_s\ B_{rs} = \int_\Omega f_r(\Delta f_s)\ dx \quad \text{(no summation on } r,\ s) \ . \tag{2.17}$$

The bound (2.15) is optimized by choosing η_r so that

$$V_r'(\eta_r) + \sum_s c_s\ B_{rs}\ \eta_s = \bar{\sigma}\ ,\quad r = 1,\ 2,\cdots,\ n\ . \tag{2.18}$$

It may be noted that

$$\sum_r c_r c_s\ B_{rs} = \int \Delta f_s\ dx = 0 \quad ,$$

from (2.17), since $\sum_r f_r(x) = 1$. Therefore, equations (2.18) imply that

$$\sum_r c_r\ V_r'(\eta_r) = \bar{\sigma} \quad , \tag{2.19}$$

exactly.

The above discussion has assumed, implicitly, that $V_r(\eta) < \infty$ (otherwise, the inequality (2.15) is true but trivial). It has thus been assumed, implicitly, that each F_r grows at least quadratically when σ is large: this will be the case in all of the examples considered. Now define

$$\sigma_r = V_r'(\eta_r) \tag{2.20}$$

and note that V_r is the convex dual of $F_r - F_0$:

$$V_r = (F_r - F_0)^* \quad . \tag{2.21}$$

It follows (see, for example, Van Tiel, 1984) that

$$\eta_r = V_r^{*}{}'(\sigma_r) = (F_r - F_0)^{**}{}'(\sigma_r) \quad .$$

Then, if $\bar{F}_r$ is defined so that

$$\bar{F}_r = (F_r - F_0)^{**} + F_0 \quad , \tag{2.22}$$

equations (2.18) can be replaced by

$$\sigma_r + \sum_s c_s B_{rs} (\bar{F}_s'(\sigma_s) - M_0\sigma_s) = \bar{\sigma} \tag{2.23}$$

and the optimal bound corresponding to (2.15) is

$$\tilde{F}(\bar{\sigma}) \geq \sum_r c_r \{\tfrac{1}{2}(\bar{\sigma} - \sigma_r)\cdot\bar{F}_r'(\sigma_r) + \bar{F}_r(\sigma_r)\} \quad . \tag{2.24}$$

It may be noted that the inequality $F_r \geq \bar{F}_r$ follows directly from (2.22) and that (2.23) also provides a bound for a composite whose rth phase has creep potential $\bar{F}_r(\sigma)$.

If $\bar{F}_r$ is replaced by F_r in (2.23) and (2.24), the right side of (2.24) provides a *variational approximation* to $\tilde{F}(\bar{\sigma})$ but it is only a bound if $\bar{F}_r = F_r$ in a neighbourhood of σ_r , for each r.

So far, no assumption has been made concerning the geometrical arrangement of the composite. When, however, the constituents are arranged in a manner that is *statistically uniform and isotropic*, it can be deduced (Willis, 1981) that

$$c_s B_{rs} = Q(\delta_{rs} - c_s) \quad , \tag{2.25}$$

for some constant tensor Q that depends only on M_0. Then, equations (2.23) reduce to

$$\sigma_r + Q[\bar{F}_r'(\sigma_r) - M_0\sigma_r] = \bar{\sigma} + Q \sum_s c_s[\bar{F}_s'(\sigma_s) - M_0\sigma_s] \quad . \tag{2.26}$$

The tensor Q can be given explicitly if M_0 is chosen to correspond to an isotropic 'comparison medium', with bulk modulus K_0 and shear modulus μ_0. Then, using the symbolic notation of Hill (1965), so that

$$M_0 = \left[\frac{1}{3K_0} \ , \ \frac{1}{2\mu_0}\right] \ , \tag{2.27}$$

and the general relation (Willis, 1981)

$$Q = L_0 - L_0 \, P \, L_0 \quad ,$$

with

$$P = \left[\frac{1}{3K_0 + 4\mu_0} \; , \; \frac{3(K_0 + 2\mu_0)}{5\mu_0(3K_0 + 4\mu_0)} \right] ,$$

gives

$$Q = \left[\frac{12K_0 \, \mu_0}{3K_0 + 4\mu_0} \; , \; \frac{2\mu_0(9K_0 + 8\mu_0)}{5(3K_0 + 4\mu_0)} \right] . \tag{2.28}$$

Specializing further, to incompressible behaviour, so that each F_r is independent of hydrostatic stress and $K_0 \to \infty$, (2.27) reduces to

$$M_0 \sim \left[0, \frac{1}{2\mu_0} \right] \tag{2.29}$$

while (2.28) gives

$$Q \sim \left[4\mu_0 \; , \; \frac{6\mu_0}{5} \right] . \tag{2.30}$$

In this case, equations (2.26) can be restricted, without loss, to a 'shear stress subspace' in which hydrostatic stress is zero. Restricted to this subspace, M_0 and Q become, respectively, $1/2\mu_0$ and $6\mu_0/5$ times the identity and equations (2.26) become

$$\sigma_r + 3\mu_0 \, \overline{F}_r'(\sigma_r) = \bar{\sigma} + 3\mu_0 \sum_s c_s \, \overline{F}_s'(\sigma_s) \; . \tag{2.31}$$

It should be noted that, although M_0 has been taken as isotropic and the *geometry* of the composite has been assumed isotropic, no assumption of isotropy has been made for F_r , and the composite may still display overall anisotropy through possession of texture. The variable σ_r has been introduced, purely mathematically, through (2.20), but it can be regarded as an *approximation* to the mean stress in material r, (2.19) guaranteeing that the variables σ_r have the correct mean value $\bar{\sigma}$. The 'primary' field $\eta(x)$ was taken as piecewise constant but there is no corresponding implication that the stress field, given by (2.11), is piecewise constant.

Before concluding this general development, it is noted that $V_r(\eta_r)$ is *always* well-defined if $M_0 = 0$, and then $\overline{F}_r = F_r$ identically. For incompressible and isotropic M_0, this corresponds to $\mu_0 \to \infty$ which implies, for (2.31), the limiting form

$$F_r'(\sigma_r) = \bar{\varepsilon} \text{ with } \Sigma\, c_r \sigma_r = \bar{\sigma} \quad , \tag{2.32}$$

for some constant tensor $\bar{\varepsilon}$. Equations (2.32) generate the 'Taylor approximation', in which a uniform strain-rate $\bar{\varepsilon}$ is imposed upon the entire composite (Taylor, 1938). It generates a bound analogous to the Voigt bound of linear elasticity.

3. The polycrystal

The polycrystalline aggregate is assumed to be composed of crystallites, each described by the same strain-rate potential $F_c(\sigma)$, relative to crystallographic axes. The constituent r is characterized by the rotation R_r that takes its crystallographic axes to their correct orientation, from some reference orientation. Thus,

$$F_r(\sigma) = F_c\ (R_r^{-1}\ \sigma) \quad , \tag{3.1}$$

where $R_r^{-1}\ \sigma$ represents the operation of the rotation *inverse* to R_r on σ. In standard tensor notation, this would become $R_r^T\ \sigma\, R_r$, the superscript T denoting transpose but the given form is preferred, in preparation for employing a matrix notation in which σ is represented as a vector in a space of 6 dimensions (or 5 when restricted to the 'shear stress space').

The texture of the polycrystal is defined, within this framework, by the distribution of orientations R_r. It has no texture, and so is isotropic, if $n \to \infty$ and all rotations are equally likely (though overall isotropic behaviour could conceivably be achieved with n finite). The behaviour of each crystallite is defined by the function $F_c(\sigma)$; this is discussed next.

It is supposed that any crystallite deforms by slip on a discrete number of slip systems, with system k characterized by a slip plane with normal $n^{(k)}$ and a direction of slip $m^{(k)}$ in the slip plane; $m^{(k)}$ and $n^{(k)}$ are taken as unit vectors. The system is specified by the second-order tensor

$$\mu_{ij}^{(k)} = \tfrac{1}{2}\,(m_i^{(k)} n_j^{(k)} + m_j^{(k)} n_i^{(k)}) \quad . \tag{3.2}$$

In terms of $\mu^{(k)}$, the resolved shear stress on the kth system is

$$\tau^{(k)} = \sigma \cdot \mu^{(k)} \quad . \tag{3.3}$$

If $\gamma^{(k)}$ denotes the shear strain-rate (engineering definition) associated with the kth system, then

$$\varepsilon = \sum_k \gamma^{(k)} \mu^{(k)} \quad . \tag{3.4}$$

It is supposed that $\gamma^{(k)}$ depends on stress only through $\tau^{(k)}$, so that

$$\gamma^{(k)} = F_c^{(k)\prime}(\tau^{(k)}) \tag{3.5}$$

for some convex function $F_c^{(k)}$; in practice, $F_c^{(k)}$ will usually be taken to be

$$F_c^{(k)}(\tau) = \frac{\alpha \tau_0^{(k)}}{n+1} \left| \frac{\tau}{\tau_0^{(k)}} \right|^{n+1} , \tag{3.6}$$

corresponding to the power-law behaviour

$$\gamma^{(k)} = \alpha \left| \frac{\tau^{(k)}}{\tau_0^{(k)}} \right|^{n-1} \left(\frac{\tau^{(k)}}{\tau_0^{(k)}} \right) . \tag{3.7}$$

It follows now that

$$F_c(\sigma) = \sum_k F_c^{(k)}(\tau^{(k)}) \tag{3.8}$$

or, in the case of power-law behaviour,

$$F_c(\sigma) = \frac{\alpha}{n+1} \sum_k \tau_0^{(k)} \left| \frac{\tau^{(k)}}{\tau_0^{(k)}} \right|^{n+1} . \tag{3.9}$$

As noted by Hutchinson (1976), the form (3.9) has several attractive features, stemming from its homogeneity. In particular, if the single-crystal constitutive law is written in the form

$$\varepsilon = F_c'(\sigma) = M_c \, \sigma \tag{3.10}$$

(in which the fourth-order tensor M_c depends on σ), then for (3.9),

$$(M_c)_{ijpq} = \sum_k (\alpha/\tau_0^{(k)}) \left|\tau^{(k)}/\tau_0^{(k)}\right|^{n-1} \mu_{ij}^{(k)} \mu_{pq}^{(k)} \tag{3.11}$$

and also

$$(M_c)_{ijpq} = \frac{1}{n} \frac{\partial^2 F_c}{\partial\sigma_{ij}\partial\sigma_{pq}} \quad . \tag{3.12}$$

This concludes the general discussion of the polycrystal, except for noting that if, in line with (3.10), material with the *modified* creep potential $\overline{F}_r$ is considered to have constitutive relation

$$\varepsilon = \overline{F}_r'(\sigma) = \overline{M}_r \, \sigma \quad , \tag{3.13}$$

then equations (2.31) can be expressed in the form

$$\left[\frac{1}{3\mu_0} I + \overline{M}_r\right] \sigma_r = \overline{\varepsilon} \quad , \quad \Sigma \, c_r \sigma_r = \overline{\sigma} \quad , \tag{3.14}$$

$\overline{\varepsilon}$ being fixed by the second of equations (3.14). The equations remain nonlinear because $\overline{M}_r$ depends on σ_r. The limiting case $\mu_0 \to \infty$, corresponding to (2.32), becomes simply

$$M_r \sigma_r = \overline{\varepsilon} \quad , \quad \Sigma \, c_r \sigma_r = \overline{\sigma} \quad , \tag{3.15}$$

where $F_r'(\sigma) = M_r \, \sigma$.

If matrix notation is employed, then from (3.1),

$$M_r = R_r \, M_c(R_r^{-1}\sigma_r) \, R_r^{-1} \quad , \tag{3.16}$$

which shows explicitly how M_r may be calculated from M_c. When F_0 is isotropic, a corresponding relation exists between $\overline{M}_r$ and $\overline{M}_c$, defined from the modified single-crystal potential

$$\overline{F}_c = (F_c - F_0)^{**} + F_0 \quad . \tag{3.17}$$

4. A relaxation of the problem

A practical difficulty with the formulation presented so far is that finding the potential $\overline{F}_c$, defined by (3.17), requires a substantial computation.

This is circumvented by introducing the auxiliary quadratic potential

$$\hat{F}_c(\sigma) = \sum_k \hat{F}_c^{(k)}(\tau^{(k)}) \quad , \tag{4.1}$$

where

$$\hat{F}_c^{(k)} = \tfrac{1}{2}a\, \tau_0^{(k)} \left| \frac{\tau^{(k)}}{\tau_0^{(k)}} \right|^2 \quad , \tag{4.2}$$

corresponding to (3.8), (3.9) with $n = 1$. Then, $\hat{F}_r(\sigma)$ is defined so that

$$\hat{F}_r(\sigma) = \hat{F}_c(R_r^{-1}\, \sigma) \quad . \tag{4.3}$$

Now

$$F_c(\sigma) = \sum_k \left[F_c^{(k)}(\tau^{(k)}) - \hat{F}_c^{(k)}(\tau^{(k)})\right] + \hat{F}_c(\sigma)$$

$$\geq \sum_k (F_c^{(k)} - \hat{F}_c^{(k)})^{**}(\tau^{(k)}) + \hat{F}_c(\sigma) \tag{4.4}$$

because $F^{**} \leq F$ for any function F. Hence, if $\bar{\bar{F}}_c(\sigma)$ is defined as

$$\bar{\bar{F}}_c(\sigma) = \sum_k (F_c^{(k)} - \hat{F}_c^{(k)})^{**}(\tau^{(k)}) + \hat{F}_c(\sigma) \quad , \tag{4.5}$$

potentials $\bar{\bar{F}}_r(\sigma)$ are given by

$$\bar{\bar{F}}_r(\sigma) = \bar{\bar{F}}_c(R_r^{-1}\, \sigma) \tag{4.6}$$

and, for any given isotropic F_0, the constant a in the definition (4.2) of $\hat{F}_c$ is chosen so that the quadratic function $\hat{F}_c - F_0$ is convex, it follows that

$$F_r(\sigma) \geq \bar{\bar{F}}_r(\sigma) \tag{4.7}$$

and

$$(\bar{\bar{F}}_r - F_0)^{**} = \bar{\bar{F}}_r - F_0 \quad . \tag{4.8}$$

The inequality (4.7) ensures that $\tilde{F}(\bar{\sigma})$ is bounded below by the overall potential of a composite whose rth phase has potential $\overline{\overline{F}}_r(\sigma)$ and this, in turn, is bounded below by (2.24), with σ_r defined by (2.23), except that in both of these $\overline{F}_r$ is replaced by $\overline{\overline{F}}_r$. The introduction of $\overline{\overline{F}}_r$ renders the inequality less sharp but it has the advantage that the functions $(F_c^{(k)} - \hat{F}_c^{(k)})^{**}$ depend only on a single variable and can be calculated explicitly. In fact, when $F_c^{(k)}(\tau)$ has the power-law form (3.6),

$$(F_c^{(k)} - \hat{F}_c^{(k)})^{**}(\tau) = F_c^{(k)}(\tau_1) - \hat{F}_c^{(k)}(\tau_1) \ , \ |\tau| \leq \tau_1$$

$$= F_c^{(k)}(\tau) - \hat{F}_c^{(k)}(\tau) \ , \ |\tau| \geq \tau_1 \ , \tag{4.9}$$

where

$$\tau_1 = \tau_0^{(k)}(a/\alpha)^{1/(n-1)} \ . \tag{4.10}$$

5. Untextured polycrystals

The overall potential $\tilde{F}(\bar{\sigma})$ of an untextured polycrystal is isotropic and hence is a symmetric function of the three principal average stresses, $\bar{\sigma}_1$, $\bar{\sigma}_2$, $\bar{\sigma}_3$ say. Equivalently, it is a function of the invariants

$$\bar{\sigma}_m = \frac{1}{3}(\bar{\sigma}_1 + \bar{\sigma}_2 + \bar{\sigma}_3) \ ,$$

$$\bar{\sigma}_e = \{ \frac{3}{2} [(\bar{\sigma}_1 - \bar{\sigma}_m)^2 + (\sigma_2 - \sigma_m)^2 + (\sigma_3 - \sigma_m)^2]\}^{1/2} \ ,$$

$$\bar{\tau} = \{ \frac{27}{2} (\bar{\sigma}_1 - \bar{\sigma}_m)(\bar{\sigma}_2 - \bar{\sigma}_m)(\bar{\sigma}_3 - \bar{\sigma}_m)\}^{1/3} \ . \tag{5.1}$$

The first invariant, $\bar{\sigma}_m$, is the mean normal stress. The second, $\bar{\sigma}_e$, is the equivalent stress, equal to $\bar{\sigma}_3$ when $\bar{\sigma}_1 = \bar{\sigma}_2 = 0$ and the third, $\bar{\tau}$, indicates the state of stress: $|\bar{\tau}| \leq \sigma_e$, with equality when the stress is uniaxial, and $\bar{\tau} = 0$ for pure shear when, for example, $\bar{\sigma}_1 = -\bar{\sigma}_3$ and $\bar{\sigma}_2 = 0$.

The polycrystals under consideration are incompressible and so independent of $\bar{\sigma}_m$; there is no loss of generality, therefore, in considering only states of stress for which $\bar{\sigma}_2 = 0$. The stresses $\bar{\sigma}_1$ and $\bar{\sigma}_3$ may then be represented in the form

$$\bar{\sigma}_1 = \bar{\sigma}_e \left\{ \cos\beta + \frac{\sin\beta}{\sqrt{3}} \right\} \ ,$$

$$\bar{\sigma}_3 = \bar{\sigma}_e \left[\cos\beta - \frac{\sin\beta}{\sqrt{3}}\right] \quad , \tag{5.2}$$

in terms of which

$$\bar{\tau} = \bar{\sigma}_e \, \{\cos\beta\,(3 - 4\cos^2\beta\}^{1/3} \quad . \tag{5.3}$$

Thus, $|\bar{\tau}|/\bar{\sigma}_e$ ranges from 1 to 0 as β ranges from 0 to $\pi/6$.

In the particular case of pure power-law (and incompressible) behaviour, $\tilde{F}(\bar{\sigma})$ becomes a homogeneous function of $\bar{\sigma}_e$ and $\bar{\tau}$ only and can be expressed, fully generally, in the form

$$\tilde{F}(\bar{\sigma}) = \frac{\alpha\bar{\tau}_0}{n+1} \left[\frac{\bar{\sigma}_e}{\bar{\tau}_0}\right]^{n+1} \quad , \tag{5.4}$$

in which the 'normalizing stress' $\bar{\tau}_0$ is a function of $(\bar{\tau}/\bar{\sigma}_e)$ only. For any chosen value of $(\bar{\tau}/\bar{\sigma}_e)$ or, equivalently, any chosen β, the lower bound (2.24) for $\tilde{F}(\bar{\sigma})$ can be translated into an upper bound for $\bar{\tau}_0$.

6. Cubic Polycrystals

The remainder of this section is devoted to the study of f.c.c. polycrystals, for which each crystallite has four slip planes of the (1, 1, 1) type and three slip directions per plane, of the (1, 1, 0) type, giving 12 slip systems in all. Power-law behaviour, as given in (3.9), will be assumed and the normalizing stress $\tau_0^{(k)}$ will be taken as the same for each system, so that $\tau_0^{(k)} = \tau_0$, say.

6.1 Elementary bounds

The simplest bound for $\tilde{F}(\bar{\sigma})$ is the upper bound given by (2.6). All twelve slip systems contribute equally to the bound, so that

$$\tilde{F}(\bar{\sigma}) \leq \frac{12\alpha}{n+1}\,\tau_0 \left\langle \left|\frac{\bar{\tau}^{(1)}}{\tau_0}\right|^{n+1} \right\rangle \quad , \tag{6.1}$$

the angled bracket here representing an average over all orientations of the slip system $k = 1$. The 'reference' orientation can be chosen so that, at this orientation,

$$m^{(1)} = (1, 0, 0) \;\text{ and }\; n^{(1)} = (0, 0, 1) \quad .$$

Then, at a general orientation characterized by the Euler angles (φ, θ, ψ),

$$m^{(1)}=(\cos\varphi\cos\psi-\sin\varphi\cos\theta\sin\psi,-\sin\varphi\cos\psi-\cos\varphi\cos\theta\sin\psi, \sin\theta\sin\psi)$$

and

$$n^{(1)} = (\sin\varphi\sin\theta, \cos\varphi\sin\theta, \cos\theta) \ .$$

The resolved average shear stress $\bar{\tau}^{(1)}$ is $\bar{\sigma}_1\, m_1^{(1)}\, n_1^{(1)} + \bar{\sigma}_3\, m_3^{(1)}\, n_3^{(1)}$, i.e.

$$\bar{\tau}^{(1)} = \bar{\sigma}_1 \ (\cos\varphi\cos\psi - \sin\varphi\cos\theta\sin\psi)\sin\varphi\sin\theta + \bar{\sigma}_3\sin\theta\sin\psi\cos\theta$$

and (6.1) requires the calculation of

$$J_n = \frac{1}{4\pi^2}\int_0^{2\pi} d\varphi \int_0^{2\pi} d\psi \int_0^{\pi/2} \sin\theta\, d\theta\, |\bar{\tau}^{(1)}|^{n+1} \ , \tag{6.2}$$

with $\bar{\sigma}_1$, $\bar{\sigma}_3$ given by (5.2). The bound (6.1) can then be translated into a lower bound for the normalizing stress $\bar{\tau}_0$:

$$\bar{\tau}_0/\tau_0 \geq (12J_n)^{-1/n} \ , \tag{6.3}$$

where J_n is evaluated with $\bar{\sigma}_e$ set equal to 1.

An elementary lower bound for $\tilde{F}(\bar{\sigma})$, which induces an upper bound for τ_0, is obtained by solving equations (3.15) and substituting the solution into (2.24), with $M_0 = 0$ so that $\bar{F}_r = F_r$. As the form of (3.15) shows, this bound is analogous to the Voigt bound of linear elasticity; in the context of plasticity it is often called the Taylor bound and this terminology will be used here. Although equations (3.15) have a solution, it is worth noting that the tensor $M_r(\sigma)$ can become singular for some stresses σ. This is because any three tensors $\mu^{(k)}$ associated with the same normal n are linearly dependent, so there exist at most 8 linearly independent $\mu^{(k)}$. If, now, s is an eigenvector of $M_c(\sigma)$ with zero eigenvalue,

$$M_c(\sigma)\, s = 0 \ \text{ implies } \ \alpha\tau_0 \sum_{k=1}^{12} \left| \frac{\mu^{(k)}\cdot\sigma}{\tau_0} \right|^{n-1} (\mu^{(k)}\cdot s)\ \mu^{(k)} = 0 \ ,$$

which can be satisfied if

$$\textit{either } \mu^{(k)}\cdot\sigma = 0 \textit{ or } \mu^{(k)}\cdot s = 0 \textit{ for each } k.$$

This is possible since (disregarding hydrostatic stress) σ and s each have 5 independent components and there are only 8 linearly independent $\mu^{(k)}$. The same argument applies to M_r, which is obtained from M_c by a rotation. It was still possible, nevertheless, to obtain a solution of equations (3.15) by Newton iteration, which is described more generally below.

6.2 *The new bound*

Improved lower bounds for $\tilde{F}(\bar{\sigma})$, which induce corresponding upper bounds for $\bar{\tau}_0$, follow by solving equations (3.14), but with $\bar{F}_r$ replaced by $\bar{\bar{F}}_r$ as given by (4.5), (4.6), and with the pair of constants a , μ_0 restricted so that $\hat{F}_c - F_0$ is convex. For the cubic polycrystal, this implies

$$4\mu_0 a \geq 3 \quad . \tag{6.4}$$

In practice, the best bound is obtained by taking equality in (6.4), and a and μ_0 were expressed in terms of the parameter τ_1, through equation (4.10), with each $\tau_0^{(k)}$ equal to τ_0. Equations (3.14) were solved by the Newton iteration scheme

$$\sigma_r + \Delta\sigma_r = \left\{\frac{1}{3\mu_0} I + \bar{\bar{F}}_r{}''(\sigma_r)\right\}^{-1} \Bigg\{ \bar{\bar{F}}_r{}''(\sigma_r)\,\sigma_r - \bar{\bar{F}}_r{}'(\sigma_r)$$

$$+ \left[\sum_s c_s \left\{\frac{1}{3\mu_0} I + \bar{\bar{F}}_s{}''(\sigma_s)\right\}^{-1}\right]^{-1}$$

$$\left[\bar{\sigma} - \sum_t c_t \left\{\frac{1}{3\mu_0} I + \bar{\bar{F}}_t{}''(\sigma_t)\right\}^{-1} \left\{\bar{\bar{F}}_t{}''(\sigma_t)\,\sigma_t - \bar{\bar{F}}_t{}'(\sigma_t)\right\}\right]\Bigg\} \; . \tag{6.5}$$

In the Taylor limit, $\mu_0 \to \infty$, $\bar{\bar{F}}_r \to F_r$ and the homogeneity of F_r reduces (6.5) to

$$\sigma_r + \Delta\sigma_r = \left[\frac{n-1}{n}\right] \sigma_r + \frac{1}{n} M_r^{-1} \left[\Sigma_s\, c_s\, M_s^{-1}\right]^{-1} \bar{\sigma} \; . \tag{6.6}$$

6.3 *Results*

Figure 1 shows plots of three bounds for the ratio $\bar{\tau}_0/\tau_0$, against $1/n$, for a loading pattern corresponding to uniaxial tension, so that $\beta = 0$ in (5.2)

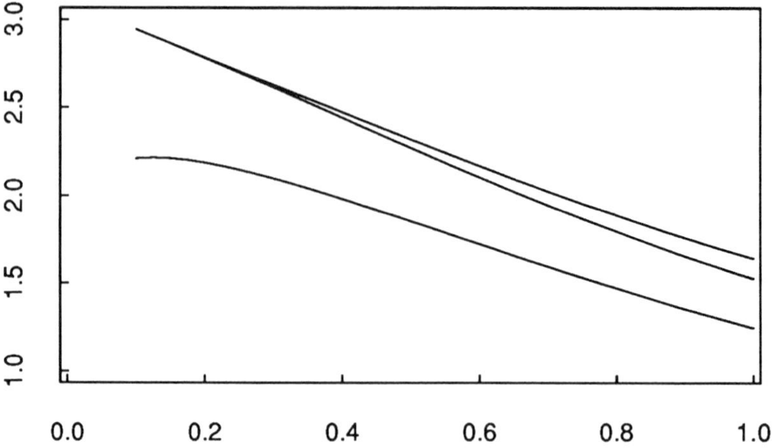

Fig. 1 Plots (in ascending order) of the elementary lower bound, the new bound and the Taylor bound, for $\bar{\tau}_0/\tau_0$, against $1/n$, for an untextured polycrystal subjected to uniaxial overall stress.

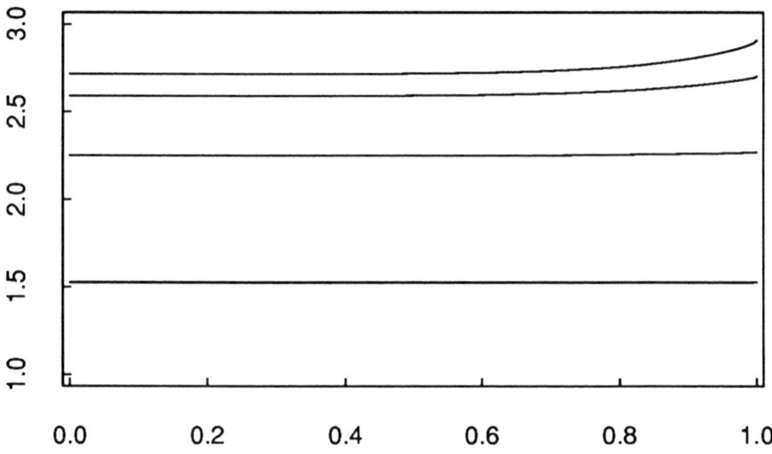

Fig. 2 Plots of the new bound for $\bar{\tau}_0/\tau_0$ against the stress state parameter $\bar{\tau}/\sigma_e$, for $n = 1, 2, 4, 8$. The lowest curve corresponds to $n = 1$ and they increase with n.

(though other values of β also yield $\bar{\tau}/\bar{\sigma}_e = 0$). The lowest curve corresponds to the lower bound (6.3) and the highest is the Taylor bound. The intermediate curve is a new upper bound, obtained as described above but with the parameter τ_1 chosen optimally. The bound is relatively insensitive to the precise value of τ_1, which varied between about 0.125 (for $n = 2$) and 0.14 (for $n = 8$). The curves shown were actually calculated for a polycrystal with 125 different crystal orientations; the precise value of τ_1 would depend on these also. In the linear case ($n = 1$), the bound reduces to the Hashin-Shtrikman value 43/28. It always lies slightly above the self-consistent estimate of Hutchinson (1976) and tends to the Taylor bound at large n.

An advantage of the present calculation over that of Hutchinson (1976) is that equations (3.14) can be solved for any pattern of loading, corresponding to any value of $|\bar{\tau}|/\bar{\sigma}_e$ between 0 and 1. Figure 2 shows plots of the new bound, for $\bar{\tau}_0/\tau_0$, against the parameter $|\bar{\tau}|/\bar{\sigma}_e$, for the four values $n = 1, 2, 4, 8$. They were obtained by varying β between 0 and $\pi/6$. The lowest curve is for $n = 1$ and shows no sensitivity to stress state, as expected. Higher curves correspond to increasing values of n and show some sensitivity, though still not much. It is emphasised that the curves provide only bounds but they nevertheless *suggest* that overall behaviour is reasonably well approximated by the form (5.4), with $\bar{\tau}_0$ taken independent of stress. Figures corresponding to Fig. 1 have been generated for values of β other than zero but, as Fig. 2 suggests, they look very much like Fig. 1.

The present formalism is also capable of generating bounds for the behaviour of polycrystals displaying texture. The simplest possibility is to solve equations (3.14), with volume fractions c_r giving different weight to different orientations; a further possibility is to relax the assumption of 'geometrical isotropy' and to generate different tensors B_{rs} from their basic definition (2.17). These options are being pursued at present.

References

Hashin, Z. and Shtrikman, S., 1962, "A variational approach to the theory of the elastic behaviour of polycrystals", *J. Mech. Phys. Solids*, Vol. 10, pp. 343-352.

Hill, R., 1965, "Continuum micromechanics of elastoplastic polycrystals", *J. Mech. Phys. Solids*, Vol. 13, pp. 89-101.

Hutchinson, J. W., 1976, "Bounds and self-consistent estimates for creep of polycrystalline materials", *Proc. R. Soc. Lond. A, Vol. 348, pp. 101-127.*

Mandel, J., 1972, "Plasticité classique et visoplasticité", *CISM Udine Courses and Lectures, no. 97*. Springer-Verlag, New York.

Ponte Castañeda, P. and Willis, J. R., 1988, "On the overall properties of nonlinearly viscous composites", *Proc. R. Soc. Lond. A, Vol. 416, pp. 217-244.*

Rice, J. R., 1971, "Inelastic constitutive relations for solids: an internal variable theory and its application to metal plasticity", *J. Mech. Phys. Solids*, Vol. 19, pp. 433-455.

Talbot, D. R. S. and Willis, J. R., 1985, "Variational principles for inhomogeneous nonlinear media", *IMA J. Appl. Math.*, Vol. 35, pp. 39-54.

Talbot, D. R. S. and Willis, J. R., 1987, "Bounds and self-consistent estimates for the overall properties of nonlinear composites", *IMA. J. Appl. Math.*, Vol. 39, pp. 215-240.

Taylor, G. I., 1938, "Plastic strain in metals", *J. Inst. Metals*, Vol. 62, pp. 307-324.

Van Tiel, J., 1984, *Convex analysis*. Wiley, New York.

Willis, J. R., 1981, "Variational and related methods for the overall properties of composites", *Advances in Applied Mechanics*, edited by C.-S. Yih, Vol. 21, pp. 1-78. Academic Press, New York.

Willis, J. R., 1983, "The overall elastic response of composite materials", *J. Appl. Mech.*, Vol. 50, pp. 1202-1209.

Willis, J. R., 1986, "Variational estimates for the overall response of an inhomogeneous nonlinear dielectric", *Homogenization and Effective Moduli of Materials and Media*, edited by J. L. Ericksen, D. Kinderlehrer, R. V. Kohn and J.-L. Lions, *The IMA Volumes in Mathematics and Its Applications*, Vol. 1, pp. 245-263. Springer-Verlag, New York.

Willis, J. R., 1989, "The structure of overall constitutive relations for a class of nonlinear composites", *IMA J. Appl. Math.*, Vol. 43, pp. 231-242.

Simplifications in the Behavior of Viscoelastic Composites with Growing Damage

R.A. Schapery*
Civil Engineering Department
Texas A&M University
College Station, TX 77843

ABSTRACT

Some analytical results for the mechanical behavior of elastic composite materials and structures with growing damage are summarized and then extended to viscoelastic media. The effect of strain rate only at crack tips is considered first; it is shown that if the crack speed is a strong function of energy release rate, the overall mechanical response is like that for an aging elastic material. Both stable crack growth and unstable crack growth followed by arrest produce this aging-like behavior. Viscoelastic behavior throughout one or more of the phases is then introduced. A simplification is used in which only one relaxation modulus characterizes the viscoelasticity, apart from that at crack tips. Upon replacing the physical displacements in the response for an elastic material by quantities called pseudo displacements, a simple model for viscoelastic composites with growing damage is obtained.

*Address effective September 1, 1990: Department of Aerospace Engineering and Engineering Mechanics, University of Texas, Austin, TX 78712.

1. Introduction

The problem of developing a realistic mathematical model of the mechanical behavior of viscoelastic composites with growing damage is a difficult one. However, it is believed that considerable simplification may be introduced in the description of both the intrinsic viscoelastic behavior and the damage, while retaining the essential elements needed for a realistic description of deformation and fracture behavior of many composites of engineering interest. The emphasis of this paper is on simplifications which appear to be applicable at least to particle and fiber reinforced polymers when the matrix is soft relative to the particles or fibers. The underlying model for elastic behavior with damage is not restricted in this way.

We shall not give a general review of the literature in this area; but, instead, contributions of the author and coworkers on the issue of simplification are emphasized. For a broader view of the subject, the reader is referred to work by Onat and Leckie (1988) and Weitsman (1988), where a tensorial description of damage is addressed. The state variables used in the present paper to characterize global deformation and the damage may be scalars or tensors, but there is no need here to identify them as such. Work on a nonlinear viscoelastic composite-like material, polycrystalline ice, is relevent to the present work. For some studies which address both viscoelasticity and damage growth see Harper (1986, 1989), Karr and Choi (1989), Schapery (1989) and Sjolind (1987).

The present discussion is based in-part on results

from Schapery (1990), as summarized in Section 2 where elastic behavior with damage growth is covered. In Sections 3 and 4 we fill in some of the details concerning effects of growing cracks which were only touched upon by Schapery (1990). In addition, some new results are obtained for the case of a body with distributed cracks which become unstable, grow dynamically, and then are arrested. Rate effects at crack tips only are considered in Sections 3 and 4, while simultaneous effects of global and crack-tip viscoelasticity are discussed in Section 5.

2. Elastic Behavior with Damage Growth

Some of the results from Schapery(1990) are collected in this section. We consider an elastic structure or material whose thermodynamic state is a function of independent generalized displacements $q_j (j = 1,2,...J)$ and internal state variables $S_m (m = 1,2,...M)$ as well as temperature or entropy; inelastic behavior arises from changes in the S_m. Generalized forces Q_j are defined in the usual way in that

$$\delta W' = Q_j \delta q_j \quad (j \text{ not summed}) \tag{1}$$

for each virtual displacement δq_j, where $\delta W'$ is the virtual work. Then, from thermodynamics

$$Q_j = \partial W / \partial q_j \tag{2}$$

where W is the Helmholtz free energy (when temperature is used as an independent state variable) or the internal energy (when entropy, instead of temperature, is an independent variable). For brevity, thermal effects will not be considered here, and therefore we shall refer to W simply as the <u>strain energy</u>. The generalized displacements q_j may be, for example, the uniform strains in a material element and Q_j the conjugate stresses, or q_j and Q_j may be, respectively, the displacements and forces applied to a structure.

The internal state variables serve here to define changes in the structure such as micro- or macro-cracking, and are called <u>structural parameters</u>. Whenever any one $\dot{S}_m \neq 0$, we specify as the evolution law,

$$f_m = \partial W_s / \partial S_m \tag{3}$$

where $W_s = W_s(S_m)$ is a state function of one or more S_m; also, f_m is the <u>thermodynamic force</u>,

$$f_m \equiv -\partial W / \partial S_m \tag{4}$$

The left side of equation (3) is the available force for producing changes in S_m, while the right side is the required force. For any specific set of processes (i.e. histories $q_j(t)$), equation (3) may not be satisfied for all M of the parameters; if it is not, those S_m will be constant. The subscript r or p will henceforth be used in place of m to designate the parameters that change, which are taken to be R in number.

The total work done on the body by Q_j during an <u>actual</u> process (i.e., a process for which parameters change in accordance with equation (3)), starting at some reference state, is denoted by W_T,

$$W_T \equiv \int Q_j dq_j \tag{5}$$

where the summation convention for repeated indices is used. From equations (2)-(5) we find

$$W_T = W + W_s \tag{6}$$

where $W = W_s = 0$ in the reference state. Thus, W_s may

be interpreted as that portion of the total work W_T which contributes to changes in the structure.

The second law of thermodynamics provides an inequality as a constraint on the changes in state,

$$\dot{W}_s = T\,S' \geq 0 \tag{7}$$

where T is absolute temperature and S' is the entropy production rate. Even if equation (3) is satisfied for any one S_r, this inequality may not allow it to change. Moreover, instantaneous values of the S_r are such that they minimize the total work when the body passes through stable states; i.e.,

$$\partial W_T/\partial S_r = 0 \tag{8}$$

$$(\partial^2 W_T/\partial S_r \partial S_p)\,\delta S_r \delta S_p \geq 0 \tag{9}$$

It is observed that equation (3) represents R equations for finding the S_r as functions of q_j. Then, $W_T = W_T(q_j, S_r(q_j), S_q)$ where the S_q are the constant parameters. From equation (5),

$$Q_j = \partial W_T/\partial q_j \tag{10}$$

showing that the body exhibits hyperelastic behavior during the time any particular set of parameters S_r undergoes change. Because the total work is a potential during inelastic processes, the incremental stiffness matrix is symmetric. Conversely, given that the stiffness matrix is symmetric when one or more S_r change, then both equations (3) and (10) follow.

If forces act on crack faces then they have to be included in the set Q_j unless they are associated with frictionless contact; in the latter case, the effect of crack opening and closing may be taken into account through the form of the strain energy function. Coulomb friction, if significant, cannot be accounted for through a work potential, and therefore the stiffness matrix is not necessarily symmetric during processes involving crack face sliding. If, however, one can use a potential to characterize the relationship between crack-face forces and relative displacements between crack faces, equation (10) may be extended to this case by including this potential (which may depend on additional structural parameters) in W_T; such a simplification is applicable with surface free-energy effects (Schapery, 1990) and was proposed by Schapery (1989) to account for crack-face friction in ice under compression.

3. Rate Effects at the Crack Tip

Here we provide the background to Section 4; in that section it is shown that a familiar equation for crack speed, discussed next, leads to equation (3) as an approximation.

Suppose that the local crack tip speed $\dot{a}$ (at an arbitrary point on the crack edge) obeys a power law in local energy release rate G,

$$\dot{a} = k\ G^q \tag{11}$$

where, for now, we assume q is a positive constant; also, $G \equiv -\partial\hat{W}/\partial A$ where $\hat{W}$ is the strain energy of a body with one or more cracks less the surface free energy, and ∂A is an increment in crack surface area. The surface free energy is usually negligible (which we assume here) so that $\hat{W} = W$. The coefficient k may change with time for various reasons, including transient temperatures, material aging, and mode-ratio effects; if the latter exist, we assume the mode ratio is constant. It will be helpful to write $k = c_1 k_1$, where c_1 is a positive dimensionless function of time and k_1 is a positive constant which has dimensions appropriate for the units used in equation (11). Now, integrate equation (11) with respect to time and then take its qth root,

$$\Delta a^{1/q} = L_G k_1^{1/q} \tag{12}$$

where $\Delta a \equiv a - a_o$, a_o is the crack size at t=0, and

$$L_G \equiv [\int_0^{\xi} G^q d\xi']^{1/q} \tag{13}$$

is the so-called Lebesgue norm of G; also,

$$\xi \equiv \int_0^t c_1(t')dt' \tag{14}$$

is <u>reduced time</u>.

If $q = \infty$, then $L_G = G_\ell$, where G_ℓ is the largest value of G up to and including the current time (Reddy and Rasmussen, 1982). For many materials $0 << q < \infty$, which leads to an approximation for L_G which is practically as simple as for $q = \infty$. Consider first the case of power-law reduced-time dependence $G \sim \xi^p$ where $p \geq 0$. Then,

$$L_G = k_2 \xi^{1/q} G \tag{15}$$

in which

$$k_2 \equiv (pq + 1)^{-1/q} \tag{16}$$

The accuracy of equation (15) was studied by Schapery (1982) (using pseudo strain rather than energy release rate as the argument of the Lebesgue norm) for cases in which the argument was a nondecreasing function of ξ, not necessarily a power law. In terms of G used here, let us define p as a logarithmic derivative,

$$p \equiv \frac{d \log G}{d \log \xi} \tag{17}$$

where $p \geq 0$. With $q \geq 4$, good agreement was reported between equation (15) and the value of L_G found by numerical integration for a variety of histories with $p = p(\xi)$. If q is large or p is not strongly time dependent, then k_2 may be taken as a constant, which we shall do here; however, even if p is negative or k_2 is not constant, approximations like that in equation (15) can be developed, and they are useful in view of numerical integration difficulties encounted when $q \gg 1$ (Schapery, 1982).

Use of equation (15) in (12) yields

$$G = (\Delta a/\xi)^{1/q}/k_2 k_1^{1/q} \tag{18a}$$

which replaces equation (11) as the means for predicting crack growth. The left side is the <u>available</u> work/area for crack growth; therefore the right side may be interpreted as the <u>required</u> work. Equation (18a) is like the crack growth equation for a brittle elastic material, where the right side is the critical fracture energy, say G_c. However, in contrast to brittle elastic behavior, G_c here varies with crack growth and time.

That the Lebesgue norm depends primarily on the current value of G when $q \gg 1$, rather than its entire history of variation, is obviously due to the integrand (G^q) being a strongly increasing function of G and the assumption $dG/d\xi \geq 0$. Thus, even when the power law equation (11) is not applicable we expect to be able to use a crack growth law of the type

$$G = G_c(\Delta a/\xi) \tag{18b}$$

to predict instantaneous values of Δa. One may arrive at equation (18b) directly by starting with

$$\dot{a} = kf(G) \tag{19a}$$

where k is such that $f(1) = 1$. The form of equation (11) results by using the exponent

$$q \equiv \log f/\log G \tag{19b}$$

(Alternatively, one could define q as a logarithmic derivative instead of a ratio.) Then, if $q \gg 1$ and $\dot{G} \geq 0$ it is anticipated that equation (18a) will be a good approximation, although its accuracy has not yet been studied. As q is now a function of G, one needs to solve for G; this yields equation (18b), in which G_c is not necessarily a power law in $\Delta a/\xi$.

Another generalization of interest is for cyclic loading when the basic growth law is like equation (11) or (19), but $\dot{a}$ and G are replaced by da/dN and the maximum value of G over a cycle, respectively. Obviously, the above results may be used in this case, but a reduced time based on N, rather than t, enters. With a small modification one may also treat in the same way the case where the amplitude of variation of G over a cycle is used in place of the maximum G (Schapery, 1990).

It should be observed that when $G < G_\ell$ and $q \gg 1$ then L_G, equation (13), is practically constant if the time period for which $G < G_\ell$ is not extremely long. This implies $\dot{a} \simeq 0$ constant and from (15) that

$$L_G = k_2 \xi_\ell^{1/q} G_\ell \tag{20}$$

where ξ_ℓ is the reduced time at which G first drops

below the <u>largest value</u> G_ℓ. This behavior is taken into account in the elastic-like model if we assume $\dot{a} = 0$ when $G < G_c$, where $G_c = G_c(\Delta a/\xi_\ell)$. If G later increases to G_ℓ, say at time ξ_L, then G_c again varies as in equation (18); but ξ should be replaced by $\xi-(\xi_L - \xi_\ell)$ for continuity of G_c. In some cases, such as for cyclic loading, it may be necessary to account for contributions to L_G when G is close to G_ℓ by modifying equation (15) (Schapery, 1982).

In arriving at equation (18) it was not necessary to specify explicitly the manner in which G varys with loading or with geometry of the one or more cracks that may exist. However, this variation certainly will affect the time-dependence of crack growth and thus determine the accuracy of equation (18) and whether or not a physically acceptable solution Δa exists. In order to illustrate this point, let us consider two special cases before discussing the connection between equations (3) and (18).

First, observe that if the energy release rate is constant in time, equation (18) is an exact result. This situation exists for some elementary delamination and transverse microcracking problems in laminates when the applied displacement is constant. A second more interesting case is that for which

$$G = aG_1 \tag{21}$$

where G_1 is a function of only the applied loads or displacements; this form may be derived by dimensional analysis for a linear or nonlinear homogeneous body with an isolated, penny-shaped crack of radius a or straight-edged, through-the-thickness crack of length a.

Equation (18a) then can be written as

$$(a - a_o)^{1/q}/\dot{a} = k_2(k_1\xi)^{1/q}G_1 \qquad (22)$$

If $\dot{G}_1 \geq 0$ this equation predicts that $\dot{a}$ is a positive function of ξ for $0<\xi<\xi_f$, where

$$\xi_f \equiv a_o(q-1)^{q-1}/k_1(k_2 q a_o G_1)^q \qquad (23)$$

Also, $\dot{a} = \infty$ at $\xi = \xi_f$ and there is no solution for $\xi > \xi_f$. The crack size at time ξ_f is

$$a = a_o q/(q-1) \qquad (24)$$

(Note that $a \simeq a_o$ if $q>>1$.) Equation (23) is the limiting time for which a physically meaningful, stable solution is obtained. Therefore ξ_f may be interpreted as the fracture time, unless the crack growth is arrested by interaction with, for example, originally remote particles or fibers.

When equation (21) is used in the original growth law (11), with $q>1$, we obtain as the exact solution

$$a/a_o = (1-I_1)^{1/(1-q)} \qquad (25)$$

where

$$I_1 \equiv (q-1)k_1 a_o^{(q-1)} \int_0^{\xi} G_1^q d\xi' \qquad (26)$$

Observe that crack size is a monotone increasing function of time and $a = \dot{a} = \infty$ at the time for which $I_1 = 1$. Denoting this fracture time by ξ_F, we find

$$\xi_F = \{(q-1)k_1 a_o^{(q-1)} G_1{}^q\}^{-1} \qquad (27)$$

if G_1 is constant. Both ξ_f and ξ_F depend on a_o and G_1 in the same way, and they are equal if we take $k_2 = (q-1)/q$; if $q>>1$, then $k_2 \simeq 1$, as in equation (16). Thus, for this case in which an instability develops, approximate equation (18a) provides essentially the same crack growth behavior as the exact solution.

4. Work Functions with Rate Effects.

Let us now combine the results in Sections 2 and 3. We assume the instantaneous geometry of all cracks in the body may be defined by the structural parameters S_m. For many particulate and fibrous composites, the cracks tend to be at or close to interfaces or, at least, to have orientations and shapes defined more by the microstructural geometry than by the loading. The orientation of a crack relative to that of the loading will of course affect its rate of growth. Elliptical delaminations, transverse cracks (which are rectangular cracks with planes parallel to fibers and normal to ply surfaces) and cracks between hard particles and a soft matrix are of this type. For these cases it is realistic to use a finite and possibly small number of parameters to define the damage state.

Stable Crack Growth: Use the right side of equation (18a) to define a crack work function for the nth crack,

$$W_n \equiv \int_n \Delta a^{1/q} dA / k_2 (k_1 \xi)^{1/q} \tag{28}$$

where the integral is taken over the area of growth of the nth crack, $A - A_o$, which may not be planar or otherwise regular; the various constants such as q may be different for different cracks. The growth Δa is a local value which is defined along a curve that is normal to the moving crack edge. If $q = \infty$ this equation is independent of the history of the crack geometry, and yields simply $W_n \sim A - A_o$. Since $1 \ll q < \infty$, the effect of history is weak, and it is therefore appropriate to use an idealization in which Δa is a single-valued function of A. For example, for cracks which can be idealized as planar, through-the-thickness cracks with straight edges, use $a = A/B$ and $a_o = A_o/B$ where B is thickness; thus

$$W_n = (A-A_o)^{\lambda}/\lambda k_2(k_1 B\xi)^{1/q} \tag{29}$$

where

$$\lambda \equiv (1+q)/q \tag{30}$$

On the other hand if the cracks are more or less elliptical with moderate aspect ratios or circular use $A = \pi a^2$ in equation (28) and find

$$W_n = 2\pi q(a-a_o)^{1/q}\{a(a-a_o)/q + (a^2 - a_o^2)\}/\lambda(1+2q)k_2(k_1\xi)^{1/q} \tag{31}$$

in which the substitution of

$$a = (A/\pi)^{1/2} \tag{32}$$

into this result gives $W_n = W_n(A,\xi)$. The area of each crack is a function of one or more of the parameters S_m, by previous assumption; if, for example, the cracks are elliptical, we may use for each crack two parameters, the major and minor axes. The total crack work function, considering all cracks, is denoted by W_s,

$$W_s \equiv \sum_n W_n \tag{33}$$

so that $W_s = W_s(S_m, \xi)$, as in equation (3), but now with time-dependence. If q is the same for all cracks, Then $W_s \sim \xi^{-1/q}$.

A crack work function based on the more general equation (18b) may be easily derived. For additional generality, also use $A = A_1 a^\alpha$, where A_1 and α are positive constants which may vary from crack-to-crack. In this case we find for the nth crack,

$$W_n = A_1 \alpha \xi^\alpha \int_0^\rho (\rho' + \frac{a_o}{\xi})^{\alpha-1} G_c(\rho') d\rho' \qquad (34)$$

where $\rho \equiv \Delta a/\xi$ and $\Delta a = (A/A_1)^{1/\alpha} - a_o$. Then equation (33) yields $W_s(S_m, \xi)$.

We may now use equation (18) to arrive at (3). Let δA be the local increment in area for a process in which one or more crack edges advance an infinitesimal amount. Multiply equation (18b) by δA and sum along the crack edges, which yields $-\delta W = \delta W_s$. We want this work equality to be valid for all changes $\delta A = (\partial A/\partial S_m)\delta S_m$ due to arbitrary changes δS_m. This requirement yields equation (3), where f_m and W_s are given, respectively, by equations (4) and (33). Inasmuch as $W_s = W_s(S_m, \xi)$, the behavior is the same as for an aging elastic material. However, it should be recalled that the aging stops during periods of "load reduction" (cf. equation (20)).

For the special case in which there are only two different q's, say one $q<\infty$ and one $q=\infty$, equations (28) and (33) provide the simple power-law time-dependence,

$$W_s = W_\infty + W_q \xi^{-1/q} \qquad (35)$$

where $W_\infty = W_\infty(S_m)$ and $W_q = W_q(S_m)$. Then, the R parameters S_r which change in a process are found from the R equations,

$$-\frac{\partial W}{\partial S_r} = \frac{\partial W_\infty}{\partial S_r} + \frac{\partial W_q}{\partial S_r} \xi^{-1/q} \qquad (36)$$

Even if the q_j are constant in time, equation (36) yields time-dependent values of S_r, and thus time-dependent stresses or forces through equation (2) or (10).

The connection between equation (3) and crack growth theory was established by considering the propagation of individual cracks. However, in modeling damage growth associated with microcracking, it is normally practical to use only a small number of averaging structural parameters which serve to approximate the actual effect of damage on overall mechanical behavior. As done previously (Schapery, 1990) let us require the approximate model to exhibit the same overall limited path-independence of work as implied by equation (10). Inasmuch as equation (3) is necessary and sufficient for such limited path-independence, this growth law should be used in the approximate model. Carrying this argument one step further, we would want the time-dependence of W_s to be essentially the same as that for the more complete representation, e.g. equation (35).

Unstable Crack Growth: The special energy release rate in equation (21) leads to unstable crack growth, which is predicted to occur at a time ξ_f, equation (23), that depends on the initial size. We shall consider a situation in which there are many parallel, planar, penny-shaped cracks with a distribution of radii a_o. It is assumed that when each crack becomes unstable, it rapidly grows in the original plane to a much larger size and then is arrested by some obstacle. By considering the effect of each crack on overall mechanical response after it reaches its final arrested size, we can develop a work potential W_T which obeys the equations of Section 2. The procedure is analogous to that used by Schapery (1990, Appendix B) for brittle elastic behavior. In this earlier work, generalized forces were used instead of displacements as independent variables. While either set could be used, for consistency with work in the previous sections, we shall use displacements.

The work potential is found to be

$$W_T = W_o - gG_1 + W_s \tag{37}$$

where $W_o = W_o(q_j)$ is the strain energy without cracks and $G_1 = G_1(q_j)$ is the function in equation (21). Also, $g=g(S)$ and $W_s = W_s(S,\xi)$, where

$$W_s = -\frac{(q-1)^Q}{k_2 q (k_1 \xi)^{1/q}} \int_S^{S_1} (S')^{-Q} \frac{dg}{dS'} dS' \tag{38}$$

in which $Q \equiv (q-1)/q$ and S_1 is the initial radius of the largest pre-existing crack. The function g is unspecified here, but depends on the details of microstructure and accounts for the effect of randomly or regularly distributed arrested crack sizes.

The stationary work condition, equation (8), is to be satisfied by the proposed W_T. From equations (37) and (38),

$$\frac{\partial W_T}{\partial S} = \{-G_1 + \frac{(q-1)^Q S^{-Q}}{k_2 q (k_1 \xi)^{1/q}}\} \frac{dg}{dS} \tag{39}$$

The quantity in braces vanishes in view of equation (23) if we consider S to be the initial radius of the crack which becomes unstable at the current time; this equation provides $S = S(G_1, \xi)$. The smaller a crack is, the longer the time is for the crack growth to become unstable, so that $\dot{S}<0$. The body is stable if $\partial^2 W_T/\partial S^2 > 0$; from equation (39) one finds that $dg/dS < 0$. This condition on g not only assures stability, but assures that the entropy production inequality, equation (7), will be satisfied when $\dot{S}<0$. If additional crack orientations are introduced, the work potential will depend in a similar way on additional structural parameters.

We conclude that the work potential based on unstable (micro)crack growth and arrest exhibits the same behavior as that based on stable growth. The power law time-dependence of W_S appears in both cases, except its physical origin is different. For stable growth, it reflects the direct effect of continuous growth, while for unstable growth it arises from the time delay for instability. The two types of growth may co-exist without changing the form of time-dependence.

5. Global Viscoelasticity

One approach to developing a model which includes viscoelastic effects throughout the matrix and fibers, besides that at crack tips, would be to introduce a rate-type evolution law for a portion of the internal state variables, say $S_\beta (\beta = 1,2,\ldots B)$, of Section 2. For example, for linear viscoelastic behavior with damage given one would use (e.g. Schapery, 1964),

$$\dot{S}_\beta = b_{\beta\gamma} f_\gamma \tag{40}$$

where $b_{\beta\gamma}$ is a symmetric matrix, and $f_\gamma = -\partial W/\partial S_\gamma$ in which W is quadratic in q_j and S_γ. The remaining internal state variables would be associated with the damage, and thus obey equation (3); $b_{\beta\gamma}$ may depend on them. The problem with this approach is that the simple crack growth theory in Section 3 and 4 is not in general applicable because there is not a simple correspondence between elastic and viscoelastic fields in the continuum. Unless considerable simplification is introduced in the description of the global viscoelastic behavior of the constituent materials, it does not appear to be possible to develop a practical analytical model. Here, we shall briefly review a manageable approach the author has used to account for linear and nonlinear viscoelasticity of the matrix; it permits the use of a slightly modified form of the crack growth theory of Sections 3 and 4.

Let us give the constitutive equation without damage and then discuss the modification needed in the elasticity theory with damage. With small strains and rotations the constitutive equation for any one of the constituent materials or phases, in terms of stresses σ_{ij} and strains $\varepsilon_{ij}(i,j = 1,2,3)$, is given as

$$\sigma_{ij} = \partial W_p / \partial \varepsilon^R_{ij} \tag{41a}$$

where $W_p = W_p(\varepsilon^R_{ij})$ and

$$\varepsilon^R_{ij} \equiv E_R^{-1} \int_{-\infty}^{t} E(t-\tau,t) \frac{\partial \varepsilon_{ij}}{\partial \tau} d\tau \tag{41b}$$

are so-called pseudo strains. The quantity $E(t-\tau,t)$ is the relaxation modulus, allowing for aging through the second argument, while E_R is a free constant which can be selected to have the units of modulus so that ε^R_{ij} is dimensionless. As discussed elsewhere (Schapery, 1981) equation (41) contains the special cases (1) linear isotropic viscoelasticity (if the Poisson's ratio is constant), (2) nonlinear elasticity ($E=E_R$) and (3) linear and nonlinear viscous theory. Inasmuch as W_p is like strain energy density, but is a function of pseudo strains, we call it pseudo strain energy density. It is easily shown that a multiphase continuum may be characterized by equations like (41) if each of the phases obeys equation (41) and all have the same relaxation modulus; phase-to-phase differences are reflected in the particular pseudo strain energy density employed. If the deformations in any phase are relatively small, so that it may be assumed rigid, then of course its relaxation modulus is not restricted to be the same as that for the other phases.

A simple correspondence exists between the mechanical state of elastic and viscoelastic bodies, with or without crack growth, when equations like (41) are applicable (Schapery, 1981, 1984). Large deformations may be taken into account by using Piola stresses and deformation gradients (in place of ε_{ij}); however, there is a basic limitation in that the pseudo strain energy density may be significantly affected by large rotations. This correspondence enables us to use all of the theory in Sections 2-4 by simply replacing q_i with generalized pseudo displacements,

$$q_i^R = E_R^{-1} \int_{-\infty}^{t} E(t-\tau,t) \frac{dq_i}{d\tau} d\tau \qquad (42)$$

while retaining the Q_i as generalized forces. The superscript R comes from use of the name "reference elastic solution" for the set of variables (q_i^R, Q_i); they may be interpreted as the displacements and forces in and on an elastic body which is identical to that of the viscoelastic body except for the relaxation modulus. Observe that when the q_i^R are used in Section 4, there will be hereditary effects due to both damage growth and viscoelasticity of the continuum.

6. Conclusions

An approach to modeling the mechanical response of viscoelastic composites with changing structure has been described. Although the rate-type evolution law used for the changing structure is that commonly identified with crack growth, equations (11) and (19a), it is not necessarily limited to crack growth. Indeed, the approach may be used for any evolution law of the form $\dot{S}_m = F_m(f_m)\hat{F}_m(S_m)$ for each m, where F_m is a strongly increasing function of the associated thermodynamic force f_m. This is a special case of the form used by Rice (1971) in a study of constitutive relations for metal and other solids; he assumed $\dot{S}_m = \dot{S}_m(f_m, S_1, \ldots, S_M)$ for each m.

Experimental verification of the viscoelastic behavior predicted by the simplified theory described herein is presently very limited. However, existing results on particle-filled rubber with constant and varying damage (Schapery, 1982) and on fiber-reinforced plastic with constant damage (Tonda and Schapery, 1987) do support the theory.

Acknowledgment

Sponsorship of this work by the Office of Naval Research is gratefully acknowledged.

References

Harper, B.D., 1986, "A Uniaxial Nonlinear Viscoelastic Constitutive Relation for Ice," Journal of Energy Resources Technology, Vol. 108, pp. 156-160.

Harper, B.D., 1989, "Some Implications of a Nonlinear Viscoelastic Constitutive Theory Regarding Interrelationships Between Creep and Strength Behavior of Ice at Failure," Journal of Offshore Mechanics and Arctic Engineering, Vol. 111, pp. 144-148.

Karr, D.G. and Choi, K., 1989, "A Three-Dimensional Constitutive Damage Model for Polycrystalline Ice,"Mechanics of Materials, Vol. 8, pp. 55-66.

Onat, E.T. and Leckie, F.A., 1988, "Representation of Mechanical Behavior in the Presence of Changing Internal Structure," Journal of Applied Mechanics, Vol. 55, pp. 1-10.

Reddy, J.N. and Rasmussen, M.L., 1982, Advanced Engineering Analysis, John Wiley & Sons, pp. 203-205.

Rice, J.R., 1971, "Inelastic Constitutive Relations for Solids: An Internal-Variable Theory and its Application to Metal Plasticity," Journal of the Mechanics and Physics of Solids, Vol. 19, pp. 433-455.

Schapery, R.A., 1964, "Application of Thermodynamics to Thermomechanical, Fracture and Birefringent Phenomena in Viscoelastic Media," Journal of Applied Physics, Vol. 35, pp. 1451-1465.

Schapery, R.A., 1981, "On Viscoelastic Deformation and Failure Behavior of Composite Materials with Distributed Flaws," 1981 Advances in Aerospace Structures and Materials, AD-Vol. 1, ASME, Edited by S.S. Wang and W.J. Renton, pp. 5-20.

Schapery, R.A., 1982, "Models for Damage Growth and Fracture in Nonlinear Viscoelastic Particulate Composites," Proc. Ninth U.S. National Congress of Applied Mechanics, Book No. H00228, ASME, pp. 237-245.

Schapery, R.A., 1984, "Correspondence Principles and a Generalized J Integral for Large Deformation and Fracture Analysis of Viscoelastic Media", International Journal of Fracture, Vol. 25, pp. 195-223.

Schapery, R.A., 1989, "Models for the Deformation Behavior of Viscoelastic Media with Distributed Damage and Their Applicability to Ice," Proc. IUTAM/IAHR Symp. on Ice/Structure Interaction, St. John's, Newfoundland, in press.

Schapery, R.A., 1990, "A Theory of Mechanical Behavior of Elastic Media with Growing Damage and Other Changes in Structure," Journal of the Mechanics and Physics of Solids, Vol. 38, pp. 215-253.

Sjolind, S.G., 1987, "A Constitutive Model for Ice as a Damaging Viscoelastic Material," Cold Regions Science and Technology, Vol. 41, pp. 247-262.

Tonda, R.D. and Schapery, R.A., 1987, "A Method for Studying Composites with Changing Damage by Correcting for the Effects of Matrix Viscoelasticity," Damage Mechanics in Composites, AD-Vol. 12, ASME, Edited by A.S.D. Wang and G.K. Haritos, pp. 45-51.

Weitsman, Y., 1988, "A Continuum Damage Model for Viscoelastic Materials," Journal of Applied Mechanics, Vol. 55, pp. 773-780.

The Effective Properties of Brittle/Ductile Incompressible Composites

Pedro Ponte Castañeda
Mechanical Engineering
The Johns Hopkins University
Baltimore, MD 21218, U.S.A.

Abstract

A new variational method for estimating the effective properties of nonlinear composites in terms of the corresponding properties of linear composites with the same microstructural distributions of phases is applied to an isotropic, incompressible composite material containing a brittle (linear) and a ductile (nonlinear) phase. More specifically, in this particular work the prescription is used to obtain bounds of the Hashin-Shtrikman type for the effective properties of the nonlinear composite in terms of the well-known linear bounds. It can be shown that in some cases the method leads to optimal bounds.

Introduction

PONTE CASTAÑEDA (1990a) has proposed a new procedure for estimating the effective properties of composite materials with phases exhibiting nonlinear constitutive behavior. The procedure, which is straightforward to implement, expresses the effective properties of the nonlinear composite in terms of the effective properties of a family of linear composites with the same distribution of phases as the nonlinear composite. Appropriate references for the linear theory of composites are given by the review article of WILLIS (1982) and by the monograph of CHRISTENSEN (1979). The new procedure was applied in the above reference to materials containing a nonlinear matrix either weakened by voids or reinforced by rigid particles. Estimates and rigorous bounds were obtained for the effective properties of such materials. The Hashin-Shtrikman bounds (obtained via the new method from the linear Hashin-Shtrikman bounds) were found to be an improvement over the corresponding bounds obtained by PONTE CASTAÑEDA and WILLIS (1988) for the same class of materials using an extension of the Hashin-Shtrikman variational principle to nonlinear problems proposed by TALBOT and WILLIS (1985). Recently, WILLIS (1990) has shown that the Hashin-Shtrikman bounds obtained via the new method can also be obtained by the method of TALBOT and WILLIS (1985) with an optimal choice of the comparison material. More generally, however, the new procedure can make use of linear higher-order bounds and estimates to yield corresponding bounds and estimates for nonlinear materials. In fact, the new procedure can be shown to yield exact results for a certain class of nonlinear composites. This is discussed in detail by PONTE CASTAÑEDA (1990b).

In this paper we apply the general procedure to a composite containing a brittle (linear) and a ductile (nonlinear) phase. We assume that the phases are perfectly bonded to each other, incompressible and isotropic. Additionally, the size of the typical heterogeneity is assumed to be small compared to the size of the specimen and the scale of variation of the applied loads. It is further assumed that the effect of the interfaces is negligible, so that the effective properties of the composite are essentially derived from the bulk behavior of the constituent phases. Both upper and lower bounds of the Hashin-Shtrikman type are given for the isotropic composite as functions of the properties and volume fractions of the phases. Specific results are given when the behavior of the nonlinear phase is linear plus power-law, including the pure power-law case. Some of the bounds are shown to be *optimal* (i.e., microstructures can be given attaining these bounds).

Effective Properties

Consider a two-phase composite occupying a region of unit volume Ω, with each phase occupying a subregion $\Omega^{(r)}$ $(r = 1, 2)$, and let the stress potential, $U(\sigma,\mathbf{x})$, be expressed in terms of the homogeneous phase potentials, $U^{(r)}(\sigma)$, via

$$U(\sigma,\mathbf{x}) = \sum_{r=1}^{2} \chi^{(r)}(\mathbf{x}) U^{(r)}(\sigma), \tag{1}$$

where

$$\chi^{(r)}(\mathbf{x}) = \begin{cases} 1 & \text{if } \mathbf{x} \in \Omega^{(r)} \\ 0 & \text{otherwise} \end{cases} \tag{2}$$

is the characteristic function of phase r. The phases are assumed to be incompressible and isotropic, so that the potentials $U^{(r)}(\boldsymbol{\sigma})$ can be assumed to depend on the stress $\boldsymbol{\sigma}$ only through the effective stress

$$\sigma_e = \sqrt{\frac{3}{2}\mathbf{S}\cdot\mathbf{S}},$$

where $\mathbf{S}$ is the deviator of $\boldsymbol{\sigma}$. Thus, we assume that there exist scalar-valued functions $f^{(r)}$ such that

$$U^{(r)}(\boldsymbol{\sigma}) = f^{(r)}(\sigma_e).$$

Then, the stress field $\boldsymbol{\sigma}$, satisfying the equilibrium equations

$$\sigma_{ij,j} = 0, \tag{3}$$

is related to the strain field $\boldsymbol{\varepsilon}$, related to the displacement field $\mathbf{u}$ via

$$\varepsilon_{ij} = \frac{1}{2}(u_{i,j} + u_{j,i}), \tag{4}$$

through the constitutive relation

$$\varepsilon_{ij} = \frac{\partial U}{\partial \sigma_{ij}}(\boldsymbol{\sigma}, \mathbf{x}). \tag{5}$$

The commas in equations (3) and (4) denote differentiation, and the summation convention has also been used in equation (3). We assume that the phases are perfectly bonded, so that the displacement is continuous across the interphase boundaries. However, the strains and, therefore, the stresses may be discontinuous across such boundaries, and hence equation (3) must be interpreted in a weak sense, requiring continuity of the traction components of the stress across the interphase boundaries.

We note that if we let $\boldsymbol{\varepsilon}$ represent the rate-of-deformation tensor and $\mathbf{u}$ the velocity field, the above equations can be used to model high-temperature creep, as well as high-rate viscoplastic deformations. Here we will present our work in the context of time-independent plasticity (deformation theory), but in view of the above comment the results could be given appropriate interpretations in nonlinear creep and viscoplasticity.

To define the effective properties of the heterogeneous material we introduce, following HILL (1963), the uniform constraint boundary condition

$$\sigma_{ij} n_j = \bar{\sigma}_{ij} n_j, \quad \mathbf{x} \in \partial\Omega, \tag{6}$$

where $\partial\Omega$ denotes the boundary of the composite, $\mathbf{n}$ is its unit outward normal, and $\bar{\boldsymbol{\sigma}}$ is a given constant symmetric tensor. Then, the average stress is precisely $\bar{\boldsymbol{\sigma}}$, i.e.

$$\overline{\boldsymbol{\sigma}} = \int_{\Omega} \boldsymbol{\sigma}(\mathbf{x}) dV \tag{7}$$

and we *define* the average strain in a similar manner by

$$\overline{\boldsymbol{\varepsilon}} = \int_{\Omega} \boldsymbol{\varepsilon}(\mathbf{x}) dV . \tag{8}$$

The effective behavior of the composite, or the relation between the average stress and the average strain then follows from the principle of minimum complementary energy, which can be stated in the form

$$\tilde{U}(\overline{\boldsymbol{\sigma}}) = \min_{\boldsymbol{\sigma} \in S(\overline{\boldsymbol{\sigma}})} \overline{U}(\boldsymbol{\sigma}), \tag{9}$$

where

$$\overline{U}(\boldsymbol{\sigma}) = \int_{\Omega} U(\boldsymbol{\sigma}, \mathbf{x}) \, dV$$

is the complementary energy functional of the problem,

$$S(\overline{\boldsymbol{\sigma}}) = \{\boldsymbol{\sigma} \,|\, \sigma_{ij,j} = 0 \text{ in } \Omega, \text{ and } \sigma_{ij} n_j = \overline{\sigma}_{ij} n_j \text{ on } \partial\Omega\}$$

is the set of statically admissible stresses, and where we have assumed convexity of the nonlinear potential $U(\boldsymbol{\sigma}, \mathbf{x})$. Thus, we have that

$$\overline{\varepsilon}_{ij} = \frac{\partial \tilde{U}}{\partial \sigma_{ij}} (\overline{\boldsymbol{\sigma}}). \tag{10}$$

Our task will be to determine bounds and estimates for $\tilde{U}(\overline{\boldsymbol{\sigma}})$, which, under the above assumptions, is known to be convex.

Bounds and Estimates

A new variational principle for determining bounds and estimates for the effective properties of nonlinear composites in terms of the effective properties of linear composites was proposed by PONTE CASTAÑEDA (1990a,b). In this section, we specialize the derivation given in PONTE CASTAÑEDA (1990b) for the case where both phases are incompressible, and phase #2 is linear so that

$$U^{(2)}(\boldsymbol{\sigma}) = \frac{1}{6\mu^{(2)}} \sigma_e^2 .$$

The new variational principle is based on a representation of the potential of the nonlinear material in terms of the potentials of a family of linear *comparison* materials. Thus, for a homogeneous nonlinear material with "stronger than quadratic" growth in its potential, $U(\boldsymbol{\sigma})$, we have that

$$U(\boldsymbol{\sigma}) \geq \max_{\mu>0}\{U_o(\boldsymbol{\sigma}) - V(\mu)\}, \tag{11}$$

where

$$V(\mu) = \max_{\boldsymbol{\sigma}}\{U_o(\boldsymbol{\sigma}) - U(\boldsymbol{\sigma})\} \tag{12}$$

and

$$U_o(\boldsymbol{\sigma}) = \frac{1}{6\mu}\sigma_e^2 \tag{13}$$

is the the potential of the comparison linear material.

To demonstrate this result, let

$$U(\boldsymbol{\sigma}) = \phi(s), \tag{14}$$

where $s = \sigma_e^2$. Then, the Legendre-Fenchel transform of the scalar-valued function ϕ is given by

$$\phi^*(\alpha) = \max_{s>0}\{\alpha s - \phi(s)\}, \tag{15}$$

where α is assumed to be positive. A well-known result in convex analysis (VAN TIEL 1984, §6.3) is that

$$\phi(s) \geq \max_{\alpha>0}\{s\alpha - \phi^*(\alpha)\}, \tag{16}$$

with equality if ϕ is a convex function of its argument. With the identifications $s = \sigma_e^2$ and $\alpha = (6\mu)^{-1}$, we can see that (11) and (12) are but simple re-statements of (16) and (15), respectively. In particular,

$$V(\mu) = \phi^*\left(\frac{1}{6\mu}\right). \tag{17}$$

To derive the new variational principle, we apply (11) to the nonlinear phase #1, and make use of the result in the complementary energy principle (9). Thus, after some manipulations, we find that

$$\tilde{U}(\overline{\boldsymbol{\sigma}}) \geq \max_{\mu^{(1)}(\mathbf{x})}\left\{\tilde{U}_o(\overline{\boldsymbol{\sigma}}) - \int_{\Omega^{(1)}} V^{(1)}(\mu^{(1)})\,dV\right\}, \tag{18}$$

where

$$\tilde{U}_o(\overline{\boldsymbol{\sigma}}) = \min_{\boldsymbol{\sigma}\in S(\overline{\boldsymbol{\sigma}})} \overline{U}_o(\boldsymbol{\sigma}), \tag{19}$$

$$U_o(\boldsymbol{\sigma},\mathbf{x}) = \sum_{r=1}^{2}\chi^{(r)}(\mathbf{x})U_o^{(r)}(\boldsymbol{\sigma}),$$

and

$$U_o^{(r)}(\boldsymbol{\sigma}) = \frac{1}{6\mu^{(r)}}\sigma_e^2.$$

Note that the comparison linear material agrees with the actual material in phase # 2 (which is linear). In the above derivation, we note that the comparison moduli $\mu^{(1)}$ are functions of $\mathbf{x}$, since the stress field $\boldsymbol{\sigma}$ will also in general be a function of $\mathbf{x}$ within phase #1. If we assume that $U^{(1)}(\boldsymbol{\sigma})$ is "strongly convex" (i.e. if ϕ is convex), then we have equality in (11), and hence, usually, equality in (18). However, if the conditions for equality are not met, relation (18) still provides a useful lower bound for $\tilde{U}(\overline{\boldsymbol{\sigma}})$. An detailed derivation of this result, discussing the precise conditions for equality, is given in PONTE CASTAÑEDA (1990b).

The variational principle described by (18) roughly corresponds to solving a completely linear problem for a heterogeneous material with arbitrary moduli variation within the nonlinear phase, and then optimizing with respect to the variations in moduli within the nonlinear phase. Thus, one can think of the nonlinear material as a "linear" material with variable moduli that are determined by prescription (18) in such a way that its properties agree with those of the nonlinear material.

This suggests that if the fields happen to be constant over the nonlinear phase, then the variable moduli $\mu^{(1)}(\mathbf{x})$ can be replaced by constant moduli $\mu^{(1)}$. More generally, however, we have the following lower bound for $\tilde{U}(\overline{\boldsymbol{\sigma}})$

$$\tilde{U}_{-}(\overline{\boldsymbol{\sigma}}) = \max_{\mu^{(1)}>0}\{\tilde{U}_{o}(\overline{\boldsymbol{\sigma}}) - c^{(1)}V^{(1)}(\mu^{(1)})\}, \tag{20}$$

where $c^{(1)}$ is the volume fraction of phase #1. The result in this form is a special case of a more general result first derived by PONTE CASTAÑEDA (1990a), when only one of the phases is nonlinear, and the other one is linear.

We note that the prescriptions (18) and (20) lead to convex expressions for the bounds and estimates of the effective potential of the nonlinear composite, provided that the corresponding bounds and estimates for the linear composite are convex. This is a desirable feature, because the effective potential of the composite is known to be convex.

Application to Hashin-Shtrikman Bounds

HASHIN and SHTRIKMAN (1962) prescribed bounds for the effective moduli of linear-elastic, isotropic composites, depending only on the volume fractions of the phases. When there are only two phases, these bounds have been shown to be optimal (i.e., microstructures can be given that simultaneously attain the bounds for the shear and bulk modulus) by FRANCFORT and MURAT (1987).

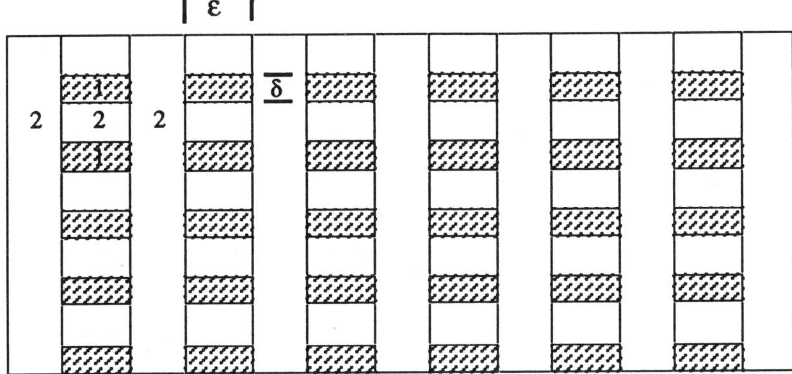

Figure 1. Rank-2 laminate.

Their construction made use of *iterated* laminates for which the effective properties can be computed exactly. Such materials are obtained by layering the two constituent phases to obtain a rank-1 laminate; the resulting material is once again layered (in an arbitrary direction) with one of the original phases in a smaller lengthscale. This procedure can obviously be iterated n times to obtain a rank-n laminate. In general such materials will be anisotropic, but by choosing appropriately the layer orientations at the different layering operations, it is possible to obtain an isotropic composite, and its properties coincide with one of the Hashin-Shtrikman (H-S) bounds depending on which constituent phase is selected to play the role of the matrix material. Figure 1 depicts a rank-2 laminate (not to scale) with phase #2 as the matrix phase.

For the special case of incompressible materials, when there is only one modulus for the composite, the H-S upper bound for the effective shear modulus can be expressed in the form

$$\tilde{\mu}_{+} = \begin{cases} \dfrac{\mu^{(1)}}{\alpha\left(\mu^{(1)},\mu^{(2)}\right)} & \text{if } \mu^{(1)} \geq \mu^{(2)} \\ \dfrac{\mu^{(2)}}{\beta\left(\mu^{(1)},\mu^{(2)}\right)} & \text{if } \mu^{(1)} \leq \mu^{(2)} \end{cases}, \tag{21}$$

where

$$\alpha\left(\mu^{(1)},\mu^{(2)}\right) = \frac{2c^{(1)}\mu^{(2)} + \left(3 + 2c^{(2)}\right)\mu^{(1)}}{\left(2 + 3c^{(2)}\right)\mu^{(2)} + 3c^{(1)}\mu^{(1)}} \tag{22}$$

and

$$\beta(\mu^{(1)},\mu^{(2)}) = \frac{2c^{(2)}\mu^{(1)} + (3+2c^{(1)})\mu^{(2)}}{(2+3c^{(1)})\mu^{(1)} + 3c^{(2)}\mu^{(2)}} \ . \tag{23}$$

The corresponding H-S lower bound is obtained by interchanging the expressions in (21) for the upper bound (and keeping the conditions on the shear moduli $\mu^{(1)}$ and $\mu^{(2)}$ fixed).

The above H-S upper bound for the effective shear modulus yields a lower bound for the potential of the linear material $\tilde{U}_o$. This information can be used in combination with prescription (20) to yield a H-S lower bound for the potential of the nonlinear material $\tilde{U}$. On the other hand, upper bounds for $\tilde{U}_o$ do not necessarily generate upper bounds for $\tilde{U}$.

The result for the lower bound on $\tilde{U}$ depends on which of the two branches of (21) is used in conjunction with (20). If $\mu^{(1)} > \mu^{(2)}$, then the average effective stress $\bar{\sigma}_e$ must be such that the condition

$$3\mu^{(2)} f'(\bar{\sigma}_e) < \bar{\sigma}_e, \tag{24}$$

is satisfied (usually when the average shear stress is small enough). Here, for simplicity, we have made the identification $f^{(1)} = f$. The corresponding form of the bound is then

$$\tilde{U}_-(\bar{\boldsymbol{\sigma}}) = \tilde{f}_1(\bar{\sigma}_e), \tag{25}$$

where

$$\tilde{f}_1(\bar{\sigma}_e) = c^{(1)} f(s) + \mu^{(2)} \left[\frac{(2+3c^{(2)})}{2} - \left(\frac{\bar{\sigma}_e}{s}\right)^2 \right] (f'(s))^2 \tag{26}$$

and s solves the equation

$$c^{(1)} + \mu^{(2)}(2+3c^{(2)}) \frac{f'(s)}{s} = \frac{5}{3} \left[\frac{(2+3c^{(2)})(s/\bar{\sigma}_e)^2 - 2}{3c^{(2)}} \right]^{-1/2} . \tag{27}$$

On the other hand, if $\mu^{(1)} < \mu^{(2)}$, then the average effective stress $\bar{\sigma}_e$ must be such that the condition

$$3\mu^{(2)} f'(\bar{\sigma}_e) > \bar{\sigma}_e, \tag{28}$$

is satisfied (i.e., when the average shear stress is large enough), and

$$\tilde{U}_-(\bar{\boldsymbol{\sigma}}) = \tilde{f}_2(\bar{\sigma}_e), \tag{29}$$

where

$$\tilde{f}_2(\overline{\sigma}_e) = c^{(1)} f(s) + \frac{\left(3+2c^{(1)}\right)\overline{\sigma}_e^2 + c^{(1)}\left(2+3c^{(1)}\right)s^2 - 10c^{(1)}s\overline{\sigma}_e}{18c^{(2)}\mu^{(2)}}, \tag{30}$$

and s solves the equation

$$9c^{(2)}\mu^{(2)} f'(s) = 5\overline{\sigma}_e - \left(2+3c^{(1)}\right)s. \tag{31}$$

The corresponding stress/strain relations have the form

$$\boldsymbol{\varepsilon} = \frac{3}{2}\tilde{f}'(\overline{\sigma}_e)\mathbf{S}, \tag{32}$$

where

$$\tilde{f}_2'(\overline{\sigma}_e) = \frac{2}{3}\frac{\left(3+2c^{(1)}\right)\overline{\sigma}_e - 5c^{(1)}s}{6c^{(2)}\mu^{(2)}}, \tag{33}$$

but $\tilde{f}_1'(\overline{\sigma}_e)$ does not have a simple expression.

In general, we do not expect the above lower bounds for $\tilde{U}$ to be optimal. In fact, expression (25) does not yield an optimal bound if condition (24) is satisfied. However, it is shown in PONTE CASTAÑEDA (1990b) that if condition (28) is satisfied, then the bound (29) is optimal. This is because the same microstructure attaining the linear bounds can be also shown to attain the nonlinear bound; the reason being that the fields are constant in the (nonlinear) inclusion phase, and hence expressions (20) and (18) are identical. Similar observations have been made by KOHN (1990) in a similar context (starting from the Talbot-Willis nonlinear variational principle) and, independently, by PONTE CASTAÑEDA (1990c) in the context of conductivity.

Conversely, in general, we do not expect that interchanging conditions (24) and (28) would turn expression (25) and (29) into upper bounds for the nonlinear potential $\tilde{U}$. This is contrary to the corresponding operation for the linear composite. All that can be said, however, is that expression (29) is an estimate for the upper bound for $\tilde{U}$ if condition (24) is satisfied and that expression (25) is an estimate for the upper bound for $\tilde{U}$ if condition (28) is satisfied. Both of these estimates are expected to get progressively better with weaker nonlinearities.

Application to Power-Law Behavior

In this section, we specialize further the calculations of the previous section by taking the constitutive behavior of the nonlinear phase to be governed by a linear plus power relation

$$f(\sigma_e)=\frac{1}{3}\left[\frac{1}{2\mu}+\left(\frac{1}{n+1}\right)\frac{1}{\eta}\sigma_e^{n-1}\right]\sigma_e^2. \tag{34}$$

Note that the case $\mu\to\infty$ corresponds to pure power-law behavior, and the limits $n\to 1$ (in addition to $\mu\to\infty$) or $\eta\to\infty$ correspond to linear behavior.

The conditions (24) and (28) determining the appropriate branch of the bound specialize to

$$\frac{\mu^{(2)}}{\mu}+\frac{\mu^{(2)}}{\eta}\overline{\sigma}_e^{n-1}<1 \tag{35}$$

and the opposite inequality, respectively. The first condition guaranteeing that (25) is a lower bound (and correspondingly that (29) is an estimate for the upper bound) corresponds to small enough average stress on the composite. Alternatively, the second condition (with > instead of <) corresponds to sufficiently large average stress. Note that, if $\mu^{(2)}/\mu>1$, condition (35) can never be satisfied and, conversely, the alternative condition is always satisfied. This condition ensures that the difference between the potential of phase #1 and that of phase #2 is convex. Here, we will consider two cases: one case, meeting this condition, with $\mu^{(2)}/\mu=2$, and the other with $\mu^{(2)}/\mu=0$, corresponding to the pure power-law case.

The results for the bounds (25) and (29) specialized to the case when (34) holds can be expressed in the form:

$$\frac{\tilde{U}(\overline{\boldsymbol{\sigma}})}{U^{(2)}(\overline{\boldsymbol{\sigma}})}=F\left\{\frac{\mu^{(2)}}{\eta}\overline{\sigma}_e^{n-1};\frac{\mu^{(2)}}{\mu},c^{(2)},n\right\}, \tag{36}$$

where the precise form of F depends on whether (25) or (29) applies, and $(\mu^{(2)}/\eta)\overline{\sigma}_e^{n-1}$ plays the role of the independent variable, with $\mu^{(2)}/\mu$, $c^{(2)}$ and n, serving as parameters.

Results for the upper and lower bounds for $\tilde{U}$ are given in Figures 2 and 3 for the case where $\mu^{(2)}/\mu=0$, and in Figure 4 for the case where $\mu^{(2)}/\mu=2$. In the first case, condition (35) determining whether (29) is an estimate for the upper bound or an optimal lower bound, and whether (25) is an estimate for the upper bound, or a non-optimal lower bound, simply reduces to the condition of whether the independent variable $(\mu^{(2)}/\eta)\overline{\sigma}_e^{n-1}$ is less or greater than unity. For that reason, we give results emphasizing the small stress and large stress domains, separately, in Figures 2 and 3, respectively.

In each plot we have three sets of curves corresponding to three values of $c^{(2)}$ (0.1, 0.5 and 0.9). Additionally, we show the limiting cases corresponding to $c^{(2)}=0$ and $c^{(2)}=1$. These limiting curves appear as straight lines, one with with variable slope depending on the value of n and $\mu^{(2)}/\mu$, and the other with

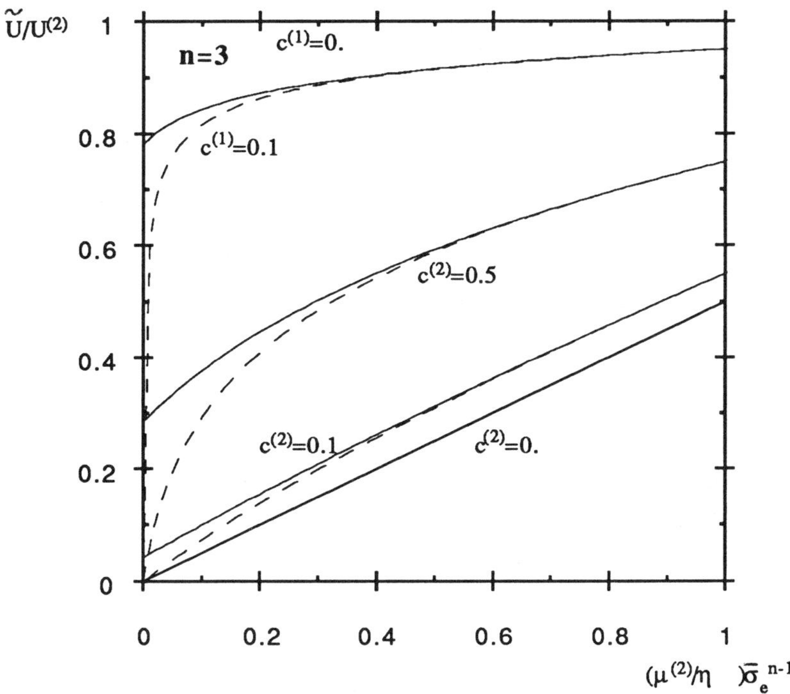

Figure 2(a). Plots of the bounds for the effective energy of the composite as functions of the average stress (appropriately normalized) for $\mu^{(2)}/\mu = 0$ and n = 3 (small stress).

zero slope (value equal to unity), respectively. The intermediate sets of curves correspond to the upper and lower bounds.

In Figure 2, depicting results for two values of the nonlinearity parameter (n = 3 and 10), the continuous line corresponds to the estimate for the upper bound (for $\tilde{U}$), and the dashed line corresponds to the rigorous lower bound. In Figure 3, showing also results for the same two values of the nonlinearity parameter, the continuous line corresponds to the optimal lower bound, and the dashed line is an estimate for the upper bound. For this value of $\mu^{(2)}/\mu$, the upper and lower bound coalesce when the value of the independent variable

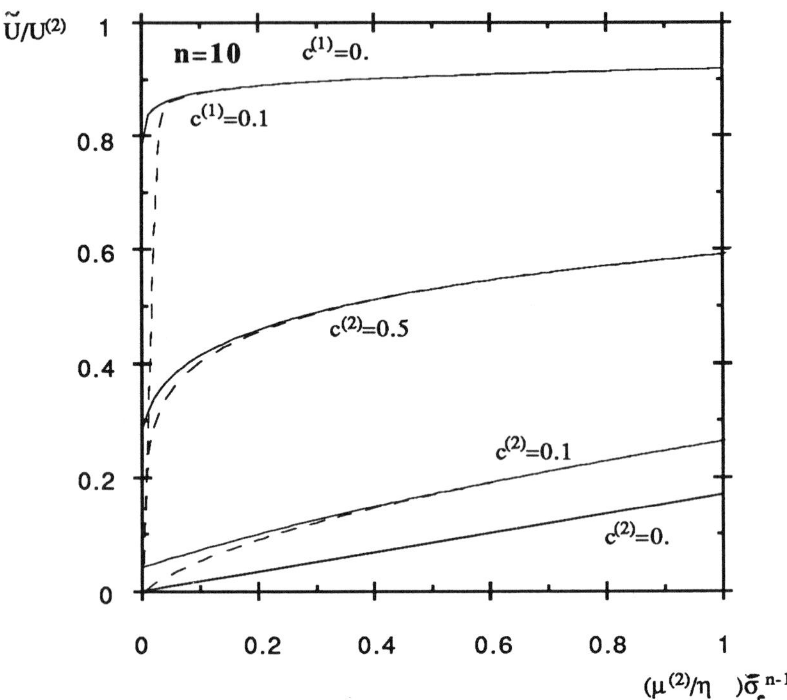

Figure 2(b). Plots of the bounds for the effective energy of the composite as functions of the average stress (appropriately normalized) for $\mu^{(2)}/\mu = 0$ and n = 10 (small stress).

$(\mu^{(2)}/\eta)\bar{\sigma}_e^{n-1}$ approaches unity. In the linear case, this behavior corresponds to the limit of the moduli of the phases approaching each other. More generally, assuming that $\mu^{(2)}/\mu$ is less than unity, there is a value of the independent variable (i.e., an average stress level) at which the bounds are equal, and hence the effective energy of the composite is known exactly. This phenomenon is related to the lack of convexity of the difference between the potentials of the nonlinear and linear phases.

In Figure 4, depicting results for the same two values of the nonlinearity parameter, the continuous line corresponds to the optimal lower bound (for $\tilde{U}$), and the dashed line corresponds to the estimate for the upper bound. In this case,

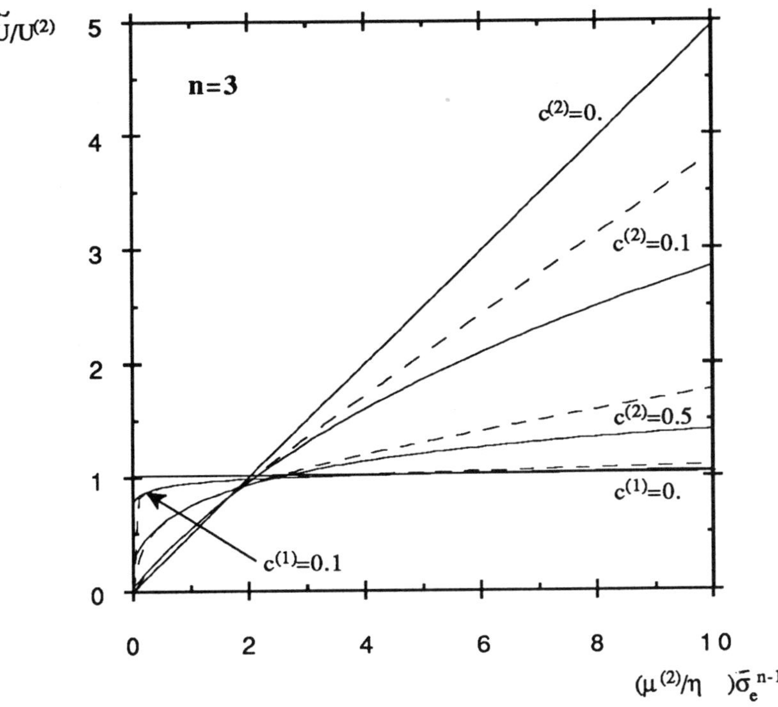

Figure 3(a). Plots of the bounds for the effective energy of the composite as functions of the average stress (appropriately normalized) for $\mu^{(2)}/\mu = 0$ and n = 3 (large stress)

with a convex difference between the nonlinear and linear potentials, there is no value of the independent variable for which the upper and lower bound are equal.

Both in Figures 3 and 4, we observe that the lower bound approaches a straight line with zero slope and the upper bound approaches a straight line with slope depending on the value of n (smaller for larger n). This is consistent with the following asymptotic behaviors for the lower and upper bounds

$$\frac{\tilde{U}(\overline{\boldsymbol{\sigma}})}{U^{(2)}(\overline{\boldsymbol{\sigma}})} \approx \frac{\left(1+\tfrac{2}{3}c^{(1)}\right)}{c^{(2)}}, \tag{37}$$

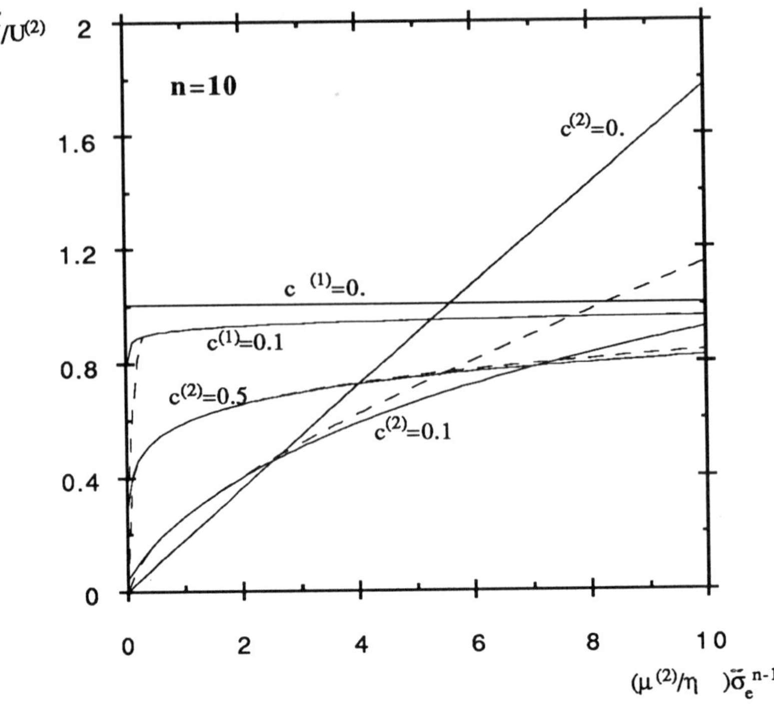

Figure 3(b). Plots of the bounds for the effective energy of the composite as functions of the average stress (appropriately normalized) for $\mu^{(2)}/\mu = 0$ and n = 10 (large stress)

and

$$\frac{\tilde{U}(\overline{\sigma})}{U^{(2)}(\overline{\sigma})} \approx \frac{2}{n+1}\frac{1}{\omega^n}\frac{\mu^{(2)}}{\eta}\overline{\sigma}_e^{n-1}, \tag{38}$$

with

$$\omega = \frac{\left(1+\frac{3}{2}c^{(2)}\right)^{\frac{n+1}{2n}}}{\left(c^{(1)}\right)^{\frac{1}{n}}} \approx 1+\left(\frac{3n+7}{4n}\right)c^{(2)} \quad \text{as } c^{(2)} \to 0 \tag{39}$$

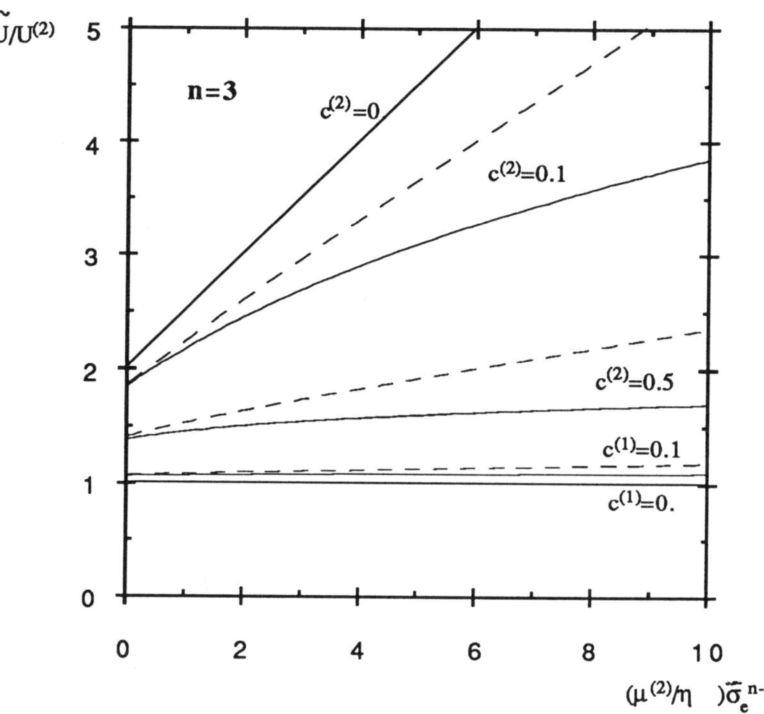

Figure 4(a). Plots of the bounds for the effective energy of the composite as functions of the average stress (appropriately normalized) for $\mu^{(2)}/\mu = 2$ and n = 3.

respectively. These two behaviors correspond physically to the cases of a linear matrix with voids and a power-law matrix with rigid inclusions (studied by PONTE CASTAÑEDA, 1990a), respectively. The reason for these behaviors is that the lower bound (for $\tilde{U}$) corresponds to putting the stiffer material in the matrix phase and the less stiff material in the inclusion phase (and viceversa for the upper bound). Clearly, for large enough stresses, the linear phase is stiffer than the nonlinear phase.

We note that accurate numerical calculations of the potential of a power-law matrix with spherical rigid inclusion have yielded results of the form (38) with

$$\omega = 1 + \left(\frac{g(n)}{n}\right) c^{(2)} \quad \text{as } c^{(2)} \to 0, \tag{40}$$

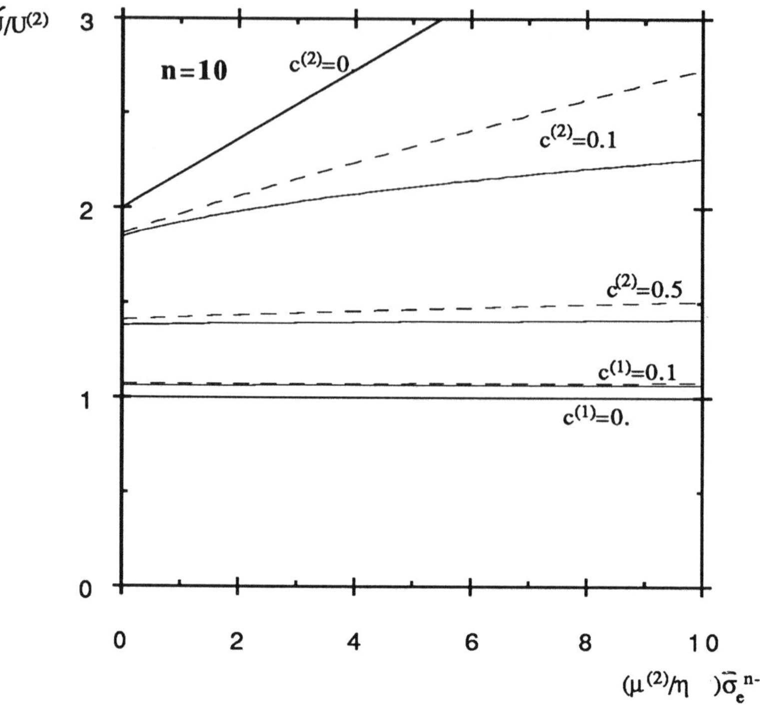

Figure 4(b). Plots of the bounds for the effective energy of the composite as functions of the average stress (appropriately normalized) for $\mu^{(2)}/\mu = 2$ and n = 10.

where $g(n)$ is such that $g(1) = 5/2$, $g(3) \approx 3.21$ and $g(10) \approx 6.09$ (LEE and MEAR, 1990), and $g(n) \rightarrow 0.38n$ as $n \rightarrow \infty$ (HUTCHINSON, 1990). These results do not compare very favorably with the corresponding results from (39): 5/2, 4.00, 9.25 and $0.75n$, but it should be recalled that these results correspond to the case for which we do not have a rigorous bound (it is simply an estimate of the bound). None the less, the results of (38) with (39) may provide reasonable estimates for larger values of the volume fraction of the linear phase.

Acknowledgements

This research was supported by the Air Force Office of Scientific Research under grant 89-0288.

References

CHRISTENSEN, R. M. (1979) *Mechanics of Composite Materials*, Wiley Interscience, New York.

FRANCFORT, G. and MURAT, F. (1987) *Archive Rat. Mech. and Analysis* **94**, 307.

HASHIN, Z. and SHTRIKMAN, S. (1962) *J. Mech. Phys. Solids* **10**, 335.

HILL, R. (1963) *J. Mech. Phys. Solids* **11**, 357.

HUTCHINSON, J.W. (1990) Private communication.

KOHN, R. V. (1990) Private communication.

LEE, B.J. and MEAR, M.E. (1990) Submitted for publication.

PONTE CASTAÑEDA, P. (1990c) To appear.

PONTE CASTAÑEDA, P. (1990b) To appear.

PONTE CASTAÑEDA, P. (1990a) *J. Mech. Phys. Solids*, in press.

PONTE CASTAÑEDA, P. and WILLIS, J. R. (1988) *Proc. R. Soc. Lond. A* **416**, 217.

TALBOT, D. R. S. and WILLIS, J. R. (1985) *IMA J. Appl. Math.* **35**, 39.

VAN TIEL, J. (1984) *Convex Analysis*, Wiley, New York.

WILLIS, J. R. (1982) In *Mechanics of Solids, The Rodney Hill 60th anniversary volume* (ed. H.G. HOPKINS & M.J. SEWELL), Pergamon Press, Oxford, 653.

WILLIS, J. R. (1990) *J. Mech. Phys. Solids*, to appear.

Inelastic Behavior II

Compressive Failure of Fibre Composites Due to Microbuckling

N.A. Fleck† and B. Budiansky*

†Cambridge University Engineering Department,
Trumpington Street, Cambridge, CB2 1PZ,
England.

*Division of Applied Sciences, Harvard University,
Cambridge, Massachusetts, 02138,
U.S.A.

ABSTRACT

The dominant compressive failure mechanism of modern fibre composites is microbuckling. This is demonstrated in the form of a fracture map. For polymer matrix composites microbuckling is a plastic event. An analysis is presented of both elastic and plastic microbuckling of unidirectional composites under remote axial and shear loading. The effects of fibre misalignment and inclination of the band are included. We find that a simple rigid-perfectly plastic analysis suffices for plastic microbuckling; it demonstrates that the axial compressive strength increases with decreasing fibre misalignment, increasing shear strength of the matrix, and decreasing remote shear stress. Finally, a calculation is performed of the remote axial and shear stress required to propagate an existing microbuckle. We find that the axial propagation stress is typically less than the shear yield stress of the matrix material.

1. INTRODUCTION

Most fibre reinforced polymer matrix composites have a compressive strength less than their tensile strength due to microbuckling of the load bearing fibres aligned with the loading direction. In many applications compressive strength is a design limiting feature. Over the past ten years significant improvements have been made to the tensile strength, impact resistance and toughness of these composites. Unfortunately, compressive strength has shown little concomitant improvement.

In this paper, previous experimental studies and theoretical models of microbuckling are reviewed. A new analysis of microbuckling is presented, based upon the kink band analysis of Budiansky (1983). The composite is subjected to remote axial compression and shear. Material inside and outside of the kink band is taken to be homogeneous but anisotropic. The kink band response is calculated for a variety of constitutive behaviours: (1) elastic, (2) rigid-perfectly plastic, and (3) elastic-perfectly plastic. An analysis is then given for the remote axial and shear stress required to propagate a microbuckle zone into undamaged material across the section of a specimen. The analysis is based upon a simple energy balance. We find remarkably low values for the propagation stress. This suggests that the compressive failure stress of large sheet structures containing a microbuckle near a stress raiser may be much less than that predicted for small undamaged specimens.

The paper deals only with the response of unidirectional unnotched composites. In many practical applications notched multi-directional composites are used. A design methodology is now emerging to deal with the effects of notches and off-axis plies (see for example, Starnes and Williams (1982), Rhodes, Mikulas and McGowan (1984), Soutis and Fleck (1990) and Soutis, Fleck and Smith (1990)).

2. PREVIOUS THEORETICAL WORK

Rosen (1965) assumed that compressive failure is by elastic microbuckling: he modelled the fibres as columns supported by an elastic foundation. He recognised that the composite plate may be a short stiff structure which does not buckle in compression on the macroscale, but the individual fibres have small diameters and buckle as slender columns on the microscale. Two possible buckling modes were distinguished, the shear mode and the extension mode. For the shear mode, shear deformation occurs in the matrix material, and the compressive strength σ_c is given by,

$$\sigma_c = \frac{G_m}{1-v_f} \tag{2.1}$$

where G_m is the shear modulus of the matrix and v_f is the fibre volume fraction. In the extension mode, matrix material suffers direct straining in a direction transverse to the fibre axis. The shear mode predicts a lower strength than the extension mode and is assumed to dominate.

The Rosen anaysis overpredicts strength typically by a factor of four. This suggests that microbuckling is a plastic rather than an elastic event. Several investigators (eg. Lager and June

(1969)) have introduced empirical correction factors in order to improve the agreement between the Rosen theory and experiment.

Argon (1972) and Budiansky (1983) identified the shear yield stress k of the matrix material and the initial misalignment angle $\bar{\phi}$ of fibres in the microbuckled band as the main factors governing the compressive strength. The misalignment angle $\bar{\phi}$, and band inclination β are defined in the insert in Fig.1. For a

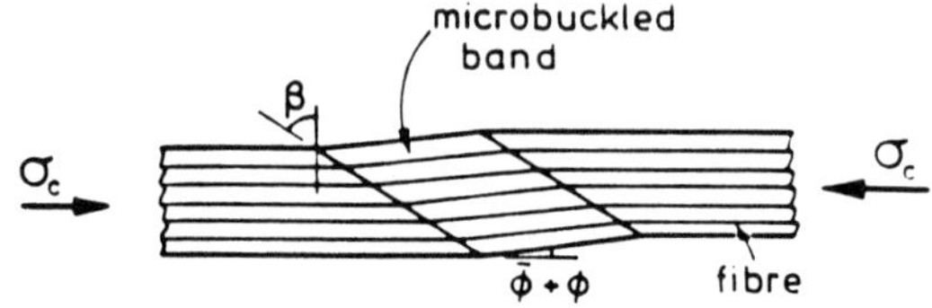

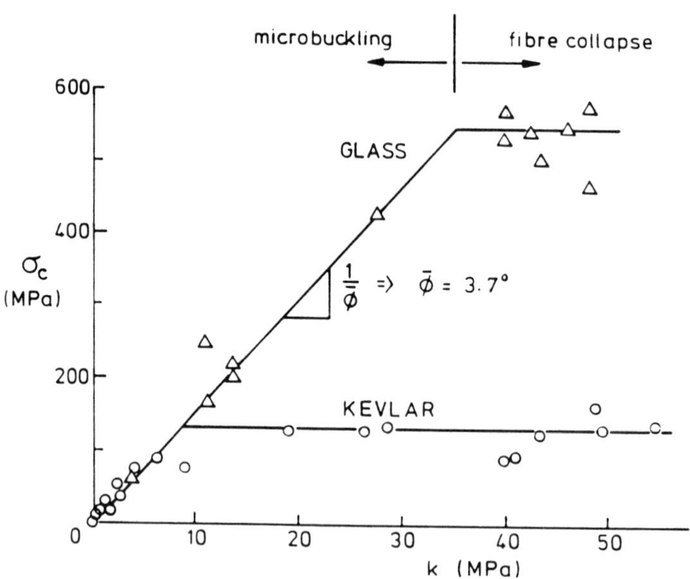

Fig.1: Effect of shear yield stress k of polyester matrix upon compressive strength σ_c of glass and Kevlar composites. Data taken from Piggott and Harris (1980).

rigid-perfectly plastic matrix material, Budiansky found that the compressive strength σ_c is given by

$$\sigma_c = \frac{k^*}{\bar{\phi}} \tag{2.2}$$

where,

$$k^* \equiv k \left(1 + \left(\frac{\sigma_{Ty}}{k}\right)^2 \tan^2\beta\right)^{1/2} \tag{2.3}$$

and σ_{Ty} is the yield stress of the composite transverse to the fibre direction.

There was little need to include fibre bending explicitly in the analysis: a kinking analysis suffices wherein material in the microbuckled band is treated as a homogeneous anisotropic solid. This approach is developed later in the present paper.

Recently, Steif (1988) has modelled the effect of fibre-matrix debonding upon the elastic microbuckling of fibre composites. The model is an adaptation of the Rosen analysis to situations where slip occurs at the fibre-matrix interface; slip begins when the interfacial shear stress attains a critical value. Interfacial shear failure is similar in many respects to shear yielding of the matrix. Steif's model gives reasonable predictions for ceramic matrix composites when the wavelength of the buckle equals the specimen length. This assumption is unrealistic.

3. AVAILABLE EXPERIMENTAL EVIDENCE

From the published literature it is apparent that unidirectional composites fail by two distinct failure mechanisms, fibre microbuckling and fibre collapse.

When the matrix yield stress is sufficiently

high, the fibres suffer compressive collapse. This is due to fibre yielding in the case of steel or Kevlar fibres (see Moncunill de Ferran and Harris (1970), Greszczuk (1972, 1975), Piggott and Harris (1980), and Piggot (1981)). Alternatively, fibre collapse is by compressive fracture from defects in the case of carbon fibres or glass fibres (see Ewins and Ham (1973), Ewins and Potter (1980), and Piggott and Harris (1980)).

Available experimental evidence for polymer matrix composites supports the hypothesis that microbuckling is a plastic rather than an elastic phenomenon. A summary of the measured compressive strengths for unidirectional, carbon fibre polymer matrix composites is given in Table 1. The first three

Composite System	Ref.	σ_c (MPa)	$\frac{G}{1-v_f}$(MPa)	k (MPa)	$\bar{\phi}$
T800/924C	Soutis (1989)	1615	6000	60	2.6°
HITEX 12K/E7jK8	U.S.Polymeric (1990)	1447	5510	40	1.4°
HITEX 46-3B/E7K8T	U.S.Polymeric (1990)	1274	4400	67	3.0°
AS4/PEEK	Jelf (1990)	1200	4000	78	3.4°
HS/MY720 (Woven)	Curtis and Bishop (1984)	400	3000	55	7.5°

Table 1: Comparison of measured compressive strength σ_c of unidirectional carbon fibre polymer matrix composites with predictions of the Rosen model, equation 2.1 and the Budiansky model, equation 2.2.

data sets refer to carbon fibre epoxy composites, the system AS4/PEEK is a carbon fibre Peek composite, and the system HS/MY720 refers to a carbon fibre epoxy 0°/90° woven layup. The table includes predicted strengths by the Rosen (1965) model, and the inferred misalignment angle $\bar{\phi}$ by substituting strength values σ_c and k into equation 2.2. For simplicity we assume $\beta = 0$ so that $k^* = k$. The error in the inferred values of $\bar{\phi}$ is at most 20% by this approximation, for typical values of E_T/G and β. We conclude from Table 1 that the Rosen model overpredicts compressive strength by a factor of approximately 4. The inferred values for $\bar{\phi}$ from Budiansky's model, equation 2.2, agree with typical measurements of $\bar{\phi}$, Jelf (1990). For polymer matrix composites, the matrix yields rather than microcracks. We are justified in viewing k as a plastic yield stress.

Direct experimental evidence to support equation 2.2 comes from measurements of the microbuckling strength of glass fibre and Kevlar fibre reinforced polyester by Piggott and Harris (1980). They varied the matrix shear yield stress by controlling the state of polyester resin cure from just jelled to fully cured. The compressive strength is proportional to k, provided that failure is by microbuckling, see Fig.1. This behaviour supports equation 2.2. The slope of the graph in Fig. 1 gives $\bar{\phi} = 3.7°$, assuming $\beta = 0$. When k is increased to sufficiently high values the glass or Kevlar fibres collapse prior to microbuckling.

Early carbon fibre epoxy composites failed by fibre collapse at test temperatures below approximately 100°C, see Fig. 2. At higher temperatures microbuckling occurred; the progressive decrease in microbuckling strength with increasing temperature T for T > 100°C is associated with the decrease in matrix shear yield stress k with increasing T, in accordance with equation 2.2. Over the last decade the compressive strength of carbon fibres has doubled, while epoxy matrices have changed little in strength

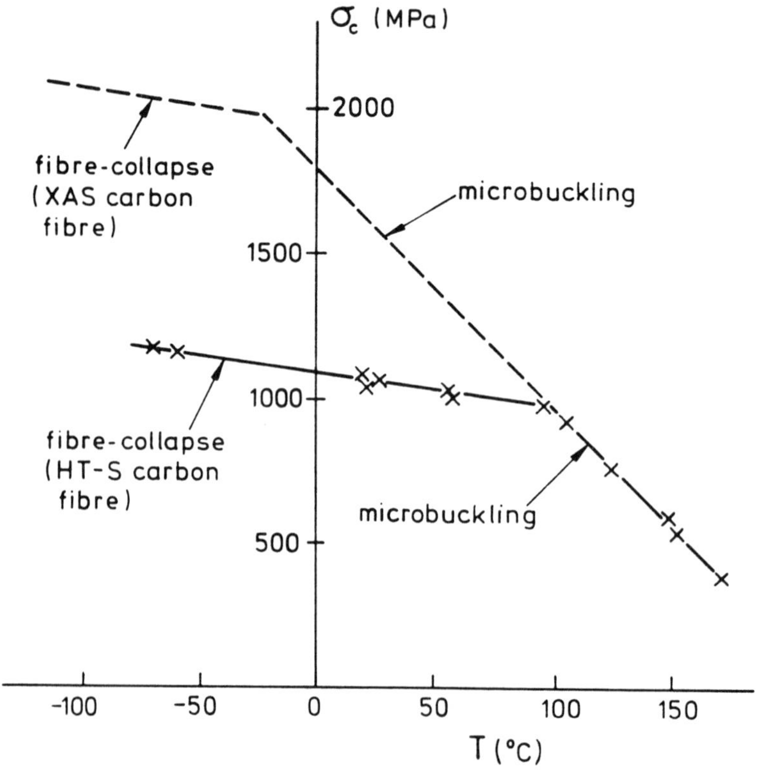

Fig. 2: Effect of temperature T upon failure strength σ_c of carbon fibre epoxy matrix composites. Experimental data x–x are taken from Ewins and Potter (1980). The dotted line gives the typical response of more recent systems.

due to demands for high impact strength and high toughness of the composite. Thus the transition temperature from microbuckling to fibre collapse has shifted from approximately 100°C to -40°C (Barker and Balasundaram (1987)), as shown in Fig. 2. Thus present day carbon fibre epoxy composites fail by microbuckling at ambient and at elevated temperatures.

The failure mechanisms exhibited by a unidirectional fibre composite may be summarised in a fracture diagram, with axes $k/\overline{\phi}$ and $G_m/(1-v_f)$, as shown in Fig. 3. Failure is by three distinct mechanisms:

1. Elastic microbuckling. Rosen's analysis predicts a microbuckling strength σ_c given by equation 2.1.

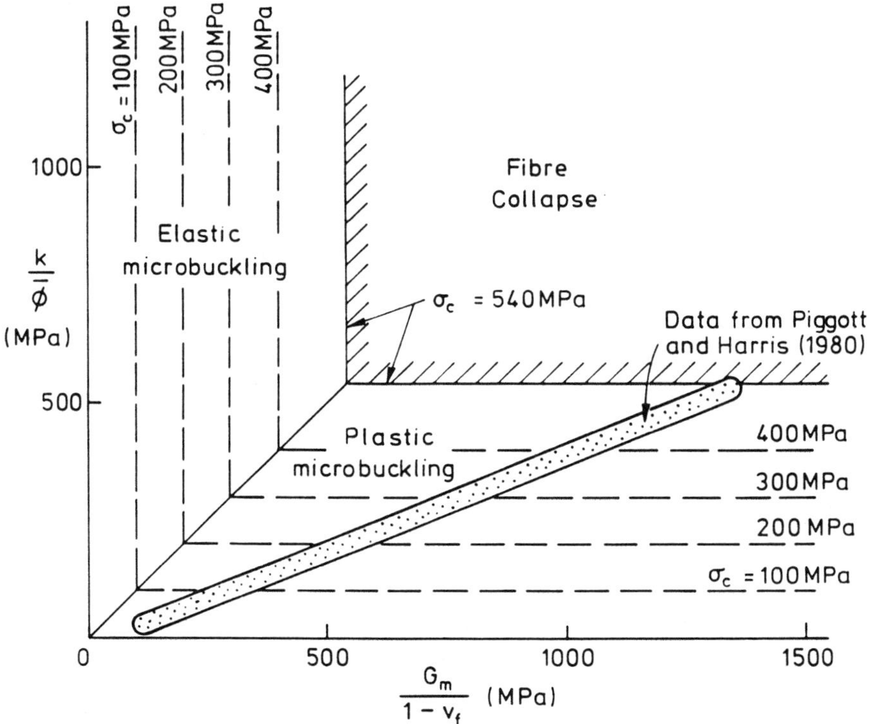

Fig.3: Fracture map for glass fibre polyester matrix composite. Data taken from Piggott and Harris (1980).

2. Plastic microbuckling. The Budiansky analysis predicts a strength given by equation 2.2. For simplicity we assume $\beta = 0$, hence $k^* = k$.

3. Fibre collapse. This occurs when the stress in the fibres attains a critical fracture value σ_f, such that

$$\sigma_c = v_f \sigma_f \tag{3.1}$$

The fracture diagram contains contours of compressive strength σ_c given by equations 2.1–2.3 and 3.1. The boundary of the fibre collapse regime depends upon fibre volume fraction v_f: otherwise the diagram is unique for a given fibre reinforcement.

Data for glass fibre reinforced polyester are included in Fig. 3, taken from the work of Piggott and Harris (1980). The data are replotted from Fig. 1. The experimental values support the common finding that the yield stress and elastic stiffness of polymer matrices scale in a linear fashion: thus the compressive strength of the fibre composite varies linearly with elastic modulus. This has led several investigators (for example Dow and Gruntfest (1960) and Rosen (1965)) to conclude erroneously that microbuckling is an elastic event for polymer matrix composites.

It is clear from the fracture diagram that the maximum attainable compressive strength is dictated by the intrinsic compressive fracture strength of the fibres. This strength is rarely achieved in practice for polymer matrix composites; requirements for high composite toughness and impact strength dictate the use of matrices with a low yield stress and high ductility.

Thus, plastic microbuckling is the usual failure mode in compression.

The ceramic matrix of ceramic fibre/ceramic matrix composites displays a non-linear response due to plasticity or to microcracking, Evans and Adler (1978). A plastic microbuckling analysis remains appropriate for such systems.

Preliminary unpublished tests by the authors show that elastic microbuckling occurs in a glass fibre/silicone rubber matrix composite. No systematic experimental investigations of elastic microbuckling in elastomeric matrix composites were found from the literature.

4. KINKING ANALYSIS

We shall analyse the behaviour of a kinked band of infinite length and finite width w, oriented at an angle β as shown in Fig. 4. First we consider the

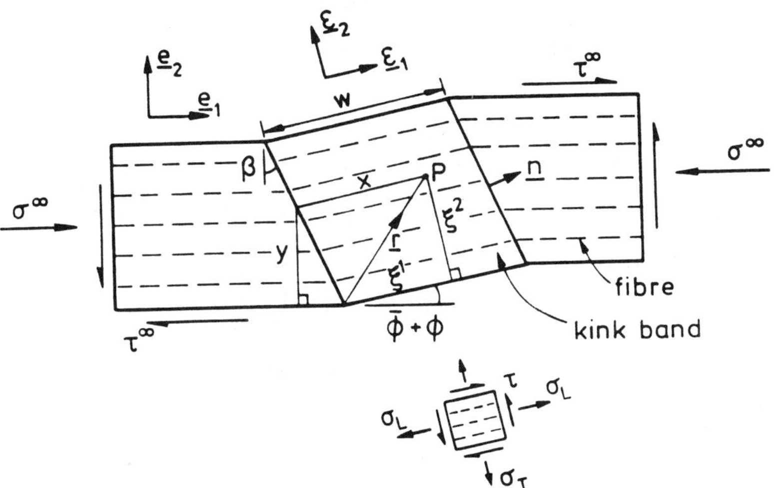

Fig.4: Detailed geometry of kink band.

kinematics and equilibrium of the band. In subsequent sections we explore the effect of the constitutive behaviour upon the buckling response.

4.1 Kinematics

We assume inextensional fibres but allow the composite to undergo direct straining transverse to the fibre direction, and shear straining parallel to the fibre direction. The fibres are assumed to have broken along the boundaries of the band. We smear out the fibres and matrix, and consider the composite to behave as a homogeneous anisotropic solid. Fibre bending is not treated explicitly; Budiansky has included the effects of fibre bending elsewhere, Budiansky (1983). He found that except for its role in setting the kink band width, fibre bending has only a small influence on the collapse response and can be neglected for most practical applications.

Consider the buckled band shown in Fig. 4. An arbitrary point P has a position vector $\underline{r}$,

$$\underline{r} = \xi_1 \, \underline{\epsilon}_1 + \xi_2 \, \underline{\epsilon}_2 \tag{4.1}$$

in terms of Cartesian co-ordinates (ξ_1, ξ_2) and fixed orthonormal base vectors $(\underline{\epsilon}_1, \underline{\epsilon}_2)$ which are instantaneously aligned with the fibre direction in the band. The velocity $\underline{v}$ of the point P is

$$\underline{v} = y \, \dot{\gamma}^{\infty} \, \underline{e}_1 + x\dot{\phi} \, \underline{\epsilon}_2 \tag{4.2}$$

where $\dot{\gamma}^{\infty}$ is the remote shear strain rate parallel to the unbuckled fibres, $\dot{\phi}$ is the rotation rate of the fibres in the band, and the fixed unit vector $\underline{e}_1$, and lengths x and y are defined in Fig. 4. We assume the remote direct strain rate transverse to the fibres equals zero.

The velocity strain rate $\dot{\underset{\sim}{\epsilon}}$ in the band is

$$\dot{\underset{\sim}{\epsilon}} = \tfrac{1}{2}\,(\underline{\nabla}\ \underline{v} + (\underline{\nabla}\ \underline{v})^T) \tag{4.3}$$

where the superscript T denotes the transpose and the gradient operator $\underline{\nabla}$ is

$$\underline{\nabla} = \underline{\epsilon}_1 \frac{\partial}{\partial \xi_1} + \underline{\epsilon}_2 \frac{\partial}{\partial \xi_2} \tag{4.4}$$

The quantitites x, y, ξ_1 and ξ_2 are related by

$$x = \xi_1 + \xi_2 \tan(\beta - \bar{\phi} - \phi)$$

and,

$$y = \xi_2 \cos\beta \sec(\beta - \bar{\phi} - \phi) \tag{4.5}$$

Here, $\bar{\phi}$ is the initial misalignment angle of fibres in the band; it serves as an imperfection.

Unit vectors $\underline{e}_1$ and $\underline{e}_2$, aligned with respect to the remote fibre direction as shown in Fig. 4, can be resolved into the $\underline{\epsilon}_1$ and $\underline{\epsilon}_2$ directions as,

$$\underline{e}_1 = \underline{\epsilon}_1 \cos(\bar{\phi} + \phi) - \underline{\epsilon}_2 \sin(\bar{\phi} + \phi)$$

$$\underline{e}_2 = \underline{\epsilon}_1 \sin(\bar{\phi} + \phi) + \underline{\epsilon}_2 \cos(\bar{\phi} + \phi) \tag{4.6}$$

We can now evaluate the strain rate via equations 4.2 - 4.7, to give

$$\dot{\underset{\sim}{\epsilon}} = (\dot{\phi}\tan(\beta-\bar{\phi}-\phi) - \dot{\gamma}^{\infty} \cos\beta \sin(\bar{\phi}+\phi) \sec(\beta-\bar{\phi}-\phi))\underline{\epsilon}_2\underline{\epsilon}_2$$
$$+ \tfrac{1}{2}(\dot{\phi}+\dot{\gamma}^{\infty}\cos\beta\cos(\bar{\phi}+\phi)\sec(\beta-\bar{\phi}-\phi))(\underline{\epsilon}_1\underline{\epsilon}_2 + \underline{\epsilon}_2\underline{\epsilon}_1) \tag{4.7}$$

But $\dot{\underset{\sim}{\epsilon}}$ equals $[\dot{e}_T\ \underline{\epsilon}_2\underline{\epsilon}_2 + \tfrac{1}{2}\ \dot{\gamma}(\underline{\epsilon}_1\underline{\epsilon}_2 + \underline{\epsilon}_2\underline{\epsilon}_1)]$ where by definition $\dot{e}_T$ is the direct strain rate transverse to the fibres in the band, and $\dot{\gamma}$ is the shear strain rate in the band. Identification of this expression for $\dot{\underset{\sim}{\epsilon}}$ with equation 4.7 gives,

$$\dot{e}_T = \dot{\phi}\tan(\beta-\bar{\phi}-\phi) - \dot{\gamma}^{\infty} \cos\beta \sin(\bar{\phi}+\phi) \sec(\beta-\bar{\phi}-\phi)$$

and, $$\dot{\gamma} = \dot{\phi} + \dot{\gamma}^{\infty} \cos\beta \cos(\bar{\phi} + \phi) \sec(\beta - \bar{\phi} - \phi) \quad (4.8)$$

For the case of vanishing remote shear, equations 4.8 may be integrated directly to give,

$$e_T = \ell n \left[\frac{\cos(\beta-\bar{\phi}-\phi)}{\cos(\beta-\bar{\phi})}\right]$$

and $$\gamma = \phi. \quad (4.9)$$

The band boundary rotates at a rate $\dot{\beta}$ which depends upon the remote shear strain rate $\dot{\gamma}^{\infty}$,

$$\dot{\beta} = - \dot{\gamma}^{\infty} \cos^2\beta \quad (4.10)$$

Integration of 4.10 yields,

$$\tan\beta = \tan\beta_o - \gamma^{\infty} \quad (4.11)$$

where β_o is the initial inclination of the band.

4.2 Equilibrium

Now consider equilibrium of the band. Equating the traction on both sides of the band boundary gives,

$$\underline{n} \cdot \underset{\sim}{\sigma}^{\infty} = \underline{n} \cdot \underset{\sim}{\sigma} \quad (4.12)$$

where $\underline{n} = \underline{e}_1 \cos\beta + \underline{e}_2 \sin\beta$ is the unit normal to the band, the remote stress $\underset{\sim}{\sigma}^{\infty}$ is,

$$\underset{\sim}{\sigma}^{\infty} = - \sigma^{\infty} \underline{e}_1 \underline{e}_1 + \tau^{\infty} (\underline{e}_1 \underline{e}_2 + \underline{e}_2 \underline{e}_1) \quad (4.13)$$

and the stress inside the band $\underset{\sim}{\sigma}$ is,

$$\underset{\sim}{\sigma} = \sigma_L \underline{\epsilon}_1 \underline{\epsilon}_1 + \sigma_T \underline{\epsilon}_2 \underline{\epsilon}_2 + \tau(\underline{\epsilon}_1 \underline{\epsilon}_2 + \underline{\epsilon}_2 \underline{\epsilon}_1) \quad (4.14)$$

Here σ_L is the direct stress in the fibre direction.

Substitution of 4.13 and 4.14 into 4.12, gives via 4.6 the two equilibrium statements,

$$- \sigma^{\infty} \cos\beta \cos(\bar{\phi} + \phi) + \tau^{\infty} \sin(\beta + \bar{\phi} + \phi)$$

$$= \tau \sin(\beta - \bar{\phi} - \phi) + \sigma_L \cos(\beta - \bar{\phi} - \phi) \qquad (4.15)$$

and $$\sigma^{\infty} \cos \beta \sin(\bar{\phi} + \phi) + \tau^{\infty} \cos(\beta + \bar{\phi} + \phi)$$

$$= \tau \cos(\beta - \bar{\phi} - \phi) + \sigma_T \sin(\beta - \bar{\phi} - \phi) \qquad (4.16)$$

The longitudinal stress σ_L along the fibre direction in the band is of limited interest (the fibres are inextensional), and we consider equation 4.15 no further. The stress components σ_T and τ are of interest, and appear explicitly in our suggested constitutive laws for the band; we shall make extensive use of equation 4.16 in the calculation of the buckling response of the kinked band.

Note that the stress rates $\dot{\sigma}_L$, $\dot{\sigma}_T$ and $\dot{\tau}$ defined with respect to the rotating fibres are objective stress rates which are not equal to the Jaumann stress rates. Nevertheless, they appear to be the natural choice.

5. ELASTIC MICROBUCKLING

In this section we calculate the buckling load and the post buckling response for an elastic composite under remote compressive axial stress σ^{∞} and remote shear stress τ^{∞}. Material inside and outside of the kinked band has a transverse stiffness E_T and a shear stiffness G, such that,

$$\tau^{\infty} = G\gamma^{\infty}, \quad \tau = G\gamma, \quad \sigma_T = E_T e_T \qquad (5.1)$$

For simplicity, we assume remote proportional loading,

$$\tau^{\infty} = e\sigma^{\infty} \tag{5.2}$$

where the dimensionless parameter e is fixed, and neglect the presence of any imperfections, $\bar{\phi} = 0$. Budiansky (1983) has argued elsewhere that imperfections produce only small knock-down factors upon the buckling load, and can be neglected.

First, we calculate the buckling load. We differentiate the equilibrium equation 4.16 with respect to ϕ and make use of equations 4.8, 4.10, 5.1 and 5.2, to give,

$$f_1(\phi)\,\frac{d(\sigma^{\infty}/G)}{d\phi} = f_2(\phi) \tag{5.3}$$

where, $f_1(\phi) \equiv \cos\beta\,\sin\phi + e\,\frac{\sigma^{\infty}}{G}\,\sin\beta\,\cos^2\beta\,\sin\phi$

$+ e\,\cos(\beta + \phi) + e^2\,\frac{\sigma^{\infty}}{G}\,\cos^2\beta\,\sin(\beta + \phi) - e\,\cos\beta\,\cos\phi$

$- e\,\frac{\tau}{G}\,\cos^2\beta\,\sin(\beta - \phi) + e\,\frac{E_T}{G}\,\cos\beta\,\sin\phi\,\tan(\beta - \phi)$

$+ e\,\frac{\sigma_T}{G}\,\cos^2\beta\,\cos(\beta - \phi)$

$$f_2(\phi) \equiv \cos(\beta - \phi) + \frac{\tau}{G}\,\sin(\beta - \phi)$$

$$+ \frac{E_T}{G}\,\sin(\beta - \phi)\,\tan(\beta - \phi) - \frac{\sigma_T}{G}\,\cos(\beta - \phi)$$

$$- \frac{\sigma^{\infty}}{G}\,\cos\beta\,\cos\phi + e\,\frac{\sigma^{\infty}}{G}\,\sin(\beta + \phi) \tag{5.4}$$

In the limit $\phi \to 0, f_1(\phi) \to 0$. Hence $f_2(0) = 0$ by equation 5.3, and the buckling load σ^{∞}_c/G is

$$\frac{\sigma^{\infty}_c}{G} = \frac{1 + \frac{E_T}{G}\,\tan^2\beta_o}{1 - 2\,e\,\tan\beta_o} \tag{5.5}$$

In equation 5.5, β_o is defined as the limit of β as $\phi \to 0$, and not the inclination of the band boundary at $\sigma^{\infty} = \tau^{\infty} = 0$.

That is, β_o is the initial inclination at buckling.

Surprisingly, we find from equation 5.5 that a negative value of remote shear stress $\tau^\infty = e\,\sigma^\infty$ reduces the buckling load. For a fixed $\tau^\infty/\sigma^\infty$ value there exists a critical initial inclination β_c such that σ_c^∞/G is a minimum; this is demonstrated in Fig. 5a.

(a)

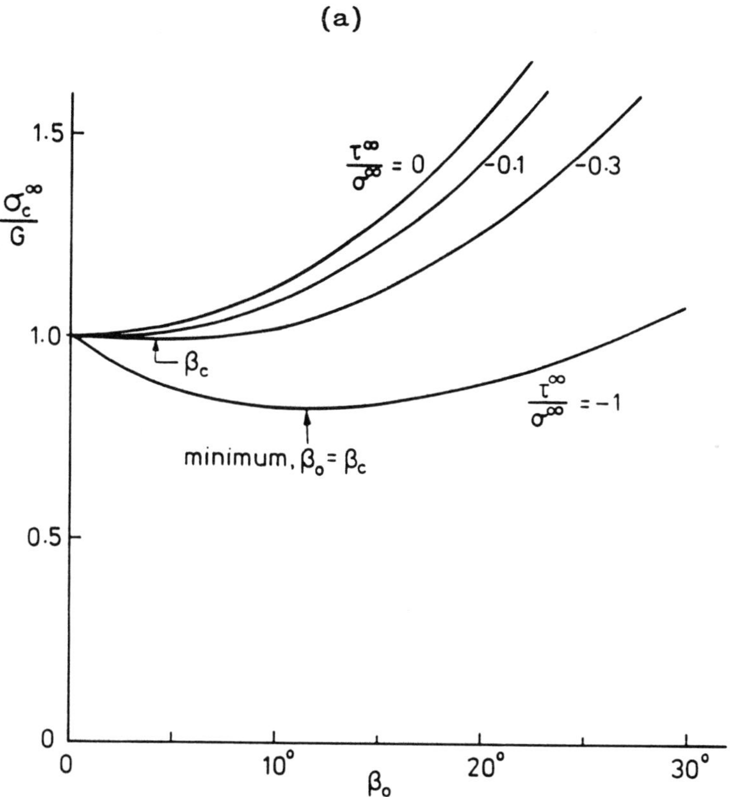

Fig.5(a): Effect of initial band inclination β_o upon the elastic buckling stress σ_c^∞, for the case $E_T/G = 4$.

An interaction diagram showing the buckling locus for $\beta_o = \beta_c$ is given in Fig. 5b. The collapse locus is sensitive to the value assumed for E_T/G, (Typically, $E_T/G \approx 4$). Equation 5.5 predicts $\beta_c = 0$ for the limit of vanishing remote shear stress. This is in disagreement with typical measurements of band angle for ceramic fibre polymer composites, where the observed angle is $\beta_o = 10° - 30°$. Budiansky (1983) argues via an elastic bending analysis that geometrical imperfections induce the onset of plastic yielding along an inclined domain at $\beta_o > 0$. Thus, in order to achieve $\beta_o > 0$ we must assume the presence of imperfections and assume the material is able to yield plastically.

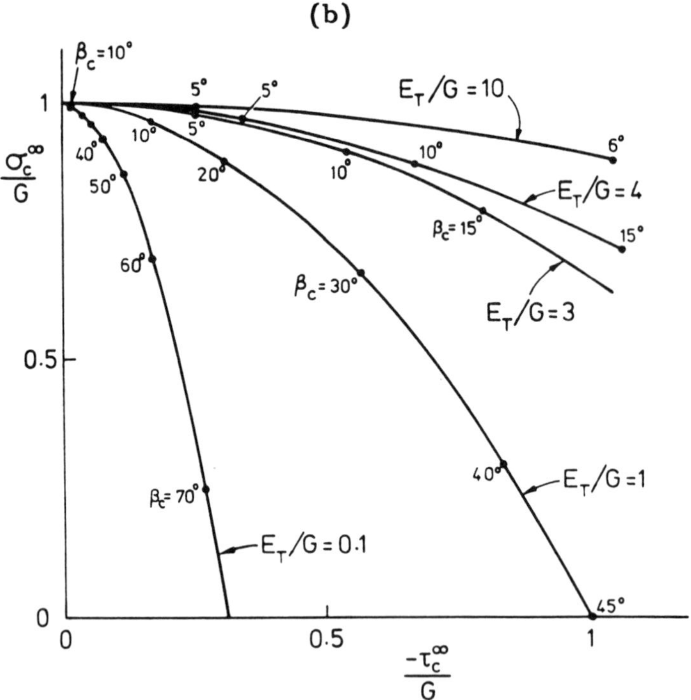

<u>**Fig.5(b):**</u> Interaction diagram for elastic microbuckling. Plot of buckling locus for weakest inclination $\beta_o = \beta_c$, for a range of E_T/G values.

For most practical cases $E_T/G > 1$ and the presence of remote shear has only a small influence on the buckling load σ_c^∞/G as shown in Fig. 5b.

5.1 Elastic Post Buckling Response

The post buckling response is determined by integrating numerically a system of 4 linear 1st order differential equations:

$$\frac{d/(\sigma^\infty/G)}{d\phi} = h_1\left(\frac{\sigma^\infty}{G}, \frac{\tau}{G}, \frac{\sigma_T}{G}, \beta, \phi\right)$$

$$\frac{d(\tau/G)}{d\phi} = h_2\left(\frac{\sigma^\infty}{G}, \frac{\tau}{G}, \frac{\sigma_T}{G}, \beta, \phi\right)$$

$$\frac{d(\sigma_T/G)}{d\phi} = h_3\left(\frac{\sigma^\infty}{G}, \frac{\tau}{G}, \frac{\sigma_T}{G}, \beta, \phi\right)$$

$$\frac{d\beta}{d\phi} = h_4\left(\frac{\sigma^\infty}{G}, \frac{\tau}{G}, \frac{\sigma_T}{G}, \beta, \phi\right) \qquad (5.6)$$

The function h_1 is given by $h_1 \equiv f_2/f_1$ from equations 5.3 and 5.4. Functions h_2, h_3 and h_4 follow naturally from equations 4.8, 4.10 and 5.1,

$$h_2 = 1 + e\,\frac{f_2}{f_1}\cos\phi\cos\beta\sec(\beta - \phi)$$

$$h_3 = \frac{E_T}{G}\tan(\beta - \phi) - e\,\frac{f_2}{f_1}\frac{E_T}{G}\sin\phi\cos\beta\sec(\beta - \phi)$$

$$\text{and } h_4 = -e\cos^2\beta\,\frac{f_2}{f_1} \qquad (5.7)$$

The system of equations 5.6 is integrated from $\phi = 0$ using a Runge-Kutta scheme. Since $f_1(\phi)$ and $f_2(\phi)$ are of order ϕ for ϕ small, care is required in evaluating $h_1 = f_2/f_1$ for small ϕ.

Typical results are shown in Fig. 6. For $\beta_o = 0$ and all $\tau^\infty/\sigma^\infty$, the post buckling response is stable: σ^∞/G increases with increasing ϕ. For $\beta_o > 0$, a softening post bifurcation response is displayed. Now consider the effect of $\tau^\infty/\sigma^\infty$ upon the buckling response, as shown in Fig. 6. When $\tau^\infty/\sigma^\infty = 0$, the minimum buckling load σ_c^∞/G is achieved at $\beta_o = 0$. When $\tau^\infty/\sigma^\infty = -1$, σ_c^∞/G is a minimum at $\beta_o = 11.7°$; the post buckling response for this critical orientation is softening initially and hardening later. A snap-through response is predicted at large values of $|\tau^\infty/\sigma^\infty|$ and β_o, such as $\tau^\infty/\sigma^\infty = -1$, $\beta_o = 30°$.

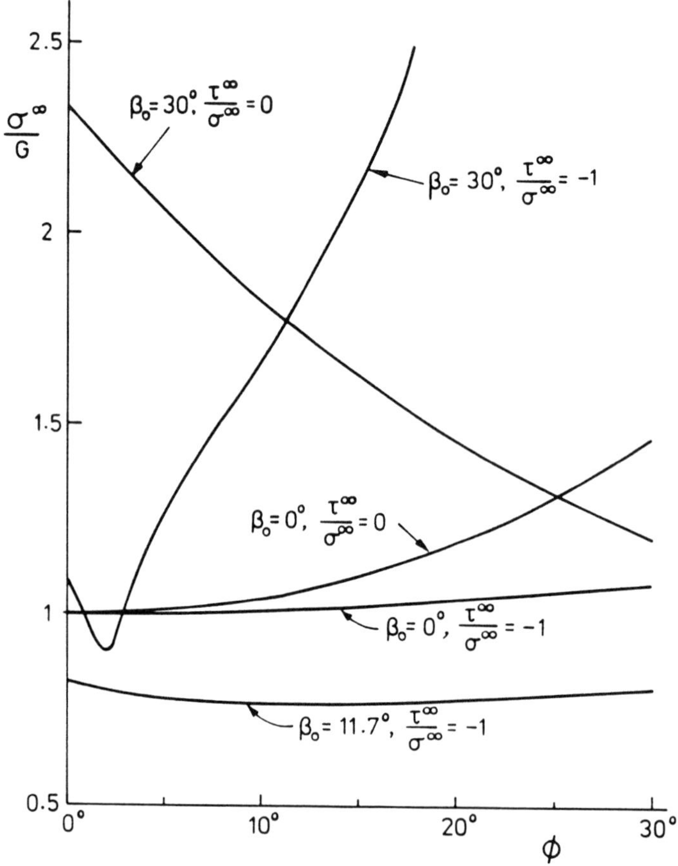

Fig.6: Elastic post buckling response, $E_T/G = 4$.

6. PLASTIC MICROBUCKLING

Polymer and metal matrix composites usually fail by plastic microbuckling. Budiansky (1983) has previously analysed plastic microbuckling by considering the response to remote axial stress σ^∞ of a kink band made from rigid-perfectly plastic material. We begin by generalising this analysis for the case of a remote stress σ^∞ with a remote shear stress τ^∞. Then, we consider microbuckling of an elastic-perfectly plastic solid under combined axial and shear stress. Plastic microbuckling in a strain hardening solid will be addressed in a future publication.

6.1 Rigid-Perfectly Plastic Solid

Consider the response of a rigid-perfectly plastic composite containing a kink band as shown in Fig. 4. The material is loaded remotely by an axial stress σ^∞ and a shear stress τ^∞. In general, the kink band is inclined at an angle β, and fibres in the kink band suffer an initial misalignment $\bar{\phi}$.

During collapse, non-proportional plastic straining occurs in the kink band. Remote material remains rigid, thus β is constant and we can drop the distinction between β and β_o. Inclined kink bands induce transverse stresses at the initiation of kinking, so that a combined-stress plasticity law must be invoked. We use arbitrarily a quadratic yield condition,

$$\left(\frac{\tau}{k}\right)^2 + \left(\frac{\sigma_T}{\sigma_{Ty}}\right)^2 = 1 \qquad (6.1)$$

where k and σ_{Ty} are the shear and tensile transverse yield stresses of the composite with respect to the fibre direction.

It appears reasonable to assume that an associated plastic flow rule applies. Then, by normality,

$$\dot{\gamma} = \frac{\tau}{k}\dot{\lambda}$$

$$\dot{e}_T = \left(\frac{k}{\sigma_{Ty}}\right)^2 \frac{\sigma_T}{k}\dot{\lambda} \tag{6.2}$$

where the non-dimensional parameter $\dot{\lambda}$ is positive for active plastic straining.

Combining equations 6.2 with equations 4.8 (recalling that $\dot{\gamma}^\infty = 0$ since remote material is rigid), gives,

$$\sigma_T = \tau \left(\frac{\sigma_{Ty}}{k}\right)^2 \tan\,(\beta - \bar{\phi} - \phi) \tag{6.3}$$

and, via equation 6.1,

$$\tau = k\left(1 + \left(\frac{\sigma_{Ty}}{k}\right)^2 \tan^2\,(\beta - \bar{\phi} - \phi)\right)^{-\frac{1}{2}} \tag{6.4}$$

We can now obtain an expression for the σ^∞ versus ϕ collapse response, by substituting the equations 6.3 and 6.4 into 4.16,

$$\sigma^\infty = \frac{k\left(1+\left(\frac{\sigma_{Ty}}{k}\right)^2 \tan^2(\beta-\bar{\phi}-\phi)\right)^{\frac{1}{2}}\cos(\beta-\bar{\phi}-\phi)-\tau^\infty\cos(\beta+\bar{\phi}+\phi)}{\cos\beta\,\sin(\bar{\phi}+\phi)} \tag{6.5}$$

For small $\bar{\phi}$ and ϕ, this simplifies to,

$$\sigma^\infty \sim \frac{k^* - \tau^\infty}{\bar{\phi} + \phi} \tag{6.6}$$

where $k^* \equiv k\,\left(1 + \left(\frac{\sigma_{Ty}}{k}\right)^2 \tan^2\,\beta\right)^{\frac{1}{2}}$ as given by equation

2.3. Equation 6.6 has been given previously by Batdorf and Ko (1987) in the limit $\beta = 0$. In the case of vanishing remote shear, equation 6.6 reduces to equation (2.2), given by Budiansky (1983).

For simplicity, we shall consider proportional remote loading with $\tau^\infty = e\sigma^\infty$. Equations 6.5 and 6.6 then reduce to,

$$\frac{\sigma^\infty \bar{\phi}}{k^*} = \left[\frac{1+(\frac{\sigma_{Ty}}{k})^2\tan^2(\beta-\bar{\phi}-\phi)}{1+(\frac{\sigma_{Ty}}{k})^2\tan^2\beta}\right]^{\frac{1}{2}} \frac{\cos(\beta-\bar{\phi}-\phi)\ \bar{\phi}}{\cos\beta\sin(\bar{\phi}+\phi)+e\cos(\beta+\bar{\phi}+\phi)} \tag{6.7}$$

and,

$$\frac{\sigma^\infty \bar{\phi}}{k^*} = (1 + \frac{e}{\bar{\phi}} + \frac{\phi}{\bar{\phi}})^{-1} \tag{6.8}$$

respectively.

Results

Equations 6.7 and 6.8 are compared in Fig. 7. We

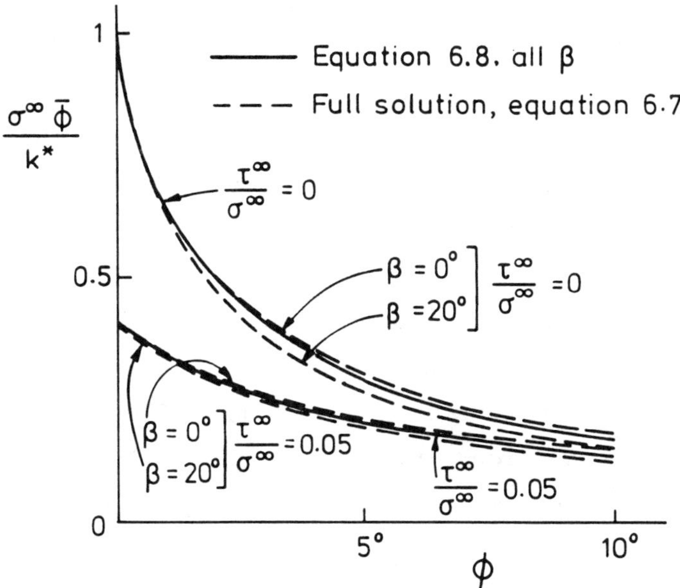

Fig.7: Accuracy of small ϕ approximation for microbuckling of a rigid-perfectly plastic solid. $\sigma_{Ty}/k = 2$, $\bar{\phi} = 2°$.

deduce that the small $(\bar{\phi} + \phi)$ approximation is adequate. Matrix failure or fibre-matrix debonding occurs at small values of ϕ (typically 3°) and the kink band then loses its load carrying capacity. Thus, equation 6.8 suffices over the range of validity of the analysis.

It is evident from equation 6.8 that the maximum value of remote stress σ^{∞} (at $\phi = 0$) is critically dependent upon the misalignment angle $\bar{\phi}$. As $\bar{\phi}$ tends to zero, σ^{∞} becomes unbounded; there is no finite bifurcation load for the perfect structure. The implication is that the materials manufacturer should arrange processing conditions to maximise fibre alignment, and thereby minimise $\bar{\phi}$.

We also deduce from equation 6.8 that a positive shear stress τ^{∞} reduces the buckling stress σ^{∞}. This contrasts with the case of elastic microbuckling where a negative value of τ^{∞} reduces the bifurcation value of σ^{∞}. To gain insight into this apparent paradox we consider next the buckling response of an elastic-perfectly plastic solid.

6.2 Elastic-perfectly Plastic Solid

We now examine the buckling response of an elastic-perfectly plastic composite, of geometry shown in Fig. 4. Consider the general case of the kink band inclined at an angle β, loaded remotely by σ^{∞} and τ^{∞}. The fibres in the band have an initial misalignment $\bar{\phi}$.

Typically, the response consists of two stages:

(a) An initial elastic response, followed by

(b) Matrix yielding and an elastic-plastic response.

We are interested in the early stages of deformation, and assume that ϕ, the various strain measures and $\bar{\phi}$ are each small, such that equations 4.8 simplify to,

$$\gamma \sim \phi + \gamma^{\infty}, \qquad e_T \sim \phi \tan \beta \tag{6.9}$$

The shear strain in the remote material γ^{∞} remains less than the shear yield strain γ_y (= 0.1%-1%) throughout the response. Thus $\beta \approx \beta_o$ by equation 4.11.

(a) Initial Elastic Response

The equilibrium equation 4.16, the constitutive law 5.1 and equation 6.9 may be combined to give the initial elastic response,

$$\sigma^{\infty} \sim \frac{(G + E_T \tan^2 \beta)}{(1 - 2 e \tan \beta)} \frac{\phi}{\bar{\phi} + \phi} \tag{6.10}$$

Here, as elsewhere, we assume remote proportional loading with $\tau^{\infty} = e\, \sigma^{\infty}$.

The matrix yields when equation 6.1 is satisfied. At this instant ϕ attains the yield value ϕ_y. Application of 5.1 and 6.9 gives,

$$\tau = G\gamma = G(\phi_y + e \frac{\sigma^{\infty}}{G})$$

$$\sigma_T = E_T e_T = E_T\, \phi_y \tan \beta \tag{6.11}$$

The value of σ^{∞} at which yield commences, $\sigma^{\infty} = \sigma^{\infty}_y$, is determined by substituting equation 6.10 and 6.11 into the yield condition 6.1, and solving for σ^∞ by the Newton-Raphson method. Predictions are compared in Fig. 8a with the buckling stresses σ_c^∞, τ_c^∞ for a rigid-perfectly plastic solid,

$$\sigma_c^\infty = \frac{k^* - \tau_c^\infty}{\bar{\phi}} \tag{6.12}$$

which is a restatement of equation 6.6 with $\phi = 0$.

It is clear from Fig. 8a that the collapse locus given by the rigid-perfectly plastic solid well

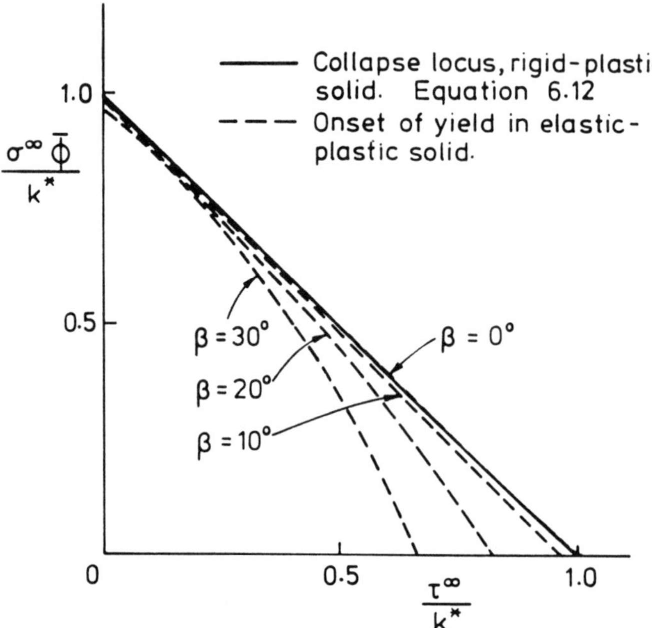

Fig.8(a): Comparison of collapse load $\sigma_c^\infty\bar{\phi}/k^*$ for rigid-perfectly plastic solid with onset of yield load $\sigma_y^\infty\bar{\phi}/k^*$ for elastic-perfectly plastic solid. $E_T/G = (\sigma_{Ty}/k)^2 = 4$, $\bar{\phi} = 2°$, $\gamma_y \equiv k/G = 0.001$.

approximates the onset of yield in the elastic-perfectly plastic solid. A positive shear stress τ^∞ decreases both the buckling stress σ_c^∞ for the rigid-perfectly plastic solid, and the stress σ_y^∞ at which the matrix yields for the elastic-perfectly plastic solid. In the limit of $e/\bar{\phi} << 1$ and $\phi/\bar{\phi} << 1$, σ_y^∞ and σ_c^∞ reduce to the same expression,

$$\sigma_y^\infty \sim \sigma_c^\infty \sim k^* \left(1 - \frac{e}{\bar{\phi}}\right) \qquad (6.13)$$

Consider the special case $\tau^\infty = 0$. Then equations 6.11 and 6.1 give,

$$\phi_y = \frac{k}{G}\left(1 + \left(\frac{k}{\sigma_{Ty}}\right)^2 \left(\frac{E_T}{G}\right)^2 \tan^2\beta\right)^{-\frac{1}{2}} \qquad (6.14)$$

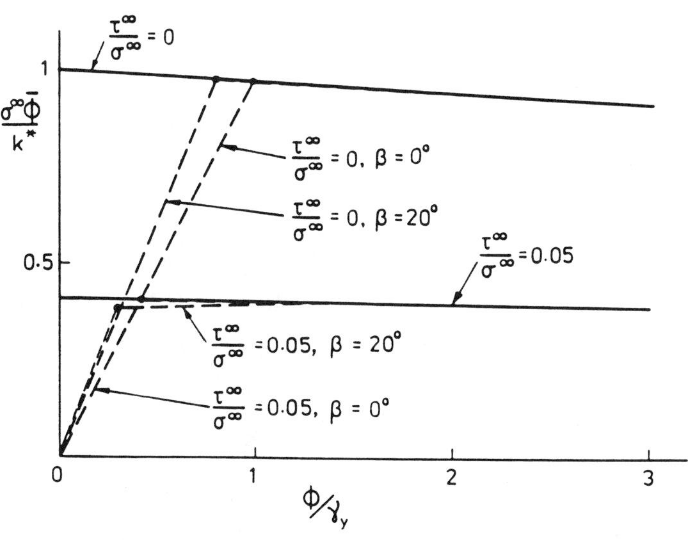

Fig.8(b): Comparison of collapse response for rigid-perfectly plastic solid with that for elastic-perfectly plastic solid. $E_T/G = (\sigma_{Ty}/k)^2 = 4$, $\bar{\phi} = 2°$, $\gamma = 0.001$.

If we assume for mathematical convenience $\frac{E_T}{G} = (\frac{\sigma_{Ty}}{k})^2$, then equations 6.10 and 6.14 reduce to,

$$\sigma_y^\infty = \frac{k^*}{\bar{\phi} + \phi_y} \qquad (6.15)$$

where $k^* \equiv k\ (1 + (\frac{\sigma_{Ty}}{k})^2\ \tan^2\ \beta)^{1/2}$, as before. This value for $(\sigma_y^\infty, \phi_y)$ lies on the σ^∞ versus ϕ response for the rigid-perfectly plastic solid given by equation 6.6.

(b) Post Yield Response

After matrix yield, the elastic-perfectly composite suffers both elastic and plastic straining in the kink band. The plastic strain rate is normal to the yield locus given by equation 6.1. A derivation of the relevant equations is given in Appendix A. Here, we describe only the results.

The pre and post yield response for the elastic-perfectly plastic solid is compared in Fig. 8b with the post buckling response for the rigid-perfectly plastic solid. We note that the matrix yields at $\phi/\gamma_y \leq 1$, where $\gamma_y = k/G$ is the shear yield strain of the matrix. The elastic-plastic response quickly approaches the rigid-perfectly plastic result, so that they are indistinguishable beyond $\phi/\gamma_y = 2$. We conclude that the rigid-perfectly plastic constitutive description is adequate for practical purposes.

In the limit of $\tau^\infty = 0$, with $\frac{E_T}{G} = (\frac{\sigma_{Ty}}{k})^2$, the post-yield elastic-perfectly plastic response coincides with the rigid-perfectly plastic response. In the limit of $\beta = 0$, material in the kink band suffers simple shear with $\tau = k$, $\sigma_T = 0$. Again, the post yield elastic-perfectly plastic response coincides with the rigid-perfectly plastic response.

7. PROPAGATION OF A MICROBUCKLE BAND

In practice, fibre microbuckling initiates at a stress raiser such as an imperfection or a hole in a sheet. The microbuckle band propagates across the remaining section of the structure. Fleck and co-workers (Soutis and Fleck (1990), Soutis, Fleck and Smith (1990)) have analysed the early stages of microbuckle propagation by treating the microbuckle as a crack with a bridging zone at its tip. This approach is reasonable if the traction is negligible across the microbuckled band, at a distance far behind the tip of the advancing microbuckle.

Here, we calculate the stress σ_p^∞ required to propagate a long microbuckle in steady state. In this limit, the rubble strength of the microbuckled material is not negligible. The geometry is shown in Fig. 9a. We assume remote proportional loading where $\tau^\infty = e\sigma^\infty$ and e is fixed. We shall use a simple energy argument to calculate σ_p^∞, and make use of the remote stress versus remote displacement response of a microbuckled band of infinite length. Chater and Hutchinson (1984) used a similar method to calculate the pressure required to propagate bulges and buckles in elastic cylinders.

The predicted remote stress σ^∞ versus ϕ response of the infinite band is shown in Fig. 9b. We assume the infinite band displays the rigid-perfectly plastic characteristic, equation 6.6, for small ϕ. When a critical fibre rotation ϕ_1 is attained, the tensile transverse strain in the band e_T equals the failure strain e_{Tf} (typically e_{Tf} = 1%) and the matrix fails, see Fig. 9b. The band strength vanishes with continued fibre rotation ϕ until e_T reduces to zero at $\phi = \phi_2$. Thereafter we imagine the fibres in the band

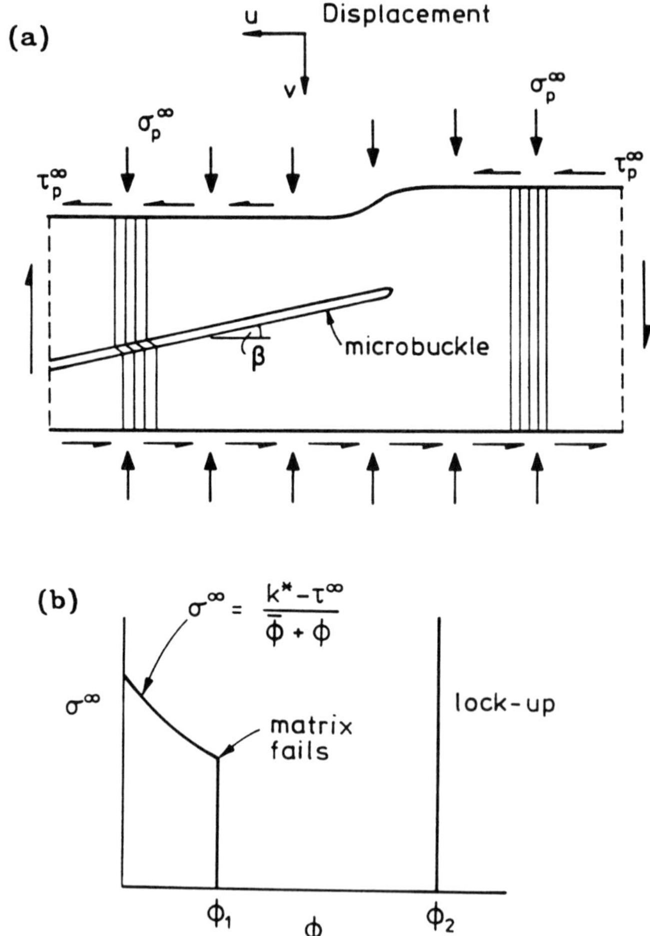

Fig.9: (a) Geometry of a propagating microbuckle. (b) Conjectured collapse response of a microbuckle.

contact each other and the band locks-up, with no further straining of the band. Lock-up is based on the idea that the composite resists compressive transverse straining in a highly stiff manner. The condition $e_T = 0$ corresponds to zero volumetric strain in the band since the fibres are considered to be inextensional. Chaplin (1977) and Evans and Adler (1978) also argue that fibre rotation stops when the volumetric strain in the band vanishes; they base their arguments on direct measurements of microbuckling.

Consider conservation of energy when the semi-infinite microbuckle shown in Fig. 9a undergoes a unit advance. For a deformation theory solid we get,

$$\sigma_p^\infty v_2 + \tau_p^\infty u_2 = \int_o^{v_1} \sigma^\infty \, dv + \int_o^{u_1} \tau^\infty \, du + G_c \sec \beta \quad (7.1)$$

where $\underset{\sim}{\sigma}_p$ is the fixed remote stress. Horizontal displacement u and vertical displacment v are defined in Fig. 9a. The first two terms on the right hand side of equation 7.1 refer to the work done by microbuckling when material is taken from a state $\phi = o$ to a state $\phi = \phi_2$, as shown in Fig. 9b. The last term on the right hand side of equation 7.1 represents the dissipation due to delamination and damage in off-axis plies.

Equation 7.1 provides a **necessary** condition for microbuckling: a detailed collapse mechanism at the tip of the advancing microbuckle would provide the **sufficient** condition.

We next calculate σ_p^∞ from equation 7.1. The matrix fails when $e_T = e_{Tf}$ at a fibre rotation ϕ_1,

$$\phi_1 = \frac{e_{Tf}}{\tan \beta - \overline{\phi} \sec^2 \beta} \quad (7.2)$$

Lock-up occurs when e_T equals zero. This is achieved at a fibre rotation ϕ_2, where by 4.9,

$$\phi_2 = 2(\beta - \bar{\phi}) \tag{7.3}$$

By kinematics, the remote displacements u and v, which form the work conjugates of τ_p^∞ and σ_p^∞ respectively, are,

$$\frac{u}{w} = \sin(\bar{\phi} + \phi) - \sin\bar{\phi}$$

$$\frac{v}{w} = \cos\bar{\phi} - \cos(\bar{\phi} + \phi) \tag{7.4}$$

When ϕ is small, these reduce to,

$$\frac{u}{w} = \phi$$

$$\frac{v}{w} = \tfrac{1}{2}\phi^2 + \bar{\phi}\phi \tag{7.5}$$

Equation 7.1 can be evaluated using equations 6.6, 7.2-7.5 and the assumption of proportional remote loading $\tau^\infty = e\,\sigma^\infty$ to give,

$$\frac{\sigma_p^\infty}{k} = \left[(1-\cos2\beta - \bar{\phi}\sin2\beta) + e(\sin2\beta - \bar{\phi}(1 + \cos2\beta))\right]^{-1}$$

$$\times \left[\phi_1 \left(1 + \left(\frac{\sigma_{Ty}}{k}\right)^2 \tan^2\beta\right)^{\frac{1}{2}} + \frac{G_c}{kw}\sec\beta\right] \tag{7.6}$$

where ϕ_1 is specified by equation 7.2.

The buckle propagation stress σ_p^∞ is plotted against band inclination β in Fig. 10, by evaluating equation 7.6. We note that σ_p^∞ increases with increasing G_c and decreasing τ^∞, as expected. A critical angle of β exists for which σ_p^∞ is a minimum, for any specified material parameters and loading ratio e. Disappointingly, the predicted values of β are in the range 45° - 75° which are larger than the values 10° - 30° typically measured. This suggests that the angle β may be set and locked-in at the **initiation** of kink propagation, by deformation patterns induced elastically via initial misalignments, as proposed by Budiansky (1983).

The kink band analysis does not provide us with the value of the kink width w. Budiansky (1983) has predicted w with reasonable success using an elastic bending analysis. He finds for perfectly brittle fibres (tensile failure strain = 0),

$$\frac{w}{d} = \frac{\pi}{4} \left(\frac{2\ k^{*}}{v_f E}\right)^{-\frac{1}{3}} \tag{7.7}$$

where d is the fibre diameter. This expression predicts correctly w/d ~ 10 for material properties typical of carbon fibre epoxy composites.

7.1 Case Study

Soutis and Fleck (1990) have examined recently the

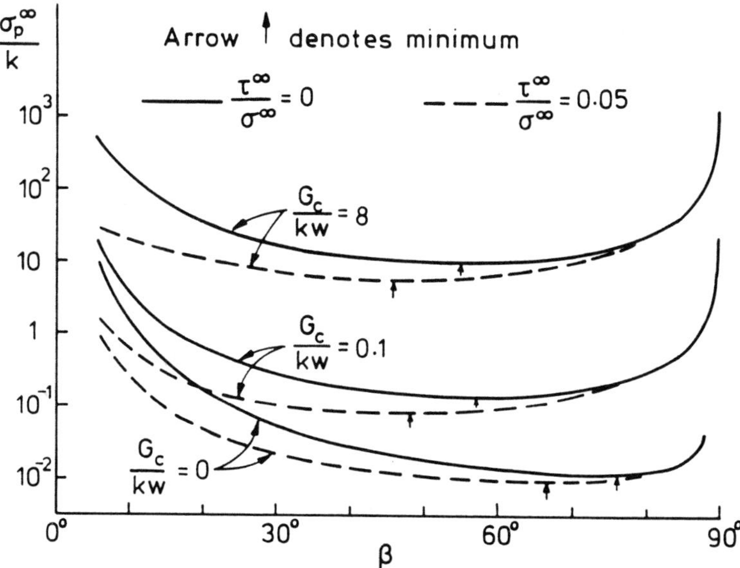

<u>Fig. 10:</u> Effect of band angle β upon propagation stress σ_p^∞/k. $\sigma_{Ty}/k = 2$, $\bar{\phi} = 2°$, $e_{Tf} = 0.01$.

compressive failure of a unidirectional and $[\pm 45°/0_2)_3]_S$ multidirectional carbon fibre epoxy composite. They measured the following material properties for the unidirectional material under remote axial loading: $\beta = 20°$, $k = 60$ MPa, $E_T/G = (\sigma_{Ty}/k)^2 = 4$, $\sigma_c = 1600$ MPa. Equation 6.6 predicts $\bar{\phi} = 2.6°$ which is consistent with the level of fibre misalignment observed. The fibre rotation after microbuckling ϕ_2 was found to satisfy equation 7.3, supporting the concept of lock-up. Soutis and Fleck were unable to measure the buckle propagation stress σ_p^∞ since unstable buckle propagation occurred once the microbuckle was initiated. This is consistent with the predicted value $\sigma_p^\infty = 13$ MPa from equation 7.6 which is much less than the measured buckling strength $\sigma_c = 1600$ MPa.

Soutis and Fleck (1990) have also measured the toughness G_c associated with splitting and delamination in the $[(\pm 45/0_2)_3]_S$ laminate. They measured $G_c = 30$ kJ/m^2 from the compressive fracture load of specimens containing central slits transverse to the loading direction. This value for G_c together with an observed value for $w = 60\mu m$, gives $G_c/kw = 8.3$, and $\sigma_p^\infty = 2900$ MPa via equation 7.6. This predicted value for σ_p^∞ is too high, as the multi-directional laminate was observed to fail unstably at a stress $\sigma_c = 810$ MPa. We conclude that more detailed modelling of damage development in the off-axis plies is required. This is not surprising, since the measured toughness G_c associated with delamination and splitting is more than two orders of magnitude greater than the dissipation due to microbuckling per unit area advance of the microbuckle.

8. CONCLUDING REMARKS

The elastic and plastic kinking analyses of microbuckling is able to account for some but not all of the experimental observations. Fibre bending must be treated explicitly in order to predict the width w of the microbuckle band and the band inclination β.

There remains a paucity of experimental evidence on the underlaying features of microbuckling. The authors are unaware of any systematic studies which examine the influences of fibre misalignment upon microbuckling strength. It is difficult to distinguish experimentally between elastic and plastic microbuckling of polymer matrix composites, as matrix yield stress usually scales linearly with matrix stiffness. Data on the shape of the yield locus for composites remain scant. It seems that few systematic experimental studies have been conducted of the elastic microbuckling of fibre composites with elastomeric matrices.

The buckle propagation analysis suggests that only a small compressive stress is required in order to propagate an existing microbuckle. No measurements of the microbuckle propagation stress were found from the literature.

ACKNOWLEDGEMENTS

This work was supported in part by DARPA University Research Initiative (Subagreement P.O. NO.

VB38639-0 with the University of California, Santa Barbara, ONR Prime Contract 00014-86-K-0753), and by the Division of Applied Sciences, Harvard University. The authors are also grateful for funding from the Procurement of Executive of the Ministry of Defence, under a joint SERC/MOD contract.

APPENDIX A: DERIVATION OF POST-YIELD ELASTIC-PERFECTLY PLASTIC RESPONSE

The post-yield elastic-perfectly plastic response is given in rate form as follows. We shall use the superscripts e and p to denote elastic and plastic, respectively. The strain rate in the kink band is,

$$\dot{\gamma} = \dot{\gamma}^{e} + \dot{\gamma}^{p}, \quad \dot{e}_{T} = \dot{e}_{T}^{e} + \dot{e}_{T}^{p} \tag{A1}$$

where,

$$\dot{\gamma}^{e} = \frac{\dot{\tau}}{G}, \quad \dot{e}_{T}^{e} = \frac{\dot{\sigma}_{T}}{E_{T}} \tag{A2}$$

in accordance with equations 5.1, and

$$\dot{\gamma}^{p} = \frac{\tau}{k}\dot{\lambda}, \qquad \dot{e}_{T}^{p} = \left(\frac{k}{\sigma_{Ty}}\right)^{2} \frac{\sigma_{T}}{k}\dot{\lambda} \tag{A3}$$

by equations 6.2; $\dot{\lambda}$ is a positive number for active plastic straining. Assume proportional remote loading with $\tau^{\infty} = e\sigma^{\infty}$. We combine equations A1-A3, with the yield condition 6.1 and the kinematic relations 4.8, to obtain $\dot{\lambda}$ and the stress rates in the kink band $\dot{\tau}$ and $\dot{\sigma}_{T}$,

$$\begin{bmatrix} \dot{\tau}/G \\ \dot{\lambda} \end{bmatrix} = \frac{1}{A} \begin{bmatrix} A_{11} & A_{12} \\ A_{21} & A_{22} \end{bmatrix} \begin{bmatrix} \dot{\phi} \\ \dot{\sigma}^{\infty}/G \end{bmatrix}$$

$$A \equiv \frac{\sigma_T}{k}\left(\frac{k}{\sigma_{Ty}}\right)^2 + \frac{\tau}{k}\frac{\tau}{\sigma_T}\left(\frac{\sigma_{Ty}}{k}\right)^2 \frac{G}{E_T}$$

$$A_{11} = \frac{\sigma_T}{k}\left(\frac{k}{\sigma_{Ty}}\right)^2 - \frac{\tau}{k}\tan\left(\beta - \bar{\phi} - \phi\right)$$

$$A_{12} = \left[\frac{\sigma_T}{k}\left(\frac{k}{\sigma_{Ty}}\right)^2 \cos(\bar{\phi}+\phi) + \frac{\tau}{k}\sin(\bar{\phi}+\phi)\right] e \cos\beta \sec(\beta-\bar{\phi}-\phi)$$

$$A_{21} = \frac{\tau}{\sigma_T}\left(\frac{\sigma_{Ty}}{k}\right)^2 \frac{G}{E_T} + \tan\left(\beta-\bar{\phi}-\phi\right)$$

$$A_{22} = \left[\frac{\tau}{\sigma_T}\left(\frac{\sigma_{Ty}}{k}\right)^2 \frac{G}{E_T}\cos(\bar{\phi}+\phi) - \sin(\bar{\phi}+\phi)\right] e \cos\beta \sec(\beta-\bar{\phi}-\phi)$$

$$\dot{\sigma}_T = -\frac{\tau}{\sigma_T}\left(\frac{\sigma_{Ty}}{k}\right)^2 \dot{\tau} \tag{A4}$$

Equations A4 simplify in an obvious manner when we assume small $(\bar{\phi}+\phi)$, and $E_T/G \equiv (\sigma_{Ty}/k)^2$. To proceed, we differentiate the equilibrium equation 4.16 with respect to time, and substitute for $\dot{\tau}$, $\dot{\sigma}_T$ using A4, and for $\dot{\beta}$ using 4.10. This results in a 1st order differential equation for $d(\sigma^\infty/G)/d\phi$ of the type given in 5.6, but with a new expression for h_1. Similarly, we obtain 1st order differential equations for $d(\tau/G)/d\phi$, $d(\sigma_T/G)/d\phi$ and for $d\beta/d\phi$. The resulting system of 1st order differential equations, analogous to equations 5.6, is integrated numerically using a Runge-Kutta routine. The starting values are given by the onset of yield condition described in part (a) of section 6.2. Results are shown in Fig. 8b.

REFERENCES

Argon, A.s. (1972). "Fracture of composites", Treatise of Materials Science and Technology, Vol. 1, Academic Press, New York.

Barker, A.J. and Balasundaram, V. (1987). "Compression testing of carbon fibre-reinforced plastic exposed to humid environments", Composites, 18(3), pp. 217-226.

Batdorf, S.G. and Ko, R.W.C. (1987). "Stress-strain behaviour and failure of uniaxial composites in combined compression and shear, Parts I and II, "Internal report, School of Engineering and Applied Science, University of California, Los Angeles, CA 90024.

Budiansky, B. (1983). "Micromechanics", Computers and Structures, 16(1), pp. 3-12.

Chaplin, C.R. (1977). "Compressive fracture in unidirectional glass-reinforced plastics", J.Mat.Sci., 12, pp. 347-352.

Chater, E. and Hutchinson, J.W. (1984). "On the propagation of bulges and buckles", J.Appl.Mech., 51, pp. 269-277.

Curtis, P.T. and Bishop, S.M. (1984). "An assessment of the potential of woven carbon fibre reinforced plastics for high temperature applications, Composites, 15(4), pp. 259-265.

Dow, N.F. and Gruntfest, I.J. (1960). "Determination of most-needed, potentially possible improvements in materials for ballistic and space vehicles". General Electric, Air Force Contract AF04(647)-269.

Evans, A.G. and Adler, W.F. (1978). "Kinking as a mode of structural degradation in carbon fiber composites", Acta Metallurgica, 26, pp. 725-738.

Ewins, P.D. and Ham, A.C. (1973). "The nature of compressive failure in unidirectional carbon fibre reinforced plastic", RAE Technical Report 73057.

Ewins, P.D. and Potter, R.T. (1980). "Some observations on the nature of fibre reinforced plastics and the implications for structural design", Phil.Trans.R.Soc.Lond., **A294**, pp. 507-517.

Greszczuk, L.B. (1972). "Failure mechanisms of composites subjected to compressive loading", AFML-TR-72107, U.S. Air Force.

Greszczuk, L.B. (1975). "Microbuckling failure of circular fibre-reinforced composites", AIAA Journal, 13 (10), pp. 1311-1318.

Jelf, M. (1990), unpublished research, Cambridge University Engineering Dept., England.

Lager, J.B. and June, R.R. (1969). "Compressive strength of boron/epoxy composites", J.Composite Material, 3(1), pp. 48-56.

Moncunill de Ferran E. and Harris, B. (1970). "Compression of polyester resin reinforced with steel wires". J.Comp.Mater. 4, p.62.

Piggott, M.R. and Harris, B. (1980). "Compression strength of carbon, glass and Kevlar-49 fibre reinforced polyester resins", J.Mat.Sci., 15, pp. 2523-2538.

Piggott, M.R. (1981). "A theoretical framework for the compressive properties of aligned fibre composites", J.Mat.Sci., 16, pp. 2837-2845.

Rosen, B.W. (1965). "Mechanics of composite strengthening", Fibre Composite materials, Am.Soc.Metals Seminar, Chapter 3.

Rhodes, M.D., Mikulas, M.M. and McGowan, P.E. (1984). "Effects of orthotrophy and width on the compression strength of graphite-epoxy panels with holes", AIAA Journal, 22(9), pp. 1283-1292.

Soutis, C. (1989). "Compressive failure of notched carbon fibre-epoxy panels", Ph.D. Thesis, Cambridge University Engineering Department, England.

Soutis, C. and Fleck, N.A. (1990). "Static compression failure of carbon fibre T800/924C composite plate with a single hole", to appear in June issue of J.Composite Materials.

Soutis, C., Fleck, N.A. and Smith, P.A. (1990). "Failure prediction technique for compression loaded carbon fibre-epoxy laminate with open holes", submitted to J.Composite Materials.

Starnes, J.H. and Williams, J.G. (1982). "Failure characteristics of graphite/epoxy structural components in compression", Mechanics of composites materials-recent advances, pp. 283-306, eds. Z. Hashin and C.T. Herakovich, Pergamon Press.

Steif, P. (1988). "A simple model for the compressive failure of weakly bonded, fiber-reinforced composites", J.Composite Materials, 22, pp.818-828.

U.S. Polymeric (1990). Data sheets on properties of carbon fibre epoxy composites, 700E Dyer Rd., Santa Ana, CA, 92707, USA.

A Critical Evaluation for a Class of Micro-Mechanics Models*

Richard M. Christensen

University of California
Lawrence Livermore National Laboratory
Livermore, California

ABSTRACT

New results are derived for the effective properties of composite materials composed of a continuous matrix phase containing a highly concentrated suspension of rigid spherical inclusions. The resulting analytical forms from several different theoretical micro-mechanics models are found to vary widely and they are assessed with respect to physical significance.

SUMMARY

Micro-mechanical analyses are required to predict the effective properties and failure characteristics of composite materials. The term micro-mechanics implies the use of a model which accounts for explicit interaction at the level of a continuous matrix phase and one or more inclusion phases. There are many theoretical micro-mechanical models which have been developed and applied to predict the effective elastic properties of composite materials, and these will be considered here. Furthermore there have been comparisons of such model predictions with data but this has primarily been done in the range where all of the more reasonable models are within the band of the experimental error and only models containing known errors or the meaningless rule of mixtures show large deviations. Few if any critical evaluations have been offered. The present work is intended to provide a critical evaluation of several of the prominent theoretical micro-mechanics models and this summary is excerpted from a recent work[1] with this objective. The many micro-mechanics models that have been developed for predicting the effective, elastic properties of composite materials admit different forms depending upon

*Work performed under the auspices of the U.S. Department of Energy by the Lawrence Livermore National Laboratory under contract number W-7405-ENG-48.

whether the case of particulate reinforcement or fiber reinforcement is being considered. Models which have a theoretical basis in either case include the following:

Differential Method
Composite Spheres (Cylinders) Model
Self Consistent Method
Generalized Self Consistent Method
(Three Phase Model)
Mori-Tanaka Method

Although there are many other micro-mechanics models they usually are either of an essentially numerical nature involving series or finite element solutions, or they are of an empirical nature, or finally they involve grossly oversimplifying assumptions. The primary references to the main theoretical models are as follows. The Differential Method has a long and interesting history, but it was most effectively developed and used by Roscoe.[2] The Composite Spheres Model is due to Hashin.[3] The Self Consistent Method is credited to Budiansky[4] and Hill.[5] The Generalized Self Consistent Method was formalized by Christensen and Lo[6] and referred to as the Three Phase Model by them. Finally, the Mori-Tanaka method has had many contributors, but the most recent and simplest derivation of it has been given by Benveniste.[7]

To proceed further it is necessary to decide whether to follow the particulate or fiber reinforced case. The present work will focus primarily upon the particulate case since it is of interest in its own right, and the conclusions found from it project over to the fiber reinforced case. In the area of particulate reinforced composite materials explicit attention will be given to the case of spherical inclusions in a continuous matrix phase. It is with this spherical inclusion case that there exist the most relevant and explicit theoretical forms, as well as the greatest liklihood of finding critical data against which to test the various theories. The case of spherical inclusions can be sub-divided into two major classes

i) Single Size Spherical Inclusions

ii) Polydisperse (Size) Spherical Inclusions

The case of single size spherical inclusions have been extensively studied by Frankel and Acrivos[8] in the form of highly concentrated suspensions of rigid inclusions, in the fluids context, and by Chen and Acrivos[9] under deformable inclusion conditions in the elastic context. The present work covers the polydisperse case of spherical particle suspensions with isotropic properties.

Within the case of polydisperse suspensions of spherical inclusions, the limiting case is specified by those suspensions which permit full packing with $c \to 1$, where c is the volume fraction of inclusions. Models which permit $c \to 1$ are of interest for at least three reasons. Micro-mechanics models which permit $c \to 1$ are particularly amenable to theoretical treatment and in fact are widely used in theoretical studies. Secondly, these micro-mechanics models giving $c \to 1$ are appropriate for use in describing practical, polydisperse suspensions involving a gradation of sizes of particles. Thirdly, models allowing $c \to 1$ can serve as reasonable approximations to the monodisperse case so long as the concentration is not too large. With the restriction that the models permit the full packing of inclusions and for purposes of rigorous comparison and evaluation, two of the previously mentioned micro-mechanics models must be excluded from consideration. These are the Composite Spheres Model and the Self Consistent Method. The Composite Spheres Model does not give a solution for the effective shear modulus but rather only for the bulk modulus. It does yield information on bounds for the shear modulus but that is not of relevance here where concern is with explicit predictions rather than bounds. Also, the shear deformation property is the primary concern here for reason to be given next. The Self Consistent Method when applied to multi-phase media does not always cover the full range of volume fraction up to $c \to 1$. This is true particularly when there is a large mis-match in properties of the two phases, which is the case of primary interest here, for reasons which also will be outlined in the following discussion. Thus, the micro-mechanics models which conform with the requirements stated here are:

Differential Method
Generalized Self Consistent Method
Mori-Tanaka Method

It is these models which will be extensively studied, compared and evaluated in the following work.

Most of the comparisons will concern the shear modulus rather than the corresponding volumetric property. This is because the shear property is much more difficult than the volumetric property to analyse. In the more physically based models the volumetric property is governed by a scalor potential whereas the shear property involves vector potentials. Accordingly it is much easier to bring spurious results into a shear modulus micro-mechanics model than it is for the volumetric property.

It is important to search for and examine the range of behavior where the differences between the various micro-mechanics models are the greatest. As already briefly mentioned, most evaluations involve comparisons of models with data in low volume fraction ranges where the differences between them are not great. To see the greatest differences between the models it is necessary to go to conditions involving concentrated suspensions. Furthermore, the greatest differences between model predictions occur for suspensions that have phases with the greatest mis-match in properties between the matrix and inclusions. The work here therefore focuses upon the suspensions involving perfectly rigid spherical particles under highly concentrated conditions in a continuous matrix phase. This case also will provide the most discrminating means of assessing the models relative to experimental data.

Brief mention should be made of the relationship of direct model predictions for properties with corresponding information upon values of upper and lower bounds for the properties. Bounds have value in several different situations. Certainly bounds are of interest when it is not possible to obtain a direct micro-mechanical solution for the effective property of interest. Also, bounds are of use in testing the direct predictions from models. If predictions from a particular model violate relevant bounds then the model is useless. If, on the other hand, the predicitons from a particular model satisfy the relevant bounds, then essentially no information is gained from the bounds relative to the model. That is the situation in operation here. No further reference need be made to bounds, other than to simply observe that all results given and interpreted here are consistent with all relevant bounds information.

The predictions from the various models are compared under high concentration conditions. This is done through the derivation of asymptotic forms appropriate in the limiting case of high concentration of the rigid spherical inclusion phase. The leading terms in the expansions give the dominant effect under high concentration conditions. The details of these lengthy derivations are given by Christensen[1] and the end results of the derivations will be given here. The dominant terms for the effective shear modulus μ obtained for the Differential Method, the Generalized Self Consistent Method and the Mori-Tanaka Method are given in Table 1 where μ_m is the shear modulus of the matrix phase and c is the volume concentration of the spherical particles.

Table 1. Effective Shear Modulus for a High Concentration Rigid Spherical Inclusion Suspension

	Compressible Matrix	Incompressible Matrix
Differential Method	$\frac{\mu}{\mu_m} \rightarrow \frac{F(\nu_m)}{(1-c)^2}$	$\frac{\mu}{\mu_m} = \frac{1}{(1-c)^{5/2}}$
Generalized Self Consistent Method	$\frac{\mu}{\mu_m} \rightarrow \frac{f(\nu_m)}{(1-c)}$	$\frac{\mu}{\mu_m} \rightarrow \frac{27}{16(1-c)^3}$
Mori-Tanaka Method	$\frac{\mu}{\mu_m} \rightarrow \frac{f(\nu_m)}{(1-c)}$	$\frac{\mu}{\mu_m} \rightarrow \frac{5}{2(1-c)}$

All results in Table 1 represent the high concentration leading terms of the corresponding expansions except the result shown for the Differential Method with the incompressible matrix phase which is an exact result valid at all concentrations. The symbols $F(\nu_m)$ and $f(\nu_m)$ and $f(\nu_m)$ in Table 1 represent functions of the Poisson's ratio, ν_m, of the matrix phase. These forms are known from the solutions given in Ref. 1. For example for $\nu_m = 1/5$ then $F(\nu_m) = 1$, $f(\nu_m) = 2.38$ and $f(\nu_m) = 2$.It is seen from Table 1 that the three micro-mechanics models give drastically different predictions of behavior under high concentration conditions. In the incompressible matrix case the exponent of the dominant (1-c) term ranges from -1 to -3 causing orders of magnitude differences in predictions under very concentrated conditions. The only cases in Table 1 where the different models give the same order of the (1-c) term is for the Generalized Self Consistent Method and the Mori-Tanaka Method in the Compressible Matrix case. However, even in this case the values of the two functions of Poisson's ratio $f(\nu_m)$ and $f(\nu_m)$ in Table 1 can be very different. This difference is shown in Table 2.

Table 2. Compressible Case, High Concentration

$$\frac{\mu}{\mu_m} \to \frac{\text{constant}}{(1-c)}$$

	$\nu_m=0$	$\nu_m=1/4$	$\nu_m=1/3$	$\nu_m=.45$	$\nu_m=1/2$
Generalized Self Consistent Method Constant	2.05	2.54	3	5.92	∞
Mori-Tanaka Method Constant	1.87	2.05	2.14	2.36	5/2

The fact the constant becomes unbounded for the Generalized Self Consistent Method as $\nu_m \to 1/2$ is simply a reflection of the fact that in the incompressible case the leading term in the expansion is of higher order than that of $1/(1-c)$ and the problem must be reformulated to get the $1/(1-c)^3$ dependence shown in Table 1.

The results shown in Tables 1 and 2 reveal the fundamental differences between the three micro-mechanics models. In Ref. 1 a detailed comparison is given of the three micro-mechanics models with experimental data. The results are more complex than can be simply stated here, because the full theoretical forms of the models must be used rather than just the asymptotic results shown in Table 1. The comparison with experimental data favors the Generalized Self Consistent Method rather than the other two models; see Ref. 1 for the conditions and limitations of the evaluation. The main qualification of these results is that of the restriction to very concentrated conditions for the suspension. Perhaps it could be argued that this limiting case is too severe and under "normal conditions" all the micro-mechanics models discussed here give reasonable predictions. The difficulty, however, is in defining the term normal conditions since there is no clear dividing line. In this sense then a model that gives reasonable behavior in some cases and unreasonable behavior in other cases is no better than an empirical model. In any case, all models recover dilute behavior adequately, and the major technical problem is to properly model the other extreme of behavior, the concentrated suspension case, as has been considered here.

REFERENCES

1. Christensen, R.M., "A Critical Evaluation for a Class of Micro-Mechanics Models", in press, J. Mech. Phys. Solids.

2. Roscoe, R. Brit, J. Appl. Phys. 3 (1952) 267.

3. Hashin, Z., J. Appl. Mech. 29 (1962), 143.

4. Budiansky, B., J. Mech. Phys. Solids 13 (1965) 223.

5. Hill, R., J. Mech. Phys. Solids, 13 (1965) 213.

6. Christensen, R. M. and Lo, K. H., J. Mech. Phys. Solids, 27 (1979), 315.

7. Benveniste, Y., Mech. of Mats. 6 (1987), 147.

8. Frankel N. A. and Acrivos, A., Chem. Eng. Sci. 22 (1967), 847.

9. Chen. H.S. & Acrivos A., Int. J. Solids Structures, 14 (1978), 349.

Experiments and Modeling in Plasticity of Fibrous Composites

George J. Dvorak, Yehia A. Bahei–El–Din,
Rahul S. Shah and Himanshu Nigam

Department of Civil Engineering
Rensselaer Polytechnic Institute
Troy, New York 12180, USA

Abstract

This paper describes a part of a continuing program in experimental verification of model predictions of the overall inelastic behavior of fibrous metal matrix composites. Measurements of initial and subsequent yield surfaces, and of plastic strains were performed on thin–walled B/Al tubes with unidirectional axial reinforcement, under torsion and internal pressure that was applied along a complex incremental path. The results were interpreted with three different models, the periodic hexagonal array (PHA) model, the bimodal plasticity theory, and a modified Mori–Tanaka scheme. The reliability of the predictions varies significantly, particularly where plastic strains are concerned. Only the PHA model appears to be of value in this regard, but the shape and position of overall yield surfaces was well predicted even by the matrix–dominated mode of the bimodal theory.

1. Introduction

Numerous analytical models have been proposed for prediction of the inelastic response of fibrous composites, an extensive bibliography appears in a recent review by Dvorak (1990). In contrast, experimental verification of such models, and the experimental work itself, have attracted much less attention. The present paper is a part of a program where such questions are addressed, both in terms of physical experiments, model development, and evaluation of the reliability of model predictions by comparison with experimental measurements of yield surfaces and plastic strains.

The experimental technique and results for incremental loading along a complex loading path are described in §2. Section 3 presents certain general relations which govern motion of subsequent yield surfaces during loading, and evaluation of plastic strains in the inelastic phases. Section 4 reviews some elements of the various models used in subsequent simulations of the experiments in §5. Some conclusions are noted in §6. In general, good predictions of overall yield surfaces and of their shape and position during plastic loading can be derived from several models. However, evaluation of plastic strains is much more difficult and less reliable even if performed with refined modeling techniques.

2. Experimental Results

The results described herein were obtained as a part of a continuing program which was first discussed by Dvorak et al. (1988). The work was performed on thin–walled tubular specimens made of a unidirectionally reinforced 6061–Al/B composite, with fibers aligned parallel to the axis of the tube. Figure 1 shows the specimen dimensions, instrumentation, and end attachments. The specimen tubes were fabricated by diffusion bonding of monolayers which were wrapped around a cylindrical madrel and then subjected to external pressure of

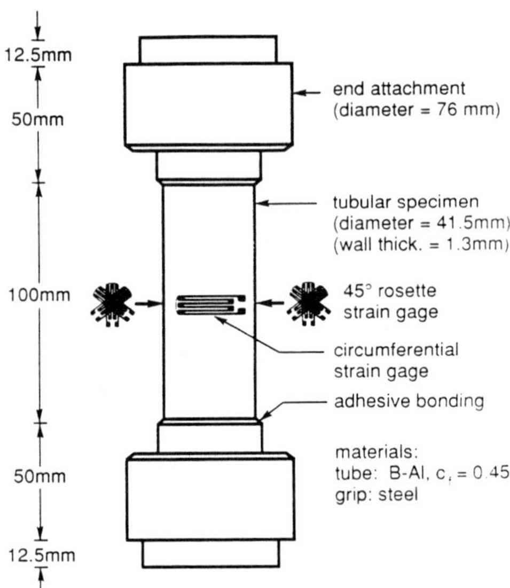

Fig. 1 Dimensions and instrumentation of composite tube specimen.

30 MPa at 500° C. In the finished form, the tube wall contained seven layers of fiber, the fiber volume fraction was $c_f = 0.45$. The specimens were annealed at 400° C for two hours, cooled at 10° C/hr to 260° C and then air cooled.

Specimen instrumentation was designed for measurement and recording of axial, hoop and longitudinal shear strains in the tube wall under incremental loading. Typical resolution was $1.0\mathrm{x}\ 10^{-6}$. Load was applied by a MTS closed–loop servo–hydraulic machine in the stress controlled mode. Independent components of axial force, torque, and internal pressure were combined to create various incremental loading paths. The machine was controlled by an IBM PC; the accuracy of load application, in terms of the average stress in the specimen wall was 0.1 MPa; the loading rate in all three directions was about 4.2 MPa/min. All tests were performed at room temperature. Additional information may be found in *op. cit.*

The purpose of the experiments was to establish overall yield surfaces and to measure plastic strain magnitudes for the selected loading sequences. The yield surfaces were constructed as loci of experimentally detected yield points on stress–strain curves found during excursions from the elastic region. As in *op. cit.*, the yield points were defined at the onset of deviation from linearity, and they were evaluated by back–extrapolation from the initial inelastic part of the stress–strain curve.

Figure 2 shows the path in the overall stress space which was selected for the present experimental work, and for the subsequent comparison of the results with several model predictions, as discussed in the sequel. The path was chosen in the $\sigma_{21}\sigma_{22}$ – stress plane, where σ_{22} denotes the normal stress transverse to the fiber, and σ_{21} is the longitudinal shear stress. The σ_{22} corresponds to application of internal pressure. No axial force was applied to compensate for the axial normal stress caused by the internal pressure, hence such stress coexists with σ_{22} and the scale on the horizontal stress axis in Fig. 2 was adjusted to reflect the resulting rotation of the stress plane actually employed in the measurements.

The path starts at point 1, and follows the sequence indicated by the arrows to point 11. The initial yield surface I was found by loading excursions from point 1 to the experimental yield points indicated. Subsequent yield surfaces

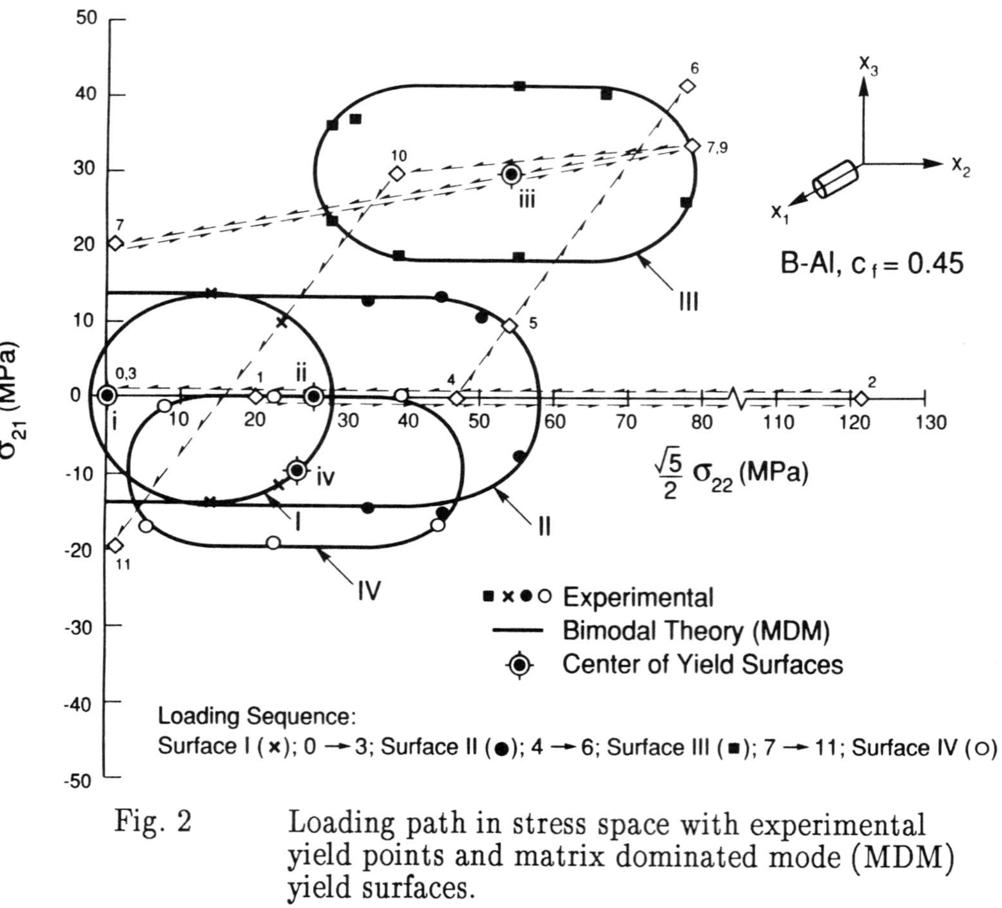

Fig. 2 Loading path in stress space with experimental yield points and matrix dominated mode (MDM) yield surfaces.

II to IV were found in a similar manner at points 4, 7, and 11. Note that points 6 and 11 lie outside their respective yield surfaces, this "piercing" phenomenon is explained below. The individual yield surfaces seen in Fig. 2 correspond to the matrix–dominated deformation mode in fibrous media (Dvorak and Bahei–El–Din 1987). The relevant equations appear next.

3. Yield surfaces and plastic strains

The origin and motion of the overall yield surfaces of the fibrous material is related to the local stress field in the phases, and to the hardening characteristics of the matrix material which alone is responsible for the inelastic deformation. The local fields may be thought of in terms of piecewise uniform distributions, such as those that may be evaluated by the finite element method. For this purpose, a representative volume of the composite material may be subdivided into $N = N_f + N_m$ subelements where the stress field is known in terms of some constant values. Under a uniform overall stress $\bar{\underset{\sim}{\sigma}}$ the local field is

$$\underset{\sim}{\sigma}_{rk} = \underset{\sim}{B}_{rk}\bar{\underset{\sim}{\sigma}} \;, \quad k = 1,2, \ldots N; \tag{1}$$

where $r = f, m$, and $\underset{\sim}{B}_{rk}$ is the stress concentration factor of subelement k. This concentration factor must be evaluated for each subelement from the solution of a boundary value problem for the representative volume of the composite microstructure under consideration, or it may be estimated by an approximate method.

The results found by Dvorak et al. (1988) indicate that the annealed 6061 aluminum matrix hardens kinematically. The equation of the matrix yield surface has the form

$$g_m(\underset{\sim}{\sigma}_m - \underset{\sim}{\alpha}_m) = 0 \tag{2}$$

and the evolution of the back stress vector $\underset{\sim}{\alpha}_m$ is reasonably well described by the Phillips (1972) rule which stipulates that

$$d\underset{\sim}{\alpha}_m = d\underset{\sim}{\sigma}_m \quad . \tag{3}$$

From the above one concludes that for each matrix subelement there is a yield surface in the overall stress space given by the equation

$$f_k(\overline{\underset{\sim}{\sigma}} - \overline{\underset{\sim}{\alpha}}_k) \equiv g_m[\underset{\sim}{B}_{mk}(\overline{\underset{\sim}{\sigma}} - \overline{\underset{\sim}{\alpha}}_k)] = 0; \; k = 1,2,...N \; . \tag{4}$$

where $\underset{\sim}{\alpha}_m$ denotes the position of the current center in the local stress space, and $\overline{\underset{\sim}{\alpha}}_k$ denotes the position of the corresponding surface in the overall stress space. The projections of the local surfaces into the overall stress space may be referred to as branches of the overall yield surface. Also, the following relation exists between the local and overall stress vectors (Hill 1967):

$$\underset{\sim}{\sigma}_{mk} - \underset{\sim}{\alpha}_{mk} = \underset{\sim}{B}_{mk}(\overline{\underset{\sim}{\sigma}} - \overline{\underset{\sim}{\alpha}}_k). \tag{5}$$

With regard to the Phillips hardening rule for the matrix (3), the motion of the loaded branches of the overall yield surface is then described as

$$d\overline{\underset{\sim}{\alpha}}_k = d\overline{\underset{\sim}{\sigma}} \quad . \tag{6}$$

The branches for subelements which currently undergo elastic deformation also experience a translation, due to the stress redistribution during plastic straining (Dvorak et al. 1988). Equation (6) suggests that regardless of the actual form of the local yield function (2), the loaded branches of the overall surface translate by an amount equal to the overall stress increment. As plastic yielding spreads through the entire matrix volume, the said translation affects all branches of the overall surface.

Evaluation of overall plastic strains in a composite material must also proceed from local constitutive equations for the matrix material. The local plastic strain is computed from the familiar formula (Ziegler 1959):

$$d\epsilon^p_{ij} = (1 + H/3G)^{-1} n_{kl} d\epsilon_{kl} n_{ij} \tag{7}$$

where $d\epsilon_{ij}$ is the total strain increment, n_{ij} is the direction of the outward normal to the yield surface at the current stress point, G is the elastic shear modulus of the matrix phase, and $H = H(\sigma_{ij})$ is its instantaneous plastic tangent modulus. This modulus was evaluated here from the two–surface theory proposed by Dafalias and Popov (1976). Their approach introduces a bounding surface, which is an isotropic expansion of the initial yield surface. The two surfaces are coaxial in the undeformed state. During deformation, the yield surface follows a given hardening rule, such as that indicated by (3), while the bounding surface undergoes the translation in which the center of this surface moves by the amount

$$d\beta_{ij} = d\alpha_{ij} - (1 - \frac{H_0}{H}) \frac{d\sigma_{mn} n_{mn}}{\mu_{kl} n_{kl}} \mu_{ij} \ . \tag{8}$$

This is illustrated in Fig. 3a which also defines the terms in (8). Note that at points s and $\bar{s}$ the two surfaces have parallel outward normals. In an uniaxial test, the two–surface approach leads to the scheme indicated in Fig. 3b. If σ is the distance between s and $\bar{s}$, then the instantaneous plastic tangent modulus H follows from the formula

$$H = H_0 + h\, \delta / (\delta_{in} - \delta) \tag{9}$$

where δ_{in} is the value of δ at the onset of yielding, H_0 is the asymptotic value of H, and h is a material parameter. Note that application of this model to a specific material requires evaluation of the parameters H_0, h, and the size of the bounding surface. After rearrangement, the above equations lead to the following formula for evaluation of the instantaneous stiffness of the matrix material:

$$L_{ijkl} = L^e_{ijkl} - \frac{2G}{1 + H/3G} n_{ij} n_{kl} \ . \tag{10}$$

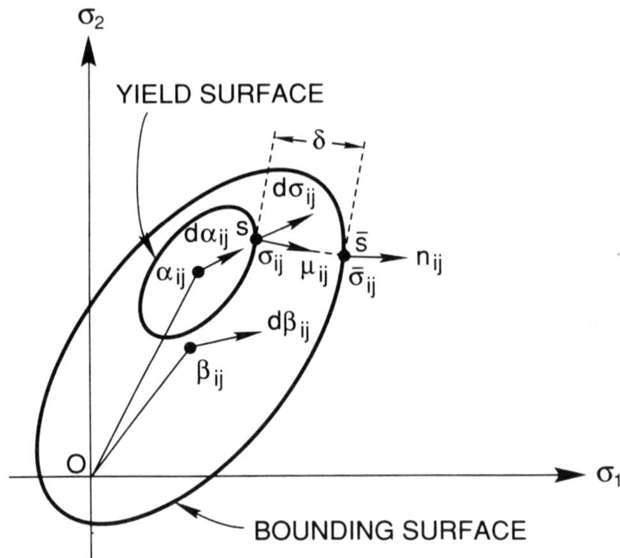

Fig. 3a Schematic representation of the yield and bounding surfaces and their motion during plastic flow.

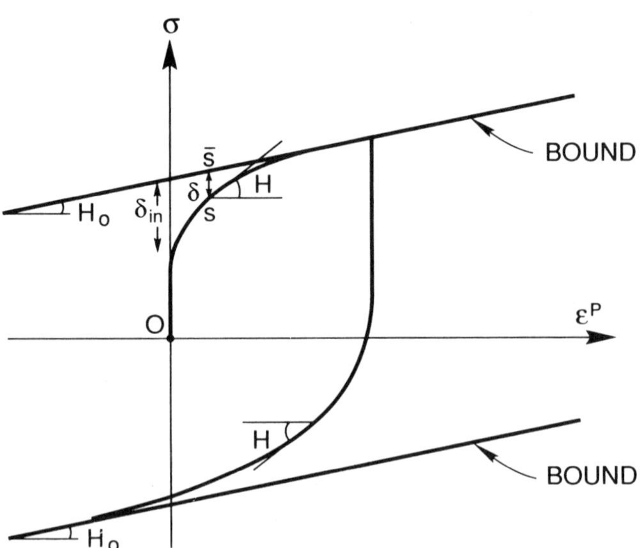

Fig. 3b Schematic representation of δ, δ_{in}, H, H_0 in a uniaxial test.

4. Composite material models

4.1 Bimodal plasticity theory

In their 1987 paper, Dvorak and Bahei–El–Din suggested that fibrous composites reinforced by aligned elastic fibers may exhibit two distinct deformation modes in response to plastic straining in the matrix phase. In the fiber mode, the composite behaves as a heterogeneous material where both fiber and matrix carry the applied load. The self–consistent, Mori–Tanaka, or other such models may be used to describe overall behavior in this mode. In contrast, the matrix dominated mode transfers all loads to the matrix and employs the fiber to restrict plastic straining in the matrix to simple shear deformation on planes that are parallel to the fiber axis. Each of the two modes is activated by a different state of applied stress, i. e., each has its own segment of the composite yield surface in the overall stress space. The matrix mode surface is independent of phase moduli, but the fiber mode surface depends on these moduli, hence the two modes are activated by different stress states in different composite systems. The fiber mode is always present, but the matrix mode is not. In general, existence of the matrix mode is favored in systems where the fiber longitudinal shear modulus is large compared to the matrix shear modulus, e.g., in the present B/Al system, but also in SiC/Ti and SiC/TiAl systems. This condition is seldom met in composites reinforced by graphite fibers, hence only the fiber mode may exist in the Gr/Al and similar systems.

The loading path of Fig. 2 activates only the matrix dominated mode in the B/Al composite used in the experiments. The corresponding yield surface was derived by Dvorak and Bahei–El–Din (1987) as

$$f_a(\underset{\sim}{\sigma}) = \left[\frac{\sigma_{21}-\alpha_{21}}{\tau_0}\right]^2 + \left[\frac{\sigma_{22}-\alpha_{22}}{\tau_0} \pm 1\right]^2 - 1 = 0 \quad \text{for } |q| \leq 1$$

$$f_b(\underset{\sim}{\sigma}) = \left[\frac{\sigma_{21}-\alpha_{21}}{\tau_0}\right]^2 - 1 = 0 \quad \text{for } |q| \geq 1 \tag{11}$$

where τ_0 is the composite yield stress in longitudinal shear, and $q = (\sigma_{21} - \alpha_{21})/(\sigma_{22} - \alpha_{22})$.

This is a straight cylinder with oval cross–section and generators parallel to the σ_{11} axis, as shown in Fig. 4 in the initial state, $\alpha_{ij} = 0$. Figure 2 contains oblique sections of this cylinder, with the yield stress τ_0 adjusted to fit the experimental yield points at each of the displayed surfaces. As observed by Dvorak et al., (1988), the matrix hardening has a small isotropic component, possibly caused by strain and age hardening, which affects the size of the experimentally detected overall surface.

The model of the matrix–dominated mode of plastic deformation also indicates a method for evaluation of plastic strains; this has been outlined in the 1987 paper.

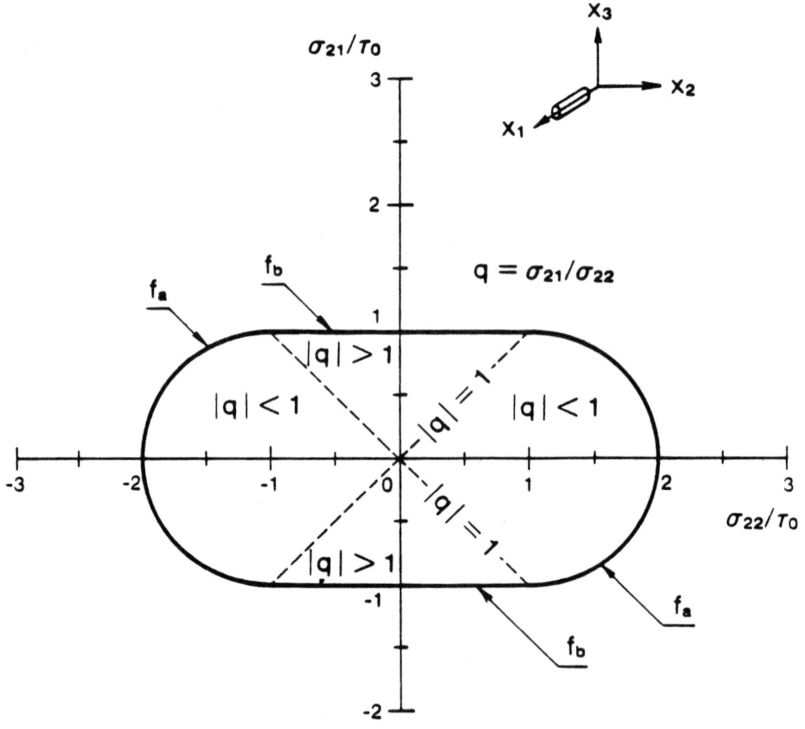

Fig. 4 Transverse cross section of the initial yield surface of the matrix–dominated mode.

4.2 The Periodic Hexagonal Array model

Dvorak and Teply (1985) and Teply and Dvorak (1988) developed this model in an attempt to obtain better estimates of the local fields, and bounds on instantaneous overall properties of elastic–plastic composite materials. The microstructural geometry is represented by a periodic distribution of the fibers in a hexagonal array, Fig. 5. The array is divided into two sets of unit cells, as indicated by the shaded and unshaded triangles in Fig. 5. Under properly prescribed periodic boundary conditions, the two sets of unit cells have related internal fields when the material is subjected to remotely applied uniform overall stresses or strains. Accordingly, only one unit cell from either set needs to be analyzed.

The actual analysis is performed by the finite element method. The unit cell is subdivided into a selected number of subelements, element material properties are prescribed as suggested by (10), and solutions are generated as the cell is subjected to prescribed incremental loading. Figure 6 indicates two possible subdivisions of the unit cell; of course the fine mesh is preferred when details of the local fields are of interest. A separate subroutine has been developed for the coarse mesh (Teply 1984), while the fine mesh has been implemented via the ABAQUS program.

4.3 The Mori–Tanaka method

This approximate method for evaluation of estimates of overall elastic properties of matrix–based heterogeneous media was originally proposed by Mori and Tanaka (1973), and more recently reformulated in a simpler form by Benveniste (1987). The essential assumption of the method is that the local field in each inclusion may be replaced by the field which exists when this inclusion resides in an large volume of matrix that is subjected to the average matrix stress or strain at infinity. In a recent paper, Lagoudas, Gavazzi and Nigam (1990) adapted this method to elastic–plastic systems. At each loading step, their approach utilizes the Eshelby solution of the inclusion problem in a homogeneous anisotropic matrix. Then, the stress and strain distributions found in the phases are replaced by their phase volume averages. These averages serve to evaluate instantaneous matrix properties from the prescribed constitutive law of the matrix, in the form suggested by (10).

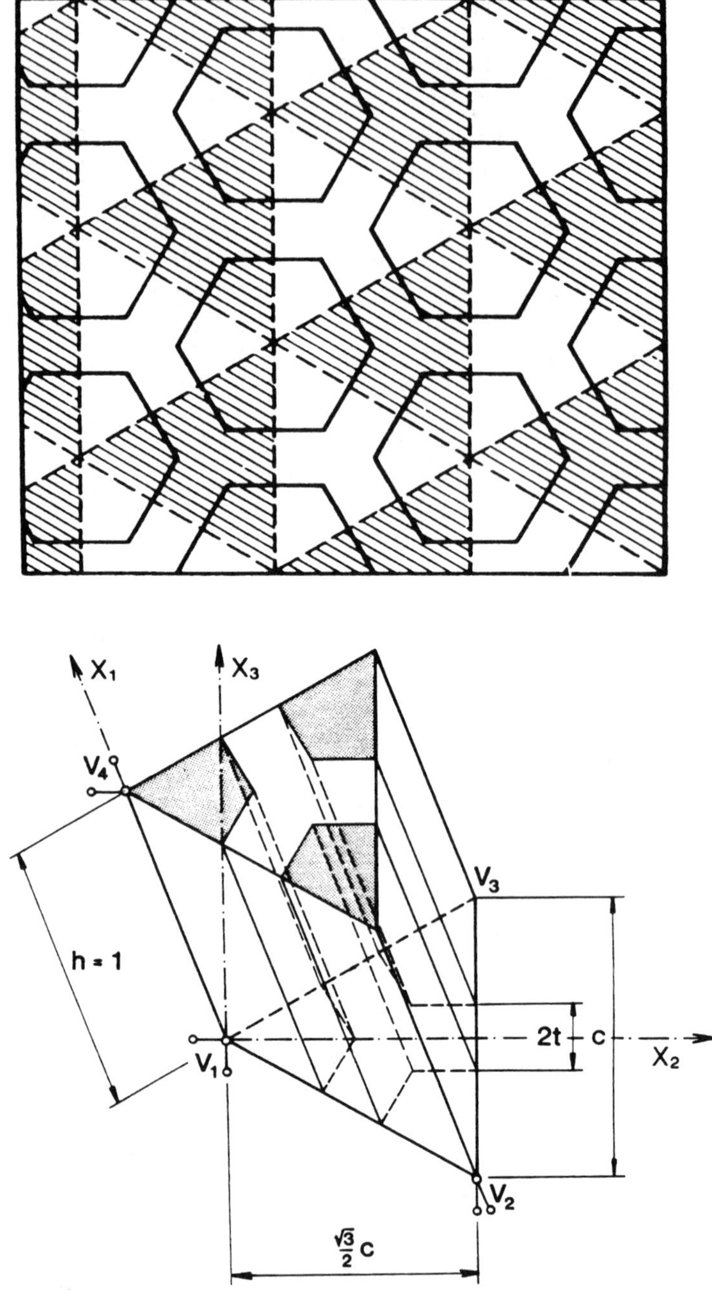

Fig. 5 Transverse cross section and unit cell of the Periodic Hexagonal Array model.

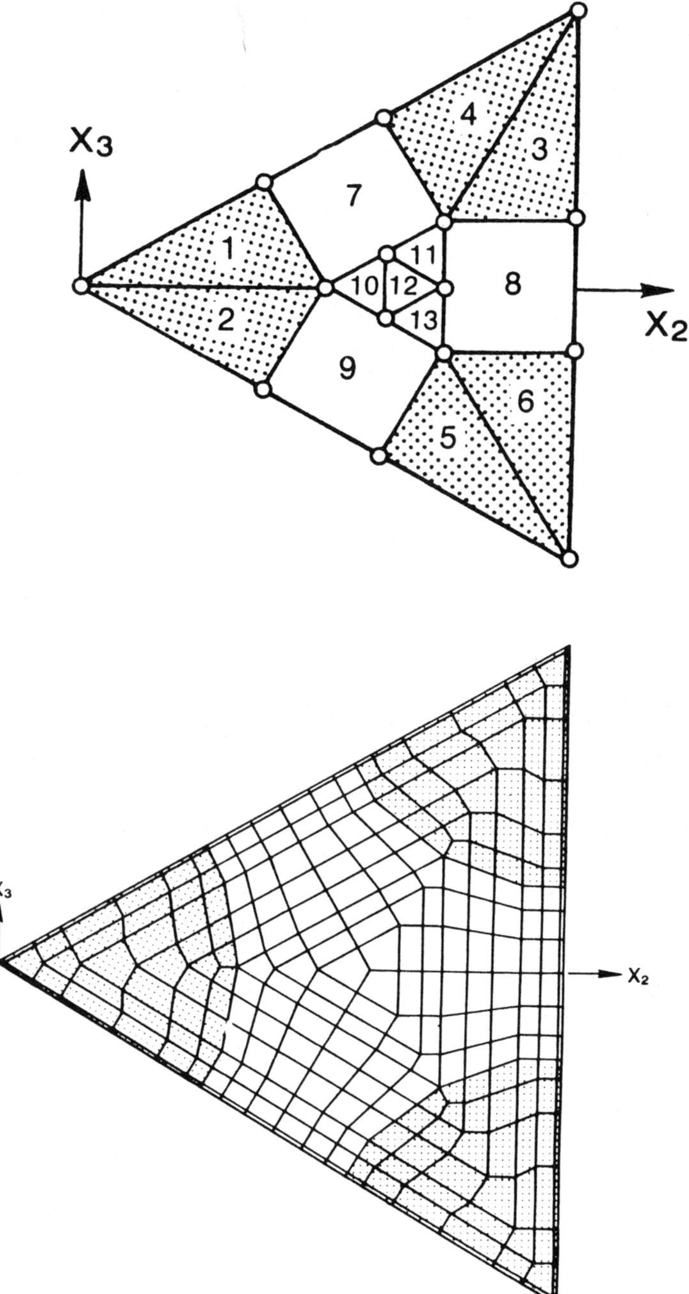

Fig. 6 Coarse and refined finite element meshes.

5. Interpretation of experimental data

We now proceed to simulate the yield surfaces and plastic strains measured along the path indicated in Fig. 2 by the three models described in §4. With reference to the figure, note that the initial part of the loading path, between points 0, 2, and 3 was used to fix the parameters in (10). This was done empirically, albeit with prior knowledge of similar parameters for the neat matrix. In particular, the initial matrix yield stress in simple tension was found to be equal to 23.64 MPa, the radius of the bounding surface in the same direction was evaluated as 69.6 MPa. The material parameters in (9) were selected as $H_0 = 1100$ MPa and $h = 80{,}000$ MPa.

Figure 7 presents the initial yield surface I from Fig. 2,with the experimentally detected yield points, together with a cluster of subelement branches of the overall yield surface derived from the fine mesh in Fig. 6, according to (4) at $\overline{\underset{\sim}{\alpha}}_k = 0$. Some of these branches fall within the MDM surface of the bimodal model, this reflects the presence of microyielding prior to a definite deviation from linearity at the yield point. Figure 8 shows the cluster after loading from the origin to point 2 of the loading path in Fig. 2. Each loaded branch translates according to (6). Since all branches pass through the loading point 2, all matrix subelements have yielded at this point. The outer normals to the innermost branches at the loading point form a cone of normals which must contain the overall plastic strain increment vector (Hill 1967). The directions of this vector found in experiments and in the PHA simulation are shown to lie within this cone.

Similar yield surface configurations were found at other loading points. Among those we display in Fig. 8 the data at point 6. Again, all elements yield, the branches form a hinge,and the cone of normals contains the experimental and numerical (PHA) evaluations of the plastic strain increment vector directions. Those are seen to lie in close proximity, but not in the direction of the normal to the MDM surface found from the bimodal theory fit to the experimentally measured yield points.

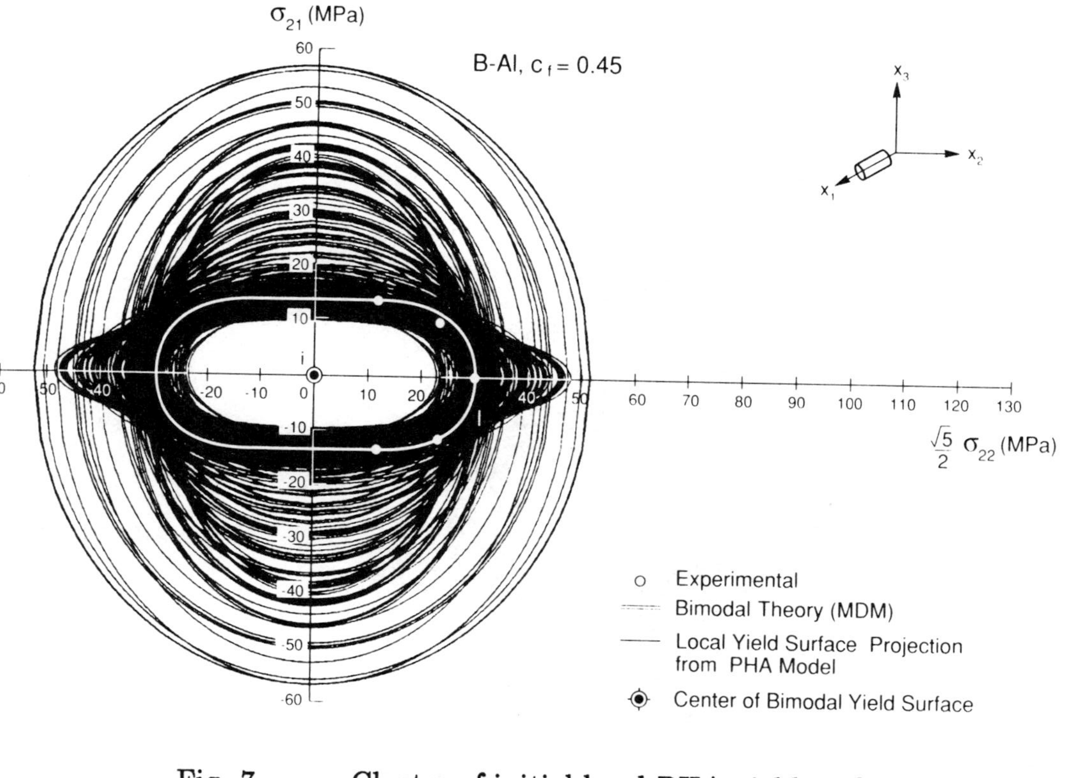

Fig. 7 Cluster of initial local PHA yield surfaces with experimental yield points and bimodal yield surface.

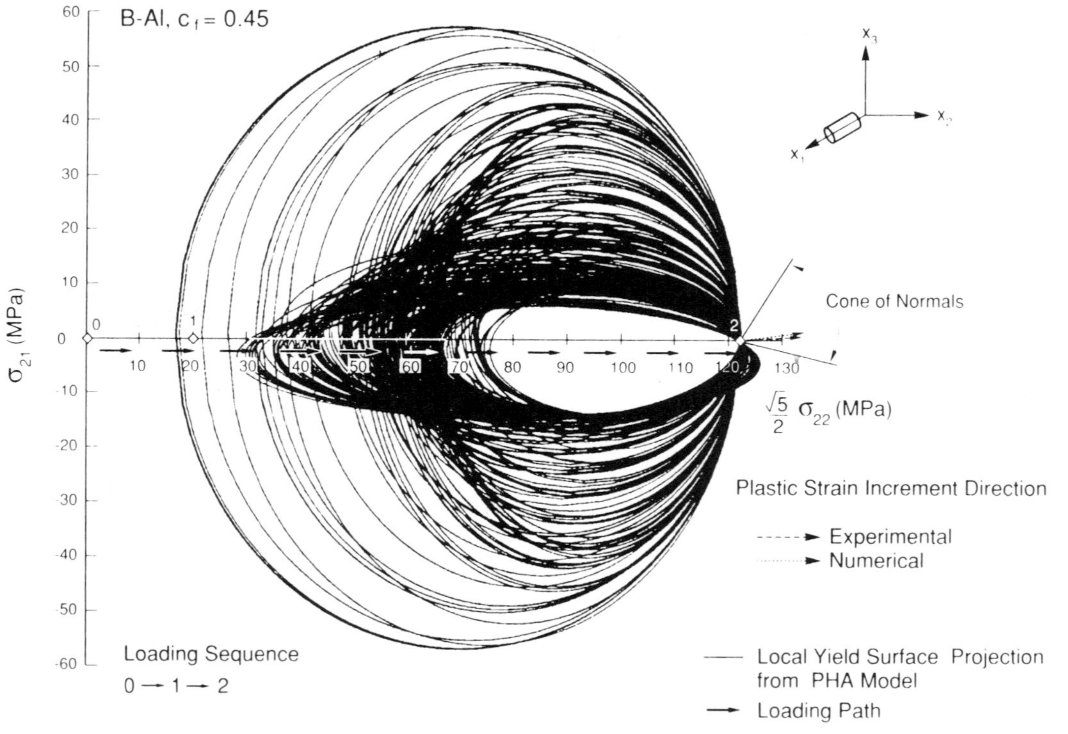

Fig. 8 Cluster of local PHA yield surfaces at loading point 2, and the plastic strain increment directions.

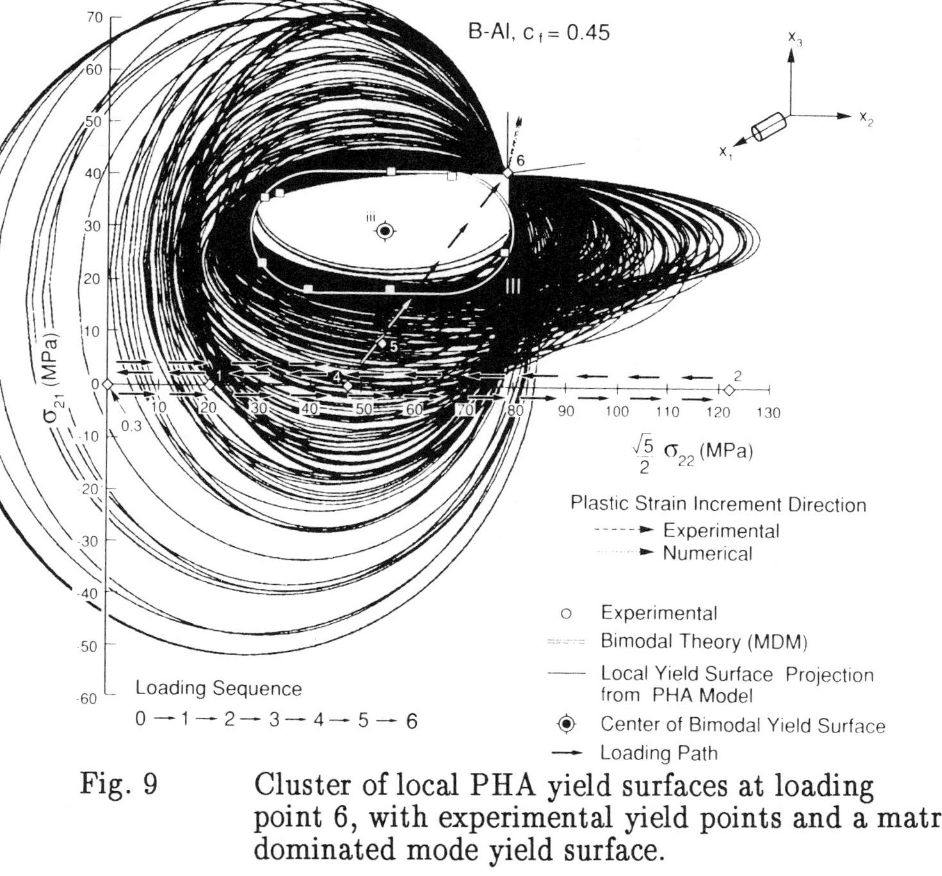

Fig. 9 Cluster of local PHA yield surfaces at loading point 6, with experimental yield points and a matrix dominated mode yield surface.

Piercing of the latter surface by the loading vector at point 6 is seen to be caused during detection of the experimental yield points. As the yield points are found after loading to point 6, the cluster of subelement surfaces undergoes a substantial rearrangement during the several loading excursions, such that the point 6 no longer lies on the final overall yield surface.

To indicate the differences in the direction of the plastic strain increment vectors that were measured experimentally and predicted by the PHA and bimodal models, we record here the respective angles, measured in degrees in the anti–clockwise direction from the σ_{22} axis:

Directions of the plastic strain increment vector

Loading point	Experimental	PHA model	Bimodal theory
2	3.21	5.55	8.17
3	175.94	174.21	171.83
5	63.43	83.03	51.34
6	83.02	72.33	51.34
8	220.58	209.49	215.00
11	244.33	253.17	NA

This suggests that both methods approximate the actual plastic strain direction with similar accuracy. Of course, it is more meaningful to compare the actual plastic strain magnitudes. This is done in Fig. 10 which shows the normal plastic strains ϵ_{22}^{p} as a function of the applied overall stress σ_{22}. Also in Fig. 11, which presents the same comparison for the longitudinal shear components.

The normal plastic strains appear to be in reasonable agreement up to point 5. There is a significant difference between the Mori–Tanaka and bimodal theory predictions and the experimentally measured plastic strains during loading between points 5 and 6 and thereafter. The PHA model predictions, while not in perfect agreement, appear to be noticeably closer to the measured strains.

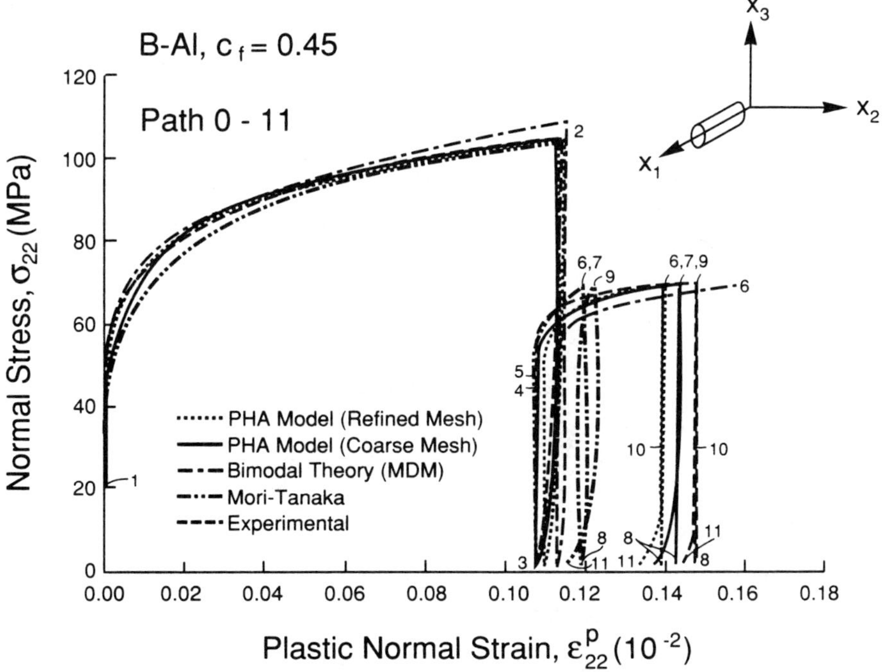

Fig. 10 Normal stress – plastic normal strain, σ_{21}–ϵ^{p}_{22}, response for loading path 0–11.

The longitudinal shear plastic strain comparisons in Fig. 11 show these differences to be even more pronounced. The Mori–Tanaka model deviates most significantly from the experiments, somewhat lesser but still large deviations are shown by the bimodal model, and even the PHA predictions are quite different from the experimental data. Note that most of these differences occurred during loading from point 5 to point 6.

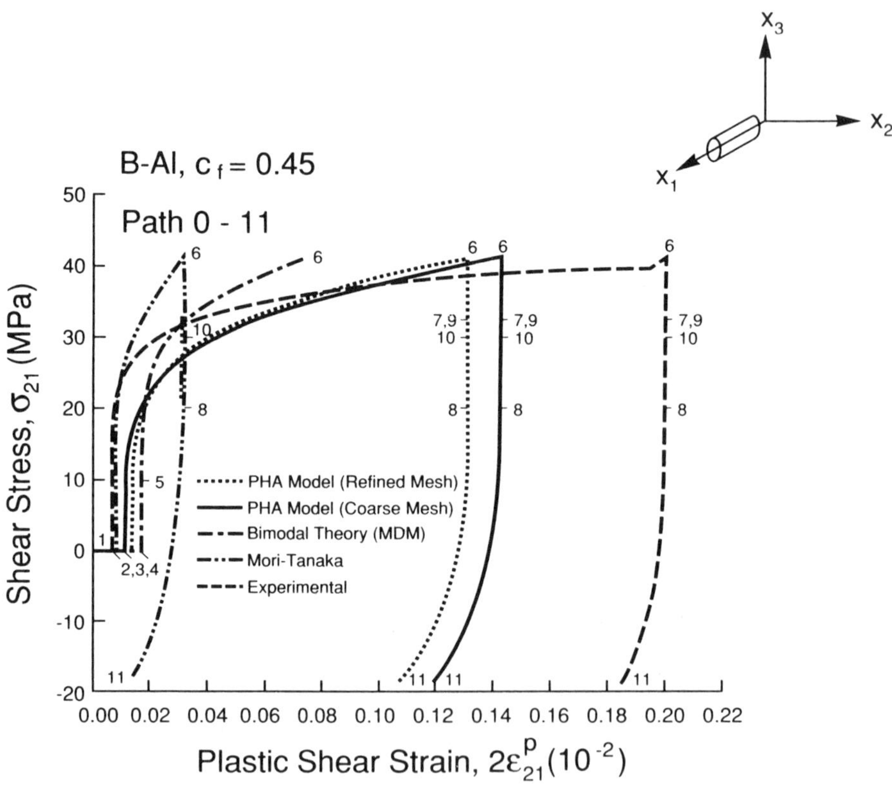

Fig. 11 Shear stress – plastic shear strain, σ_{21}–$2\epsilon^{p}_{21}$, response for loading path 0–11.

A final comparison, in Fig. 12, shows the plastic strain path followed by the actual B/Al composite material and by the respective models during the entire loading program. Here, the PHA model comes relatively close to the experimental data, whereas both the Mori–Tanaka and bimodal predictions are very far away.

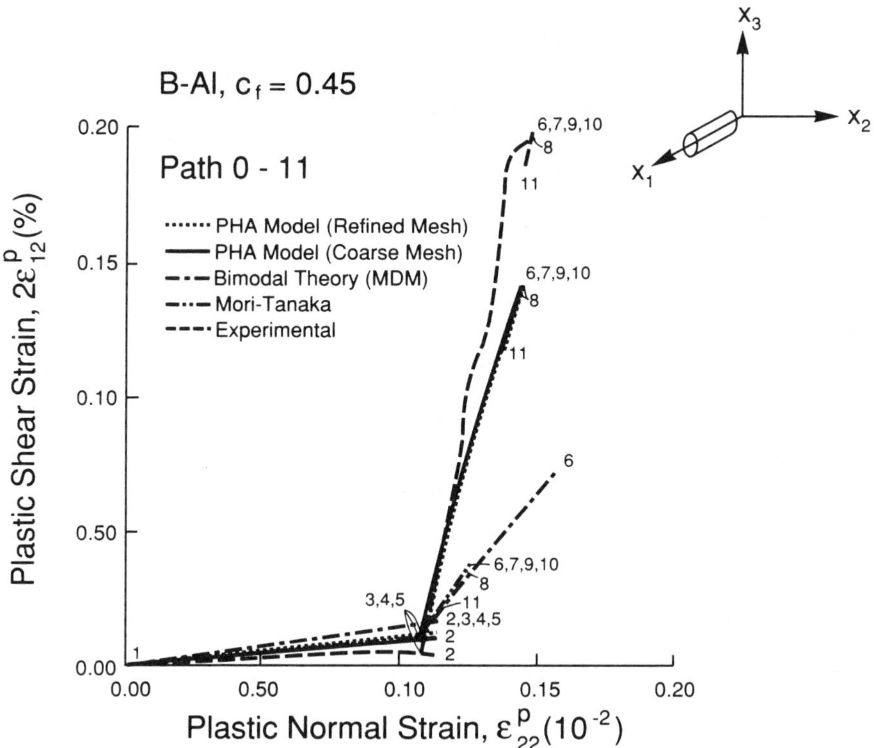

Fig. 12 Plastic normal stress – plastic shear strain, ϵ^p_{21}–$2\epsilon^p_{21}$, response for loading path 0–11.

6. Conclusions

The results suggest that heterogeneous media posses a number of yield surfaces both initially and at subsequent loading points. Each point of the inelastic phase or phases may be associated with a branch of the composite yield surface in the overall space. These branches form a cluster of yield surfaces, the internal envelope of this cluster delineates the purely elastic region devoid of any inelastic deformation. In contrast, the experimentally detected surface which connects the yield points that signify a definite deviation from linearity on observed stress–strain diagrams in various loading directions, is actually a locus of vertices formed in the above cluster by the loading excursions. As suggested by Dvorak et al, (1988), the experimental surface should be regarded as a penetration envelope of the rearranged cluster of subelement surfaces. This serves to emphasize the fact that the normality rule applies only within a cone of normals, and not with respect to the surface drawn through the vertices.

It is remarkable that the initial and subsequent experimental surfaces are so well fitted by the bimodal model, and specifically in the present case by the matrix–dominated surface of Fig. 4. Also, it is surprising that the rigid–body translation according to (6), which actually applies only to the currently loaded branches of the cluster, is exhibited by the entire experimental surface. In any event, it appears that the bimodal model, in conjunction with (6) for a matrix that follows the Phillips hardening law (3), is rather useful in evaluation of subsequent yield surfaces along a general loading path in the overall stress space. Our related work shows that equally good predictions are obtained from the PHA model, albeit with much greater effort.

Apart from the kinematic hardening which clearly dominates overall behavior, there are isotropic hardening components. In part, they are probably caused by age and strain hardening. However, numerical simulation of certain experimental loading sequences with the PHA model has revealed that overall isotropic hardening may take place in the PHA model material even if the matrix hardens only kinematically, as demanded by (3). At the present time, reliable modeling of the isotropic effects appears to be beyond reach, and the effects themselves may not be very significant. The same applies to time–dependent effects at the loading rate employed herein. However, significant rate effects have been detected in the present system at very low rates.

Evaluation of plastic strains appears to be the most difficult goal to reach in modeling. Both the Mori–Tanaka and bimodal models give unreliable predictions. This is no doubt caused by the reliance of these models on normality of the plastic strain increment vector to model approximations of a single current yield surface. The experiments and their simulations show that normality holds within the cone of normals, but not with respect to either the experimentally detected surface, or to its approximations. The PHA model appears to be more reliable in this regard, but it is also susceptible to significant deviations from measured plastic strains. Clearly, the problem is due, in part, to the unexact nature of plastic strains predictions in the matrix material; this is inherent in all available models, including the Dafalias–Popov scheme used herein.

Acknowledgement

This work was sponsored by the Mechanics Division of the Office of Naval Research. Dr. Yapa Rajapakse served as program monitor.

References

Benveniste, Y., 1987, "A New Approach to the Application of Mori–Tanaka's Theory in Composite Materials," *Mechanics of Materials*, Vol. 6, pp.. 147–157.

Dafalias, Y.F. and Popov, E.P., 1976, "Plastic Internal Variables Formalism of Cyclic Plasticity," *ASME Journal of Applied Mechanics*, Vol. 43, pp. 645–651.

Dvorak, G.J., and Teply, J.L., 1985, "Periodic Hexagonal Array Models for Plasticity of Composite Materials," *Plasticity Today: Modeling, Methods and Applications*, A. Saczuk and V. Bianchi, eds., Elsevier, Amsterdam, pp. 623–642.

Dvorak, G.J. and Bahei–El–Din, Y.A., 1987, "A Bimodal Plasticity Theory of Fibrous Composite Materials," *Acta Mechanica*, Vol. 69, pp. 219–241.

Dvorak, G.J., Bahei–El–Din, Y.A., Macheret, Y., and Liu, C.H., 1988, "An Experimental Study of Elastic–Plastic Behavior of a Fibrous Boron–Aluminum Composite," *Journal of the Mechanics and Physics of Solids*, Vol. 36, pp. 665–687.

Dvorak, G.J., 1990, "Plasticity Theories for Fibrous Composite Materials," in *Metal Matrix Composites*, Vol. 2, *Mechanisms and Properties*, ed. by R.K. Everett and R.J. Arsenault, *Treatise on Materials Science and Technology*, Academic Press, Boston, v. 33, pp. 1–77.

Lagoudas, D.C., Gavazzi, A. and Nigam, H., 1990, "Elastoplastic Behavior of Metal Matrix Composites Based on Incremental Plasticity and the Mori–Tanaka Averaging Scheme," Submitted to *Journal of Computational Mechanics.* (See also paper in this volume by the authors).

Hill, R., 1967, "The Essential Structure of Constitutive Laws for Metal Composites and Polycrystals," *Journal of the Mechanics and Physics of Solids*, Vol. 15, pp. 79–95.

Mori, T. and Tanaka, K., 1973, "Average Stress in Matrix and Average Elastic Energy of Materials with Misfitting Inclusions," *Acta Metallurgica*, Vol. 21, pp. 571–574.

Phillips,A., Liu, C.S., and Justusson, J.W., 1972, "An Experimental Investigation of Yield Surfaces at Elevated Temperatures," *Acta Mechanica*, Vol. 14, pp. 119–146.

Teply, J.L., 1984, "Periodic Hexagonal Array Models for Plasticity Analysis of Composite Materials," Ph.D. Dissertation, The University of Utah, Salt Lake City.

Teply, J.L. and Dvorak, G.J., 1988, "Bounds on Overall Instantaneous Properties of Elastic–Plastic Composites," *Journal of the Mechanics and Physics of Solids*, Vol. 36, pp. 29–58.

Ziegler, H., 1959, "A Modification of Prager's Hardening Rule," *Quarterly of Applied Mathematics*, Vol. 17, pp. 55–65.

Computational Methods

The Effect of Superposed Hydrostatic Stress on the Mechanical Response of Metal-Matrix Composites

T. Christman[†], J. Llorca[‡], S. Suresh[*] and A. Needleman[*]

[†]Division of Engineering and Applied Science, California Institute of Technology, Pasadena, CA 91125, U.S.A.

[‡]Department of Materials Science, Polytechnic University of Madrid, Madrid, Spain

[*]Division of Engineering, Brown University, Providence, RI 02912, U.S.A.

Abstract

The effects of a superposed hydrostatic stress on the deformation and failure behavior of whisker reinforced metal-matrix composites are analyzed numerically. The applied loading path consists of the imposition of a hydrostatic stress followed by tension along the fiber axis. Matrix cavitation is the sole failure mechanism analyzed and an elastic-viscoplastic material model is used that accounts for ductile fracture by the nucleation and subsequent growth of voids to coalescence. The effect of the distribution of the whiskers on failure is illustrated. A superposed hydrostatic stress is found to have a much greater effect on ductility when the whiskers are clustered than when they are uniformly distributed in the matrix. The predicted variations in ductility for tensile and compressive superposed hydrostatic stress, and the presence of zones which show highly localized strains, are in qualitative agreement with available experimental results.

Introduction

Whisker-reinforced ductile matrix composites represent a broad class of advanced structural materials which often possess excellent specific strength and stiffness values. The primary disadvantage of many particle-reinforced materials is that they exhibit low ductility and poor fracture toughness. In order to improve these macroscopic characteristics, it is essential to develop an understanding of the deformation mechanisms on the local or micro-level. A survey of published experimental work indicates that ductile failure in the matrix by the nucleation, growth and coalescence of cavities is a dominant failure mode in some composite materials with strongly bonded interfaces [McDanels, 1984; You *et al.*, 1987; Christman *et al.*, 1989]. Llorca *et al.* [1990] have carried out a parametric numerical study of the effects of matrix porosity on deformation and failure behavior using a phenomenological porous-plastic constitutive relation to characterize the matrix material.

It has long been appreciated that stress triaxiality plays an important role in ductile failure by void nucleation, growth and coalescence. Recent finite element simulations have shown that significant stress triaxiality develops in a whisker reinforced composite matrix as a consequence of constrained deformation, Christman *et al.* [1989]. For the particular case of metals reinforced by whiskers that are oriented in the direction of tensile loading, tensile hydrostatic stresses develop at fiber ends while compressive hydrostatic stresses are predicted for the matrix material located in the regions between the ends of the fibers. The tensile hydrostatic stresses give rise to an apparent increase in strength. It has also been shown, Christman, Needleman and Suresh, [1989] and Tvergaard [1990], that particle distribution plays an important role in the development of local field quantities, in particular on the development of stress triaxiality.

Here, the framework in Christman, Needleman and Suresh [1989] and Llorca *et al.* [1990] is used to investigate the effect of superposed hydrostatic stress on composite deformation and failure. The applied loading path consists of the imposition of a hydrostatic stress (either tension or compression) followed by uniaxial tension.

Problem Formulation

The formulation of the boundary value problem follows that in Christman, Needleman and Suresh [1989] and Llorca *et al.* [1990], where further details and further references can be found. A convected coordinate Lagrangian formulation of the field equations is employed with the initial unstressed state taken as a reference.

All field quantities are considered to be functions of convected coordinates, y^i, which serve as particle labels, and time, t. Attention is confined to quasi-static deformations and, with body forces neglected, the rate form of the principal of virtual work is written as

$$\Delta t \int_V [\dot{\tau}^{ij}\delta E_{ij} + \tau^{ij}\dot{u}^k_{,i}\delta u_{k,j}]\, dV = \\ = \Delta t \int_S \dot{T}^i \delta u_i \, dS - \left[\int_V \tau^{ij}\delta E_{ij}\, dV - \int_S T^i \delta u_i \, dS \right]. \quad (1)$$

Here, τ^{ij} are the contravariant components of the Kirchhoff stress ($\boldsymbol{\tau} = J\boldsymbol{\sigma}$, where $\boldsymbol{\sigma}$ is the Cauchy stress and J is the ratio of current to reference volume of a material element) on the deformed convected coordinate net. The quantities V and S are the volume and surface, respectively, of the body in the reference configuration and $(\dot{\,}) = \partial()/\partial t$ at fixed y^i. The second term on the right hand side represents an equilibrium correction term that is used in the numerical procedure to reduce drift from the equilibrium path due to the finite time step.

The nominal traction components, T^i, and the Lagrangian strain components, E_{ij}, are given by

$$T^i = (\tau^{ij} + \tau^{kj} u^i_{,k})\nu_j \tag{2}$$

$$E_{ij} = \frac{1}{2}(u_{i,j} + u_{j,i} + u^k_{,i} u_{k,j}) \tag{3}$$

where $\boldsymbol{v}$ is the surface normal in the reference configuration, u_j are the components of the displacement vector on base vectors in the reference configuration, and $(\ )_{,i}$ denotes covariant differentiation in the reference frame.

The composite material is idealized in terms of a periodic array of identical cells. In some calculations, an axisymmetric cylindrical cell is used that can be regarded as an approximation to a three dimensional array of hexagonal cylinders [Tvergaard, 1982]. In other calculations, in order to investigate clustering effects, a plane strain cell model is used. Consideration is restricted to deformations for which the straight lines bounding each cell remain straight after deformation (this is a stronger constraint than required by periodicity). Also, attention is confined to deformations that preserve the mirror symmetry of the array so that straight lines connecting the centers of the cells remain straight.

In the axisymmetric analyses a cylindrical coordinate system (r, θ, z) is used and the identifications $y^1 = r$, $y^2 = z$ and $y^3 = \theta$ are made, while in the plane strain analyses the y^1-y^2 plane is the plane of deformation. In either case, the region analyzed numerically is $0 \le y^1 \le w_0$ and $0 \le y^2 \le b_0$, and the boundary conditions on this region are

$$\dot{u}^2 = 0 \qquad \dot{T}^1 = 0 \qquad \text{on} \qquad y^2 = 0 \tag{4}$$

$$\dot{u}^2 = \dot{U}_2 = \dot{\epsilon}_{ave}(b_0 + U_2) \qquad \dot{T}^1 = 0 \qquad \text{on} \quad y^2 = b_0 \tag{5}$$

$$\dot{u}^1 = \dot{U}_1 \qquad \dot{T}^2 = 0 \qquad \text{on} \quad y^1 = w_0. \tag{6}$$

Here, $\dot{\epsilon}_{ave}$ is a prescribed constant while $\dot{U}_1$ is determined from the specification of the transverse stressing. Defining,

$$\Sigma_1 = \frac{1}{\bar{A}_1} \int_{A_1} T^1 dA_1 \tag{7}$$

$$\Sigma_2 = \frac{1}{\bar{A}_2} \int_{A_2} T^2 dA_2 \tag{8}$$

where A_i is the reference area of the cell face with normal in the y^i direction and $\bar{A}_i$ is the corresponding current area. The quantity $\dot{U}_1$ is determined from the condition that

$$\dot{\Sigma}_1 = \rho \dot{\Sigma}_2. \tag{9}$$

The loading history consists of two stages. In the first stage, the cell is subject to overall hydrostatic stressing (in-plane hydrostatic stressing in the plane strain) and $\rho = 1$. The value of $\Sigma_1 = \Sigma_2$ at the end of the first stage of loading is denoted by σ_H and is referred to as the superposed hydrostatic stress. In the second stage, tensile loading parallel to the whisker axes is imposed with the overall transverse stress (in plane strain, the in-plane overall transverse stress) held constant, $\rho = 0$. Within the cell the fibers are assumed to remain perfectly bonded to the matrix and traction and displacement continuity conditions are maintained across the interface.

The material is characterized as an elastic-viscoplastic Gurson-type solid [Gurson, 1975 and 1977]. The total rate of deformation, $\mathbf{D}$, is written as the sum of an elastic part, $\mathbf{D}^e$, and a plastic part, $\mathbf{D}^p$, so that

$$\mathbf{D} = \mathbf{D}^e + \mathbf{D}^p. \tag{10}$$

The flow rule is

$$\mathbf{D}^p = \dot{\Lambda}\frac{\partial \Phi}{\partial \boldsymbol{\sigma}} \tag{11}$$

where

$$\Phi = \frac{\sigma_e^2}{\bar{\sigma}^2} + 2q_1 f^* \cosh\left(\frac{3q_2\sigma_h}{2\bar{\sigma}}\right) - 1 - {q_1}^2 {f^*}^2 = 0 \tag{12}$$

and

$$\dot{\Lambda} = \frac{(1-f)\bar{\sigma}\dot{\bar{\epsilon}}}{\boldsymbol{\sigma} : \frac{\partial \Phi}{\partial \boldsymbol{\sigma}}}. \tag{13}$$

Here, $\bar{\sigma}$ is the average strength of the matrix material and

$$\sigma_e^2 = \frac{3}{2}\boldsymbol{\sigma} : \boldsymbol{\sigma} \qquad \sigma_h = \frac{1}{3}\boldsymbol{\sigma} : \mathbf{I}. \tag{14}$$

The parameters q_1 and q_2 were introduced by Tvergaard [1981, 1982] to bring predictions of the model into closer agreement with full numerical analyses of a periodic array of voids. Values of q_1=1.25 and q_2=1 have given good agreement between the predictions of the modified Gurson model and the results of a numerical study of void coalescence in isotropically hardening materials [Koplik and Needleman, 1988]. The function $f^*(f)$, which was introduced by Tvergaard and Needleman [1984] to account for the effects of rapid void coalescence at failure, has the form

$$f^* = \begin{cases} f & f \le f_c\,; \\ f_c + \dfrac{f_u^* - f_c}{f_f - f_c}(f - f_c) & f \ge f_c. \end{cases} \tag{15}$$

The constant f_u^* is the value of f^* at zero stress in (12), i.e. $f_u^* = 1/q_1$. As $f \to f_f$, $f^* \to f_u^*$ and the material loses all stress carrying capacity. In the analyses of Koplik and Needleman [1988], it was found that the value of f_c varies slowly with stress triaxiality and matrix strain hardening, but there is a rather strong dependence on the initial void volume fraction. For the volume fraction of void nucleating particles considered in this study (see next section for details), the values of f_c=0.10 and f_f=0.25 were used in the calculations. In general the evolution of the void volume fraction results from growth of existing voids and nucleation of new voids,

$$\dot{f} = (1-f)\mathbf{D}^p : \mathbf{I} + \dot{f}_{nucleation}. \tag{16}$$

The first term on the right hand side of (16) gives the rate of increase of void volume fraction due to the growth of existing voids and, in the present calculations, the nucleation of new voids is given by,

$$\dot{f}_{nucleation} = \mathcal{D}\dot{\bar{\epsilon}}. \tag{17}$$

As suggested by Chu and Needleman [1980], void nucleation is taken to follow a normal distribution, so that for plastic strain-controlled nucleation,

$$\mathcal{D} = \frac{f_N}{s_N\sqrt{2\pi}} \exp\left[-\frac{1}{2}\left(\frac{\bar{\epsilon}-\epsilon_N}{s_N}\right)^2\right] \tag{18}$$

The strain and strain rate hardening of the matrix material is described by

$$\dot{\bar{\epsilon}} = \dot{\epsilon}_0[\bar{\sigma}/g(\bar{\epsilon})]^{1/m} \qquad g(\bar{\epsilon}) = \sigma_0(\bar{\epsilon}/\epsilon_0+1)^N \qquad \epsilon_0 = \sigma_0/E \tag{19}$$

where $\bar{\epsilon} = \int \dot{\bar{\epsilon}}\, dt$.

The function $g(\bar{\epsilon})$ represents the effective stress versus effective strain response in a tensile test carried out at a strain-rate such that $\dot{\bar{\epsilon}} = \dot{\epsilon}_0$. Also, σ_0 is a reference strength and N and m are the strain hardening exponent and strain rate hardening exponent, respectively.

Standard kinematic relations are used to express the constitutive law as a relation between the contravariant components (on the current base vectors) of the convected rate of Kirchhoff stress, $\hat{\tau}^{ij}$, and Lagrangian strain rate, $\dot{E}_{ij}$. When $f \equiv 0$, this constitutive relation reduces to that for an elastic-viscoplastic Mises solid.

The deformation history is calculated in a linear incremental manner and, in order to increase the stable time step, the rate tangent modulus method of Peirce *et al.* [1984] is used. This is a forward gradient method based on an estimate of the plastic strain rate in the interval between t and $t+\Delta t$. The incremental boundary value problem is solved using a combined finite element Rayleigh-Ritz method [Tvergaard, 1976].

Material model

The material chosen as a model system for the finite element analysis is a 2124 aluminum alloy (powder metallurgy matrix) reinforced with 13 volume percent of SiC whiskers. The evolution of microstructure in this system in response to aging treatments has been well characterized using transmission electron microscopy [Christman and Suresh, 1988]. The properties of the matrix material selected for the analysis represent the peak-aged condition of the matrix. The choice of matrix properties is such that the occurrence of accelerated aging in the composite matrix due to the presence of the ceramic whiskers is accounted for (see Christman and Suresh [1988] and Christman, Needleman and Suresh [1989] for details). The mechanical properties of the matrix material are: $E = 70$ GPa, $\nu = 0.33$, $\sigma_0 = 290$ MPa, $N = 0.13$, $m = 0.004$. Both the whisker aspect ratio and the cell aspect ratio were chosen equal to 5, based on optical microscopy results. The elastic constants of the SiC whiskers are taken to be $E = 450$ GPa and $\nu = 0.17$. On the basis of prior experimental evidence, the interface between the SiC whisker and the matrix was assumed to be perfectly bonded. The whiskers were also taken to be aligned along the extrusion direction. The nucleation parameters in (18), namely the volume fraction of void nucleating particles f_N, void nucleation strain ϵ_N and standard deviation s_N, were chosen equal to 0.05, 0.05 and 0.01, respectively. These values are estimated from the experimental measurements performed by Van Stone *et al.* [1974] on 2xxx series aluminum alloys.

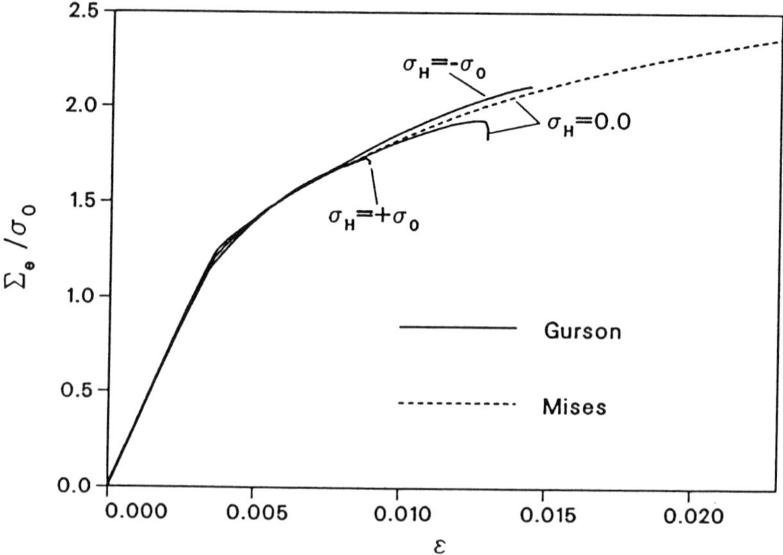

Figure 1. Effect of superposed hydrostatic stress on monotonic deformation. Axisymmetric model.

Results

Axisymmetric model

The results for the axisymmetric calculations are shown in Fig. 1, where the overall effective stress $\Sigma_e = |\Sigma_2 - \Sigma_1|$ (normalized by the reference stress σ_0) is plotted against the axial strain. The numerical results show that the main effect of the superposed hydrostatic stress is on the ductility of the composite; the stress-strain behavior prior to failure is not significantly affected by cavitation in the matrix. This behavior can be rationalized by the following arguments. Void nucleation is taken to be controlled by the effective strain. At the onset of deformation, the effective strains in the matrix are below the critical nucleation strain and no voids are present in the matrix. Under these conditions, the Mises and Gurson models are identical and the effective strain does not depend on the hydrostatic stress level. Once a critical strain is reached, voids are nucleated in the metallic matrix. Under the influence of a far-field tensile load (applied along the fiber direction), a high level of tensile mean normal stress, σ_h defined in (14), builds up at the ends of the fibers. This positive triaxiality (generated by the constrained flow of the ductile matrix between the brittle fibers), when superposed on externally imposed positive hydrostatic stresses, produces rapid void growth in the ductile matrix. Consequently, failure takes place shortly after the nucleation of the voids.

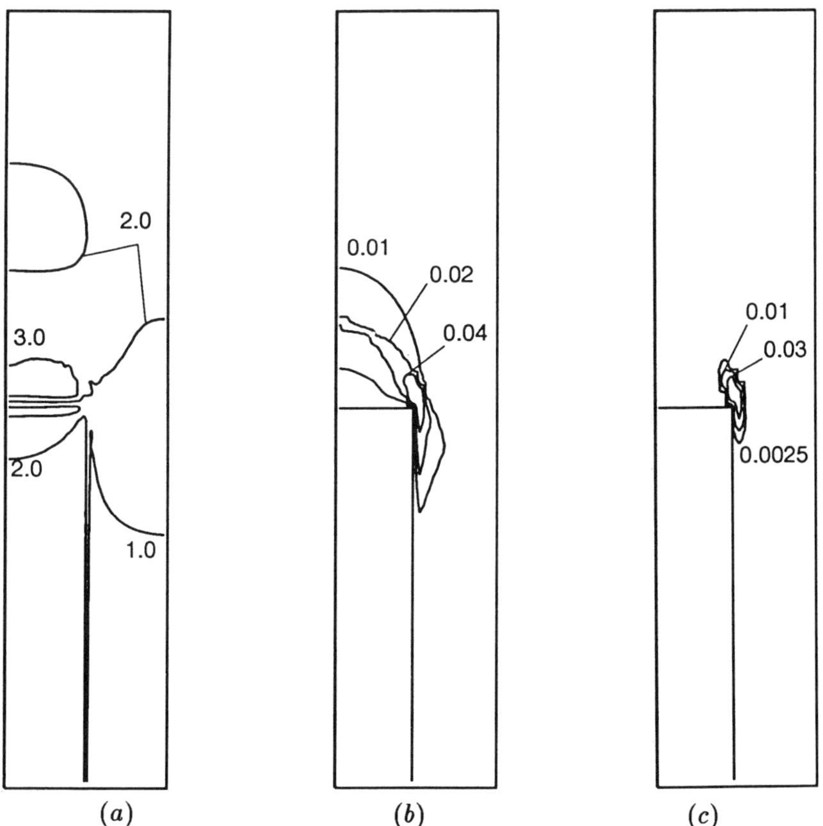

Figure 2. Contours of (a) normalized mean normal stress, σ_h/σ_0, (b) plastic strain, $\bar{\epsilon}$, and (c) void volume fraction, f at $\epsilon_{ave} = 0.01$. $\sigma_H = +\sigma_o$.

Figure 2 shows contours of constant mean normal stress, plastic strain and void volume fraction in the aluminum matrix at a far-field tensile strain of 0.01 and for an externally imposed hydrostatic stress of $+\sigma_o$. The superposition of the internally generated matrix triaxiality (due to constrained flow) with the remotely applied positive hydrostatic stresses causes very large tensile mean stresses to develop in the matrix, especially close to the fiber ends. Conversely, when compressive hydrostatic stresses are imposed externally, the conditions become less favorable for the growth of voids as the severity of tensile hydrostatic stresses in the matrix is partially offset by the external hydrostatic stress, Fig. 3. As a result, the maximum extent of matrix cavitation is also noticeably diminished, as shown in Fig. 3*c*. It is worth noting that failure occurs in a very localized zone around the whisker ends when a tensile hydrostatic stress is superposed; the application of compressive hydrostatic stresses promotes more distributed damage (compare Figs. 2 and 3).

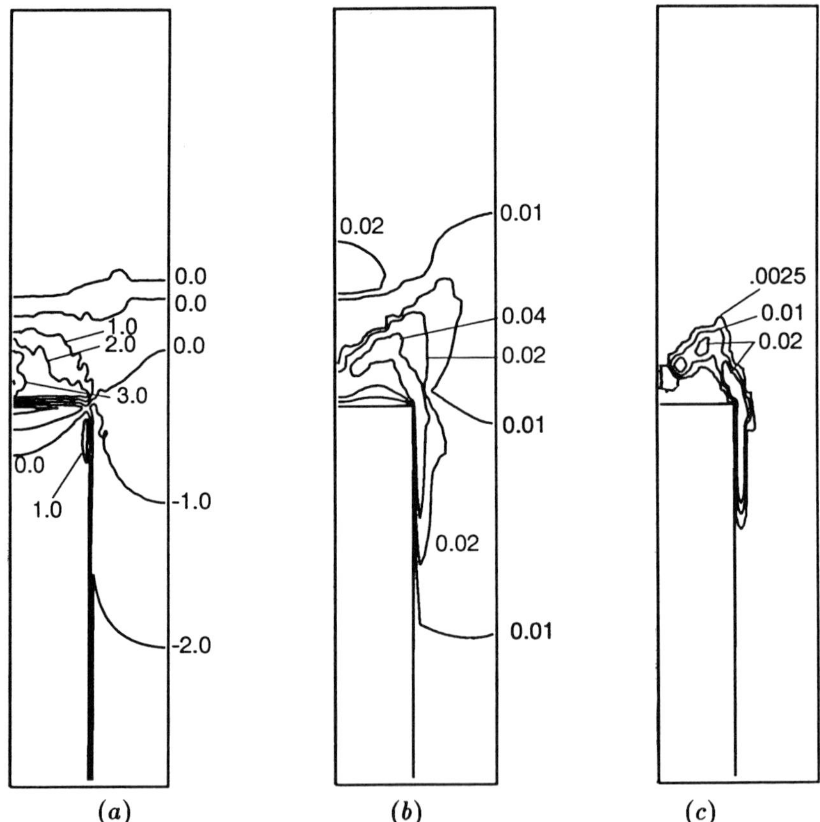

Figure 3. Contours of (a) normalized mean normal stress, σ_h/σ_0, (b) plastic strain, $\bar{\epsilon}$, and (c) void volume fraction, f at $\epsilon_{ave} = 0.013$. $\sigma_H = -\sigma_o$.

Plane Strain Model

Numerical studies by Christman, Needleman and Suresh [1989] have shown that the distribution of the brittle fibers in the ductile matrix has a strong influence on the overall constitutive response. One of the principal causes for this geometrical effect is the change in matrix triaxiality caused by the fiber distribution. The highest levels of matrix triaxiality develop under a far-field tensile stress, when the fibers are arranged in a perfectly aligned and uniform manner throughout the matrix. Relatively shifting the fibers either in the direction of the fiber axis or in the transverse direction leads to a reduction in the average level of triaxiality due to constrained flow. This effect is more pronounced for the former case of "vertical clustering" than for the latter "horizontal clustering". For this reason, finite element analyses of the effect of imposed hydrostatic stresses on tensile deformation were carried out in this work for a specific case of vertical clustering.

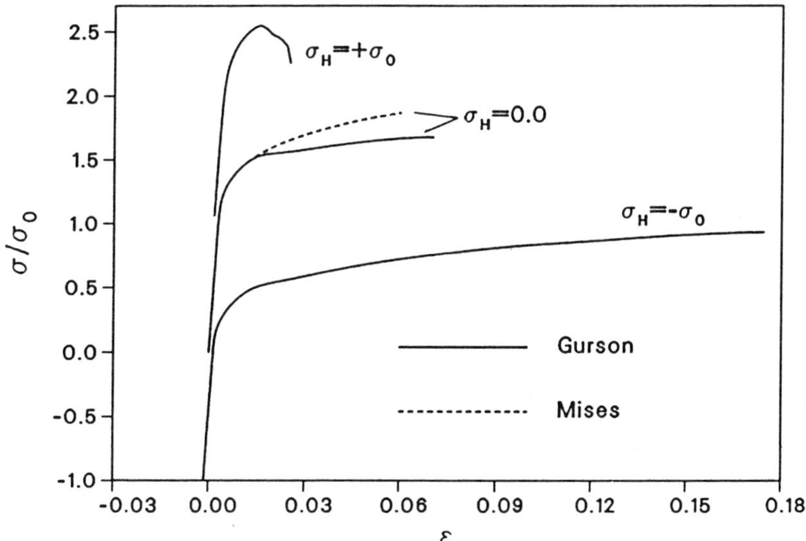

Figure 4. Effect of superposed in-plane hydrostatic stress on monotonic deformation. Plane strain model.

Detailed investigations of the effect of fiber clustering on the constitutive response within the context of an axisymmetric unit cell formulation require a full three-dimensional analysis. However, as shown by Christman, Needleman and Suresh [1989], the basic features of the role of fiber distribution in influencing deformation can be captured (with significantly less computational resources) by plane strain models. In the plane strain model adopted in this study, the whiskers were clustered vertically by 50% as compared to the perfectly periodic and uniform distribution. If the shortest distance between the ends of two coaxial fibers (aligned one beneath the other) is a in the uniform and periodic distribution, and if this distance is reduced to a lower value a' for the vertically clustered arrangement, then percent of vertical clustering is given by $[(a - a')/a] \times 100$ (see Christman, Needleman and Suresh [1989] for full details on vertical clustering). It should be noted that in the plane strain model, the external hydrostatic stresses are imposed only in the plane of the unit cell.

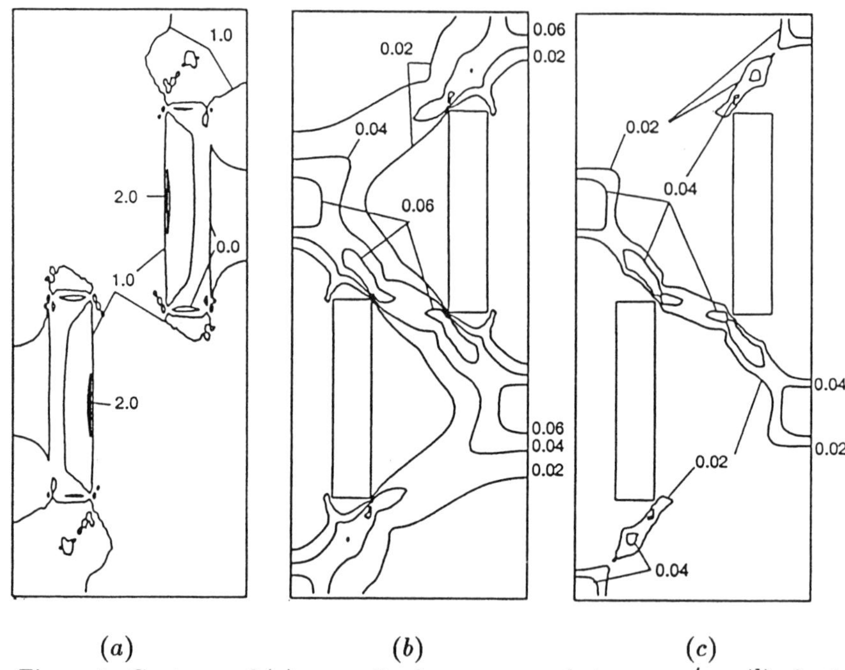

(a) (b) (c)

Figure 5. Contours of (a) normalized mean normal stress, σ_h/σ_0, (b) plastic strain, $\bar{\epsilon}$, and (c) void volume fraction, f at $\epsilon_{ave} = 0.02$. $\sigma_H = 0$.

Figure 4 shows the numerically predicted tensile stress-strain response for the composite material where the effect of imposed in-plane hydrostatic stress is illustrated for the plane strain model with 50% vertical clustering. The role of the void growth process in influencing the overall stress-strain behavior of the composite is evident from the results shown for $\sigma_H = 0$. The lower strength and lower strain hardening exponent exhibited by the Gurson matrix material (as compared to the Mises matrix) are a consequence of the softening arising from the nucleation and growth of voids. The superposition of a negative in-plane hydrostatic stress leads to a strain-to-failure value which is three times higher than that without any superposed hydrostatic stress and seven times higher than that with a far-field tensile hydrostatic stress of the same magnitude. Although the two-dimensional nature of the model precludes any direct comparison with experimental results, it is worth noting that the ductility improvements predicted by the present analysis for imposed negative hydrostatic stress compare favorably with experimental observations reported in the literature for 2xxx aluminum alloys reinforced with SiC whiskers and particulates [Vasudevan *et al.*, 1989; Liu *et al.*, 1989].

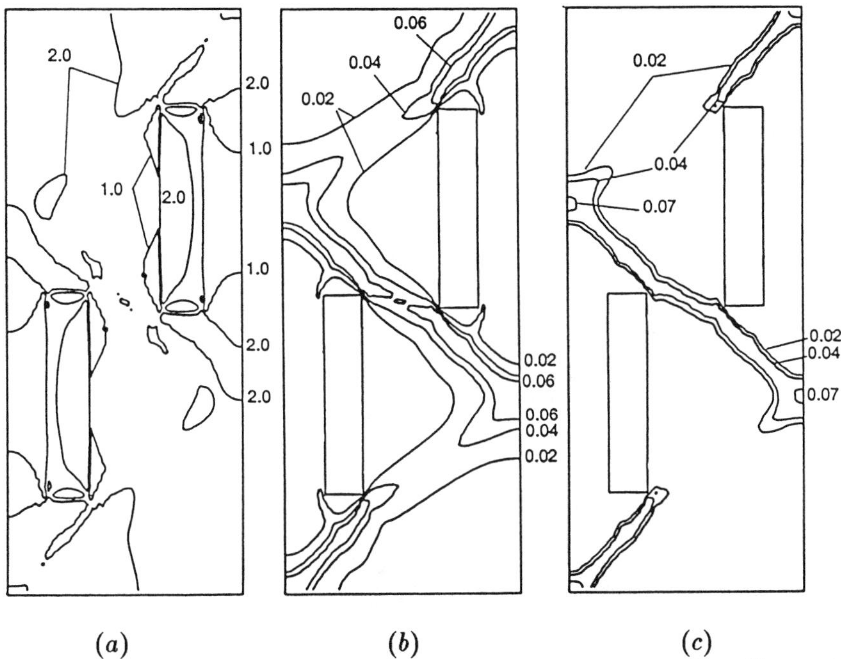

(*a*) (*b*) (*c*)

Figure 6. Contours of (*a*) normalized mean normal stress, σ_h/σ_0, (*b*) plastic strain, $\bar{\epsilon}$, and (*c*) void volume fraction, f at $\epsilon_{ave} = 0.02$. $\sigma_H = +\sigma_o$.

Figures 5 and 6 display contours of constant hydrostatic stress, plastic strain and void volume fraction in the matrix for numerical simulations pertaining to superposed in-plane hydrostatic stresses of 0 and $+\sigma_o$ at a far-field tensile strain of 0.02. Note that the higher level of tensile triaxiality for externally imposed tensile hydrostatic stresses (Fig. 6*c*) causes a larger void volume fraction to develop than in the case of zero external triaxiality (Fig. 5*c*). In both cases, void nucleation is especially pronounced between the ends of adjacent whiskers where large plastic strains and mean stresses are produced. For positive superposed hydrostatic stress, the void volume fraction reaches a critical value of 0.1 at an axial strain of 0.025 and the composite fails. Under compressive superposed hydrostatic stress, void growth is suppressed and the matrix undergoes a larger amount of plastic

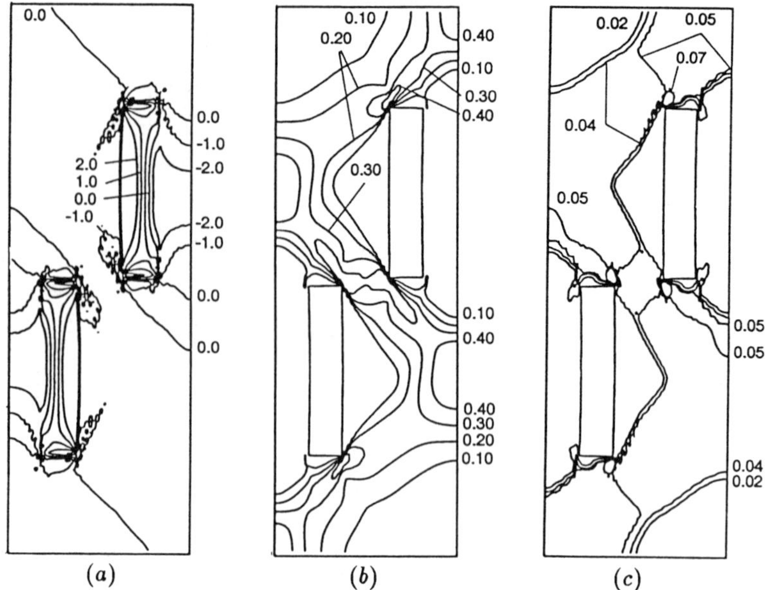

Figure 7. Contours of (a) normalized mean normal stress, σ_h/σ_0, (b) plastic strain, $\bar{\epsilon}$, and (c) void volume fraction, f at $\epsilon_{ave} = 0.12$. $\sigma_H = -\sigma_o$.

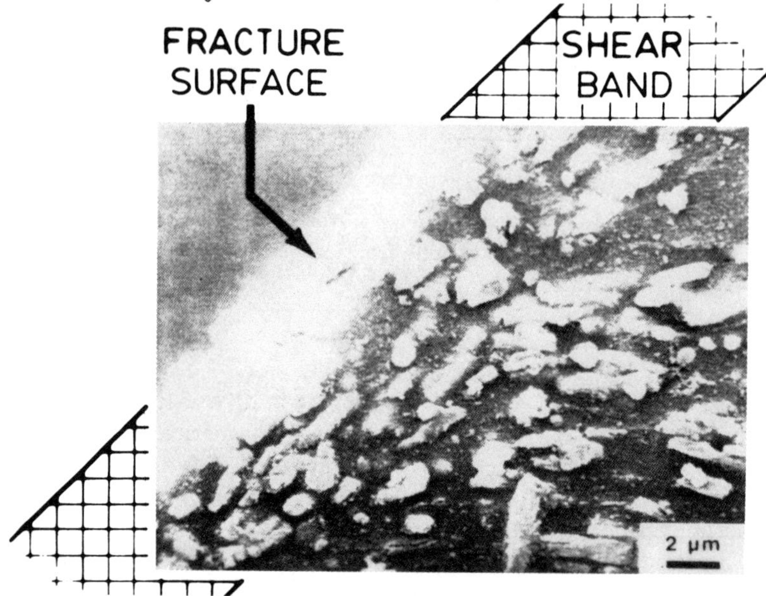

Figure 8. Experimental observation of strain localization in 2124 Al-13 vol. % SiC whisker composite under an imposed negative hydrostatic stress. (From Vasudevan *et al.*, 1989)

deformation before failure. This trend is illustrated in Fig. 7 with contours of constant hydrostatic stress, plastic strain and void volume fraction in the matrix for an externally imposed negative in-plane hydrostatic stress of $-\sigma_o$ at a far-field tensile strain of 0.12.

In Fig. 7, note that the porosity extends over a significant fraction of the unit cell and that the void volume fraction is lower than the critical value. The plastic strain contours reveal that strains of up to 0.4 occur in the regions populated by the voids. The material confined within the ends of adjacent whiskers undergoes the greatest extent of plastic flow. In this analysis of deformation for 50% vertical clustering, the localization of strain under negative imposed hydrostatic stress is clearly evident along the line linking the whisker ends. Such strain localization under a negative imposed hydrostatic stress has been documented experimentally for a 2124 aluminum alloy reinforced with 13.2% SiC whiskers, Fig. 8 [Vasudevan *et al.*, 1989].

Concluding Remarks

Recent finite element simulations have shown that significant hydrostatic stress levels develop in the matrix of whisker reinforced composites as a consequence of constrained plastic deformation, Christman, Needleman and Suresh [1989]. The level of this constraint-induced hydrostatic stress is sensitive to the distribution and shape of the reinforcement particles. Additionally, numerical studies of porosity evolution in whisker-reinforced metal-matrix composites [Llorca *et al.*, 1990] show that the tensile hydrostatic stresses which develop near the whisker ends promote failure due to matrix void nucleation, growth and coalescence. In the calculations reported here, a positive superposed hydrostatic stress enhances this effect, while a negative superposed hydrostatic stress counteracts it. The effect of the superposed hydrostatic stress on ductility is found to be much greater than on the deformation response. Vertically clustered (overlapping) whiskers in plane strain exhibit a seven fold increase in the strain-to-failure when the superposed hydrostatic pressure changes from $+\sigma_o$ to $-\sigma_o$, while the effect is much smaller for an axisymmetric model of a uniform distribution of whiskers aligned with the tensile axis. This is because of the larger hydrostatic stresses that develop due to constrained plastic flow in the aligned fibers.

The only failure mechanism modelled in this investigation is matrix cavitation. A superposed hydrostatic stress would be expected also to affect failure by fiber decohesion and by fiber cracking.

Acknowledgements

J.Ll. expresses his gratitude to the Fulbright Commission for the fellowship support provided to carry out this research. A.N. is grateful for support provided by the Office of Naval Research through grant N00014-89-J-3054. S.S. acknowledges the support of the Materials Research Group on Plasticity and Fracture which is funded by the National Science Foundation.

Computations reported here were carried out on the Cray YMP at the Pittsburgh Supercomputer Center.

References

Christman, T., Needleman, A., Nutt, S. and Suresh, S., 1989, "On the Microstructural Evolution and Micromechanical Modelling of Deformation of a Whisker-Reinforced Metal-Matrix Composite," *Materials Science and Engineering*, Vol. A107, pp. 49-61.

Christman, T., Needleman, A. and Suresh, S., 1989, "An Experimental and Numerical Study of Deformation in Metal-Ceramic Composites," *Acta Metallurgica*, Vol. 37, pp. 3029-3050. See also Corrigenda, 1990, Vol. 38, pp. 879-880.

Christman, T. and Suresh, S., 1988, "Microstructural Development in an Aluminum Alloy-SiC Whisker Composite," *Acta Metallurgica*, Vol. 36, pp. 1691-1704.

Chu, C. C. and Needleman, A., 1980, "Void Nucleation Effects in Biaxially Stretched Sheets," *Journal of Engineering Materials and Technology*, Vol. 102, pp. 249-256.

Gurson, A. L., 1975, "Plastic Flow and Fracture Behavior of Ductile Materials Incorporating Void Nucleation, Growth and Interaction," Ph. D. Thesis, Brown University.

Gurson, A. L., 1977, "Continuum Theory of Ductile Rupture by Void Nucleation and Growth: Part I - Yield Criteria and Flow rules for Porous Ductile Media," *Journal of Engineering Materials and Technology*, Vol. 99, pp. 2-15.

Koplik, J. and Needleman, A., 1988, "Void Growth and Coalescence in Porous Plastic Solids," *International Journal of Solids and Structures*, Vol. 24, 835-853.

Liu, L. S., Manoharan, M. and Lewandowski, J.J., 1989, "Effects of Microstructure on the Behavior of an Aluminum Alloy and an Aluminum Matrix Composite Tested under Low Levels of Superimposed Hydrostatic Pressure," *Metallurgical Transactions*, Vol. 20A, pp. 2409-2417.

Llorca, J., Needleman, A. and Suresh, S., 1990, "An Analysis of the Effects of Matrix Void Growth on Deformation and Ductility of Metal-Ceramic Composites," to be published.

McDanels, D. L., 1984, "Analysis of Stress-Strain, Fracture, and Ductility Behavior of Aluminum Matrix Composites Containing Discontinuous Silicon Carbide Reinforcement," *Metallurgical Transactions*, Vol. 16A, pp. 1105-1115.

Pierce, D., Shih, C. F. and Needleman, A., 1984, "A Tangent Modulus Method for Rate Dependent Solids," *Computers and Structures*, Vol. 18, pp. 875-887.

Tvergaard, V., 1976, "Effect of Thickness Inhomogeneities in Internally Pressurized Elastic-Plastic Spherical Shells" *Journal of the Mechanics and Physics of Solids*, Vol. 24, pp. 291-304.

Tvergaard, V., 1981, "Influence of Voids on Shear Band Instabilities under Plane Strain Conditions," *International Journal of Fracture*, Vol. 17, pp. 389-407.

Tvergaard, V., 1982, "On Localization in Ductile Materials Containing Spherical Voids," *International Journal of Fracture*, Vol. 18, pp.237-252.

Tvergaard, V., 1990, "Analysis of Tensile Properties of a Whisker-Reinforced Metal-Matrix Composite," *Acta Metallurgica et Materialia*, Vol. 38, pp. 185-192.

Tvergaard, V. and Needleman, A., 1984, "An Analysis of the Cup-Cone Fracture in a Round Tensile Bar," *Acta Metallurgica*, Vol. 32, pp. 157-169.

Van Stone, R. H., Merchant, R. H. and Low, J. R., 1974, "Investigation of the Plastic Fracture of High-Strength Aluminum Alloys," *ASTM STP 556*, American Society for Testing and Materials, pp. 93-124.

Vasudevan, A. K., Richmond, O., Zok, F. and Embury, J. D., 1989, "The Influence of Hydrostatic Pressure on the Ductility of Al-SiC Composites," *Materials Science and Engineering*, Vol. A107, pp. 63-69.

You, C. P., Thompson, A. W. and Bernstein, I. M., 1987, "Proposed Failure Mechanism in a Discontinuously Reinforced Aluminum Alloy," *Scripta Metallurgica*, Vol. 21, pp. 181-185.

Micromechanical Modeling of Plasticity and Texture Evolution in Semi-Crystalline Polymers

DAVID M. PARKS and SAID AHZI
Department of Mechanical Engineering
Massachusetts Institute of Technology
Cambridge, MA 02319

IUTAM Symposium on
Inelastic Deformation of Composite Materials
RPI, May 1990

Abstract

The microstructure of many semi-crystalline polymers consists of rather broad, thin composite inclusions of crystalline and amorphous domains. A micromechanically based composite model is proposed to study plasticity and texture evolution in these composite materials. Special consideration is given to molecular chain inextensibily within the crystalline phase. Molecular alignment within the amorphous phase is accounted for by introducing a back stress tensor in the corresponding flow rule. Interface compatibility and traction equilibrium are considered within each composite inclusion. A modified Taylor-like localization law is proposed to deal with the singularity problems related to kinematic deficiency in some inclusions. Applications of this model are made to predict the behavior and deformation texture evolution in initially isotropic High Density Polyethylene. Comparisons of our predictions to experimental data are given.

1. Introduction

Semi-crystalline polymers are a class of naturally two-phase composite materials. The morphology of these composites consists of co-existing crystalline domains and amorphous domains associated with each other in a fine plate-like morphological structure. In an undeformed state, semi-crystalline polymers are generally spherulitic. The material element representative of the spherulite consists of a two-phase composite inclusion lying in the radial direction of of this spherulite (Figure 1). Under plastic straining, morphological and crystallographic textures evolve with plastic deformation; the evolution of these textures affects the macroscopic behavior. For instance, a strong textural hardening is observed in tension but not in simple shear of Polyethylene (G'Sell and Jonas [1]; G'Sell, Boni and Shrivastava, [2]).

The aim of this work is to develop a micromechanically-based model to describe the plasticity of semi-crystalline polymers and to predict the evolution of textural anisotropy and the macroscopic behavior under different loads. Possible plastic deformations mechanisms in semi-crystalline polymers were reviewed by Bowden and Young [3] and by Haudin [4]. Like other crystalline materials, the crystalline phase of the polymers deforms by crystallographic slip, twinning and martensitic transformation. However, all of these mechanisms leave the crystallographic chain direction inextensible, and therefore provide less than five independent systems. The plasticity of the amorphous domain can be described by the double kink model proposed by Argon [5]. The molecular alignment within this phase can be represented by a back stress tensor of the type used in the work of Boyce, Parks and Argon [6].

We propose a viscoplastic composite model in which we neglect elasticity and pressure sensitivity but account for the contributions of both crystalline and amorphous phases to large plastic straining and rotation. Incompressibility is assumed in both phases. In this model, we solve for the kinematics and equilibrium within a two-phase composite inclusion embedded in an infinite matrix. The composite inclusion consists of a crystalline lamella and its corresponding amorphous layer (Figure 1). Applications of this composite model have been made to predict crystallographic and morphological textures and stress-strain behavior of High Density Polyethylene (HDPE) under tension and simple shear. Predicted results are compared to experimental data.

2. Constitutive relations

2.1 Crystalline phase

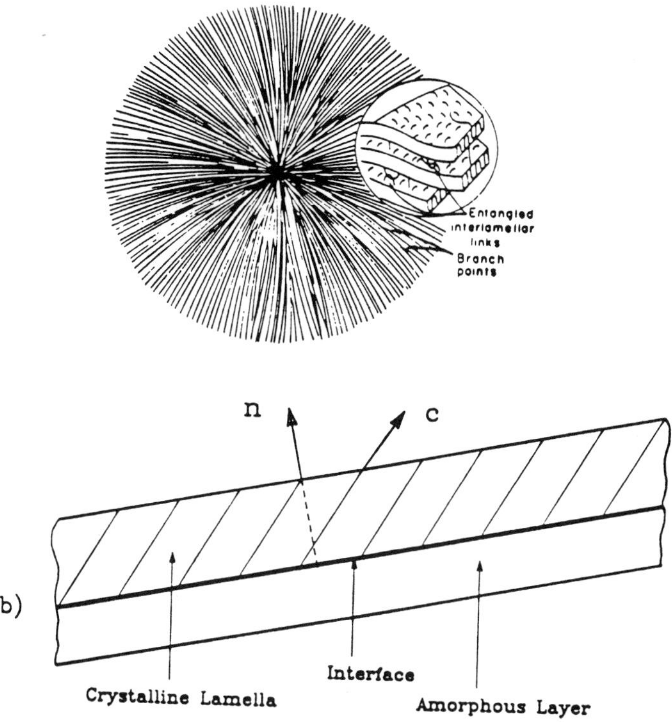

Figure 1: a) schematic reperesentation of the spherulite (HOFFMAN et al., [14])
b) composite inclusion.

	slip system	normalized resistance
Chain slip	(100)[001]	1.
	(010)[001]	1.
	{110}[001]	1.
Transverse slip	(100)[010]	1.1
	(010)[100]	1.4
	$\{110\} < 1\bar{1}0 >$	1.8

Table 1: Slip systems of polyethylene and estimates of their corresponding normalized initial critical resolved shear stresses.

In this study, we consider plasticity by slip. The available slip systems in the othorhombic unit cell, within the crystalline phase of polyethylene, are summarized in Table 1. Estimates of initial deformation resistances of these systems are normalized to that of chain slip and are also shown in Table 1.

Each of the K slip systems, in a given crystalline lamella, is characterized by the couple ($\mathbf{s}^\alpha$,$\mathbf{n}^\alpha$) where $\mathbf{s}^\alpha$ and $\mathbf{n}^\alpha$ are the unit vectors representing the slip direction and slip plane normal of the slip system α, respectively. The slip rate $\dot{\gamma}^\alpha$ of each slip system can be related to its conjugate stress measure τ^α (resolved shear stress) via a power law relation (Hutchinson [7]). If we denote by g^α the shear strength associated with the system α and by $\dot{\gamma}_0$ a reference shear rate identical to all slip systems, the power law relation can be expressed as follows:

$$\dot{\gamma}^\alpha = \dot{\gamma}_0 \frac{\tau^\alpha}{g^\alpha} \left|\frac{\tau^\alpha}{g^\alpha}\right|^{n-1}, \tag{1}$$

where n is the inverse rate sensitivity coefficient. g^α may evolve with strain hardening.

When elasticity is neglected, the strain rate $\mathbf{D}^c$ within the crystalline lamella is generated by the shear rates of all slip systems. According to relation (1), the shear rates of non-active systems are negligible. Due to chain inextensibility in the crystalline lamella, less than five independent slip systems are available in crystalline polymers. Slip systems of polyethylene, given by Table 1, comprise four independent systems. Therefore, the missing degree of freedom can be expressed as a constraint on the the strain rate:

$$\mathbf{C}' \cdot \mathbf{D}^c = 0. \tag{2}$$

$\mathbf{C}' \equiv \mathbf{c} \otimes \mathbf{c} - \frac{1}{3}\mathbf{1}$ is the deviatoric part of the dyadic $\mathbf{c} \otimes \mathbf{c}$, with $\mathbf{c}$ representing the the crystallographic unit vector in the chain direction (constrained direction), and $\mathbf{1}$ is the second order identity tensor.

Let us denote by $\mathbf{S}^c$ the deviatoric Cauchy stress within the crystalline lamella. This traceless tensor can be decomposed as a sum of a four-dimensional stress $\mathbf{S}^{c,*}$ perpendicular to $\mathbf{C}'$ and the fifth component S^c_C in the constrained direction $\mathbf{C}'$. Parks and Ahzi [8] show that the resolved shear stress τ^α, on the slip system α, is independent on the stress component S^c_C:

$$\tau^\alpha = \mathbf{S}^{c,*} \cdot \mathbf{R}^\alpha. \tag{3}$$

where $\mathbf{R}^\alpha = \frac{1}{2}\{\mathbf{s}^\alpha \otimes \mathbf{n}^\alpha + \mathbf{n}^\alpha \otimes \mathbf{s}^\alpha\}$ is the symmetric part of Schmid tensor associated with slip system α.

A constitutive law for the constrained single crystal is obtained by expressing the strain rate $\mathbf{D}^c$ as a sum of slip system contributions to the rate of plastic shearing ($\mathbf{D}^c = \sum_{\alpha=1}^{K} \dot{\gamma}^\alpha \mathbf{R}^\alpha$). By the use of (1) and (3), this constitutive law can be expressed as follows:

$$\mathbf{D}^c = \left\{ \dot{\gamma}_0 \sum_{\alpha=1}^{K} \frac{1}{g^\alpha} \left(|\frac{\mathbf{S}^{c,*} \cdot \mathbf{R}^\alpha}{g^\alpha}| \right)^{n-1} \mathbf{R}^\alpha \otimes \mathbf{R}^\alpha \right\} [\mathbf{S}^{c,*}]. \tag{4}$$

Further detail on constrained crystal plasticity may be found in the work of Parks and Ahzi [8].

The spin tensor $\mathbf{W}^c$ of the lamella can be decomposed as a sum of plastic spin $\mathbf{W}^p$ and lattice spin $\mathbf{W}^*$. The lattice spin controls the rate of change of crystallographic axes; it is expressed as:

$$\begin{aligned} \mathbf{W}^* &= \mathbf{W}^c - \mathbf{W}^p \\ &= \mathbf{W}^c - \sum_{\alpha=1}^{N} \dot{\gamma}^\alpha \mathbf{A}^\alpha, \end{aligned} \tag{5}$$

where $\mathbf{A}^\alpha = \frac{1}{2}(\mathbf{s}^\alpha \otimes \mathbf{n}^\alpha - \mathbf{n}^\alpha \otimes \mathbf{s}^\alpha)$ is the skew part of the Schmid tensor associated with the slip system α. We note that $\mathbf{D}^c$ and $\mathbf{W}^c$ are the symmetric part and skew part, respectively, of the velocity gradient $\mathbf{L}^c$: $\mathbf{L}^c = \mathbf{D}^c + \mathbf{W}^c$.

In the constrained hybrid (CH) model that Parks and Ahzi [8] applied to an idealized 100% crystalline polyethylene, the stress component S^c_C in the constrained direction was estimated to equal the corresponding macroscopic stress component. This estimation is not needed for the current composite model since the local equilibrium provides, in general, a closed form solution for this stress component (Parks and Ahzi [9]).

2.2 Amorphous phase

The amorphous phase of polymers is a disordered macromolecular solid. The plastic deformation in this phase occurs by thermally activated molecular segment rotations (Argon [5]). Under plastic straining, molecular alignment (amorphous texturing) develops within this phase. Boyce, Parks and Argon [6] introduced a back stress tensor in the flow rule to account for the plastic resistance increase accompanying the molecular alignment.

To model the plastic deformation within the amorphous domains of semi-crystalline polymers, we assume a power law relation between the shear rate $\dot{\gamma}^a$ and the corresponding shear stress of each amorphous domain:

$$\dot{\gamma}^a = \dot{\gamma}_0 \, \frac{\tau^a}{a\tau_0} \left(\frac{\tau^a}{a\tau_0}\right)^{n-1}, \tag{6}$$

where $\dot{\gamma}_0$ and τ_0 are reference strain rate and reference stress respectively and n is the rate exponent. These material parameters can be chosen equal to those of the crystalline lamella, with $\tau_0 = g^{[001]}$ representing the initial shear strength on the easiest slip system (chain slip). The coefficient a characterizes the relative softness of the amorphous domain. For an amorphous phase with flow strength of the same order to that of the crystalline one (glassy amorphous), the coefficient a is expected to be of the order of unity. In the other hand, for an amorphous phase with a flow strength smaller than the crystalline one (rubbery amorphous), the coefficient a should be considerably smaller than unity.

Following the work of Boyce, Parks and Argon [6], we account for molecular alignment effect, in the amorphous domain, by introducing a back stress tensor $\mathbf{B}^a$ in the flow rule. If we denote by $\mathbf{S}^a$ the deviatoric Cauchy stress, the driving stress within the amorphous domain is defined as: $\mathbf{S}^a - \mathbf{B}^a$. The resolved shear stress τ^a is defined as a norm of the driving stress. Introducing (5) in the flow rule leads to the following three dimensional constitutive law for the amorphous domain:

$$\mathbf{D}^a = \dot{\gamma}_0 \left(|\frac{\mathbf{S}^a - \mathbf{B}^a}{a\tau_0}| \right)^{n-1} \left(\frac{\mathbf{S}^a - \mathbf{B}^a}{a\tau_0} \right). \tag{7}$$

The back stress principal components B_i^a evolve with the principal components V_i^p of the plastic stretch tensor according to the following relation (Boyce, Parks and Argon [6]):

$$B_i^a = C^R \frac{\sqrt{N}}{3} \left[V_i^p \mathcal{L}^{-1} \left(\frac{V_i^p}{\sqrt{N}} \right) - \frac{1}{3} \sum_{j=1}^{3} V_j^p \mathcal{L}^{-1} \left(\frac{V_j^p}{\sqrt{N}} \right) \right], \tag{8}$$

where N is the number of rigid links between entanglements and is equal to the tensile locking stretch squared ($N = \lambda_L^2$). C^R is approximately the rubbery modulus and $\mathcal{L}$ is the Langevin function defined by $\mathcal{L}(\beta_i) = coth(\beta_i) - 1/\beta_i = V_i^p/\sqrt{N}$. We note that the plastic stretch tensor $\mathbf{V}^p \equiv \mathbf{V}$ (elasticity neglected) is obtained from the deformation gradient tensor $\mathbf{F}^a$ of the amorphous phase: $\mathbf{V}^p \equiv \mathbf{V} = (\mathbf{F}^a \mathbf{F}^{aT})^{\frac{1}{2}}$. Figure 2 shows simulated stress strain curves for tension and compression of an amorphous phase deformed homogeneously according to relation (7). Further detail on this constitutive law is given by Parks and Ahzi [9].

3. Kinematics and Equilibrium within the Composite Inclusion

The basic element that constitutes the spherulites of semi-crystalline polymers consists of a thin, broad composite inclusion of crystalline lamella an amorphous layer (Figure 1). These composite inclusions are modeled as infinitely extended 'sandwiches' with planar crystalline-amorphous interfaces. In what follows, we denote by $\mathbf{n}^I$ the inclusion normal and by ψ^I the angle between $\mathbf{n}^I$ and the crystallographic chain direction $\mathbf{c}$.

3.1 Inclusion volume-averaging

Let us denote by $\mathbf{L}^I$, $\mathbf{L}^c$ and $\mathbf{L}^a$ the average velocity gradients of the inclusion, crystalline lamella and amorphous layer, respectively. The inclusion-average velocity gradient can be expressed as:

$$\mathbf{L}^I = f_a\, \mathbf{L}^a + (1 - f_a)\, \mathbf{L}^c, \tag{9}$$

where f_a represents the relative thickness of the amorphous layer (amorphous volume fraction). Similarly, the inclusion-average stresses can be written as:

$$\mathbf{S}^I = f_a\, \mathbf{S}^a + (1 - f_a)\, \mathbf{S}^c, \tag{10}$$

$$p^I = f_a\, p^a + (1 - f_a)\, p^c, \tag{11}$$

where $\mathbf{S}^I$ is the average deviatoric stress within the composite inclusion. p^I, p^c and p^a are the pressure terms of the inclusion, crystalline lamella and amorphous layer respectively.

3.2 Jump conditions

The first jump condition demands continuous velocities across the lamella-amorphous interface. If we denote by ε the permutation tensor and by ${n_p}^I$ the components of the interface normal $\mathbf{n}^I$, the continuity conditions are given by the following relation:

$$\varepsilon_{ipq} \Delta L_{jq} {n_q}^I = \varepsilon_{ipq}(L^c_{jq} - L^a_{jq}){n_q}^I = 0. \tag{12}$$

The second jump condition consists of traction equilibrium across the interface. It is expressed as:

$$(\mathbf{T}^c - \mathbf{T}^a)\, \mathbf{n}^I = (\mathbf{S}^c - \mathbf{S}^a)\, \mathbf{n}^I - (p^c - p^a)\, \mathbf{1}\, \mathbf{n}^I = \mathbf{0}. \tag{13}$$

$\mathbf{T}^c$ and $\mathbf{T}^a$ are the Cauchy stresses in the crystalline and amorphous phases of the inclusion, respectively. Since incompressibility is assumed and pressure sensitivity is neglected, only the two equations of (13) not involving pressure terms are considered. The third equilibrium equation is assumed.

To solve equilibrium equations for local kinematics, Parks and Ahzi [9] considered three different cases: 1) no constraint in either of the two phases, 2) one constraint in the chain direction of the lamella, and 3) two constraints in the lamella. Here we consider only the case of one constraint, which applies to

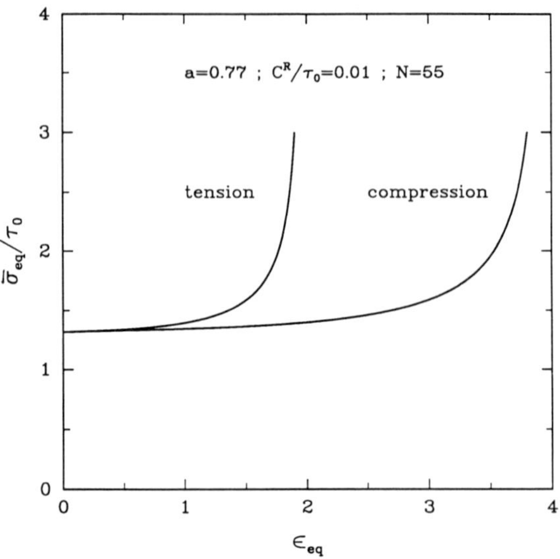

Figure 2: stress–strain response of the amorphous phase in tension and compresion.

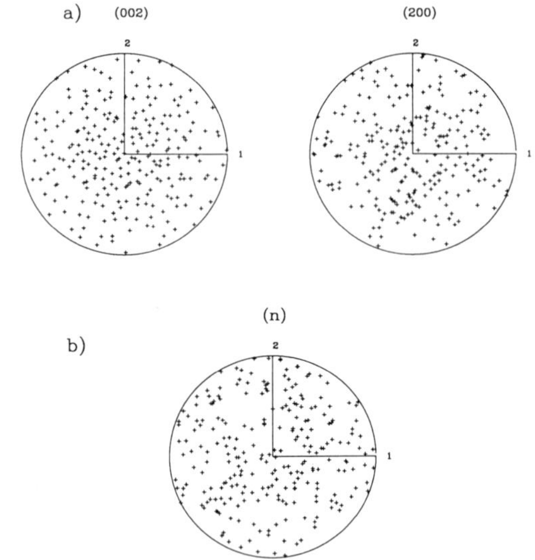

Figure 3: a) (002) and (200) pole figures of the initial isotropic texture (244 orientations); b) isotropic distribution of the lamellae normals.

polyethylene crystals, and assume the most general situation in which the lamella normal $\mathbf{n}^I$ is not parallel to the chain direction $\mathbf{c}$ ($\psi^I \neq 0, \pi$). To prescribe the inclusion velocity gradient $\mathbf{L}^I$, a localization law should be imposed on the composite inclusion in the manner of existing models (Taylor, Constrained Hybrid, Self-Consistent, Sachs, ...).

4. Localization Law and Texture Update

4.1 Localization law

The simplest an the mostly used localization law is known in crystal plasticity as the Taylor model. This model is based on the suggestion of Taylor [10] that, in a pollycrystalline aggregate, local deformation is approximately uniform and equal to the macroscopic uniform one imposed on the aggregate. When the representative element of the aggregate is a single crystal, five independent slip systems are needed to accommodate this uniform deformation. For crystals lacking five independent systems, Parks and Ahzi [8] proposed a modification of Taylor model and applied it to several constrained crystals. In the present study, the representative element of the aggregate (semi-crystalline polymer) is a two-phase composite inclusion comprising a crystalline lamella lacking five independent slip systems. When $\mathbf{n}^I$ is parallel to $\mathbf{c}$ ($\psi^I = 0$), the entire inclusion I is constrained in $\mathbf{n}^I$ direction.

The most general case, for initially spherulitic polyethylene, is that the normals $\mathbf{n}^I$ and their corresponding chain axes $\mathbf{c}$ do not coincide. The angle ψ^I between these two directions has been measured by several authors (Basset and Hodge [11]; Varnell, et al., [12]). The reported mean value varies from 17° to 40°. However, a general deformation can be accommodated by the composite inclusion, and the Taylor model (strain rate continuity) can be applied. On the other hand, under plastic straining the angle ψ^I may decrease for some inclusions and become close to zero. When this occurs, due to both the non-deformability of the lamella in the $\mathbf{c}$ direction and incompressibility, the deformation within the entire inclusion in $\mathbf{c}$ or $\mathbf{n}^I$ directions will also decrease, and the stresses predicted by the Taylor model become unreasonably high. The applicability of the Taylor model then becomes an issue for all inclusions having a small angle between $\mathbf{n}^I$ and $\mathbf{c}$ (quasi-constrained inclusions). To solve this numerical problem, we propose a phenomenological modification of the Taylor model to be applied to the quasi-constrained inclusions ($0 < \psi^I < \psi_0$), where ψ_0 is a suitably small angle. This modification is a local relaxation of the imposed strain rate component in the $\mathbf{n}^I$ direction without any loss in global compatibility. The relaxation will depend on the angle ψ^I . The stain rate $\mathbf{D}^I$ within each inclusion I, can then be related to the macroscopic strain rate $\bar{\mathbf{D}}$ by the following relations:

$$\mathbf{D^I} = \mathcal{P} < \mathcal{P} >^{-1} \bar{\mathbf{D}} \quad for\ \psi^I < \psi_0,$$
$$\mathbf{D^I} = \bar{\mathbf{D}} \quad for\ \psi^I \geq \psi_0; \tag{14}$$

where

$$\mathcal{P} = \mathcal{I} - \frac{3}{2}(1 - (\psi^I/\psi_0)^m)\mathbf{N}' \otimes \mathbf{N}'. \tag{15}$$

$\mathcal{I}$ is the fourth order identity tensor and $\mathbf{N}' = \mathbf{n}^I \otimes \mathbf{n}^I - \frac{1}{3}\mathbf{1}$ is a tensorial representation of the lamella normal direction. ψ_0 is a critical angle, m is an arbitrary power coefficient and $< . >$ designates the volume average over all quasi-constrained inclusions. The main effect of this local relaxation is to limit strain rate in the $\mathbf{n}^I$ and $\mathbf{c}$ directions in those inclusions with small ψ^I values; in turn, much more modest stress levels are generated within each inclusion. When the shape effect is neglected, the inclusion spin $\mathbf{W}^I$ can approximately be equated to the macroscopic spin $\bar{\mathbf{W}}$:

$$\mathbf{W}^I = \bar{\mathbf{W}}. \tag{16}$$

4.2 Texture update

The rate of change of lattice orientation is given by the spin $\mathbf{W}^*$ defined by (5). Following Asaro and Rice [14], the rate of change of the crystallographic axes, for instance chain axis $\mathbf{c}$, can be expressed as:

$$\dot{\mathbf{c}} = \mathbf{W}^* \mathbf{c}. \tag{17}$$

Since the distribution of lamellae normals (morphological texture) is, in general, noncoincident with that of any particular crystallographic direction within the crystalline regions, it is important to update the lamella normals independently. For this, we use the simple concept based on convected material coordinates. Let $\delta\mathbf{x}_1$ and $\delta\mathbf{x}_2$ be two infinitesimal vectors in the reference inclusion plane at time "0", emanating from a common origin. At generic time t, when deformation gradient is $\mathbf{F}(t)$, these vectors are convected to $\delta\mathbf{h}_1 = \mathbf{F}\delta\mathbf{x}_1$ and $\delta\mathbf{h}_2 = \mathbf{F}\delta\mathbf{x}_2$, respectively. The crystal-amorphous interface is material, however, so the interface normal at time $t = 0$ is

$$\mathbf{n}^I(0) = \frac{\delta\mathbf{x}_1 \times \delta\mathbf{x}_2}{|\delta\mathbf{x}_1 \times \delta\mathbf{x}_2|}, \tag{18}$$

while at time t, it is

$$\mathbf{n}^I(t) = \frac{\mathbf{F}\delta\mathbf{x}_1 \times \mathbf{F}\delta\mathbf{x}_2}{|\mathbf{F}\delta\mathbf{x}_1 \times \mathbf{F}\delta\mathbf{x}_2|}. \tag{19}$$

We update $\mathbf{F}$ for each inclusion according to $\dot{\mathbf{F}} = \mathbf{LF}$ or, in incremental form, $\mathbf{F}(t + \Delta t) = \exp(\mathbf{L}\Delta t)\mathbf{F}(t)$. Although $\mathbf{L}^a \neq \mathbf{L}^c$, so that $\mathbf{F}^c \neq \mathbf{F}^a$ in general, the compatibility requirements on $\mathbf{L}^a$ and $\mathbf{L}^c$ ensure that the normal $\mathbf{n}^I(t)$, as calculated using either $\mathbf{F}^a$ or $\mathbf{F}^c$, will be identical. However, an updated Lagrangian

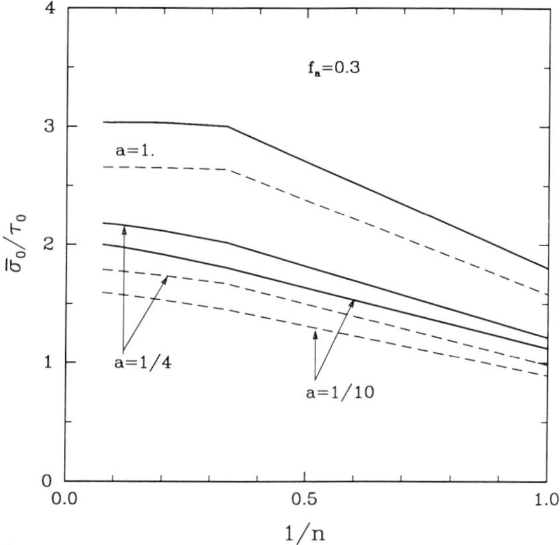

Figure 4: Normalized tensile reference stress.
Solid lines for slip system resistances in Table 1;
dashed lines for slip systems with the same resistance.

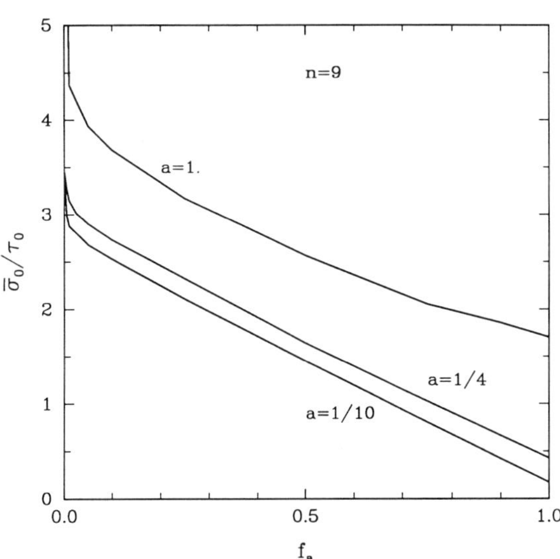

Figure 5: Normalized tensile reference stress versus f_a.

scheme (current configuration $\equiv$ reference configuration) is convenient for updating $\mathbf{n}^I$. For this, we use $\mathbf{F}^c$ with $\mathbf{F}^c(t+\Delta t) = \exp(\mathbf{L}^c\Delta t)$. The molecular alignment (amorphous phase texturing) is accounted for by the back stress $\mathbf{B}^a$. To update $\mathbf{B}^a$, it is necessary to use the Lagrangian scheme (initial configuration $\equiv$ reference configuration). This can be obtained by using $\mathbf{F}^a(t+\Delta t) = \exp(\mathbf{L}^a\Delta t)\mathbf{F}^a(t)$, with $\mathbf{F}^a(t)$ the cumulative deformation gradient from time '0' to time 't'.

5. Results and Conclusions

5.1 Results

We applied the present composite model to an initially isotropic (quasi-spherulitic) high density polyethylene (HDPE) with $f_a = 0.3$. The initial distribution of chain axes $\mathbf{c}$ is shown by Figure 3a for 244 orientations. A random distribution of lamellae normals $\mathbf{n}^I$, shown by Figure 3b, is generated by assuming an initial angel ψ^I of about 30° for all inclusions. The values of $m = 5$ and $\psi_0 = 15^o$ have been chosen in the projection operator $\mathcal{P}$ defined by (15). No crystallographic strain hardening is considered in the following applications. Our first application consists of predicting the uniaxial reference stress $\bar{\sigma}_0$ in the following power law creep equation:

$$\bar{D}^{eq} = \dot{\gamma}_0(\bar{\sigma}^{eq}/\bar{\sigma}_0)^n, \tag{20}$$

where $\bar{D}^{eq} = \sqrt{\frac{2}{3}\bar{\mathbf{D}} : \bar{\mathbf{D}}}$ and $\bar{\sigma}^{eq} = \sqrt{\frac{3}{2}\bar{\mathbf{S}} : \bar{\mathbf{S}}}$ are the macroscopic equivalent uniaxial tensile strain rate and the equivalent uniaxial tensile stress, respectively. Figure 4 shows the normalized tensile reference stress as function of the rate sensitivity coefficient $1/n$ for $a = 1$, $a = 1/4$ and $a = 1/10$. Also shown in Figure 4 for each value of a is an indication of the sensitivity of model results to the relative strengths of the slip systems. The lower dashed curves for each case represent an assumption of equal slip system resistances. The influence of the amorphous phase volume fraction on the tensile reference stress is shown in Figure 5. For the extreme cases of $f_a = 0$ and $f_a = 1$, the tensile reference stresses are calculated by the 100% crystalline model (CH model) and by 100% amorphous model (equation (7)), respectively. The singularity shown by figure 5, for $a = 1$ when f_a goes to zero, is due to high stresses predicted by Taylor-like localization law. Whereas, for soft amorphous domains ($a = 1/4$; $a = 1/10$), more modest stresses are generated within the amorphous domains even for very small amorphous volume fraction. As a second application, a uniaxial tension test is simulated for HDPE. The predicted stress-strain curve is compared to data of G'Sell, et al., [1,2], in Figure 6. The predicted texture (crystallographic and lamellae normals distributions) are shown by pole figures in Figure 7. The final application corresponds to simulation of a simple shear test. For the same parameters chosen to fit the tensile stress-strain curve to experimental data, the predicted macroscopic behavior in simple shear is again compared to the data of G'Sell, et al., [2], (Figure 8). The

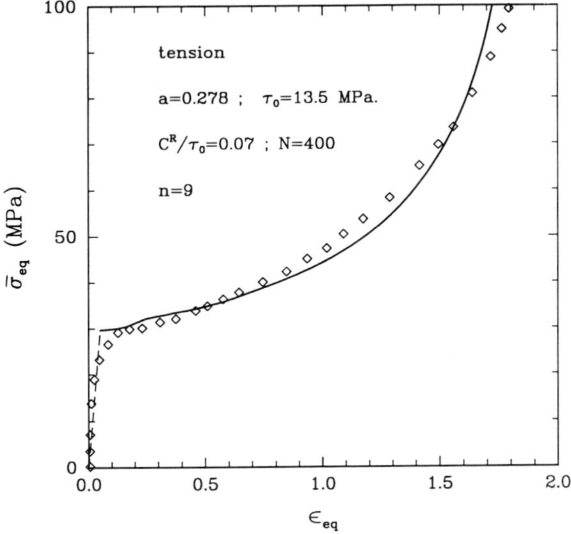

Figure 6: Predicted stress–strain behavior (solid line) for tension in comparisons to G'Sell's data.

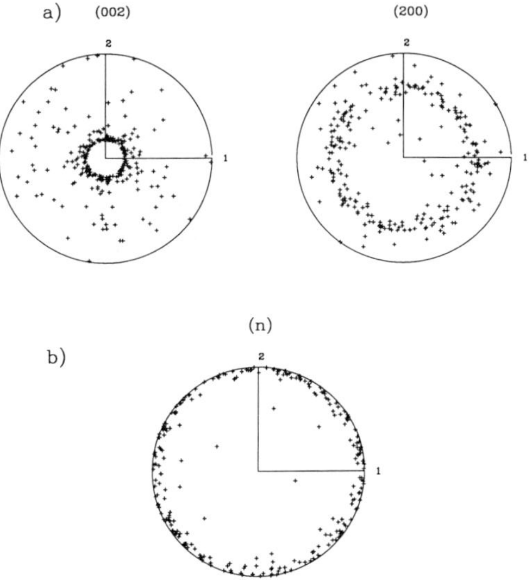

Figure 7: a) Predicted (002) and (200) pole figures for tension at ϵ_{eq}=180% (tension axis = 3)

b) Distribution of the lamellae normals (ϵ_{eq}=180%).

predicted textures in simple shear are represented in Figure 9. We note that the difference in the values of τ_0 chosen the fit tension and simple shear curves is due to the results reported by G'Sell, et al., [2]. In an unpublished work of G'Sell, this difference was corrected and in fact the the von-Mises yield stresses in tension and simple shear coincide.

The influence of texture can be seen by the hardening difference in tension and simple shear. The fiber crystallographic texture developed in tension in conjunction with molecular alignment induces high stresses because of the lack of deformation mechanisms at large plastic straining. On the other hand, for simple shear, the texture development evolves so as to continue to permit easy mechanisms (chain slip) to operate. Therefore, negligible textural hardening is involved. The discrepancy between our predictions and experimental data for simple shear (Figure 8) is probably due to the upper bound nature of the Taylor model that we apply to the composite inclusions.

5.2 Conclusions

We developed a micromechanically-based composite model for semi-crystalline polymers which accounts for the co-existence of both amorphous and crystalline domains. Constitutive equations accounting for kinematic deficiency in the crystalline phase are given. The molecular alignment and the relative softness of the amorphous phase are included in a proposed power law constitutive relation for this phase. Results of a Taylor-type localization law applied to the sandwich inclusions of HDPE are encouraging and satisfactory. Other applications of this model are under investigation (Parks and Ahzi [9]).

ACKNOWLEGMENT

This work was supported by the DARPA U.R.I. program under ONR contract No. N00014-86-K-0768. We are pleased to acknowledge discussions with Professors A. S. Argon, R. E. Cohen and C. G'Sell.

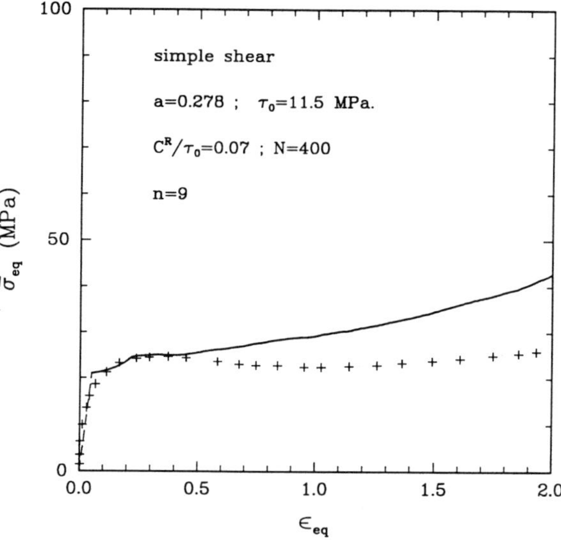

Figure 8: Predicted stress–strain behavior (solid line) for simlpe shear in comparison to G'Sell's data.

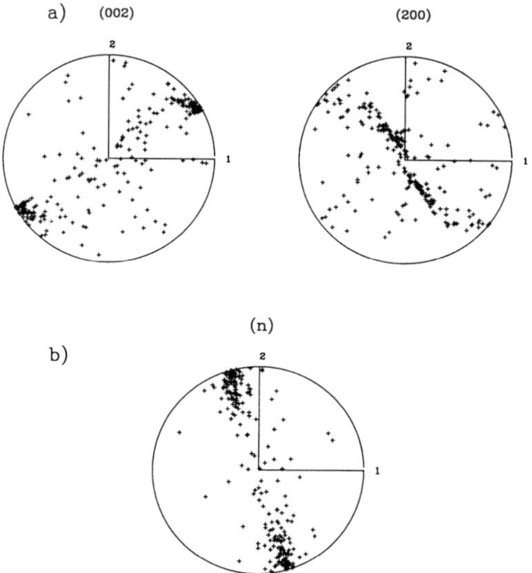

Figure 9: a) (002) and (200) pole figures of the for simple shear at γ=3.46

b) Distribution of the lamellae normals (γ=3.46).

REFERENCES

[1] G'Sell, C. and Jonas, J. J., *J. Mat. Sci.*, **14**, 583, (1979).
[2] G'Sell, C., Boni, S. and Shrivastava, S., *J. Mat. Sci.*, **18**, 903, (1983).
[3] Bowden, P. B. and Young, R. J., *J. Mat. Sci.* **9**, 2034, (1974).
[4] Haudin, J. M., *Plastic Deformation of Amorphous and Semi–Crystalline Materials*, p. 291, Eds. B. ESCAIG and C. G'SELL Les Editions de Physique, Les Ulis, (1982).
[5] Argon, A. S., *Phil. Mag.*, **28**, 39, (1973).
[6] Boyce, M. C., Parks, D. M. and Argon, A. S., *Mech. Materials*, **7**, 17, (1988)
[7] Hutchinson, J. W., *Proc. R. Soc. Lond.* **A348**, 101, (1976).
[8] Parks, D. M. and Ahzi, S., *J. Mech. Phys. Solids*, in press, (1990).
[9] Parks, D. M. and Ahzi, S., ms. in preparation, (1990).
[10] Taylor, G.I., *J. Inst. Metals* **62**, 307, (1938).
[11] Bassett, D. C. and Hodge, A. M., *Proc. Roy. Soc. London,* **A377**, 25, (1981)
[12] Varnell, W. D., Ryba, E., *J. Macromol. Sc.* -Phys., **B26**, (2), 135, (1987)
[13] Asaro, R. J. and Rice, J. R., *J. Mech. Phys. Solids*, **25**, 309, (1977).
[14] Hoffman, J. D., Davis, G. T. and Laurizten J. I., *Treatise on Solid State Chemistry* , **3**, Edited by HANNAY N. B., Plenum, N.Y., p.497, (1976).

A Unified Formulation of Micromechanics Models of Fiber-Reinforced Composites

J. L. Teply* and J. N. Reddy**

1. Introduction

In a micromechanics constitutive model the overall instantaneous properties of fibrous composites are defined by relations between overall stress and strain averages. Such averaging techniques are also known as 'homogenization'. The models may account for fiber, matrix and fiber-matrix interface properties and their interactions (see [1-5]). Various micromechanics approaches are used to calculate overall stress and strain fields using different representative micro-geometries (i.e. unit cells); see, for example, the self-consistent method of Hill [1], the variational formulation of Hashin [2], the vanishing fiber diameter model of Dvorak and Bahei-El-Din [3], periodic rectangular array of Aboudi [4], and the periodic hexagonal array (PHA) model of Teply and Dvorak [5]. Although various models use different representative geometries of the unit cell and different approximations of the displacements and/or stresses to obtain the overall properties, they all share certain common basis. The present paper has the objectives of finding the common basis so that the models can be related to each other.

The finite element formulations based on the principle of virtual displacements, the principle of virtual forces or a mixed variational principle are used to develop the unified formulation of micromechanics models. In order to develop the unified formulation for the overall averages of a model, the local interpolation of displacement and stress fields are developed such that the local fields of the model are matched. The interface-continuity of displacements and tractions and overall boundary conditions are enforced either point-wise or in a variational (i.e. integral) sense, as required by the original model. The relationship between the overall stiffness properties of the unified model and the original model is usually established via the comparison of unit cell energies. The unified finite-element formulation enables

*Senior Technical Specialist, Alcoa Laboratories, Alcoa Center, PA 15069.
**Clifton C. Garvin Professor, Department of Engineering Science and Mechanics, Virginia Polytechnic Institute, Blacksburg, VA 24061.

us to study relative convergence, accuracy characteristics and the effect of the approximation of fiber geometry and boundary conditions on the solution. The effect of the unit cell geometry can also be evaluated. The micromechanics models of Aboudi [4] and Teply and Dvorak [5] are considered in the present study for detailed discussion.

The model of Teply and Dvorak [5] is already developed in the form of a finite element model. The Aboudi model [4], which is presented as a boundary-value problem of elasticity, is reformulated using the finite element formulation with independent approximations of displacements and stresses. It is shown that linear approximation of the displacements and piece-wise constant (i.e., constant within each subcell) approximation of stresses naturally results in Aboudi's original model.

2. Formulation

The microstructure of Aboudi's model consists of square fibers surrounded by the square cylinders of the matrix material (see Figure 1). It is assumed that a volume Ω with boundary Γ is filled by a large number of square fibers. It means that the smallest typical dimension of the volume, h_v, is much larger than max $(h_1, h_2, \ell_1, \ell_2)$. There are several ways to fill the volume by a periodic microstructure. Aboudi selected a rectangular unit cell which consists of a square or rectangular fiber at one of the unit cell vertices and three rectangular cylinders of the matrix material at the other three vertices. To fill the volume, a simple translational transformation is applied to the unit cell. The boundary Γ is approximated as shown in Figure 2.

The Hellinger-Riessner variational principle for the entire volume Ω can be written as follows (see Reddy [6]):

$$\Pi = \int_{\Omega} \sigma_{ij} \frac{1}{2} (u_{i,j} + u_{j,i}) d\Omega - \frac{1}{2} \int_{\Omega} \sigma_{ij} M_{ijk\ell} \sigma_{k\ell} d\Omega - \int_{\Gamma_t} \overset{\star}{t}_i u_i d\Gamma - \int_{\Gamma_u} \sigma_{ij} n_j \overset{\star}{u}_i d\Gamma \quad (1)$$

where σ_{ij} and u_i are the stress and displacement components, $M_{ijk\ell}$ are the components of the tensor of material compliances, $\overset{\star}{t}_i$ are the components of the boundary traction vector defined on Γ_t, $u_i^\star$ are the components of the specified displacement vector on Γ_u, and $\Gamma = \Gamma_t + \Gamma_u$. The tensor components $M_{ijk\ell}$ are piecewise constant functions with discontinuities along the fiber/matrix boundaries. In what follows, it will be shown that Aboudi's displacement and equilibrium equations can

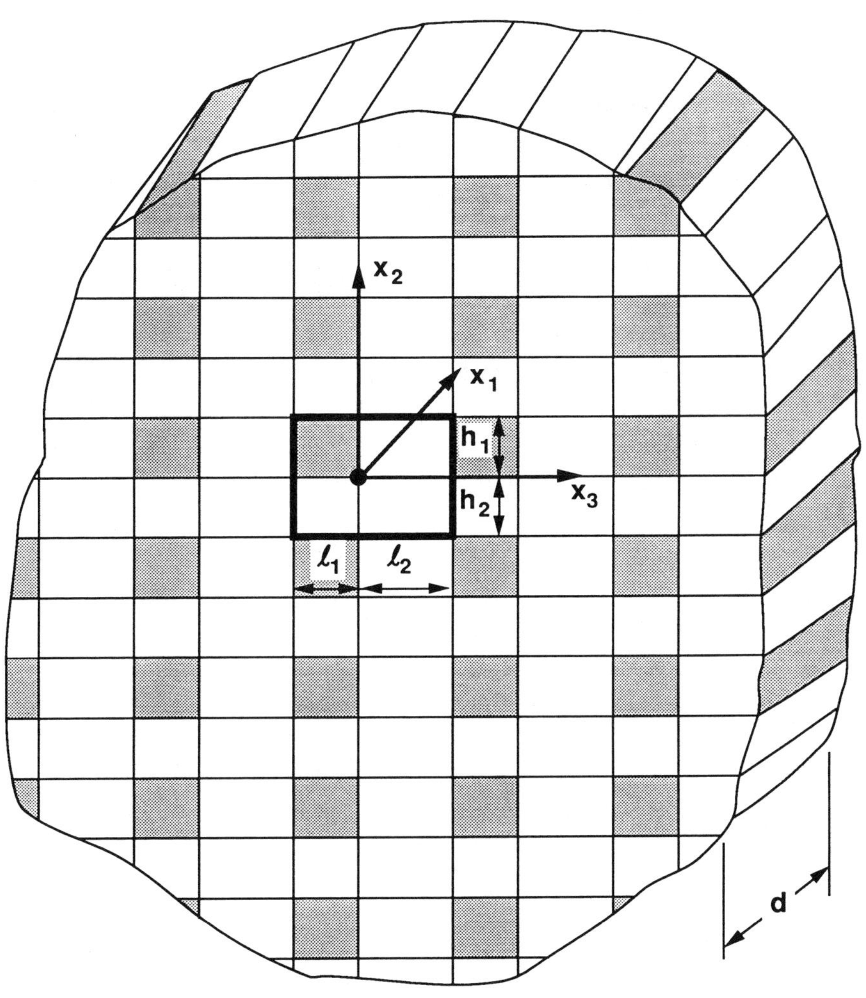

Periodic Microstructure of Aboudi
Figure 1

be derived when piecewise linear and piecewise constant finite-element interpolation functions are applied to the two independent fields u_i and σ_{ij}, respectively, and Γ_u is equal to zero. Of course, the interpolation functions must be periodic in Ω with the periodicity interval $(h_1 + h_2)$ in the x_2-direction and $(\ell_1 + \ell_2)$ in the x_3-direction.

In order to facilitate our proof, we shall use Aboudi's notation for the unit cell. The unit cell subcells and subcell coordinate systems are shown in Figure 3. In the present paper the subcells will be sometimes referred to as the fiber or matrix finite elements. The unit cell has length d in the x_1-direction. Length d is arbitrary because the fiber and matrix deformation in the x_1-direction is assumed to be uniform. Due to this uniform deformation no local $x_1^{(\beta\gamma)}$ need to be considered in the subcells.

The finite element interpolation of u_i, which is equivalent to Aboudi's displacement field, is of the form

$$u_i^{(\beta\gamma)} = \sum_j^4 u_{ij}^{(\beta\gamma)} \theta_j^{(\beta\gamma)} \quad i = 1,2,3 \tag{2}$$

where $u_{ij}^{(\beta\gamma)}$ denotes the i-th displacement component at j-th node, $\theta_j^{(\beta\gamma)}$ are the linear Lagrange interpolation functions [7], and $u_i^{(\beta\gamma)}$ are the displacement component of an arbitrary subcell $(\beta\gamma)$. In view of the displacement continuity across subcell boundaries, the nodal values $u_{ij}^{(\beta\gamma)}$ satisfy the following relations:

$$\begin{aligned} u_{i2}^{(11)} &= u_{i3}^{(21)} \quad , \quad u_{i3}^{(11)} = u_{i2}^{(21)} \\ u_{i2}^{(12)} &= u_{i3}^{(22)} \quad , \quad u_{i3}^{(12)} = u_{i2}^{(22)} \end{aligned} \tag{3}$$

and

$$\begin{aligned} u_{i1}^{(11)} &= u_{i2}^{(12)} + u_{i3}^{(12)} - u_{i1}^{(12)} \\ u_{i1}^{(12)} &= u_{i2}^{(11)} + u_{i3}^{(11)} - u_{i1}^{(11)} \\ u_{i1}^{(21)} &= u_{i2}^{(22)} + u_{i3}^{(22)} - u_{i1}^{(22)} \\ u_{i1}^{(22)} &= u_{i2}^{(21)} + u_{i3}^{(21)} - u_{i1}^{(21)} \end{aligned} \tag{4}$$

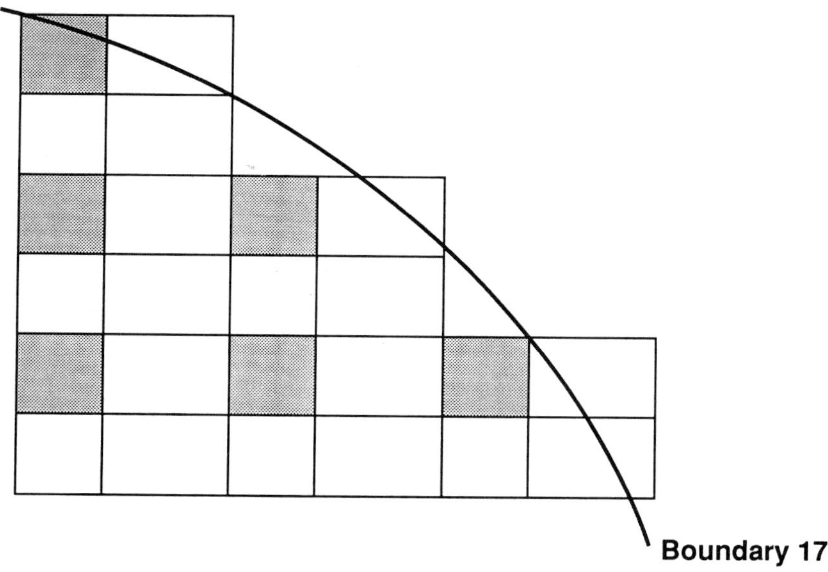

Approximation of Boundary by Unit Cells
Figure 2

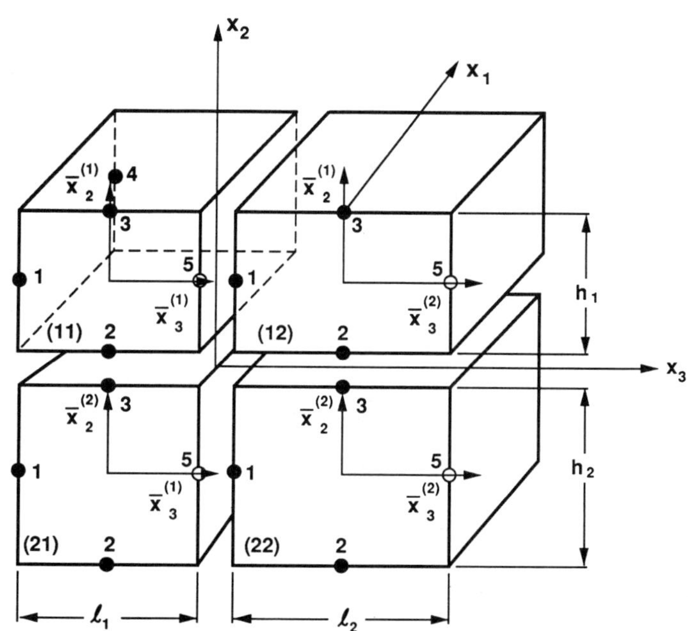

Aboudi's Notation in Finite Element Concept
Figure 3

Note that continuity conditions (3) are automatically introduced during the finite-element assembly procedure (see [7]) whereas equations (4) are not a part of the assembly procedure and must be introduced via linear combinations for unknown $u_{ij}^{(\beta\gamma)}$ $(j = 1,2,3)$; u_{i4} in equation 2 corresponds to the nodal value at distance d above nodal displacement u_{i1} in the x_1-direction. The shape functions $\theta_i^{(\beta\gamma)}$ are equal to

$$\theta_1^{(\beta\gamma)} = -\frac{2}{\ell_j}\bar{x}_3^{(\gamma)}$$

$$\theta_2^{(\beta\gamma)} = \frac{1}{2} - \frac{1}{2d}x_1 - \frac{1}{h_\beta}\bar{x}_2^{(\beta)} + \frac{1}{\ell_\gamma}\bar{x}_3^{(\gamma)}$$

$$\theta_3^{(\beta\gamma)} = \frac{1}{2} - \frac{1}{2d}x_1 + \frac{1}{h_\beta}\bar{x}_2^{(\beta)} + \frac{1}{\ell_\gamma}\bar{x}_3^{(\gamma)} \tag{5}$$

$$\theta_4^{(\beta\gamma)} = \frac{1}{d}x_1$$

We define functions $W_i^{(\beta\gamma)}(x_1)$, $\Phi_i^{(\beta\gamma)}(x_1)$ and $\Psi_i^{(\beta\gamma)}(x_1)$ by

$$W_i^{(\beta\gamma)} = \frac{1}{2}(u_{i2}^{(\beta\gamma)} + u_{i3}^{(\beta\gamma)}) + \frac{1}{d}[u_{i4}^{(\beta\gamma)} - \frac{1}{2}(u_{i2}^{(\beta\gamma)} + u_{i3}^{(\beta\gamma)})]x_1$$

$$\Phi_i^{(\beta\gamma)} = \frac{1}{h_\beta}(u_{i3}^{(\beta\gamma)} - u_{i2}^{(\beta\gamma)}) \tag{6}$$

$$\psi_i^{(\beta\gamma)} = \frac{1}{\ell_\gamma}(u_{i3}^{(\beta\gamma)} + u_{i2}^{(\beta\gamma)} - 2u_{i1}^{(\beta\gamma)})$$

Then Eq. (2) can be cast in the following form

$$u_i^{(\beta\gamma)} = W_i^{(\beta\gamma)} + \Phi_i^{(\beta\gamma)}\bar{x}_2^{(\beta)} + \psi_i^{(\beta\gamma)}\bar{x}_3^{(\gamma)} \tag{7}$$

which is Aboudi's form of the displacement functions in individual subcells. The derivation of Aboudi's periodic displacement field can be now completed by the transformation of equations (3) and (4) to Aboudi's continuity conditions. To accomplish the transformation, we first express nodal displacements $u_{ij}^{(\beta\gamma)}$ in terms of $W_i^{(\beta\gamma)}$, $\Phi_i^{(\beta\gamma)}$ and $\Psi_i^{(\beta\gamma)}$.

From (6), one can readily derive

$$
\begin{aligned}
u_{i1}^{(\alpha\beta)} &= W_{i0}^{(\beta\gamma)} - \frac{\ell_\beta}{2}\,\Psi_i^{(\beta\gamma)} \\
u_{i2}^{(\beta\gamma)} &= W_{i0}^{(\beta\gamma)} - \frac{h_\beta}{2}\,\Phi_i^{(\beta\gamma)} \\
u_{i3}^{(\beta\gamma)} &= W_{i0}^{(\beta\gamma)} + \frac{h_\beta}{2}\,\Phi_i^{(\beta\gamma)}
\end{aligned}
\tag{8}
$$

where $W_{i0}^{(\beta\gamma)}$ is equal to $W_i^{(\beta\gamma)}$ when x_1 is equal to zero. This restriction on $W_i^{(\beta\gamma)}$ does not affect the generality of our derivation because equations (3) to (7) remain unchanged with respect to a translation of the $x_1 = 0$ plane along the x_1-axis.

$$
\bar{x}_1 = x_1^1 + x_1 \tag{9}
$$

where x_1^1 is an arbitrary coordinate from Ω.

Substitution of (8) into (3) yields

$$
\begin{aligned}
W_{i0}^{(11)} - \frac{h_1}{2}\,\Phi_i^{(11)} - W_{i0}^{(21)} - \frac{h_2}{2}\,\Phi_i^{(21)} &= 0 \\
W_{i0}^{(11)} + \frac{h_1}{2}\,\Phi_i^{(11)} - W_{i0}^{(21)} + \frac{h_2}{2}\,\Phi_i^{(21)} &= 0 \\
W_{i0}^{(12)} - \frac{h_1}{2}\,\Phi_i^{(12)} - W_{i0}^{(22)} - \frac{h_2}{2}\,\Phi_i^{(22)} &= 0 \\
W_{i0}^{(12)} + \frac{h_1}{2}\,\Phi_i^{(12)} - W_{i0}^{(22)} + \frac{h_2}{2}\,\Phi_i^{(22)} &= 0
\end{aligned}
\tag{10}
$$

Similarly, the substitution of (8) into (4) yields

$$
\begin{aligned}
W_{i0}^{(11)} - \frac{\ell_1}{2}\,\Psi_i^{(11)} - W_{i0}^{(12)} - \frac{\ell_2}{2}\,\Psi_i^{(12)} &= 0 \\
W_{i0}^{(11)} + \frac{\ell_1}{2}\,\Psi_i^{(11)} - W_{i0}^{(12)} + \frac{\ell_2}{2}\,\Psi_i^{(12)} &= 0 \\
W_{i0}^{(21)} - \frac{\ell_1}{2}\,\Psi_i^{(21)} - W_{i0}^{(22)} - \frac{\ell_2}{2}\,\Psi_i^{(22)} &= 0 \\
W_{i0}^{(21)} + \frac{\ell_1}{2}\,\Psi_i^{(21)} - W_{i0}^{(22)} + \frac{\ell_2}{2}\,\Psi_i^{(22)} &= 0
\end{aligned}
\tag{11}
$$

Equations (7), (10) and (11) confirm that the finite element interpolation function (2), (3) and (4) correctly describe the kinematics of Aboudi's model.

The interpolation functions for the stress components $\sigma_{ij}^{(\beta\gamma)}$ can be selected to be constant in each matrix or fiber element

$$\sigma_{ij}^{(\beta\gamma)} = S_{ij}^{(\beta\gamma)} \tag{12}$$

The stress components $S_{ij}^{(\beta\gamma)}$ are constant in each $(\beta\gamma)$ element; $S_{ij}^{(\beta\gamma)}$ are defined inside and on the boundary of the $(\beta\gamma)$-element, and they are zero anywhere else.

The mixed variational principle (1) is now used to derive Aboudi's equilibrium equations for $\sigma_{ij}^{(\beta\gamma)}$. Let us substitute (2), (3), and (4) into (1)

$$\Pi = \int_{\Omega} \frac{1}{2} [\sigma_{ij}^{(\beta\gamma)}(\theta_{k,j}u_{ik}^{(\beta\gamma)} + \theta_{k,i}u_{jk}^{(\beta\gamma)}) - \sigma_{ij}^{(\beta\gamma)}M_{ijk\ell}S_{k\ell}^{(\beta\gamma)}]d\Omega$$

$$- \int_{\Gamma_t} t_i^{\star}\theta_k u_{ik}^{(\beta\gamma)} d\Gamma \tag{13}$$

Integration by parts can be applied to the first term in (13). Since $u_i^{(\beta\gamma)}$ are discontinuous along interelement boundaries, the integration of (13) must include surface integrals on these boundaries:

$$\Pi = \int_{\Omega} \frac{1}{2} [\sigma_{ij,j}^{(\beta\gamma)}\theta_k u_{ik}^{(\beta\gamma)} + \sigma_{ij,i}^{(\beta\gamma)}\theta_k u_{jk}^{(\beta\gamma)} + \sigma_{ij}^{(\beta\gamma)}M_{ijk\ell}\sigma_{k\ell}^{(\beta\gamma)})d\Omega$$

$$+ \int_{\hat{\Gamma}} \frac{1}{2} [\sigma_{ij}^{(\beta\gamma)}n_j^{(\beta\gamma)}\theta_k u_{ik}^{(\beta\gamma)} + \sigma_{ij}^{(\beta\gamma)}n_i^{(\beta\gamma}\theta_k u_{jk}^{(\beta\gamma)}]d\Gamma^{(\beta\gamma)} \tag{14}$$

$$- \int_{\Gamma_t} \{t_i^{\star}\theta_k u_{ik}^{(\beta\gamma)} - \frac{1}{2} [\sigma_{ij}^{(\beta\gamma)}n_j^{(\beta\gamma)}\theta_k u_{ik}^{(\beta\gamma)} + \sigma_{ij}^{(\beta\gamma)}n_i^{(\beta\gamma)}\theta_k u_{jk}]\}d\Gamma^{(\beta\gamma)}$$

where $\hat{\Gamma}$ is the sum of all interelement boundaries, $n_i^{(\beta\gamma)}$ is the i-th component of the unit vector perpendicular to $\Gamma^{(\beta\gamma)}$ boundary:

$$n_i^{(\beta\gamma)} = \text{cosine of the angle between the unit normal vector } \hat{n}^{(\beta\gamma)} \text{ and the } x_i\text{-axis}$$

$$= \cos(\hat{n}^{(\beta\gamma)}, x_i) \tag{15}$$

Since $\sigma_{ij} = \sigma_{ji}$, equation (14) is equal to

$$\Pi = \int_{\Omega} [(\sigma_{ij,j}^{(\beta\gamma)}\theta_k)u_{ik}^{(\beta\gamma)} + \frac{1}{2}\sigma_{ij}^{(\beta\gamma)}M_{ijk\ell}\sigma_{k\ell}^{(\beta\gamma)}]d\Omega$$

$$+ \int_{\hat{\Gamma}} \sigma_{ij}^{(\beta\gamma)} n_j^{(\beta\gamma)} \theta_k u_{ik}^{(\beta\gamma)} d\hat{\Gamma} - \int_{\Gamma_t} (t_i^* - \sigma_{ij}^{(\beta\gamma)} n_j^{(\beta\gamma)}) \theta_k u_{ik}^{(\beta\gamma)} d\Gamma \quad (16)$$

In order to use the Hellinger-Riessner variational principle (16) in the derivation of overall properties of composite materials, it is necessary to establish relations between the independent variables σ_{ij} and u_i. A simple way to establish this relationship is to transform the mixed formulation (16) to the equilibrium formulation. The equilibrium formulation is based on the complementary energy principle for linear elastic materials, or minimum stress rate variational principle for elastic-plastic or viscoplastic materials. The functional (16) is transformed to a complementary energy functional when the first term in volume integral and the interelement boundary integral are set equal to zero. Since $u_{ik}^{(\beta\gamma)}$ are arbitrary nodal variables the first term in the volume integral must be zero:

$$\int_{\Omega} \sigma_{ij,j}^{(\beta\gamma)} \theta_k d\Omega = 0 \quad (17)$$

Evaluation of (17) yields Aboudi's equations of equilibrium in terms of stress moments (see [4]) if Legendre polynomial are used to approximate $\sigma_{ij}^{(\beta\gamma)}$.

In all Aboudi's calculations, only the constant terms of Legendre polynomials are used in $\sigma_{ij}^{(\beta\gamma)}$ approximation as suggested by Eq. (12). Under these conditions, the first term in (16) is automatically equal to zero, i.e., equations of equilibrium are trivially satisfied inside each $(\beta\gamma)$-element. The second term in (16) introduces interelement continuity conditions for $\sigma_{ij}^{(\beta\gamma)}$. For example, consider the interelement boundary between elements $\beta\gamma = 11$ and $\beta\gamma = 12$. The interelement boundary integral has the form

$$- \int_{-h_1/2}^{h_n/2} \int_0^d \{\sigma_{i2}^{(11)} [-u_{i1}^{(11)} + (1 - \frac{x_1}{2d} - \frac{\bar{x}_2^{(1)}}{h_1}) u_{i2}^{(11)} + (1 - \frac{x_1}{2d} + \frac{\bar{x}_2^{(1)}}{h_1}) u_{i3}^{(11)}$$

$$+ \frac{x_2}{d} u_{i4}^{(11)}] - \sigma_{i2}^{(12)} [u_{i1}^{(12)} - (\frac{x_1}{2a} + \frac{\bar{x}_2^{(1)}}{h_1}) u_{i2}^{(11)} \quad (18)$$

$$+ (- \frac{x_1}{2d} + \frac{x_2^{(1)}}{h_1}) u_{i3}^{(12)} + \frac{x_1}{d} u_{i4}^{(12)}]\} dx_1 d\bar{x}_2^{(i)} = 0$$

After substituting $\sigma_{i1}^{(11)}$ and $\sigma_{i1}^{(22)}$ from (12), the evaluation of (17) yields

$$S_{i2}^{(11)} dh_\beta (- u_{i1}^{(11)} + \frac{3}{4} u_{i2}^{(11)} + \frac{3}{4} u_{i3}^{(11)} + \frac{1}{2} u_{i4}^{(11)})$$

$$- S_{i2}^{(12)} dh_\beta (u_{i1}^{(12)} - \frac{1}{4} u_{i2}^{(12)} - \frac{1}{4} u_{i3}^{(12)} + \frac{1}{2} u_{i4}^{(12)}) = 0 \qquad (19)$$

The expressions in parentheses are the nodal displacement at points $P^{(11)}(\frac{d}{2}, 0, \frac{\ell_1}{2})$ and $P^{(12)}(\frac{d}{2}, 0, -\frac{\ell_2}{2})$ [see Eqs. (5) and (2)]. Since the strains ε_{11} in fiber and matrix elements are the same, the nodal variable $u_{i2}^{(\beta\gamma)}$, $u_{i3}^{(\beta\gamma)}$ and $u_{i4}^{(\beta\gamma)}$ must be related as follows:

$$u_{i4}^{(11)} - \frac{1}{2} (u_{i2}^{(11)} + u_{i3}^{(11)}) = u_{i4}^{(21)} - \frac{1}{2} (u_{i2}^{(21)} + u_{i3}^{(21)})$$

$$= u_{i4}^{(12)} - \frac{1}{2}(u_{i2}^{(12)} + u_{i3}^{(12)}) = u_{i4}^{(22)} - \frac{1}{2} (u_{i2}^{(21)} + u_{i3}^{(21)}) \qquad (20)$$

The multiplication of (20) by $\frac{1}{2}$ and addition of the first and third term of (20) to the right hand and left hand side of the second connectivity equation (4), respectively, yields

$$u_{i1}^{(12)} - \frac{1}{4} u_{i2}^{(12)} - \frac{1}{4} u_{i3}^{(12)} + \frac{1}{2} u_{i4}^{(12)}$$

$$= - u_{i1}^{(11)} + \frac{3}{4} u_{i2}^{(11)} + \frac{3}{4} u_{i3}^{(11)} + \frac{1}{2} u_{i4}^{(11)} \qquad (21)$$

Then, for arbitrary $u_{ij}^{(11)}$ or $u_{ij}^{(12)}$, the condition (14) yields a strong continuity for stress components

$$S_{i2}^{(11)} - S_{i2}^{(12)} = 0. \qquad (22)$$

Similar results can be readily obtained for the rest of the interelement boundaries. In summary we may write the following set of interelement equilibrium conditions

$$S_{i2}^{(11)} - S_{i2}^{(21)} = 0 \quad , \quad S_{i2}^{(21)} - S_{i2}^{(22)} = 0$$

$$S_{i3}^{(21)} - S_{i3}^{(11)} = 0 \quad , \quad S_{i3}^{(22)} - S_{i3}^{(12)} = 0 \qquad (23)$$

and

$$S_{i1}^{(\beta\gamma)} u_{i4} - \frac{1}{2} S_{i1}^{(\beta\gamma)} (u_{i2} + u_{i3}) = 0$$

These are identical to the stress continuity relations developed by Aboudi [4] when element-wise constant approximation is used for the stress field. Note from Eq. (16) that the traction boundary conditions

$$\sigma_{ij}^{(\beta\gamma)} n_j^{(\beta\gamma)} = t_i^*$$

are satisfied in the strong sense.

3. Finite Element Formulation

The equation (16) for a subcell $(\beta\gamma)$, in view of equation (17), reduces to

$$\Pi^{(\beta\gamma)} = -\frac{1}{2}\int_{\Omega^{(\beta\gamma)}} \sigma_{ij}^{(\beta\gamma)} M_{ijk\ell}^{(\beta\gamma)} \sigma_{k\ell}^{(\beta\gamma)} d\Omega + \int_{\Gamma^{(\beta\gamma)}} \sigma_{ij}^{(\beta\gamma)} n_j^{(\beta\gamma)} \theta_k u_{ik}^{(\beta\gamma)} d\Gamma \qquad (24)$$

Equation (24) can be readily cast in a matrix form when the following vectors are defined:

$$\{\sigma^{(\beta\gamma)}\} = \{\sigma_{11}^{(\beta\gamma)}, \sigma_{22}^{(\beta\gamma)}, \sigma_{33}^{(\beta\gamma)}, \sigma_{12}^{(\beta\gamma)}, \sigma_{13}^{(\beta\gamma)}, \sigma_{23}^{(\beta\gamma)}\}^T \qquad (25)$$

and

$$\{u^{(\beta\gamma)}\} = \{u_{11}^{(\beta\gamma)}, u_{21}^{(\beta\gamma)}, u_{31}^{(\beta\gamma)}, u_{41}^{(\beta\gamma)}, u_{21}^{(\beta\gamma)}, u_{22}^{(\beta\gamma)}, \ldots, u_{34}^{(\beta\gamma)}\}^T$$

The additional index variables, $M_{ijk\ell}$, n_j and θ_k can be also organized in matrix forms in accordance with definitions (25). The matrix form of (25) for constant approximation of $\sigma_{ij}^{(\beta\gamma)}$ equal to $S_{ij}^{(\beta\gamma)}$ has the form

$$\Pi^{(\beta\gamma)} = -\frac{1}{2}\,\Omega^{(\beta\gamma)}\{S^{\beta\gamma}\}^T[M^{(\beta\gamma)}]\{S^{(\beta\gamma)}\} + \{S^{(\beta\gamma)}\}^T[H^{(\beta\gamma)}]\{u^{(\beta\gamma)}\} \qquad (26)$$

where $[M^{\beta\gamma}]$ is a (6 x 6) matrix of the compliance moduli of the $(\beta\gamma)$ element and $[N^{\beta\gamma}]$ is a (6 x 12) matrix of

$$\int_{\Gamma^{(\beta\gamma)}} n_j^{(\beta\gamma)} \theta_k^{(\beta\gamma)} d\Gamma^{(\beta\gamma)}$$

coefficients.

Since the components $S_{ij}^{(\beta\gamma)}$ in the vector $\{S^{(\beta\gamma)}\}$ are internal to each $(\beta\gamma)$ element, they can be eliminated from the mixed varational principle (16) by a variation of the functional $\Pi^{(\beta\gamma)}$ with respect to $S_{ij}^{(\beta\gamma)}$ in each element:

$$\delta\Pi^{(\beta\gamma)} = -\,\Omega^{(\beta\gamma)}[M^{(\beta\gamma)}]\{S^{(\beta\gamma)}\} + [H^{\beta\gamma}]\{u^{(\beta\gamma)}\} = 0 \qquad (27)$$

For invertible $[M^{(\beta\gamma)}]$, we can immediately write

$$\{S^{(\beta\gamma)}\} = \frac{1}{\Omega^{(\beta\gamma)}} [M^{(\beta\gamma)}]^{-1}[H^{(\beta\gamma)}]\{u^{(\beta\gamma)}\} \tag{28}$$

Equation (28) establishes the relation between the originally independent fields $\sigma_{ij}^{(\beta\gamma)}$ and $u_i^{(\beta\gamma)}$ in (13) and (16). Note that an explicit form of the relations between $\sigma_{ij}^{(\beta\gamma)}$ and $u_i^{(\beta\gamma)}$ can be obtained when one organizes the components of the matrix $[H^{(\beta\gamma)}]$ back to the subscript notation before the integration along $\Gamma^{(\beta\gamma)}$ is applied to $[H^{\beta\gamma}]$. It is easy to show that for the constant approximation of $\sigma_{ij}^{(\beta\gamma)}$ the relation is as follows:

$$\sigma_{ij}^{(\beta\gamma)} = (M_{ijk\ell}^{(\beta\gamma)})^{-1} \frac{1}{2} (u_{i,j}^{(\beta\gamma)} + u_{j,i}^{(\beta\gamma)}) \tag{29}$$

This result confirms the fact that, for constant $\sigma_{ij}^{(\beta\gamma)}$ the zeroing of the first volume integral and the interelement surface integrals in (16) transforms the mixed formulation (16) into a complementary energy or minimum stress rate variational formulation.

4. Homogenization

In this section, the finite element homogenization procedure developed by Teply and Dvorak ([5] and [8]) is used to imply Aboudi's solution of overall (homogenized) properties with the solution of the overall properties via the mixed finite element functional (16). The Teply-Dvorak homogenization procedure is based on the comparison of unit cell energies. The energies considered in the comparison are the energies of the periodic and homogenized microstructures. Using the finite element method, both unit cell energies are expressed in the form of quadratic functions of contact point displacements. The quadratic functions are compared and homogenized properties calculated from the equivalence of the contact point displacements. The homogenization procedure has been developed for the Periodic Hexagonal Array (PHA) model described in [5]. A detailed explanation of the two concepts can be found in [8].

The contact points are defined as the points in Ω at which the displacements of the homogenized material are equal to the displacement of the periodic microstructure. At the same time the boundary Γ is subjected to the conditions corresponding to a uniform stress or strain field. For the variational principle (16), where $\Gamma = \Gamma_t$, the boundary tractions are

$$t_i^* = \sigma_{ij}^0 n_j \tag{30}$$

where σ_{ij}^o are the components of the uniform stress field in Ω. It will be shown that the selection of the subcell centroids as the contact points yields Aboudi's homogenization equations for displacements.

Suppose that Ω is a homogenized medium subjected to the boundary conditions (30). The displacement components u_i^o are then linear functions of x_1, x_2, and x_3. The linear functions u_i^o can be readily expressed in terms of the finite element interpolation functions when, for example, the contact points are selected as the nodes for u_i^o. This approach has been used in [8]. However, in the present study it is advantageous to select four vertices of a unit cell (see Figure 4) as the interpolation points, and derive the displacements at the contact points (the centroids of the $\beta\gamma$-elements) from a linear combination of the four vertex displacements u_{ik}^o. Note that the contact points are also not nodal points in the finite element mesh for the periodic microstructure.

Assuming the four vertices of the unit cell to be the nodal points, the interpolation functions for u_i^o can be written as

$$u_i^o = \theta_k^o u_{ki}^o \qquad i = 1,2,3 \tag{31}$$

where

$$\begin{aligned}
\theta_1^o &= \frac{h_1\ell_2 - h_2\ell_1}{(h_1 + h_2)(\ell_1 + \ell_2)} - \frac{1}{d}x_1 - \frac{1}{h_1 + h_2}x_2 - \frac{1}{\ell_1 + \ell_2}x_3 \\
\theta_2^o &= \frac{\ell_1}{\ell_1 + \ell_2} + \frac{1}{\ell_1 + \ell_2}x_3 \\
\theta_3^o &= \frac{h_2}{h_1 + h_2} + \frac{1}{h_1 + h_2 x_2} \\
\theta_4^o &= \frac{1}{d}x_1
\end{aligned} \tag{32}$$

The origin of the coordinate system (x_1,x_2,x_3) is taken at the intersections of the four subcells (see Figure 4).

The equality of displacements u_1^o and $u_i^{(\beta\gamma)}$ at the contact points is readily obtained when we substitute

$$x_1 = 0,\ x_2 = (-1)^{(\beta-1)}h_\beta/2 \text{ and } x_3 = (-1)^\gamma \ell_\gamma/2$$

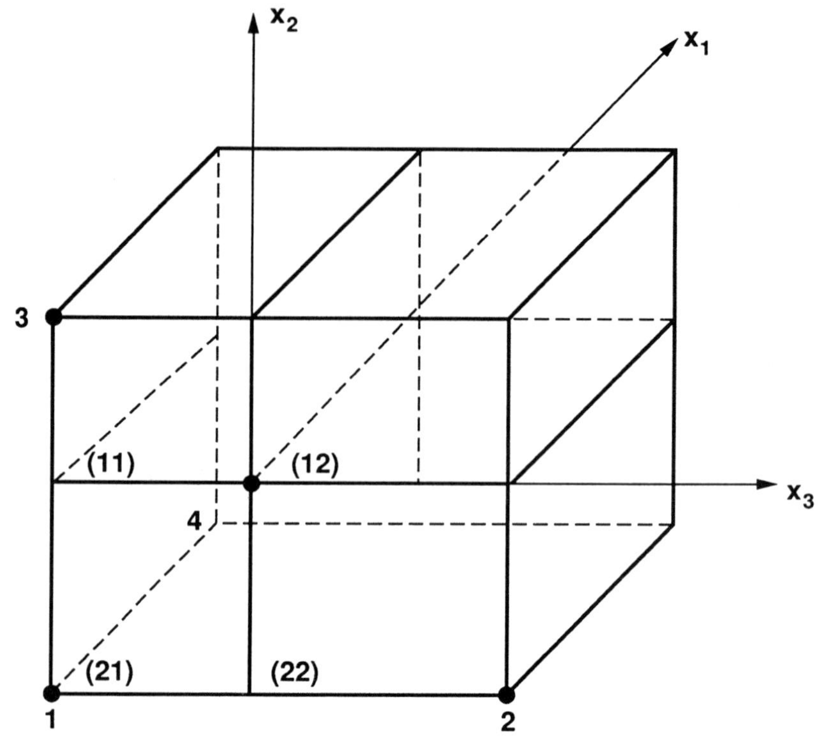

Nodal Points for Homogenized Material
Figure 4

in u_i^o [see equation (31)], and

$$x_1 = 0,\ \bar{x}_2^{(\beta)} = 0 \text{ and } \bar{x}_3^{(\gamma)} = 0$$

in $u_i^{(\beta\gamma)}$ [see equation (2)]. The equations of the displacement equivalence at the four contact points are:

$$\left[\frac{h_1\ell_2 - h_2\ell_1}{(h_1 + h_2)(\ell_1 + \ell_2)} - \frac{h_1}{2(h_1 + h_2)} + \frac{\ell_1}{2(\ell_1 + \ell_2)}\right]u_{i1}^i$$

$$+ \frac{\ell_1}{2(\ell_1 + \ell_2)} u_{i2}^o + \frac{2h_2 + h_1}{2(h_1 + h_2)} u_{i3}^o = \frac{1}{2} (u_{i2}^{(11)} + u_{i3}^{(11)})$$

$$\left[\frac{h_1\ell_2 + h_2\ell_1}{(h_1 + h_2)(\ell_1 + \ell_2)} + \frac{h_2}{2(h_1 + h_2)} + \frac{\ell_1}{2(\ell_1 + \ell_2)}\right]u_{i1}^o$$

$$+ \frac{\ell_1}{2(\ell_1 + \ell_2)} u_{i2}^o + \frac{h_2}{2(h_1 + h_2)} u_{i3}^o = \frac{1}{2} (u_{i2}^{(21)} + u_{i3}^{(21)})$$

$$\left[\frac{h_1\ell_2 - h_2\ell_1}{(h_1 + h_2)(\ell_1 + \ell_2)} - \frac{h_1}{2(h_1 + h_2)} - \frac{\ell_2}{2(\ell_1 + \ell_2)}\right]u_{i1}^o$$

$$+ \frac{2\ell_1 + \ell_2}{2(\ell_1 + \ell_2)} u_{i2}^o + \frac{2h_2 + h_1}{2(h_1 + h_2)} u_{i3}^o = \frac{1}{2} (u_{i2}^{(12)} + u_{i3}^{(12)})$$

$$\left[\frac{h_1\ell_2 - h_2\ell_1}{(h_1 + h_2)(\ell_1 + \ell_2)} + \frac{h_2}{2(h_1 + h_2)} - \frac{\ell_2}{2(\ell_1 + \ell_2)}\right]u_{i1}^o$$

$$+ \frac{2\ell_1 + \ell_2}{2(\ell_1 + \ell_2)} u_{i2}^o + \frac{h_2}{2(h_1 + h_2)} u_{i3}^o = \frac{1}{2} (u_{i2}^{(22)} + u_{i3}^{(22)}) \tag{33}$$

Equations (33) can be cast into Aboudi's notation when the following relations are considered.

$$w_i = u_i^o$$

$$\frac{\partial w_i}{\partial x_2} = \frac{1}{h_1 + h_2} (u_{3i}^o + u_{1i}^o) \tag{34}$$

$$\frac{\partial w_i}{\partial x_3} = \frac{1}{\ell_1 + \ell_2} (u_{2i}^o - u_{1i}^o)$$

Since displacement w_i is continuous in Ω, we can define the projection of w_i on a subcell $(\beta\gamma)$.

$$w_i^{\beta\gamma} = w_i, \tag{35}$$

when x_2 and x_3 in (22) belong to $\Omega^{(\beta\gamma)}$ and its boundary $\Gamma^{(\beta\gamma)}$. Equation (35) is a natural condition for the assembly of the homogenized unit cells in Ω. The displacements $w_i^{\beta\gamma}$ must be continuous in Ω and can be derived from a single function w_i. Equation (35) is a mere consequence of the u_i^o definition - u_i^o is defined continuous and linear in Ω. Nevertheless we include Eq. (35) in our paper to establish a complete equivalence between the contact point approach and Aboudi's formulation. Note that in Aboudi's work (35) represent the constraint equations for the homogenized displacement $w_i^{\beta\gamma}$ to introduce the continuity of w_i in Ω. The Aboudi's constraints for the derivative of $w_i^{\beta\gamma}$ (34) can be derived from the equivalence of contact point displacements (33). For example, subtracting the second equation (33) from the first one, we obtain

$$u_{i3}^o - u_{i1}^o = u_{i2}^{(11)} + u_{i3}^{(11)} - u_{i2}^{(21)} - u_{i3}^{(21)} \tag{36}$$

where $u_{i2}^{(11)}$ and $u_{i3}^{(21)}$ are nodal displacements at the common node between (11) and (21) elements. Notice that displacements $u_{i3}^{(11)}$ and $u_{i2}^{(21)}$ are the displacements at two different nodes; the former is the displacement at the midpoint in the upper edge of the (11)-element and the latter is the displacement at the midpoint on the lower edge of the (21)-element (see Figure 2). Hence, the connectivity equations (3) can be applied only to $u_{i2}^{(11)}$ and $u_{i3}^{(21)}$. The application of the first connectivity equation (3) yields

$$u_{i3}^o - u_{i1}^o = u_{i3}^{(11)} - u_{i2}^{(11)} + u_{i3}^{(21)} - u_{i2}^{(21)} \tag{37}$$

which, with the aid of Eqs. (34) and (6), can be cast in Aboudi's form

$$\frac{\partial w_i}{\partial x_2} = \frac{h_1}{h_1 + h_2} \Phi_i^{(11)} + \frac{h_2}{h_1 + h_2} \Phi_i^{(21)}. \tag{38}$$

Similar result is obtained when the fourth equation (33) is subtracted from the third one to yield,

$$\frac{\partial w_i}{\partial x_2} = \frac{h_1}{h_1 + h_2} \Phi_i^{(12)} + \frac{h_2}{h_1 + h_2} \Phi_i^{(22)}. \tag{39}$$

The derivative of w_i in the x_3-direction is determined when the third equation (33) is subtracted from the first one. The equivalence of the contact point displacements yields:

$$u_{i2}^{o} - u_{i1}^{o} = \frac{1}{2}\,(u_{i2}^{(12)} + u_{i3}^{(12)}) - \frac{1}{2}\,(u_{i2}^{(11)} + u_{i3}^{(11)}) \tag{40}$$

Expanding the righthand side of Eq. (40) by $2u_{i1}^{(12)} - 2u_{i1}^{(12)}$ and using the second connectivity equation (4), equation (40) takes on the following form

$$u_{i2}^{o} - u_{i1}^{o} = u_{i2}^{(11)} + u_{i3}^{(11)} - 2u_{i1}^{(11)} + u_{i2}^{(12)} + u_{i3}^{(12)} - 2u_{i1}^{(12)} \tag{41}$$

Equation (41) can be changed to Aboudi's constraint equation when Eqs. (34) and (6) are used:

$$\frac{\partial w_i}{\partial x_3} = \frac{\ell_1}{\ell_1 + \ell_2}\,\psi_i^{(11)} + \frac{\ell_2}{\ell_1 + \ell_2}\,\psi_i^{(12)}. \tag{42}$$

The last constraint equation is obtained when the second and fourth equations (33) are considered:

$$\frac{\partial w_i}{\partial x_3} = \frac{\ell_1}{\ell_1 + \ell_2}\,\psi_i^{(21)} + \frac{\ell_2}{\ell_1 + \ell_2}\,\psi_i^{(22)} \tag{43}$$

In summary, we can conclude that the derivation of equations (35), (38), (39), (42) and (43) has shown that the kinematics of the finite element formulation is identical to the kinematic equations of Aboudi's model when the concept of contact points is applied at the subcell centroids and linear shape functions (2) and (31) are used for the displacement functions in the subcells and homogenized unit cell. Using Eq. (33) it can also be shown that the overall strains in Aboudi's model are simple weighted averages of two or three local strains, $\frac{1}{2}\,(u_{i,j}^{(\beta\gamma)} + u_{j,i}^{(\beta\gamma)})$. Although the derivation of weighted averages for the strains in homogenized material is straightforward, it involves tedious algebra which falls outside the scope of this paper. For that reason only the final results are summarized. From equations (2), (3), (4), (20), (28), (31), (32) and (33), and assuming the following definitions

$$\begin{aligned} \varepsilon_{ij}^{(\beta\gamma)} &= H_{ijk\ell}^{(\beta\gamma)} u_{\xi\ell}^{(\beta\gamma)} \\ \varepsilon_{ij}^{o} &= \frac{1}{2}\,(u_{i,j}^{o} + u_{j,i}^{o}) \end{aligned} \tag{44}$$

where $H^{(\beta\gamma)}_{ijk\ell}$ are the components of matrix $[H^{(\beta\gamma)}]$ defined in (28), one can calculate ε^{o}_{ij} as follows

$$\varepsilon^{o}_{11} = \varepsilon^{(\beta\gamma)}_{11}$$

$$\varepsilon^{o}_{22} = \frac{h_1}{h_1 + h_2}\,\varepsilon^{(11)}_{22} + \frac{h_2}{h_1 + h_2}\,\varepsilon^{(21)}_{22} = \frac{h_1}{h_1 + h_2}\,\varepsilon^{(12)}_{22} + \frac{h_2}{h_1 + h_2}\,\varepsilon^{(22)}_{22}$$

$$\varepsilon^{o}_{33} = \frac{\ell_1}{\ell_1 + \ell_2}\,\varepsilon^{(11)}_{33} + \frac{\ell_2}{\ell_1 + \ell_2}\,\varepsilon^{(12)}_{33} = \frac{\ell_1}{\ell_1 + \ell_2}\,\varepsilon^{(21)}_{33} + \frac{\ell_2}{\ell_1 + \ell_2}\,\varepsilon^{(22)}_{33}$$

$$\varepsilon^{o}_{12} = \frac{h_1}{h_1 + h_2}\,\varepsilon^{(11)}_{12} + \frac{h_2}{h_1 + h_2}\,\varepsilon^{(21)}_{12} = \frac{h_1}{h_1 + h_2}\,\varepsilon^{(12)}_{12} + \frac{h_2}{h_1 + h_2}\,\varepsilon^{(22)}_{12}$$

$$\varepsilon^{o}_{13} = \frac{\ell_1}{\ell_1 + \ell_2}\,\varepsilon^{(11)}_{13} + \frac{\ell_2}{\ell_1 + \ell_2}\,\varepsilon^{(12)}_{33} = \frac{\ell_1}{\ell_1 + \ell_2}\,\varepsilon^{(21)}_{13} + \frac{\ell_2}{\ell_1 + \ell_2}\,\varepsilon^{(22)}_{13}$$

(45)

and for $h_2\ell_1 = h_1\ell_2$, we have

$$\varepsilon^{o}_{23} = \frac{h_1\ell_1}{(h_1 + h_2)(\ell_1 + \ell_2)}\,\varepsilon^{(11)}_{23} + \frac{h_2\ell_2}{(h_1 + h_2)(\ell_1 + \ell_2)}\,\varepsilon^{(21)}_{23}$$

$$+ \frac{\ell_1 h_2}{(h_1 + h_2)(\ell_1 + \ell_2)}\,(\varepsilon^{(11)}_{23} + \varepsilon^{(21)}) = \frac{h_1\ell_1}{(h_1 + h_2)(\ell_1 + h_2)}\,\varepsilon^{(21)}_{23}$$

$$+ \frac{h_2\ell_2}{(h_1 + h_2)(\ell_1 + \ell_2)}\,\varepsilon^{(22)}_{23} + \frac{\ell_1 h_2}{(h_1 + h_2)(\ell_1 + \ell_2)}\,(\varepsilon^{(21)}_{23} + \varepsilon^{(22)}_{23})$$

The requirement $h_2\ell_1 = h_1\ell_2$ may not be generally satisfied in Aboudi's model, but for most model applications the rectangular fibers and unit cells can be arranged such that the above requirement is satisfied. For example, when the ratio of fiber sides, ℓ_1 and h_1, is equal to the ratio of the unit cell sides, $(\ell_1 + \ell_2)$ over $(h_1 + h_2)$, the geometry requirement is satisfied. Equations (45) are special forms of strain volume averages specific to Aboudi's geometry. The integral form of the strain volume averages can be readily derived from (45). For example, consider the second equation in (45) and multiply its right hand side with $(\ell_1 + \ell_2)/(\ell_1 + \ell_2)$

$$\varepsilon_{22}^{o} = \frac{h_1 \ell_2}{(h_1 + h_2)(\ell_1 + \ell_2)} \varepsilon_{22}^{(12)} + \frac{h_2 \ell_2}{(h_1 + h_2)(\ell_1 + \ell_2)} \varepsilon_{22}^{(22)}$$

$$+ \frac{\ell_1}{\ell_1 + \ell_2} \left(\frac{h_1}{(h_1 + h_2)} \varepsilon_{22}^{(12)} + \frac{h_2}{(h_1 + h_2)} \varepsilon_{22}^{(22)}\right) \qquad (46)$$

The term in parenthesis is equal to the lefthand side of the second equation (45). Substitution of the lefthand side from (45) yield the integral form of the strain volume average

$$\varepsilon_{22}^{o} = \frac{h_2 \ell_1}{(h_1 + h_2)(\ell_1 + \ell_2)} \varepsilon_{22}^{(11)} + \frac{h_2 \ell_1}{(h_1 + h_2)(\ell_1 + \ell_2)} \varepsilon_{22}^{(21)}$$

$$+ \frac{h_1 \ell_2}{(h_1 + h_2)(\ell_1 + \ell_2)} \varepsilon_{22}^{(22)} + \frac{h_2 \ell_2}{(h_1 + h_2)(\ell_1 + \ell_2)} \varepsilon_{22}^{(22)} \qquad (47)$$

Similar expressions can be obtained for the other strain components. Strain volume averages (45) suggest that the subcell strain components $\varepsilon_{ij}^{(\beta\gamma)}$ are not independent. It can be seen that three strain components, for example $\varepsilon_{ij}^{(11)}$, $\varepsilon_{ij}^{(21)}$ and $\varepsilon_{ij}^{(12)}$, are sufficient to calculate the overall average ε_{ij}^{o} and also to determine the fourth strain $\varepsilon_{ij}^{(21)}$. This feature of Aboudi's model is specific to the selection of the model geometry - rectangular unit cells containing square or rectangular fibers. The strain constraints (45) are also the result of the small number of subcells (four) in the unit cell. It will be shown that the constraints (45) can be relaxed when fiber cross section is approximated by a n-sided polygon and the number of subcells is increased.

Having established the identity between the kinematics of the mixed finite element formulation (29) and Aboudi's model, we shall now derive the equivalence between overall properties calculated via Aboudi's and the finite element approach. For that purpose we shall use the above derived kinematic relations, and apply them to the derivation of the contact point displacement quadratic forms of the unit cell energies for the homogenized and periodic microstructures. In the course of our derivations, it will be shown that the Aboudi's and the finite element model yield identical equilibrium equation in terms of the global and local displacements u_i^{o} and $u_i^{(\beta\gamma)}$, respectively.

First, let us consider a unit cell of the periodic microstructure. In order to develop the unit cell energy quadratic form, unit cell boundary conditions must be derived. In accordance with the overall boundary conditions (30) and variational principle (16), the stress components, $\sigma_{ij}^{(\beta\gamma)}$, of the unit cells located on the boundary Γ of Ω must be equal to

$$\sigma_{ij}^{(\beta\gamma)} = \sigma_{ij}^{o} \tag{48}$$

Substituting the stress approximation coefficients $S_{ij}^{(\beta\gamma)}$ in (46), and assuming the strong continuity of $S_{ij}^{(\beta\gamma)}$ cross the interelement boundary $\hat{\Gamma}$ [see Eqs. (22) and (23)], one can immediately write the unit cell boundary condition in the form of the subcell stress coefficients

$$S_{ij}^{(\beta\gamma)} = \sigma_{ij}^{o} \tag{49}$$

Substitution of Eq. (49) in (28) introduces the overall boundary tractions (30) in the finite element formulation

$$\{\sigma_o\} = \frac{1}{\Omega^{(\beta\gamma)}} [M^{(\beta\gamma)}]^{-1} [H^{(\beta\gamma)}]\{u^{(\beta\gamma)}\} \tag{50}$$

The last equations can be, with the aid of Eqs. (6) and (35), cast in Aboudi's notation. First, let us use Eqs. (6) and (35) to define a matrix $[G^{(\beta\gamma)}]$ by

$$[G^{(\beta\gamma)}] = [H^{(\beta\gamma)}]\{u^{(\beta\gamma)}\} = \begin{bmatrix} \frac{\partial w_1^{(\beta\gamma)}}{\partial x_1} & 0. & 0 \\ 0 & \Phi_2^{(\beta\gamma)} & 0 \\ 0. & 0 & \Psi_3^{(\beta\gamma)} \\ \Phi_1^{(\beta\gamma)} & \frac{\partial w_2^{(\beta\gamma)}}{\partial x_1} & 0 \\ \Psi_1^{(\beta\gamma)} & 0. & \frac{\partial w_3^{(\beta\gamma)}}{\partial x_1} \\ 0 & \Psi_2^{(\beta\gamma)} & \Phi_3^{(\beta\gamma)} \end{bmatrix} \tag{51}$$

Substitution of Eq. (51) in (50) yields the final transformation equation between the finite element model and Aboudi's formulation

$$\{\sigma^o\} = \frac{1}{\Omega^{(\beta\gamma)}} [M^{(\beta\gamma)}]^{-1}[G^{(\beta\gamma)}] \tag{52}$$

This equation relates the overall stress components σ^{o}_{ij} to Aboudi's displacement functions $w^{(\beta\gamma)}_{i}$, $\Phi^{(\beta\gamma)}_{2}$ and $\Psi^{(\beta\gamma)}_{i}$. In Aboudi's model, Equations (52) together with (35), (38), (39), (42) and (43) are used in the selection of a set of solution equations to determine a specific elastic stiffness modulus or the stress-strain curve for a specific nonlinear loading path. For example, assuming $\{\sigma^{o}\}$ equal to $\{0,0,0,\sigma_{12},0,0\}$, a set of equations can be selected from the matrix equations (52) and the above mentioned kinematic constraints (35), (38), (39), (42) and (43) to calculate either the longitudinal shear modulus or longitudinal shear stress-strain curve of fiber reinforced composites. A comprehensive review of various solutions of (25), (32), (39), (42), (43) and (52) is given in [9].

It has been shown [see Eq. (28)], that Eqs. (50) and (52) are two different forms of the stationary condition for the mixed functional (16). The stationary condition (28) is calculated with respect to the δS_{ij} variation of the functional. It can be also shown that for the homogenized material, the kinematic constraints (35), (38), (39), (42) and (43) are the stationary conditions of the mixed functional (16) with respect to δu_{ij} variations. It means that the solution of Aboudi's equations is equivalent to the stationary points of the mixed functional (16) when the linear interpolation functions are used for $u^{(\beta\gamma)}_{i}$, and u^{o}_{ij} and $\sigma^{(\beta\gamma)}_{ij}$ are approximated by an elementwise constant functions $S^{(\beta\gamma)}_{1i}$. This concludes the unification of the finite element and Aboudi formulations.

In Aboudi's formulation, different equations are used to solve the stress-strain curves for two different loading paths. Such arrangement can become inconvenient for solutions of nonlinear problems with a nonproportional loading path in the σ^{o}_{ij}-space. Therefore, it is advantageous to continue our derivations and use Teply-Dvorak's method to derive the instantaneous overall stiffness moduli from the equivalence of two contact point displacement quadratic forms.

In order to develop the quadratic form for the periodic microstructure, it is necessary to construct transformation matrices between nodal displacements $u^{(\beta\gamma)}_{ik}$ and contact point displacements $a^{(\beta\gamma)}_{i\ell}$ (ℓ = 1,2,3,4). The transformation can be written in a matrix form when the following vectors of nodal and contact point displacements are defined

$$\{a\} = \{a^{(21)}_{11}, a^{(21)}_{21}, a^{(21)}_{31}, a^{(22)}_{12}, a^{(22)}_{22}, a^{(22)}_{32}, a^{(11)}_{13}, a^{(11)}_{33}, a^{(21)}_{14}$$

$$a_{24}^{(21)}, a_{34}^{(21)}\}^T$$

and

$$\{u\} = \{u_{11}^{(11)}, u_{21}^{(11)}, u_{31}^{(11)}, u_{11}^{(21)}, u_{21}^{(21)}, u_{31}^{(31)}, u_{31}^{(11)}, u_{32}^{(11)}, u_{33}^{(11)},$$
$$u_{21}^{(11)}, u_{22}^{(11)}, u_{23}^{(11)}, u_{21}^{(21)}, u_{22}^{(21)}, u_{23}^{(21)}, u_{31}^{(12)}, u_{32}^{(12)}, u_{33}^{(12)},$$
$$u_{21}^{(12)}, u_{22}^{(12)}, u_{23}^{(12)}, u_{21}^{(22)}, u_{22}^{(22)}, u_{23}^{(22)}, u_{41}^{(21)}, u_{42}^{(21)}, u_{43}^{(21)}\}^T \tag{53}$$

where $a_{1i}^{(\beta\gamma)}, a_{2i}^{(\beta\gamma)}, a_{3i}^{(\beta\gamma)}$ $(i = 1,2,3)$ are the displacements at the centroids of the bottom surfaces $(x_1 = 0)$ of elements (21), (22) and (11), respectively; displacements $a_{4i}^{(21)}$ are at the centroid top surfaces $(x_1 = d)$ of the element (21) (see Figure 5). Notice that according to the connectivity equations (3) and (4) different nodal displacements $u_{ik}^{(\beta\gamma)}$ can be used to define the same vector $\{u\}$. Assuming that the rigid body motion of a unit cell is restricted, then there exist matrices $[T]$ and $[T^{(\beta\gamma)}]$ such that

$$\{u\} = [T]\{a\}$$

$$\{u^{(\beta\gamma)} = [T^{(\beta\gamma)}]\{a\}. \tag{54}$$

where, according to the second definition (25) and equations (3), (4) and (20), $\{u^{(\beta\gamma)}\}$ are vectors from subspaces of the vector space $\{u\}$. It means that $[T^{(\beta\gamma)}]$ is a certain submatrix of $[T]$.

Matrices $[T]$ and $[T^{(\beta\gamma)}]$ can be derived by several means. Thesimpliest approach is illustrated in Figure 5. The influence function, $\{a_{11}^{(\beta\gamma)}\} = \{1,0,0,\ldots,0\}^T$, is constructed to derive the first columns of $[T]$ and other four matrices $[T^{(\beta\gamma)}]$ from the displacements at the connectivity points. Similar procedures can be used to calculate the coefficients of the other eleven columns in $[T]$ and $[T^{(\beta\gamma)}]$. Such displacement functions $u_i^{(\beta\gamma)}$ are linear and they satisfy the connectivity equations (3), (4) and (20). A subspace of the vector space $\{u\}$ can be found such that $[T]$ and $[T^{(\beta\gamma)}]$ are uniquely defined and invertible. For example, the subspace can be defined by the following vectors

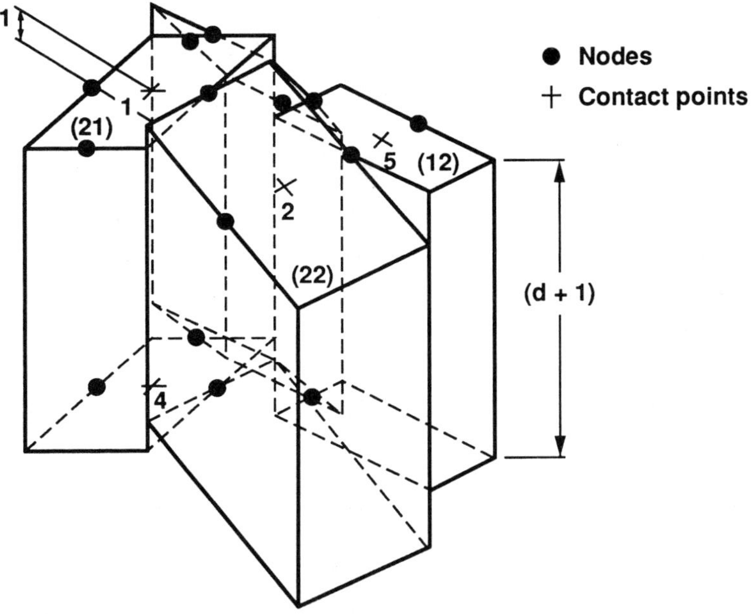

Influence Function {a} = {1, 0, 0, ..., 0}
Figure 5

$$\{u_a\} = \{u_{31}^{(11)}, u_{32}^{(11)}, u_{33}^{(11)}, u_{21}^{(21)}, u_{22}^{(11)}, u_{23}^{(11)},$$

$$u_{21}^{(12)}, u_{22}^{(12)}, u_{23}^{(12)}, u_{41}^{(21)}, u_{42}^{(21)}, u_{43}^{(21)}\}^T \quad (55)$$

The remaining coordinates of the vector space $\{u\}$ define vectors $\{u_b\}$ from the subspace which are complemenetary to $\{u_a\}$ in the vector space $\{u\}$. It can be shown that $\{u_a\}$ $\{u_F\}$ and the first equation (54) is equal to

$$\begin{Bmatrix} u_a \\ u_b \end{Bmatrix} = \begin{bmatrix} t \\ 0 \end{bmatrix} \{a\} \quad (56)$$

In accordance with the definition of the subspace $\{u_a\}$, the rest of the transformation equations (54) can be written in the following form:

$$\begin{Bmatrix} u_a^{(\beta\gamma)} \\ u_b^{(\beta\gamma)} \end{Bmatrix} = \begin{bmatrix} t^{(\beta\gamma)} \\ 0 \end{bmatrix} \{a\} \quad (57)$$

where $[t^{(\beta\gamma)}]$ are certain submatrices of $[t]$.

Having derived relations between $\{u_a^{(\beta\gamma)}\}$, $\{u_b^{(\beta\gamma)}\}$ and $\{a\}$, we may now approach the final steps in the derivation of the energy quadratic form $\Pi^{(\beta\gamma)}$ [see (26)].

First, we substitute $\{u_a^{(\beta\gamma)}\}$ and $\{u_b^{(\beta\gamma)}\}$ in (28), and use the result to cast the matrix of $\Pi^{(\beta\gamma)}$ in the form of a quadratic function of the nodal displacements, $\{u_a^{(\beta\gamma)}\}$ and $\{u_b^{(\beta\gamma)}\}$.

$$\Pi^{(\beta\gamma)} = \frac{1}{2} (\{u_a^{(\beta\gamma)}\}^T \{u_b^{(\beta\gamma)}\}^T)[K_0^{(\beta\gamma)}] \begin{Bmatrix} \{u_a^{(\beta\gamma)}\} \\ \{u_b^{(\beta\gamma)}\} \end{Bmatrix} \quad (58)$$

The functional (58) can be varied with respect to the nodal displacements $\{u_b^{(\beta\gamma)}\}$. The variation yields:

$$\delta\Pi^{(\beta\gamma)} = [K^{(\beta\gamma)}] \begin{Bmatrix} \{u_a^{(\beta\gamma)}\} \\ \{u_b^{(\beta\gamma)}\} \end{Bmatrix} = \{0\} \quad (59)$$

where $[K^{(\beta\gamma)}]$ is a submatrix of $[K_o^{(\beta\gamma)}]$.

Note that $\{u_a^{(\beta\gamma)}\}$ and $\{u_b^{(\beta\gamma)}\}$ in (59) satisfy the boundary equilibrium condition (28) and the connectivity eqations (3), (4) and (20). The free nodal displacements $\{u_b^{(\beta\gamma)}\}$ can be determined from (59) as functions of $\{u_a^{(\beta\gamma)}\}$:

$$\{u_b^{(\beta\gamma)}\} = [R^{(\beta\gamma)}]\{u_a^{(\beta\gamma)}\} \tag{60}$$

This result can be substituted in Eq. (58) to reduce $\Pi^{(\beta\gamma)}$ to the quadratic form containing only $\{u_a^{(\beta\gamma)}\}$ displacements. The substitution of $\{a\}$ for $\{u_a^{(\beta\gamma)}\}$ from (57) yields the final form of $\Pi^{(\beta\gamma)}$:

$$\Pi^{(\beta\gamma)} = \frac{1}{2}\{a\}^T[K_a^{(\beta\gamma)}]\{a\} \tag{61}$$

Note that $[K_a^{(\beta\gamma)}]$ in (61) represents the energy contribution of the $(\beta\gamma)$-element to the entire assembly of a unit cell because the connectivity equations together with the transformation (57) represent the typical finite element assembly procedure for the $(\beta\gamma)$-elements.

The total energy for the unit cell is readily obtained when the sum of $\Pi^{(11)}$,$\Pi^{(21)}$,$\Pi^{(12)}$ and $\Pi^{(22)}$ is calculated

$$\Pi = \frac{1}{2}\{a\}^T[K_a]\{a\} \tag{62}$$

The procedure similar to that used in the derivation of (62) can be used to derive the energy Π_o of the homogenized unit cell in the terms of the contact point displacements $\{a\}$.

The simplest way to derive the quadratic form of Π_o is to consider the entire unit cell as a single finite element and apply the results of Section 2 and the displacement interpolation functions (31). Without going into the detials of required algebra, the matrix form of the homogeneized unit cell energy Π_o can be written as follows:

$$\Pi_o = \frac{1}{2}\{a\}^T[H_o]^T[M_o]^{-1}[H_o]\{a\} \tag{63}$$

Vector $\{a\}$ in (62) and (63) consists of the twelve contact point displacements [see definition (53)]. These displacement characterize general motion of the unit cell, including the rigid body motion. The rigid body motion must be eliminated from the unit cell energies Π and Π_0, in order to obtain a unique solution of $[M_0]^{-1}$. To accomplish the rigid body elimination, the unit cell can be supported in three points in such a way that six displacements $a_i^{\beta\gamma}$ from $\{a\}$ are equal to zero. It has been shown by Teply and Dvorak [5] that the rigid body motion elimination yields an invertible matrix $[H_0]$. After the elimination of the rigid body motion, the equivalence of (62) and (63) carries a simple formula for the calculation of overall moduli.

$$[M_0]^{-1} = [H_0]^{T^{-1}}[K][H_0]^{-1} \tag{64}$$

For boundary conditions (30), $[M_0]^{-1}$ is equal to overall stiffness matrix $[L_0]$. Note that a different $[L_0]$ would be obtained when displacement boundary conditions are assumed, see [5]. For nonlinear problems, energy increment $\Delta\Pi$ and $\Delta\Pi_0$ must be used in derivation of (64). It is simple to show that the comparison of $\Delta\Pi$ and $\Delta\Pi_0$ yield the same result (64) when instantaneous local moduli are considered in the derivation of $[K]$. In order to calculate the instantaneous local moduli in $\beta\gamma$-elements, it is necessary to establish relation between traction increment Δt_i^* and displacement increments $\Delta u_{ik}^{(\beta\gamma)}$, and $\Delta a_{ik}^{(\beta\gamma)}$. The former is readily available when (50) is transformed to an incremental form. The latter is obtained when (60) and (57) are first transformed in incremental forms and then substituted in the incremental form of (50). Finally, the incremental form of (49) must be used to obtain:

$$\{\Delta S^{(\beta\gamma)}\} = \frac{1}{\Omega^{(\beta\gamma)}} [M^{(\beta\gamma)}]^{-1}[H^{(\beta\gamma)}]\{\Delta u^{b\gamma}\}$$

or

$$\{\Delta S^{(\beta\gamma)}\} = \frac{1}{\Omega^{(\beta\gamma)}} [M^{(\beta\gamma)}]^{-1}[H^{(\beta\gamma)}][R^{\beta\gamma}][t^{\beta\gamma}]\{\Delta a\} \tag{65}$$

where $[M^{(\beta\gamma)}]^{-1}$ is the matrix of instantaneous stiffness moduli in the $(\beta\gamma)$-element, and

$$\Delta t_{ij}^* = \Delta\sigma_{ij}^o n_j = \Delta S_{ij}^{(\beta\gamma)} n_j \tag{66}$$

Equation (64) and (65) are extensions of Aboudi's model to arbitrary, nonproportional loading paths for elastic-plastic and viscoplastic composite materials. Equation (64) can also be used to show that Aboudi's model supplies lower bounds to the actual stiffness moduli.

5. Discussion of Results

It is shown that, for Aboudi's representative cell and microstructure, the finite element formulation results in the overall stiffness moduli $[M_0]^{-1}$ that is identical to that of Aboudi. It has also been shown that the use of the Teply-Dvorak's concept of the equivalence of the contact point displacements [5] yields the solution of $[L_0]$ or $[M_0]^{-1}$ in the same matrix form as it is used in the Periodic Hexagonal Model.

Since the Hellinger-Reissner variational principle is also used in the PHA model to establish bounds on elastic and elastic-plastic instantaneous overall moduli, it is possible to determine the relationship between Aboudi's approximation of $[M_0]^{-1}$ and the actual stiffnesses $[M^*]^{-1}$ of the periodic microstructure. In both models, PHA and Aboudi's, the two independent variables - stress and displacement - are related in such a way that the Hellinger-Reissner principle is transformed in the complementary energy or minimum stress rate principles. The relationship between stress and displacement fields in the Aboudi's model is shown in (28). It is shown in [5] that for the minimum stress rate principle and the boundary conditions (30), the following inequality is valid for the coefficient of diagonalized matrices $[M_0]$ and $[M^*]$

$$\text{diag } m_i^* \le \text{diag } m_{0i} \tag{67}$$

where m_i^* are ordered eigenvalues of $[M^*]$ and m_{0i} are ordered eigenvalues of $[M_0]$. For detail derivation of (67) see [5]. Equation (67) shows that the diagonal coefficients of Aboudi's compliance matrix, $\text{diag}[M_0]$, are always higher than the coefficients of the actual compliances $\text{diag}[M^*]$. If overall stiffness $[L_0]$ is equal to $[M_0]^{-1}$, one can readily relate the stiffness moduli $\text{diag}[L_0]$ to $\text{diag}[L^*]$,

$$\text{diag } \ell_{0i} \le \text{diag } \ell_i^* \tag{68}$$

In summary, it has been shown that the stiffness and compliance moduli calculated via Aboudi's model can be used as lower and upper estimates of the actual moduli of fibrous composite.

It is also shown in [5] that bounds on elastic or elastic-plastic instantaneous moduli can depend on the model microstructural properties such as fiber shape and fiber arrangement in a periodic array. It has been shown in [10] and [11] that fiber shape and fiber arrangement have a pronounced influence on the estimates of overall stiffness or compliance moduli $[L_0]$ or $[M_0]$. For example, it is shown in [10] that different elastic-plastic solutions are obtained when hexagonal, dodecagonal and higher n x 6-sided polygons are used to approximate circular fibers in the PHA model. Similar results are expected for square or rectangular fiber arrays used by Aboudi's.

In order to estimate the effect of fiber shape approximation on Aboudi's estimates of overall mechanical properties $[M_0]$ and $[L_0]$, Aboudi's unit cell has to be redefined. The change in the selection of the unit cell is necessary because the rounding of the fiber element (11-element) would ask for different connectivity equations (3) and (4) to account for the periodicity of the displacement in Aboudi's model. The periodic conditions for the displacement can be preserved and the approximation of circular fibers by n x 4-sided polygons readily introduced to Aboudi's model when the unit cell vertices are shifted to the fiber centroids as shown in Figure 6. It is probably obvious to the reader from the derivations of (3), (4) and (33) that these kinematic equations remain unchanged for the new unit cells. Additionally, the boundary and interelement equilibrium equations (30) and (22) are also preserved in this unit cell. Since both kinematic and equilibrium equations are preserved, matrices $[K]$, $[H_0]$ and most importantly $[M_0]^{-1}$ must be identical for Aboudi's and the new unit cell.

Having established the identity between Aboudi's and the unit cell defined in Figure 6, we may now approach the approximation of circular fibers by n x 4 sided polygon or even by arbitrary n-sided polygon. The simplest extension of Aboudi's square or rectangular fibers is to octagonal fibers. The unit cell with octagonal fibers is shown in Figure 7. For this unit cell, the connectivity equations (3), (4) and (20) remain unchanged; the displacement compatibility at the contact points (33) and the boundary interelement equilibrium equations (30) and (22) must be reformulated in terms of new shape functions for $(\beta\gamma)$-elements. The reformulation is simple and straightforward. Note that matrices $H^{(\beta\gamma)}$ must be changed, but $[H_0]$ remain the same as for Aboudi's unit cell.

The addition of more sides to the polygonal approximation of the fiber cross section is asking for more subcells in the unit cell. The subdivision of the unit cell into more than four subcells leads us to the outside of the framework of Aboudi's model and bring us closer to the finite element convergence studies used for the PHA model in [10]. In such convergence studies, Aboudi's model or its extension to

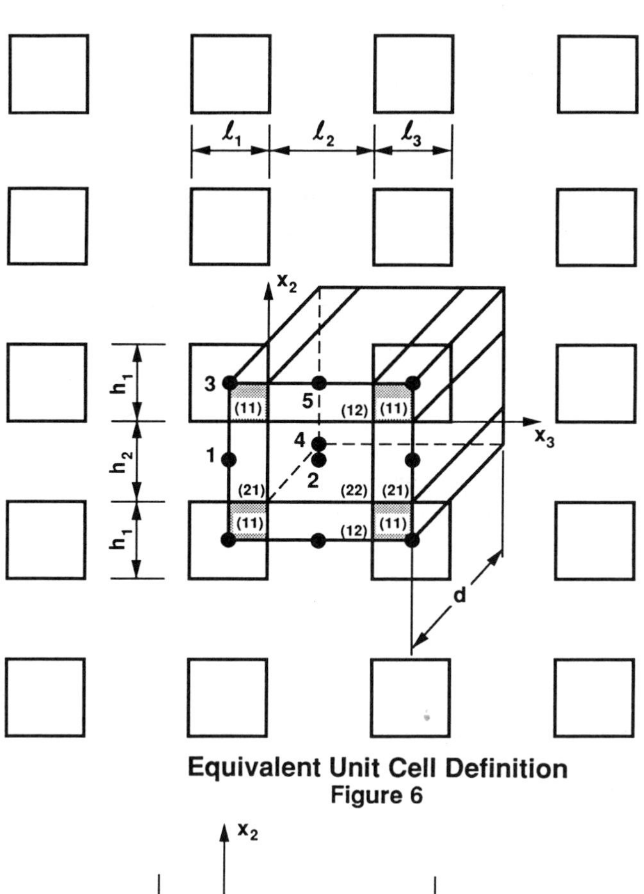

Equivalent Unit Cell Definition
Figure 6

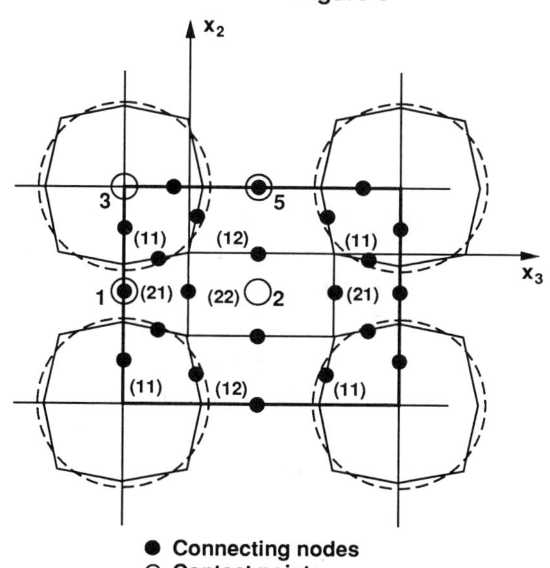

Unit Cell with Octagonal Fibers
Figure 7

octagonal fibers may serve as first estimates for elastic-plastic or viscoplastic instantaneous moduli.

References

1. Hill, R., "Theory of Mechanical Properties of Fiber-Strengthened Materials," *J. Mech. Phys. Solids*, Vol. 13, pp. 189 (1965).
2. Hashin, Z. and Shtrikman, S., "A Variational Approach to the Elastic Behavior of Multiphase Materials," *J. Mech. Phys. Solids,* Vol. 10 p. 343 (1962).
3. Dvorak, G. J. and Bahei-El-Din, Y. A., "Elastic-Plastic Behavior of Fibrous Composites," *J. Mech. Phys. Solids*, Vol. 27, p. 51 (1979).
4. Aboudi, J., "A Continuum Theory for Fiber-Reinforced Elastic-Viscoelastic Composites," *Int. J. Engng. Sci.*, Vol. 20, p. 605 (1982).
5. Teply, J. L., and Dvorak, G. J., "Bounds on Overall Instantaneous Properties of Elastic-Plastic Composites," *J. Mech. Phys. Solids*, Vol. 36, No. 1, pp. 29-58 (1988).
6. Reddy, J. N., *Energy and Variational Methods in Applied Mechanics*, John Wiley, New York (1984).
7. Reddy, J. N., *An Introduction to the Finite Element Method,* McGraw-Hill, New York (1984).
8. Teply, J. L., "Periodic Hexagonal Array Model for Plasticity Analysis of Composite Materials," The University of Utah, Ph.D. Dissertation, 1984, University Microfilm International-8409545.
9. Arenburg, T. and Reddy, J. N., "Elastoplastic Analysis of Metal Matrix Composite Structures," Virginia Polytechnic Institute and State University Report, CCMS-89-02, February 1989.
10. Shah, R. and Teply, J. L., "A Final Element Convergence Study of the Periodic Hexagonal Array Model," Alcoa Division Report 57-89-32, Alcoa Laboratories, PA 15069, November 1989.
11. Brokenbrough, J. R. and H. A. Wienecke, "Constitutive Response of Continuous Fiber-Reinforced Metal-Matrix Composites: Effect of Fiber Shape and Spatial Distribution," Alcoa Division Report 57-89-35 Alcoa Laboratories, PA 15069, December 1989.

Inelastic Behavior III

A Micromechanical Composite Yield Model Accounting for Residual Stresses

C. T. Herakovich, J. Aboudi[1] and J. L. Beuth, Jr.
Civil Engineering Department
University of Virginia
Charlottesville, VA 22903

Abstract

An analytical micromechanical model is used to predict yielding in continuous-fiber unidirectional metal-matrix composite materials. The von Mises criterion is used to predict yielding of the composite matrix based on (1) the average stresses in the matrix, and (2) the largest of the average stresses in each of the modelled matrix subcells. Two-dimensional yield surfaces are generated under thermomechanical loading conditions for two metal matrix composites, boron/aluminum and silicon carbide/titanium. Results indicate that, depending on the material, temperature excursions typically experienced in processing may cause matrix yielding at zero far-field applied stress. The analysis shows that thermal stresses distort and shift the yield surface based upon subcell stresses. Thus the importance of micromechanics is demonstrated.

1. Introduction

The ability to use metal matrix composites at high temperatures is one of their important advantages over resin matrix composites. Since the metal matrix is an elastoplastic material, it appears that the prediction of the overall yield surface of the composite is a fundamental step toward the study of its behavior. Yielding of the composite is caused by the yielding of its metal matrix. The prediction of the initial yield surfaces of metal matrix composite in the absence of thermal effects was presented by Pindera and Aboudi (1988). It was shown that yield surfaces generated on the basis of the average matrix behavior generally underestimate initial yielding as compared with predictions based on local matrix stresses and that the results obtained on the basis of local matrix stresses correlate very well with finite element predictions of Dvorak et al (1973). The approach presented by Pindera and Aboudi (1988) is based on the micromechanical model of periodic array of fibers which was recently reviewed by Aboudi (1989). This micromechanical approach is analytical and requires minimal computational effort, while offering the ability to model generalized

[1]Visiting from Tel Aviv University, Tel Aviv, Israel

loading conditions. In the case of metal matrix composites, the prediction of initial yield surfaces can be easily carried out with minimal effort.

In the present paper, initial yield surfaces of metal matrix composites at elevated temperatures are generated on the basis of the above micromechanical model (the method of cells) under a variety of thermomechanical loading conditions. The effects of the presence of residual stresses on the initial yield surfaces of the composite are studied. It is shown that depending on the applied stress state and temperature loading, the residual stresses will cause a translation of yield surfaces coupled with changes in size and shape. Furthermore, depending on the material combination and temperature excursion experienced in processing, the matrix may yield at zero far-field applied stress. Predictions based on the average stresses in the entire ductile matrix are also given for comparison.

2. Micromechanical Model

The geometry of the micromechanical model analyzed by the method of cells is illustrated in Fig. 1. The fiber of the composite extends along the 1-direction. A representative volume element of the composite consists of four subcells. Subcell ($\beta = 1, \gamma = 1$) represents a single material fiber. Subcells ($\beta, \gamma = 1, 2$; $\beta + \gamma \neq 2$) represent the surrounding matrix material. The overall behavior of the composite is obtained by imposing continuity of displacements and tractions between subcells and neighboring cells on an average basis. Assuming perfectly elastic fiber and matrix materials, the following thermomechanical constitutive equation can be established for the overall behavior of the unidirectional fibrous composite (Herakovich et al, 1988)

$$\bar{\boldsymbol{\sigma}} = \mathbf{E}\,\bar{\boldsymbol{\varepsilon}} - \mathbf{U}\,\Delta T \tag{1}$$

where $\bar{\boldsymbol{\sigma}}$ and $\bar{\boldsymbol{\varepsilon}}$ are the average stresses and strains in the composite, ΔT is the temperature deviation from a reference temperature at which the composite is stress-free when the strains are zero. In (1), $\mathbf{E}$ is the matrix of the effective stiffnesses of the composite which represents its overall transversely isotropic behavior, and $\mathbf{U}$ is a vector which incorporates the thermal effects. All material properties are considered temperature independent in the present analysis. The explicit form of $\mathbf{E}$ and $\mathbf{U}$ can be found in Aboudi (1989).

It can be easily verified that the predicted effective moduli are in excellent agreement with those computed from elasticity analysis for glass/epoxy and graphite/epoxy systems (Pickett 1968, Chen and Chang 1970, Behrens 1971).

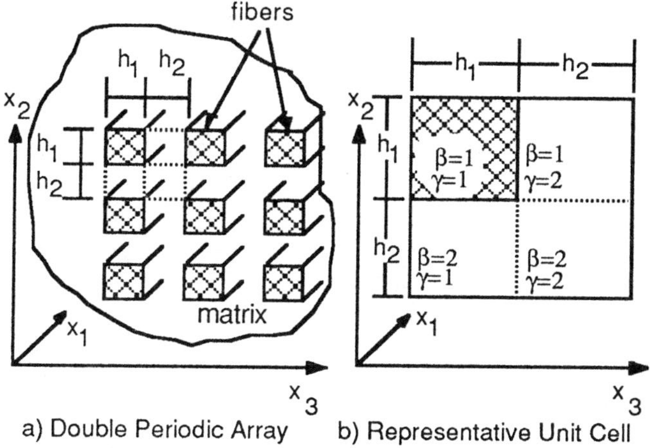

Fig. 1. Method of cells model

Such comparisons are presented in Figs. 2 & 3. Furthermore, as shown in Fig. 4, the effective transverse shear modulus of a unidirectional glass/polyester composite as predicted from the present micromechanical approach, coincides with the corresponding modulus computed from the three phase model (Christensen 1979). Also shown in the figure is the lower bound as predicted by Hashin & Shtrikman (1963) for arbitrary phase geometry. These comparisons are significant since the prediction of the transverse shear modulus or, equivalently, the transverse Young's modulus, of a unidirectional composite provides a critical check for the validity of a micromechanical model.

Let $S_{ij}^{(\beta\gamma)}$ represent the average stresses in the subcell $(\beta\gamma)$. It is possible to relate $S_{ij}^{(\beta\gamma)}$ to the average stresses $\bar{\sigma}_{ij}$ and the temperature deviation ΔT. The resulting equation can be represented in the form

$$\mathbf{S}^{(\beta\gamma)} = \mathbf{B}^{(\beta\gamma)}\, \bar{\boldsymbol{\sigma}} - \mathbf{V}^{(\beta\gamma)}\, \Delta T \tag{2}$$

The elements of the concentration matrix $\mathbf{B}^{(\beta\gamma)}$ and vector $\mathbf{V}^{(\beta\gamma)}$ can be readily determined from the micromechanics analysis. Assuming that yielding of the

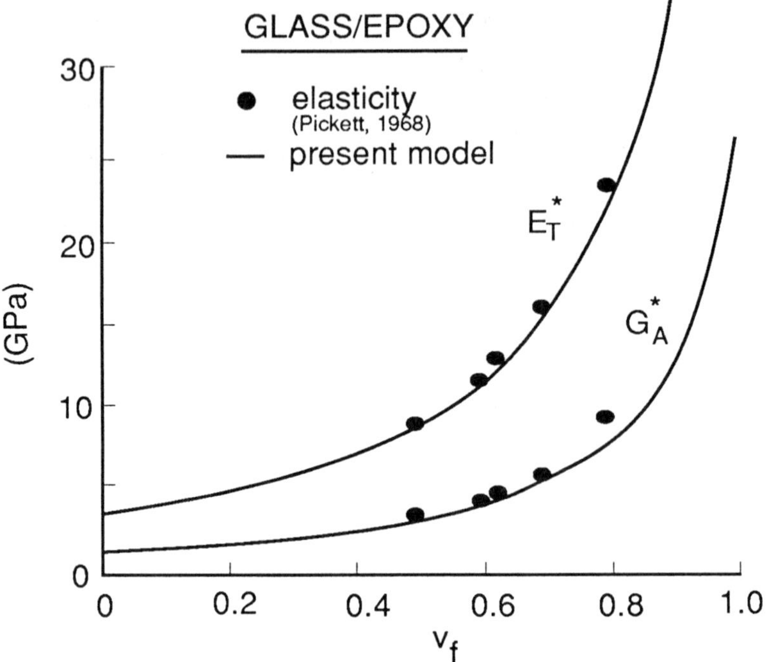

Fig. 2. Theoretical comparisons for elastic properties of unidirectional glass/epoxy.

matrix is governed by the von Mises criterion, it follows that yielding of the composite occurs when the von Mises condition in any given matrix subcell is fulfilled. This requires that

$$f(S_{ij}^{(\beta\gamma)}) = \frac{1}{2}\, \hat{S}_{ij}^{(\beta\gamma)}\, \hat{S}_{ij}^{(\beta\gamma)} - \frac{1}{3}\, Y^2 = 0 \qquad \beta + \gamma \neq 2 \tag{3}$$

where $\hat{S}_{ij}^{(\beta\gamma)}$ is the deviator of $S_{ij}^{(\beta\gamma)}$, and Y is the yield stress of the matrix in simple tension. When the average stress in the entire ductile matrix is employed for the prediction of the composite initial yield surface, $S_{ij}^{(\beta\gamma)}$ in (2) - (3) should be replaced by the matrix average stresses. These are given by

$$\bar{\sigma}_{ij}^{(m)} = \left[h_1\, h_2\, (S_{ij}^{(12)} + S_{ij}^{(21)}) + h_2^2\, S_{ij}^{(22)} \right] / \, (2h_1\, h_2 + h_2^2) \tag{4}$$

Consequently, eqn. (2) will take, in this case, the form

$$\bar{\boldsymbol{\sigma}}^{(m)} = \mathbf{B}^{(m)}\, \bar{\boldsymbol{\sigma}} - \mathbf{V}^{(m)}\, \Delta T \tag{5}$$

and yielding of the composite occurs when

$$f(\overline{\sigma}_{ij}^{(m)}) = \frac{1}{2}\,\hat{\sigma}_{ij}^{(m)}\,\hat{\sigma}_{ij}^{(m)} - \frac{1}{3}\,Y^2 = 0 \tag{6}$$

where $\hat{\sigma}_{ij}^{(m)}$ is the deviator of $\overline{\sigma}_{ij}^{(m)}$.

3. Results

Initial yield surface predictions are given for unidirectional metal matrix composites at room and elevated temperatures. The yield surfaces are generated on the basis of the largest of the average stresses in the matrix subcells as well as on the average stresses in the entire matrix. Two types of metal matrix composites are considered, boron/aluminum and silicon carbide/titanium (SCS-6/ Ti-6Al-4V). The properties of the four constituents are given in Tables 1 and 2. The fiber volume fraction, v_f, was equal to 0.3 for all calculations. All results were obtained using a user friendly, interactive PC software package (Herakovich, et al, 1988).

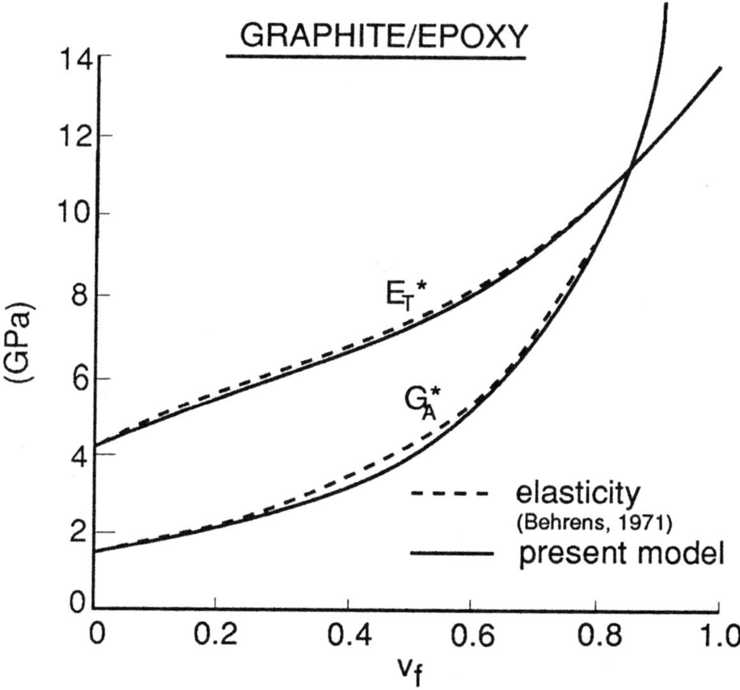

Fig. 3. Theoretical comparisons for elastic properties of unidirectional graphite/epoxy.

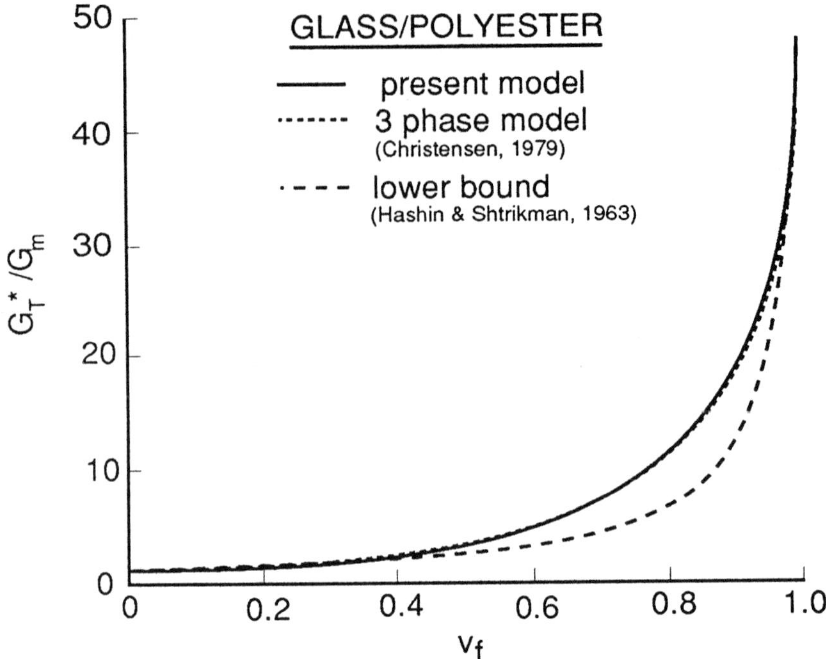

Fig. 4. Theoretical comparisons for transverse shear modulus of unidirectional glass/polyester.

For the given material constants and fiber volume fraction, the micromechanics model readily provides the temperature changes which cause yielding of the matrix phase of the unidirectional traction-free composites. Table 3 presents values for the temperature increase from the stress free state to

Table 1. Material properties of boron fibers and aluminum matrix.

	E (GPa) Young's modulus	**ν Poisson's ratio**	**Y (MPa) yield stress**	**α (10^{-6} / °C) coefficient thermal exp.**
Boron	**413.6**	**0.21**	**---**	**8.1**
Aluminum	68.9	0.33	262	23.4

Table 2. Material properties of SCS-6 fibers and Ti-6Al-4V matrix.

SCS-6	**E_A(GPa)+**	**ν_A**	**E_T(GPa)**	**ν_T**
	388.8	0.133	394.3	0.147
	G_A(GPa)	**Y (MPa)**	**$\alpha_A(10^{-6}/°C)$**	**$\alpha_T(10^{-6}/°C)$**
	142	---	5	4.6
Ti-6Al-4V	**E_A(GPa)**	**ν_A**	**E_T(GPa)**	**ν_T**
	110.3	0.31	110.3	0.31
	G_A(GPa)	**Y (MPa)**	**$\alpha_A(10^{-6}/°C)$**	**$\alpha_T(10^{-6}/°C)$**
	42.1	903	8.69	8.69

+ (subindices A and T denote axial and transverse directions)

cause yielding as predicted by both methods employed and for both materials under consideration. It can be readily noted that the predicted values provided by the two methods are quite different. This leads to the conclusion that the use of the matrix average stress for the prediction of yielding of metal matrix composites at elevated temperatures may lead to significant error. Clearly, some composite materials may have significant residual thermal stresses at room temperature. Depending upon the constituents, these residual stresses may actually cause yielding of the matrix material. Specifically, the ratio of fiber and matrix coefficients of thermal expansion and the ratio of the fiber and matrix elastic moduli play strong roles, as does the yield stress of the matrix. For example, the results in Table 3 indicate that B/Al yields at a much lower temperature than $SCS_6/Ti{-}6A{-}4V$. Indeed, the temperature change required to cause yielding in the two material systems differs by an order of magnitude.

Table 3. Temperature changes for first yielding of unidirectional composites ($v_f = 0.3$).

	ΔT (°C) subcell stress	ΔT (°C) average matrix stress
Boron / aluminum	241	382
Silicon carbide / titanium	2350	4330

Results of yield surfaces for unidirectional boron / aluminum are given in Figs. 5 - 8 under zero thermal loading conditions $\Delta T = 0$ and thermomechanical loading with $\Delta T = 222°C$. The different figures correspond to different biaxial applied stress states with all other components of applied stress being identically zero. It can be readily observed that for $\Delta T = 0$, (Figs. 5a, 6a & 7a) the difference between the two methods of prediction is not substantial. The shapes of the yield surfaces are similar for all cases, however, the matrix subcell stress method generally provides more conservative predictions. For yielding due to axial and transverse normal stresses, (Fig. 5a) the subcell yield surface is interior to the matrix average yield surface except at the two points $\bar{\sigma}_{11} = \pm Y$. The subcell yield surface for the combination of two transverse normal stresses is entirely within the matrix average yield surface (Fig. 6a), but the two yield surfaces are essentially identical for the case of transverse shear $\bar{\sigma}_{23}$ and axial normal $\bar{\sigma}_{11}$ loadings. This latter result indicates that the microlevel stress states associated with these two loadings are nearly uniform throughout the matrix. It is noted that the "corners" in the subcell predictions are associated with yielding of the individual subcells and are not indicative of the actual material response. The results in this study (for $\Delta T = 0$) agree with those of Pindera & Aboudi (1988).

For thermal mechanical loading (Figs. 5b, 6b, & 7b) there are substantial differences between the two predictive methods. The subcell yield surface is both

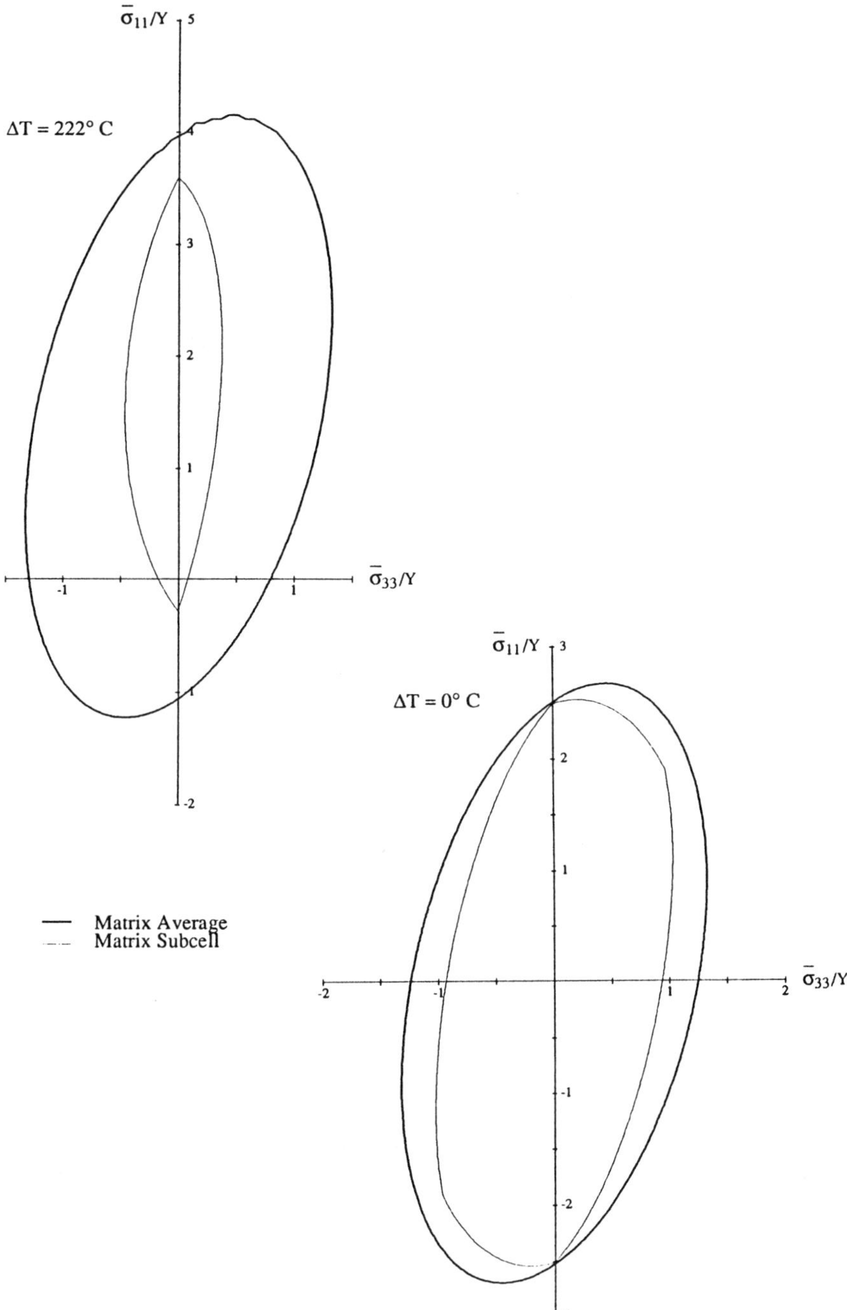

Fig. 5. Yield surface of a unidirectional boron / aluminum in $\bar{\sigma}_{11} - \bar{\sigma}_{33}$ stress space.

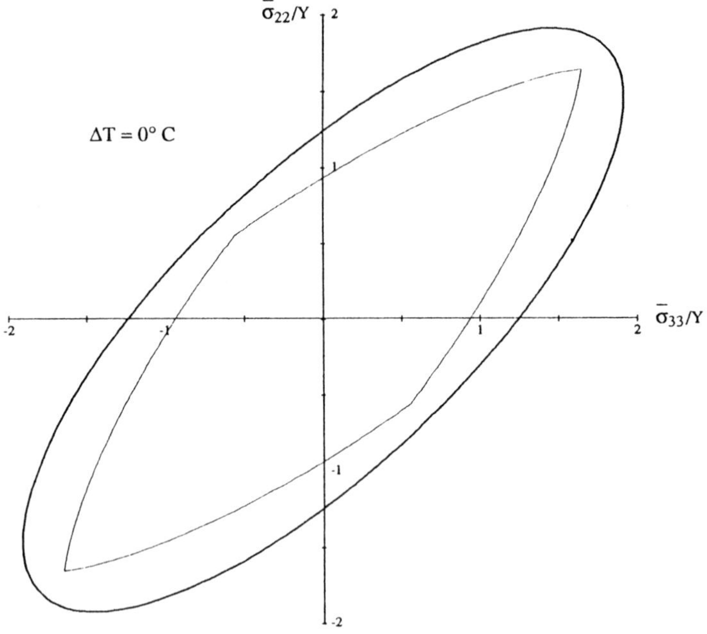

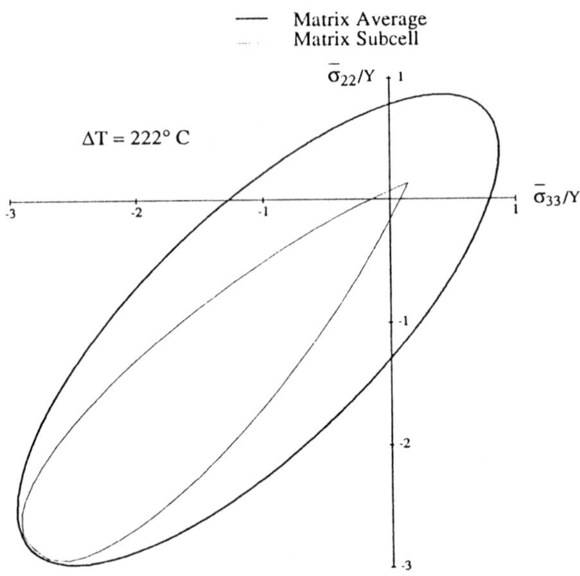

Fig. 6. Yield surface of a unidirectional boron / aluminum in $\bar{\sigma}_{22} - \bar{\sigma}_{33}$ stress space.

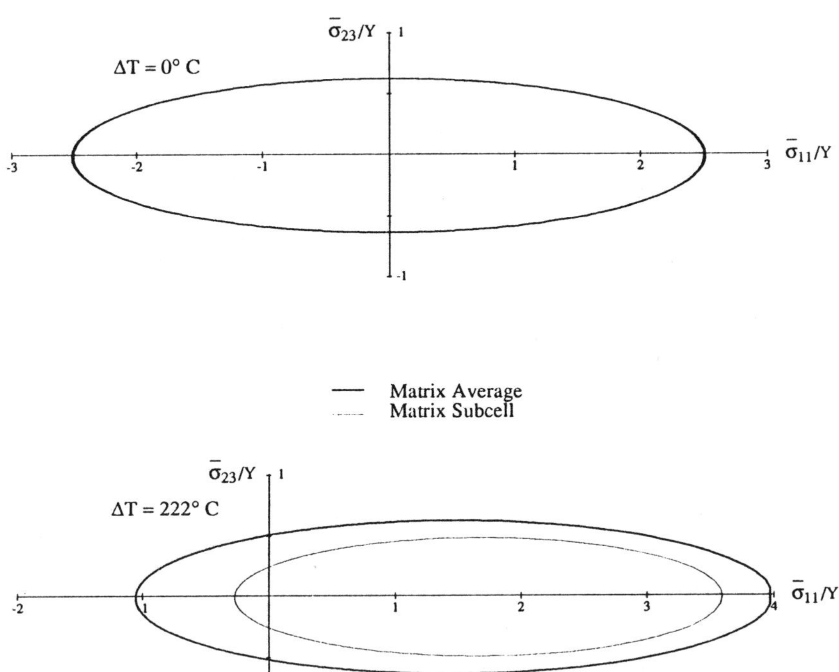

Fig. 7. Yield surface of a unidirectional boron / aluminum in $\bar{\sigma}_{23} - \bar{\sigma}_{11}$ stress space.

distorted and shifted by the presence of thermal stresses. In contrast, the matrix average yield surface experiences nearly pure shifting. For the combination of $\bar{\sigma}_{11}$ and $\bar{\sigma}_{33}$ loading (Fig. 5b) the yield surfaces translates along the $\bar{\sigma}_{11}$ axis. In the case of $\bar{\sigma}_{22}$ and $\bar{\sigma}_{33}$ (Fig. 6b), the translation is along the $\bar{\sigma}_{22} = \bar{\sigma}_{33}$ line and for $\bar{\sigma}_{23} - \bar{\sigma}_{11}$ loading (Fig. 7b) the translation is along the $\bar{\sigma}_{11}$ axis.

The subcell yield surfaces in Figs. 5b, 6b & 7b approach the origin at one point since the temperature change of 222°C is close to the ΔT required for thermal yielding (241°C). Temperature changes greater than 241°C would not cause the yield surface to translate beyond the origin; the initial yield surface for mechanical loading loses its significance in such a case since the material has already yielded.

Figure 8 shows a comparison of the subcell yield surfaces in the $\bar{\sigma}_{23} - \bar{\sigma}_{33}$ stress space. This figure clearly shows a strong effect of thermal stresses on the

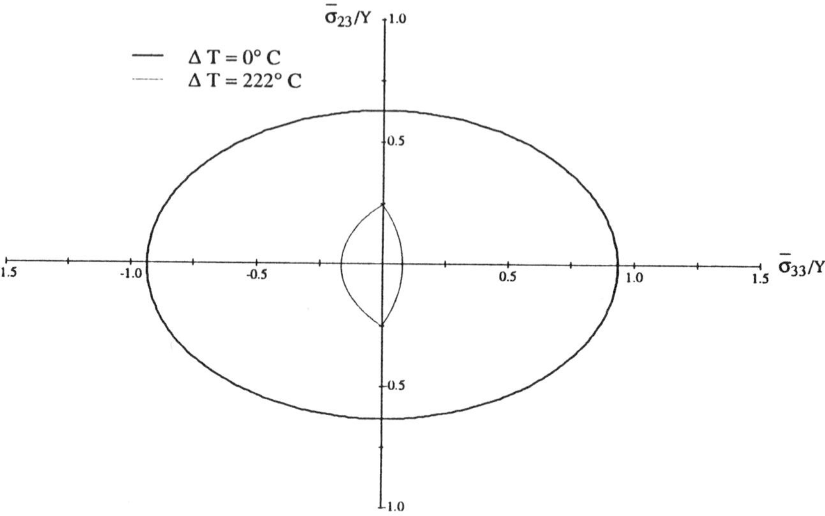

Fig. 8. Yield surface of a unidirectional boron / aluminum in $\bar{\sigma}_{23} - \bar{\sigma}_{33}$ stress space.

yield surface. At the elevated temperature the yield surface is distorted and shrunk considerably. A larger ΔT would shrink the yield surface even farther. The substantial differences noted in the predictions for the matrix subcell and average matrix stress methods for the various thermomechanical cases highlights the importance of modeling local matrix stresses. The use of average matrix stresses in analyzing inelastic response of composites may be misleading. Subcell yield surfaces for $SCS-6/Ti-6A-4V$ are presented for ΔT=0 and ΔT=1000°C in Figs. 9, 10, &11. This temperature change is only 42% of that required for yielding due to pure thermal loading. This is quite different from the temperature change used for the B/Al which was 92% of that required for thermal yielding. Although the ΔT for the $SCS-6/Ti-6A-4V$ is well below that required for yielding it nevertheless represents a large ΔT in terms of service environment and is 4.5 times the ΔT considered for B/Al. The results in $\bar{\sigma}_{23} - \bar{\sigma}_{33}$ stress space (Fig. 9) show a reduction in size of the yield surface and shift in the negative $\bar{\sigma}_{33}$ direction at elevated temperature. The results in $\bar{\sigma}_{13} - \bar{\sigma}_{22}$ stress space (Fig. 10) indicate an unusual phenomenon. The yield surface at elevated temperature is shifted in the negative $\bar{\sigma}_{22}$ direction and also distorted such that the elevated temperature yield surface is actually outside the room temperature yield surface

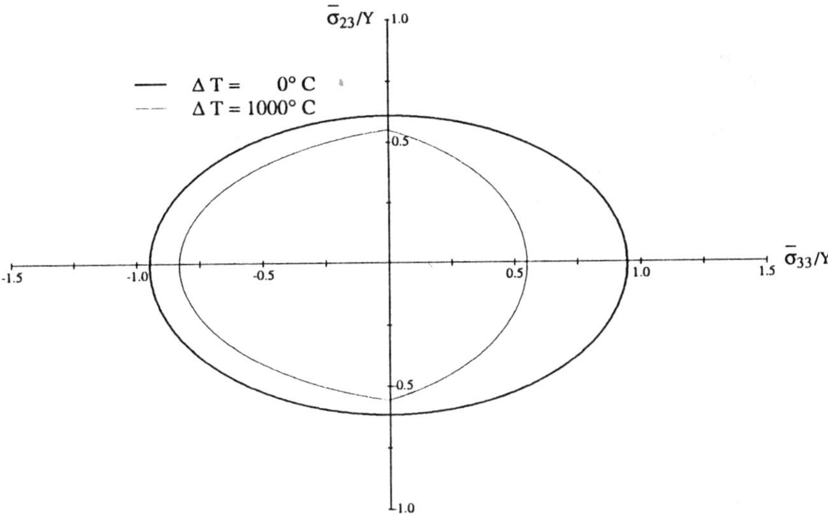

Fig. 9. Yield surface of a unidirectional *SCS–6/Ti–6Al–4V* in $\bar{\sigma}_{23} - \bar{\sigma}_{33}$ stress space.

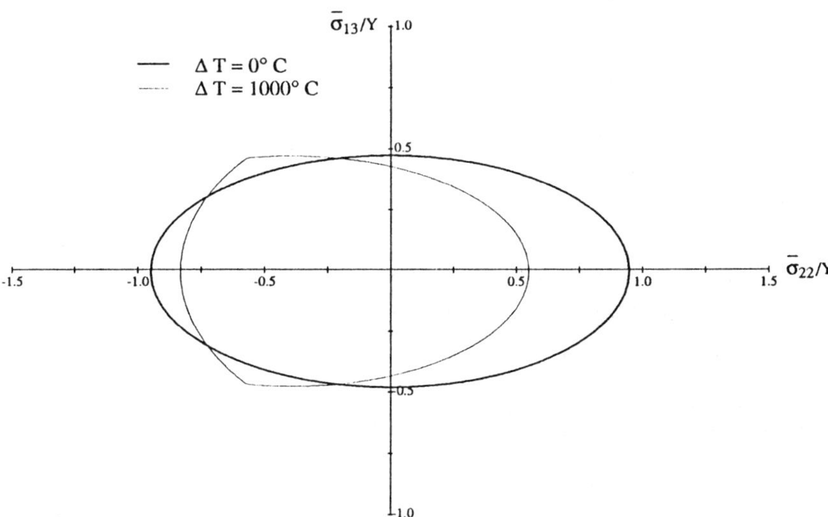

Fig. 10. Yield surface of a unidirectional *SCS–6/Ti–6Al–4V* in $\bar{\sigma}_{13} - \bar{\sigma}_{22}$ stress space.

over a range of negative $\bar{\sigma}_{22}$ values. The $\bar{\sigma}_{22} - \bar{\sigma}_{33}$ stress space results in Fig. 11 are very similar to those for *B/Al* in Fig. 6b. The elevated temperature yield surface has shifted along the $\bar{\sigma}_{22}=\bar{\sigma}_{33}$ line and distorted. The results indicate that thermal stresses have a more severe effect on the yield surface of the $SCS_6/Ti\text{–}6A\text{–}4V$ than on the *B/Al*.

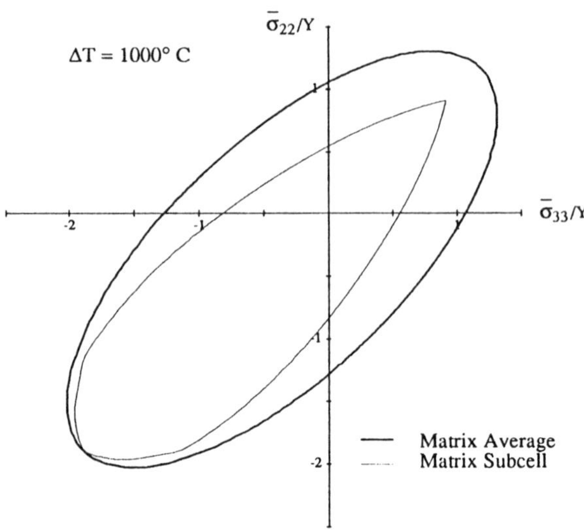

Fig. 11. Yield surface of a unidirectional $SCS\text{–}6/Ti\text{–}6Al\text{–}4V$ in $\bar{\sigma}_{22} - \bar{\sigma}_{33}$ stress space.

4. Conclusions

A micromechanical method that requires minimal computational effort has been used for the prediction of initial yield surfaces of metal matrix composite subjected to a variety of thermomechanical loadings. The two methods which have been employed in the generation of yield surfaces are based on the average stress in the matrix and subcell matrix stress.

Application of the von Mises criterion results in the prediction that some metal matrix composites may experience yielding at zero load due to relatively small temperature excursions during thermal processing. The temperature change required to cause yielding is a function of the mismatch between fiber and matrix materials, and the magnitude of matrix yield stress.

The study has shown the importance of a micromechanics analysis for the prediction of yield surfaces in the presence of thermal-mechanical loading. It has been shown that the subcell yield surface is both distorted and shifted due to thermal stresses. In contrast, the yield surface based upon the average matrix stresses is predicted to shift but not distort significantly due to thermal effects. Both the shifting and the distortion must be included in problems involving thermal mechanical yielding.

5. Acknowledgment

The authors are greatly appreciative of the support for this effort provided by the General Electric Company, the Center for Innovative Technology, and the University of Virginia's Center for Academic Enhancement.

6. References

Aboudi, J., 1989, "Micromechanical Analysis of Composites by the Method of Cells", *Appl. Mech. Rev. 42*, pp. 193-221.

Behrens, E., 1971, "Elastic Constants of Fiber-reinforced Composites with Transversely Isotropic Constituents", *J. Appl. Mech. 38*, pp. 1062-1065.

Chen, C. H. and Chang, S., 1970, "Mechanical Properties of Anisotropic Fiber-reinforced Composites", *J. Appl. Mech. 37*, pp. 186-189.

Christensen, R. M., 1979, *Mechanics of Composite Materials*, Wiley, New York.

Dvorak, G. J., Rao, M. S. M. and Tarn, J. Q., 1973, "Yielding in Unidirectional Composites Under External Loads and Temperature Changes", *J. Composite Mat. 7*, pp. 194-216.

Hashin, Z. and Shtrikman, S., 1963, "A Variational Approach to the Theory of the Elastic Behavior of Multiphase Materials", *J. Mech. Phys. Solids*, vol. 11, pp. 127-140.

Herakovich, C. T., Aboudi, J. and Beuth,Jr. J. L., 1988, "Nonlinear Composite Materials Micromechanics Model Phase II", Univ of Virginia, AM-88-01.

Pickett, G., 1968, "Elastic Moduli of Fiber Reinforced Plastic Composites", in *Fundamental Aspects of Fiber Reinforced Plastic Composites*, (R. T. Schwartz and H. S. Schwartz eds.) Interscience.

Pindera, M. J. and Aboudi, J., 1988, "Micromechanical Analysis of Yielding of Metal Matrix Composites", *Int. J. of Plasticity 4*, pp. 195-214.

The Effects of Thermal, Plastic and Elastic Stress Concentrations on the Overall Behavior of Metal Matrix Composites

F. Corvasce, P. Lipinski, M. Berveiller

Laboratoire de Physique et Mécanique des Matériaux, CNRS, ISGMP,
University of Metz (France)

Abstract

This study deals with the modelling of the inelastic thermomechanical behavior of Metal Matrix Composites considered as microinhomogeneous and macrohomogeneous materials. The theory is developed within the framework of the classical thermoelastoplasticity without damage. At first, the local constitutive equation is recalled. Introducing stress and (or) strain concentration tensors for mechanical and thermal fields, the behavior of the equivalent homogeneous medium is deduced in a compact form. In order to determine these concentration tensors, both for the reinforcements and the grains of the polycrystalline matrix, a general integral equation is proposed. Some various approximation methods for solving such an equation are listed.

The self-consistent approach is developed for a granular medium in order to determine the thermoelastic (linear) as well as the thermoelastoplastic behavior.

Results concerning monotonic loading are presented. A special attention is given to residual stresses. For complex loading pathes (biaxial loading), initial and subsequent yield surfaces are shown.

1. Introduction

The evaluation of the linear thermoelastic behavior of microinhomogeneous materials such as polycrystals or composites was studied by many authors : Kröner (1958), Levin (1967), Hill (1967)... In particular, Levin (1967) has shown that the overall thermal expansion coefficients of a composite medium can be calculated from the thermoelastic-constants of the constituents and the overall elastic moduli. The determination of the thermal stress concentration tensor is not necessary in order to find the overall thermal behavior.

In the past, many models based on the self consistent theory of Kröner (1958, 1961) have been proposed to evaluate the macroscopic behavior of heterogeneous media from the properties of the constituents and the interactions between them : Hutchinson (1970), Walpole (1981), Laws (1973), Dvorak (1986), Budiansky (1970), Berveiller and Zaoui (1979). More recently, the integral equation formulation for linear properties (Dederichs and Zeller (1973), Kröner (1977), Korringa (1973)) has been developed and unified variational treatment (Willis, 1977) generate estimations of overall properties.

Such an integral formulation was extended to elastoplasticity of polycrystals (Berveiller and Zaoui (1984), Lipinski and Berveiller (1989)). Thus, for linear behavior and weakly inhomogeneous elastoplasticity, one can consider that the formal solution of such kind of problem is known.

The question becomes different when dealing with thermoelastoplastic strongly inhomogeneous materials like metal-matrix composites (M.M.C.). At first, plastic strain in the matrix is a supplementary source of internal stress but simultaneously it can reduce internal stress (for example thermal) by plastic accomodation. Thus, even if the local elastic and plastic behavior is temperature independent, thermal, elastic and plastic internal stresses are strongly coupled which leads to a complex overall behavior. The second difficulty is associated with the nonlinearity of the plastic constitutive equation which needs an incremental formulation.

Partial solution of problems of this kind has been limited to simple microstructure and (or) restricted loading situations : De Silva and Chadwick (1969), Dvorak and Rao (1976). The most general development was proposed by Dvorak (1986) for fibrous composite with transversely isotropic phases.

It was shown that the overall thermal stress and strain vectors can be obtained through superposition of certain uniform field in the phases and local fields caused by uniform stress or strain. The elastoplastic behavior of the matrix is assumed to be piecewise linear.

This paper contains four major parts. In section 2, the local constitutive equations for thermoelastoplastic behavior are reviewed. Perfect bounding between reinforcements and matrix and between grains is assumed. The plasticity of the matrix is described through the glide on crystallographic slip systems inside the grains of the polycrystalline matrix. This allows to take into account some crystallographic (and) or morphologic textures of the matrix due to forming processes.

By analogy with the elastic behavior of microinhomogeneous material, (Kröner (1977), Dederichs and Zeller (1973)) the metal matrix composite is considered as a continuum medium with microstructure. Using equilibrium and compatibility conditions, the partial differential equations of the problem are transformed into an integral equation in section 3. A modified self consistent scheme, taking into account the influence of the reinforcement shape on the local fields inside the matrix is proposed in section 4. Applications and results are presented in the last part.

2. Local and overall constitutive equations

The infinite microinhomogeneous medium of volume V undergoes a thermomechanical loading defined by the state of overall stress Σ, the temperature θ and the corresponding rates $\dot{\Sigma}$ and $\dot{\theta}$.

At a point r of the medium, the total strain rate $\dot{\varepsilon}^T(r)$ is the sum of the elastic $\dot{\varepsilon}^e(r)$, the plastic $\dot{\varepsilon}^p(r)$ and the thermal $\dot{\varepsilon}^{th}(r)$ parts

$$\dot{\varepsilon}^T(r) = \dot{\varepsilon}^e(r) + \dot{\varepsilon}^p(r) + \dot{\varepsilon}^{th}(r) \tag{1}$$

For each grain of the matrix, the elastoplastic behavior is first specified when the elementary plastic deformation mechanism is the crystallographic glide (plastic resolved shear strain rate $\dot{\gamma}^g$) on (g) slip systems with n^g and m^g the unit slip plane and slip direction vectors, respectively. The corresponding plastic strain rate $\dot{\varepsilon}^p_{ij}$ is given by

$$\dot{\varepsilon}^p_{ij} = \sum_{g=1}^{N} R^g_{ij}\,\dot{\gamma}^g \tag{2}$$

with

$$R^g_{ij} = \frac{1}{2}\,(m^g_i\, n^g_j + m^g_j\, n^g_i) \tag{3}$$

The resolved shear stress rate $\dot{\tau}^g$ on (g) for a given stress rate $\dot{\sigma}_{ij}$ with respect to the crystal lattice is :

$$\dot{\tau}^g = R^g_{ij}\,\dot{\sigma}_{ij} \tag{4}$$

If a linear strain hardening behavior is assumed (H^{mn} = hardening matrix) so that the critical shear stress rate on (g), $\dot{\tau}^g_c$, is given by

$$\dot{\tau}^g_c = \sum_{h=1}^{N} H^{gh}\,\dot{\gamma}^h \tag{5}$$

the following formal constitutive equation may be written

$$\dot{\varepsilon}^p_{ij} = \sum_{g,h} R^g_{ij}\, K^{gh}\, R^h_{kl}\,\dot{\sigma}_{kl} \tag{6}$$

where $K = H^{-1}$ for the active system only.

The thermoelastoplastic strain rate for a grain of the matrix is now given by

$$\dot{\varepsilon}^{T}_{ij} = \left\{ s_{ijkl} + \sum_{g,h} R^{g}_{ij} K^{gh} R^{h}_{kl} \right\} \dot{\sigma}_{kl} + n_{ij} \dot{\theta} \tag{7}$$

where $s = c^{-1}$ is the crystal elastic compliance tensor (c = elastic moduli) and n are the local thermal expansion coefficients. Equation (7) is only valid if the local yield conditions are fullfilled (see Berveiller and Lipinski, 1989). A more general case of a constitutive equation including phase transformation was given by Patoor et al. (1989).

Now (7) is written in the form

$$\dot{\varepsilon}^{T}_{ij}(r) = g_{ijkl}(r)\, \dot{\sigma}_{kl}(r) + n_{ij}(r)\, \dot{\theta} \tag{8}$$

or by inversion as

$$\dot{\sigma}_{ij}(r) = l_{ijkl}(r)\, \dot{\varepsilon}^{T}_{kl}(r) - m_{ij}(r)\, \dot{\theta} \tag{9}$$

For the linear thermoelastic behavior of reinforcements, the constitutive equation has the same form as in (8) or (9) but with g = s and l = c.
The overall thermomechanical constitutive equation takes a form analogous to (8) or (9)

$$\dot{E}^{T}_{ij} = G_{ijkl}\, \dot{\Sigma}_{kl} + N_{ij}\, \dot{\theta} \tag{10}$$

or

$$\dot{\Sigma}_{ij} = L_{ijkl}\, \dot{E}^{T}_{kl} - M_{ij}\, \dot{\theta} \tag{11}$$

with

$$\begin{aligned} l &= g^{-1} \\ m &= l : n \\ L &= G^{-1} \\ M &= L : N \end{aligned} \tag{12}$$

The overall stress and strain rates $\dot{\Sigma}$ and $\dot{E}^{T}$ may be deduced from the local ones by the usual averaging operations

$$\dot{\Sigma}_{ij} = \frac{1}{V} \int_{V} \dot{\sigma}_{ij}(r)\, dV \equiv \overline{\dot{\sigma}_{ij}(r)} \tag{13}$$

$$\dot{E}^T_{ij} = \frac{1}{V}\int_V \dot{\varepsilon}^T_{ij}(r)\, dV \equiv \overline{\dot{\varepsilon}^T_{ij}}(r) \tag{14}$$

Following Hill (1977), thermal and mechanical concentration tensors relating local and overall fields are introduced for the local stress and strain tensors

$$\dot{\sigma}_{ij}(r) = B_{ijkl}(r)\, \dot{\Sigma}_{kl} + b_{ij}(r)\, \dot{\theta} \tag{15}$$

$$\dot{\varepsilon}^T_{ij}(r) = A_{ijkl}(r)\, \dot{E}^T_{kl} + a_{ij}(r)\, \dot{\theta} \tag{16}$$

By (13) (14) (15) and (16), one has

$$\bar{B} = I \quad , \quad \bar{A} = I \quad , \quad \bar{b} = 0 \quad , \quad \bar{a} = 0 \tag{17}$$

where I is the fourth order unit tensor.

Now the overall thermomechanical behavior follows from (9) (16) and (13), (14), as

$$\dot{\Sigma}_{ij} = \overline{l_{ijkl}(r)\, A_{klmn}(r)}\, \dot{E}^T_{mn} + \overline{l_{ijkl}(r)\, a_{kl}(r) - m_{ij}(r)}\, \dot{\theta} \tag{18}$$

The overall instantaneous moduli L and thermal stress coefficients M of (11) can be identified comparing with (18)

$$\begin{aligned} L_{ijmn} &= \overline{l_{ijkl}(r)\, A_{klmn}(r)} \\ M_{ij} &= \overline{m_{ij}(r) - l_{ijkl}(r)\, a_{kl}(r)} \end{aligned} \tag{19}$$

At a current loading state, local stress and plastic strain depending on the thermomechanical loading path are supposed to be known. Then, the local moduli l and m can be determined from (7) or (9) but the determination of L and M requires the knowledge of the concentration tensors A and a. They can be calculated by the self consistent method or using some other hypothesis. Nevertheless, all the models of the literature constitute an approximation of the integral equation proposed in the next section.

3. Integral equation formulation

We consider the inhomogeneous solid as a continuum medium with a microstructure satisfying

* the static equilibrium equation

$$\dot{\sigma}_{ij,j}(r) = 0 \tag{20}$$

* the thermomechanical constitutive relations (9)

$$\dot{\sigma}_{ij}(r) = l_{ijkl}(r)\, \dot{\varepsilon}^T_{kl}(r) - m_{ij}(r)\, \dot{\theta}$$

* the kinematic relations

$$\dot{\varepsilon}^T_{ij} = \frac{1}{2}(\dot{u}_{i,j} + \dot{u}_{j,i}) \tag{21}$$

where $\dot{u}$ is the velocity field.
For the sake of brevity, only the displacement boundary value problem, which requires the condition

$$\dot{u}_i(r) = \dot{u}^o_i \quad = \dot{E}^T_{ij}\, x_j \qquad (r \in \partial v) \tag{22}$$

is considered.
$\dot{\theta}$ is supposed uniform through the solid.
The question is now the calculation of $\dot{\varepsilon}^T(r)$ from $\dot{E}^T$ and $\dot{\theta}$ which allows to find A(r) and a(r). As it was done by Dederichs and Zeller (1973) or Kröner (1977) for an elastic behavior, it is convenient to introduce a homogeneous fictitious medium with tangent moduli L° and M°.
The spatial variation of l(r) and m(r) are now contained in δl and δm defined by

$$\begin{aligned} \delta l\,(r) &= l\,(r) - L^o \\ \delta m\,(r) &= m\,(r) - M^o \end{aligned} \tag{23}$$

Equations (9), (20) and (21) take the form

$$L^o_{ijkl}\, \dot{u}^T_{k,lj} + \left[\delta l_{ijkl}(r)\, \dot{\varepsilon}^T_{kl}(r) - \delta m_{ij}(r)\, \dot{\theta}\right]_{,j} = 0 \tag{24}$$

The solution of these equations may be obtained thanks to the Green tensor G° of the homogeneous fictitious medium through the integral equation

$$\dot{\varepsilon}^T_{ij}(r) = \dot{E}^T_{ij} + \int_V \Gamma_{ijkl}(r - r') \left[\delta l_{klmn}(r') \, \dot{\varepsilon}^T_{mn}(r') - \delta m_{kl}(r') \, \dot{\theta} \right] dV' \tag{25}$$

with

$$\Gamma_{ijkl}(r - r') = \frac{1}{2} \left(G^o_{ik,jl'} + G^o_{jk,il'} \right) \tag{26}$$

Here l', instead of l, indicates a derivation with respect to r'.

For a medium composed of grains inside the matrix and reinforcements, the first approximation of (25) may be obtained assuming an uniform thermoelastoplastic behavior within each constituent. In consequence, the strain rate $\dot{\varepsilon}^T$ is also assumed to be uniform at the grain or reinforcement level.

The deviation parts of the instantaneous moduli and thermal coefficients are now

$$\delta l\,(r') = \sum_N \Delta l^N \, H^N\,(r') \tag{27}$$

$$\delta m\,(r') = \sum_N \Delta m^N \, H^N\,(r') \tag{28}$$

where

$$H^N\,(r') = \begin{cases} 0 & \text{if } r' \notin V_N \\ 1 & \text{if } r' \in V_N \end{cases} \tag{29}$$

and V_N is the volume of constituent with the label N.
Similarly, by denoting the average value of the strain rate over a constituent N as

$$\dot{\varepsilon}^N = \frac{1}{V_N} \int_{V_N} \dot{\varepsilon}^T\,(r)\, dV \tag{30}$$

the strain rate field can be expressed in the form

$$\dot{\varepsilon}^T\,(r) = \sum_N \dot{\varepsilon}^N \, H^N\,(r) \tag{31}$$

so that the quantity $\delta l\,(r')\,\dot{\varepsilon}^T(r') - \delta m\,(r')\,\dot{\theta}$ is simply equal to

$$\left[\delta l\,(r')\,\dot{\varepsilon}^T(r') - \delta m\,(r')\,\dot{\theta}\right] = \sum_N \left[\Delta l^N \dot{\varepsilon}^N - \Delta m^N \dot{\theta}\right] H^N(r') \tag{32}$$

Substituting eq. (32) in the integral equation (25), one finds after taking the mean value on both side

$$\dot{\varepsilon}^N_{ij} = \dot{E}^T_{ij} + \sum_M T^{NM}_{ijkl} \left\{ \Delta l^M_{klmn}\, \dot{\varepsilon}^M_{mn} - \Delta m^M_{kl}\, \dot{\theta} \right\} \tag{33}$$

$$\text{with } T^{NM}_{ijkl} = \frac{1}{V_N} \int_{V_N} \int_{V_M} \Gamma_{ijkl}\,(r - r')\, dV\, dV' \tag{34}$$

For ellipsoïdal inclusions, and when M = N, the tensor T^{MM} is linked to the Eshelby's tensor S^M by the expression

$$S^M_{ijmn} = T^{MM}_{ijkl}\, L^o_{klmn}$$

For $M \neq N$, the tensor T^{MN} describes the interactions between two inclusions M and N. This tensor was introduced by Berveiller and al. (1987) where a method of calculation of T^{MN} in the case of a pair of ellipsoïdal inclusions embedded within an anisotropic matrix is also given.

Equation (33) contains directly the Eshelby's equivalence between inelastic strain ($\Delta m\,\dot{\theta}$) and an inhomogeneous mechanical strain ($\delta l\,\dot{\varepsilon}^T$) which is the basis of the Levin approach. The relation between thermal and mechanical concentrations tensors are contained in (33).

For example, for a single inclusion, equation (33) gives

$$\dot{\varepsilon}^N = \dot{E}^T + T^{NN}\,(\Delta l^N\, \dot{\varepsilon}^N - \Delta m^N\, \dot{\theta}) \tag{36}$$

The mechanical strain tensor A^N is given by

$$A^N = (I - T^{NN}\,\Delta l^N)^{-1} \tag{37}$$

and the thermal strain tensor a^N follows from (36) and (37)

$$a^N = (I - A^N)^{-1}\, \Delta l^{N-1}\, \Delta m^N \tag{38}$$

This formulae is analogous to the equation proposed recently by Benveniste and Dvorak (1990) by a more complex treatment.

For a real composite, the central problem is to solve the integral equation (25) in order to obtain, the concentration tensors A and a. Due to the strong inhomogeneous behavior of the MMCs and their particular two phase microstructure, a general acceptable approach is difficult to obtain.

In the next section, we propose a two stage self consistent approach which allows to take into account the special mechanical coupling between reinforcements and grains.
Another possibility lies in a multisite self consistent approach proposed in the case of elasticity by Fassi-Fehri and al. (1989) but this modelling is not investigated in this work.

4. A two stages self consistent approximation

The summation over all the N constituents which appears in the integral equation (33) makes difficult the determination of the concentration tensors. On the other hand, the total microstructure is unknown in general. These complex questions justifie the introduction of raisonnable approximations. The self consistent approach is well known in elasticity or plasticity of weakly inhomogeneous materials.
This approach has to be modified in the case of MMC for which a two stage localisation is necessary in order to take into account the granular structure of the matrix and the presence of reinforcement.
In view of the above relation between thermal and mechanical concentration tensors, we restrict ourselve to the discussion for the case where $\theta = 0$.
In this case, we have to find two types of localisation tensors : one A^{gM} for a grain g of the matrix and one A^I for the Ith reinforcement.
A direct one site self consistent scheme would give from (33) :

$$\dot{\varepsilon}^{gM} = \dot{E}^T + \overset{S}{T}{}^{gM}_{eff}\, \Delta l^{gM}_{eff}\, \dot{\varepsilon}^{gM} \tag{39}$$

$$\dot{\varepsilon}^I = \dot{E}^T + T^I_{eff}\, \Delta l^I\, \dot{\varepsilon}^I \tag{40}$$

where

$\dot{\varepsilon}^{gM}$ is the strain rate in the gth grain of the matrix

$\overset{s}{T}{}^{gM}_{eff}$ is defined by the integral (34) and calculated for a spherical inclusion located inside the infinite homogeneous medium with $L^o = L^{eff}$ of the elastoplastic tangent moduli Δl^{gM} ($\Delta l^{gM} = l^{gM} - L^{eff}$)

Here, we have supposed that the real shape of the grain of the matrix is spherical. It is clear from (39) that such a direct localisation for the grains of the matrix dont take into account the long range stress field in the neighbourhood of the reinforcement.

Instead of doing directly (39) and (40) it seems to be better and more precise to relate the strain rate of a grain g^M of the matrix to the mean strain rate ε^M of the matrix by the relation

$$\dot{\varepsilon}^{gM} = \dot{\varepsilon}^{M} + \overset{S}{T}{}^{gM}_{M} \, \Delta l^{gM}_{M} \, \dot{\varepsilon}^{gM} \tag{41}$$

where now

$\overset{s}{T}{}_M{}^{gM}$ is calculated from (34) for a spherical grain located inside the matrix with L^M for elasto plastic tangent moduli.

$$\Delta l^{gM}_{M} = l^{gM} - L^{M} \tag{42}$$

From the average theorem

$$f \, \dot{\varepsilon}^{I} + (1\text{-}f) \, \dot{\varepsilon}^{M} = \dot{E}^{T} \tag{43}$$

and the localisation equation (40), the relation between $\dot{\varepsilon}^M$ and $\dot{E}^T$ is now :

$$\dot{\varepsilon}^{M} = \frac{I - f\,(I - T^{I}_{eff} \, \Delta l^{I}_{eff})^{-1}}{(1 - f)} \, \dot{E}^{T} \tag{44}$$

From (41) and (44), one obtains the localisation tensor for a grain in the MMC

$$\dot{\varepsilon}^{gM} = \left(I - \overset{S}{T}{}^{gM}_{M} \, \Delta l^{gM}_{M}\right)^{-1} \left(\frac{I - f\left(I - T^{I}_{eff} \, \Delta l^{I}_{eff}\right)^{-1}}{1 - f}\right) \dot{E}^{T} \tag{45}$$

The so introduced localisation tensor is very different from those obtained directly from (39) which has the form

$$\dot{\varepsilon}^{gM} = \left(I - \overset{S}{T}{}^{gM}_{eff} \, \Delta l^{gM}_{eff}\right)^{-1} \dot{E}^{T} \tag{46}$$

It is now possible to come back to a classical one site self consistent scheme in which we are looking for an equivalent ellipsoïdal inclusion with unknown virtual shape "e" in order to ob-

tain the equivalence between (45) and (46).
We can write

$$\left(I - \overset{e}{T}{}^{gM}_{eff}\,\Delta l^{gM}_{eff}\right)^{-1} = \left(I - \overset{S}{T}{}^{gM}_{M}\,\Delta l^{gM}_{M}\right)^{-1} \left(\frac{I - f\left(I - T^{I}_{eff}\,\Delta l^{I}_{eff}\right)^{-1}}{1 - f}\right) \qquad (47)$$

from which the tensor $\overset{e}{T}{}^{gM}{}_{eff}$ can be calculated.

It is now easy to evaluate the effect of the long range stress field of the reinforcement on the mechanical localisation of the grains of the matrix.
For usual MMC, the heterogeneity of the matrix is small compared with those between matrix and reinforcement

$$\Delta l^{gM}_{M} << \Delta l^{gM}_{eff}$$

Neglecting the granular heterogeneity of the matrix ($\Delta l^{gM}{}_{M} \sim 0$) and supposing that the tensors $T_{eff}\,\Delta l$ are small in comparison with the unit tensor, a first order approximation of (47) gives

$$I + \overset{e}{T}{}^{gM}_{eff}\,\Delta l^{gM}_{eff} \approx \frac{I - f\left(I + T^{I}_{eff}\,\Delta l^{I}_{eff}\right)}{1 - f} \qquad (48)$$

or

$$(1 - f)\,\overset{e}{T}{}^{gM}_{eff}\,\Delta l^{gM}_{eff} = -\,f\,T^{I}_{eff}\,\Delta l^{I}_{eff} \qquad (49)$$

This expression shows a strong influence of the shape of the reinforcement on the equivalent inclusion for the grains of the matrix.
Introducing more simplifications, we assume that the L^{eff} can be approximated by the Voigt's hypothesis : $L^{eff} \sim f\,l^{I} + (1 - f)\,l^{M}$.

Equation (49) is now reduced to

$$\overset{e}{T}{}^{gM}_{eff} \approx T^{I}_{eff} \qquad (50)$$

This equation shows that the best choice for the equivalent inclusions describing the grains of the matrix is to assume that their shape are those of the reinforcements.
This model has been developed and some results are presented in the following section. These results confirm clearly the above conclusions.

5. Applications and results

The simulated MMC corresponds globaly to an Al-SiC composite. The material properties are summarized in Table II. Both phases are supposed to have an isotropic thermoelastic behavior.

The first application concerns the elasto plastic properties of the composite with the spherical reinforcement (a = b =c) and for several volume fractions (f = 0, 0.1, 0.2) loaded in tension at constant temperature. The evolution of the initial yield point and the hardening modulus is depicted in Fig. (1) as a function of volume fraction.

The next figure (Fig. 2) shows, for the same composite (a = b = c = 1, f = 0.1) the effect of the plastic offset on the shape of the initial yield surface in the Σ_{11} - Σ_{22} subspace. In this case, the isotropic response is observed as expected. For the elongated and parallel reinforcements one can expect a highly anisotropic behavior. Fig. 3 shows the initial yield surfaces for a composite with ellipsoïdal reinforcements (a = b = 1, c = 7, f = 0.1).
For purely mechanical loadings, it was verified that the shape of the equivalent grain of the matrix has a rather negligible influence on the overall behavior (fig. 4) for a tensile test.

On the other hand, during a thermomechanical loading the effect of the shape of the equivalent inclusion is much more important. The next figure (Fig. 5) presents the 33-component of the deviatoric part of the thermal stress concentration tensor b'_{33} for any grain of the matrix as a function of the aspect ratio c'/a' of the matrix grains ($b'_{33} = b_{33} - 1/3\ b_{kk}$)
These results have been obtained for ellipsoïdal reinforcements (c/a = c/b = 7) oriented in 3 direction and for three volume fractions (f = 0.1, 0.2, 0.3).
The rapid evolution of b'_{33} from low values corresponding to spherical matrix inclusions toward some asymptotical values for c'/a' = c/a can be observed.
The anisotropy of the thermomechanical behavior of the composite is visible for spherical grains of the matrix (c'/a' = 1) but this anisotropy has a smaller effect than the mechanical coupling between elongated reinforcements and matrix.
This phenomenon is also observed during a mechanical loading on the composite with preliminarily introduced thermal stresses.
Figures 6a, 6b and 6c present tensile and compression curves at two temperatures for three aspect ratios of the equivalent matrix grains. The influence of thermal stresses is very small when dealing with spherical grains and becomes more and more pronounced if the aspect ratios of grains tend to those of the reinforcements.

Table II

		Reinforcements	Matrix	Unit
Elasticity	μ	27 559	2 692	daN/mm^2
	ν	0.27	0.3	-
Thermal expansion coefficient	α	4.10^{-6}	21.10^{-6}	K^{-1}
Plasticity	slip systems	-	12 x (111)<110>	
	critical shear stress	-	$\tau^\circ = 14$	daN/mm^2
	hardening matrix	-	$H_1 = \mu/250$ $H_2 = A \times H_1$	daN/mm^2
Microstructure	Volumic fraction	f	(1 - f)	
	shape and orientation	ellipsoïdal a, b, c	real shape : spherical effective inclusion : ellipsoïdal a', b', c'	

The last figure (Fig. 7) has been obtained for a non proportional thermomechanical loading path. The composite has been first loaded in compression under the yield point and then its temperature has been increased. The thermaly induced plastic strain depends also strongly on the aspect ratio of the matrix'grains.

6. Discussion and conclusions

A general thermomechanical integral equation relating the local strain rate to the overall loading parameters has been proposed. This equation contains directly the relations between the thermal and mechanical strain concentration tensors.

All previous models can be deduced from this equation. In the case of one site self consistent approximation of the above integral equation for MMC materials, the attention must be focused on the mechanical coupling between the grains of the matrix and the effective medium.

By introducing a fictitious shape of the matrix grains, a new two stage self consistent approach has been developed.

Some applications concerning Al-SiC composites has been presented and the mentionned mechanical coupling effect was demonstrated.

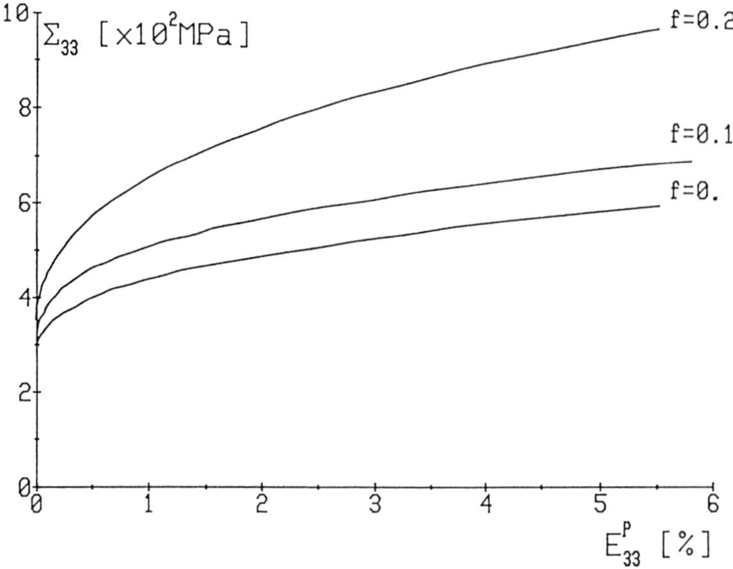

Fig. 1 : Uniaxial tensile curves for several volume fraction of spherical reinforcements (a = b = c).

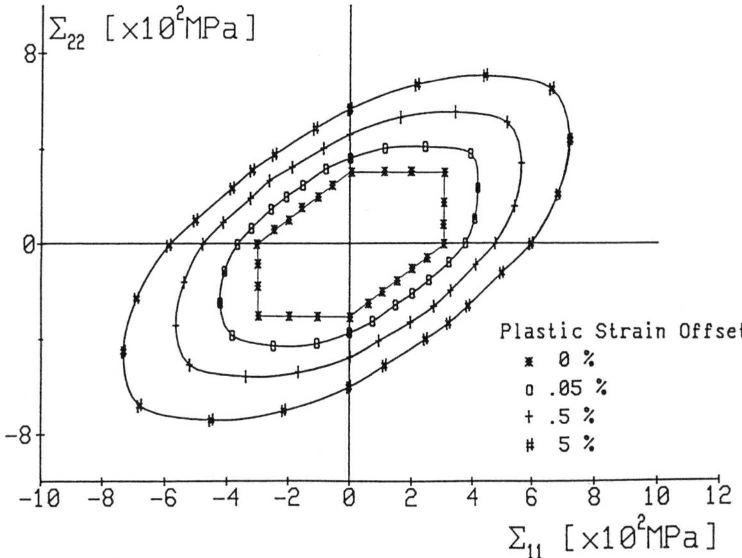

Fig. 2 : Initial yield surfaces for different offset of plastic strain. MMC with spherical reinforcements and 10% volume fraction.

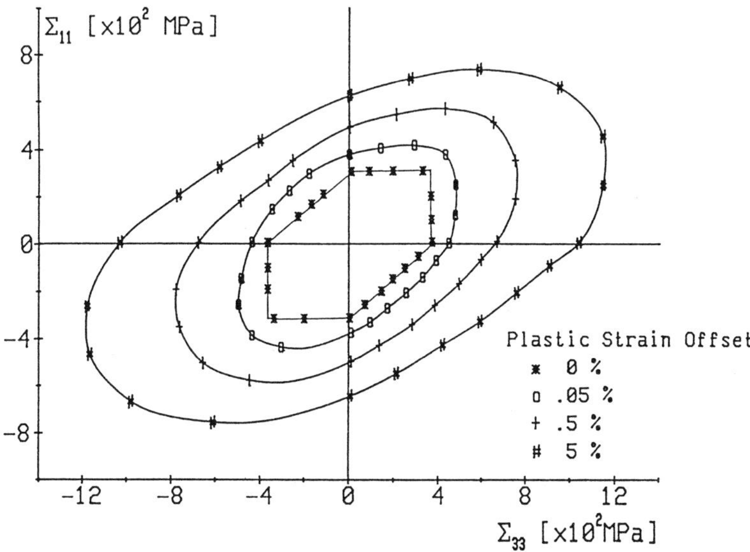

Fig. 3 : Initial yield surfaces for elongated reinforcements (10a = 10b = c) and 10% volume fraction.

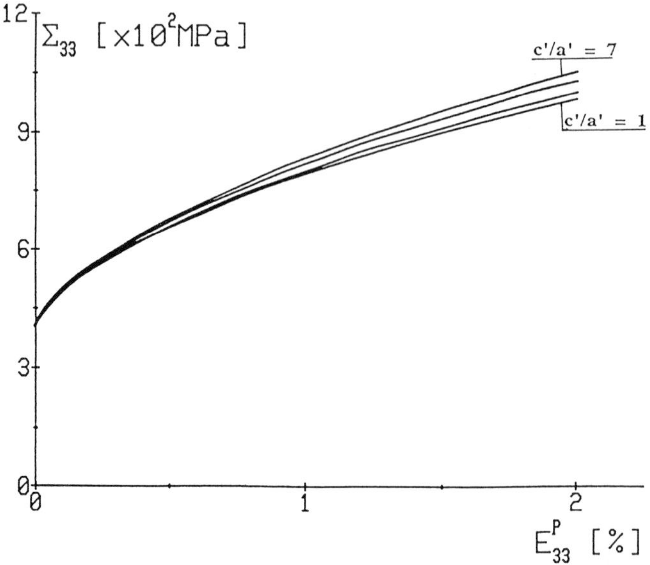

Fig. 4 : Effect of the equivalent grain shape on the tensile curve. MMC with 10% fibers and aspect ratio a = b = c/7.

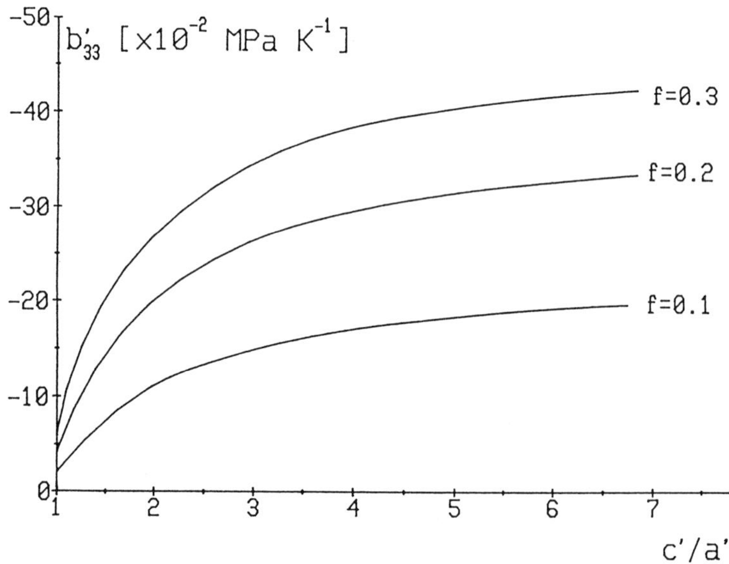

Fig. 5 : Evolution of the grain deviatoric stress concentration component b'_{33} as a function of the shape of equivalent grains. MMC with 10%, 20%, 30% volume fraction of parallel reinforcements whose aspect ratio is c/a = c/b = 7.

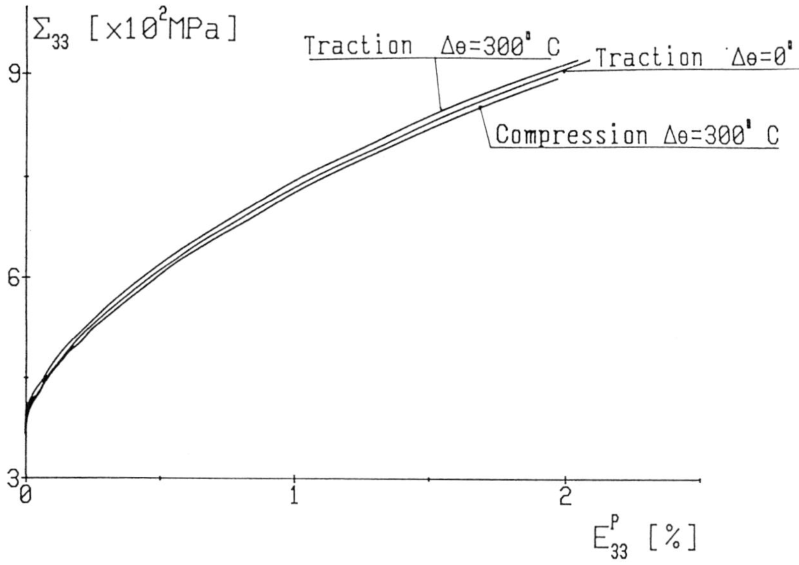

Fig. 6 : Tensile and compression curves with two temperature increments : $\Delta\theta = 0$ and $\Delta\theta = 300°C$. MMC with 10% volume fraction of ellipsoïdal (c/a = 7) reinforcements

(a) equivalent grains are spherical (a' = b' = c')

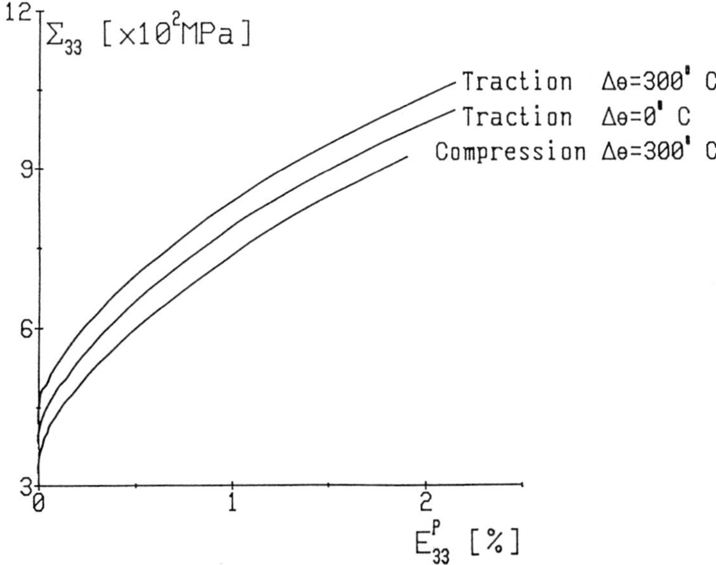

(b) equivalent grains are ellipsoïdal (c'/a' = c'/b' = 2)

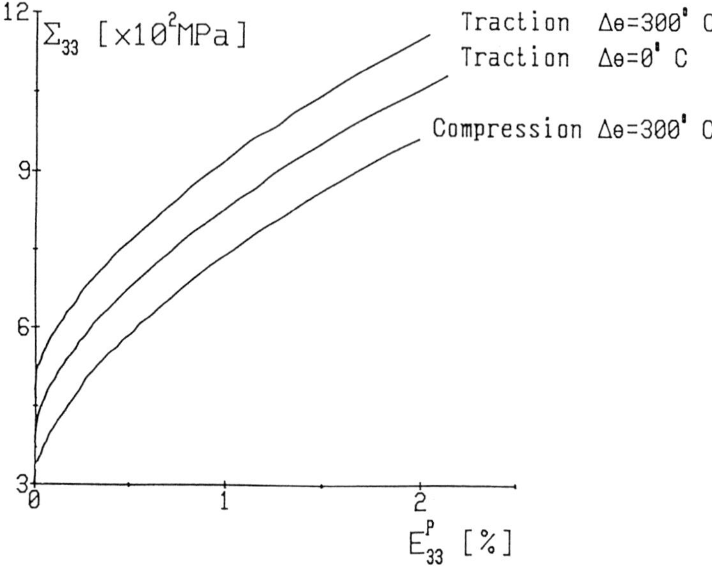

(c) equivalent grains are ellipsoïdal (c'/a' = c'/b' = 7).

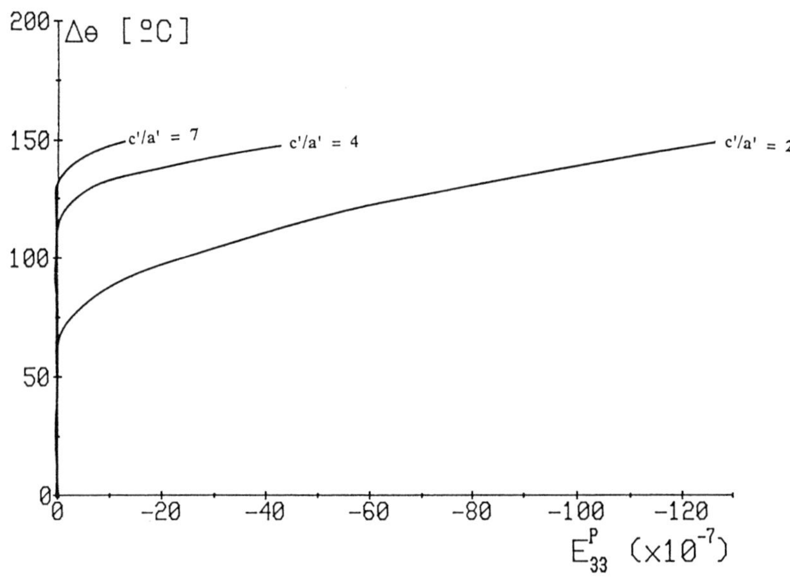

Fig. 7 : Thermal induced plastic strain at constant compression stress (Σ_{33} = - 365 MPa) MMC with 10% volume fraction of ellipsoïdal (c/a = 7) reinforcements and different shape for the equivalent inclusion describing the grain of the matrix.

References

Benveniste, Y., Dvorak, G.J., 1990, "On a correspondance between Mechanical and Thermal effects in Two Phase Composites", *Micromechanics and Inhomogeneity*, G.J. Weng and al. eds, Springer Verlag, New York.

Berveiller, M. and Zaoui, A., 1979, "An Extension of the self consistent scheme to plastically flowing polycrystals", *J. Mech. Phys. Sol.*, Vol 26, pp 325-339.

Berveiller, M. and Zaoui, A., 1984, "Modelling of the plastic behavior of inhomogeneous media", *J. of Engng. Mat. and Tech.*, Vol 106, pp 295-298.

Berveiller, M., Fassi-Fehri, O., Hihi, A., 1987, "The problem of two plastic and heterogeneous inclusions in an anisotropic medium", *Int. J. Engineering Sciences*, Vol 25, n° 6, pp 691-709.

Budiansky, B., 1965, "On the elastic moduli of some heterogeneous materials", *J. Mech. Phys. Solids*, Vol 13, pp 201-203.

Dederichs, P.H. and Zeller, R., 1973, "Variational treatment of the elastic constants of disordered materials", *Z. Phys.*, Vol 259, pp 103-116.

De Silva, A.R.T. and Chadwick, G.A., 1969, "Thermal Stresses in Fiber reinforced composites", *J. of Mech. Phys. of Sol.*, Vol 17, pp 387-403.

Dvorak, G.J. and Rao, M.S.M., 1976, "Thermal stresses in heat-treated fibrous composites", *ASME J. of Applied Mechanics*, Vol 98, pp 619-624.

Dvorak, G.J., 1986, "Thermal expansion of elastic-plastic composite materials", *ASME J. of Applied Mechanics*, Vol 53, pp 737-743.

Fassi-Fehri, O., Hihi, A., Berveiller, M., 1989, "Multiple site self consistent scheme", *Int. J. Engng Sci.*, Vol 27, n° 5, pp 495-502.

Hill, R., 1967, "The essential structure of constitutive laws for metal composites and polycrystals", *J. Mech. Phys. of Solids*, Vol 15, pp 79-89.

Hutchinson, J.W., 1970, "Elastic-plastic behavior of polycrystalline metals and composites", *Proc. Roy. Soc.*, Vol A319, pp 247-272.

Korringa, J., 1973, "Theory of elastic constants of heterogeneous media", *J. Math. Phys.*, Vol 14, pp 9-17.

Kröner, E., 1958, "Berechnung der elastischen Konstanten des Vielkristalls aus den Konstanten des Einkristalls", *Z. Phys.*, Vol 151, pp 504-518.

Kröner, E., 1961, "Zur plastischen Verformung des Vielkristalls", *Act. Metall.*, Vol 9, pp 155-161.

Kröner, E., 1977, "Bounds for effective elastic moduli of disordered materials", *J. Mech. Phys. Solids*, Vol 21, pp 9-17.

Laws, N., 1973, "On the thermostatics of composite materials", *J. Mech. Phys. Solids*, Vol 21, pp 9-17.

Levin, V.M., 1967, "Thermal expansion coefficients of heterogeneous materials", *Mekhanika Tverdogo Tela*, n° 2, pp 88-94.

Lipinski, P. and Berveiller, M., 1989, "Elastoplasticity of micro-inhomogeneous metals at large strains", *Int. J. Plasticity*, Vol 5, pp 149-172.

Patoor, E., Eberhardt, A., Berveiller, M., 1989, "Thermomechanical behavior of shape memory alloys", *Arch. of Mechanics*, Vol 40 (In press).

Walpole, L.J., 1981, "Elastic Behavior of Composite Materials : Theoretical fondations", *Adv. in Appl. Mechanics*, Vol 21, pp 169-242.

Willis, J.R., 1977, "Bounds and self consistent estimates for the overall properties of anisotropic composites, *J. Rech. Phys. Solids*, Vol 25, pp 185-202.

Inelastic Behavior IV

Residual Stresses in Fibrous Metal Matrix Composites: A Thermoviscoplastic Analysis

Erhard Krempl and Nan–Ming Yeh
Mechanics of Materials Laboratory
Rensselaer Polytechnic Institute
Troy, N. Y. 12180–3590

ABSTRACT

The vanishing fiber diameter model together with the thermoviscoplasticity theory based on overstress are used to analyze the thermomechanical rate (time)–dependent behavior of unidirectional fibrous metal–matrix composites. For the present analysis the fibers are assumed to be transversely isotropic thermoelastic and the matrix constitutive equation is isotropic thermoviscoplastic. All material functions and constants can depend on current temperature. Yield surfaces and loading/unloading conditions are not used in the theory in which the inelastic strain rate is solely a function of the overstress, the difference between stress and the equilibrium stress, a state variable of the theory. Assumed but realistic material elastic and viscoplastic properties as a function of temperature which are close to Gr/Al and B/Al composites permit the computation of residual stresses arising during cool down from the fabrication. These residual stresses influence the subsequent mechanical behavior in fiber and transverse directions. Due to the viscoplasticity of the matrix time–dependent effects such as creep and change of residual stresses with time are depicted. For Gr/Al residual stresses are affecting the free thermal expansion behavior of the composite under temperature cycling. The computational results agree qualitatively with scarce experimental results.

INTRODUCTION

Metal matrix composites consist of a ductile, usually low strength matrix reinforced with elastic, brittle and strong fibers. Ideally, the strength of the fiber and the ductility of the matrix combine to provide a new material with superior properties. Selecting the best combinations of fiber and matrix materials is a difficult task which involves conflicting demands and many compromises. To prevent self stresses from developing during cool down from the manufacturing temperature it is desirable to have the same coefficient of thermal expansion for fiber and matrix. This ideal, however, is seldom achieved as other considerations but the coefficient of thermal expansion have priority in selecting the constituent materials.

It is known that the residual stresses have an influence on the mechanical behavior, Cheskis and Heckel [1970], Dvorak and Rao [1976], Min and Crossman [1982]. Moreover, the thermal expansion behavior of metal matrix composites is shown to be influenced by the residual stresses, Garmong [1973], Kural and Min [1984] and Tompkins and Dries [1988]. In precision applications the exact thermal expansion behavior is of great interest as it influences the performance.

It is the purpose of this paper to provide a comparatively simple and approximate means of calculating the residual stresses in a unidirectional metal matrix composite during cool down from the manufacturing temperature and to assess their influence on subsequent mechanical behavior as well as on the thermal expansion of the composite under uniform temperature changes. To accomplish this task the vanishing fiber diameter model of Dvorak and Bahei–El–Din [1982] is combined with the thermoviscoplasticity theory based on overstress (TVBO) of Lee and Krempl [1990]. TVBO is a "unified" theory which does not separately postulate constitutive laws for creep and plasticity but models all inelastic deformation as rate dependent. Experiments with modern servocontrolled testing machines have shown rate dependence even at room temperature for engineering alloys, e.g stainless steels, Krempl [1979], 6061–T6 Al alloy, Krempl and Lu [1983], and Titanium alloys, Kujawski and Krempl [1981]. The transition from low to high homologous temperature behavior is usually characterized by a decrease in strength and an increase in rate dependence with an increase in temperature. This behavior can be modeled easily by TVBO by making certain constants depend on temperature. It is not necessary to postulate different laws in different temperature regimes.

First the governing equations are stated. They are represented by a system of first order, nonlinear, coupled differential equations which must be solved for a given boundary condition and loading/temperature history. Base data for 6061–T6 Al alloy for which viscoplastic material properties were determined by Yao and Krempl [1985]. Plausible changes of these properties with temperature were postulated and the system of differential equations was integrated to depict the properties of the model. Of special interest is the influence of the residual stresses set up during cooling from the manufacturing temperature of 660 °C. Owing to the viscoplastic nature of the matrix constitutive model the residual stresses redistribute while the composite is at ambient temperature. For the material properties chosen in the numerical experiment this redistribution slows down rapidly with time at ambient temperature and after 30 days a nearly constant residual stress state is reached. Since the subsequent response of the composite is affected by the residual stresses an influence of time spent at room temperature on the subsequent behavior is predicted by this analysis. The influence of residual stresses on the subsequent isothermal mechanical behavior and on the thermal expansion behavior of a composite subjected to thermal cycling is investigated by numerical experiments. The computations agree qualitatively with scarce experimental results reported by others.

THE COMPOSITE MODEL. THERMOVISCOPLASTICITY THEORY BASED ON OVERSTRESS (TVBO) AND THE VANISHING FIBER DIAMETER MODEL (VFD)

For the representation of the equations, the usual vector notation for the stress tensor components σ and the small strain tensor components ϵ are used. Boldface capital letters denote 6x6 matrices.

Stresses and strains without a superscript designate quantities imposed on the composite as a whole. Superscripts ^{f} and ^{m} denote fiber and matrix, respectively. The fiber volume fraction is c^f and c^m denotes the matrix volume fraction with $c^f + c^m = 1$.

A unidirectional fibrous composite element is assumed where the fiber is transversely isotropic thermoelastic, the matrix is isotropic and thermoviscoplastic and represented by TVBO. Fiber orientation in the 3–direction is postulated.

For the VFD model, Dvorak and Bahei–El–Din [1982], the following constraint equations hold

$$\begin{aligned} \dot{\sigma}_i &= \dot{\sigma}_i^f = \dot{\sigma}_i^m \quad \text{for } i \neq 3 \\ \dot{\sigma}_3 &= c^f\, \dot{\sigma}_3^f + c^m\, \dot{\sigma}_3^m \\ \dot{\epsilon}_i &= c^f\, \dot{\epsilon}_i^f + c^m\, \dot{\epsilon}_i^m \quad \text{for } i \neq 3 \\ \dot{\epsilon}_3 &= \dot{\epsilon}_3^f = \dot{\epsilon}_3^m. \end{aligned}$$

When they are combined with the TVBO equations by Lee and Krempl [1990] the composite is characterized by the following set of equations: (details can be found in Yeh and Krempl [1990])

$$\dot{\epsilon} = \overline{\mathbf{C}}^{-1}\dot{\sigma} + (\mathbf{K}^m)^{-1}\mathbf{X}^m + (\dot{\mathbf{R}}^f)^{-1}\sigma^f + (\dot{\mathbf{R}}^m)^{-1}\sigma^m + \overline{\boldsymbol{\alpha}}\dot{\mathbf{T}} \tag{1}$$

together with a separate growth law for the σ_3^m component of the matrix

$$\dot{\sigma}_3^m = \frac{E^m}{E_{33}}\dot{\sigma}_3 - \frac{c^f}{E_{33}} L(\dot{\sigma}_1 + \dot{\sigma}_2) - \frac{c^f E_{33}^f E^m}{E_{33}} \left\{\frac{1}{K^m k^m[\Gamma^m]}\left[X_3^m - 0.5(X_1^m + X_2^m)\right]\right\}$$

$$- \frac{c^f E_{33}^f E^m}{E_{33}} \left\{\left[\frac{1}{(E_{33}^f)^2}(\dot{\nu}_{31}^f E_{33}^f - \nu_{31}^f \dot{E}_{33}^f) - \frac{1}{(E^m)^2}(\dot{\nu}^m E^m - \nu^m \dot{E}^m)\right](\sigma_1 + \sigma_2) + \frac{\dot{E}_{33}^f}{(E_{33}^f)^2}\sigma_3^f - \frac{\dot{E}^m}{(E^m)^2}\sigma_3^m\right\}$$

$$- \frac{c^f E_{33}^f E^m}{E_{33}}(\alpha^m - \alpha_3^f)\dot{T}. \tag{2}$$

In addition growth laws for the two state variables of TVBO, the matrix equilibrium stress $\mathbf{g}^m$ and the kinematic stress $\mathbf{f}^m$, are given as

$$\dot{\mathbf{g}}^m = q^m[\Gamma^m]\dot{\sigma}^m + \dot{T}\frac{\partial q^m[\Gamma^m]}{\partial T}\sigma^m + \left\{q^m[\Gamma^m] - \theta^m\left[q^m[\Gamma^m] - p^m(1 - q^m[\Gamma^m])\right]\right\}\frac{\mathbf{X}^m}{k^m[\Gamma^m]} \tag{3}$$

$$\dot{\mathbf{f}}^m = \frac{p^m}{k^m[\Gamma^m]}\mathbf{X}^m. \tag{4}$$

with

$$
\begin{aligned}
(\Gamma^m)^2 &= (\mathbf{X}^m)^t\mathbf{H}(\mathbf{X}^m) \\
(\theta^m)^2 &= \frac{1}{(A^m)^2}(\mathbf{Z}^m)^t\mathbf{H}(\mathbf{Z}^m) \\
\mathbf{X}^m &= \sigma^m - \mathbf{g}^m \\
\mathbf{Z}^m &= \mathbf{g}^m - \mathbf{f}^m
\end{aligned} \tag{5}
$$

In the above $\bar{\mathbf{C}}^{-1}$ is the symmetric overall compliance matrix whose components are functions of the elastic properties of fiber and matrix. The viscosity matrix $(\mathbf{K}^m)^{-1}$ is not symmetric and its components together with those of $\bar{\mathbf{C}}^{-1}$ are listed in Appendix I. The matrices $(\dot{\mathbf{R}}^f)^{-1}$ and $(\dot{\mathbf{R}}^m)^{-1}$ contain time derivatives of the elastic constants of the fiber and the matrix, respectively. Both matrices are not symmetric. Their components are listed in Appendix I. These matrices represent the "additional" terms which can play a significant role in modeling thermomechanical behavior, see [Lee and Krempl 1990a]. The viscosity function $k^m[\Gamma^m]$ and the dimensionless shape function $q^m[\Gamma^m]$ are decreasing ($q^m[0] < 1$ is required) and control the rate dependence and the shape of the stress–strain diagram, respectively. (Square brackets following a symbol denote "function of".) The quantity p^m represents the ratio of the tangent modulus E_t^m at the maximum inelastic strain of interest to the viscosity factor K^m. It sets the slope of stress–inelastic strain diagram at the maximum strain of interest. $\bar{E}_{33}$, L, $\bar{\boldsymbol{\alpha}}$ together with the components of the dimensionless matrix $\mathbf{H}$ and other material properties are defined in Appendix I. An explanation of TVBO is given by Lee and Krempl [1990] and the derivation of the above equations can be found in Yeh and Krempl [1990].

Eq. (1) shows that the overall strain rate is the sum of the overall elastic strain rate, the inelastic strain rate of the matrix and the overall thermal strain rate in the case of constant elastic properties. If temperature dependent elastic properties are assumed then two additional terms contribute to the overall strain rate. They insure that the elastic behavior is path independent, see Lee and Krempl [1990, 1990a].

Eq. (2) is used to calculate the instantaneous axial matrix stress which can not be obtained from the overall boundary conditions directly. σ_3^m is affected by mechanical and thermal loadings and their loading paths. For instance for the isothermal case when $\dot{T} = 0$, matrix stresses in the fiber direction (σ_3^m, g_3^m, f_3^m) can evolve in unidirectional transverse loading, or may evolve in unidirectional shear loading provided the initial value of X_3^m is nonzero. For pure thermal loading (overall stresses are zero), σ_3^m together with g_3^m, f_3^m will develop due to the difference in the coefficients of thermal expansion of fiber and matrix; these matrix stresses in the fiber direction cause coupling between the mechanical and thermal loading in the inelastic range.

NUMERICAL SIMULATION

Eqs. (1) – (5) constitute the model which must now be applied. The boundary conditions must be specified in addition to the uniform temperature history. Also material properties must be known as a function of temperature. For the purposes of this paper two metal matrix systems, Gr/Al and B/Al are simulated. The matrix viscoplastic properties for 6061–T6 Al alloy are known at room temperature from experiments reported by Yao and Krempl [1985]. Since no experiments were available at other temperatures a plausible temperature dependence was postulated. The elastic properties and the coefficient of thermal expansion for the Gr and B fibers are listed in Table 1. They are assumed to be independent of temperature for simplicity. The matrix properties which are close to 6061–T6 Al alloy are listed in Table 2. They yield the matrix stress–strain diagrams at a strain rate of 10^{-4} s^{-1} depicted in Fig. 1. A decrease in modulus, flow stress and the asymptotic tangent modulus with increasing temperature is modeled.

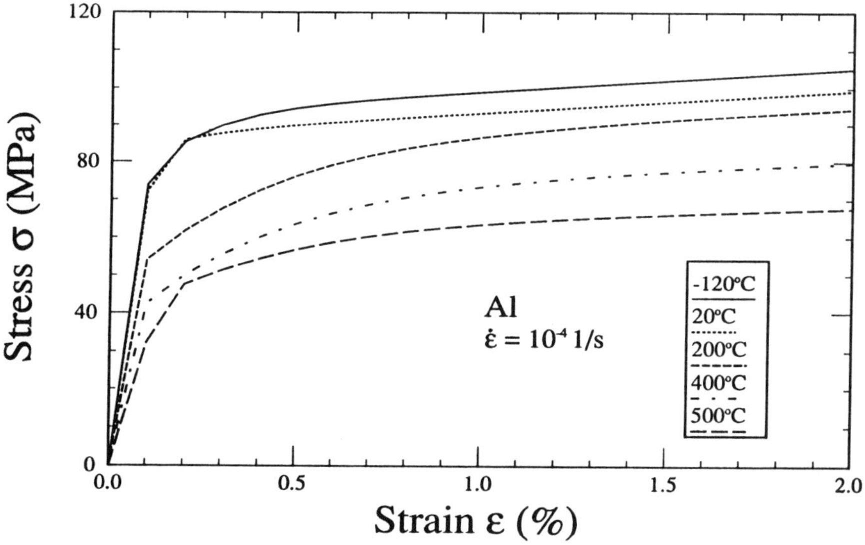

Fig. 1. *Stress–strain diagrams of matrix material at various temperatures.*

Table 1.

Elastic Properties for Boron and Graphite Fibers

Properties	B		Gr (****)
E^f_{33} (MPa)	413400	(*)	689650
ν^f_{31}	0.21	(*)	0.41
G^f_{44} (MPa)	170830	(**)	15517
α^f_3 (m/m/°C)	6.3E–6	(***)	–1.62E–6
E^f_{11} (MPa)	413400	(*)	6069
G^f_{66} (MPa)	170830	(**)	2069
α^f_1 (m/m/°C)	6.3E–6	(***)	1.08E–5

* Kreider and Prewo [1974]

** Estimate

*** Tsirlin [1985]

**** Wu, et al [1989]

Table 2.

Thermoelastic and Thermoviscoplastic Properties of the Matrix

---

$E^m = 74657[1 - (\frac{T}{933})^3]$ (MPa) (*)

$\nu^m = 0.33$ (**)

$G^m = 28066[1 - (\frac{T}{933})^3]$ (MPa) (*)

$\alpha^m = 2.35E{-}5 + 2.476E{-}8(T - 273)$ (m/m/°C) (**)

$q^m[\Gamma^m] = \Psi^m[\Gamma^m]/E^m, \quad p^m = E_t^m/K^m$

Viscosity function $k^m[\Gamma^m] = k_1(1+ \frac{\Gamma^m}{k_2})^{-k_3}$

$k_1 = 314200$ (s), $k_2 = 71.38$ (MPa) (***)

$k_3 = 53 - 0.05(T{-}273)$ (**)(***)

Viscosity Factor $K^m = E^m$

$E_t^m = 619[1 - (\frac{T}{933})^3]$ (MPa) (**)

$A^m = 72.24[1 - (\frac{T}{933})^3]$ (MPa) (**)

Shape function $\Psi^m[\Gamma^m] = c_1 + (c_2{-}c_1)\exp(-c_3\Gamma^m)$

$c_1 = 16511[1 - (\frac{T}{933})^3]$ (MPa) (**)

$c_2 = 73910[1 - (\frac{T}{933})^3]$ (MPa) (**)

$c_3 = 8.43E{-}2 + 1.06E{-}4(T{-}273) + 1.914E{-}6(T{-}273)^2 + 5.304E{-}9(T{-}273)^3$ (MPa^{-1}) (**)

Inelastic Poisson's Ratio: 0.5

$T = {}^oK$, $153^oK < T < 933^oK$

(*) Estimate. Temperature dependence due to Hillig [1985]

(**) Estimate

(***) Yao and Krempl [1985]

For the integration of the coupled set of differential equations the IMSL routine DGEAR is used.

Residual Stresses upon Cool–Down from Manufacturing Temperature

Overall stresses are assumed to be zero and the temperature is decreased at a constant rate of 0.033 ^{0}C/s. It is assumed that the composite is stress free at 660 ^{0}C and that perfect bonding starts at that temperature. Since the coefficient of thermal expansion is larger for the matrix than for the fibers tensile matrix stresses develop as shown in Fig. 2a for B/Al and in Fig. 2b for Gr/Al. Owing to the assumed fiber volume fraction

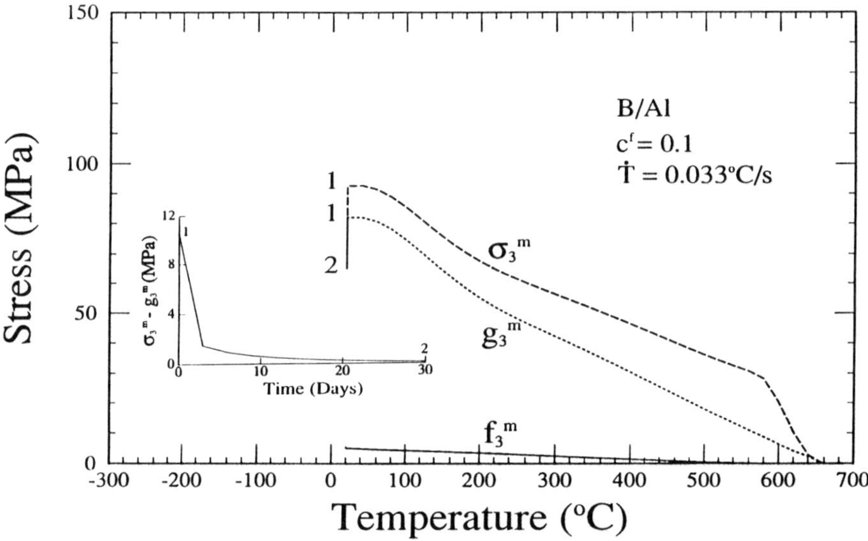

Fig. 2a. *Development of matrix stress σ_3^m, matrix equi–librium g_3^m and kinematic stress f_3^m during cool down from manufacturing temperature. The inset shows the decrease of the overstress during the room temperature hold 1 — 2. Boron/Aluminum.*

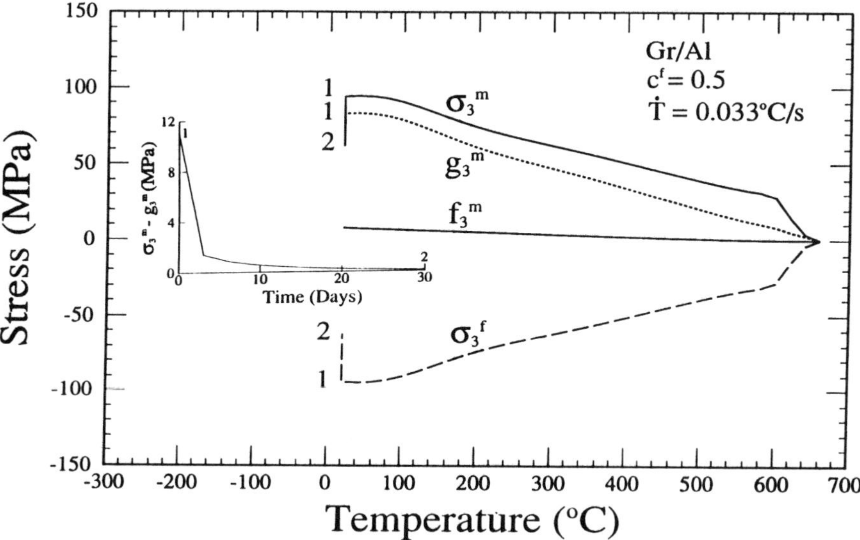

Fig. 2b. *Same as Fig. 2a except that material is Graphite/ Aluminum. The fiber stress σ_3^f is also shown.*

of 0.5 the fiber stresses are equal and opposite in Fig. 2b. They are not shown in Fig. 2a where the fiber volume fraction is 0.1. The VFD assumption listed previously yields $\sigma_3^f = -9\sigma_3^m$. At point 1 room temperature is reached. Due to the viscoplastic nature of the matrix the stresses relax to point 2 with time. The inset shows the overstress $\sigma_3^m - g_3^m$, which "drives" the inelastic deformation, rapidly decreasing with time. All residual stresses enter as initial conditions for simulations of subsequent tests. They can affect the modeled behavior and therefore time appears to influence it. After 30 days the residual stress state is nearly constant. Then the model predicts that the subsequent response becomes independent of the rest time at room temperature. On the scale of this graph the kinematic variable f_3^m does not appear to change with time. However, the digital output confirms the slight increase predicted by Eq (4).

Influence of Residual Stresses on Room Temperature Mechanical Behavior

In this case a B/Al composite with $c^f = 0.1$ is considered and uniaxial tensile tests in the fiber and the transverse directions are performed at a strain rate of 10^{-4} s^{-1}. When a strain of 0.5% is reached the overall stress is kept constant to allow creep deformation to evolve during a short period of 300 s.

Fig. 3 shows the mechanical behavior for tests in the fiber (3) – direction. The overall stress, the matrix stress and equilibrium stress are plotted vs. overall strain for 3 cases. Fig. 3a shows the behavior without residual stresses, Fig. 3b has the residual stress state at point 1 in Fig. 2 as initial conditions. This is called Case 1 and simulates a tensile test performed immediately after the composite reached room temperature. The relaxed residual state of stress represented by point 2 in Fig. 2 forms the set of initial condition for Case 2. The mechanical behavior with this set of initial conditions is given in Fig. 3c.

By comparing the figures the significant influence of residual stresses on the overall stress–strain diagram can be clearly ascertained. It can be seen that the initial slope, the stress level at which the transition to another slope takes place and the overall appearance of the composite stress–strain diagram are significantly affected by the residual stresses. Owing to a nearly zero overstress in Case 2 the initial slope seems to be

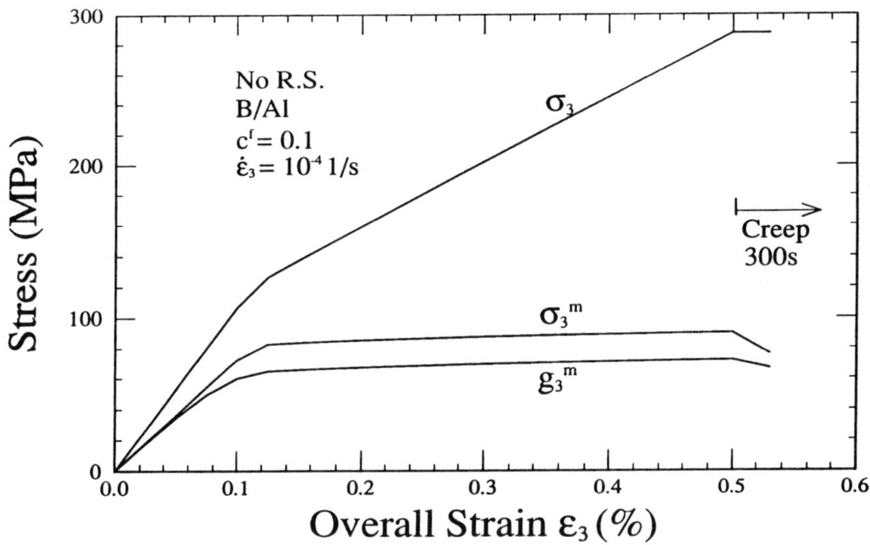

Fig. 3a. *Stresses vs strain in fiber direction at room temperature with a 300 s creep period at the maximum stress.*

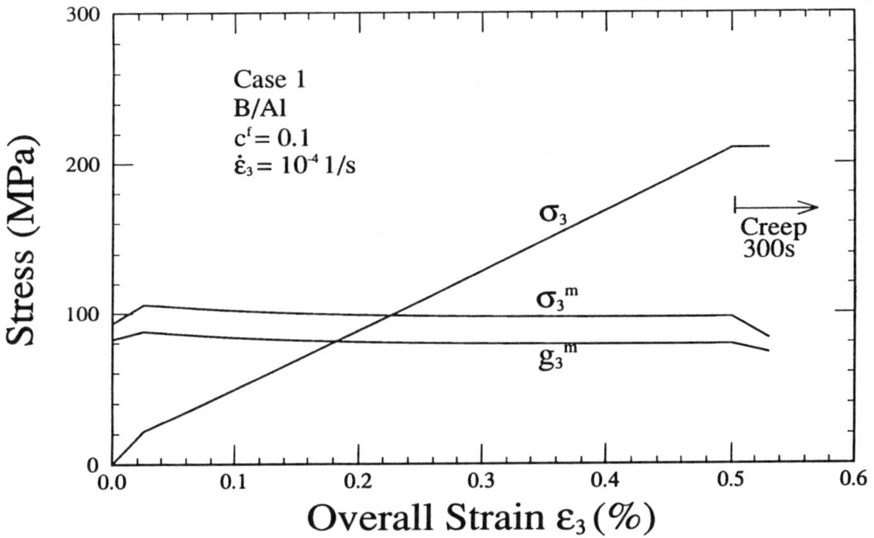

Fig. 3b. *Same as Fig. 3a for Case 1.*

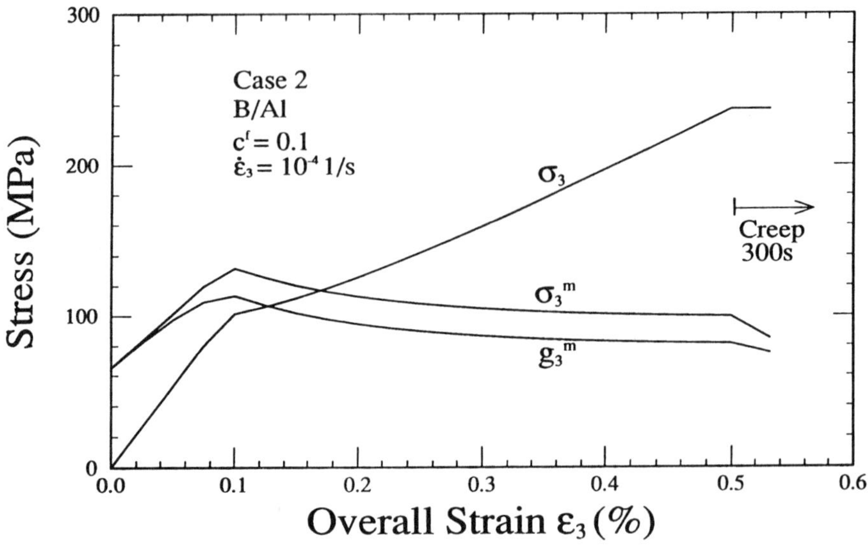

Fig. 3c. *Same as Fig. 3a for Case 2.*

identical to the stress–strain diagram with no residual stresses. The level of the overall stress is considerably lower for Case 2 than for the case without residual stresses. Since it is unlikely that a tensile test will be performed right after reaching room temperature and since the overstress decreases rapidly with time, see inset in Fig.2, an experiment would yield the results of Case 2. The residual stress state has an influence on the relation between the strain in the fiber direction and the transverse strain as shown in Fig. 4. From these curves the actual Poisson's ratio based on total strain could be calculated. The simulation of a tensile test in the transverse direction is shown in Fig. 5a and the relation between the transverse strain ϵ_1 and the two perpendicular strains ϵ_2 and ϵ_3 are shown in Fig. 5b and Fig. 5c, respectively. A significant influence of the residual stress state is evident, especially in Figs. 5a and 5b.

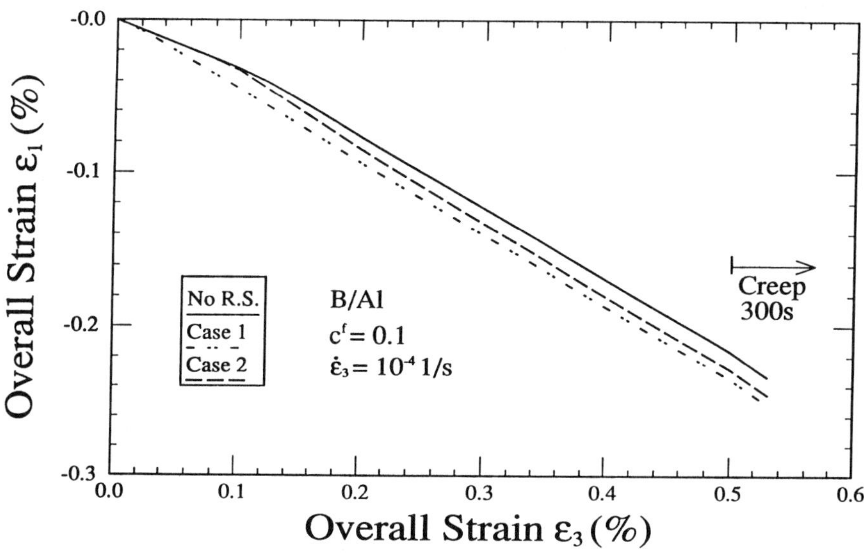

Fig. 4. *The development of the transverse strain during the tests shown in Fig. 3.*

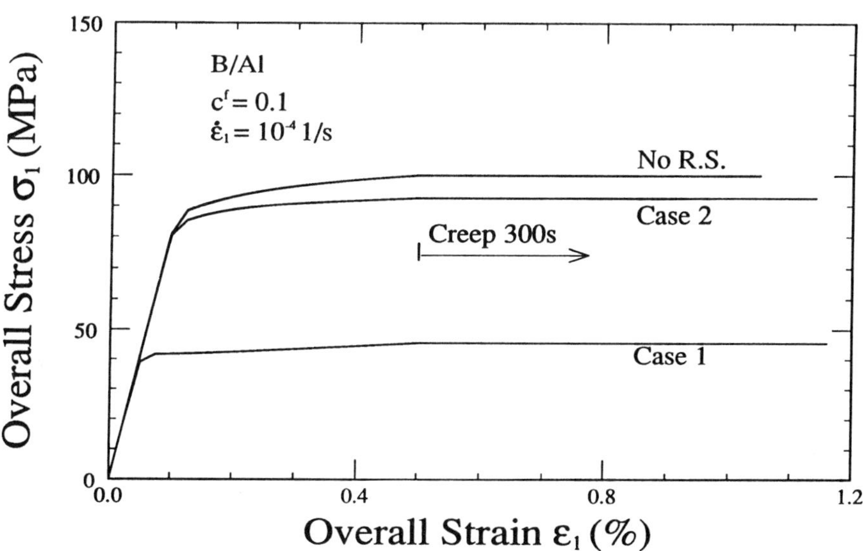

Fig. 5a. *Simulation of a transverse tensile test at room temperature as a function of the residual stresses; transverse stress–strain diagram.*

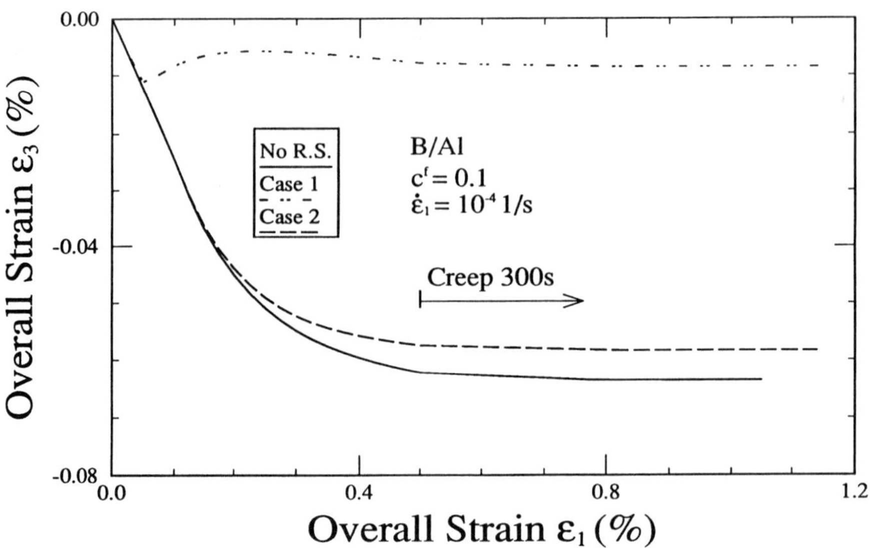

Fig. 5b. *The development of the transverse strain in the 3–direction.*

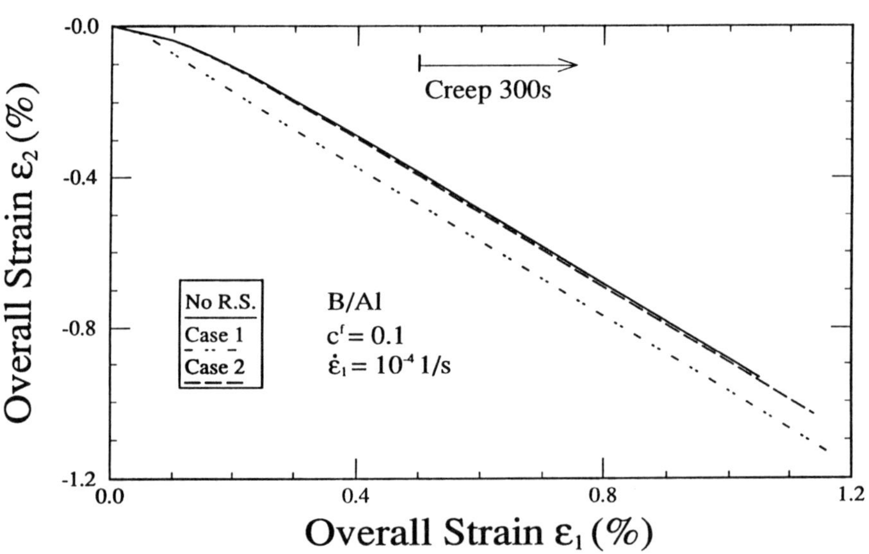

Fig. 5c. *The development of the transverse strain in the 2–direction.*

In Figs. 3 through 5 the behavior during the 300 s creep period is specially marked. As expected the total creep strain accumulated is very significant in the matrix dominated transverse mode, see Fig. 5a. It is small for the fiber direction as shown in Figs. 3. In each case primary creep is modeled with a rapidly decreasing rate. This is shown in Fig. 5d for the transverse case. This corresponds to the so–called "cold creep" phenomenon found at room temperature for ductile engineering alloys. For the strain vs. strain curves , Figs. 4, 5b and 5c, the creep periods do not differ significantly from the periods under increasing stress. Only a slight break in slope is noticeable at the outset of the creep period.

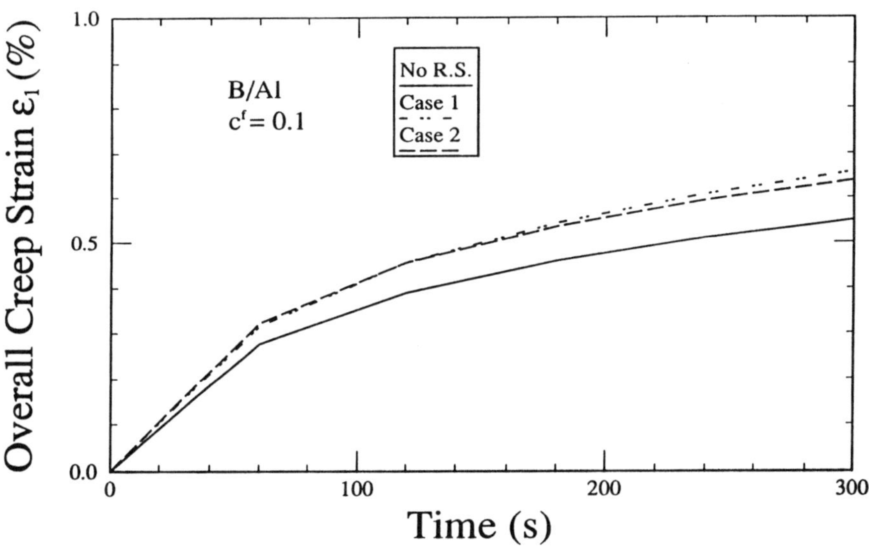

Fig. 5d. *Transverse creep strain during the 300s creep period.*

The Influence of Residual Stresses on the Thermal Cycling Behavior of Gr/Al Composite.

The thermal cycling behavior of Gr/Al is of special interest due to the negative axial CTE of Graphite, see Table 1. It gives rise to some unusual expansion behavior, see Wu et al [1989] and Tompkins and Dries [1988]. In this paper we simulate that the composite is free to expand (overall stresses are zero) and is subjected to a temperature cycle starting from room temperature to ± 120 ^{0}C at a rate of 0.033 ^{0}C/s.

The resulting strain in the fiber direction – temperature hysteresis loop is depicted in Fig. 6a. It is seen that the composite expands on the segment 0–1 but then contracts with increasing temperature, segment 1–2. Upon decrease of temperature from 120^{0}C the composite shrinks as expected but

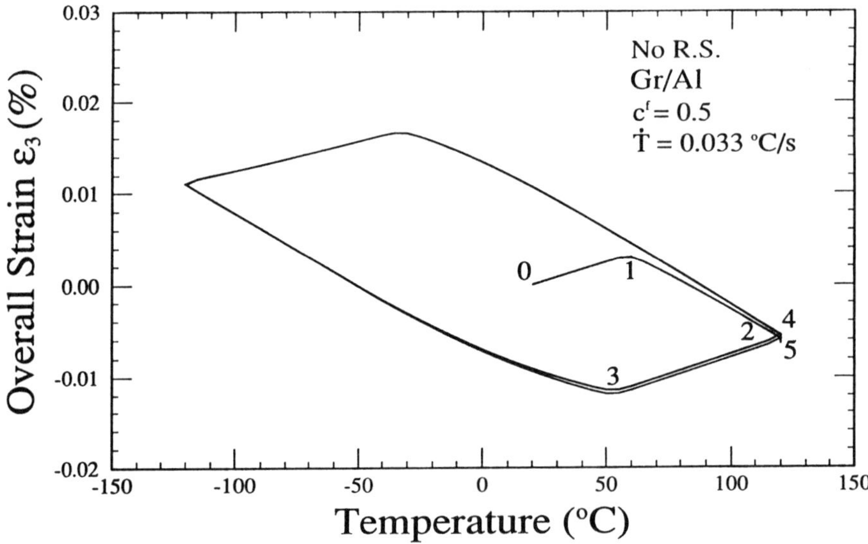

Fig. 6a. *Temperature–strain in the fiber direction loop during temperature cycling of Gr/Al composite.*

expands at point 3 although the temperature continues to decrease. This pattern continues in the subsequent reversals. At point 4 a 600 s temperature hold is introduced and the strain decreases by a small amount, the composite "creeps" under zero external load and the creep curve is shown in Fig. 6b. To demonstrate that the temperature rate has an influence the calculation was repeated with a rate of 0.1 ^{0}C/s. There is very little influence on the temperature/strain curve, but creep during the temperature hold period is accelerated as shown in Fig. 6b.

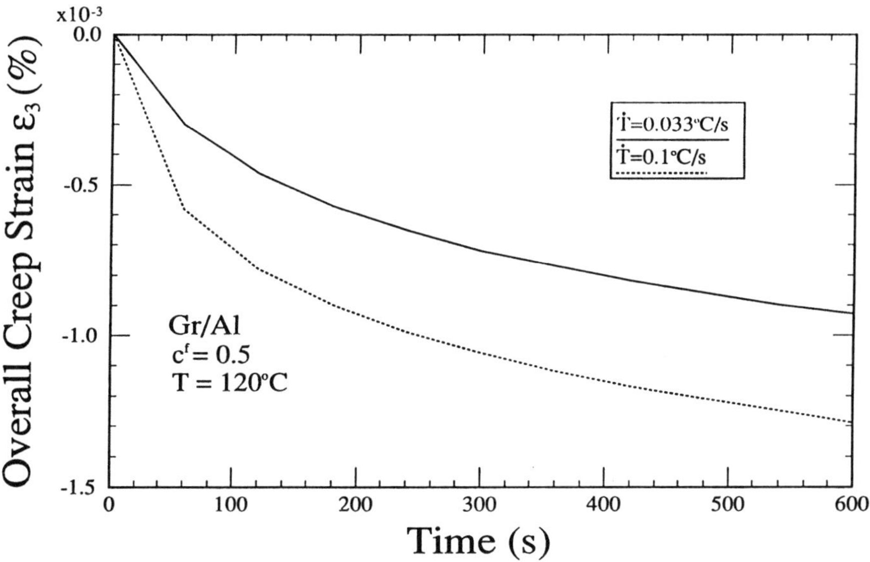

Fig. 6b. *Creep curves during temperature hold at 120° C, see points 4, 5 in Fig. 6a.*

The explanation of this unusual behavior can be found in the development of the fiber and matrix stresses during cycling as shown in Fig. 6c. It is seen that a temperature–stress hysteresis loop develops and that the matrix starts yielding at

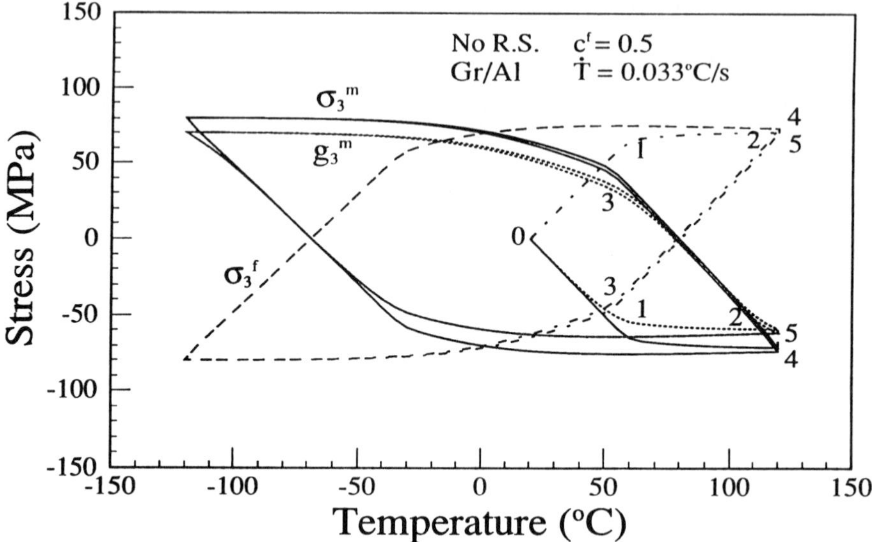

Fig. 6c. *"Internal stresses" which develop during temperature cycling shown in Fig. 6a.*

points 1 and 3 where the breaks in Fig. 6a occur. The unusual behavior is due to the matrix yielding. In the inelastic range the stiffness of the matrix is low and the overall behavior is dominated by the fiber which has a negative axial CTE.

To show the influence of residual stresses Cases 1 and 2 are simulated in Fig. 7a. and Fig. 7b, respectively. Cooling down takes place on 0–1. While the composite rests free of overall stresses at room temperature, see Fig. 2b, the overall strain increases on path 1–2, see Fig. 7b (this portion is absent in Fig. 7a which depicts Case 1). At 2 temperature cycling begins, the composite expands first, 2–3, but starts to shrink, 3–4 and then the pattern of Fig. 6a continues. However, this time the first part of the first cycle 2–5 is not inside the

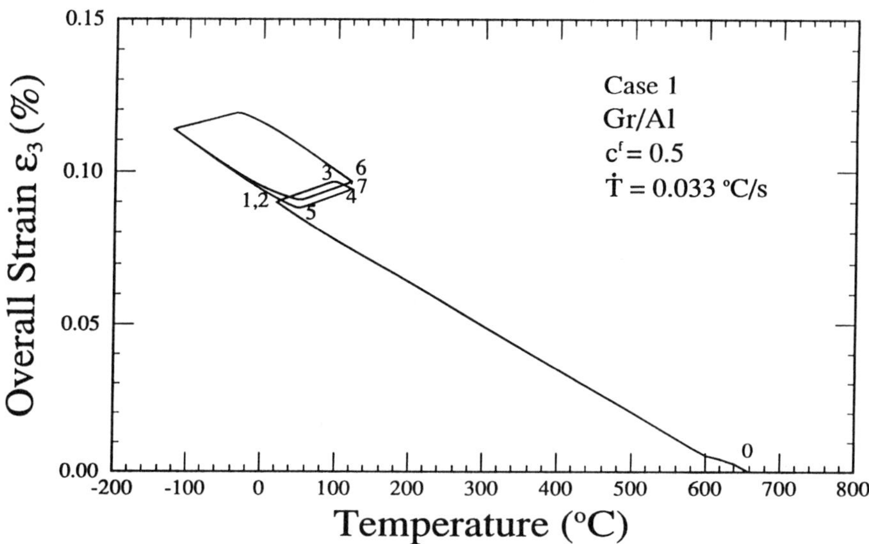

Fig. 7a. *Temperature–strain graph during cool down from 660° C and subsequent cycling as in Fig. 6a. Case 1.*

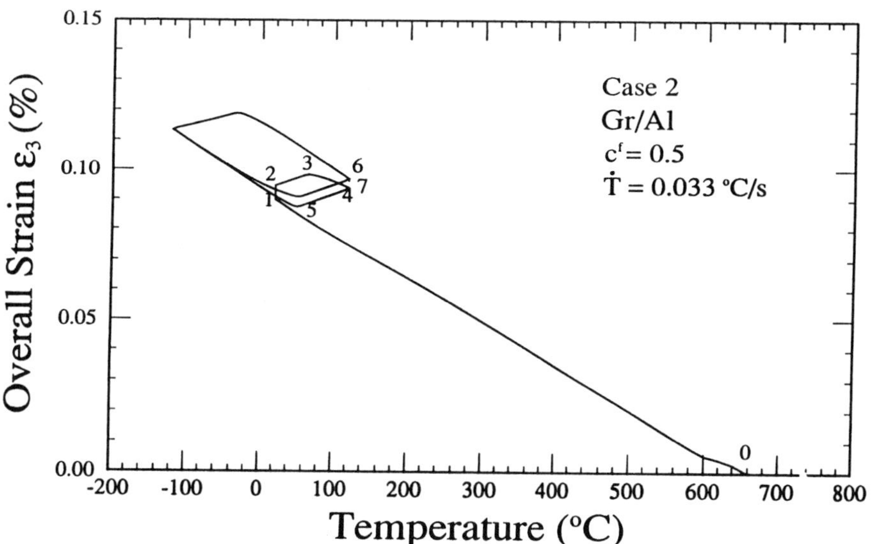

Fig. 7b. *Same as Fig. 7a for Case 2.*

subsequent loop as it was the case for Fig. 6a, see segment 0–3. Rather the first segment is shifted and the shift depends on the case considered. The residual matrix and fiber stresses have altered the cycle pattern. Their development during cycling (the cool–down portion 0–2 is omitted) is depicted in Fig. 7c for Case 2. For the identification the same numbering scheme has been used as in Fig. 7a and in Fig. 7b. It can again be ascertained that the "breaks" in the expansion behavior are coinciding with the onset of inelastic deformation of the matrix.

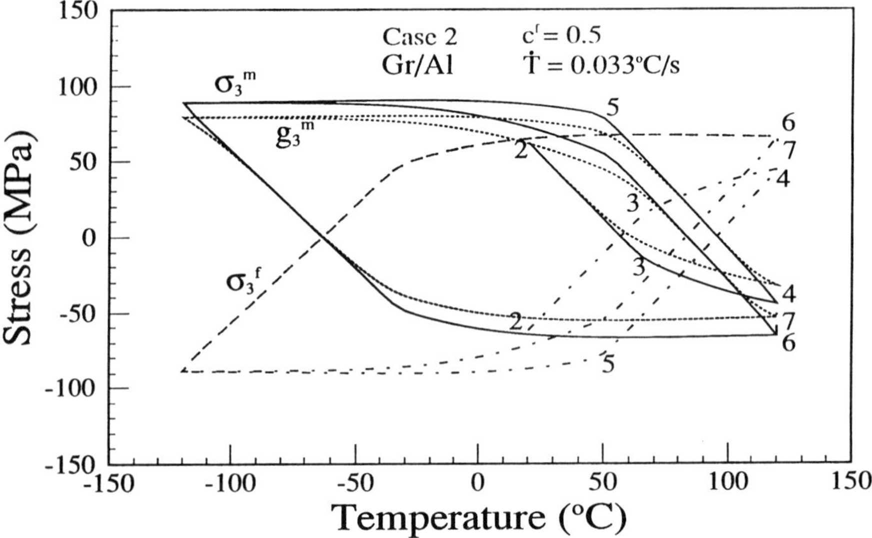

Fig. 7c. *The "internal stresses" developed during temperature cycling for Case 2. Curves start at room temperature, point 2 in Fig. 7b.*

DISCUSSION

A "unified" viscoplastic constitutive model for composite analysis, the thermoviscoplasticity theory based on overstress, was used in conjunction with the vanishing fiber diameter model in a simple analysis of the influence of fiber/matrix residual stresses on the mechanical and thermal cycling behavior. Realistic but assumed material properties permitted the execution of numerical experiments. The stress–strain diagrams reported in Figs. 3a–3c correspond qualitatively with those reported by Cheskis and Heckel [1970]. In both cases a break in the slope of the overall stress–strain diagram is observed when the matrix starts to deform inelastically in an appreciable manner. The presence of residual stresses shift the location of this break point, see Figs. 3a–3b.

Another feature exhibited by the present theory is the manifestation of the influence of rate dependence on the behavior. The first example was the redistribution of the residual stresses while the composite element was sitting stress free at room temperature after cool–down from manufacturing temperature. The theory predicts that this redistribution will nearly come to an end after some time which depends on material constants, especially the viscosity function used. In the present application the redistribution is almost finished after 30 days. While the stresses redistribute the time at room temperature appears to have an influence on the subsequent behavior.

For Gr/Al the residual stresses were shown to affect the free thermal expansion of the composite. The results of Figs. 7a and 7b suggest that residual stresses are responsible for the special shape of the first part of the first cycle of Figs. 6 and 7 of Tompkins and Dries [1988]. In comparing their figures with Figs. 7a and 7b it has to be kept in mind that the presently used theory models only cyclic neutral behavior whereas real matrix alloys may exhibit cyclic hardening or softening. These aspects could be added to the present theory in a refined approach.

The present paper intends to show the capabilities in principle. For the exact modeling of a metal matrix composite various refinements are possible. Included are the determination of matrix and fiber properties as a function of temperature and the use of other micromechanical models.

Acknowledgment

This research was supported by DARPA/ONR Contract N00014–86–K0770 with Rensselaer Polytechnic Institute. Discussions with Dr. Y. A. Bahei–El–Din are acknowledged with thanks.

REFERENCES:

Cheskis, H. P. and Heckel, R. W., 1970, "Deformation Behavior of Continuous–Fiber Metal–Matrix Composite Materials," Metallurgical Transactions, Vol 1, pp. 1931–1942.

Dvorak, G. J. and Rao, M. S. M., 1976, "Thermal Stresses in Heat–Treated Fibrous Composites," ASME Journal of Applied Mechanics, pp. 619–624.

Dvorak, G. J. and Bahei–EL–Din, Y. A., 1982, "Plasticity Analysis of Fibrous Composites," ASME Journal of Applied Mechanics, Vol. 49, pp. 327–335.

Garmong, G., 1974, "Elastic–Plastic Analysis of Deformation Induced by Thermal Stress in Eutectic Composites:1 Theory," Metallurgical Transactions, Vol. 5, pp. 2183–2190.

Hillig, W. B., 1985, "Prospects for Ultra–High–Temperature Ceramic Composites," Report No. 85CRD152, General Electric Research and Development Center.

Krempl, E.,1979, "An Experimental Study of Room–Temperature Rate Sensitivity, Creep and Relaxation of Type 304 Stainless Steel," Journal of the Mechanics and Physics of Solids, Vol. 27, pp. 363–375.

Krempl, E. and Lu, H., 1983, "Comparison of the Stress Responses of an Aluminum Alloy Tube to Proportional and Alternate Axial and Shear Strain Paths at Room Temperature," Mechanics of Materials, Vol. 2, pp. 183–192.

Kreider, K. G. and Prewo, K. M., 1974, "Boron–Reinforced Aluminum," Composite Materials, Vol. 4, Metallic Matrix Composites, Edited by Kenneth G. Kreider, Academic Press.

Kujawski, D., Krempl, E., 1981, "The Rate (Time)–Dependent Behavior of Ti–7Al–2Cb–1Ta Titanium Alloy at Room Temperature Under Quasi–Static Monotonic and Cyclic Loading," ASME Journal of Applied Mechanics, Vol. 48, pp. 55–63.

Kural, M. K. and Min, B. K., 1984, "The Effects of Matrix Plasticity on the Thermal Deformation of Continuous Fiber Graphite/Metal Composites," Journal of Composite Materials, Vol. 18, pp. 519–535.

Lee, K. D. and Krempl, E., 1990, "An Orthotropic Theory of Viscoplasticity Based on Overstress for Thermomechanical Deformations," to appear in International Journal of Solids and Structures.

Lee, K. D. and Krempl, E., 1990a, "Uniaxial thermo–mechanical loading. Numerical experiments using the thermal viscoplasticity theory based on overstress," MML Report 90–1, Rensselaer Polytechnic Institute, March.

Min, B. K. and Crossman, F. W. ,1982, "History–Dependent Thermo–mechanical Properties of Graphite/Aluminum Unidirectional Composites," Composite Materials: Testing and Design (Sixth Conference), **ASTM STP** 787, I. M. Daniel, Ed, American Society for Testing and Materials, pp. 371–392.

Tompkins, S. S. and Dries, G. A., 1988, "Thermal Expansion Measurement of Metal Matrix Composites," **ASTM STP** 964, P. R. DiGiovanni and N. R. Adsit, Editors, American Society for Testing and Materials, Philadelphia, pp. 248–258.

Tsirlin, A. M., 1985, "Boron Filaments," Handbook of Composites, **Volume** 1, "Strong Fibers," Editors: Watt, W. and Perov, B. V., North–Holland.

Wu, J. F., Shephard, M. S., Dvorak, G. J. and Bahei–EL–Din, Y. A., 1989, "A Material Model for the Finite Element Analysis of Metal–Matrix Composites," Composites Science and Technology, Vol. 35, pp. 347–366.

Yao, D. and Krempl, E., 1985, "Viscoplasticity Theory Based on Overstress. The Prediction of Monotonic and Cyclic Proportional and Nonproportional Loading Paths of an Aluminum Alloy," Int. Journal of Plasticity, Vol. 1, pp. 259–274.

Yeh, N. M. and Krempl, E., 1990, " Thermoviscoplastic Analysis of Fibrous Metal–Matrix Composites," MML Report 90–2, Rensselaer Polytechnic Institute, March.

APPENDIX I

For the transversely isotropic (fiber) and the isotropic (matrix) elastic properties the usual designations are employed. For convenience the following quantities are defined and used

$$\bar{E}_{33} = c^f E^f_{33} + c^m E^m$$
$$L = \nu^f_{31} E^m - \nu^m E^f_{33}$$
$$\bar{\nu}_{31} = c^f \nu^f_{31} + c^m \nu^m.$$

The components of the overall elastic compliance matrix $\bar{C}^{-1}$ are

$$(\bar{C}^{-1})_{11} = \frac{c^f}{E^f_{11}} + \frac{c^m}{E^m} - \frac{c^f c^m L^2}{E^f_{33} E^m \bar{E}_{33}} = (\bar{C}^{-1})_{22}$$

$$(\bar{C}^{-1})_{12} = -(c^f \frac{\nu^f_{12}}{E^f_{11}} + c^m \frac{\nu^m}{E^m} + \frac{c^f c^m L^2}{E^f_{33} E^m \bar{E}_{33}}) = (\bar{C}^{-1})_{21}$$

$$(\bar{C}^{-1})_{13} = \frac{-\bar{\nu}_{31}}{\bar{E}_{33}} = (\bar{C}^{-1})_{23} = (\bar{C}^{-1})_{31} = (\bar{C}^{-1})_{32}$$

$$(\bar{C}^{-1})_{33} = \frac{1}{\bar{E}_{33}}$$

$$(\bar{C}^{-1})_{44} = \frac{c^f}{G^f_{44}} + \frac{c^m}{G^m} = (\bar{C}^{-1})_{55}$$

$$(\bar{C}^{-1})_{66} = \frac{c^f}{G^f_{66}} + \frac{c^m}{G^m}$$

with all other $(\bar{C}^{-1})_{ij} = 0$.

The viscosity matrix $(\mathbf{K}^m)^{-1}$ is given by the components (the argument of the viscosity function k^m is omitted)

$$(K^m)^{-1}_{11} = \frac{c^m}{K^m k^m}(1 + 0.5\frac{c^f L}{E_{33}}) = (K^m)^{-1}_{22}$$

$$(K^m)^{-1}_{12} = \frac{-c^m}{2K^m k^m}(1 - \frac{c^f L}{E_{33}}) = (K^m)^{-1}_{21}$$

$$(K^m)^{-1}_{13} = \frac{-c^m}{K^m k^m}(0.5 + \frac{c^f L}{E_{33}}) = (K^m)^{-1}_{23}$$

$$(K^m)^{-1}_{31} = \frac{-c^m E^m}{2E_{33} K^m k^m} = (K^m)^{-1}_{32}$$

$$(K^m)^{-1}_{33} = \frac{c^m E^m}{E_{33} K^m k^m}$$

$$(K^m)^{-1}_{66} = \frac{3c^m}{K^m k^m} = (K^m)^{-1}_{44} = (K^m)^{-1}_{55}.$$

All other $(K^m)^{-1}_{ij} = 0$.

The components of the "extra terms" $(\dot{\mathbf{R}}^{\mathrm{f}})^{-1}$ and $(\dot{\mathbf{R}}^{\mathrm{m}})^{-1}$ are

$$(\dot{R}^f)^{-1}_{11} = -c^f\frac{\dot{E}^f_{11}}{(E^f_{11})^2} - \frac{c^f c^m L}{(E^f_{33})^2 \overline{E}_{33}}(\dot{\nu}^f_{31}E^f_{33} - \nu^f_{31}\dot{E}^f_{33})$$

$$= (\dot{R}^f)^{-1}_{22}$$

$$(\dot{R}^f)^{-1}_{12} = -\left[\frac{c^f}{(E^f_{11})^2}(\dot{\nu}^f_{12}E^f_{11} - \nu^f_{12}\dot{E}^f_{11}) + \frac{c^f c^m L}{(E^f_{33})^2\overline{E}_{33}}\right.$$

$$\left.(\dot{\nu}^f_{31}E^f_{33} - \nu^f_{31}\dot{E}^f_{33})\right] = (\dot{R}^f)^{-1}_{21}$$

$$(\dot{R}^f)^{-1}_{13} = \frac{c^f}{E^f_{33}\overline{E}_{33}}(\overline{\nu}_{31}\dot{E}^f_{33} - \dot{\nu}^f_{31}\overline{E}_{33}) = (\dot{R}^f)^{-1}_{23}$$

$$(\dot{R}^f)^{-1}_{31} = \frac{c^f}{E^f_{33}\overline{E}_{33}}(\nu^f_{31}\dot{E}^f_{33} - \dot{\nu}^f_{31}E^f_{33}) = (\dot{R}^f)^{-1}_{32}$$

$$(\dot{R}^f)^{-1}_{33} = -\frac{c^f\dot{E}^f_{33}}{E^f_{33}\overline{E}_{33}}$$

$$(\dot{R}^f)^{-1}_{44} = -c^f\frac{\dot{G}^f_{44}}{(G^f_{44})^2} = (\dot{R}^f)^{-1}_{55}$$

$$(\dot{R}^f)^{-1}_{66} = -c^f\frac{\dot{G}^f_{66}}{(G^f_{66})^2}$$

with all other $(\dot{R}^f)^{-1}_{ij} = 0$.

$$(\dot{R}^m)^{-1}_{11} = -c^m \frac{\dot{E}^m}{(E^m)^2} + \frac{c^f c^m L}{(E^m)^2 E_{33}}(\dot{\nu}^m E^m - \nu^m \dot{E}^m) = (\dot{R}^m)^{-1}_{22}$$

$$(\dot{R}^m)^{-1}_{12} = \frac{c_m}{(E^m)^2}(\nu^m \dot{E}^m - \dot{\nu}^m E^m)(1 - \frac{c^f L}{E_{33}}) = (\dot{R}^m)^{-1}_{21}$$

$$(\dot{R}^m)^{-1}_{13} = \frac{c^m}{E^m E_{33}}(\overline{\nu}_{31} \dot{E}^m - \dot{\nu}^m E_{33}) = (\dot{R}^m)^{-1}_{23}$$

$$(\dot{R}^m)^{-1}_{31} = \frac{c^m}{E^m E_{33}}(\nu^m \dot{E}^m - \dot{\nu}^m E^m) = (\dot{R}^m)^{-1}_{32}$$

$$(\dot{R}^m)^{-1}_{33} = -\frac{c^m \dot{E}^m}{E^m E_{33}}$$

$$(\dot{R}^m)^{-1}_{66} = -c^m \frac{\dot{G}^m}{(G^m)^2} = (\dot{R}^m)^{-1}_{44} = (\dot{R}^m)^{-1}_{55}.$$

All other $(\dot{R}^m)^{-1}_{ij} = 0$.

The overall coefficient of thermal expansion vector $\overline{\boldsymbol{\alpha}}$ is represented by

$$(\overline{\boldsymbol{\alpha}})_1 = c^f \alpha^f_1 + c^m \alpha^m - \frac{c^f c^m L}{E_{33}}(\alpha^m - \alpha^f_3) = (\overline{\boldsymbol{\alpha}})_2$$

$$(\overline{\boldsymbol{\alpha}})_3 = (c^f \alpha^f_3 E^f_{33} + c^m \alpha^m E^m)/E_{33}$$

$$(\overline{\boldsymbol{\alpha}})_4 = (\overline{\boldsymbol{\alpha}})_5 = (\overline{\boldsymbol{\alpha}})_6 = 0.$$

Finally

$$\mathbf{H} = \begin{bmatrix} 1 & -0.5 & -0.5 & 0 & 0 & 0 \\ & 1 & -0.5 & 0 & 0 & 0 \\ & & 1 & 0 & 0 & 0 \\ \text{sym} & & & 3 & 0 & 0 \\ & & & & 3 & 0 \\ & & & & & 3 \end{bmatrix}.$$

Elasto-Plastic Analysis for Cracked Fibrous Composites under Axial and Thermal Loads

K.P. Herrmann and Y.Q. Wang

Paderborn University, Laboratorium für Technische Mechanik
D-4790 Paderborn, Federal Republic of Germany

ABSTRACT

This investigation is concentrated on the analysis of existing matrix cracks in a brittle fiber-ductile matrix composite unit cell under tensile and/or cooling loads. An analytical formula for predicting the microstresses of a unit cell is presented. Longitudinal radial cracks and penny-shaped cracks as well located orthogonal to the fiber direction are studied.

1 INTRODUCTION

Metal matrix composites represent a very important class of fibrous composites and they are at present widely used in modern engineering structures. Thereby a successful application of these composites depends on a sound understanding of the fiber-matrix interactions as well as on the inherent defect structure, particularly cracks, produced by the fabrication process and/or the loading conditions. It has been found that the damage growth in composite materials is usually caused by the growth of existing microcracks(Sendeckyj et al, 1978; Nuismer et al, 1982), and the growth of these microcracks depends significantly on the microstresses of the composite elements. Hashin and Rosen(1964) firstly introduced the corresponding cylindrical composite unit cell in order to study the macromechanical properties of composites. Ebert et al(1968) studied the stress-strain response of a composite cylinder under axial loading. Dvorak et al(1976) developed an elasto-plastic solution concerning the microstresses of a unit cell under thermal load. Furthermore, from the standpoint of fracture mechanics, Herrmann et al(1978,1986,1988) studied the small initial cracks in composite microcomponents with elastic or elasto-plastic matrices subjected to well-defined macroscopic thermal stress fields and also interface cracks in more complicated material models. Taya and Chou(1982) calculated the energy release rates of various penny-shaped cracks in order to predict the non-linear region of the stress-strain curve. Dvorak and Bahei-

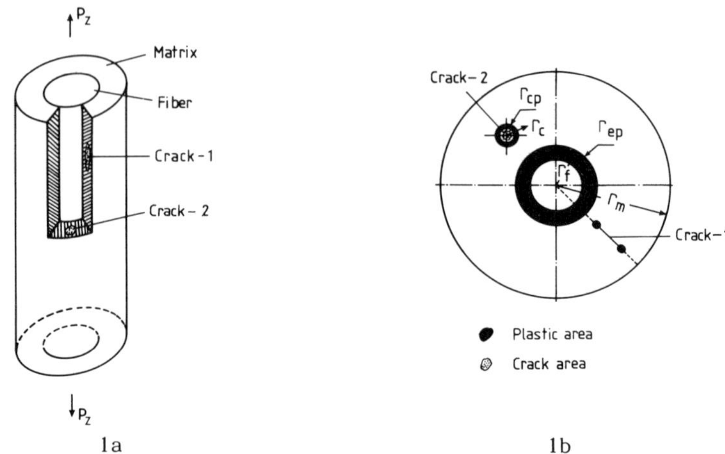

Fig.1 A unit cell of a fibrous composite with two different crack types

El-Din(1989) studied the fracture properties of metal matrix composites and Law and Dvorak(1987) found a loss of stiffness due to penny-shaped cracks associated with fiber breaks. Nuismer and Tan(1982) studied the role of matrix cracking and the resulting stiffness loss.

This paper concentrates on the analysis of existing matrix cracks in a brittle fiber-ductile matrix composite microcomponent under tensile and/or cooling loads. For most metal matrix composites, the inequality $\alpha_f < \alpha_m$ is valid for the coefficients of the linear thermal expansion of the fiber and the matrix, respectively. Shrinkage occurs over the fiber-matrix interface which results in compressive radial stresses over the material interface and tensile circumferential stresses and axial stresses within the matrix. Therefore, the radial longitudinal cracks and the penny-shaped cracks orthogonal to the axial direction(cf.Fig.1) are especially important for the cohesive strength of a fibrous composite. In the present study, the stress distribution characters in the matrix plastification process are analyzed by means of numerical and analytical methods, respectively, in representative elements of different brittle fiber-ductile matrix composites(material combinations: Glass/Al, Si/Steel, Boron/Al). Cracks located far-away from a fiber as well as near to a fiber are studied.

2 STRESS DISTRIBUTION OF THE UNCRACKED UNIT CELL

2.1 Basic Formulations and Assumptions

The model of a composite unit cell without any crack can be described by an axial symmetrical boundary-value problem. There often exists

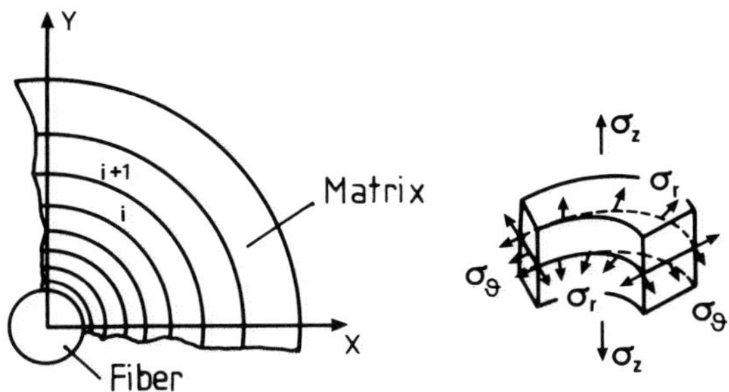

Fig.2 FEM model

a plastic zone around a fiber(cf.Fig.1b) because of the stress concentration in the vicinity of the fiber-matrix interface. Therefore, the unit cell is divided into three parts: fiber, plastic zone around the fiber, and elastic area of the matrix, respectively. The boundary and transition conditions are

$$\sigma_r(r_m) = 0 \tag{1a}$$

$$\sigma_r(r_{ep})\big|_{\text{plastic zone}} = \sigma_r(r_{ep})\big|_{\text{elastic zone}} \tag{1b}$$

$$u_r(r_{ep})\big|_{\text{plastic zone}} = u_r(r_{ep})\big|_{\text{elastic zone}} \tag{1c}$$

$$\sigma_r(r_f)\big|_{\text{fiber}} = \sigma_r(r_f)\big|_{\text{matrix}} \tag{1d}$$

$$u_r(r_f)\big|_{\text{fiber}} = u_r(r_f)\big|_{\text{matrix}} \tag{1e}$$

Because the unit cell is very long in comparison with the dimensions of its cross section, the research is based on a generalized plane strain assumption, that means ϵ_z keeps constant through the cross section. Further, the von Mises yield condition together with the associated flow rule and the Prandtl-Reuss theory are applied for the plastic analysis.

2.2 The Finite Element Analysis

Fig.2 shows the considered model which is composed of a fiber and the surrounding matrix. The incremental method is used in the finite element calculation. For the sake of convenience of the analysis the axial strain ϵ_z is regarded as a nodal displacement. Then, there are three nodal displacements for every element: $u_i, u_{i+1}, \epsilon_{zi}$. Because ϵ_z keeps constant through the cross section of the composite microcomponent, ϵ_{zi}-values of all elements are the same. Therefore, they can be regarded as one degree of freedom and put at the same position in the global displacement matrix.

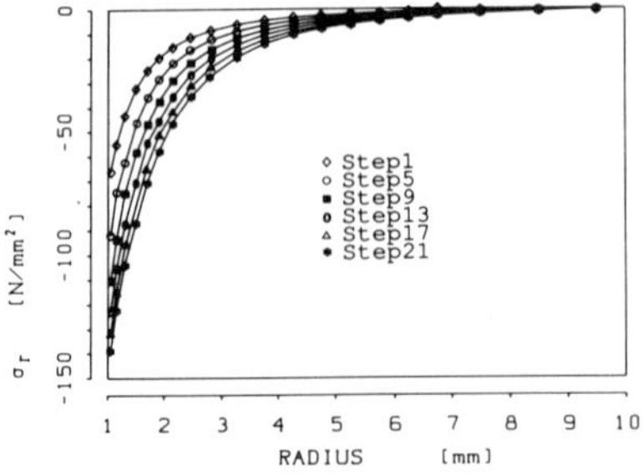

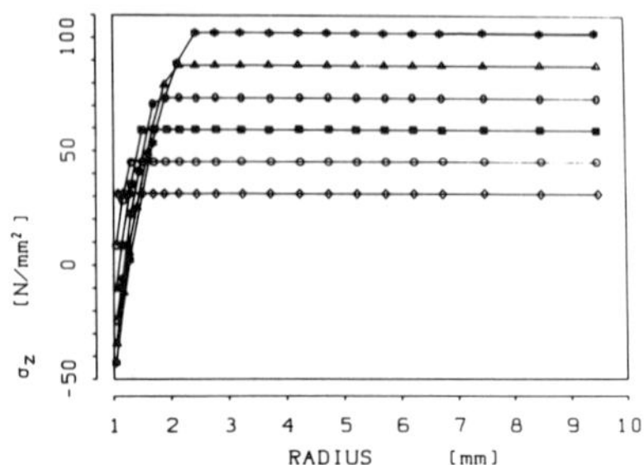

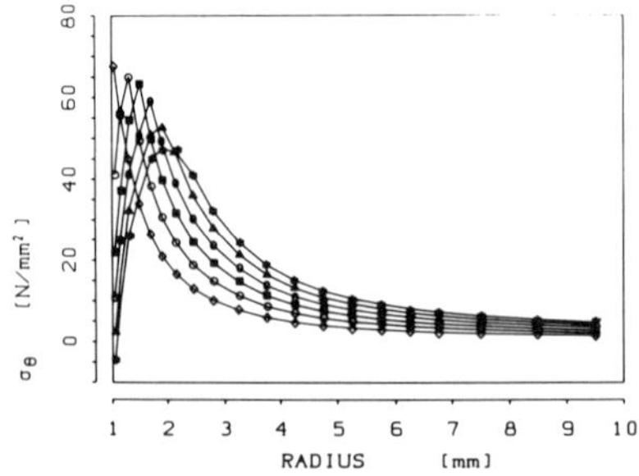

Fig.3 Stress distributions of model 1 under cooling and tensile loads $P_z/\Delta T$ = -100 [N/°K]

Twelve models have been considered. The geometrical parameters and material properties of them are shown in Table 1, where the subscripts f and m mean the fiber and the matrix, respectively, ν is Poisson's ratio, E is Young's modulus and σ_y is the yield stress of the matrix.

Fig.3 shows the numerical results of model 1 under thermal and axial loads as well obtained by means of the FEM program. The thermal load is ΔT = -300°K and the axial tensile load is P_z = 30000 N. The first element yields for ΔT = -92.84°K and P_z = 9284 N. In every loading step the temperature increment is $\Delta_i T$ = -10.36°K and the axial load increment is $\Delta_i P_z$ = 1036 N, respectively. The calculation was finished after 21 loading steps. The numerical results show that if the fiber volume fraction is relatively small, then

a) The local deviatoric stresses,i.e.,the stress differences($\sigma_\theta - \sigma_r, \sigma_r - \sigma_z$), keep basically constant and the hydrostatic stresses only increase with increasing load after yielding (see Table 2).

b) The existence of a small plastic zone surrounding the fiber has little influence on the stress distribution in the elastic zone. It can be seen from Fig.5 that graphs from an elastic analysis and an elasto-plastic

Table 1 Material properties

Model	r_f/r_m	α_f $[K^{-1}]$	α_m $[K^{-1}]$	E_f $[N/mm^2]$	E_m $[N/mm^2]$	σ_y $[N/mm^2]$	ν_f	ν_m
1	0.1	$6.75\cdot10^{-6}$	$2.39\cdot10^{-5}$	$8.40\cdot10^{4}$	$7.20\cdot10^{4}$	120.	0.21	0.34
2	0.2	$6.75\cdot10^{-6}$	$2.39\cdot10^{-5}$	$8.40\cdot10^{4}$	$7.20\cdot10^{4}$	120.	0.21	0.34
3	0.3	$6.75\cdot10^{-6}$	$2.39\cdot10^{-5}$	$8.40\cdot10^{4}$	$7.20\cdot10^{4}$	120.	0.21	0.34
4 Glass/Al	0.4	$6.75\cdot10^{-6}$	$2.39\cdot10^{-5}$	$8.40\cdot10^{4}$	$7.20\cdot10^{4}$	120.	0.21	0.34
5	0.1	$7.0\cdot10^{-6}$	$1.11\cdot10^{-5}$	$1.13\cdot10^{5}$	$2.10\cdot10^{5}$	520.	0.3	0.3
6	0.2	$7.0\cdot10^{-6}$	$1.11\cdot10^{-5}$	$1.13\cdot10^{5}$	$2.10\cdot10^{5}$	520.	0.3	0.3
7	0.3	$7.0\cdot10^{-6}$	$1.11\cdot10^{-5}$	$1.13\cdot10^{5}$	$2.10\cdot10^{5}$	520.	0.3	0.3
8 Si/Steel	0.4	$7.0\cdot10^{-6}$	$1.11\cdot10^{-5}$	$1.13\cdot10^{5}$	$2.10\cdot10^{5}$	520.	0.3	0.3
9	0.1	$4.5\cdot10^{-6}$	$2.39.10^{-5}$	$3.85\cdot10^{5}$	$7.20\cdot10^{4}$	120.	0.13	0.34
10	0.2	$4.5\cdot10^{-6}$	$2.39.10^{-5}$	$3.85\cdot10^{5}$	$7.20\cdot10^{4}$	120.	0.13	0.34
11	0.3	$4.5\ 10^{-6}$	$2.39.10^{-5}$	$3.85\cdot10^{5}$	$7.20\cdot10^{4}$	120.	0.13	0.34
12 Boron/Al	0.4	$4.5\cdot10^{-6}$	$2.39.10^{-5}$	$3.85\cdot10^{5}$	$7.20\cdot10^{4}$	120.	0.13	0.34

Table 2 Stress differences in the matrix $[N/mm^2]$

$\sigma_r - \sigma_\theta$

r_m/r_f	Step1	Step3	Step5	Step7	Step9	Step11	Step13	Step15	Step17	Step19	Step21
1.05	-134.0	-133.2	-132.9	-132.8	-132.9	-133.2	-133.4	-133.7	-133.9	-134.1	-134.4
1.15	-111.6	-131.4	-130.8	-130.5	-130.4	-130.7	-130.8	-131.1	-131.4	-131.6	-131.8
1.3	-88.0	-107.7	-127.5	-128.5	-128.2	-128.3	-128.3	-128.6	-128.8	-129.0	-129.3
1.5	-65.8	-80.5	-95.3	-109.9	-121.7	-121.7	-121.6	-121.8	-122.1	-122.4	-122.7
1.7	-51.1	-62.5	-73.9	-85.3	-96.6	-107.7	-113.7	-114.0	-114.3	-114.6	-115.0
1.9	-40.8	-49.9	-59.1	-68.1	-77.1	-86.1	-94.9	-103.7	-104.3	-104.8	-105.3
2.15	-32.0	-39.1	-46.3	-53.4	-60.5	-67.5	-74.4	-81.3	-87.9	-91.7	-92.4
2.45	-24.5	-30.0	-35.5	-41.0	-46.4	-51.8	-57.1	-62.4	-67.5	-72.5	-76.8
2.80	-18.8	-23.0	-27.2	-31.4	-35.6	-39.7	-43.8	-47.8	-51.8	-55.6	-59.2
3.25	-14.0	-17.1	-20.2	-23.3	-26.4	-29.5	-32.5	-35.5	-38.4	-41.3	-44.0
3.75	-10.4	-12.8	-15.1	-17.4	-19.7	-22.0	-24.3	-26.5	-28.7	-30.9	-32.9
4.25	-8.1	-9.9	-11.7	-13.5	-15.3	-17.1	-18.9	-20.6	-22.3	-23.9	-25.5
4.75	-6.4	-7.9	-9.3	-10.8	-12.2	-13.6	-15.1	-16.4	-17.8	-19.1	-20.4
5.25	-5.3	-6.4	-7.6	-8.8	-10.0	-11.1	-12.3	-13.4	-14.5	-15.6	-16.6
5.75	-4.4	-5.4	-6.3	-7.3	-8.3	-9.3	-10.2	-11.2	-12.1	-13.0	-13.8
6.25	-3.7	-4.5	-5.4	-6.2	-7.0	-7.8	-8.6	-9.4	-10.2	-11.0	-11.7
6.75	-3.2	-3.9	-4.6	-5.3	-6.0	-6.7	-7.4	-8.1	-8.7	-9.4	-10.0
7.50	-2.6	-3.1	-3.7	-4.3	-4.9	-5.5	-6.0	-6.6	-7.1	-7.6	-8.2
8.50	-2.0	-2.4	-2.9	-3.3	-3.8	-4.2	-4.7	-5.1	-5.5	-5.9	-6.3
9.50	-1.6	-1.9	-2.3	-2.7	-3.0	-3.4	-3.7	-4.1	-4.4	-4.7	-5.0

$\sigma_z - \sigma_r$

r_m/r_f	Step1	Step3	Step5	Step7	Step9	Step11	Step13	Step15	Step17	Step19	Step21
1.05	97.4	99.5	100.3	100.4	100.2	99.5	98.9	98.2	97.6	97.1	96.5
1.15	86.2	101.4	102.8	103.4	103.6	103.1	102.8	102.2	101.7	101.2	100.6
1.30	74.3	90.9	107.6	109.0	109.5	109.4	109.3	109.0	108.6	108.2	107.6
1.50	63.3	77.4	91.6	105.8	117.3	117.3	117.3	117.1	116.8	116.5	116.1
1.70	55.9	68.4	81.0	93.5	106.0	118.6	125.4	125.2	124.9	124.7	124.4
1.90	50.8	62.1	73.6	85.0	96.3	107.8	119.2	130.6	131.1	130.9	130.6
2.15	46.4	56.7	67.2	77.6	88.0	98.4	108.9	119.3	129.7	135.8	135.6
2.45	42.6	52.2	61.8	71.4	81.0	90.6	100.2	109.9	119.5	129.1	137.9
2.80	39.8	48.7	57.6	66.6	75.5	84.6	93.6	102.6	111.6	120.6	129.6
3.25	37.3	45.7	54.1	62.5	71.0	79.4	87.9	96.4	104.9	113.4	121.9
3.75	35.6	43.6	51.6	59.6	67.6	75.7	83.8	92.0	100.1	108.2	116.4
4.25	34.4	42.1	49.9	57.6	65.4	73.3	81.1	89.0	96.9	104.8	112.7
4.75	33.6	41.1	48.7	56.3	63.9	71.6	79.2	86.9	94.7	102.4	110.2
5.25	33.0	40.4	47.8	55.3	62.8	70.3	77.8	85.4	93.0	100.6	108.3
5.75	32.6	39.9	47.2	54.6	61.9	69.4	76.8	84.3	91.8	99.3	106.9
6.25	32.2	39.5	46.7	54.0	61.3	68.7	76.0	83.4	90.9	98.3	105.8
6.75	32.0	39.1	46.3	53.6	60.8	68.1	75.4	82.8	90.1	97.5	105.0
7.50	31.7	38.8	45.9	53.0	60.2	67.5	74.7	82.0	89.3	96.7	104.0
8.50	31.4	38.4	45.5	52.6	59.7	66.8	74.0	81.3	88.5	95.8	103.1
9.50	31.2	38.2	45.2	52.2	59.3	66.4	73.5	80.7	88.0	95.2	102.5

Note: Underline _ means that the point has been yielded

analysis, respectively, are very close in the elastic zone but separate in the interface between the plastic and the elastic zone.

The results from the other models show similar attributes. Therefore, in the following an analytical approach is presented based on those essential features of this elasto-plastic boundary value problem.

2.3 The Analytical Approach

2.3.1 Basic Assumptions

An approximate analytical solution is presented here based on the following assumptions:

a) The local principal stress differences keep constant after yielding.

b) The existence of a plastic zone surrounding the fiber has no influence on the stress distribution in the elastic zone.

2.3.2 The Stresses in the Elastic Zone

Therefore, according to assumption 2.3.1.b), the stresses in the elastic zone can be estimated by an elastic analysis without consideration of any plastic area as the following:

$$\sigma_r = (a_1 \Delta T + a_2 P_z) \frac{r_f^2}{r_m^2 - r_f^2} \left(1 - \frac{r_m^2}{r^2}\right) \tag{2}$$

$$\sigma_\theta = (a_1 \Delta T + a_2 P_z) \frac{r_f^2}{r_m^2 - r_f^2} \left(1 + \frac{r_m^2}{r^2}\right) \tag{3}$$

$$\sigma_z = E_m(a_3 \Delta T + a_4 P_z) - \alpha_m \Delta T E_m + \frac{2\nu_m r_f^2 (a_1 T + a_2 P_z)}{r_m^2 - r_f^2} \tag{4}$$

By using the definition

$$a_i = \frac{a_i'}{b} \quad (i = 1,2,3,4) \tag{5}$$

it follows

$$a_1' = \left[E_m(r_m^2 - r_f^2)(1+\nu_f) + E_f r_f^2 (1+\nu_m)\right](\alpha_m - \alpha_f) \tag{6a}$$

$$a_2' = (\nu_f - \nu_m)/\pi \tag{6b}$$

$$a_3' = 2\left[(1+\nu_f)\alpha_f - (1+\nu_m)\alpha_m\right](\nu_m - \nu_f) r_f^2 - \left[\alpha_m E_m (r_m^2 - r_f^2) + \alpha_f E_f r_f^2\right] \cdot \frac{(1-\nu_m - 2\nu_m^2) r_f^2 + (1+\nu_m) r_m^2}{E_m (r_m^2 - r_f^2)} + (1-\nu_f - 2\nu_f^2)/E_f \tag{6c}$$

$$a_4' = -(1-\nu_f - 2\nu_f^2)/E_f/\pi - \frac{(1-\nu_m - 2\nu_m^2) r_f^2 + (1+\nu_m) r_m^2}{\pi E_m (r_m^2 - r_f^2)} \tag{6d}$$

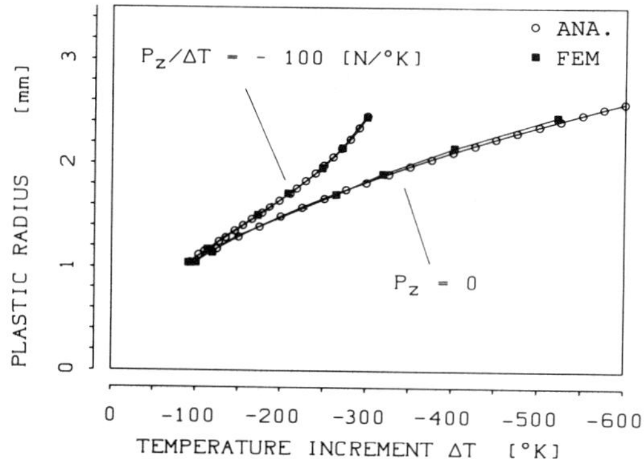

Fig.4 The size of the plastic zones under different loading

$$b = \left[4\nu_f\nu_m - 2 + \nu_m + \nu_f + \frac{E_m}{E_f}(1-\nu_f-2\nu_f^2) - \frac{E_f}{E_m}(1+\nu_m)\right] r_f^2$$

$$- \left[1+\nu_m + (1-\nu_f-2\nu_f^2)\frac{E_m}{E_f}\right] r_m^2 - \frac{2\, E_f r_f^4(1-\nu_m^2)}{E_m(r_m^2 - r_f^2)} \tag{6e}$$

2.3.3 <u>The Plastic Radius</u>

If the load is large enough there will be a plastic zone near to the fiber on the side of the matrix(Fig.1b). The yielding area can be estimated by using the von Mises yield condition:

$$\left[(\sigma_r-\sigma_\theta)^2+(\sigma_r-\sigma_z)^2+(\sigma_\theta-\sigma_z)^2\right] \geq 2\,\sigma_y^2 \tag{7}$$

where σ_y is the equivalent yield stress of the matrix. By substituting the equations (2)-(4) into the inequality (7), the plastic zone size can be obtained as:

$$r_f \leq r \leq r_{ep} \tag{8}$$

with

$$r_{ep} = \frac{r_f\sqrt{f_1\,\Delta T}}{\sqrt[4]{\sigma_y^2 - f_2^2(\Delta T)^2}} \tag{9}$$

and with the definitions

$$f_1 = \frac{\sqrt{3}\; r_m^2(a_1+a_2P_z/\Delta T)}{(r_m^2 - r_f^2)} \tag{10}$$

$$f_2 = E_m(a_3+a_4P_z/\Delta T) - \alpha_m E_m + \frac{(2\nu_m-1)r_f^2(a_1+a_2P_z/\Delta T)}{(r_m^2 - r_f^2)} \tag{11}$$

Further, eq.(9) can be rewritten as

$$\Delta T = \frac{\pm \sigma_y r_{ep}^2}{\sqrt{f_2^2 r_{ep}^4 + f_1^2 r_f^4}} \tag{12}$$

In the following negative temperature increments only have been considered. Fig.4 gives the comparison between the results obtained by eq.(9) and the FEM program, respectively.

2.3.4 Stresses in the Plastic Zone

According to assumption 2.3.1.a), the local stress differences keep constant after yielding. Therefore, the stress differences of a point in the plastic zone can be estimated by those stress differences by which the point starts yielding. From eq.(12) the yielding temperature load at this moment can be calculated. Then, by substituting the result into eqs.(2)-(4),

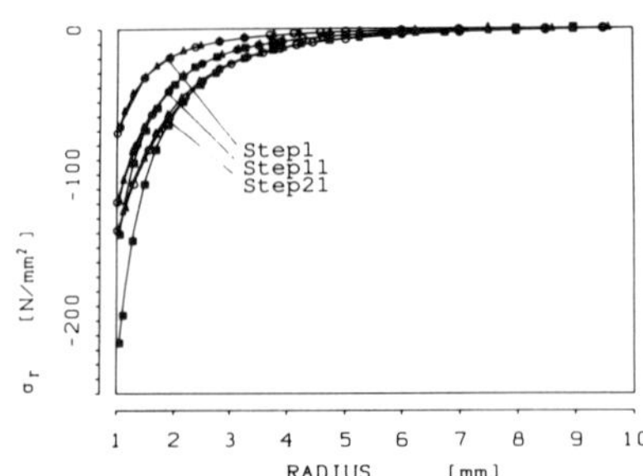

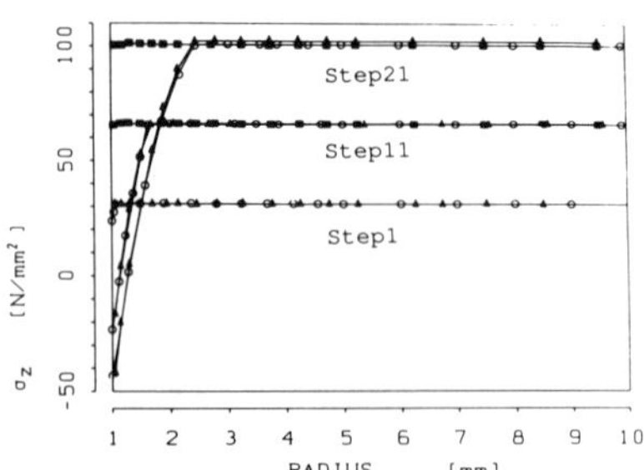

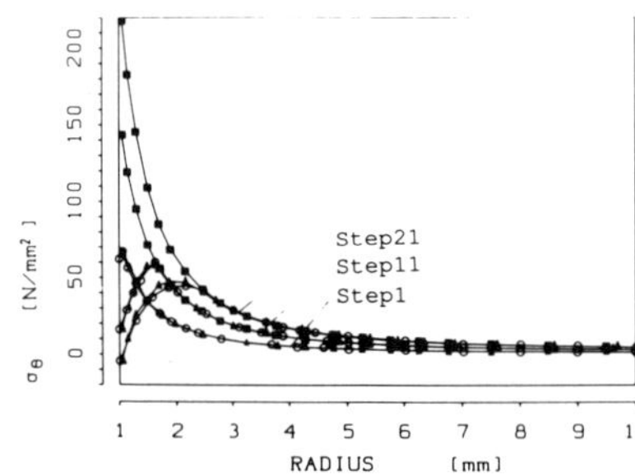

Notes:

- ▴ - Data obtained from an elasto-plastic analysis by the aid of a FEM program
- ▪ - Data obtained from an elastic analysis by the aid of a FEM program
- ∘ - Data obtained from an elasto-plastic analysis by the aid of an analytical approach

Fig.5 Comparison between results obtained by FEM and by an analytical approach

the stress differences in the plastic zone can be obtained as

$$\sigma_r - \sigma_\theta = \frac{2\sigma_y f_1 r_f^2}{\sqrt{3(f_2^2 r^4 + f_1^2 r_f^4)}} \tag{13}$$

$$\sigma_r - \sigma_z = \frac{\sigma_y (\sqrt{3}\, f_2 r^2 + f_1 r_f^2)}{\sqrt{3(f_2^2 r^4 + f_1^2 r_f^4)}} \tag{14}$$

Moreover, from the equilibrium condition follows

$$\frac{d\sigma_r}{dr} = - \frac{\sigma_r - \sigma_\theta}{r} = \frac{-2\sigma_y f_1 r_f^2}{r\sqrt{3(f_2^2 r^4 + f_1^2 r_f^4)}} \tag{15}$$

Then, the solution of the differential equation (15) reads

$$\sigma_r = - \frac{\sigma_y}{\sqrt{3}} \ln \frac{|f_1|\, r_f^2 + \sqrt{f_1^2 r_f^4 + f_2^2 r^4}}{|f_2|\, r^2} + c \tag{16}$$

where c is a constant which can be obtained from the following boundary condition:

$$\sigma_r = (a_1 \Delta T + a_2 P_z) \frac{r_f^2}{r_m^2 - r_f^2} \left(1 - \frac{r_m^2}{r^2}\right) \text{ for } r = r_{ep} \tag{17}$$

Further, the other two stress components can be obtained from the eqs.(13)-(14). Thus, the stress distribution in the plastic zone reads

$$\sigma_r = - \frac{\sigma_y}{\sqrt{3}} \ln \left(\frac{|f_1| r_f^2 + \sqrt{f_1^2 r_f^4 + f_2^2 r^4}}{|f_1| r_f^2 + \sqrt{f_1^2 r_f^4 + f_2^2 r_{ep}^4}} \frac{r_{ep}^2}{r^2} \right) + $$

$$+ (a_1 \Delta T + a_2 P_z) \frac{r_f^2 (r_{ep}^2 - r_m^2)}{r_{ep}^2 (r_m^2 - r_f^2)} \tag{18}$$

$$\sigma_\theta = \sigma_r - \frac{2\sigma_y f_1 r_f^2}{\sqrt{3(f_2^2 r^4 + f_1^2 r_f^4)}} \tag{19}$$

$$\sigma_z = \sigma_r - \frac{\sigma_y \left(\sqrt{3}\, f_2 r^2 + f_1 r_f^2\right)}{\sqrt{3(f_2^2 r^4 + f_1^2 r_f^4)}} \tag{20}$$

where r_{ep} can be obtained from eq.(9). Fig.5 shows the comparison of the results obtained by the FEM program and by the eqs.(18)-(20), respectively.

3 DUGDALE CRACK ANALYSIS FOR A CRACK TIP FAR-AWAY FROM A FIBER

It is clear that if a crack tip is situated far-away from a fiber and if its size is small in comparison to the size of the matrix, the effect of the fiber-matrix interface on the crack can be neglected. Then, the stress field without the crack delivers the boundary conditions for the calculation of the crack tip stresses from a corresponding crack boundary value problem. It is also clear from the stress distribution characters that the dangerous crack types concerning the strength of a fibrous composite are linear cracks in the radial direction of a composite microcomponent and the plane cracks orthogonal to the axial direction of the unit cell(cf.Fig.1a and Fig.1b). In this section, the Dugdale model for these two kinds of cracks are considered.

3.1 Dugdale Model for a Radial Longitudinal Crack

Because the stress distribution in the elastic zone of the matrix is not remarkably influenced by a small plastic zone around a single fiber, the Dugdale crack model presented by Herrmann(1978) can be used also for the situation that a thin plastic zone exists.

Modifying Herrmann's solution, the lengths $s_i(i=l,r)$ of the plastic zones at the two crack tips are the roots of the following equation where the plus and minus signs correspond to the left and right crack tip,

$$\frac{a}{2} + \frac{r_m^2 \left[r_0 \pm (a/2+s)\right]}{\left[r_0^2 - (a/2+s)^2\right]^{3/2}} = \frac{2\,(r_m^2 - r_f^2)\,\sigma_y}{(a_1 \Delta T + a_2 P_z)\, r_f^2} \arccos\left(\frac{a/2}{a/2+s}\right) \tag{21}$$

and the crack opening displacement reads

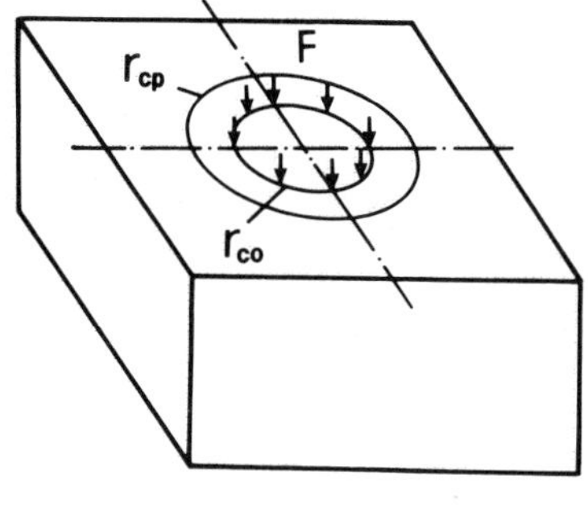

Fig.6 Virtual concentrated ring load

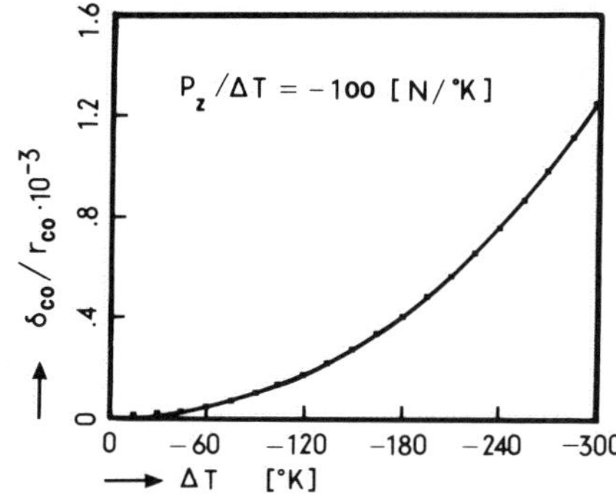

Fig.7 Crack opening displacement

$$\delta_{co} = \frac{8\,(1-\nu_m^{\,2})\ \sigma_y}{E_m}\sqrt{as + s^2}\ \text{arc cos}\left(\frac{a/2}{a/2+s}\right) \tag{22}$$

where r_0 is the coordinate of the position of the crack center and a is the crack length.

3.2 Dugdale Model for a Penny-Shaped Crack

3.2.1 Basic problem

Similarly to the Dugdale crack model for a linear crack it is assumed that the plastic zone at the crack border consists of a thin annular region. Then an appropriate Dugdale model for a penny-shaped crack is obtained by consideration of the following equivalent elastic crack boundary value problem(cf.Fig.1b and Fig.6):

$$\sigma_z^{\,c} = \begin{cases} -\sigma_z & \text{for } r_c \leq r_{co} \\ -\sigma_z + \sigma_y & \text{for } r_{co} < r_c < r_{cp} \end{cases} \tag{23}$$

where r_{co} is the radius of the circular crack and $r_{co} < r_c < r_{cp}$ represents the plastic zone at the crack border(cf. Fig.1b). Further, the stress σ_z is given by eq.(4). According to the theory of the Dugdale crack model the size of the plastic zone can be determined by use of the following condition at the fictitious crack border

$$K_I = 0 \tag{24}$$

where K_I is the stress intensity factor.

According to Kassir and Sih(1975) the K_I-value for a penny-shaped crack under normal load reads

$$K_I = -\frac{2}{\sqrt{\pi r_{cp}}} \int_0^{r_{cp}} \frac{r_c \sigma_z^c}{\sqrt{r_{cp}^2 - r_c^2}} dr_c$$

$$= \frac{2}{\sqrt{\pi r_{cp}}} \left[\sigma_z r_{cp} - \sigma_y \sqrt{r_{cp}^2 - r_{co}^2} \right] \qquad (25)$$

By substituting eq.(25) into eq.(24), the size of the plastic area at the crack border can be calculated as follows

$$r_{cp} = \frac{\sigma_y r_{co}}{\sqrt{\sigma_y^2 - \sigma_z^2}} \qquad (26)$$

3.2.2 The crack tip opening displacement

According to Fig.6 virtual concentrated ring loads F in $r_c = r_{co}$ are introduced in order to get the crack tip opening displacement.

By applying Castigliano's theorem the displacement of the ring should be equal to the partial derivative of the strain energy with respect to F. If F approaches zero, the displacement of the ring will be the crack tip opening displacement. It reads

$$\delta \Big|_{r_c = r_{co}} = \lim_{F \to 0} \frac{\partial E}{\partial F} \qquad (27)$$

where E is the strain energy. The energy release rate G_I for a mode I-crack in an elastic region is defined as

$$G_I = \frac{\partial E}{\partial A} \qquad (28)$$

where A is the area of the crack. Obviously, E depends on σ_z^c, F, and A. Therefore

$$E(\sigma_z^c, F, A) = E(\sigma_z^c, F, 0) + \int_0^{r_{cp}} G_I \, dA \qquad (29)$$

It is easy to prove that the relationship between G_I and K_I in the present problem is the same as that for a plane strain problem. Therefore, eq.(29) can be rewritten as

$$E = E(\sigma_z^{\,c},F,0) + \frac{1-\nu_m^{\,2}}{E_m}\int_0^{r_{cp}} (K_I^{\sigma_z^c} + K_I^{\,F})^2 \, dA \tag{30}$$

By introducing eq.(30) into eq.(27) one obtains

$$\delta_{co} = \lim_{F\to 0} \frac{2(1-\nu_m^{\,2})}{E_m}\int_0^{r_{cp}} K_I^{\sigma_z^c} \frac{\partial K_I^{\,F}}{\partial F}\, dA \tag{31}$$

According to Kassir and Sih(1975) it valids

$$K_I^{\,F} = \begin{cases} 0 & \text{for } r_c \leq r_{co} \\ \dfrac{F}{\pi^{3/2}\sqrt{r_c(r_c^{\,2}-r_{co}^{\,2})}} & \text{for } r_c > r_{co} \end{cases} \tag{32}$$

and $K_I^{\sigma_z^c}$ can be obtained from eq.(25). Then it follows

$$\delta_{co} = \frac{4\,(1-\nu_m^{\,2})}{\pi^2 E_m}\int_{r_{co}}^{r_{cp}} \frac{(\sigma_z r_c - \sigma_y\sqrt{r_c^{\,2} - r_{co}^{\,2}})}{r_c\sqrt{r_c^{\,2} - r_{co}^{\,2}}}\, dA$$

$$= \frac{8(1-\nu_m^{\,2})}{\pi E_m}\left[\sigma_z\sqrt{r_{cp}^{\,2} - r_{co}^{\,2}} - \sigma_y\,(r_{cp} - r_{co})\right] \tag{33}$$

By introducing eq.(26) into eq.(33) the crack-tip opening displacement can be expressed as:

$$\delta_{co} = \frac{8\,(1-\nu_m^{\,2})r_{co}}{\pi E_m}\left(\sigma_y - \sqrt{\sigma_y^{\,2} - \sigma_z^{\,2}}\right) \tag{34}$$

A numerical evaluation of this equation is given in Fig.7.

4 ANALYSIS FOR A CRACK TIP NEAR TO A FIBER

If the crack tip is near to a fiber the effect of the fiber-matrix interface on the crack has to be considered. Especially, if the crack tip contacts the plastic zone around the fiber the above mentioned Dugdale model can not be applied in order to estimate the fracture resistance of a composite microcomponent. Thereby, many investigations have pointed out that the interface properties are very important for a determination of the fracture behaviour of a fibrous composite.

It can be seen from Fig.3 that the circumferential and the axial stresses, respectively, will decrease rapidly in the vicinity of the material interface if there is a plastic zone around the fiber. Therefore, associated with the plastic energy dissipation at the crack tip the existence of the plastic zone will improve the fracture resistance of fibrous composites. It is possible to sustain a stable crack growth or arrest a crack in the plastic zone. Further, the stable crack growth is described here by means of the energy release rate. A special finite element program was developed for the numerial calculation of the energy release rate. Based on these analyses the effects of plastic zones around fibers on the fracture resistance of fibrous composites are discussed.

4.1 Energy Release Rate

If a crack extends by a small amount $\Delta_i a$, the stress σ_y along the new extended crack surface will be released and there will be an opening displacement v. For simplicity this process can be regarded as a continuous process as shown in Fig.8. The crack tip opens when the stress σ_y along $\Delta_i a$ is gradually released. Obviously, the released energy is the volume covered by the σ_y-stress surface and the energy release rate can be expressed as

$$G = \frac{2}{\Delta_i a} \int_0^{\Delta_i a} dx \int_0^{v(0,x)} \sigma_y(x,v)\, dv \tag{35}$$

The process can be modeled by the FEM-model shown in Fig.9. The nodal forces at the nodes 1 and 2 are released in steps and the energy

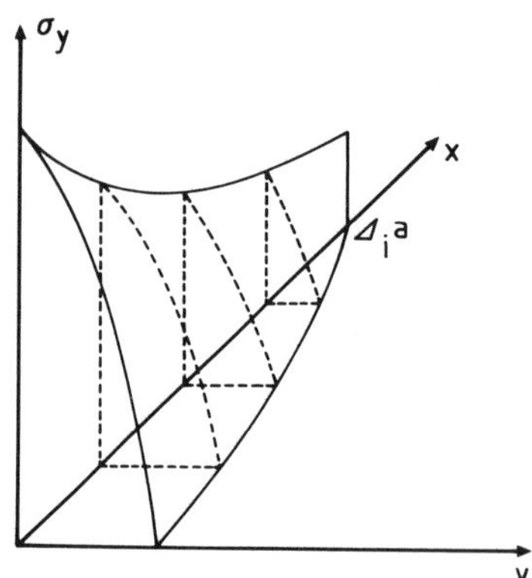

Fig.8 Energy release

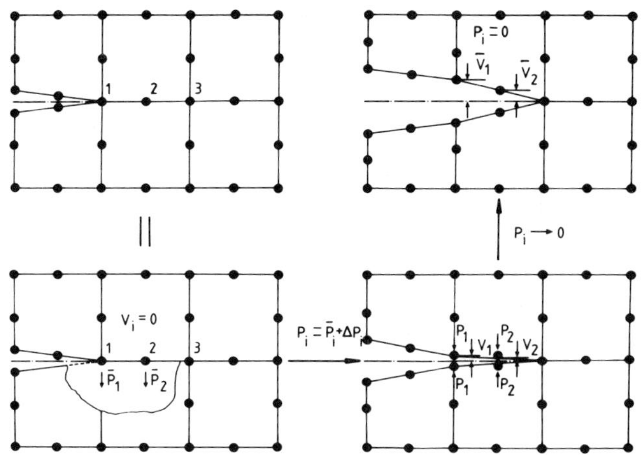

Fig.9 FE-model for an energy releasing process

release rate G can be obtained by means of the following integral

$$G = \frac{2}{\Delta_i a} \left(\int_0^{\bar{v}_1} P_1 \, dv_1 + \int_0^{\bar{v}_2} P_2 \, dv_2 \right) \tag{36}$$

4.2 Numerical Calculation

Fig.10 shows a finite element model for the numerical analysis of a cracked composite unit cell. Because of the symmetry of the cell only one half of the unit cell was considered. For the sake of convenience of data preparation this unit cell was divided into several substructures. The thick lines in Fig.10 represent the boundaries of the substructures. The crack length is a and the crack propagates towards the fiber. The largest crack propagation length is Δ a = 1/3 a. The stable crack growth for model 3 in Table 1 with four different positions of the crack tip under at least four different thermal loads (ΔT = -100°K, -200°K, -300°K, -400°K) and tensile loads, respectively, was calculated. The positions of the crack tip near to a fiber(cf.Fig.10) are given by r_a = 3.3, 4.3, 4.7, and 5.1mm, respectively. The crack length a (cf.Fig.10) is 1.5mm.

Fig.11 shows the energy release rate for r_a = 4.7mm and r_a= 3.3mm, respectively. It can be seen from Fig.11a that the energy release rate increases firstly with increasing thermal load until ΔT = -200°K and decreases gradually for higher thermal load. This phenomenon can be

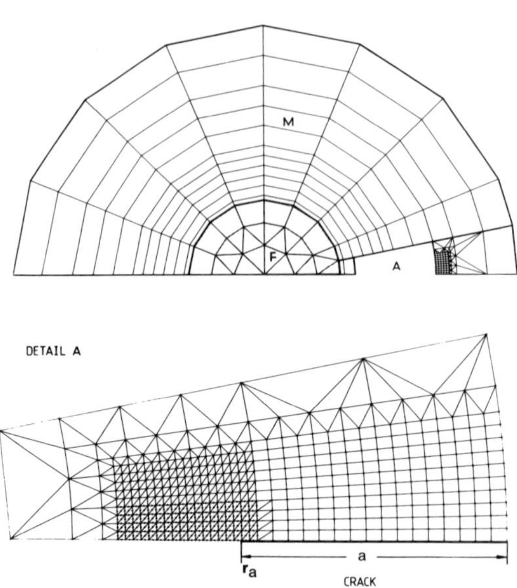

Fig.10 FE-model of a cracked composite microcomponent

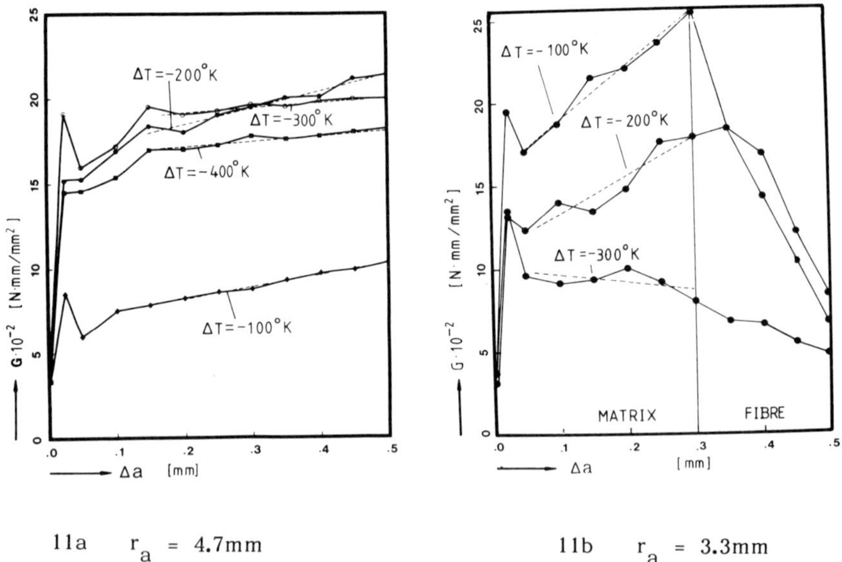

11a r_a = 4.7mm 11b r_a = 3.3mm

Fig.11 Energy release rate G under different thermal loads

explained by the wake of the plastic zone at the crack tip shown in Fig.12 which gives the plastic area in region A (see Fig.10). For $\Delta T = -100°K$ both plastic zones around the fiber and around the crack tip are small and they don't contact each other. The circumferential stresses near to the crack tip will increase with increasing thermal load and then the energy release rate will increase too. But for $\Delta T = -200°K$ both plastic zones start to contact and get together gradually. Because the circumferential stresses decrease rapidly in the plastic zone around the fiber (cf.Fig.3), the stresses near to the crack tip are improved and the energy release rate will not increase for the higher load. This can also be seen clearly from Fig.13 in which the dotted lines represent that thermal load at which the two plastic zones contact. Fig.11b shows energy release rates for a crack tip very near to a fiber where both plastic zones join when the thermal load is rather small. This picture shows the situation when the crack penetrates into the fiber from the matrix side and the energy release rate will decrease very quickly after this penetration.

Fig.14 gives the profiles of a stable crack extension. The crack becomes much sharper after the onset of growth than it was in the initial state. Therefore, the energy release rates are always much smaller at the initial crack growth than those for further stable crack growth.

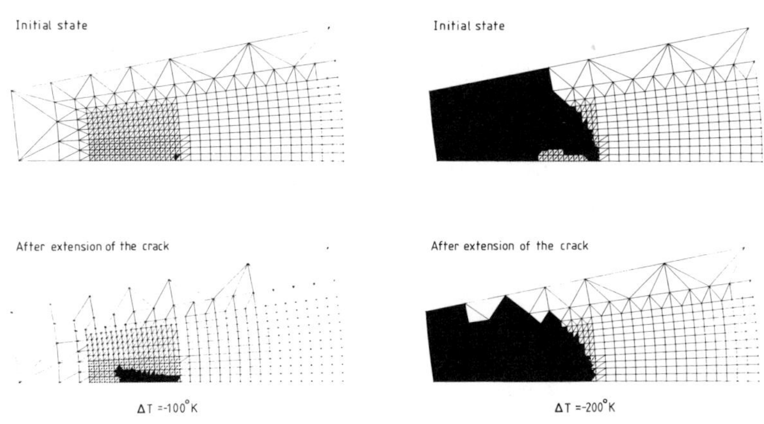

- Plastic zone

Fig.12 The wakes of the plastic zones for stable crack growth

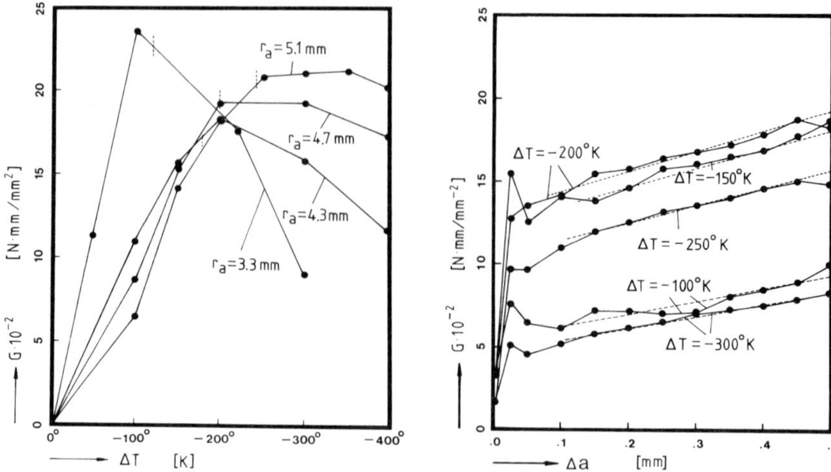

Fig.13 Energy release rate G in dependence on temperature

Fig.15 Energy release rate for an initial crack tip at r_a = 4.7mm and $P_z/\Delta T$ = -100 N/°K

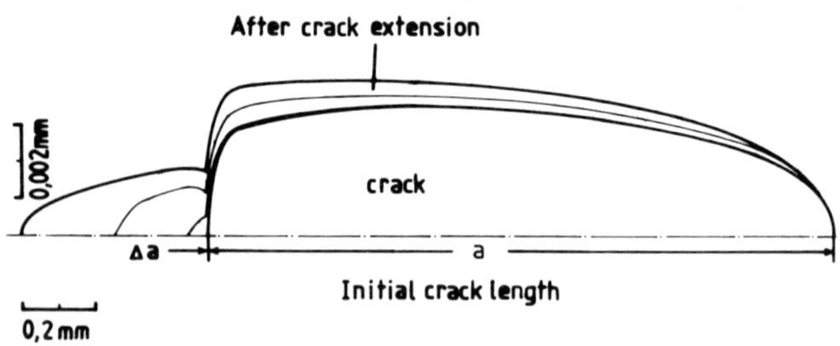

Fig.14 The profiles of a stable crack extension

Fig.15 shows the energy release rate at a crack tip under combined thermal and tensile load. It can be seen that the fracture resistance is improved in comparison with Fig.11a because of an acting tensile load which makes the plastic zone around the fiber of a thermally loaded unit cell larger(cf.Fig.5).

Similar to the distribution of the circumferential stresses the axial stresses decrease rapidly in the plastic zones around fibers. Therefore, the existence of a plastic zone may also improve the fracture resistance to penny-shaped cracks located in unit cells which are submitted to tensile loads. Because the plastic zones around fibers are mainly caused by thermal loads, it means that the fracture resistance to penny-shaped cracks of a unit cell under both tensile and thermal loads may be better than that of a cell under tensile load only. A detailed investigation is presently on the way.

ACKNOWLEDGEMENT

The support of the Alexander von Humboldt-Foundation for one of the authors(Y.Q.Wang) is gratefully acknowledged.

REFERENCES

Achenbach,J.D. and Zhu,H.,1989, "Effect of interfacial zone on mechanical behaviour and failure of fiber reinforced composites", Journal of the Mechanics and Physics of Solids Vol. 37, pp. 381-393.

Bahei-El-Din,Y.A., Dvorak,G.J. and Wu,J.F.,1989, "Fracture of fibrous metal matrix composites-II. Modeling and numerical analysis", Engineering Fracture Mechanics Vol. 34, pp. 105-123.

Christensen,R.M.,1982, "Mechanical Properties of Composite Materials", in Zvi Hashin and Carl T. Herakovich(Eds.), Mechanics of Composite Materials: Recent Advances, Pergamon Press, pp. 1-16.

Dvorak,G.J. and Rao,M.S.M.,1976, "Thermal stresses in heat-treated fibrous composites", Journal of Applied Mechanics Vol. 43, pp. 619-624.

Dvorak,G.J., Bahei-El-Din Y.A. and Bank,L.C.,1989, "Fracture of fibrous metal matrix composites-I. Experimental results", Engineering Fracture Mechanics Vol. 34, pp. 87-104.

Ebert, L.J, Hecker,S.S and Hamilton,C.H.,1968, "The stress-strain behaviour of concentric composite cylinders", Journal of Composite Materials Vol. 2, pp. 458-476.

Hashin,Z. and Rosen,B.W.,1964, "The elastic moduli of fiber reinforced materials", Journal of Applied Mechanics Vol. 31, pp. 223-232.

Herrmann,K.,1978, "Interaction of cracks and self-stresses in a composite structure", In J.W. Provan(Ed.), SM Study 12; Continuum Models of Discrete Systems, University of Waterloo Press, pp. 313-338.

Herrmann,K. and Pawliska,P.,1986, "Finite element analysis of thermal crack growth in self-stressed fiber reinforced composites with partially plastified matrices", International Journal of Fracture 31, R11-R16.

Herrmann,K. and Ferber,F.,1988, "Numerical and experimental investigations of branched thermal crack systems in self-stressed models of unidirectionally reinforced fibrous composites", in S.N. Atluri(Ed.), Computational Mechanics 88, Springer, 3.V.1-8.V.4.

Herrmann,K.P. and Mihovsky,I.M.,1988, "Approximate analytical investigation of the elastic-plastic behaviour of fibrous composites", XVII International Congress of Theoretical and Applied Mechanics, Grenoble, France.

Kassir,M.K. and Sih,G.C.,1975, Three-Dimensional Crack Problems, Mechanics of Fracture 2, Noordhoff International Publishing Leyden.

Laws,N and Dvorak,G.J.,1987 "The effect of fiber breaks and aligned penny-shaped cracks on the stiffness and energy release rates in unidirectional composites", International Journal of Solids and Structures Vol.23, pp. 1269-1283.

McCullough,R.L.,1982, "Influence of microstructure on the thermoelastic and transport properties of particulate and short fiber composites", in Zvi Hashin and Carl T. Herakovich(Eds.), Mechanics of Composite Materials: Recent Advances,Pergamon Press, pp.17-29.

Nuismer,R.J. and Tan,S.C.,1982, "The role of matrix cracking in the continuum constitutive behaviour of a damaged composite ply", In Zvi Hashin and Carl T. Herakovich(Eds.), Mechanics of Composite Materials: Recent Advances,Pergamon Press, pp. 437-448.

Olster,E.F. and Jones,R.C.,1974, "Effect of interface on fracture", in Arthur G. Metcalfe(Ed.), Interfaces in Metal Composites, Academic Press, pp. 245-284.

Sendeckyj,G.P., Maddux,G.E. and Tracy,N.A.,1978, "Comparison of holographic, radiographic and ultrasonic techniques for damage defection in composite materials", Proceedings of 2nd International Conference on Composite Materials, Toronto, pp. 1037-1056.

Taya,M. and Chou,T.W.,1982, "Energy release rates of various microcracks in short-fiber composites", In Zvi Hashin and Carl T. Herakovich, Mechanics of Composite Materials: Recent Advances, Pergamon Press pp. 457-472.

Incremental Elastoplastic Behavior of Metal Matrix Composites Based on Averaging Schemes

Dimitris C. Lagoudas and **Andres C. Gavazzi***

Institute Center for Composite Materials and Structures and
Department of Civil Engineering
Rensselaer Polytechnic Institute, Troy, N.Y. 12180

*On leave from Technical University of Milan, Milano, Italy.

Abstract

The applicability of averaging schemes, with emphasis on the Mori–Tanaka method, for the prediction of the response of binary composites in the plastic range is discussed. The applied loading is subdivided into small increments and the Eshelby solution for the inhomogeneity problem is used in conjunction with an averaging scheme to obtain the load increments in the various phases. Since the Eshelby solution depends on the instantaneous matrix material properties and these are updated at the end of each load increment, using the backward difference scheme, an iterative procedure is necessary for the calculation of the correct load increments in the phases (concentration factors). The performance of Mori–Tanaka and self–consistent schemes is compared with results obtained using the modified periodic hexagonal array (PHA) finite element model for a SiC_w–$A\ell$ particulate composite.

Introduction

Averaging methods based on the Eshelby solution of the ellipsoidal inhomogeneity (Eshelby (1957)) have been used extensively for the evaluation of the effective elastic properties of fibrous or particulate composites. The most prominent ones are the self–consistent method and recently the Mori–Tanaka method, both summarized in the review paper by Dvorak (1990). An attempt to extend the self–consistent method to composites undergoing elastoplastic deformations has been made by Hutchinson (1970) (following Hill's (1965a) formulation) for polycrystals and by Dvorak and Bahei–El–Din (1979) for fibrous composites under axisymmetric mechanical loading and uniform thermal loading, among others. The extension of the Mori–Tanaka method to deformation plasticity has been attempted by Tandon and Weng (1988) for particulate composites.

In this work, averaging schemes are formulated for binary composites with an elastic phase dispersed in an elastoplastic matrix modeled by a time independent incremental plasticity constitutive law. In the first section, the average elastic stress and strain concentration factors are obtained for binary composites, starting with the equivalence principle of Eshelby and following the Mori–Tanaka (1973) assumption of an average stress in the matrix. An extension to include elastoplastic matrix behavior is then attempted and history dependent concentration factors are defined. In the second section, a numerical scheme is described for the integration of the combined system of matrix plasticity equations together with the incremental equations that define the concentration factors. Explicit evaluations are given for a Mises yield criterion and kinematic hardening according to Phillips hardening rule. The third section contains comparisons of the predictions of Mori–Tanaka and self–consistent methods with numerical simulations using a periodic hexagonal array model (PHA) (Dvorak and Teply (1985)) for a SiC_W–Al composite.

1. Concentration Factors

1.1 Elastic Concentration Factors by the Mori–Tanaka Method

Let a composite material consist of two elastic phases and be loaded by a macroscopically uniform stress, $\bar{\sigma}$, or strain, $\bar{\epsilon}$. The uniform applied fields are assumed to extend over a representative volume V. The representative volume forms a characteristic sample of the microstructure and consists of a sufficiently large number of inclusions of each phase, yet its size is small compared to the macroscopic length scale (Hill (1963)). To explicitly evaluate the phase average stress and strain concentration factors for a binary composite we assume that the stress and strain fields in the dispersed phase can be represented by a single inhomogeneity embedded in an infinite medium. According to the self–consistent method (Budiansky (1965), Hill (1965b)), the interactions among the inhomogeneities for finite volume fractions are approximately taken into account by embedding the inhomogeneity in a medium with the effective elastic properties. The above methodology was initially developed to model the average constitutive behavior of polycrystals but was later applied to composites. The basic difference between composites and polycrystals is the existence of a distinct matrix phase in composites which supports the various dispersions. This realization established a new trend in evaluating effective properties for composites. The ellipsoidal inhomogeneity representing the dispersed phase is embedded in an infinite matrix material instead of an effective medium and the interactions are taken into account by appropriately modifying the average stress (strain) field in the matrix from the applied one. The idea of modifying the stress field in the matrix instead of altering its material properties was originally proposed by Mori and Tanaka (1973) for the transformation problem and subsequently modified for the inhomogeneity problem by Taya and Mura (1981), Weng (1984) and recently reconsidered by Benveniste (1987). Following the above principle we will briefly go through a derivation of stress and strain concentration factors.

Assume first that a macroscopically uniform strain $\bar{\boldsymbol{\epsilon}}$ is applied to the representative volume V. Due to the interactions of the inclusions, the average strain in the matrix is perturbed by an amount $\tilde{\boldsymbol{\epsilon}}$, while the strain in the inhomogeneity has an additional perturbation $\boldsymbol{\epsilon}^{pt}$ (Weng (1984)). Based on Eshelby's (1957) equivalence principle, the stress field in the inhomogeneity can be made equal to the stress field of an equivalent inclusion that has the matrix material properties and undergoes a transformation strain $\boldsymbol{\epsilon}^{t}$, thus compensating for the change in the material properties. The equivalence principle has the explicit form

$$\mathbf{L}_f(\bar{\boldsymbol{\epsilon}} + \tilde{\boldsymbol{\epsilon}} + \boldsymbol{\epsilon}^{pt}) = \mathbf{L}_m(\bar{\boldsymbol{\epsilon}} + \tilde{\boldsymbol{\epsilon}} + \boldsymbol{\epsilon}^{pt} - \boldsymbol{\epsilon}^{t}) \ , \tag{1}$$

where the subscript f refers to the inhomogeneity and the subscript m to the matrix material. We also follow Hill's notation in representing by bold face capital letters fourth order tensors and by bold face lower–case letters second order tensors. In the above, and also in the sequel, **L** denotes the stiffness tensor, while **M** denotes the compliance tensor, with appropriate subscripts to indicate the phases. The relation between the perturbation strain ϵ^{pt} and the transformation strain ϵ^{t} is given by Eshelby's solution through the **S** tensor, namely $\epsilon^{pt} = \mathbf{S}\epsilon^{t}$. The strain $\tilde{\epsilon}$ can be expressed in terms of the transformation strain by using the requirement that the volume average strain, $<\epsilon>$, must be equal to the macroscopically applied strain, that is

$$<\epsilon> = c_f(\bar{\epsilon} + \tilde{\epsilon} + \epsilon^{pt}) + c_m(\bar{\epsilon} + \tilde{\epsilon}) = \bar{\epsilon} \ , \tag{2}$$

from which we deduce that $\tilde{\epsilon} = -c_f\epsilon^{pt} = -c_f\mathbf{S}\epsilon^{t}$. This result allows us to write the average strain in the inhomogeneity ϵ_f in terms of the applied and the transformation strains by the formula

$$\epsilon_f = \bar{\epsilon} + \tilde{\epsilon} + \epsilon^{pt} = \bar{\epsilon} + c_m\mathbf{S}\epsilon^{t} \ . \tag{3}$$

To be able to evaluate the strain concentration factor for the inhomogeneity $(\epsilon_f = \mathbf{A}_f\bar{\epsilon})$, we should eliminate the transformation strain ϵ^{t}. Writing the equivalence principle (1) in the form $\mathbf{L}_f\epsilon_f = \mathbf{L}_m\epsilon_f - \mathbf{L}_m\epsilon^{t}$ we easily obtain

$$\epsilon^{t} = \mathbf{M}_m\,(\mathbf{L}_m - \mathbf{L}_f)\,\epsilon_f \ , \tag{4}$$

and, substituting the above result back into equation (3), we finally have the strain concentration factor for the inhomogeneity given by

$$\mathbf{A}_f = [\mathbf{I} + c_m\mathbf{S}\,\mathbf{M}_m(\mathbf{L}_f - \mathbf{L}_m)]^{-1} = [\mathbf{I} + c_m\mathbf{M}_m(\mathbf{I} - \mathbf{T})(\mathbf{L}_f - \mathbf{L}_m)]^{-1} \ . \tag{5}$$

In a similar way the inhomogeneity stress concentration factors defined by $\sigma_f = \mathbf{B}_f\bar{\sigma}$ are obtained, with the result being

$$\mathbf{B}_f = [\mathbf{I} + c_m\mathbf{L}_m(\mathbf{I} - \mathbf{S})(\mathbf{M}_f - \mathbf{M}_m)]^{-1} = [\mathbf{I} + c_m\mathbf{T}\,\mathbf{L}_m(\mathbf{M}_f - \mathbf{M}_m)]^{-1} \ . \tag{6}$$

We stress that **S** denotes the Eshelby tensor for an ellipsoidal inclusion with the shape of the inhomogeneity and the matrix material properties. The computation of the Eshelby tensor for a general anisotropic material has been carried out numerically (see Appendix), while analytical expressions can be found in Mura (1987) for various special cases of geometry and inclusion shapes. We also report the dual formulae of the concentration factors in equations (5) and (6) using the **T** tensor (**T**, defined by $\sigma^{pt}=\mathbf{T}\sigma^{t}$, is the dual to **S**). These alternative expressions are obtained using Hill's (1965b) formulae which connect the **S** and **T** tensors, namely $\mathbf{S}\,\mathbf{M}_m = \mathbf{M}_m(\mathbf{I}-\mathbf{T})$ and $\mathbf{T}\,\mathbf{L}_m = \mathbf{L}_m(\mathbf{I}-\mathbf{S})$. Similar expressions to the ones given in equations (5) and (6) for the concentration factors have been derived by Weng (1984) and Benveniste (1987). Since they are all based on Eshelby's equivalence principle, they are all equivalent.

The **average** concentration factors for the matrix are found by using the well known relations

$$c_f\mathbf{A}_f + c_m\mathbf{A}_m = \mathbf{I}\ ,\ c_f\mathbf{B}_f + c_m\mathbf{B}_m = \mathbf{I}\ , \tag{7}$$

while the overall effective properties are evaluated from (Hill (1965b))

$$\mathbf{L} = c_f\mathbf{L}_f\mathbf{A}_f + c_m\mathbf{L}_m\mathbf{A}_m\ ,\ \mathbf{M} = c_f\mathbf{M}_f\mathbf{B}_f + c_m\mathbf{M}_m\mathbf{B}_m\ . \tag{8}$$

1.2 Concentration Factors in Composites with Elastoplastic Phases

Assume that one or both of the phases behave elastoplastically. The local fields in the phases are not only non–uniform due to the complicated geometry but they are also functionals of the prescribed loading history, that is,

$$\sigma_r(\mathbf{x},t) = \sigma_r[\mathbf{x};\bar{\sigma}(t)] \quad \text{and} \quad \epsilon_r(\mathbf{x},t) = \epsilon_r[\mathbf{x};\bar{\sigma}(t)]\ , \tag{9}$$

where the brackets denote general functionals of the loading history ($\bar{\sigma}(t)$ is replaced by $\bar{\epsilon}(t)$ for strain history). The subscript r indicates the phases, which for the purposes of this work will be limited to two, namely r=f, denoting the inhomogeneity phase with the convex geometry (ellipsoidal shape), and r=m, denoting its complementary shape which is occupied by the matrix material. If we increment the loading parameter by an infinitesimal amount, dt, inducing an infinitesimal load increment, $d\bar{\sigma} = \partial\bar{\sigma}(t)/\partial t\ dt$, the increments in the local stress and strain fields are given by

$$d\sigma_r(\mathbf{x},t) = \sigma_r[\mathbf{x};\bar{\sigma}(t)+d\bar{\sigma}(t)] - \sigma_r[\mathbf{x};\bar{\sigma}(t)] \equiv \hat{\mathbf{B}}_r[\mathbf{x},\bar{\sigma}(t)]\, d\bar{\sigma}(t) \ ,$$

$$d\epsilon_r(\mathbf{x},t) = \epsilon_r[\mathbf{x};\bar{\sigma}(t)+d\bar{\sigma}(t)] - \epsilon_r[\mathbf{x};\bar{\sigma}(t)] \equiv \hat{\mathbf{A}}_r[\mathbf{x},\bar{\sigma}(t)]\, d\bar{\epsilon}(t) \ , \tag{10}$$

with similar formulas whenever a strain history is prescribed. Notice that $d\bar{\epsilon} = \mathbf{M}[\mathbf{x};\bar{\sigma}(t)]\, d\bar{\sigma}$, where $\mathbf{M}$ is the fourth order instantaneous effective compliance tensor.

To proceed further, we make the usual assumption in plasticity that the history dependence is expressed at every point $\mathbf{x}$ by the stress state and additional internal state variables (i.e., back stresses) which we will collectively denote by $\boldsymbol{\alpha}$. In that case, equations (10) reduce to

$$d\sigma_r(\mathbf{x},t) = \hat{\mathbf{B}}_r[\mathbf{x};\sigma_f(\mathbf{x}),\boldsymbol{\alpha}_f(\mathbf{x}),\sigma_m(\mathbf{x}),\boldsymbol{\alpha}_m(\mathbf{x})]\, d\bar{\sigma}(t) \ ,$$

$$d\epsilon_r(\mathbf{x},t) = \hat{\mathbf{A}}_r[\mathbf{x};\sigma_f(\mathbf{x}),\boldsymbol{\alpha}_f(\mathbf{x}),\sigma_m(\mathbf{x}),\boldsymbol{\alpha}_m(\mathbf{x})]\, d\bar{\epsilon}(t) \ , \tag{11}$$

where $\sigma_r(\mathbf{x})$ indicates the stress state and $\boldsymbol{\alpha}_r(\mathbf{x})$ indicates the internal state variables carrying information about the loading history at each point $\mathbf{x}$ of the phase volume V_r. The implicit dependence on the spatial position $\mathbf{x}$ of the material points in V will be influenced by the specific geometry of the microstructure. $\hat{\mathbf{B}}_r$ and $\hat{\mathbf{A}}_r$ will be called the instantaneous local stress and strain concentration factors.

The volume averages of the local fields are defined in the usual way by

$$\langle d\sigma_r \rangle = \frac{1}{V_r}\int_{V_r} d\sigma_r \, dV = \frac{1}{V_r}\int_{V_r} \hat{\mathbf{B}}_r[\mathbf{x};\sigma_f(\mathbf{x}),\boldsymbol{\alpha}_f(\mathbf{x}),\sigma_m(\mathbf{x}),\boldsymbol{\alpha}_m(\mathbf{x})]\, d\bar{\sigma}\, dV$$
$$\equiv \mathbf{B}_r[\sigma_f(\mathbf{x}),\boldsymbol{\alpha}_f(\mathbf{x}),\sigma_m(\mathbf{x}),\boldsymbol{\alpha}_m(\mathbf{x})]\, d\bar{\sigma} \ ,$$

$$\langle d\epsilon_r \rangle = \frac{1}{V_r}\int_{V_r} d\epsilon_r \, dV = \frac{1}{V_r}\int_{V_r} \hat{\mathbf{A}}_r[\mathbf{x};\sigma_f(\mathbf{x}),\boldsymbol{\alpha}_f(\mathbf{x}),\sigma_m(\mathbf{x}),\boldsymbol{\alpha}_m(\mathbf{x})]\, d\bar{\epsilon}\, dV \tag{12}$$
$$\equiv \mathbf{A}_r[\sigma_f(\mathbf{x}),\boldsymbol{\alpha}_f(\mathbf{x}),\sigma_m(\mathbf{x}),\boldsymbol{\alpha}_m(\mathbf{x})]\, d\bar{\epsilon} \ ,$$

where $V_f + V_m = V$. The average concentration factors $\mathbf{A}_r$, $\mathbf{B}_r$ are general functions of the stress state and the internal variables in V at the instance of the application of the load increment. The evaluation of the concentration factors therefore requires the knowledge of the deformation history pointwise for the whole representative volume, which can be obtained by numerical solutions (i.e., Dvorak and Teply (1985)).

To be able to utilize the averaging techniques and to extend the methods used in elasticity, we further assume that **the average concentration factors are approximately equal to the concentration factors based on the phase averages of the state variables**, that is, they are functions of the average stresses $\sigma_r \equiv <\sigma_r(\mathbf{x})>$ and the internal state variables $\alpha_r \equiv <\alpha_r(\mathbf{x})>$ in the phases. We thus have the final form of equations (12):

$$
\begin{aligned}
<d\sigma_r> &= \mathbf{B}_r[\sigma_f(\mathbf{x}),\alpha_f(\mathbf{x}),\sigma_m(\mathbf{x}),\alpha_m(\mathbf{x})]\, d\bar{\sigma} \cong \mathbf{B}_r(\sigma_f,\sigma_m,\alpha_f,\alpha_m)\, d\bar{\sigma} \;, \\
<d\epsilon_r> &= \mathbf{A}_r[\sigma_f(\mathbf{x}),\alpha_f(\mathbf{x}),\sigma_m(\mathbf{x}),\alpha_m(\mathbf{x})]\, d\bar{\epsilon} \cong \mathbf{A}_r(\sigma_f,\sigma_m,\alpha_f,\alpha_m)\, d\bar{\epsilon} \;.
\end{aligned} \tag{13}
$$

It is evident at this point that the elastoplastic response of the r–th phase will depend on the average elastoplastic response of both constituents (the same way as the elasticity solution in a phase of a heterogeneous medium depends on the elastic constants of all phases). The above assumption is a drastic one because it replaces the average response to non–uniform fields of inherently inhomogeneous phases (due to plastic deformations) by the response to the average fields of instantaneously homogeneous phases. It requires that all instantaneous phase moduli $\mathbf{L}_r[\mathbf{x};\sigma_f(\mathbf{x}),\alpha_f(\mathbf{x}),\sigma_m(\mathbf{x}),\alpha_m(\mathbf{x})]$ be replaced by $\mathbf{L}_r(\sigma_f,\alpha_f,\sigma_m,\alpha_m)$. The validity of this assumption will be tested in Section 3 by comparing the analysis results with experiments and numerical simulations. The basic consequence of the above assumption is that the Eshelby solution for the ellipsoidal inhomogeneity now holds since only some average change in the instantaneous properties is taken into account and the phases are considered to be instantaneously homogeneous. Therefore, the analysis presented in Section 1.1 applies for the calculation of the average concentration factors, and, in particular, equations (5) and (6) can be used if the material parameters are taken to be the instantaneous ones, depending on the phase average stresses σ_r and the average internal state variables α_r.

2. Constitutive Description for Plastically Deforming Phases

The time independent elastoplastic response of the r–th phase in a composite can be characterized by the following set of equations

$$d\epsilon_r = d\epsilon_r^e + d\epsilon_r^p = \mathbf{M}_r^e\, d\sigma_r + d\epsilon_r^p \ , \tag{14}$$

$$d\epsilon_r^p = d\lambda_r\, \mu_r(\sigma_r,\alpha_r),\ d\alpha_r = d\lambda_r\, \nu_r(\sigma_r,\alpha_r),\ d\lambda_r \geq 0 \ , \tag{15}$$

$$\phi_r(\sigma_r,\alpha_r) \leq 0 \ , \ \phi_r\, d\lambda_r = 0 \ , \ d\phi_r\, d\lambda_r = 0 \ , \tag{16}$$

where $d\epsilon_r$ is the phase average of the total strain increment and $d\epsilon_r^e$, $d\epsilon_r^p$ represent its decomposition into an elastic increment and a plastic increment; σ_r is the average of Cauchy stress and α_r a set of internal variables. Equation (14) represents the commonly assumed additive decomposition of the total strain increment into an elastic and a plastic part, where the elastic part is derived from the elastic compliance by the generalized Hooke's law. The evolution equations for the plastic strains in terms of a general non–associative flow rule and internal state variables are given by equations (15). ϕ_r is the yield function of the material of the r–th phase and equations (16) require that when $\phi_r<0$ then necessarily $d\lambda_r = 0$, i.e., elastic response, and when $d\lambda_r > 0$ then $\phi_r= 0$, $d\phi_r=0$, i.e., plastic loading. The last of these equations (16) assures that the loading point remains in contact with the yield surface during infinitesimal loading.

In the case of homogeneous materials, the loading history is given either in terms of strains or stresses as a function of time. The constitutive equations can then be integrated, assuming that ϵ_r or σ_r are known functions of time. In the case of composite materials, the loading history is given in the overall space and hence equations (13), relating overall to local quantities through the concentration factors, must be utilized. Since the concentration factors depend on the instantaneous material properties of the constituents, ϵ_r or σ_r are not only determined by the applied loading, as is the case for homogeneous materials, but they also depend on the elastoplastic solution (i.e., the current moduli of the elastoplastic phases). As a result, equations (14–16) have to be integrated simultaneously for all elastoplastic phases since they are coupled through the concentration factors in equations (13).

To simplify the analysis, and also realistically model most of the fibrous composites in engineering applications, we will assume that the fiber phase remains always linearly elastic. This will result in the average matrix concentration factors being functions of $\boldsymbol{\sigma}_{\mathrm{m}}$ and $\boldsymbol{\alpha}_{\mathrm{m}}$, that is

$$d\boldsymbol{\sigma}_{\mathrm{m}} = \mathbf{B}_{\mathrm{m}}(\boldsymbol{\sigma}_{\mathrm{m}},\boldsymbol{\alpha}_{\mathrm{m}})\, d\bar{\boldsymbol{\sigma}} \ , \ d\boldsymbol{\epsilon}_{\mathrm{m}} = \mathbf{A}_{\mathrm{m}}(\boldsymbol{\sigma}_{\mathrm{m}},\boldsymbol{\alpha}_{\mathrm{m}})\, d\bar{\boldsymbol{\epsilon}} \ , \tag{17}$$

thus decoupling the elastoplastic problem in the matrix from that in the fibers. The concentration factors for the elastic fibers are evaluated by equations (7).

The integration of equations (14–16) can be carried out by different integration schemes, as discussed in the work by Ortiz and Popov (1985). The same methodology can also be applied here, with the necessary modification due to the additional equations (17) that define the loading path for the matrix through the instantaneous concentration factors. Referring to the work by Ortiz and Popov, it can be easily proven that both the generalized trapezoidal and the generalized midpoint integration schemes applied in our case provide first order accuracy for an arbitrary selection of the parameter α and second order accuracy for $\alpha = 1/2$ (α denotes the weight of the contribution of the end point of an integration interval, i.e., α=1 corresponds to the backward difference). The numerical stability analysis for our case is more complicated and no general conclusions can be drawn because the stability will depend on the micromechanical model used for the evaluation of the concentration factors. For the backward difference method (α=1) absolute stability is obtained.

The backward difference method has been used as the integration scheme for equations (14–17) in this work, and, by doing so, it is equivalent to having neglected the path dependence of the matrix material within every step increment. In order to maintain good accuracy, small time steps in the loading history have been used. For the specific examples considered, the following assumptions have been used for the constitutive behavior of the matrix material:
i) Associative flow rule,
ii) Mises yield surface, and
iii) Kinematic hardening (the internal state variable is the center of the yield surface) with Phillips hardening rule (Phillips (1972)).
The Mises criterion reads:

$$\phi=(\sigma_m-\alpha_m)^T C(\sigma_m-\alpha_m)-Y_m^2 \leq 0 \tag{18}$$

where $\mathbf{C}$ is a 6×6 symmetric matrix, with non zero entries $c_{11}=c_{22}=c_{33}=1$, $c_{12}=c_{13}=c_{23}=-1/2$ and $c_{44}=c_{55}=c_{66}=3$, and Y_m is the matrix yield stress in uniaxial tension. The flow rule is given by the equation

$$d\epsilon_m^p = d\lambda\, \mathbf{n}_m = d\lambda\, \partial\phi_m/\partial\sigma_m = d\lambda\, \mathbf{C}(\sigma_m-\alpha_m) \ . \tag{19}$$

The Phillips hardening with Ziegler assumption leads to the evolution of the internal variable α_m (Dvorak (1990))

$$d\alpha_m = d\lambda\, [(c\mathbf{n}_m^T \mathbf{n}_m^*)/d\sigma_m^T \mathbf{n}_m]\, d\sigma_m = d\sigma_m \ , \tag{20}$$

where c=2H/3, with H being the instantaneous tangent modulus, and $n_k^*=n_k$ for k=1,2,3, $n_k^*=n_k/2$ for k=4,5,6. The last equality in (20) results from the consistency condition. For this particular case of Phillips hardening the backward difference scheme leads to a set of expressions:

$$\Delta\epsilon_m^e = \mathbf{M}_m^e\, \Delta\sigma_m \ , \quad \Delta\epsilon_m^p = \frac{\mathbf{n}_m\mathbf{n}_m^T}{c\mathbf{n}^T\mathbf{n}^*}\, \Delta\sigma_m \ , \quad \Delta\alpha_m = \Delta\sigma_m \ , \tag{21}$$

$$\Delta\sigma_m = \mathbf{B}_m\, \Delta\bar{\sigma} \ , \quad \mathbf{M}_m = \frac{\mathbf{n}_m\mathbf{n}_m^T}{(c\mathbf{n}_m^T\mathbf{n}_m^*)} + \mathbf{M}_m^e \ , \tag{22}$$

where $\mathbf{n}_m$, c, $\mathbf{B}_m$ are all evaluated at the point $(\sigma_m+\Delta\sigma_m, \alpha_m+\Delta\sigma_m)$ and $\mathbf{B}_m=(\mathbf{I}-c_f\mathbf{B}_f)/c_m$ ($\mathbf{B}_f$ is given by (6)).

The matrix stress increment is found by solving the implicit system of equations $\Delta\sigma_m = \mathbf{B}_m(\sigma_m+\Delta\sigma_m, \alpha_m+\Delta\sigma_m)\, \Delta\bar{\sigma}$. The iterative procedure consists of the calculation of the elastic predictor given by

$$\Delta\sigma_m^j = \mathbf{B}_m^e\, \Delta\bar{\sigma} \ , \qquad j=0 \ , \tag{23}$$

which provides a first estimate of the stress increment in each phase. The concentration factors are evaluated using the elastic material properties of both phases. The above estimate is then corrected by

$$\Delta\sigma_m^{(j+1)} = \mathbf{B}_m^j \, \Delta\bar{\sigma} \;, \qquad j=1,2,\ldots \;, \tag{24}$$

where the stress concentration factors $\mathbf{B}_m^j$ are evaluated using the material properties at stress $(\sigma_m + \Delta\sigma_m^j)$. The matrix material moduli are updated by performing the elastoplastic analysis described above for matrix stress increment $\Delta\sigma_m^j$. At this point of the analysis only the new stiffness or compliance matrices are necessary (equation 22). Using the updated matrix moduli $\mathbf{B}_m^j$ is calculated. The convergence criterion is given by

$$\frac{\| \Delta\sigma_m^{(j+1)} - \Delta\sigma_m^j \|}{\| \Delta\sigma_m^j \|} = e_m \leq \gamma_m \tag{25}$$

where γ_m is a given tolerance. If the tolerance criterion is satisfied, then using the final values of the local stress increments $\Delta\sigma_m = \Delta\sigma_m^{(j+1)}$ the local quantities are evaluated from equations (21). The total strain increment is given by

$$\Delta\bar{\epsilon} = c_f \, \mathbf{M}_f \, \mathbf{B}_f \, \Delta\bar{\sigma} + c_m(\Delta\epsilon_m^p + \Delta\epsilon_m^e) \;, \tag{26}$$

where $\mathbf{B}_f$ is calculated from (6) based on the evaluation of matrix moduli from (22). The plastic and elastic strain increments in the matrix are computed from equations (21). This concludes the calculations in the current stress increment. If a new stress increment $\Delta\bar{\sigma}$ is applied, the above procedure is repeated until the loading path is completed.

In the case of Phillips hardening there is only one system of equations to be solved, namely equations (22), for the matrix stress increments, since the evolution of hardening is exactly equal with the evolution of the matrix stress $(\Delta\alpha_m = \Delta\sigma_m)$. For a general hardening model another set of differential equations needs to be solved for the evolution of the internal state variables and, if the backward difference method is employed, it will require the solution of an additional nonlinear system of algebraic equations. In general, therefore, for the backward difference integration scheme, the following algebraic equations should be solved simultaneously

$$\Delta\sigma_m = \mathbf{B}_m(\sigma_m + \Delta\sigma_m, \boldsymbol{\alpha}_m + \Delta\boldsymbol{\alpha}_m)\,\Delta\bar{\sigma}\ ,$$
$$\Delta\boldsymbol{\alpha}_m = \Delta\lambda\ \nu_m(\sigma_m + \Delta\sigma_m, \boldsymbol{\alpha}_m + \Delta\boldsymbol{\alpha}_m)\ , \qquad (27)$$

together with the consistency condition $\phi(\sigma_m + \Delta\sigma_m, \boldsymbol{\alpha}_m + \Delta\boldsymbol{\alpha}_m) = 0$ for the evaluation of $\Delta\sigma_m, \Delta\boldsymbol{\alpha}_m$ and $\Delta\lambda$. The iterative solution procedure described previously can also be used for the above system with initial values based on the elastic predictor, namely $\Delta\sigma_m = \Delta\sigma_m^e = \mathbf{B}_m^e \Delta\bar{\sigma}$, $\Delta\boldsymbol{\alpha}_m = 0$, $\Delta\lambda = 0$ (the loading path is prescribed in the overall stress space).

We briefly mention that if the loading path is prescribed in the strain space, the differential equations to be integrated during plastic loading (i.e. whenever $\phi(\sigma_m + d\sigma_m^e) > 0$) are the following:

$$d\sigma_m = \mathbf{L}_m^e\,(d\epsilon_m - d\lambda\ \mu_m(\sigma_m, \boldsymbol{\alpha}_m))\ ,\ \ d\boldsymbol{\alpha}_m = d\lambda\ \nu_m(\sigma_m, \boldsymbol{\alpha}_m)\ , \qquad (27)$$

together with the consistency condition and the definition of the matrix strain increment $d\epsilon_m = \mathbf{A}_m(\sigma_m, \boldsymbol{\alpha}_m)\, d\bar{\epsilon}$. The elastic predictor is given by $d\sigma_m^e = \mathbf{L}_m^e \mathbf{A}_m^e d\bar{\epsilon}$.

3. Comparison with Numerical Simulations – Particulate Composites

The prediction of the Mori–Tanaka method discussed above has been compared with numerical results performed on a particulate SiC_w–$A\ell$ composite. A composite system with aluminum matrix and silicon carbide short fibers, periodically placed, has been considered. The results of the averaging methods have been compared with those obtained using a periodic hexagonal array (PHA) model developed by Dvorak and Teply (1985) and modified for particulate composites by Dvorak (1990) and Lin and Dvorak (1990). The PHA is a micromechanical model based on a finite element discretization of the representative volume element (RVE). The geometry of the whisker reinforcement is taken to be similar to a hexagonal prism. A volumetrically equivalent ellipsoid to the given prism was obtained assuming two of its principal axes equal and using the conditions that both the prism and the ellipsoid have the same volume and they correspond to the same volume fractions. The following properties are chosen for the numerical calculations: Aluminum: E_m=72 GPa, ν_m=.33, Y_m=70 MPa (initial yield stress in tension), H_m=14.5 GPa (linear hardening). Silicon Carbide: E_f=430 GPa, ν_f=.25. The whisker reinforcement was considered elastic until failure. For the matrix material Phillips hardening rule was adopted to model the kinematic hardening, while the yield surface was defined by the Mises yield criterion.

Figures 1 and 2 show the numerical results of the PHA model together with the results from the averaging techniques for multiaxial proportional loading ($\bar{\sigma}_{33}=2\bar{\sigma}_{13}$) for the SiC_w/Al short fiber composite. In addition to the Mori–Tanaka averaging scheme for the evaluation of the concentration factors, the self–consistent (Hill (1965), Budiansky (1965)) scheme is also plotted for comparison purposes. For monotonic loading the Mori–Tanaka predictions are closer to the numerical results than the self–consistent predictions, which underestimate the total strains. In addition to its better accuracy, the convenience of using the Mori–Tanaka method, with respect to the self–consistent, becomes more evident when the computer time is considered. In fact, because of its implicit nature, the self–consistent model requires, on the average, five times more CPU time than the Mori–Tanaka method.

It is of interest to compare the yield surfaces predicted by the averaging techniques with the PHA results. Figure 3 shows the initial yield surfaces as predicted by the PHA numerical scheme and the Mori–Tanaka and self–consistent averaging schemes. The initial yield surfaces are not very different and the Mori Tanaka yield surface is a better approximation than the self–consistent yield surface, which predicts higher initial yield stresses. Figure 4 shows the yield surfaces at the end of the non–proportional path (95,0)→(95,40) in the overall ($\bar{\sigma}_{33},\bar{\sigma}_{13}$) stress space. The PHA surface is obtained as the envelope of the yield surfaces of each numerical integration point of the finite element mesh (Dvorak (1990), Lin and Dvorak (1990)) and, even though locally the matrix behaves purely kinematically, the overall response of the composite presents a macroscopic isotropic softening due to different motion of the local yield surfaces. On the contrary, the averaging techniques are not able to produce a change in the yield surface size if only kinematic hardening is considered for the matrix. This remark is general and is

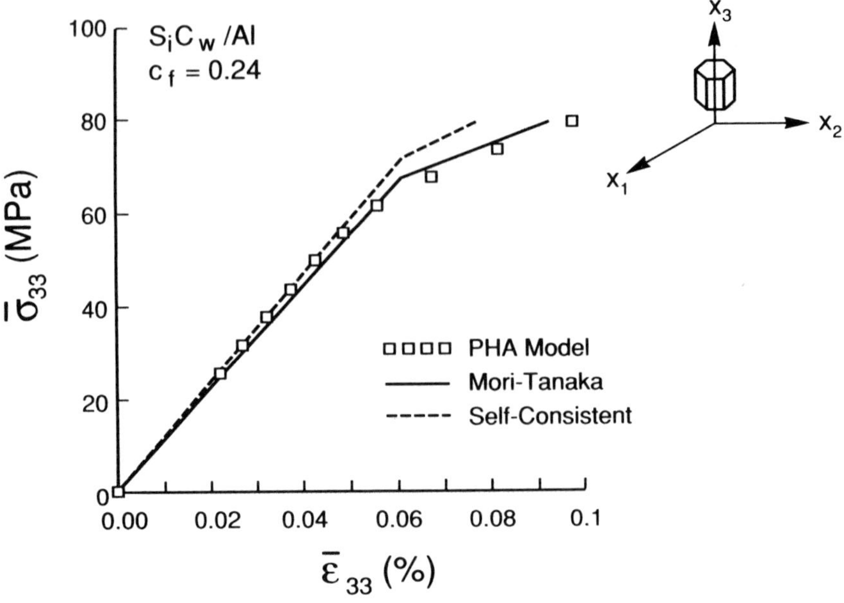

Fig. 1. Axial Stress vs. Axial Strain in a Silicon Carbide Whisker Reinforced Composite under Bi-axial Loading $(\bar{\sigma}_{33} = 2\bar{\sigma}_{13})$

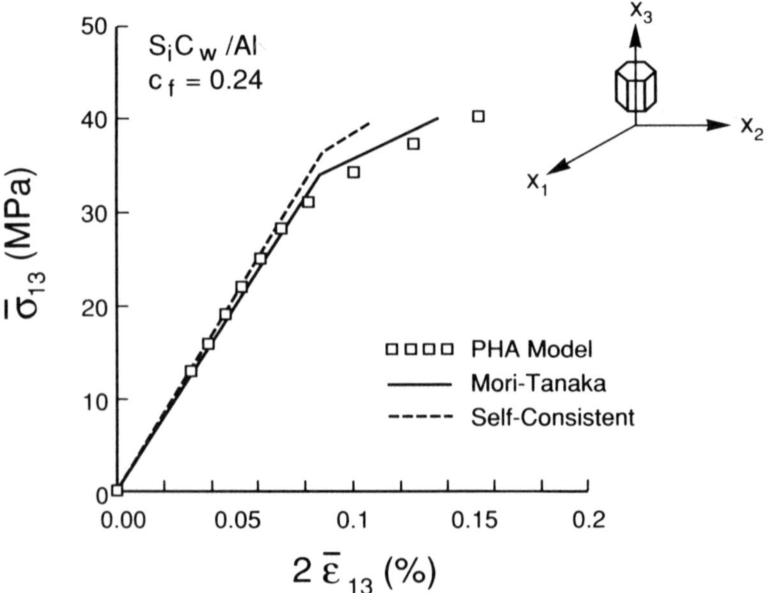

Fig. 2. Longitudinal Shear Stress vs. Longitudinal Shear Strain in a Silicon Carbide Whisker Reinforced Composite under Bi-axial Loading $(\bar{\sigma}_{33} = 2\bar{\sigma}_{13})$

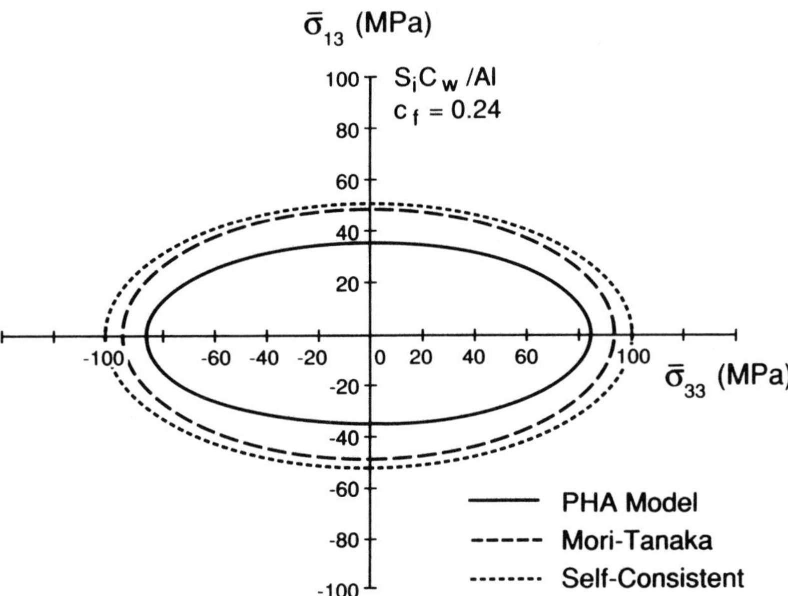

Fig. 3. Initial Yield Surface for a Silicon Carbide Whisker Reinforced Composite as Predicted by the PHA Model and the Averaging Schemes

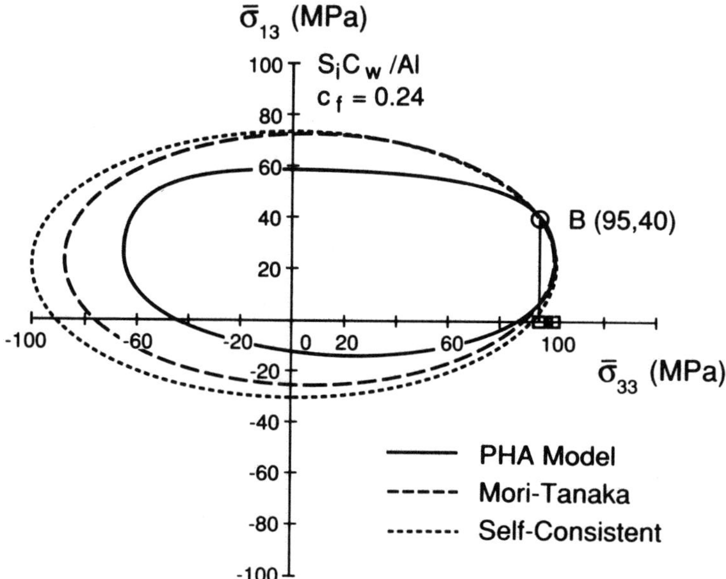

Fig. 4. Subsequent Yield Surface for a Silicon Carbide Whisker Reinforced Composite as Predicted by the PHA Model and the Averaging Schemes

valid for every averaging model based on a single matrix yield surface and assuming that the matrix is incrementally homogeneous. Even though subsequent yield surfaces show larger differences between PHA evaluations and averaging models in their sizes, it can be observed from Figure 4 that, so long as unloading does not occur, the plastic strains predicted by the three models will be close because the yield surfaces coincide at the contact point with almost identical normal vectors.

4. Discussion

The predictions of the Mori–Tanaka method and, to a lesser extent, of the self–consistent method for the constitutive behavior of a whisker reinforced composite are close to the prediction of numerical simulations. In general, good estimates are expected for particle reinforced composites with elastoplastic matrix where the matrix is free to yield in all activated slip planes. For fibrous composites reinforced with stiff fibers, such as boron fibers, and for loading paths that induce matrix dominated mode plasticity, the averaging schemes might not yield accurate results. This is the subject of an ongoing investigation to be published elsewhere (Lagoudas et al. (1990)).

A remark concerning the validity of the use of averaging techniques to model the constitutive behavior of elastoplastic composites based on the Eshelby solution for the inhomogeneity problem is appropriate. The main advantage of Eshelby's solution is the prediction of uniform stress and strain fields for the inhomogeneity when embedded into an infinite matrix under the application of uniform stress or strain fields. But the fields outside the inhomogeneity are not uniform and, for most engineering applications, the matrix yields at applied loads well below the design limit. The nonuniformity of the matrix stress field induces a nonuniform plastic strain field, which destroys the homogeneity of the instantaneous moduli of the matrix and renders the Eshelby solution inside the inhomogeneity invalid. The basic assumption, therefore, of the averaging methods (equation 13), to replace the average instantaneous concentration factors by those induced by the average phase fields, might not be accurate for complex non–proportional paths.

Acknowledgement

The authors acknowledge the many useful discussions and help they received from Professors G.J. Dvorak and Y. Bahei–El–Din during the course of this work.

References

Bahei–El–Din, Y.A. and Dvorak, G.J., 1989, "A Review of Plasticity Theory of Fibrous Composite Materials," **Metal Matrix Composites: Testing, Analysis and Failure Modes, ASTM STP 1032**, pp. 103–129.

Benveniste, Y., 1987, "A New Approach to the Application of Mori–Tanaka's Theory in Composite Materials," **Mech. of Materials**, Vol. 6, pp. 147–157.

Budiansky, B., 1965, "On the Elastic Moduli of some Heterogeneous Materials," **J. Mech. Phys. Solids**, Vol. 13, pp. 223–227.

Dvorak, G.J., and Bahei–El–Din, Y.A., 1979, "Elastic–Plastic Behavior of Fibrous Composites," **J. Mech. Phys. Solids**, Vol. 27, pp. 51–72.

Dvorak, G.J., and Teply, 1985, "Periodic Hexagonal Array Models for Plasticity Analysis of Composite Materials," **Plasticity Today: Modeling, Methods and Applications**, A.Sawczuk, ed., Elsevier, London.

Dvorak, G.J., and Bahei–El–Din, Y.A., 1987, "A Bimodal Plasticity Theory of Fibrous Composite Materials," **Acta Mechanica**, Vol. 69, pp. 219–241.

Dvorak, G.J., 1990, "Plasticity Theories for Fibrous Composite Materials," **Metal Matrix Composites, Vol. 2: Mechanisms and Properties**, R.K. Everett and R.J. Arsenault, eds, Academic Press, Boston.

Eshelby, J.D., 1957, "The Determination of the Elastic Field of an Ellipsoidal Inclusion, and Related Problems," **Proc. Roy. Soc. Lond. A**, Vol. 241, pp. 376–396.

Gavazzi, A.C., and Lagoudas, D.C., 1990, "On The Numerical Evaluation of Eshelby's Tensor and its Application to Elastoplastic Fibrous Composites," **Computational Mechanics**, in print.

Ghahremani, F., 1977, "Numerical Evaluation of the Stresses and Strains in Ellipsoidal Inclusions in an Anisotropic Elastic Material," **Mech. Res. Comm.**, Vol. 4, pp. 89–91.

Hill, R., 1963, "Elastic Properties of Reinforced Solids: Some Theoretical Principles," **J. Mech. Phys. Solids**, Vol. 11, pp. 357–372.

Hill, R., 1965a, "Continuum Micro–Mechanics of Elastoplastic Polycrystals," **J. Mech. Phys. Solids**, Vol. 13, pp. 89–101.

Hill, R., 1965b, "A Self Consistent Mechanics of Composite Materials," **J. Mech. Phys. Solids**, Vol. 13, pp. 213–222.

Hutchinson, J.W., 1970, "Elastic–plastic Behaviour of Polycrystalline Metals and Composites," **Proc. Roy. Soc. Lond. A**, Vol. 319, pp. 247–272.

Lagoudas, D.C., Gavazzi, A.C., and Nigam, H., 1990, "Elastoplastic Behavior of Metal Matrix Composites Based on Incremental Plasticity and the Mori–Tanaka Averaging Scheme," **Computational Mechanics**, submitted.

Lin, J.I., Dvorak, G.J., 1990, "Local and Overall Elastic–Plastic Response of Particle–Reinforced Metal Matrix Composites," to appear.

Mura, T., 1987, "**Micromechanics of Defects in Solids,**" Martinus Nijhoff, Dordrecht.

Mori, T. and Tanaka, K., 1973, "Average Stress in Matrix and Average Energy of Materials with Misfitting Inclusions," **Act. Metall.**, Vol. 21, pp. 571–574.

Ortiz, M., Popov, E.P., 1985, "Accuracy and Stability of Integration Algorithms for Elastoplastic Constitutive Relations," **Int. J. Num. Met.Eng.**, Vol. 21, pp. 1561–1576.

Phillips, A., Liu, C.S., and Justusson, J.W. (1972), An Experimental Investigation of Yield Surfaces at Elevated Temperatures, **Acta Mechanica 14,** pp. 119–146.

Tandon, G.P. and Weng, G.J., 1988, "A Theory of Particle–Reinforced Plasticity," **J. Appl. Mech**, Vol. 55, pp. 126–135.

Taya, M., and Mura, T., 1981, "The Stiffness and Strength of an Aligned Short–Fiber Reinforced Composite Containing Fiber–End Cracks Under Uniaxial Applied Stress," **J. Appl. Mech**, Vol. 43, pp. 361–367.

Weng, G.J., 1984, "Some Elastic Properties of Reinforced Solids, with Special Reference to Isotropic ones Containing Spherical Inclusions," **Int. J. Engng Sci.**, Vol. 22, pp. 845–856.

Appendix

Numerical Evaluation of Eshelby's S Tensor

Consider an infinitely extended homogeneous linearly elastic medium with an ellipsoidal inclusion of volume V and of the same material imbedded in it. A rectangular Cartesian coordinate system x_i, i=1,2,3 is introduced such that the origin coincides with the center of the ellipsoid and the three principal axes of the ellipsoid, a_1, a_2, a_3 are aligned with the Cartesian axes x_i. If the ellipsoidal region undergoes a uniform, stress–free transformation strain ϵ^t_{ij}, the resulting strain field in V is uniform and is given in terms of the well known Eshelby tensor **S** (Eshelby (1957))

$$\epsilon^c_{ij} = S_{ijkl}\epsilon^t_{kl} \quad , \tag{A1}$$

where S_{ijkl} are the Cartesian components of **S** and they are in general a function of the stiffness tensor $\mathbf{L}_0$ of the infinite medium and the shape of the inclusion.

For a generic anisotropic material the Eshelby tensor is given by the following surface integral, parameterized on the surface of the unit sphere (Mura (1987)):

$$S_{ijkl} = \frac{1}{8\pi} L_{0_{mnkl}} \int_{-1}^{+1} d\zeta_3 \int_0^{2\pi} \{G_{imjn}(\overline{\xi})+G_{jmin}(\overline{\xi})\}\, d\omega\,, \tag{A2}$$

where

$$\begin{aligned}
&G_{ijkl}(\overline{\xi}) = \overline{\xi}_k \overline{\xi}_l N_{ij}(\overline{\xi})/D(\overline{\xi}) \\
&\overline{\xi}_i = \zeta_i/a_i \\
&\zeta_1 = (1-\zeta_3^2)^{1/2}\cos\omega \\
&\zeta_2 = (1-\zeta_3^2)^{1/2}\sin\omega \\
&\zeta_3 = \zeta_3 \\
&D(\overline{\xi}) = \epsilon_{mnl} K_{m1} K_{n2} K_{l3} \\
&N_{ij}(\overline{\xi}) = \frac{1}{2}\epsilon_{ikl}\,\epsilon_{jmn}\, K_{km}\, K_{ln} \\
&K_{ik} = L_{0_{ijkl}} \overline{\xi}_j \overline{\xi}_l
\end{aligned} \tag{A3}$$

with ϵ_{ijk} being the permutation tensor and $L_{0_{ijkl}}$ the components of the stiffness tensor $\mathbf{L}_0$.

In some special cases, such as for isotropic and transversely isotropic materials and for different values of the parameters a_i, the quadrature in (A2) has been performed in closed form and a review of these results can be found in Mura (1987). For the case of a fully anisotropic material, no explicit formula for S_{ijkl} has yet been obtained, and the only result available is a computer program by Ghahremani (1977), which evaluates numerically the components of **S**. Ghahremani used for the numerical evaluation of **S** a formula different from (A2) with a different parameterization of the unit sphere and, according to the author, and to our direct experience, the program's accuracy depends on the values of the aspect ratios a_1/a_2 and a_1/a_3 for given material anisotropy.

In this work we develop a numerical scheme for the evaluation of S_{ijkl} based on (A2). The double integration in (A2) is performed using the following Gaussian quadrature formula

$$S_{ijkl} = \frac{1}{8\pi} \sum_{p=1}^{M} \sum_{q=1}^{N} L_{0_{mnkl}} \{G_{imjn}(\omega_q, \zeta_{3p})\, G_{jmin}(\omega_q, \zeta_{3p})\} W_{pq} \,, \tag{A4}$$

where M and N refer to the points used for the integration over ζ_3 and ω, respectively, and W_{pq} are the Gaussian weights. M and N are not fixed numbers in the program but they are selected according to the ratios a_1/a_2 and a_1/a_3 so that the maximum error in the evaluation of the components of S occurs after the sixth significant digit.

In order to demonstrate the capabilities of the computer program, some selected results and comparisons with available exact evaluations of S are given in the following. Notice that to simulate cylindrical inclusions and penny shape cracks a_1/a_3 was set equal to 10^{-40} and 10^{40}, respectively.

i) Isotropic materials.

For different values of the Poisson ratio ν of the matrix material, three types of inclusions were chosen so that they represent most of the common shapes of inclusions in composites.

a) Spherical inclusion ($a_1=a_2=a_3$). For M=4 and N=16 the exact evaluation of S (using double accuracy) is obtained.

b) Penny–shape crack ($a_3 \to 0$). In this case it is sufficient to consider M=2,N=2 in order to obtain the exact evaluation.

c) Cylindrical inclusion ($a_3 \to \infty$). Table 1 shows the number of integration points necessary to achieve accuracy up to six significant digits and makes a comparison with the results of Ghahremani's program, which always uses 32 x 32 = 1024 Gaussian points.

a_1/a_2	M	N	Accuracy (# of significant digits) Present Program	Ghahremani (1977)
1	2	16	exact	exact
2	2	46	6	6
5	2	100	6	3
10	2	160	6	2
20	2	280	6	2
50	2	400	6	2

Table 1. Accuracy in the evaluation of **S** for cylindrical fibers with elliptic cross–section for an isotropic material.

ii) Transversely isotropic materials.

In the case of a cylindrical inclusion with circular cross section the evaluation of **S** is exact if M=2 and N=16 for different combinations of the four independent ratios of the five material parameters.

iii) Anisotropic Materials.

For the extreme case of a cylindrical inclusion the values of M and N, necessary to give the required accuracy of six significant digits, depend on the value of the ratio a_1/a_2, as well as the values of the material parameters. Table 2 shows the number of required Gaussian points. Since for these cases no exact results are available for comparison, the values of M and N in Table 2 must be regarded as the minimum values of M and N, such that the entries of the Eshelby's tensor up to the first six digits remain unchanged if M and N increase. More details for the case of anisotropic matrix materials can be found in Gavazzi and Lagoudas (1990).

a_1/a_2	M	N
1	2	100
2	2	100
5	2	160
10	2	300
20	2	600
50	2	1000

Table 2. Gaussian points needed for the exact evaluation of Eshelby's **S** tensor up to six significant digits for a cylindrical inclusion in an anisotropic matrix material.

Inelastic Behavior V

Global and Internal Time Dependent Behaviour of Polymer Matrix Composites

A.H. Cardon*

Introduction

Polymeric materials exhibit specific time dependent effects under mechanical and environmental loading histories. The fibers used for the reinforcement of the polymeric matrix are generally elastic or elasto-plastic, but may even be viscoelastic. The polymer matrix composite is a multiphase continuum in which elastic and visco-elastic phases interact over interphase regions. Even with a high volume fraction of elastic components (between 0.55 and 0.65), this multiphase continuum has a basic viscoelastic behaviour, oriented or not.

The most specific viscoelastic behaviour is observed on a unidirectional reinforced polymer composite lamina under in plane loading in the transverse direction. The viscoelastic behaviour changes in relation to the angle between the loading direction and the fiber orientation becoming elastic if those directions coincide.

In a laminate, built up with lamina reinforced in alternated directions, the characteristic global response times for the viscoelastic behaviour are shifted to longer periods so that an elastic behaviour over a limited time range is observed.

---

* Professor, Composite Systems and Adhesion Research Group of the Free University of Brussels - COSARGUB
VUB (TW - KB), Pleinlaan, 2, B-1050 Brussels - Belgium.

For the study of this global behaviour a laminate analysis can be performed over the individual lamina with their anisotropic viscoelastic-elastic in plane characteristics.

Related to this global behaviour are the internal stress transferts between the elastic and viscoelastic components. Those stress transferts occur not only in a lamina between matrix and fibers, but also between lamina, function of the respective fiber orientations. For the analysis of these stress transferts the hypothesis of perfect adhesion between fibers and matrix, and between lamina is a necessary assumption. Corrections on this behaviour for rupture of adhesion between fibers and matrix, and for delaminations may later be introduced.

The global viscoelastic behaviour combined with the internal stress transferts influences the long term behaviour of the polymer matrix composite and must be taken into account for the durability analysis of the composite component under mechanical and environmental loading histories. For such a durability analysis the relation between short period tests under different extreme loading conditions and the long term behaviour under normal loading conditions must be established and verified experimentally.

A general thermodynamic based viscoelasticity theory, as proposed by R. Schapery and co-workers, is an interesting framework for such extrapolations to long term behaviour from short time test results.

1. Creep of unidirectional reinforced laminates

On specimen of 8 layers, cut out at different angles from unidirectional reinforced plates, with thickness of 1 mm (± 0.04 mm), transverse dimension of 25 mm (± 0.2 mm) and active length of 165 mm (± 1 mm) creep tests were performed at different stress levels and at different temperatures. Some examples are given below.

Figure 1.a gives the obtained creep results on APC-2 over 8 hours for stress levels of 0.6, 0.65, 0.7, 0.75 and 0.8 of the rupture stress for $\vartheta = 15°$, ($\sigma_r = 333$ MPa) at room temperature.

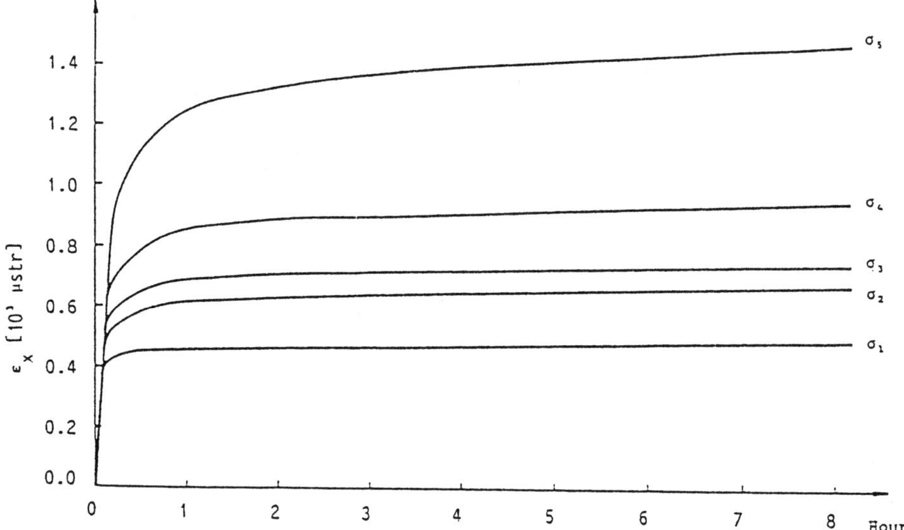

Figure 1.b gives the same creep results at the same relative stress levels for $\theta=45°$, ($\sigma_r = 125$ MPa) at room temperature.

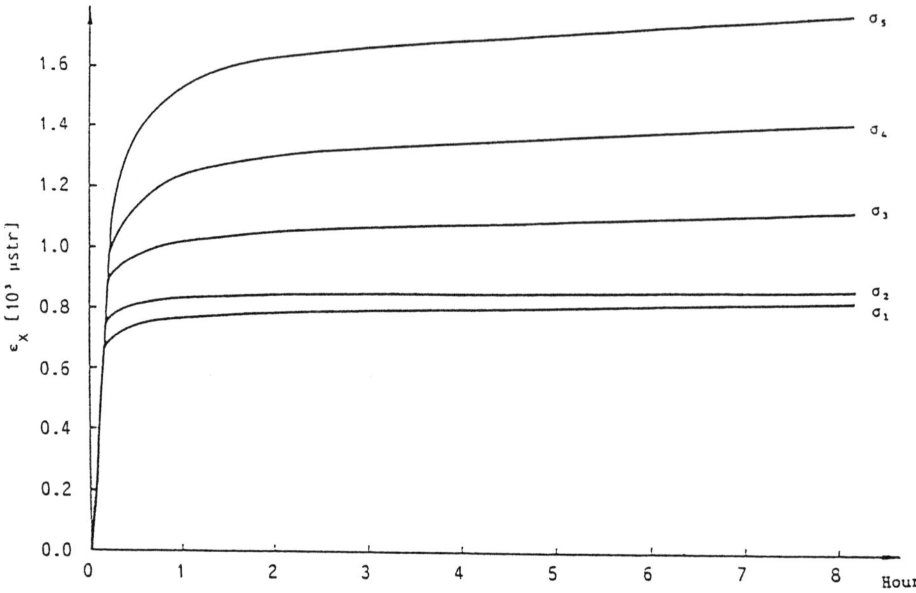

Figure 1.c gives the creep and creeep recovery results obtained on a graphite-epoxy specimen of the same dimensions under 0.6 and 0.7 of the rupture stress for $\theta=15°$ at 95°C for 120 hours creep time and 120 hours recovery time.

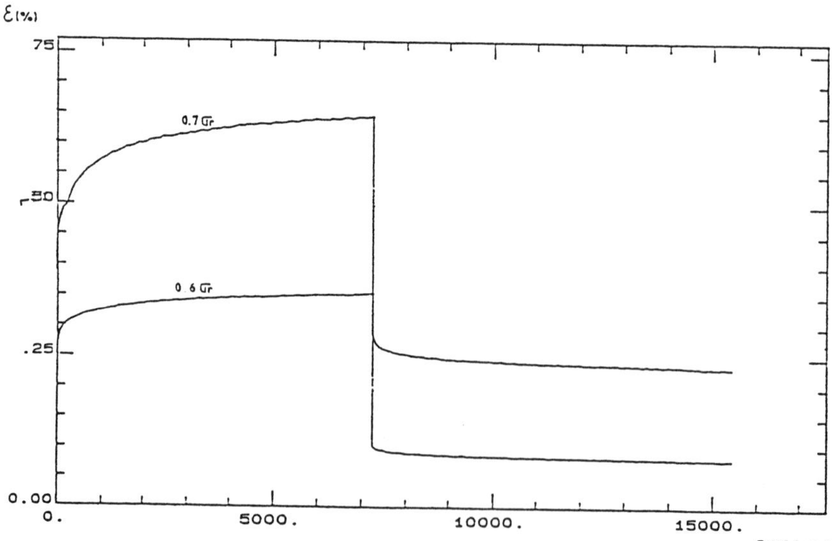

2. Creep of general laminates

For a laminate with alternated stacking sequences the global creep and creep recovery effects are reduced in a proportion that can be estimated by the numer of 0° lamina, constraining the viscoelastic behaviour of the other layers. The creep behaviour can be predicted by laminate programs as was described by M. Tuttle and H. Brinson, [1]. At higher temperature levels the creep is accelerated. If part of the mechanical loading is of oscillating type there is interaction between the two type of loadings with an important increase of the creep behaviour as was shown by J. Bax, [2], and later by J. Lee and F.X. de Charentenay, [3].

Within the linear viscoelastic range creep and relaxation characteristics can be obtained by dynamical testing for the measurement of the real and imaginary part of the complex modulus. This can be done over limited frequency ranges at different temperature levels so that by application of the time temperature superposition principle the results are obtained over a large frequency range for a reference temperature as was suggested by H. Brinson, [4].

3. Durability analysis

The stress level can be used as accelerator for the evolution of the behaviour of the polymer matrix composite laminate by a combined Time-Temperature-Stress-Superposition principle proposed by H. Brinson and co-workers, [4], [5].

If the short term tests are carried out at high stress levels the viscoelastic behaviour of the composite becomes nonlinear and in order to analyse the results a set of nonlinear viscoelastic constitutive equations are needed.

For this analysis the nonlinear viscoelastic theory, based on irreversible thermodynamics, proposed by R. Schapery, [6],[7],[8] is very well adapted.

The possibility to use this Schapery formulation for long term predictions was studied by H. Brinson and D. Dillard, [9], and by M. Tuttle and H. Brinson, [1]. Further specific aspects were developed by C. Hiel, [10],[11], by R. Brouwer, [12], for biaxial stress fields and by X. Xiao, [13],[14] for thermoplastic resin composites.

As example Figure 2 gives the variation of the shear compliance as function of time and shear stress obtained on APC-2 by X. Xiao, [13].

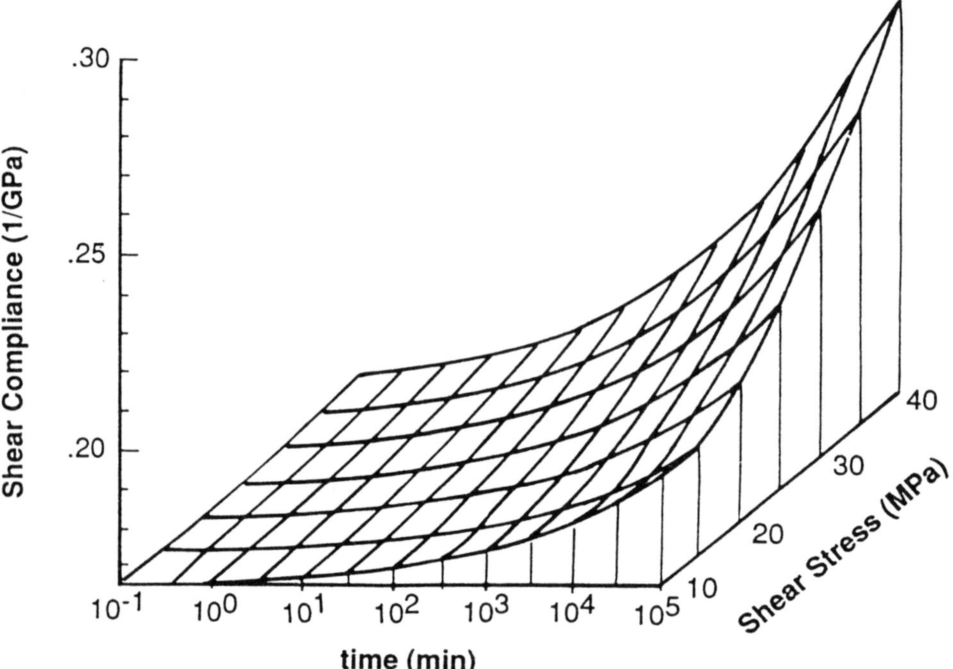

4. Internal stress transferts in a polymer matrix composite

If there is a perfect adhesion between matrix and fibers the strain being the same there is an instantaneous stress distribution between fibers and matrix under a given in plane stress program followed by a transfer of the stress from the matrix, (viscoelastic, presenting relaxation efforts), to the fibers.

The same type of stress transfer happens between adjacent lamina with non coinciding elastic directions.

In order to analyse in a simple way this aspect of the viscoelastic behaviour of the polymer matrix composite we consider a sandwich beam element where the two outer layers are elastic, the central core being viscoelastic.

For simplicity we consider a one-dimensional case and we assume the viscoelastic layer of the type 3-parameter standard solid material :

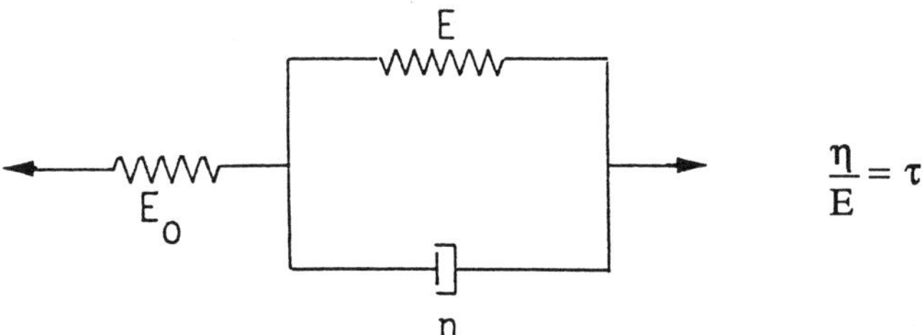

$$\frac{\eta}{E} = \tau$$

$$\frac{d\varepsilon}{dt} + \frac{\varepsilon}{\tau} = \frac{1}{E_o}\frac{d\sigma}{dt} + \frac{1}{\tau}\left(\frac{1}{E_o} + \frac{1}{E}\right)\sigma \qquad (4.1)$$

Starting from the following relations :

$$E_e \varepsilon_e = \sigma_e \qquad (4.2.a)$$

$$\frac{d\varepsilon_v}{dt} + \frac{1}{\tau}\varepsilon_v = \frac{1}{E_o}\frac{d\sigma_v}{dt} + \frac{1}{\tau}\left(\frac{1}{E_o} + \frac{1}{E}\right)\sigma_v \qquad (4.2.b)$$

$$S_e\sigma_e + S_v\sigma_V = \sigma(S_e + S_V) \tag{4.2.c}$$

$$\varepsilon_e = \varepsilon_v = \varepsilon \tag{4.2.d}$$

we obtain the constitutive differential relation for the sandwich element as :

$$\frac{d\varepsilon}{dt} + \frac{1}{\tau}\left\{\frac{S_eE_e(\frac{1}{E_o} + \frac{1}{E}) + S_v}{S_e\frac{E_e}{E_o} + S_v}\right\}\varepsilon = \frac{S_e + S_v}{S_e\frac{E_e}{E_o} + S_v}\left[\frac{1}{E_o}\frac{d\sigma}{dt} + \frac{1}{\tau}(\frac{1}{E_o} + \frac{1}{E})\,\sigma\right] \tag{4.3}$$

or

$$\frac{d\varepsilon}{dt} + \frac{1}{\tau}\{1+\alpha\psi\}\,\varepsilon = \phi\,\frac{d\sigma}{dt} + \frac{1}{\tau}(1+\alpha)\,\phi\sigma\ , \tag{4.4}$$

with

$$\frac{E_o}{E} = \alpha\ ;\ S_eE_e + S_vE_o = A\ ;\ \frac{S_eE_e}{A} = \psi\ ;\ \frac{S_e + S_v}{A} = \phi \tag{4.5}$$

If we impose $\sigma = \sigma_o H(t)$, we obtain the creep function k(t) as :

$$k(t) = \frac{\phi}{1 + \alpha\psi}\left[1 + \alpha + \alpha(\psi - 1)\,e^{-(1+\alpha\psi)\frac{t}{\tau}}\right] \tag{4.6}$$

With this creep function it is now possible to obtain $\varepsilon(t)$ under a given stress variation $\sigma(t)$ and with the conditions (4.2) we obtain $\sigma_e(t)$ and $\sigma_v(t)$.

This analysis can be generalised for a lamina and afterwards between lamina when the fiber orientations are rotated over an angle α.

Particularly this analysis shows that even under constant loading conditions stresses are transferred between matrix and fibers in every lamina and between lamina.

Under changing loading conditions these local stress variations can act as local sources for damage propagations.

Numerical predictions and experimental verifications of these local stress changes and their consequences on damage evolution are worked out on the basis of symmetric laminates $(0,90)_n$.

Conclusions

The viscoelastic nature of the polymer matrix induces time effects on the global behaviour of the polymer composite. As consequence the thermomechanical characteristics of the polymer composite will change during time. These variations are accelerated by temperature, stress level and moisture diffusion.

As consequence there is a need for the prediction of the long term behaviour of the thermomechanical characteristics of the polymer matrix composite. The thermodynamic based model of Schapery seems a good basis for this long term behaviour predictions.

A general durability analysis must combine the effect of different types of loadings, mechanical and environmental ones, their interaction and the consequent limit behaviour of a composite element. The most promising methodology is based on a damage analysis integrating the different specific results of the different loadings and their interactions.

The juxtaposition of elastic and viscoelastic elements in a lamina and the presence of different elastic orientations in adjacent lamina in a laminate results in a local permanent redistribution of stresses between matrix and fibers, and between lamina. The analysis of these local changes is important for the study of the damage propagation in a polymer matrix composite element.

Acknowledgements

We wish to thank the Belgian National Science Fund (NFWO-FKFO) and the Research Council of the Free University Brussels (OZR-VUB) for their financial support.

Many thanks also to the experimental polymer matrix group of COSARGUB, especially to D.Van Hemelrijck, L. Schillemans and F. Boulpaep.

An important help was also obtained from A. Vrijdag and nothing could be realized without the permanent support of M. Bourlau.

References

Tuttle, M., and Brinson, H.F., 1986, "Prediction of the Long-term Creep Compliance of General Composite Laminates", *Experimental Mechanics*, March, pp. 89-102.

Bax, J., 1970, "Deformation Behaviour and Failure of Glassfibre-Reinforced Resin Material", *Plastics and Polymers*, pp. 27-30.

Lee, J., and de Charentenay, F.X., 1986, "Modelisation of Cumulative Damage on an Unnotched Carbon Epoxy Laminate $(45°)_{2s}$, submitted to Fatigue and Creep Fatigue Loading".

Brinson, H.F., 1981, "Experimental Mechanics Applied to the Accelerated Characterization of Polymer Based Composites", *New Trends in Experimental Mechanics*, J.T. Pindera, ed., Springer Verlag, Vienna, pp. 1-69.

Brinson, H.F., 1985, "Viscoelastic Behaviour and Life-time (Durability) Predictions", *Mechanical Characterisation of Load Bearing Fibre Composite Laminates*, A.H. Cardon & G. Verchery, eds., Elsevier Applied Science Publ., London, pp. 3-20.

Schapery, R.A., 1964, "Application of Thermodynamics to Thermomechanical, Fracture and Birefringent Phenomena in Viscoelastic Media", *Journal of Applied Physics*, Vol. 35, No. 5, pp. 1451-1465.

Schapery, R.A., 1966, "A Theory of Nonlinear Thermoviscoelasticity based on Irreversible Thermodynamics", Proceedings of the 5th US National Congress of Applied Mechanics, ASME, pp. 511-530.

Schapery, R.A., 1970, "On a Thermodynamic Constitutive Theory and Its Application to Various Nonlinear Materials", Themoinelasticity, B.A. Boly, ed., Springer Verlag, Wien-New York, pp. 259-285.

Brinson, H.F., and Dillard D.A., 1982, "The Prediction of Long-term Viscoelastic Properties of Fiber Reinforced Plastics", Proceedings of ICCM IV, Tokyo.

Hiel, C.C., 1983, "The Nonlinear Viscoelastic Response of Resin Matrix Composite Materials, PhD-Thesis, Free University Brussels, Belgium.

Hiel, C.C., Cardon, A.H., and Brinson, H.F., 1984, "The Nonlinear Viscoelastic Response of Resin Matrix Composite Laminates", NASA Contractor Report 3772.

Brouwer, H.R., 1986, "Nonlinear Viscoelastic Characterization of Transversely Isotropic Fibrous Composites under Biaxial Loading", PhD-Thesis, Free University Brussels, Belgium.

Xiao, X., 1987, "Viscoelastic Characterization of Thermoplastic Matrix Composites", PhD-Thesis, Free University Brussels, Belgium.

Xiao, X., 1989, "Studies on the Viscoelastic Behaviour of a Thermoplastic Resing Composite", *Composite Science and Technology*, No. 34, pp. 163-182.

A Local-Field Theory for the Overall Creep of Fiber-Reinforced Metal-Matrix Composites

Y.M. Wang and G.J. Weng
Department of Mechanics and Materials Science
Rutgers University
New Brunswick, NJ 08903, USA

Abstract

A local-field theory is developed to calculate the development of the overall creep strain of a fiber-reinforced metal-matrix composite, where the creep rate of the constituents may depend nonlinearly on the stress and both phases may undergo the primary as well as the secondary creep. The theory is constructed with the combination of Eshelby's equivalent inclusion principle, Kroner's elastic constraint, Mori and Tanaka's mean-field concept, and Luo and Weng's local solution of a three-phase cylindrically concentric solid; it is intended for the low to moderate fiber concentration within the small creep-strain range. The theory thus developed also serves to evaluate the accuracy of the simpler mean-field theory, which is found to be reliable enough for both the axial and the plane-strain, biaxial tensile creep, but generally less so for the transverse tensile creep, and the transverse and axial shear creep. As compared to the longitudinal tensile creep data of a Borsic/aluminum composite, the theoretical prediction, though slightly lower, appears to lie within an acceptable range of accuracy.

1. Introduction

This paper is concerned with the development of a theoretical principle for the determination of the overall creep strain of a fiber-reinforced composite under any constant, combined stress. Our focus here is on the metallic constituents, whose creep rate depends nonlinearly on the stress and, due to work-hardening, both primary and secondary creep may be present. We take the fibers and the matrix to be both capable of creeping, and are perfectly bonded together. The circular fibers are aligned and homogeneously dispersed in the matrix so that the composite as a whole appears to be transversely isotropic.

Under a constant external stress, the initial response of the composite is elastic and the overall elastic strain can be determined from its effective elastic moduli. The initial elastic response is then followed by the time-dependent creep, whose local creep-rate depends on the local stress and creep strain according to the constitutive equations. Such a time-dependent deformation leads to a continuous stress transfer between the constituents and, even within each phase, the stress field is also constantly changing as a result of heterogeneous creep. A precise analysis of such a nonlinear, time-dependent, and heterogeneous problem is not a simple one and, as it stands now, no exact solution seems to exist for this problem. The method to be proposed is an approximate one; it is primarily based upon the combination of Zhu and Weng's (1987,89) mean-field theory for time-dependent creep and Luo and Weng's (1989) solution for the stress field in a three-phase cylindrically concentric solid due to a "stress-free" transformation strain in the central fiber region. Zhu and Weng's theory in turn was derived using Eshelby's (1957) equivalent inclusion principle, Mori and Tanaka's (1973) concept of average stress in the matrix, and Kröner (1961) and Budiansky and Wu's (1962) elastic constraint approximation, in the treatment of average-stress redistribution between the constituents during a time-dependent creeping flow. In this mean-field approach, the average creep-rate of a phase is calculated directly from its mean stress and mean creep strain according to the constitutive equation; the information regarding the local stress and strain inhomogeneity within each phase is therefore not required. To be rigorous, however, the mean creep-rate of a phase ought to be calculated from the mean of its local creep-rate, each determined at the level of its local stress and local creep-strain. If the creep rate were to depend on the stress and creep strain linearly (as in many polymeric materials), these two approaches would have yielded an identical result. For the metallic constituents, the stress exponent n usually lies between 3 to 7, and therefore, the mean-field approach will invariably underestimate the average creep-strain of the phases and consequently of the composite. At a dilute concentration, however, the stress field in both phases tends to be more uniform and therefore justifies the "mean-field" approach. Zhu and Weng's theory (1987,89) thus developed was indeed directed towards such a condition, where particle concentration of 5% was specifically investigated. The incorporation here of Luo and Weng's (1989) solution for the heterogeneous stress field is aimed at a somewhat improved concentration, though still not yet at the "finite" level. In this same spirit a local theory for the creep of a particle-reinforced composite has also been developed recently by Zhu and Weng

(1990), using Luo and Weng's (1987) solution of a three-phase spherically concentric solid. It must be borne in mind though that the adoption of Kroner's (1961) assumption of an elastic constraint for a creeping (or plastic) matrix implies that the theory thus constructed is suitable only for the small creep-strain range; applications to a large creep strain (say one or two orders of magnitude higher than the elastic strain) can result in a significant underestimate for the overall creep strain.

2. Constitutive Equations

We shall refer to the matrix as phase 0 and fibers as phase 1. Both phases will be taken to be elastically isotropic for simplicity, with the bulk and shear moduli denoted by κ_r and μ_r, respectively, for the r-th phase. The corresponding volume fraction will be denoted by c_r, such that $c_0 + c_1 = 1$.

Under a constant tensile stress σ, the steady and primary creep rates of the r-th phase usually can be written as

$$\dot{\epsilon}_s^{c(r)} = a_r \cdot (\sigma^{(r)})^{n_r},$$

$$\dot{\epsilon}_p^{c(r)} = b_r \cdot [d_r \cdot (\sigma^{(r)})^{n_r} - \epsilon^{c(r)}], \quad r = 0, 1 \tag{2.1}$$

where a_r, n_r, b_r and d_r are the material constants for the r-th phase, which can be readily determined from two tensile creep curves. This set of constitutive equations is seen to be general enough to cover those of a viscoelastic solid (n=1) or nonlinear steady creep only (b = 0). The total creep rate at any instant is given by the sum of the two $\dot{\epsilon}^c = \dot{\epsilon}_s^c + \dot{\epsilon}_p^c$, where the primary creep rate will decrease with increasing creep strain (as a result of strain hardening) and is set to zero numerically when ϵ^c reaches the critical value $d \cdot \sigma^n$.

To account for the multiaxial stress and strain state, von Mises' effective stress and strain are introduced

$$\sigma^{*(r)} = (\frac{3}{2} \sigma'^{(r)}_{ij} \sigma'^{(r)}_{ij})^{1/2},$$

$$\epsilon^{*c(r)} = (\frac{2}{3} \epsilon^{c(r)}_{ij} \epsilon^{c(r)}_{ij})^{1/2}, \tag{2.2}$$

where σ'_{ij} is the deviatoric part of σ_{ij}. Equation (2.1) is then modified to

$$\dot{\epsilon}_s^{*c(r)} = a_r \cdot (\sigma^{*(r)})^{n_r}$$

$$\dot{\epsilon}_p^{*c(r)} = b_r \cdot [d_r \cdot (\sigma^{*(r)})^{n_r} - \epsilon^{*c(r)}] , \qquad r=0, 1 \tag{2.3}$$

with $\dot{\epsilon}^{*c} = \dot{\epsilon}_s^{*c} + \dot{\epsilon}_p^{*c}$. The creep-rate components then follow from the Prandtl-Reuss relation

$$\dot{\epsilon}_{ij}^{c(r)} = \frac{3}{2} \frac{\dot{\epsilon}^{*c(r)}}{\sigma^{*(r)}} \sigma'^{(r)}_{ij} , \qquad r = 0, 1 . \tag{2.4}$$

Thus at a given stress σ_{ij} the effective stress σ^* can be calculated from (2.2) and the effective initial creep rates from (2.3). The creep-rate components are then available from (2.4) which, over a time increment dt, gives rise to an incremental creep strain $d\epsilon_{ij}^c = \dot{\epsilon}_{ij}^c \, dt$. This is to be added to whatever the existing $\epsilon_{ij}^c(t)$ to yield a new $\epsilon_{ij}^c(t+dt)$, leading to a new ϵ^{*c} for the next time increment.

3. The Initial Elastic State

The Mori-Tanaka (1973) method will be applied to determine the initial strain of the composite and the mean stress of the constituents. The equivalent transformation strain derived also allows us to use Luo and Weng's (1989) solution to evaluate the heterogeneous stress field in the matrix which, together with the mean stress, permit us to calculate the initial local creep rate at its field point.

To place the theory in perspective, we recall from the Mori-Tanaka theory that when the composite is subjected to an external boundary traction giving rise to a uniform stress σ, the strain of the comparison material, with the elastic moduli tensor L_0, is given in terms of the familiar symbolic notations by (see Weng (1984), for instance)

$$\epsilon^0 = L_0^{-1} \bar{\sigma} , \tag{3.1}$$

and the average stress of the matrix can be written as

$$\bar{\sigma}^{(0)} = \bar{\sigma} + \tilde{\sigma} = L_0 \, \epsilon^{(0)} = L_0(\epsilon^0 + \tilde{\epsilon}) , \tag{3.2}$$

where $\tilde{\epsilon}$ is its mean variation from ϵ^0, such that when $c_1 \to 0$, $\tilde{\epsilon} \to 0$. Henceforth it is to be noted that whenever σ and ϵ appear without the indicial components, they represent the general stress and strain tensors, respectively. The average stress of the fibers, in terms of its own moduli tensor and of Eshelby's (1957) equivalent inclusion principle, can be written as

$$\bar{\sigma}^{(1)} = \bar{\sigma} + \tilde{\sigma} + \sigma^{pt} = L_1 \bar{\epsilon}^{(1)} = L_1(\epsilon^0+\tilde{\epsilon}+\epsilon^{pt}) = L_0(\epsilon^0+\tilde{\epsilon}+\epsilon^{pt}-\epsilon^*) , \tag{3.3}$$

where ϵ^* is Eshelby's equivalent transformation strain (or Kroner's (1958) polarization strain, Mura's (1987) eigenstrain) introduced into the fiber regions so that L_1 can be replaced by L_0 to yield the same $\bar{\sigma}^{(1)}$. Furthermore, the average perturbed strain in the fibers is assumed to be connected to ϵ^* through

$$\epsilon^{pt} = S\,\epsilon^* , \tag{3.4}$$

in terms of Eshelby's S-tensor, and the condition $\bar{\sigma} = \Sigma c_r \bar{\sigma}^{(r)}$ also leads to

$$\tilde{\epsilon} = -c_1(\epsilon^{pt}-\epsilon^*) = c_1(I-S)\epsilon^* , \tag{3.5}$$

where I is the symmetric, fourth-rank identity tensor. This set of equations provides the equivalent transformation strain

$$\epsilon^* = A^*\epsilon^0 , \qquad A^* = -[(L_1+L_0)(c_1 I+c_0 S)+L_0]^{-1}(L_1-L_0) . \tag{3.6}$$

The total strain of the composite is given by $\bar{\epsilon} = \Sigma c_r \bar{\epsilon}^{(r)}$, leading to

$$\bar{\epsilon} = \epsilon^0 + c_1\epsilon^* = (I+c_1 A^*)\epsilon^0 . \tag{3.7}$$

Then recalling the effective stress-strain relation $\bar{\sigma} = L\bar{\epsilon}$, one arrives at the effective moduli

$$L = L_0(I+c_1 A^*)^{-1} , \tag{3.8}$$

for the system.

With the circular fibers the non-vanishing components of the S-tensor are

$$S_{2222} = S_{3333} = \frac{5-4\nu_0}{8(1-\nu_0)}, \qquad S_{2233} = S_{3322} = \frac{4\nu_0-1}{8(1-\nu_0)},$$

$$S_{2211} = S_{3311} = \frac{\nu_0}{2(1-\nu_0)}, \qquad S_{2323} = \frac{3-4\nu_0}{8(1-\nu_0)}, \tag{3.9}$$

$$S_{1212} = S_{1313} = \frac{1}{4},$$

where ν_0 is the Poisson ratio of the matrix, direction 1 is the fiber axis, and plane 2-3 the cross section. With these components the five independent elastic moduli of the composite are known to coincide with the expressions of Hill's (1964) and Hashin's (1965) lower bounds if the matrix is the softer phase (see Zhao et al. 1989):

$$E_{11} = c_1E_1 + c_0E_0 + \frac{4c_1c_0(\nu_1-\nu_0)^2}{c_1/\bar{\kappa}_0+c_0/\bar{\kappa}_1+1/\mu_0},$$

$$\nu_{12} = c_1\nu_1 + c_0\nu_0 + \frac{c_1c_0(\nu_1-\nu_0)(1/\bar{\kappa}_0-1/\bar{\kappa}_1)}{c_1/\bar{\kappa}_0+c_0/\bar{\kappa}_1+1/\mu_0},$$

$$\kappa_{23} = \bar{\kappa}_0 + \frac{c_1}{1/(\bar{\kappa}_1-\bar{\kappa}_0)+c_0/(\bar{\kappa}_0+\mu_0)},$$

$$\mu_{12} = \mu_0 + \frac{c_1\mu_0}{\mu_0/(\mu_1-\mu_0)+c_0/2},$$

$$\mu_{23} = \mu_0 + \frac{c_1\mu_0}{\mu_0/(\mu_1-\mu_0)+c_0(\bar{\kappa}_0+2\mu_0)/[2(\bar{\kappa}_0+\mu_0)]}, \tag{3.10}$$

where E_{11}, ν_{12}, κ_{23}, μ_{12}, μ_{23} are the longitudinal Young's modulus, major Poisson ratio, plane strain bulk modulus, axial and transverse shear moduli, respectively, and κ_r is the plane-strain bulk modulus of the r-th phase. This set of moduli allows us to determine the initial strain of the composite $\bar{\epsilon}$ under a given $\bar{\sigma}$.

Although the stress field in the matrix is heterogeneous, that in the fiber is generally less so and therefore the mean-field approach can be reasonably used to calculate its mean creep-rate. From (3.2), (3.5) and (3.6), one has the mean stress for the matrix

$$\bar{\sigma}^{(0)} = B_0\bar{\sigma}\ , \qquad B_0 = I + c_1 L_0(I-S)A^* L_0^{-1}, \tag{3.11}$$

and, with which, the mean stress of the fibers follow from

$$\bar{\sigma}^{(1)} = \frac{1}{c_1}\,(\bar{\sigma} - c_0\bar{\sigma}^{(0)})\ , \tag{3.12}$$

which is identical to (3.3). Thus central to the determination of $\bar{\sigma}^{(0)}$ and $\bar{\sigma}^{(1)}$ is ϵ^*, which is also needed for the determination of the heterogeneous stress field in the matrix.

For circular fibers aligned along direction 1, the ϵ^* in (3.6) can be expanded into the following components

$$\epsilon^*_{11} = a^*_1\epsilon^0_{11} + a^*_2(\epsilon^0_{22} + \epsilon^0_{33})\ ,$$

$$\epsilon^*_{22} = a^*_3\epsilon^0_{11} + a^*_4\epsilon^0_{22} + a^*_5\epsilon^0_{33},$$

$$\epsilon^*_{33} = a^*_3\epsilon^0_{11} + a^*_5\epsilon^0_{22} + a^*_4\epsilon^0_{33},$$

$$\epsilon^*_{12} = a^*_6\epsilon^0_{12}, \qquad \epsilon^*_{13} = a^*_6\epsilon^0_{13}\ , \qquad \epsilon^*_{23} = (a^*_4 - a^*_5)\epsilon^0_{23}, \tag{3.13}$$

where

$$a^*_1 = -\frac{1}{c_0 p}\left[\frac{2\nu_0 - 1}{2(1-\nu_0)} + \frac{1}{3c_0}\left(\frac{2\kappa_1}{\kappa_1 - \kappa_0} + \frac{\mu_1}{\mu_1 - \mu_0}\right)\right],$$

$$a^*_2 = \frac{1}{3c_0^2 p}\left(\frac{\kappa_1}{\kappa_1 - \kappa_0} - \frac{\mu_1}{\mu_1 - \mu_0}\right),$$

$$a_3^* = \frac{1}{c_0 p}\left[\frac{\nu_0}{2(1-\nu_0)} + \frac{1}{3c_0}\left(\frac{\kappa_1}{\kappa_1-\kappa_0} - \frac{\mu_1}{\mu_1-\mu_0}\right)\right],$$

$$a_4^* = -\frac{1}{2}\ \frac{1}{\mu_1/(\mu_1-\mu_0)-c_0/[4(1-\nu_0)]} - \frac{1}{2c_0 p}\left[-1 + \frac{1}{3c_0}\left(\frac{\kappa_1}{\kappa_1-\kappa_0} + \frac{2\mu_1}{\mu_1-\mu_0}\right)\right],$$

$$a_5^* = \frac{1}{2}\ \frac{1}{\mu_1/(\mu_1-\mu_0)-c_0/[4(1-\nu_0)]} - \frac{1}{2c_0 p}\left[-1 + \frac{1}{3c_0}\left(\frac{\kappa_1}{\kappa_1-\kappa_0} + \frac{2\mu_1}{\mu_1-\mu_0}\right)\right],$$

$$a_6^* = -\frac{1}{\mu_1/(\mu_1-\mu_0)-c_0/2}, \tag{3.14}$$

and

$$p = (a+b)\left(\frac{1}{2(1-\nu_0)} + a + 2b\right) - 2b\left(\frac{\nu_0}{2(1-\nu_0)} + b\right),$$

$$a = \frac{\mu_1}{c_0(\mu_1-\mu_0)} - 1, \qquad b = \frac{1}{3c_0}\left(\frac{\kappa_1}{\kappa_1-\kappa_0} - \frac{\mu_1}{\mu_1-\mu_0}\right). \tag{3.15}$$

Similarly the mean stress of the matrix in (3.11), now written in its deviatoric (primed) form for ready calculation of the creep rate, has the components

$$\bar{\sigma}'^{(0)}_{11} = \frac{1}{c_0}\bar{\sigma}'_{11} - \frac{c_1}{c_0^2 p}\frac{\mu_1}{\mu_1-\mu_0}\left[\left(\frac{1}{c_0}\frac{\kappa_1}{\kappa_1-\kappa_0} - \frac{2}{3}\frac{1-2\nu_0}{1-\nu_0}\right)\bar{\sigma}'_{11} + \frac{1}{9}\frac{1-2\nu_0}{1-\nu_0}\bar{\sigma}_{kk}\right],$$

$$\bar{\sigma}'^{(0)}_{22} = \frac{1}{c_0}\bar{\sigma}'_{22} - \frac{c_1}{c_0^2 p}\left\{\frac{\mu_1}{\mu_1-\mu_0}\left[\left(\frac{1}{c_0}\frac{\kappa_1}{\kappa_1-\kappa_0} - \frac{2}{3}\frac{1-2\nu_0}{1-\nu_0}\right)\frac{\bar{\sigma}'_{22}+\bar{\sigma}'_{33}}{2} - \frac{1}{18}\frac{1-2\nu_0}{1-\nu_0}\bar{\sigma}_{kk}\right]\right.$$

$$\left. + \frac{c_0 p}{1-c_0(\mu_1-\mu_0)/[4\mu_1(1-\nu_0)]}\ \frac{\bar{\sigma}'_{22}-\bar{\sigma}'_{33}}{2}\right\},$$

$$\bar{\sigma}'^{(0)}_{33} = \frac{1}{c_0}\bar{\sigma}'_{33} - \frac{1}{c_0^2 p}\{\frac{\mu_1}{\mu_1-\mu_0}[(\frac{1}{c_0}\frac{\kappa_1}{\kappa_1-\kappa_0} - \frac{2}{3}\frac{1-2\nu_0}{1-\nu_0})\frac{\bar{\sigma}'_{22}+\bar{\sigma}'_{33}}{2} - \frac{1}{18}\frac{1-2\nu_0}{1-\nu_0}\bar{\sigma}_{kk}]$$

$$+ \frac{c_0 p}{1-c_0(\mu_1-\mu_0)/[4\mu_1(1-\nu_0)]}\frac{\bar{\sigma}'_{33}-\bar{\sigma}'_{22}}{2}\},$$

$$\bar{\sigma}^{(0)}_{kk} = \frac{1}{c_0}\bar{\sigma}_{kk} - \frac{c_1}{c_0^2 p}\frac{\kappa_1}{\kappa_1-\kappa_0}[\frac{1+\nu_0}{2(1-\nu_0)}\bar{\sigma}'_{11} + (\frac{\mu_1}{c_0(\mu_1-\mu_0)} - \frac{5-4\nu_0}{6(1-\nu_0)})\bar{\sigma}_{kk}],$$

$$\bar{\sigma}^{(0)}_{12} = \frac{\mu_1+\mu_0}{(1+c_1)\mu_1+c_0\mu_0}\bar{\sigma}_{12}, \qquad \bar{\sigma}^{(0)}_{13} = \frac{\mu_1+\mu_0}{(1+c_1)\mu_1+c_0\mu_0}\bar{\sigma}_{13},$$

$$\bar{\sigma}^{(0)}_{23} = \frac{\{(3-4\nu_0)/[4(1-\nu_0)]\}(\mu_1-\mu_0)+\mu_0}{\{c_1+c_0(3-4\nu_0)/[4(1-\nu_0)]\}(\mu_1-\mu_0)\}+\mu_0}\bar{\sigma}_{23}. \tag{3.16}$$

The mean stress of the fibers follows from (3.12), which can be used in its constitutive equations to calculate its initial creep rate.

It may be noted in passing that when both phases share the same elastic moduli, the following relations can be used to simplify the foregoing relation

$$\frac{\kappa_1}{p(\kappa_1-\kappa_0)} = 0, \qquad \frac{\mu_1}{p(\mu_1-\mu_0)} = 0,$$

$$c_0^2 p[\kappa_1/(\kappa_1-\kappa_0)]\,[\mu_1/(\mu_1-\mu_0)] = 1. \tag{3.17}$$

To determine the initial local creep-rate in the matrix, we now construct its initial heterogeneous stress field associated with ϵ^*.

4. Local Stress Distribution in the Matrix Due to An Eigenstrain ϵ^*_{ij} in the Central Fiber

It must be noted at the outset that the solutions to be listed in this section not only serve to determine the initial stress distribution in the matrix, but also the change of its stress field during the subsequent creep under the assumption of Kroner's (1961) elastic constraint.

The displacement field in a three-phase cylindrically concentric solid due to an eigenstrain ϵ^* in the central fiber was recently derived by Luo and Weng (1989), where the circular fiber is embedded in the matrix, which is further embedded in an infinitely extended medium with the property of the composite. The radii of the central fiber and the intermediate matrix were denoted by a and b, respectively, so that $c_1 = (a/b)^2$. One is reminded that though both moduli L_1 and L_0 appear explicitly in the following auxiliary solutions, the values of L_1 should be replaced by those of L_0 in actual detemrination of the local strain field of the matrix (in accordance with the equivalence principle). Luo and Weng's (1989) notations -- including a, b and p below -- are kept here; they bear no relations with those appearing elsewhere.

4.1 *Strain field due to* ϵ^*_{11}

In this case the nonvanishing components of the strain field are, in cylindrical coordinates

$$\epsilon^{(0)}_{rr} = (a_2 - a_3 \frac{b^2}{r^2})\, \epsilon^*_{11}, \qquad \epsilon^{(0)}_{\theta\theta} = (a_2 + a_3 \frac{b^2}{r^2})\epsilon^*_{11}, \tag{4.1}$$

where, in terms of Lame constant λ_r of the r-th phase,

$$a_2 = -c_1\lambda_1(\mu_{23}-\mu_0)/p\,, \qquad a_3 = c_1\lambda_1(\bar{\kappa}_0+\mu_{23})/p\,,$$

$$p = 2[(\bar{\kappa}_1+\mu_0)(\bar{\kappa}_0+\mu_{23})-c_1(\bar{\kappa}_1-\bar{\kappa}_0)(\mu_{23}-\mu_0)]\,. \tag{4.2}$$

(Thus in application of the equivalence principle λ_1 and $\bar{\kappa}_1$ here must be replaced by λ_0 and $\bar{\kappa}_0$, respectively, as stated above).

4.2 Strain field due to biaxial stretching $\epsilon^*_{22} = \epsilon^*_{33}$

The results are the same as in (4.1), but with a_2, a_3 and ϵ^*_{11} replaced by b_2, b_3 and ϵ^*_{22}, respectively, such that

$$b_2 = -2c_1\bar{\kappa}_1(\mu_{23}-\mu_0)/p, \qquad b_3 = 2c_1\bar{\kappa}_1(\mu_0+\mu_{23})/p . \tag{4.3}$$

4.3 Strain field due to an axial shear $\epsilon^*_{12} = \epsilon^*_{21}$

The results are

$$\epsilon^{(0)}_{zr} = \frac{1}{2}(h_2 - h_3\frac{b^2}{r^2})\epsilon^*_{12}\cos\theta ,$$

$$\epsilon^{(0)}_{z\theta} = -\frac{1}{2}(h_2 + h_3\frac{b^2}{r^2})\epsilon^*_{12}\sin\theta , \tag{4.4}$$

where

$$h_2 = -2c_1\mu_1(\mu_{12}-\mu_0)/q ,$$

$$h_3 = 2c_1\mu_1(\mu_{12}+\mu_0)/q ,$$

$$q = (\mu_1+\mu_0)(\mu_0+\mu_{12}) - c_1(\mu_1-\mu_0)(\mu_{12}-\mu_0) , \tag{4.5}$$

and θ is measured from the 2 to the 3-axis.

4.4 Strain field due to a transverse shear $\epsilon^*_{22} = -\epsilon^*_{33} = e^*$

This is a more complicated problem and the solution was found with the aid of Christensen and Lo's (1979) formulation. The results are

$$\epsilon^{(0)}_{rr} = [d_3 + 3d_4(\eta_0-3)\frac{r^2}{b^2} - 3d_5\frac{b^4}{r^4} - d_6(\eta_0+1)\frac{b^2}{r^2}]e^*\cos2\theta ,$$

$$\epsilon^{(0)}_{\theta\theta} = [-d_3+3d_4(\eta_0+1)\frac{r^2}{b^2} + 3d_5\frac{b^4}{r^4} - d_6(\eta_0-3)\frac{b^2}{r^2}]e^* \cos 2\theta , \tag{4.6}$$

$$\epsilon^{(0)}_{r\theta} = [-d_3+6d_4\frac{r^2}{b^2} - 3d_5\frac{b^4}{r^4} - 2d_6\frac{b^2}{r^2}]e^* \sin 2\theta ,$$

where $\eta_0 = 1+2\mu_0/\bar{\kappa}_0$. The coefficients d_i were found to be

$$d_i = E^{-1}_{ij}F_j, \qquad i=1,2,\ldots,8; \quad j=1,2,\ldots,8 , \tag{4.7}$$

where for the four d_i in (4.6), the E_{ij} components ($i = 3,4,5,6$) are

$$E_{ij} = \begin{bmatrix} \mu_1/\mu_0 & 0 & -1 & 0 & 3/c_1^2 & 4/c_1 & 0 & 0 \\ -\mu_1/\mu_0 & 6\mu_1/\mu_0 & 1 & -6c_1 & 3/c_1^2 & 2/c_1 & 0 & 0 \\ 0 & 0 & 1 & \eta_0-3 & 1 & \eta_0+1 & -1 & -(\eta_c+1) \\ 0 & 0 & -1 & \eta_0+3 & 1 & -(\eta_0-1) & -1 & (\eta_c-1) \end{bmatrix} , \tag{4.8}$$

with $\eta_c = 1+2\mu_{23}/\kappa_{23}$ for the composite, and for the F_j its transpose is

$$F_j^T = [0,\ 0,\ \mu_1/\mu_0,\ -\mu_1/\mu_0,\ 0,\ 0,\ 0,\ 0], \quad j=1,2,\ldots,8. \tag{4.9}$$

With these four types of eigenstrains, any ϵ_{ij} derived in Section 3 or in the next section for the subsequent creep can be expressed by a proper superposition of these pairs. These strain fields are given in the cylindrical coordinate and it must be transformed back to the Cartesian one; then the corresponding local stress field follows as

$$\sigma^{(0)}(x) = L_0\, \epsilon^{(0)}(x) , \tag{4.10}$$

in the matrix.

The volume average of $\sigma^{(0)}(x)$ in (4.10) does not always give rise to a mean $\bar{\sigma}^{(0)}$ which is identical to (3.15); to ensure such a self-consistency the local stress field in the matrix is calculated as

$$\sigma^{(0)}(x) = \bar{\sigma}^{(0)} + L_0(\epsilon^{(0)}(x) - <\epsilon^{(0)}(x)>) , \tag{4.11}$$

where $<\cdot>$ denotes the volume average of the said quantity. The stress field in (4.11) then can be used in the constitutive equations (2.2) to (2.4) to calculate the local initial creep rate of the matrix $\dot{\epsilon}^{c(0)}$ (x, t=0).

5. Stress Variation in the Fibers and Matrix Following an Incremental Creep

As in the elastic case we first consider the mean-stress variation in both phases and then the local-stress change in the matrix.

Once the creep rates have been calculated, the mean incremental creep strains of the constituents over a time increment dt are

$$d\bar{\epsilon}^{(1)} = \dot{\bar{\epsilon}}^{c(1)}dt , \qquad \text{for the fibers}$$

$$\tag{5.1}$$

$$d\bar{\epsilon}^{c(0)} = <\dot{\epsilon}^{c(0)}(x)>dt \qquad \text{for the matrix,}$$

where at any generic time $\dot{\bar{\epsilon}}^{c(1)}$ and $\dot{\epsilon}^{c(0)}(x)$ are calculated at the corresponding levels of $\bar{\sigma}^{(1)}$ and $\bar{\epsilon}^{c(1)}$, and of $\sigma^{(0)}(x)$ and $\epsilon^{c(0)}(x)$, respectively, according to their constitutive equations.

To find the average stress variations in both phases we first fictitiously take the fibers out of the matrix and let both creep separately by $d\bar{\epsilon}^{c(1)}$ and $d\bar{\epsilon}^{c(0)}$ without any constraint. This is a truly "stress-free" process (incrementally) in the sense of Eshelby (1957), and, now to place the crept fibers into the deformed holes, the misfit strain is identified as

$$d\epsilon^* = d\bar{\epsilon}^{c(1)} - d\bar{\epsilon}^{c(0)} . \tag{5.2}$$

Under the assumption of the elastic constraint in the matrix, the mean stress variations can be found by the process similar to (3.1) to (3.5). This time, however, the incremental boundary condition is $d\sigma = c_1 d\bar{\sigma}^{(1)} + c_0 d\bar{\sigma}^{(0)} = 0$. In parallel to (3.2) and (3.3), we have the mean-stress variations

$$d\sigma^{(0)} = d\tilde{\sigma} = L_0 d\epsilon^{(0)} = L_0 d\tilde{\epsilon} , \tag{5.3}$$

in the matrix, and remembering that the fibers now possess a misfit strain $d\epsilon^*$,

$$d\bar{\sigma}^{(1)} = d\tilde{\sigma} + d\sigma^{pt} = L_1 d\bar{\epsilon}^{(1)} = L_1(d\tilde{\epsilon}+d\epsilon^{pt}-d\epsilon^*)$$

$$= L_0(d\tilde{\epsilon}+d\epsilon^{pt}-d\epsilon^*-d\epsilon^{**}) , \quad (5.4)$$

where $d\epsilon^{**}$ is the incremental equivalent transformation strain, and $d\epsilon^*+d\epsilon^{**}$ together forms the total value of the transformation strain during the incremental process.

The perturbed strain $d\epsilon^{pt}$ is now given by

$$d\epsilon^{pt} = S(d\epsilon^*+d\epsilon^{**}) , \quad (5.5)$$

and the boundary condition $d\bar{\sigma} = 0$ leads to

$$d\tilde{\epsilon} = c_1(I-S)(d\epsilon^*+d\epsilon^{**}) . \quad (5.6)$$

The analysis can be carried out in a similar fashion, giving rise to

$$d\epsilon^* + d\epsilon^{**} = [(L_1-L_0)(c_1 I+c_0 S)+L_0]^{-1} L_1 d\epsilon^* , \quad (5.7)$$

where $d\epsilon^*$ on the right is readily available from (5.2). To find the overall creep strain of the composite we remember that a prior homogeneous $d\bar{\epsilon}^{c(0)}$ had already taken place before the crept fibers were placed back to the deformed holes; thus

$$d\bar{\epsilon}^c = d\bar{\epsilon}^{c(0)} + c_0 d\tilde{\epsilon} + c_1(d\tilde{\epsilon}+d\epsilon^{pt}) = d\bar{\epsilon}^{c(0)} + c_1(d\epsilon^*+d\epsilon^{**}) . \quad (5.8)$$

The average-stress variation in the matrix then follows from (5.3) and (5.6)

$$d\bar{\sigma}^{(0)} = c_1 L_0(I-S)(d\epsilon^*+d\epsilon^{**}) , \quad (5.9)$$

and for the fibers

$$d\bar{\sigma}^{(1)} = -(c_0/c_1) d\bar{\sigma}^{(0)} , \quad (5.10)$$

which can also be found identically from (5.4).

The total transformation strain $d\epsilon^*+d\epsilon^{**}$ is now called for, which is also needed for the determination of the local stress variation in the matrix. From (5.7), one has, with the constants p. a and b in (3.15)

$$d\epsilon^{*}_{11} + d\epsilon^{**}_{11} = \frac{1+a}{p}\left(\frac{1}{2(1-\nu_0)} + a + 3b\right)\left(d\bar{\epsilon}^{c(1)}_{11} - d\bar{\epsilon}^{c(0)}_{11}\right),$$

$$d\epsilon^{*}_{22} + d\epsilon^{**}_{22} = \frac{1+a}{2p}\left(\frac{\nu_0}{1-\nu_0} + a + 3b\right)\left[\left(d\bar{\epsilon}^{c(1)}_{22} + d\bar{\epsilon}^{c(1)}_{33}\right) - \left(d\bar{\epsilon}^{c(0)}_{22} + d\bar{\epsilon}^{c(0)}_{33}\right)\right]$$

$$+ \frac{1+a}{2\{(3-4\nu_0)/[4(1-\nu_0)]+a\}}\left[\left(d\bar{\epsilon}^{c(1)}_{22} - d\bar{\epsilon}^{c(1)}_{33}\right) - \left(d\bar{\epsilon}^{c(0)}_{22} - d\bar{\epsilon}^{c(0)}_{33}\right)\right],$$

$$d\epsilon^{*}_{33} + d\epsilon^{**}_{33} = \frac{1+a}{2p}\left(\frac{\nu_0}{1-\nu_0} + a + 3b\right)\left[\left(d\bar{\epsilon}^{c(1)}_{22} + d\bar{\epsilon}^{c(1)}_{33}\right) - \left(d\bar{\epsilon}^{c(0)}_{22} + d\bar{\epsilon}^{c(0)}_{33}\right)\right]$$

$$+ \frac{1+a}{2\{(3-4\nu_0)/[4(1-\nu_0)]+a\}}\left[\left(d\bar{\epsilon}^{c(1)}_{33} - d\bar{\epsilon}^{c(1)}_{22}\right) - \left(d\bar{\epsilon}^{c(0)}_{33} - d\bar{\epsilon}^{c(0)}_{22}\right)\right],$$

$$d\epsilon^{*}_{12} + d\epsilon^{**}_{12} = \frac{2(1+a)}{1+2a}\left(d\bar{\epsilon}^{c(1)}_{12} - d\bar{\epsilon}^{c(0)}_{12}\right),$$

$$d\epsilon^{*}_{13} + d\epsilon^{**}_{13} = \frac{2(1+a)}{1+2a}\left(d\bar{\epsilon}^{c(1)}_{13} - d\bar{\epsilon}^{c(0)}_{13}\right),$$

$$d\epsilon^{*}_{23} + d\epsilon^{**}_{23} = \frac{1+a}{(3-4\nu_0)/[4(1-\nu_0)]+a}\left(d\bar{\epsilon}^{c(1)}_{23} - d\bar{\epsilon}^{c(0)}_{23}\right). \tag{5.11}$$

The mean-stress variations then follow from (5.9) and (5.10) for both phases.

To evaluate the corresponding local stress variation in the matrix with an equivalent transformation strain $d\epsilon^{*}+d\epsilon^{**}$ in the central fiber, the formula given in Section 4 can be used again, but now with the following substitutions

$$\epsilon^{*} \rightarrow d\epsilon^{*}+d\epsilon^{**}, \quad \epsilon^{(0)}(x) \rightarrow d\epsilon^{(0)}(x), \quad \langle\epsilon^{(0)}(x)\rangle \rightarrow \langle d\epsilon^{(0)}(x)\rangle . \tag{5.12}$$

In addition, the heterogeneous creep behavior in the ductile matrix itself will also result in a further stress redistribution. This is reminiscent of the heterogeneous creep of a polycrystalline aggregate among its constitient grains where each grain is taken to be elastically isotropic but due to the different crystallographic orientations each will experience its own creep activity (see Brown, 1970 and Weng, 1981). Within the assumed framework of elastic constraint the associated stress redistribution is most conveniently described by Kroner's (1961) original self-consistent relation

$$d\sigma(x) = -2\mu_0(1-\beta_0)(d\epsilon^{c(0)}(x) - <d\epsilon^{c(0)}(x)>)\,, \tag{5.13}$$

where $\beta_0 = (2/15)(4-5\nu_0)/(1-\nu_0)$.

Thus summing up all the major contributions, the local stress variation at a field point x can be expressed as

$$d\sigma^{(0)}(x) = d\bar{\sigma}^{(0)} + L_0(d\epsilon^{(0)}(x) - <d\epsilon^{(0)}(x)>)$$

$$- 2\mu_0(1-\beta_0)(d\epsilon^{c(0)}(x)-<d\epsilon^{c(0)}(x)>)\,, \tag{5.14}$$

where once again, the first one is given by (5.9), the second by (5.12) in conjunction with Section 4, and the last by (5.13).

Now that the local-field theory has been established, we recall that, in retrospect, if the simpler mean-field theory of Zhu and Weng (1987,89) were to be used, only the first term in (4.11) for the initial elastic state and also the first one in (5.14) for the subsequent creep would have been all needed in the calculation.

6. Numerical Results, Comparison with Experiments, and Assessment of the Elastic Constraint

The local-field theory thus developed is now applied to calculate the creep behavior of a practical system, for which we choose the Borsic-fiber/aluminum-matrix composite at room temperature. The elastic moduli of both phases are (Ericksen, 1973 and Allred et al., 1974)

$$\begin{aligned} &\text{Borsic filaments:} \quad E_1 = 392.7\ \text{GPa}, \quad \nu_1 = 0.15 \\ &\text{Aluminum matrix:} \quad E_0 = 68.9\ \text{GPA}, \quad \nu_0 = 0.33\,. \end{aligned} \tag{6.1}$$

The creep properties of both phases have been derived by simulation of the experimental data of Erickson (1973), as

$$\begin{aligned} &\text{Borsic filaments:} \quad a_1 = 1.81\text{x}10^{-25}, \quad n_1 = 6.1, \quad b_1 = 25 \\ &\qquad \text{and } d_1 = 3.7\text{x}10^{-23} \\ &\text{Aluminum matrix:} \quad a_0 = 8\text{x}10^{-13}, \quad n_0 = 4, \quad b_0 = 8, \\ &\qquad \text{and } d_0 = 7.7\text{x}10^{-11}\,, \end{aligned} \tag{6.2}$$

where stress, creep strain and time are in the units of MPa, m/m, and hour, in turn. The original test data and the simulated creep curves with these constants for the aluminum matrix and the Borsic fibers are shown in Figs. 1 and 2, respectively. The creep strain of the Borsic filament is seen to be about one order of magnitude smaller than that of the alumimum matrix, despite its much higher applied stress (about 15 times higher).

Since the composite is transversely isotropic, five types of creep curves have been calculated, all under the same external von Mises effective stress $\bar{\sigma}^* = 100$ MPa and each at three fiber -concentration levels $c_1 = 5\%$, 10% and 20%. To place the mean -field approximation in light of the more accurate local-field one, both types of results are plotted as dashed and solid lines, respectively, in Figs. 3 to 7. Since the stress exponent n is greater than one, the mean-field approach generally provides a lower estimate for the overall creep strain. Under the axial tension, however, the discrepancy between the two, as shown in Fig. 3, is small and therefore it should serve as a good approximation. This is attributed to the fact that the stress field in the matrix is rather uniform under this loading condition. Such is not the case under a transverse tension, whose results are depicted in Fig. 4. The percentage error by the mean-field theory, as measured by $(\bar{\epsilon}^c_{22(\mathrm{local})} - \bar{\epsilon}^c_{22(\mathrm{mean})})/\bar{\epsilon}^c_{22(\mathrm{local})}$, tends to increase with increasing fiber concentration (from about 10% at $c_1 = 0.05$ to 20% at $c_1 = 0.2$ after 12 hours of creep). Comparing Fig. 4 to Fig. 3, the transverse creep is seen to be about one order of magnitude higher than the axial one.

Both the axial shear and the transverse shear creep curves, shown in Fig. 5 and 6, respectively, suggest the necessity -- especially at a higher c_1 -- to use the local-field theory. The shear creep strain of the composite is slightly larger in the transverse direction, but the difference between the two is not as pronounced as in the corresponding tensile cases displayed in Figs. 3 and 4.

The fifth calculation is for the plane-strain biaxial tension, under the boundary condition $\bar{\sigma}_{22} = \bar{\sigma}_{33} = \bar{\sigma}$ and $\bar{\epsilon}_{11}(= \bar{\epsilon}^e_{11} + \bar{\epsilon}^c_{11}) = 0$. The initial $\bar{\sigma}$ has been chosen to provide $\bar{\sigma}^* = 100$ MPa, which, under the plane-strain condition, requires $\bar{\sigma}(0) = \bar{\sigma}^*/(1-2\nu_{12})$ initially. It must be noted that the subsequent creep will generate $d\bar{\epsilon}^c_{11}$ and, to maintain the plane-strain condition, $d\bar{\epsilon}^e_{11} = -d\bar{\epsilon}^c_{11}$ must be applied for each time increment. Thus $d\bar{\epsilon}^c_{11}$ is accompanied by a stress relaxation $d\bar{\sigma}_{11} = -E_{11}\, d\bar{\epsilon}^c_{11}$, which results in an additional $d\bar{\sigma}^{(0)}$ and $d\bar{\sigma}^{(1)}$, calculated in the elastic context of (3.11) and (3.12).

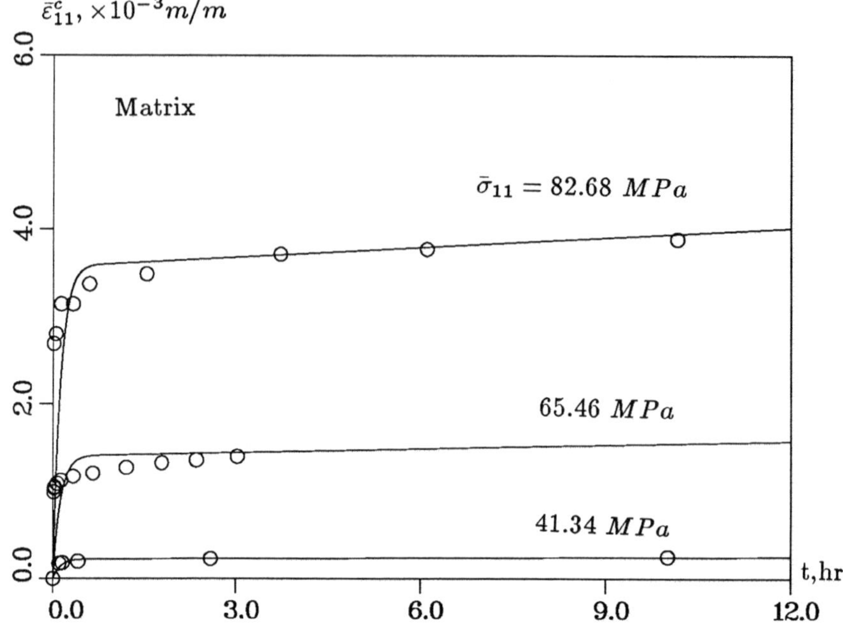

Fig. 1. Creep strains of the Al 1100 matrix at room temperature

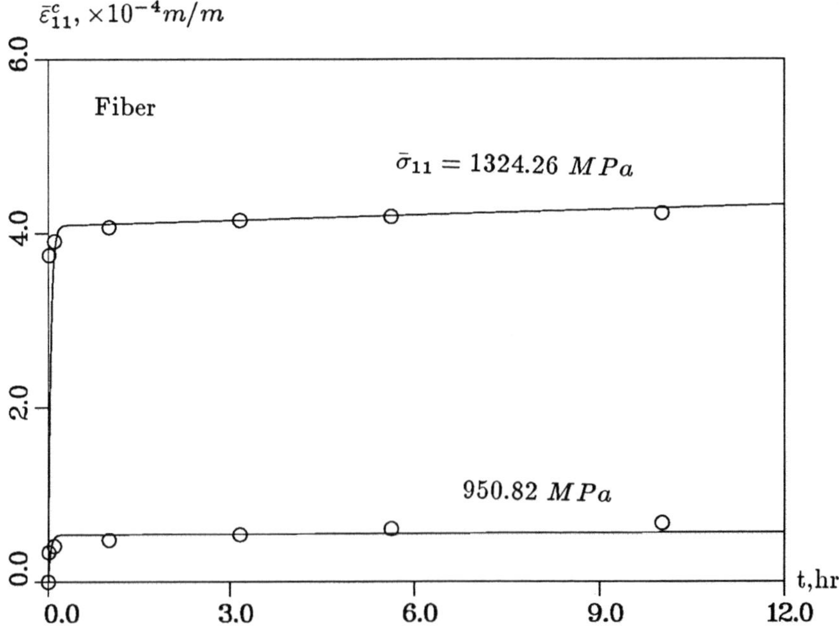

Fig. 2. Creep strains of the Borsic fibers at room temperature

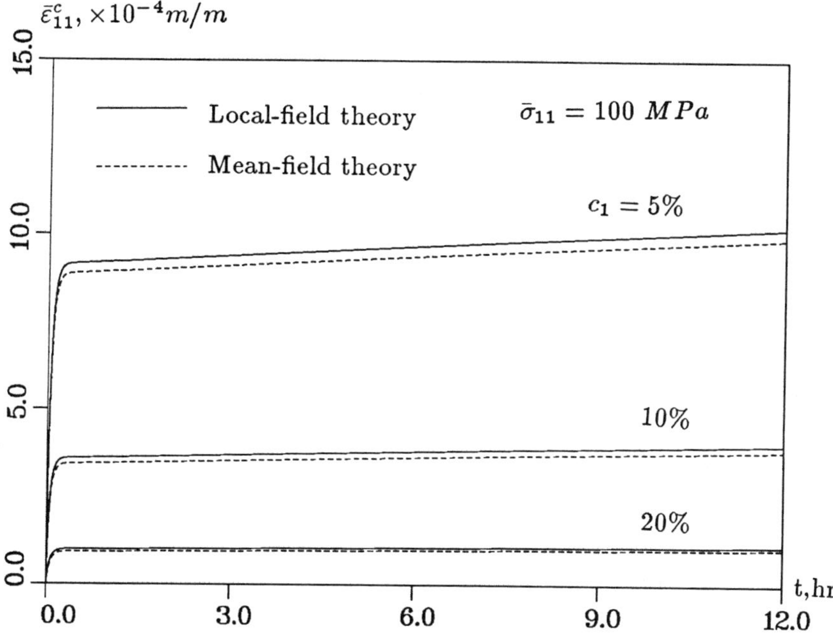

Fig. 3. Axial tensile creep strains of the composite

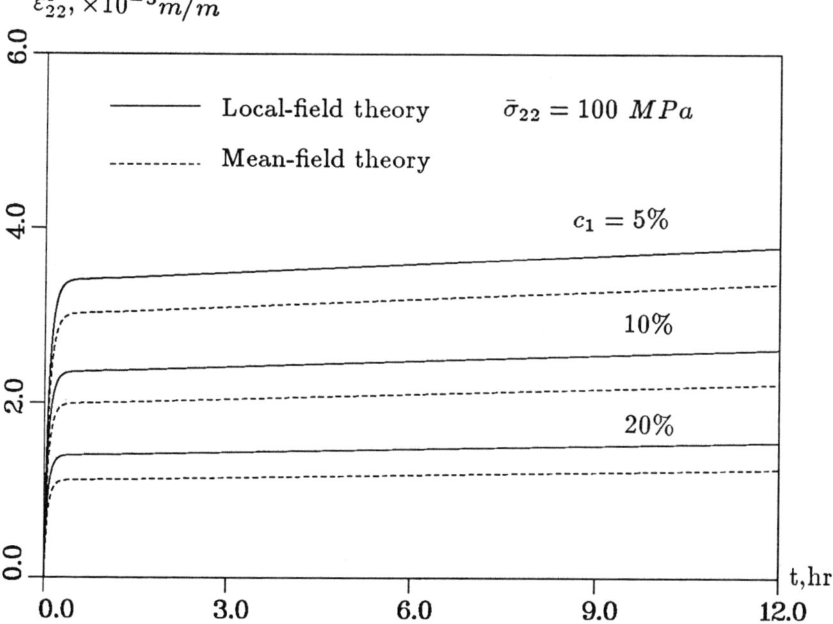

Fig. 4. Transverse tensile creep strains of the composite

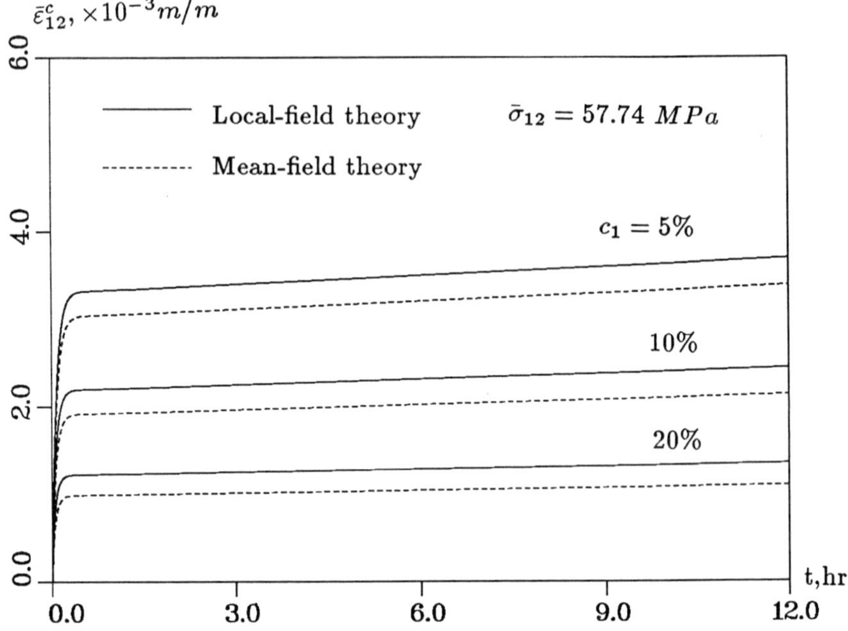

Fig. 5. Axial shear creep strains of the composite

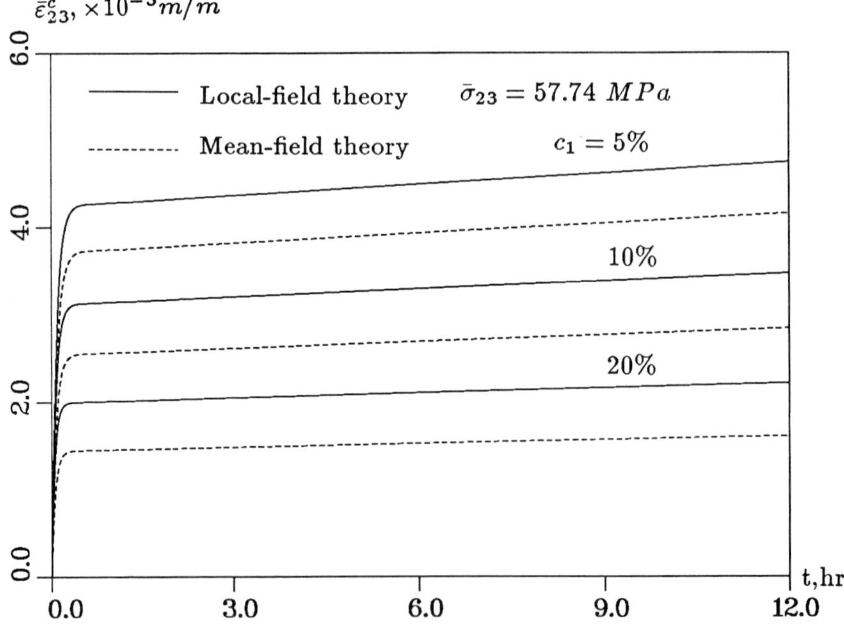

Fig. 6. Transverse shear creep strains of the composite

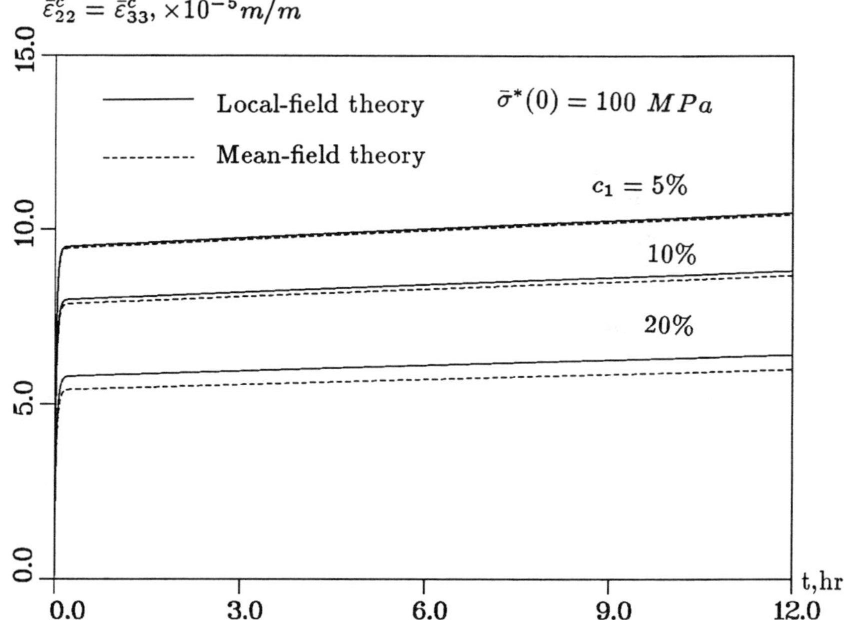

Fig. 7. Plane-strain, biaxial tensile strains of the composite

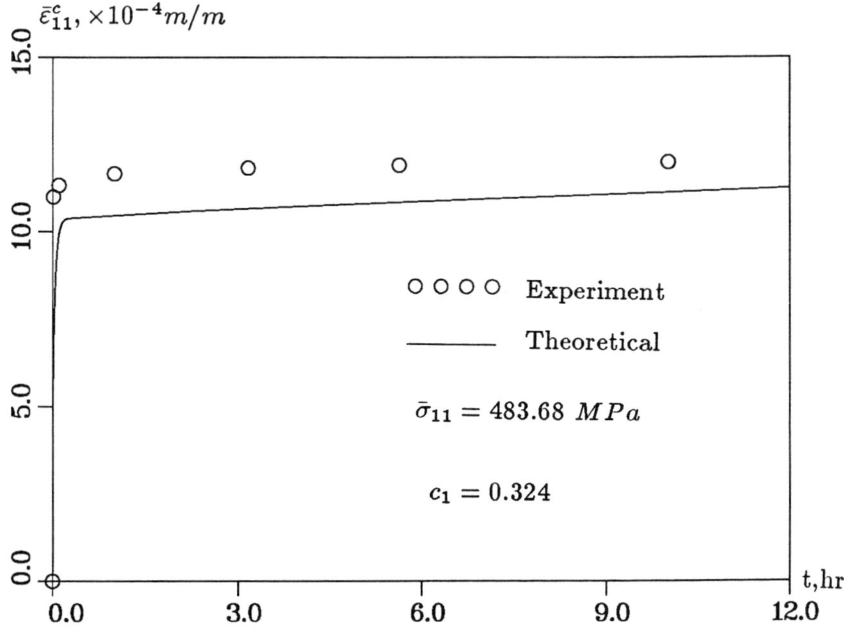

Fig. 8. Comparison between the theoretical prediction and experimental data for axial creep of the composite

The strain associated by the application of this $d\bar{\sigma}_{11}$ is time-dependent and should be included in the overall creep strain. The corresponding biaxial creep curves of the composite are shown in Fig. 7; they are one order of magnitude lower than the axial tensile creep and two orders less than the transverse one. In this case the mean-field approach is seen to be a good approximation to the local one again.

It is now desirable to see where the theoretical prediction stands as compared to the experimental data. Although the range of fiber concentration is now improved over the mean-field one, the application of the local theory is perhaps still safer in the lower range of concentration. The lowest fiber concentration where we have been able to locate the experimental data was c_1 = 32.4%, again from Ericksen (1973), under an axial stress $\bar{\sigma}_{11}$ = 483.68 MPa for the same Borsic/aluminum composite. The experimental data are shown in Fig. 8 as open circles and the theoretical prediction as a solid line. While the prediction appears to lie within a tolerable range of accuracy, its somewhat lower estimate is evident, and is believed to be caused by the assumption of elastic constraint of the elastic-creeping matrix.

Finally, in order to examine the accuracy of the assumption of elastic constraint of an elastic-creeping matrix, we note that, although more rigorous theoretical models involving an elastic-nonlinear creep matrix do not seem to exist at present, those involving a linear viscoelastic matrix are available and can be called for. In this context, Hashin (1965,66) has developed a correspondence principle between an elastic and a viscoelastic composite. For the fiber-reinforced solid an explicit formula with a Maxwell matrix was also provided for the overall axial shear creep (Hashin, 1966). To compare these two estimates numerically we used the same elastic moduli given in (6.1), neglected the creep of the Borsic filament and the primary creep of the aluminum matrix, and set a_0 = 2.24×10^{-7}, n_0 = 1 for the steady creep of aluminum. (This a_0 was chosen so that it would provide the same steady creep rate for aluminum at σ = 65.46 MPa -- the middle curve in Fig. 1 -- with this new n_0 value). The results for both theories (now the mean-field theory and the local one coincide) are shown in Fig. 9 for the total strain ($\bar{\epsilon}^e_{12} + \bar{\epsilon}^c_{12}$), where Hashin's results are plotted as solid lines and the present ones as dashed lines, with both predicting an identical initial response for the composite. The present theory generally provides a lower creep strain for the composite; this is an indication that the assumption of elastic constraint for an elastic-creeping matrix is generally

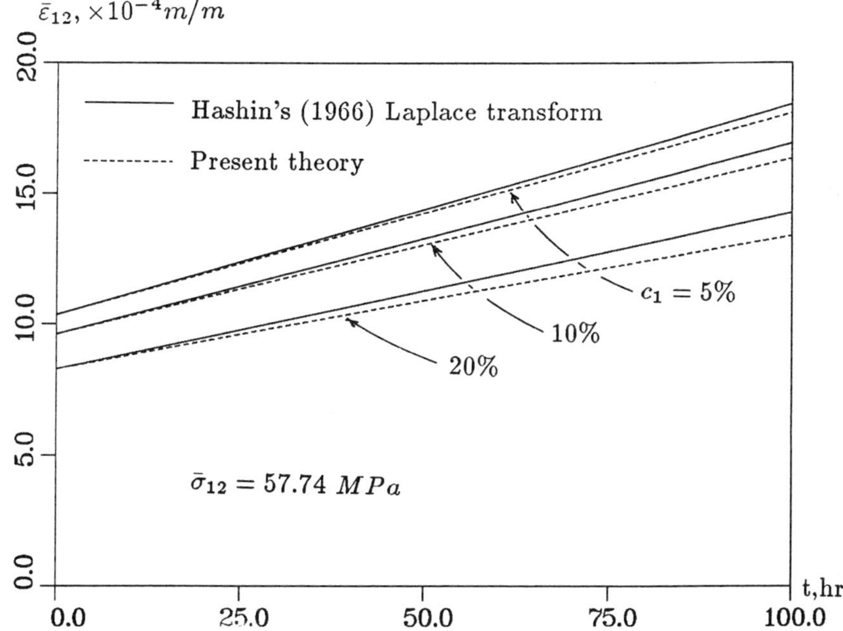

Fig. 9. Comparison between the present elastic-constraint approximation and Hashin's Laplace-transform approach

stronger than what it actually imposes on the inclusion. However, when the creep strain is small -- say in the order of the elastic strain as shown here -- the constraint power is predominantly elastic (in fact it is exactly elastic at t=0) and it is then a reasonable approximation. This is even more so when the fiber (or inclusion) concentration is low, where the creep strain can be one order of magnitude higher. Our calculations for the particle -reinforced composite also show a similar trend and, indeed, even with a three-parameter standard solid, the nature of the comparison remains essentially unaltered. Another important factor is the elastic modular ratio of the constituents; when the modular ratio is not high (say 1 to 6 as in most metal-matrix composite), the range of creep strain can go as indicated. With an excessively higher modular ratio (say rigid inclusions or voids) the range of permissible creep strain will become slightly lower.

In this light, and in view of the fact that the present theory is capable of accounting for the nonlinear stress-dependence of the creep rate and both the primary and the secondary creep of the constituents, the local theory thus developed is believed to be a valuable approximation for the estimate of the time-dependent overall creep strain of a fiber-reinforced metal-matrix composite.

Acknowledgement

This work was supported by the U.S. National Science Foundation, Solid and Geo-Mechanics Program, under Grants MSM-8614151 and MSS-8918235.

References

Allred, R.E., Hoover, W.R., and Horak, J.A., 1974, "Elastic-Plastic Poisson's Ratio of Borsic-Aluminum," *Journal of Composite Materials*, Vol. 8, pp. 15-28.

Brown, G.M., 1970, "A Self-Consistent Polycrystalline Model for Creep Under Combined Stress State," *Journal of the Mechanics and Physics of Solids*, Vol. 18, pp. 367-381.

Budiansky, B. and Wu, T.T., 1962. "Theoretical Prediction of Plastic Strains of Polycrystals," *Proceedings of the 4th U.S. National Congress of Applied Mechanics*, pp. 1175-1185.

Christensen, R.M. and Lo, K.H., 1979, "Solutions for Effective Shear Properties in Three Phase Sphere and Cylinder Models," *Journal of the Mechanics and Physics of Solids*, Vol. 27, pp. 315-330.

Eshelby, J.D., 1957, "The Determination of the Elastic Field of an Ellipsoidal Inclusion, and Related Problems," *Proceedings of the Royal Society*, London, Vol. A241, pp. 376-396.

Ericksen, R.H., 1973, "Room Temperature Creep of Borsic - Aluminum Composites," *Metallurgical Transactions*, Vol. A4, pp. 1687-1693.

Hashin, Z., 1965, "On Elastic Behavior of Fiber Reinforced Materials of Arbitrary Transverse Phase Geometry," *Journal of the Mechanics and Physics of Solids*, Vol. 13, pp. 119-134.

Hashin, Z., 1965, "Viscoelastic Behavior of the Heterogeneous Media," *Journal of Applied Mechanics*, Vol. 32, pp. 630-636.

Hashin, Z., 1966, "Viscoelastic Fiber Reinforced Materials," *AIAA Journal*, Vol. 4, pp. 1411-1417.

Hill, R., 1964, "Theory of Mechanical Properties of Fiber - Strengthened Materials: I. Elastic Behavior," *Journal of the Mechanics and Physics of Solids*, Vol. 12, pp. 199-212.

Kroner, E., 1958, *Kontinuumstheorie der Versetzungen und Eigenspannungen*, Springer, Berlin.

Kroner, E., 1961, "Zur Plastischen Verformung des Vielkristalls," *Acta Metallurgica*, Vol. 9, pp. 155-161.

Luo, H.A. and Weng, G.J., 1987, "On Eshelby's Inclusion Problem in a Three-Phase Spherically Concentric Solid, and a

Modification of Mori-Tanaka's Method," *Mechanics of Materials*, Vol. 6, pp. 347-361.

Luo, H.A. and Weng, G.J., 1989, "On Eshelby's S-Tensor in a Three-Phase Cylindrically Concentric Solid, and the Elastic Moduli of Fiber-Reinforced Composites," *Mechanics of Materials*, Vol. 8, pp. 77-88.

Mori, T. and Tanaka, K., 1973, "Average Stress in the Matrix and Average Elastic Energy of Materials with Misfitting Inclusions," *Acta Metallurgica*, Vol. 21, pp. 571-574.

Mura, T., 1987, *Micromechanics of Defects in Solids*, Martinus Nijhoff, Dordrecht, The Netherlands.

Weng, G.J., 1981, "Self-Consistent Determination of Time - Dependent Behavior of Metals," *Journal of Applied Mechanics*, Vol. 48, pp. 41-46.

Weng, G.J., 1984, "Some Elastic Properties of Reinforced Solids, with Special Reference to Isotropic ones Containing Spherical Inclusions," *International Journal of Engineering Science*, Vol. 22, pp. 845-856.

Zhao, Y.H., Tandon, G.P. and Weng, G.J., 1989, "Elastic Moduli for a Class of Porous Materials," *Acta Mechanica*, Vol. 76, pp. 105-130.

Zhu, Z.G. and Weng, G.J., 1987, "Micromechanics of Time - Dependent Deformation in a Dispersion-Hardened Polycrystal," *Acta Mechanica* (The Aris Phillips Memorial Volume), Vol. 69, pp. 295-313.

Zhu, Z.G. and Weng, G.J., 1989, "Creep Deformation of Particle -Strengthened Metal-Matrix Composites," *ASME Journal of Engineering Materials and Technology*, Vol. 111, pp. 99-105.

Zhu, Z.G. and Weng, G.J., 1990, "A Local Theory for the Calculation of Overall Creep Strain of Particle-Reinforced Composites," *International Journal of Plasticity*, (in press).

The Overall Behaviour of a Nonlinear Fibre Reinforced Composite

D. R. S. Talbot[1] and J. R. Willis[2]

[1]Department of Mathematics,
Coventry Polytechnic,
Coventry, CV1 5FB.

[2]School of Mathematical Sciences,
University of Bath,
Bath, BA1 7AY.

Abstract

A nonlinear fibre reinforced composite is considered whose local behaviour is described by a complementary energy function. The composite is regarded as nonlinearly elastic but subject to geometrically linear strains and attention is restricted to fields which are independent of the coordinate in the direction of the generators of the fibres. The overall complementary energy is defined and a generalization of the Hashin-Shtrikman variational structure is applied to the resulting problem. The results take the form of bounds for the complementary energy and results are presented for a random distribution of linearly elastic fibres embedded in a nonlinear matrix.

1. Introduction

This work is concerned with describing the overall behaviour of a composite comprising a matrix reinforced by a random distribution of long aligned fibres firmly bonded to the matrix. If the composite were made of linearly elastic constituents the overall behaviour would also be linearly elastic, characterized by an overall tensor of moduli, and a number of methods exist for estimating and bounding these moduli. Here, the constituents are taken to be nonlinear so that the functional form of the overall response is not known a priori. The approach adopted is to apply a generalization of the Hashin-Shtrikman variational structure due to Willis (1983), Talbot and Willis (1985) and Willis (1986).

The local behaviour of each constituent phase is described by a complementary energy function and the composite is subjected only to infinitesimal deformations. The approach generates bounds for the "overall complementary energy function", even though its functional form is unknown, which make allowance for the two-point statistics of the medium. However, in contrast to the linear problem it has not so far proved possible to obtain both upper and lower bounds. In the present work a lower bound only is derived.

The plan of the remainder of this work is as follows. First, general definitions of overall properties for any composite are given and then specialized to a fibre reinforced material. A novel feature of the present work is the introduction of an auxiliary energy function which describes a plane problem for a cross section perpendicular to the fibre direction and allows for a strain in the fibre direction. The nonlinear Hashin-Shtrikman principle is then applied. In the particular case of a two phase composite some explicit nonlinear equations are derived and solved when linearly elastic incompressible fibres are embedded in an incompressible nonlinear matrix. Finally, some specimen results are presented.

2. Energy Relations

Although a fibre reinforced composite will be considered in the sequel, the general definitions given initially apply to any inhomogeneous medium. A medium is considered which occupies a domain Ω, with boundary $\partial\Omega$. The medium is taken to have constitutive equation

$$\sigma_{ij} = \frac{\partial W}{\partial e_{ij}} (e_{kl} \; ; x) \quad , \quad x \in \Omega$$

or symbolically,

$$\sigma = W'(e; x) \; , \tag{2.1}$$

where σ is Cauchy stress and W is a convex function of the infinitesimal strain e. The overall response of the medium is defined, following Hill (1963, 1972), by reference to a boundary value problem for which the displacement u is prescribed as

$$u_i = \bar{e}_{ij} x_j \tag{2.2}$$

over $\partial\Omega$. The tensor $\bar{e}$ is constant and symmetric and it follows that the mean strain in the medium is exactly $\bar{e}$. The minimum energy principle gives

$$\widetilde{W}(\bar{e}) = \inf_{e'} \frac{1}{|\Omega|} \int_\Omega W(\bar{e} + e'; x) \, dx \tag{2.3}$$

for the mean strain energy density where the infimum is taken over strain fields e' derived from displacement fields u' which vanish on $\partial\Omega$.

For the work to follow it is more convenient to use an alternative characterization of $\widetilde{W}(\bar{e})$ given by Willis (1989). If $W^*(\sigma; x)$ denotes the complementary energy density at x, then since W is convex

$$W^*(\sigma;\, x) = \sup_e \{\sigma \cdot e - W(e;\, x)\} \ . \tag{2.4}$$

It now follows from (2.3) that

$$\tilde{W}(\bar{e}) = \inf_{e'} \frac{1}{|\Omega|} \int_\Omega \sup_\sigma \{\sigma \cdot (\bar{e} + e') - W^*(\sigma;\, x)\}\, dx \tag{2.5}$$

and, interchanging the operations of supremum and infimum then evaluating the infimum, it follows that

$$\tilde{W}(\bar{e}) \geq \sup \frac{1}{|\Omega|} \int_\Omega \{\sigma \cdot \bar{e} - W^*(\sigma;\, x)\}\, dx \ . \tag{2.6}$$

The supremum in (2.6) is taken over fields for which div $\sigma = 0$, since the infimum over e' is $-\infty$ otherwise. This result may be written

$$\tilde{W}(\bar{e}) \geq \sup_{\bar{\sigma}} \{\bar{\sigma} \cdot \bar{e} - \tilde{W}_*(\bar{\sigma})\} \ , \tag{2.7}$$

where

$$\tilde{W}_*(\bar{\sigma}) = \inf \frac{1}{|\Omega|} \int_\Omega W^*(\sigma;\, x)\, dx \ , \tag{2.8}$$

and the infimum is taken over fields σ for which div $\sigma = 0$ and which have mean value $\bar{\sigma}$. Equality is usually attained in (2.7) and in particular will be when $W(e;\, x)$ is a smooth function of e; then $\tilde{W}_* = \tilde{W}^*$ and it follows that

$$\bar{\sigma} = \tilde{W}'(\bar{e}) \ , \quad \bar{e} = \tilde{W}^{*\prime}(\bar{\sigma}) \ .$$

A complete discussion is given by Toland and Willis (1989) in the context of nonlinear electrostatics. It will be assumed in the sequel that $\tilde{W}^* = \tilde{W}_*$ although (2.7) provides a lower bound even when equality is not attained.

The remainder of this work is devoted to obtaining bounds for $\tilde{W}^*(\bar{\sigma})$ for a fibre reinforced composite. The medium envisaged comprises a cylinder with generators parallel to the x_3 axis containing a distribution of circular fibres whose axes are also parallel to the x_3 axis. The cross section of the cylinder perpendicular to the x_3 direction will be denoted by A with boundary ∂A. Units of length are chosen so that A has unit area.

Attention is restricted to fields which are functions of x_1 and x_2 only. For a cylinder of finite length l in the x_3 direction

$$\widetilde{W}^*(\overline{\sigma}) \leq \inf \frac{1}{|\Omega|} \int_{\Omega} W^*(\sigma; x)\, dx,$$

where the infimum is now taken over fields $\sigma(x_1, x_2)$. However, if l is very much greater than a typical dimension, a say, in either the x_1 or x_2 direction, then σ will be a function of x_1 and x_2 only except close to the ends and in the limit $l \to \infty$ the inequality tends to equality. Hence defining

$$\widetilde{W}_2^*(\overline{\sigma}) = \inf \int_A W^*(\sigma; x)\, dx, \tag{2.9}$$

where the infimum is taken over such fields, it is reasonable to seek bounds on $\widetilde{W}_2^*(\overline{\sigma})$.

To complete the formulation, first note that σ_{33} is unrestricted by the requirement div $\sigma = 0$. In order to evaluate the infimum over σ_{33} let

$$s_{\alpha j} = s_{j\alpha} = \sigma_{\alpha j} \quad , \tag{2.10}$$

where greek suffixes take the values 1 and 2 and define

$$W^{\times}(s, e_{33}) = \inf_{\sigma_{33}} (W^*(\sigma) - \sigma_{33} e_{33}) \; . \tag{2.11}$$

Here, and in the sequel, $W^{\times}$ and W^* depend also on x but for brevity this dependence is not shown explicitly. Also let

$$\widetilde{W}^{\times}(\overline{s}, \overline{e}_{33}) = \inf \int_A (W^*(\sigma) - \sigma_{33}\overline{e}_{33})\, dx, \tag{2.12}$$

where the infimum is taken over fields σ such that div $\sigma = 0$ and $\overline{s}_{\alpha j} = \overline{s}_{j\alpha} = \overline{\sigma}_{\alpha j}$ is prescribed but $\overline{\sigma}_{33}$ is unrestricted. Then

$$\widetilde{W}^{\times}(\overline{s}, \overline{e}_{33}) = \inf_{\overline{\sigma}_{33}} \{\inf \int_A (W^*(\sigma) - \overline{\sigma}_{33}\overline{e}_{33})\, dx\},$$

where $\overline{\sigma}_{33}$ is prescribed for the inner infimum. Hence

$$\widetilde{W}^{\times}(\overline{s}, \overline{e}_{33}) = \inf_{\overline{\sigma}_{33}} (\widetilde{W}_2^*(\overline{\sigma}) - \overline{\sigma}_{33}\overline{e}_{33})$$

and it follows that

$$\widetilde{W}_2^*(\overline{\sigma}) = \sup_{\overline{e}_{33}} (\overline{\sigma}_{33}\overline{e}_{33} + \widetilde{W}^{\times}(\overline{s}, \overline{e}_{33})) \; . \tag{2.13}$$

Finally, note that σ_{33} depends on x_1 and x_2 only and so is unconstrained by the requirement div $\sigma = 0$. Thus,

$$\tilde{W}^{\times}(\bar{s}, \bar{e}_{33}) = \inf_{s} \int_A W^{\times}(s, \bar{e}_{33})\, dx \ . \tag{2.14}$$

The problem of finding bounds for $\tilde{W}_2^*$ is now reduced to finding bounds for $\tilde{W}^{\times}$ which is an easier problem than working with (2.9) directly. It defines a two-dimensional problem for stresses (other than σ_{33}) in a fibre-reinforced material subjected to the prescribed axial strain $\bar{e}_{33}$.

3. Hashin-Shtrikman structure

The results of this section are derived from a generalization of the Hashin-Shtrikman variational principles to nonlinear problems developed by Willis (1983, 1986), Talbot and Willis (1985) and Willis (1990). The idea is to introduce a convex comparison function, W_0, with dual function W_0^*. Then for any symmetric second order tensor η

$$(W^* - W_0^*)^*(\eta;\, x) = \sup_{\sigma} \{\sigma{\cdot}\eta - W^*(\sigma;\, x) + W_0^*(\sigma;\, x)\}. \tag{3.1}$$

It is convenient to define

$$V(\eta;\, x) = (W^* - W_0^*)^*(\eta;\, x) \ . \tag{3.2}$$

It now follows that

$$W^*(\sigma;\, x) \geq \sigma{\cdot}\eta + W_0^*(\sigma;\, x) - V(\eta;\, x) \tag{3.3}$$

for any σ, η and hence from (2.12)

$$\tilde{W}^{\times}(\bar{s}, \bar{e}_{33}) \geq \inf_{s,\sigma_{33}} \int_A (\sigma{\cdot}\eta - \sigma_{33}\bar{e}_{33} + W_0^*(\sigma;x) - V(\eta;x))\, dx. \tag{3.4}$$

Next, evaluating the infimum over σ_{33} yields

$$\tilde{W}^{\times}(\bar{s}, \bar{e}_{33}) \geq \inf_{s} \int_A (s{\cdot}\eta + W_0^{\times}(s, \bar{e}_{33}-\eta_{33}) - V(\eta;x))\, dx, \tag{3.5}$$

where $W_0^{\times}$ is defined via (2.11). The inequality (3.5) provides, using (2.13), a lower bound for W_2^* for any choice of η. It is finite provided the right side of (3.1) is finite. This is the case so long as $W^*(\sigma; x) - W_0^*(\sigma; x)$ tends to infinity faster than $||\sigma||$ as $||\sigma|| \to \infty$.

To make further progress, W_0^* is taken to be quadratic in σ and independent of x. Symbolically

$$W_0^* = \tfrac{1}{2}\, \sigma \cdot M_0 \sigma, \tag{3.6}$$

where M_0 is a fourth order 'tensor of compliances '. Its inverse is the 'tensor of moduli' L_0. The restriction on the growth of $W^* - W_0^*$ limits the usefulness of the form (3.6) to functions W^* which grow at least quadratically with σ when $||\sigma||$ is large. However, this will be the case for the examples treated later.

In order to calculate $W_0^\times$ explicitly it is necessary to have knowledge of the tensor L_0. To be definite, L_0 is now taken to be isotropic, with bulk and shear moduli K_0, μ_0 respectively; in the notation of Hill (1965)

$$L_0 = (3K_0, 2\mu_0) \ . \tag{3.7}$$

It then follows that

$$M_0 = \left[\frac{1}{3K_0}, \frac{1}{2\mu_0}\right] \tag{3.8}$$

and a little algebra shows that

$$W_0^\times(s, e_{33}) = \frac{1}{2}\left\{(a - \frac{1}{2}b)\, s_{\alpha\alpha}^2 + 2b s_{\alpha\beta}\, s_{\alpha\beta} + 4b s_{\alpha 3}^2 - d\, s_{\alpha\alpha}\, e_{33} - c\, e_{33}^2\right\}, \tag{3.9}$$

where

$$a = \frac{1}{8\mu_0}\left[\frac{5\mu_0 - 3K_0}{3K_0 + \mu_0}\right], \quad b = \frac{1}{4\mu_0}, \quad c = \frac{9\mu_0 K_0}{3K_0 + \mu_0}, \quad d = \frac{3K_0 - 2\mu_0}{3K_0 + \mu_0}. \tag{3.10}$$

With the further definitions

$$W_0^{*(2)}(s) = \frac{1}{2}\left\{(a - \frac{1}{2}b)\, s_{\alpha\alpha}^2 + 2b s_{\alpha\beta}\, s_{\alpha\beta}\right\}, \tag{3.11}$$

$$\eta^*_{\alpha j} = \eta^*_{j\alpha} = \eta_{\alpha j} - \frac{1}{2}\,\delta_{\alpha j}\, d(\bar{e}_{33} - \eta_{33}) \tag{3.12}$$

it follows from (3.5) that

$$\tilde{W}^{\times}(\bar{s},\, \bar{e}_{33}) \geq \inf_{s} \int_A \left[s\cdot\eta^* + W_0^{*(2)}(s) + 2bs_{\alpha 3}^2 - V(\eta;\, x) - \frac{1}{2}\, c(\bar{e}_{33} - \eta_{33})^2 \right] dx. \tag{3.13}$$

The infimum is attained when

$$e_{\alpha\beta} := \frac{\partial W_0^{*(2)}(s)}{\partial s_{\alpha\beta}} + \eta^*_{\alpha\beta}$$

$$e_{\alpha 3} := 2bs_{\alpha 3} + \eta^*_{\alpha 3} \tag{3.14}$$

is compatible with a displacement field u satisfying a boundary condition of the form (2.2) on ∂A for some $\bar{e}$. Rewriting (3.14) as

$$e = M_0^* s + \eta^*$$

so that

$$s = L_0^*(e - \eta^*) \tag{3.15}$$

which define M_0^* and L_0^*, the problem now is to find u satisfying (2.2) on ∂A such that s and e are related through (3.15) and $\operatorname{div} s = 0$. The value of $\bar{e}$ in (2.2) is then chosen so that s has mean value $\bar{s}$. Thus it is necessary to solve

$$\operatorname{div}\, [L_0^*(e - \eta^*)] = 0 \quad , \quad x \in A \quad , \tag{3.16}$$

subject to the boundary condition (2.2). This uncouples into equations for the plane and antiplane components. A formal solution is

$$e = \bar{e} - \Gamma(L_0^* \eta^*) \quad , \tag{3.17}$$

where Γ is a linear operator whose kernel contains two derivatives of the Green function for the plane or antiplane problem as appropriate. Calculating s through (3.15) and choosing $\bar{e}$ appropriately gives

$$s = \bar{s} - \Delta\eta^* \quad , \tag{3.18}$$

where the linear operator Δ is defined by

$$\Delta\eta^* = L_0^*(\eta^* - \bar{\eta}^*) - L_0^*\Gamma L_0^*\eta^* \quad . \tag{3.19}$$

An elementary property of Γ is that

$$\Gamma L_0^* \Gamma = \Gamma$$

and it follows that

$$\Delta M_0^* \Delta = \Delta \quad , \tag{3.20}$$

see, for example, Willis (1981). It follows now on substituting (3.18) in (3.13) and using (3.20) that

$$\tilde{W}^\times(\bar{s}, \bar{e}_{33}) \geq \int_A \{\bar{s}\cdot\eta^* - \frac{1}{2}\eta^*\cdot\Delta\eta^* - V(\eta; x) - \frac{1}{2}c(\bar{e}_{33} - \eta_{33})^2\}\, dx$$
$$+ W_0^{*(2)}(\bar{s}) + 2b\bar{s}_{\alpha 3}^2 \quad . \tag{3.21}$$

4. Application to composites

No immediate advantage is obtained by specializing to a fibre-reinforced composite. Instead the body is taken to be a mixture of n firmly bonded phases. The complementary energy function of the rth phase is written W_r^* so that

$$W^*(\sigma; x) = \sum_{r=1}^{n} W_r^*(\sigma)\, f_r(x) \quad , \tag{4.1}$$

where $f_r(x)$ has value 1 if x lies in phase r and has value zero otherwise. It follows from (3.2) that

$$V(\eta; x) = \sum_{r=1}^{n} V_r(\eta)\, f_r(x) \quad , \tag{4.2}$$

where

$$V_r(\eta) = (W_r^* - W_0^*)^*(\eta) \quad . \tag{4.3}$$

The right side of (3.21) provides a bound for any η; the procedure now is to choose

$$\eta(x) = \Sigma\, \eta_r\, f_r(x) \quad , \tag{4.4}$$

where the η_r are constants, substitute into (3.21) and (2.13) and then maximise the right side with respect to the variables η_r to produce a bound for $\widetilde{W}_2^*$. Substituting (4.4) into (3.21) gives

$$\widetilde{W}^{\times}(\overline{s}, \overline{e}_{33}) \geq \Sigma\, c_r \left[\overline{s}\cdot\eta_r^* - V_r(\eta_r) - \frac{1}{2} c(\overline{e}_{33} - (\eta_r)_{33})^2 \right]$$
$$- \frac{1}{2}\, \Sigma\Sigma\, \eta_r^*\cdot B_{rs}\eta_s^* + W_0^{*(2)}(\overline{s}) + 2b\overline{s}_{\alpha 3}^2 \; , \tag{4.5}$$

where c_r is the volume fraction of phase r, given by

$$c_r = \int_A f_r(x)\, dx \quad , \tag{4.6}$$

the tensors B_{rs} are given by

$$B_{rs} = \int_A f_r(\Delta f_s)\, dx \tag{4.7}$$

and η_r^* is found from η_r using (3.12).

For the particular case of a two phase composite ($n = 2$), $f_1 + f_2 = 1$ and

$$B_{rs} = c_r(\delta_{rs} - c_s)Q \text{ (no summation on } r) \quad , \tag{4.8}$$

where

$$c_1 c_2 Q = \int_A f_1(\Delta f_1)\, dx \quad . \tag{4.9}$$

Using this in (4.5) and then substituting into (2.13) now gives

$$\widetilde{W}_2^*(\overline{\sigma}) \geq \sup_{\overline{e}_{33}} \left\{ \overline{\sigma}_{33}\overline{e}_{33} + \Sigma\, c_r \left[\overline{s}\cdot\eta_r^* - V_r(\eta_r) - \frac{1}{2} c(\overline{e}_{33} - (\eta_r)_{33})^2 \right] \right.$$
$$\left. - \frac{1}{2}\, \Sigma\, c_r(\eta_r^* - \overline{\eta}^*)\cdot Q(\eta_r^* - \overline{\eta}^*) \right\} + W_0^{*(2)}(\overline{s}) + 2b\overline{s}_{\alpha 3}^2 \; , \tag{4.10}$$

where $\bar{\eta} = \Sigma\, c_r \eta_r$ and $\bar{s}_{\alpha j} = \bar{\sigma}_{\alpha j}$. The right hand side is maximized with respect to the variables η_r when they satisfy

$$\zeta_{\alpha j} + (Q(\eta^* - \bar{\eta}^*))_{\alpha j} = \bar{s}_{\alpha j} \quad , \tag{4.11}$$

$$\zeta_{33} - c(\bar{e}_{33} - \eta_{33}) + \frac{d}{2} \left[Q(\eta^* - \bar{\eta}^*) \right]_{\alpha\alpha} = \frac{1}{2}\, d\, \bar{s}_{\alpha\alpha} \quad , \tag{4.12}$$

where ζ is either ζ_1 or ζ_2 and ζ_r is a subgradient of V_r at η_r:

$$\zeta_r \in \partial V_r(\eta_r) \quad . \tag{4.13}$$

Finally the supremum over $\bar{e}_{33}$ is attained when

$$c(\bar{e}_{33} - \bar{\eta}_{33}) = \bar{\sigma}_{33} - \frac{1}{2}\, d\, \bar{s}_{\alpha\alpha} \tag{4.14}$$

and adding this to (4.12) gives

$$\zeta_{33} + c(\eta_{33} - \bar{\eta}_{33}) + \frac{d}{2} \left[Q(\eta^* - \bar{\eta}^*) \right]_{\alpha\alpha} = \bar{\sigma}_{33} \ . \tag{4.15}$$

5. Example: an array of incompressible fibres in an incompressible matrix

The remainder of this work is devoted to considering an array of incompressible fibres which exhibit linear behaviour embedded, with transversely isotropic symmetry, in an incompressible nonlinear matrix. Specifically the matrix (phase 1) is taken to be characterized by a complementary energy density

$$W_1^*(\sigma) = \int_0^{\sigma_e} F(t)\, dt \quad , \tag{5.1}$$

where the equivalent stress $\sigma_e = (3\sigma'_{ij}\sigma'_{ij}/2)^{\frac{1}{2}}$ is defined in terms of the deviatoric stress $\sigma'_{ij} = \sigma_{ij} - \delta_{ij}\,\sigma_m$ and the mean stress $\sigma_m = \sigma_{ii}/3$. The function $F(t)$ is taken to be convex. The fibres (phase 2) are taken to have energy density

$$W_2^*(\sigma) = \frac{1}{6\mu_2}\, \sigma_e^2 \quad . \tag{5.2}$$

The comparison medium is taken to have the corresponding form with modulus μ_0:

$$W_0^*(\sigma) = \frac{1}{6\mu_0}\,\sigma_e^2 \quad . \tag{5.3}$$

It follows by taking the incompressible limit of results given by Walpole (1969) that

$$\begin{aligned}(Q\eta^*)_{\alpha\beta} &= \mu_0\eta_{\alpha\beta}^* + \frac{1}{2}\,\mu_0\eta_{\gamma\gamma}^*\,\delta_{\alpha\beta} \ , \\ (Q\eta^*)_{\alpha 3} &= (Q\eta^*)_{3\alpha} = \mu_0\eta_{\alpha 3}^* \ .\end{aligned} \tag{5.4}$$

Also, in this limit, $a = -\,\mu_0/8$, $c = 3\mu_0$, $d = 1$ and the polarizations η satisfy $\eta_{kk} = 0$. With the further definition

$$\eta_{\alpha j}'' = \eta_{\alpha j} - \frac{1}{2}\,\delta_{\alpha j}\eta_{\gamma\gamma} \tag{5.5}$$

it follows that (4.11) and (4.15) can be written

$$\zeta_{\alpha j} + \mu_0(\eta_{\alpha j}'' - \bar{\eta}_{\alpha j}'') = \bar{\sigma}_{\alpha j} \ , \tag{5.6}$$

$$\zeta_{33} + 3\mu_0(\eta_{33} - \bar{\eta}_{33}) = \bar{\sigma}_{33} \ , \tag{5.7}$$

where $\bar{s}_{\alpha j}$ has been replaced by $\bar{\sigma}_{\alpha j}$. From (5.6) $\zeta_{\gamma\gamma} = \bar{\sigma}_{\gamma\gamma}$ so it follows that

$$\zeta_{\alpha j}'' + \mu_0(\eta_{\alpha j}'' - \bar{\eta}_{\alpha j}'') = \bar{\sigma}_{\alpha j}'' \ , \tag{5.8}$$

$$\zeta_{33}' + 2\mu_0(\eta_{33}' - \bar{\eta}_{33}') = \bar{\sigma}_{33}' \ , \tag{5.9}$$

where ζ' and η' are defined like σ' and ζ'' is defined like η''.

Now

$$\zeta_2' = \left[\frac{1}{2\mu_2} - \frac{1}{2\mu_0}\right]^{-1} \eta_2' \tag{5.10}$$

and, provided F is smooth enough and $W_1^* - W_0^*$ is convex,

$$\eta_1' = \frac{1}{\mu_0}\,\varphi(\zeta_{1e})\,\zeta_1' \ , \tag{5.11}$$

where

$$\varphi(\zeta) = \frac{\mu_0}{\zeta}\left[\frac{3}{2}\,F(\zeta) - \frac{\zeta}{2\mu_0}\right] . \tag{5.12}$$

If $W_1^* - W_0^*$ is not convex, (5.11) still holds while $\zeta_1\ \varphi/\mu_0$ is, in general, a subgradient of $(W_1^* - W_0^*)^{**}(\zeta)$. The solution to (5.8), (5.9) now follows by eliminating ζ_2, η_2 and then η_1 from the equations to give

$$(\zeta_1)_{\alpha j}'' = \frac{\lambda_2}{(\lambda_1 + c_2\varphi)}\ \bar{\sigma}_{\alpha j}'' \tag{5.13}$$

and

$$(\zeta_1)_{33}' = \frac{\lambda_4}{(\lambda_3 + 2c_2\varphi)}\ \bar{\sigma}_{33}' \ , \tag{5.14}$$

where the constants λ_1 , $\cdots$, λ_4 are

$$\begin{aligned}
\lambda_1 &= 1 + c_1\ \mu_0\left[\frac{1}{2\mu_2} - \frac{1}{2\mu_0}\right] ,\\
\lambda_2 &= 1 + \mu_0\left[\frac{1}{2\mu_2} - \frac{1}{2\mu_0}\right] ,\\
\lambda_3 &= 1 + 2\mu_0\ c_1\left[\frac{1}{2\mu_2} - \frac{1}{2\mu_0}\right] ,\\
\lambda_4 &= \mu_0/\mu_2 \ .
\end{aligned} \tag{5.15}$$

An equation for ζ_{1e} then follows as

$$\frac{2}{3}\,\zeta_{1e}^2 = \frac{\lambda_2^2}{(\lambda_1 + c_2\varphi)^2}\ \tau^2 + \frac{3}{2}\ \frac{\lambda_4^2}{(\lambda_3 + 2c_2\varphi)^2}\ \bar{\sigma}_{33}'^2 \ , \tag{5.16}$$

where

$$\tau^2 = \bar{\sigma}_{\alpha\beta}''\ \bar{\sigma}_{\alpha\beta}'' + 2\bar{\sigma}_{\alpha 3}\ \bar{\sigma}_{\alpha 3} \ . \tag{5.17}$$

The solution to (5.16) for ζ_{1e} determines $(\zeta_1)''_{\alpha j}$, $(\zeta_1)'_{33}$ from (5.13), (5.14), η'_1 from (5.11) and, noting that $\bar{\zeta} = \bar{\sigma}$ from (5.6), (5.7), ζ_2 can be found and η_2 follows from (5.10).

Finally, $\zeta_r \in \partial(W_r^* - W_0^*)^*(\eta_r)$ and hence $\eta_r \in \partial(W_r^* - W_0^*)^{**}(\zeta_r)$. It now follows that

$$V_r(\eta_r) = \zeta_r \cdot \eta_r - (W_r^* - W_0^*)^{**}(\zeta_r) \tag{5.18}$$

and some algebra shows that (4.10) becomes

$$\tilde{W}_2^*(\bar{\sigma}) \geq \Sigma\, c_r \left[\frac{1}{2}\bar{\sigma}_{ij}(\eta_r)_{ij} - \frac{1}{2}(\zeta_r)_{ij}(\eta_r)_{ij} + (W_r^* - W_0^*)^{**}(\zeta_r)\right]$$
$$+ W_0^{*(2)}(\bar{\sigma}) + \frac{1}{2\mu_0}\,\bar{\sigma}_{\alpha 3}\bar{\sigma}_{\alpha 3} + \frac{3}{8\mu_0}\,\bar{\sigma}_{33}'^2 \,. \tag{5.19}$$

In deriving (5.18), use has been made of (4.14), (5.8) and (5.9).

The right side of (5.19) provides a non-trivial bound for any $\mu_0 \geq \mu_2$ and the best bound is found by optimizing over μ_0 with this restriction. A procedure previously adopted by Ponte Castañeda and Willis (1988), Willis (1989) and Willis and Talbot (1989) would restrict μ_0 to values for which $W_r^* - W_0^* = (W_r^* - W_0^*)^{**}$ at the ζ_r of interest. This would certainly produce a bound but it need not be optimal. See Willis (1990) for further discussion.

6. Results

Before proceeding, it is convenient to write the bound in (5.19) as a function of ζ_{1e} only. Upon using the relations (5.10), (5.11), (5.13), (5.14) and eliminating ζ_2 using $\bar{\zeta} = \bar{\sigma}$ the right side of (5.19) becomes, after some algebra,

$$W_L^*(\bar{\sigma}) = c_1(W_1^* - W_0^*)^{**}(\zeta_{1e})$$
$$+ \frac{c_2}{2}\left[\frac{1}{2\mu_2} - \frac{1}{2\mu_0}\right]\left\{\left[\frac{1+\varphi}{\lambda_1 + c_2\varphi}\right]^2 \tau^2 + \frac{3}{2}\left[\frac{1+2\varphi}{\lambda_3 + 2c_2\varphi}\right]^2 \bar{\sigma}_{33}'^2\right\}$$
$$+ \frac{1}{\mu_0}\, c_1 c_2 \left[\mu_0\left[\frac{1}{2\mu_2} - \frac{1}{2\mu_0}\right] - \varphi\right]^2 \left\{\frac{1}{2}\,\frac{\tau^2}{(\lambda_1 + c_2\varphi)^2} + \frac{3}{2}\,\frac{\bar{\sigma}_{33}'^2}{(\lambda_3 + 2c_2\varphi)^2}\right\}$$

$$+ \frac{1}{4\mu_0} \tau^2 + \frac{3}{8\mu_0} \bar{\sigma}'^2_{33} \ . \tag{6.1}$$

The particular form of F for which results are presented is

$$F(t) = e_0 \left\{ \frac{t}{\sigma_0} + \left[\left(\frac{t}{\sigma_0} \right)^m - \left(\frac{\sigma_y}{\sigma_0} \right)^m \right] H(t - \sigma_y) \right\} . \tag{6.2}$$

Then, writing $3\mu_1 = \sigma_0 / e_0$, two cases need to be considered: firstly $\mu_1 < \mu_2$ and secondly $\mu_1 > \mu_2$. In the first case the fibres are stiffer than the matrix and $W_1^* - W_0^*$ and $W_2^* - W_0^*$ are convex for $\mu_0 \geq \mu_2$. The bound (6.1) is optimized by choosing $\mu_0 = \mu_2$, φ has the form (5.12), $W_1^* - W_0^* = (W_1^* - W_0^*)^{**}$ and the bound follows by solving (5.16) and substituting the result in (6.1).

In the second case, the matrix is stiffer than the fibres for low levels of applied stress and although $W_2^* - W_0^*$ is convex for $\mu_0 \geq \mu_2$, $W_1^* - W_0^*$ need not be. If σ_c denotes the value of σ_e for which $W_1^* - W_0^*$ is minimized, that is σ_c is the solution to

$$\frac{\sigma_c}{3\mu_0} = F(\sigma_c) \quad , \tag{6.3}$$

with the constraint $\sigma_c \geq \sigma_y$, it is easy to show that

$$(W_1^* - W_0^*)^{**}(\sigma) = \int_0^{\sigma_c} F(t)\, dt - \frac{\sigma_c^2}{6\mu_0} \quad , \quad \sigma_e \leq \sigma_c$$

$$= \int_0^{\sigma_e} F(t)\, dt - \frac{\sigma_e^2}{6\mu_0} \quad , \quad \sigma_e \geq \sigma_c \ , \tag{6.4}$$

and the function φ is now given by

$$\varphi(\zeta) = 0 \quad , \quad \zeta \leq \sigma_c$$

$$= \frac{\mu_0}{\zeta} \left[\frac{3}{2} F(\zeta) - \frac{\zeta}{2\mu_0} \right] , \quad \zeta \geq \sigma_c \ . \tag{6.5}$$

For low levels of applied stress it is convenient to treat σ_c and μ_0 as the primary variables; for then, $\zeta < \sigma_c$, $\varphi = 0$ and it can be shown that

$$W_L^*(\bar{\sigma}) = \int_0^{\sigma_c} F(t)dt - \frac{\sigma_c^2}{6\mu_0} + \frac{1}{4\mu_0}\left[\frac{2\lambda_2 - \lambda_1}{\lambda_1}\tau^2 + \frac{3}{2}\frac{\lambda_4}{\lambda_3}\bar{\sigma}_{33}'^2\right] . \tag{6.6}$$

The best bound is obtained when

$$\frac{\partial W_L^*}{\partial \mu_0} = 0 \quad , \tag{6.7}$$

with the constraint $\mu_0 \geq \mu_2$ and σ_c given by the solution to (6.3) with the constraint $\sigma_c \geq \sigma_y$. The strategy is to solve this problem for μ_0 and σ_c ignoring the constraint $\mu_0 \geq \mu_2$. If the value of μ_0 so obtained is greater than μ_2 the bound is given by (6.6). At a certain level of applied stress $\mu_0 = \mu_2$ is obtained corresponding to the least value of $\bar{\sigma}$ for which $W_1^* - W_0^* = (W_1^* - W_0^*)^{**}$ and the problem then reverts to solving (5.16) with φ given by (5.12) and $\mu_0 = \mu_2$, and substituting the result into (6.1).

A computer program has been written to evaluate (6.1) and sample results are presented for

$$\sigma_y/\sigma_0 = 0.2 \text{ and } c_2 = 0.3 \ .$$

Fig. 1 shows the lower bound $W_L^*(\bar{\sigma})/(e_0\sigma_0)$ plotted against τ/σ_0 for $\bar{\sigma}_{33}'/\sigma_0 = 0.1$ and $\mu_1/\mu_2 = 2$. The dotted curve corresponds to $m = 3$ while the solid curve corresponds to $m = 10$. The bound W_L^* is a function of $\bar{\sigma}$ through τ, $\bar{\sigma}_{33}'$ and is a lower bound for $\tilde{W}^*(\bar{\sigma})$. Although it does not follow that $\partial W_L^*/\partial\bar{\sigma}_{ij}$ is a bound for $\bar{e}_{ij}$, Fig. 2 nevertheless shows plots of τ/σ_0 against $\dfrac{\partial W_L^*}{\partial \tau}/e_0$; this provides *approximations* for the mean strain components $\bar{e}_{\alpha\beta}$, $\bar{e}_{\alpha 3}$. Some checks have been made and the slope of the straight portion of the curves at low values of τ/σ_0 agrees with the Hashin-Shtrikman upper bound for the shear modulus of a composite comprising linearly elastic fibres with modulus μ_2 embedded in a linearly elastic matrix with modulus μ_2. Fig. 3 shows plots of $W_L^*(\bar{\sigma})/(e_0\sigma_0)$ against τ/σ_0 for $\bar{\sigma}_{33}'/\sigma_0 = 0.1$ and $\mu_1/\mu_2 = 0.5$. No new features emerge and the comments above about the behaviour of the bound as $\tau/\sigma_0 \to 0$ still obtain.

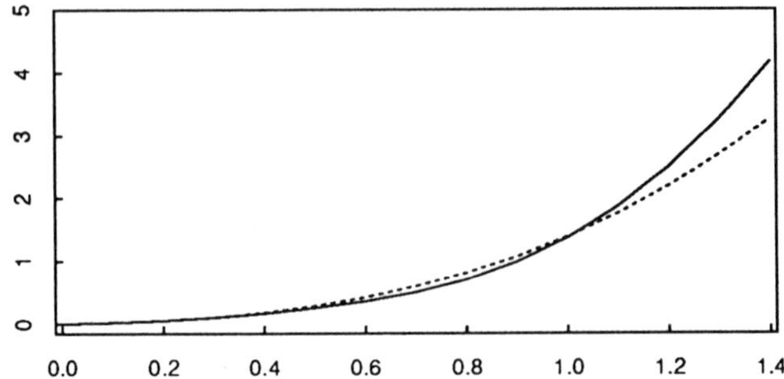

Fig.1 The lower bound $W_L^*(\bar{\sigma})/(e_0\sigma_0$ versus τ/σ_0, for $\bar{\sigma}'_{33}/\sigma_0 = 0.1$, $\mu_1/\mu_2 = 2$ and $m = 3$ (dotted curve), $m = 10$ (solid curve).

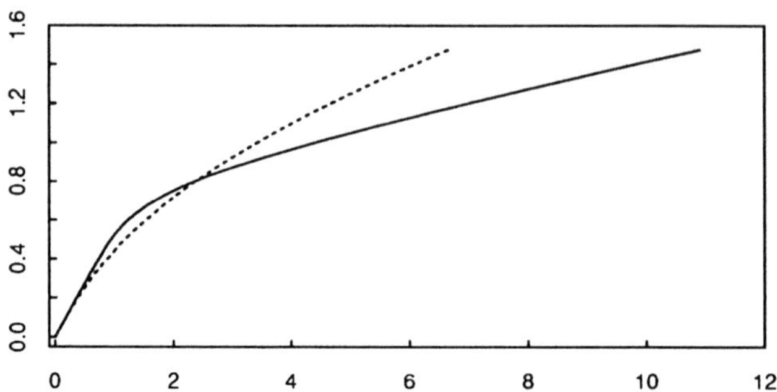

Fig.2 Plots of τ/σ_0 against $(\partial W_L^*/\partial\tau)/e_0$, for the same parameter values as in Fig. 1.

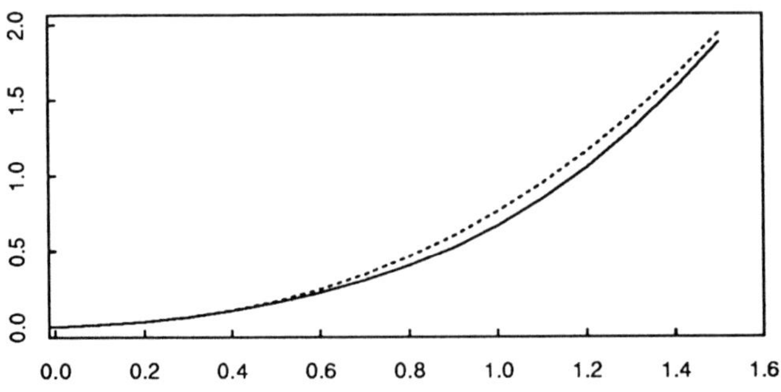

Fig.3 The lower bound $W_L^*(\bar{\sigma})/(e_0\sigma_0)$ versus τ/σ_0, for $\bar{\sigma}'_{33}/\sigma_0 = 0.1$, $\mu_1/\mu_2 = 0.5$ and $m = 3$ (dotted curve), $m = 10$ (solid curve).

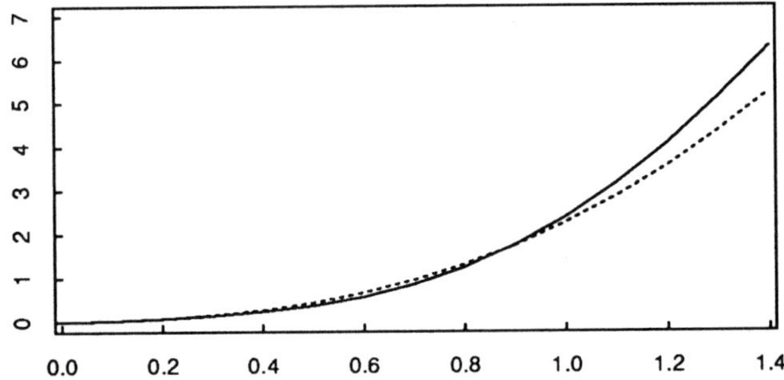

Fig.4 The lower bound $W_L^*(\bar{\sigma})/(e_0\sigma_0)$ versus $\bar{\sigma}_{33}'/\sigma_0$, for $\tau/\sigma_0 = 0.1$, $\mu_1/\mu_2 = 2$ and $m = 3$ (dotted curve), $m = 10$ (solid curve).

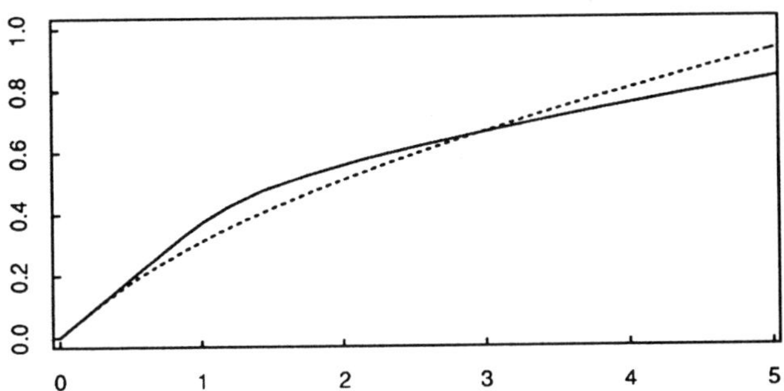

Fig.5 Plots of $\bar{\sigma}_{33}'/\sigma_0$ against $(\partial W_L^*/\partial\bar{\sigma}_{33}')/e_0$ for the same parameter values as in Fig. 4.

Figs. 4 and 5 show plots of $W_L^*(\bar{\sigma})/(e_0\sigma_0)$ against $\bar{\sigma}_{33}'/\sigma_0$ and $\bar{\sigma}_{33}'/\sigma_0$ against $\dfrac{\partial W_L^*}{\partial\bar{\sigma}_{33}'}/e_0$ for $\tau/\sigma_0 = 0.1$ and $\mu_1/\mu_2 = 2$. The behaviour of the bounds at low values of $\bar{\sigma}_{33}'/\sigma_0$ now agrees with the Voigt bound for the shear modulus of a linear composite. The behaviour of the bounds in Figs. 1, 3 and 4 for high levels of applied stress is also of interest. In this case $\mu_0 = \mu_2$, (5.16) becomes

$$\left[\frac{\zeta_{1e}}{\sigma_0}\right]^{2m} \simeq \frac{6}{c_2^2}\,\frac{\mu_1^2}{\mu_2^2}\left[\left[\frac{\tau}{\sigma_0}\right]^2 + \frac{3}{8}\left[\frac{\bar{\sigma}_{33}'}{\sigma_0}\right]^2\right], \tag{6.8}$$

and (6.1) is

$$W_L^*(\bar{\sigma})/(e_0\sigma_0) \simeq \frac{c_1}{2}\left[1 - \frac{\mu_1}{\mu_2}\right]\left[\frac{\zeta_{1e}}{\sigma_0}\right]^2 + \frac{c_1}{m+1}\left[\frac{\zeta_{1e}}{\sigma_0}\right]^{m+1} - c_1\left[\frac{\sigma_y}{\sigma_0}\right]^m \frac{\zeta_{1e}}{\sigma_0}$$

$$+ \frac{3\mu_1}{\mu_2}\left\{\frac{1}{4}\frac{(1+c_1)}{c_2}\left[\frac{\tau}{\sigma_0}\right]^2 + \frac{3}{8}\frac{1}{c_2}\left[\frac{\bar{\sigma}_{33}'}{\sigma_0}\right]^2\right\}. \qquad (6.9)$$

These show that for $\tau/\sigma_0 \gg 1$ or $\bar{\sigma}_{33}'/\sigma_0 \gg 1$ the bound is approximately a quadratic function. This is borne out by the shape of the curves in Figs. 2 and 5. This form for the bound agrees with intuition for axial loading (Fig. 4) but clearly underestimates the energy with the loading in Figs. 1 and 3. In this latter case τ can be interpreted as an applied shear and it is expected that once the matrix has yielded it will continue to flow and that the overall energy should obey some sort of power law. The inability of the results presented here to model this behaviour probably reflects the lack of sensitivity of the correlation functions to which phase constitutes the matrix and which the fibres. This lower bound would model more accurately a linearly elastic matrix containing nonlinear fibres.

References

Hill, R., 1963, "Elastic Properties of Reinforced Solids: Some Theoretical Principles", *J. Mech. Phys. Solids*, Vol. 11, pp. 357-372.

Hill, R., 1965, "Continuum Micro-Mechanics of Elastoplastic Polycrystals", *J. Mech. Phys. Solids*, Vol. 13, pp. 89-101.

Hill, R., 1972, "On Constitutive Macro-Variables for Heterogeneous Solids at Finite Strain", *Proc. R. Soc.*, A326, pp. 131-147.

Ponte Castañeda, P., and Willis, J. R., 1988, "On the Overall Properties of Nonlinearly Viscous Composites", *Proc. R. Soc., A416, pp. 217-244.*

Talbot, D. R. S., and Willis, J. R., 1985, "Variational Principles for Inhomogeneous Nonlinear Media", *IMA J. Appl. Math.*, Vol. 35, pp. 39-54.

Toland, J. F., and Willis, J. R., 1989, "Duality for Families of Natural Variational Principles in Nonlinear Electrostatics", *SIAM J. Math. Anal.*, Vol. 20, pp. 1283-1292.

Walpole, L. J., 1969, "On the Overall Elastic Moduli of Composite Materials", *J. Mech. Phys. Solids*, Vol. 17, pp. 235-251.

Willis, J. R., 1981, "Variational and Related Methods for the Overall Properties of Composites", *Advances in Applied Mechanics*, edited by C.-S. Yih, Vol. 21, pp. 1-78, Academic Press, New York.

Willis, J. R., 1983, "The Overall Elastic Response of Composite Materials", *J. Appl. Mech.*, Vol. 50, pp. 1202-1209.

Willis, J. R., 1986, "Variational Estimates for the Overall Response of an Inhomogeneous Nonlinear Dielectric", *Homogenization and Effective Moduli of Materials and Media*, edited by J. L. Ericksen, D. Kinderlehrer, R. V. Kohn and J.-L.

Lions, *The IMA Volumes in Mathematics and Its Applications*, Vol. 1, pp. 245-263, Springer, New York.

Willis, J. R., 1989, "Variational Estimates for the Overall Behaviour of a Nonlinear Matrix-Inclusion Composite", *Micromechanics and Inhomogeneity - The Toshio Mura Anniversary Volume*, edited by G. J. Weng, M. Taya and H. Abe, pp. 581-597, Springer-Verlag, New York.

Willis, J. R., and Talbot, D. R. S., 1989, "Variational Methods in the Theory of Random Composite Materials", *Proceedings of the 6th Symposium on Continuum Models and Discrete Systems*, Dijon, edited by G. A. Maugin, (to appear).

Willis, J. R., 1990, "On Methods for Bounding the Overall Properties of Nonlinear Composites", *J. Mech. Phys. Solids*. (to appear).

Damage and Failure

A Continuum Model for Damage Evolution in Laminated Composites

D.C. Lo
D.H. Allen

Aerospace Engineering Department, Texas A&M University
College Station, Texas, USA 77843

and

C.E. Harris

Mechanics of Materials Branch, NASA Langley Research Center
Hampton, Virginia, USA 23665

Abstract

The accumulation of matrix cracking is examined using continuum damage mechanics lamination theory. A phenomenologically based damage evolutionary relationship is proposed for matrix cracking in continuous fiber reinforced laminated composites. The use of material dependent properties and damage dependent laminate averaged ply stresses in this evolutionary relationship permits its application independently of the laminate stacking sequence. Several load histories are applied tocrossply laminates using this model and the results are compared to published experimental data. The stress redistribution among the plies during the accumulation of matrix damage is also examined. It is concluded that characteristics of the stress redistribution process could assist in the analysis of the progressive failure process in laminated composites.

Introduction

A unique property of composite materials is their evolutionary failure characteristic. The inhomogeneity of the microstructure provides numerous paths in which loads can be redistributed around the damaged region. Thus, the integrity and response of the component are affected more by the collective effects of the accumulated subcritical damage than by any single damage event. For laminated composites, this subcritical damage takes the form of matrix cracks, delaminations, debonding, and fiber fractures. Because the transfer of load away from the damaged area influences the damage evolution in the adjacent areas, the stresses at the ply level play an important role in the evolution of damage and the ultimate failure of the structure. Thus, this unique failure process that makes composites an attractive engineering material has also limited their efficient use in structures. The primary reason has been the shortage of analytical means to model the complex events occurring within the laminate and the prediction of damage evolution.

A review of the current literature indicates that three approaches have been taken to solve this problem. The first approach is the phenomenological methodology. Empirical relationships are developed from a body of experimentally measured data. Statistical theory is frequently employed to correlate this data. Models of this type tend to be restricted to the stacking sequence utilized to construct them. The second approach, called the crack propagation method, identifies damage as dominant cracks and fracture mechanics is applied to predict crack growth. The physical significance of each damage mode is retained with this approach. Unfortunately, the damage state in composite materials contains a multitude of interacting defects. Analysis in this manner is complex and perhaps unmanageable. For example, if there are hundreds of cracks, then the finite element method would require tens of thousands of elements. The third and most recent approach is the use of internal state variables in a continuum damage mechanics framework to model the damage accumulation. The averaged effects of the damage are represented through the internal state variables. This theory provides a thermodynamically rigorous characterization of the continuum under examination. The continuum damage mechanics approach entails the formulation of constitutive relationships, damage variable descriptions, and the damage evolution laws. The damage evolution laws may either be phenomenological, mechanistic, or even some combination of both. Because this approach is capable of accounting for the stacking sequence, it differs considerably from the phenomenological approach. Furthermore, because the effects of each crack are treated in the constitutive equations rather than via fracture mechanics, the computational solution is simpler than the crack propagation approach.

In the current paper, the continuum damage mechanics lamination theory proposed by Allen, et al. (1987a,b) is reviewed. Also a matrix crack damage evolutionary law is developed and this phenomenologically based evolutionary relationship is used to examine the accumulation of damage and the accompanying stress redistribution among the plies in laminates subjected to fatigue loading conditions.

Review of the Damage Dependent Lamination Model

The continuum damage mechanics approach used herein is based on the thermodynamics of irreversible processes. It is postulated that the state of a material point in a system undergoing a dissipative process can be characterized by a set of observable and internal state variables if this process is sufficiently close to the equilibrium state. Through the application of constraints from the fundamental principles of thermodynamics as proposed by Coleman and Mizel (1966) and the assumption of interdependence among these state variables, Coleman and Gurtin (1967) showed that constitutive equations for the material can be constructed in terms of the strain tensor, temperature, and internal state variables. The internal state variables may represent any dissipative process occurring in the medium. For the current application to distributed matrix cracks in laminated polymeric composites, Allen, et al. (1987a) selected a second order tensor value internal state variable to represent the kinematics of the cracks. This tensor was first defined by Vakulenko and Kachanov (1971) to be

$$\alpha_{ij} = \frac{1}{V_L} \int_S u_i n_j dS, \tag{1}$$

where α_{ij} are the components of the internal state variable tensor, V_L is a local volume in which statistical homogeneity can be assumed, u_i and n_j are the crack face displacement and normals, respectively, and S is the crack surface area.

The thermomechanical response of an elastic material with damage was found to be related to the damage dependent Helmholtz free energy as follows,

$$\sigma_{Lij} = \frac{\partial h_L}{\partial \varepsilon_{Lij}} \tag{2}$$

where σ_{Lij} are the components of the locally averaged stress tensor, h_L is the volume averaged Helmholtz free energy and ε_{Lij} are components of the infinitesimal locally averaged strain tensor. The subscript L will be used herein to denote volume averaged quantities. The Helmholtz free energy, h_L, for an elastic material with distributed damage is defined by

$$h_L = h_{EL} + u_L^c, \tag{3}$$

where h_{EL} is the Helmholtz free energy of an equivalent undamaged elastic body and u_L^c is the energy associated with the damage effect on the material. These two energy quantities are expressed by second order Taylor series expansions of the corresponding state variables. If the higher ordered and residual effects are neglected and isothermal conditions are assumed, The ply level thermomechanical constitutive relationship of the damaged material is,

$$\sigma_{ij} = C_{Lijkl}\varepsilon_{kl} + I_{Lijkl}\alpha_{kl}. \tag{4}$$

where C_{Lijkl} is the undamaged material modulus tensor and I_{Lijkl} is denoted the damage modulus tensor. The components of this damage modulus tensor were shown by Allen, et al. (1987b) to be related to the modulus tensor as follows

$$I_{Lijkl} \approx -C_{Lijkl}. \tag{5}$$

Another result of the Coleman-Mizel formulation is the entropy production by the distributed damage in the absence of thermal gradients,

$$f_{ij}^T \dot{\alpha}_{ij} \geq 0 \tag{6}$$

where the thermodynamic force, f_{ij}^T, associated with the evolution of α_{ij} is denoted by

$$f_{ij}^T = -\frac{\partial h}{\partial \alpha_{ij}}. \tag{7}$$

Inequality (6) will admit only those processes that yield non-negative rates of entropy production.

The response of a multilayered laminate with matrix damage is obtained by inplane averaging the ply level constitutive relationship show in equation (4) and imposing the Kirchhoff hypothesis through the thickness to obtain the following modified set of laminate equations,

$$\begin{aligned}\{N\} = \sum_{k=1}^{n} [Q]_k (z_k - z_{k-1}) \{\varepsilon^o\} - \frac{1}{2} \sum_{k=1}^{n} [Q]_k (z_k^2 - z_{k-1}^2) \{\kappa\} \\ - \sum_{k=1}^{n} [Q]_k (z_k - z_{k-1}) \{\alpha^M\}_k,\end{aligned} \tag{8}$$

where $\{N\}$ is the resultant force per unit length vector, $[Q]_k$ is the elastic modulus matrix for the kth ply in laminate coordinates, z_k is the distance from the midplane to the kth ply, ε^o and κ are the midplane strains and curvatures, respectively; and $\{\alpha^M\}$ are the damage variables corresponding to the matrix cracks in the kth ply. The effects of the matrix cracking on the resultant forces are contained in the last group of terms on the right hand side of equation (8). An expression for the resultant moments can be written in a similar form. The stiffness loss of composite crossply laminates with matrix cracks has been successfully predicted with this model by Allen, et al. (1987b, 1988) for specified damage states using equation (8). In addition, although it will not be discussed in this paper, the model has also been extended to include the effects of delamination by Harris, et al. (1987, 1988). A damage evolution relationship for this model will be introduced in the following section.

Evolutionary Equation for Internal State Variables

In order to characterize the development of the internal damage, the conditions for the initiation and progression of damage as well as a means to quantify changes in the damage state must be established. Concepts from linear elastic fracture mechanics can often be employed to assist in the development of the evolutionary equations for brittle damage. The thermodynamic force, f_{ij}^T, as defined by equation (7), represents the available free energy that can be delivered by the system for a small change in the internal state variable. However, this change will occur only if the energy delivered is equal to or greater than the energy required to produce this change in the internal state. Thus, the following initiation criterion based on the relative magnitudes of the available and required thermodynamic forces, $(f^T_{ij})_{req.}$ can be used for this model,

$$f^T_{ij} \geq (f^T_{ij})_{req.}. \tag{9}$$

The implementation of this criterion will require the determination of the required thermodynamic forces. The required thermodynamic forces are most likely to change with the damage state. This criterion is analogous to comparing the strain energy release rate to the critical value in fracture mechanics. Had the internal state variable been defined as the crack surface area, the thermodynamic force would be identical to the strain energy release rate.

One approach to the formulation of the internal state variable evolutionary relationships is through micromechanical considerations. However, this approach is dependent on the availability of micromechanical solutions that can model the essential physical characteristics of the damage state. For the problem of matrix cracks embedded in an orthotropic medium that is layered between two other orthotropic media, the solutions that are currently available are applicable only to very specific loading conditions and damage geometries. Therefore, the evolutionary equation proposed herein is phenomenological in nature. The form of the damage evolutionary relationship employed in this paper is based on the observation made by Wang, et al. (1984) that for some materials the rate of damage surface evolution per load step, $\frac{dS}{dN}$, follows the power law as shown below, in which the strain energy release rate, G, and a material parameter, n, serves as the basis and exponent, respectively.

$$\frac{dS}{dN} = PG^n \tag{10}$$

where P is a material constant. To develop internal state variable evolutionary equations in the form of equation (10), α_{ij} has to be related to the damage surface area. Since α_{ij}, as defined by equation (1), represents the kinematics of the crack faces, the damage surface area alone will not be sufficient to describe the crack face displacements. Assuming that each crack in the material volume shares a common geometry and orientation, then the specification of the far field strains will complete this description. The rate of change of the internal state variable can therefore be expressed as follows,

$$\frac{d\alpha_{ij}}{dN} = \frac{d\alpha_{ij}}{dS}\frac{dS}{dN} \tag{11}$$

where the far field strains are reflected in the term $\frac{d\alpha_{ij}}{dS}$, which relates the changes in the internal state variable during the development of damage surfaces. Thus, using equations (10) and (11), the stable evolution of the internal state variable is defined by

$$d\alpha_{ij} = \frac{d\alpha_{ij}}{dS} k_1 G^n dN \tag{12}$$

where $\frac{d\alpha_{ij}}{dS}$ can be determined analytically for simplified damage geometries and loading conditions and k_1 is a material parameter.

It will now be shown that the strain energy release rate can be determined from the thermodynamic forces. The Helmholtz free energy for a body with distributed damage was defined earlier to be the sum of the Helmholtz free energy in an equivalent undamage body, h_{EL}, and the energy of damage creation, u_L^c. Since h_{EL} is independent of the internal state variables, the thermodynamic force can be expressed in terms of u_L^c as

$$f_{ij}^T = -\frac{\partial u_L^c}{\partial \alpha_{ij}}. \tag{13}$$

Allen, et al. (1987a) defined u_L^c as the mechanical energy of a continuum due to equivalent crack surface tractions acting on the crack faces. This energy encompasses the energy available for crack extension and the energy loss due to the apparent stiffness reduction caused by existing cracks. An association between the energy for crack extension and the strain energy release rate, G, is defined by

$$u_L^c = \frac{1}{V_L}\int_S G dS. \tag{14}$$

The strain energy release rate is obtained from equation (14) by differentiation with respect to the crack surface area. Thus,

$$G = V_L \frac{du_L^c}{dS}. \tag{15}$$

If the process is restricted to isothermal conditions, u_L^c depends only on the strain tensor, ε_{ij}, and the internal state variable, α_{ij}, so that equation (15) becomes

$$G = V_L \frac{\partial u_L^c}{\partial \varepsilon_{ij}}\frac{d\varepsilon_{ij}}{dS} + V_L \frac{\partial u_L^c}{\partial \alpha_{ij}}\frac{d\alpha_{ij}}{dS}. \tag{16}$$

The relationship between the strain energy release rate and the available thermodynamic force is obtained by permitting crack extension under fixed grip conditions and using the definition of the thermodynamic force in equation (13),

$$G = -V_L \frac{d\alpha_{ij}}{dS} f_{ij}^T. \tag{17}$$

The internal state variable evolutionary equation shown in (12) can thus be expressed in terms of the available thermodynamic force as follows:

$$d\alpha_{ij} = \frac{d\alpha_{ij}}{dS} k_1 \left[-\frac{d\alpha_{mn}}{dS} f_{mn}^T \right]^n dN \qquad \text{when} \quad f_{ij}^T \geq \left(f_{ij}^T\right)_{req.}. \tag{18}$$

For crack propagation under a single fracture mode, the crack surface kinematics for a thin laminate can be characterized by a single component of the internal state variable tensor. For example, consider the matrix crack damage state shown in Fig. 1, where all the crack planes are flat and oriented perpendicular to the plane formed by the lamina and parallel to the fibers, a pure opening mode (mode I) would correspond to the crack faces moving in a direction parallel to the crack face normals. Thus, in reference to the ply level cartesian coordinate system, the internal state variable tensor as defined by equation (1) will be zero except for α_{22}. Likewise, for the pure shearing mode (mode II), the only non-zero component of the internal state variable tensor would be α_{12}.

Matrix Crack Development in Multi-Ply Laminates

The evolutionary relationship shown in equation (18) provides a description of the damage development at the ply level. In laminates containing multiple plies of different orientation, the development of damage is influenced by the adjacent plies. This is attributed to the different stress states found in the neighboring plies and the redistribution of load among the plies that occurs during the accumulation of damage. The effects of this interaction on the damage development are included implicitly in the evolutionary equation through the thermodynamic forces and $\frac{d\alpha_{ij}}{dS}$ because laminate averaged damage dependent ply responses calculated from equation (8) are used to determine these quantities. Since the adjacent ply constraining effects are reflected through the laminate equations, the evolutionary equation can be applied independent of the laminate stacking sequence.

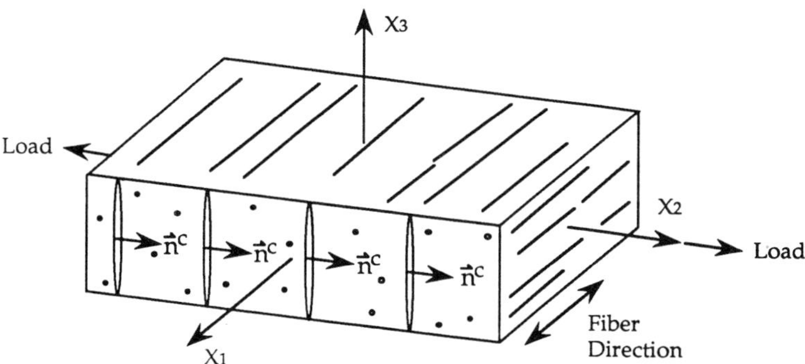

Figure 1. Idealized configuration of ply with matrix cracking.

The term $\frac{d\alpha_{ij}}{dS}$, found in equation (18), reflects the changes to the internal state variable with respect to changes to the damage surfaces. $\frac{d\alpha_{ij}}{dS}$ can be obtained analytically from relationships describing the kinematics of the crack surfaces for given damage states and loading conditions, should such solutions exist. For transverse matrix cracks in crossply laminates, the average crack face displacement in the pure opening mode can be approximated by a solution obtained by Lee, et al. (1989) for a medium containing an infinite number of alternating 0° and 90° plies. Thus $\frac{d\alpha_{22}}{dS}$ can be determined for crossply laminates subjected to uniaxial loading conditions. It has been found that for typical continuous fiber reinforced Graphite/Epoxy systems $\frac{d\alpha_{22}}{dS}$ can be assumed to be constant for a given applied load until the damage state has reached an advanced stage of development. This assumption has facilitated the determination of the material parameters k_1 and n. The damage state at any point in the loading history can now be determined by the integration of equation (18) using the laminate averaged ply responses. This integration is performed numerically because of the nonlinearity of the damage evolutionary equation. The fourth order Runge-Kutta method has been found to be suitable for this application.

The development of the matrix crack damage state in crossply laminates subjected to uniaxial fatigue loading is examined using the proposed damage evolution equation. To maintain the thermodynamic admissibility of the fatigue loading process, it is assumed that the values of the internal state variables remain constant during the unloading portion of the load cycle. It was further assumed that the required thermodynamic force is very small compared to the available thermodynamic force, thus α_{22} will change at the onset of load application. The material properties for AS4/3501-6 Graphite/Epoxy are used in the calculations to enable the comparison of model prediction to experimental measurements made by Chou, et al. (1982). The material parameters for this polymeric composite system have been found to be

$$k_1 = 4.42 \qquad \text{and} \qquad n = 6.39 \quad . \tag{19}$$

The damage histories for two crossply layups have been predicted using the model. One layup consists of four consecutive 90° plies laminated between 0° layers of two ply thickness. The other crossply laminate contains six consecutive 90° plies in the center. The model predictions of the damage state in the $[0_2/90_2]_s$ laminates fatigue loaded at a maximum stress amplitude of 38 ksi and 43 ksi are shown in Figs. 2 and 3, respectively. The lower stress amplitude is equivalent to eighty percent of the monotonic crack initiation stress, while the higher stress amplitude is equal to ninety percent of the initiation stress. The experimentally measured damage states were originally measured in terms of the crack density. However, the corresponding α_{22} for each damage level can be approximated by the relationship proposed by Lee, et al. (1989).

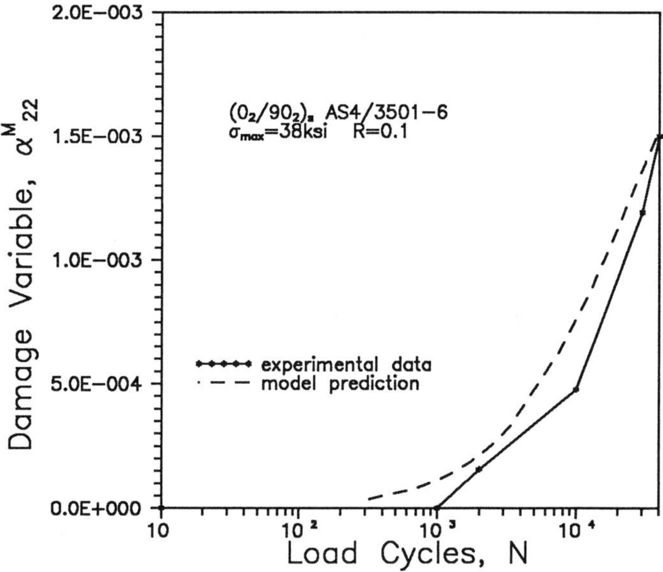

Figure 2. Matrix crack damage in the 90° plies of a $[0_2/90_2]_s$ AS4/3501-6 laminate loaded at a stress amplitude of 38ksi .

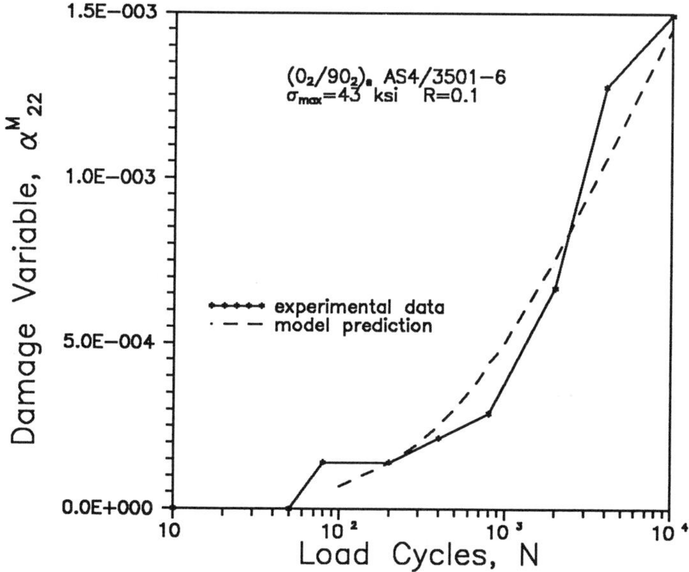

Figure 3. Matrix crack damage in the 90° plies of a $[0_2\ 90_2]_s$ AS4/3501-6 laminate loaded at a stress amplitude of 43ksi .

Good agreement is found between the model predictions and the experimental results. The damage evolution for the thicker $[0_2/90_3]_s$ laminate is shown in Fig. 4. This laminate was loaded at a maximum stress amplitude of 26 ksi. This amplitude corresponds to eighty percent of the quasi-static matrix crack initiation stress. The results for this load case indicated good agreement with the experimental data. The effect of the load redistribution on the damage evolution is apparent in this load case. A marked decrease in the rate of damage evolution after fifty thousand load cycles was indicated by the model. On the other hand, the experimental data showed this decrease to occur after only ten thousand load cycles. Since the evolution of delaminations were not included in the analysis, the predicted decrease in the rate of damage evolution is attributed only to the matrix crack induced transfer of load from the 90° plies to the adjacent 0° plies and the resulting decrease in the available thermodynamic force. The measured values of the damage state, however, may have been influenced by the formation of delaminations along the free edges and in the interior. Such occurrence can drastically affect the stress distribution among the plies and the available thermodynamic force for damage evolution.

To examine the amount of stress redistribution that occurs during the damage accumulation, the model was used to determine the axial stress in the 90° plies of the $[0_2/90_3]_s$ laminate fatigue loaded at three different stress amplitudes. Fig. 5 shows that for the stress amplitude of 38 ksi, the axial stress in the 90° plies after forty thousand cycles was less than fifty percent of the original stress level in the undamaged laminate. Therefore, the rate of damage evolution is expected to be relatively low during the latter stages of the loading history. This is observed in Fig. 6, which shows the corresponding values of the internal state variable, α_{22}, for the three load cases. The 26 ksi stress amplitude load case, on the other hand, produced only gradual changes in the axial stress and damage state as compared to the other two stress amplitudes. The percentage decrease from the original undamage stress level increased with the fatigue stress amplitude. These results demonstrate that the stress redistribution characteristics among the plies in the laminate are dependent on the loading conditions. These redistribution characteristics will affect the manner in which damage develops in the surrounding plies as well as eventual failure of the laminate.

Conclusion

A damage evolutionary relationship for the accumulation of matrix cracks has been presented. This phenomenologically based relationship operates as part of a continuum damage mechanics model developed to analyze the response of laminated composites. The utilization of material dependent quantities and damage dependent laminate averaged ply responses in the evolutionary relationship has enable it to function independent of the lam-

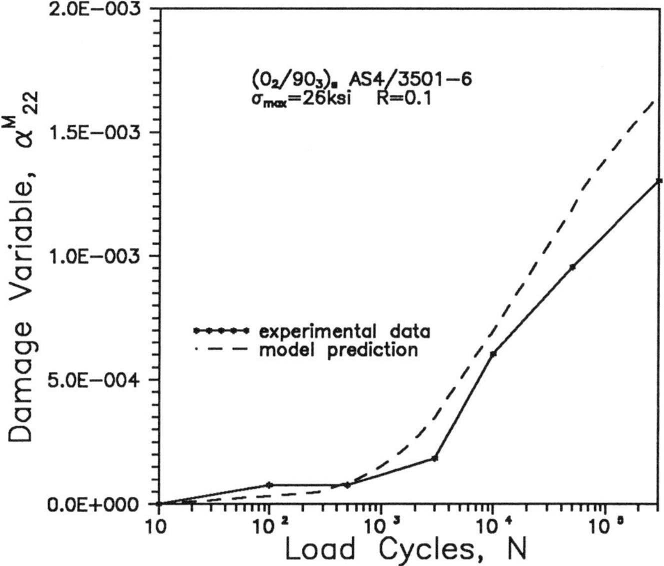

Figure 4. Matrix crack damage in the 90° plies of a $[0_2/90_3]_s$ AS4/3501-6 laminate loaded at a stress amplitude of 26ksi .

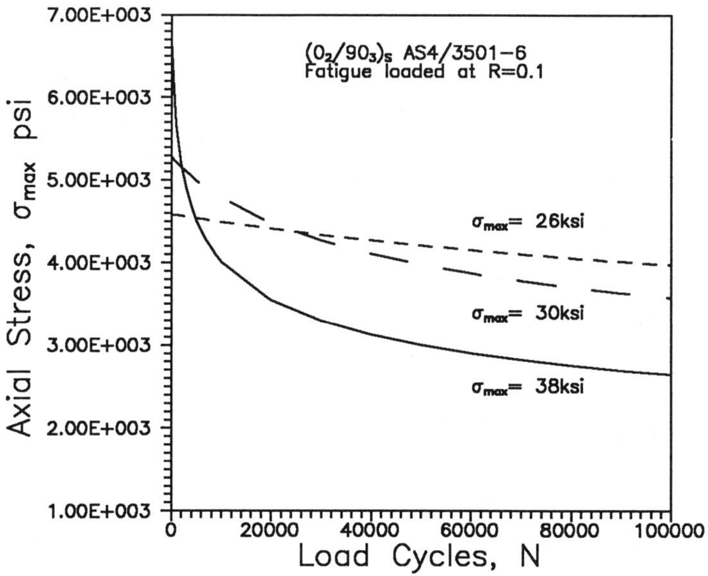

Figure 5. Damage induced stress redistribution in the 90° plies of $[0_2/90_3]_s$ laminates subjected to constant stress amplitude fatigue loading.

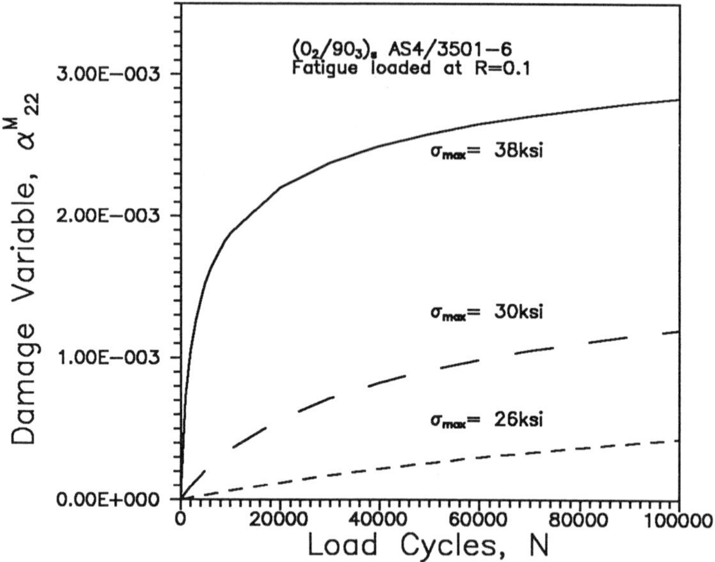

Figure 6. The accumulation of damage in the 90° plies of $[0_2/90_3]_s$lamin -ates subjected to constant stress amplitude fatigue loading.

inate stacking sequence. The capability to predict the evolution of matrix cracks in crossply laminates subjected to fatigue loading conditions has been demonstrated. The evolutionary relationship has also been used to examined stress redistribution among the plies during the damage history of the laminate. It is found that the stress distribution behavior is dependent on the load amplitude. Higher applied loads result in rapid changes in the axial stress and damage state during the initial portion of the loading history. This is then followed by low rates of change for the remainder of the loading history. For lower load amplitudes, the stress redistribution process and damage accumulation proceed in a gradual manner.

The existence of other types of damage, such as delaminations, will alter the redistribution of stress among the plies and thus the evolution of the damage state. An evolutionary relationship for delamination is currently under development. The inclusion of this type of damage should provide better insight into the complex events occurring within the laminate. The information obtained from the stress redistribution histories could enhance the knowledge of failure characteristics in laminated composites and aid in the development of models for this progressive failure process.

Acknowledgments

This research was supported by NASA Langley Research Center under contract no. NAG-1-979 and NGT-50262.

References

Allen,D.H., Harris,C.E., and Groves,S.E., 1987a, "A Thermomechanical Constitutive Theory for Elastic Composites With Distributed Damage–I. Theoretical Development," Int. Journal of Solids and Structures, Vol. 23, pp. 1301-1318.

Allen,D.H., Harris,C.E., and Groves,S.E., 1987a, "A Thermomechanical Constitutive Theory for Elastic Composites With Distributed Damage–II. Application to Matrix Cracking in Laminated Composites," Int. Journal of Solids and Structures, Vol. 23, pp. 1319-1338.

Allen,D.H., Harris,C.E., Groves,S.E., and Norvell,R.G., 1988, "Characteristics of Stiffness Loss in Crossply Laminates With Curved Matrix Cracks," Journal of Composite Materials, Vol. 22, pp. 71-80.

Chou,P.C., Wang,A.S.D., and Miller,H., 1982, "Cumulative Damage Model for Advanced Composite Materials, AFWAL-TR-82-4083".

Coleman,B.D., and Mizel,V.J., 1966, "Thermodynamics and Departures from Fourier's Law of Heat Conduction," Arch. Rational Mech. Analysis, 23, pp. 245-261.

Coleman,B.D., and Gurtin,M.E., 1967, "Thermodynamics With Internal State Variables," The Journal of Chemical Physic, Vol. 47, pp. 597-613.

Harris,C.E., Allen,D.H., and Nottorf,E.W., 1987, "Damage–Induced Changes in the Poisson's Ratio of Cross-Ply Laminates: An Application of a Continuum Damage Mechanics Model for Laminated Composites," Damage Mechanics in Composites, ASME, AD-Vol. 12, pp. 17-23.

Harris,C.E., and Allen,D.H., 1988, "A Continuum Damage Model of Fatigue–Induced Damage in Laminated Composites," SAMPE Journal, Vol. 24, No. 4 , pp. 43-51.

Lee,J.W., Allen, D.H., and Harris,C.E., 1989, "Internal State Variable Approach for Predicting Stiffness Reductions in Fibrous Laminated Composites With Matrix Cracks," Journal of Composite Materials, Vol. 23 ,pp. 1273-1291.

Vakulenko,A.A., and Kachanov,M.L., 1971, "Continuum Theory of Cracked Media," Mekh. Tverdogo Tela, 6, p.159.

Wang,A.S.D., Chou,P.C., and Lei,S.C., 1984, "A Stochastic Model for the Growth of Matrix Cracks in Composite Laminates," Journal of Composite Materials, Vol. 18, pp.239-254.

Lower and Upper Bound Estimates for the Macroscopic Strength Criterion of Fiber Composite Materials

PATRICK DE BUHAN*, JEAN SALENÇON*, ALBERTO TALIERCIO**

* Laboratoire de Mécanique des Solides, Ecole Polytechnique
91128 Palaiseau Cedex, France

** Dipartimento di Ingegneria Strutturale, Politecnico di Milano
20133 Milano, Italy

Abstract : The formulation of a homogenization procedure within the framework of the yield design theory makes it possible to derive a strength criterion for a fiber composite material in a rigorous way, from the only definitions of the strength properties of the constituents (matrix and fibers) and of their geometrical, structural and voluminal arrangement. Making use of both yield design static and kinematic approaches, quite simple analytical lower and upper bound estimates are obtained for a unidirectional fiber composite. A detailed analysis of those estimates is carried through in the specific case when the composite is subjected to a uniaxial solicitation or to plane strain conditions parallel to the fibers direction .

---

1. Introduction

Among the various theoretical investigations being carried out for analysing the inelastic behaviour of fiber composite materials, some are more specifically concerned with the determination of their ultimate strength properties. They are generally based on a "micro-macro" approach which aims at relating the overall behaviour of the composite to the strength properties of its constituents taking into account their structural arrangement (e.g. respective voluminal proportions, geometry of the fibers etc.) (Saurindranath and Mc Laughlin, 1975 ; Sawicki, 1981 ; Dvorak and Bahei-el-Din, 1982). When the fibers are laid out throughout the matrix material following a regular pattern, a homogenization method formulated within the yield design theory (de Buhan and Salençon, 1987) provides a rigorous mechanical framework for solving such a problem.

This paper is devoted to the application of that method to a unidirectional fiber composite material. The general static and kinematic definitions of the corresponding macroscopic strength criterion are given, and lower and upper bounds are derived from the implementation of the yield design static and kinematic approaches respectively. These bounds can be expressed by means of explicit analytical formulae in case of simple solicitations such as uniaxial or plane strain loading conditions. Their major interest lies in providing easily usable estimates for the overall strength of the composite, which remain valid for any voluminal proportion of the fibers.

2. *General definitions of the strength criterion*

2.1. *Working hypotheses and notations*

Consider a composite material made of one single array of long parallel fibers (unidirectional composite) periodically distributed throughout a homogeneous matrix material in such a way that a rectangular parallepiped of dimensions a_1, a_2, a_3 along three orthogonal directions of a $Ox_1x_2x_3$ coordinate system may be exhibited as a representative volume (Fig. 1), called a "unit cell" and denoted by $\mathcal{a}$. This unit cell consists of a cylindrical, not necessarily circular, volume $\mathcal{a}_f$, of fiber material embedded in the matrix material which occupies the remaining part $\mathcal{a}_m$ of the cell.

The constituents are supposed to obey von Mises strength conditions, namely :

$$F_m(\underline{\underline{\sigma}}) = [\,J_2(\underline{\underline{\sigma}})\,]^{1/2} - K_m/\sqrt{3} \leq 0 \qquad (1)$$

for the matrix, and

$$F_f(\underline{\underline{\sigma}}) = [\,J_2(\underline{\underline{\sigma}})\,]^{1/2} - K_f/\sqrt{3} \leq 0 \qquad (2)$$

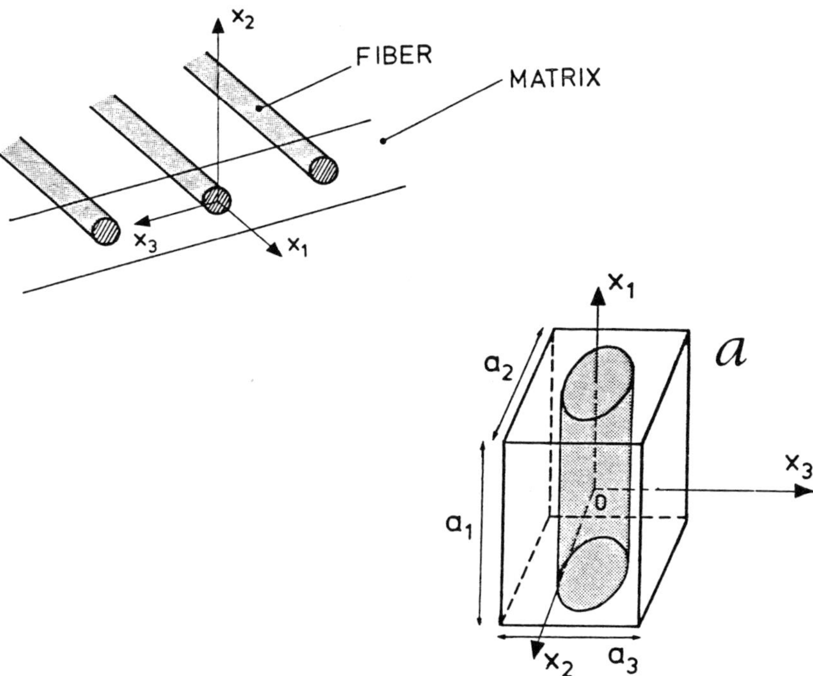

Fig. 1 : Representative unit cell for a unidirectional fiber composite.

for the fiber material, where $J_2(\underline{\underline{\sigma}})$ denotes the second invariant of the deviatoric stress $(J_2(\underline{\underline{\sigma}}) = 1/2 \quad \underline{\underline{s}} : \underline{\underline{s}}^{(\star)})$, with K_m (resp. K_f) representing the uniaxial strength of the matrix (resp. fiber) material.

Furthermore, failure by lack of adherence between the fiber and the surrounding matrix will be discarded in the present analysis, thereby assuming perfect bonding between the constituents.

($\star$) $1/2 \; \underline{\underline{s}} : \underline{\underline{s}} = 1/2 \; s_{ij} \, s_{ji}$, *with summation over the repeated subscripts.*

2.2. Static definition

According to the yield design homogenization theory (Suquet, 1985) the macroscopic strength condition of the above described composite, considered as a homogeneous material, writes :

$$F_{hom}(\underline{\underline{\Sigma}}) \leqslant 0 \tag{3}$$

where the macroscopic stress $\underline{\underline{\Sigma}}$ is computed as the volume average over $\mathcal{Q}$ of a stress field $\underline{\underline{\sigma}}$ which must comply with the following conditions :

a) *boundary periodicity*, i.e. the origin O being taken at the center of the parallelepiped

$$\sigma_{ij}\left(x_j = -\frac{a_j}{2}\right) = \sigma_{ij}\left(x_j = +\frac{a_j}{2}\right) \tag{4-a}$$

whatever i and j ;

b) *equilibrium* :

div $\underline{\underline{\sigma}} = 0$ (and $[\![\underline{\underline{\sigma}}]\!] . \underline{n}$ across possible stress discontinuity surfaces) ; (4-b)

c) *strength requirements* :

$$F_m(\underline{\underline{\sigma}}(\underline{x})) \leqslant 0 \quad (\text{resp. } F_f(\underline{\underline{\sigma}}(\underline{x})) \leqslant 0) \tag{4-c}$$

whatever $\underline{x}$ belonging to $\mathcal{Q}_m$ (resp. $\mathcal{Q}_f$).

2.3. Kinematic definition

Introducing the support function π_{hom} of the macroscopic strength criterion, defined as (Salençon, 1983) :

$$\pi_{hom}(\underline{\underline{D}}) = \sup_{\underline{\underline{\Sigma}}} \{\underline{\underline{\Sigma}} : \underline{\underline{D}} \ ; \ F_{hom}(\underline{\underline{\Sigma}}) \leqslant 0\} \tag{5}$$

where $\underline{\underline{D}}$ represents any symmetric second order tensor, it can be proved (Suquet, 1985 ; de Buhan, 1986) that :

$$\pi^{hom}(\underline{\underline{D}}) = \inf_{\underline{v}} \{<\pi(\underline{\underline{d}})> \; ; \; <\underline{\underline{d}}> = \underline{\underline{D}}\} \tag{6}$$

where :

• $\underline{v}$ denotes any velocity field giving a strain rate field $\underline{\underline{d}}$ periodic over $\mathcal{a}$

• $<\underline{\underline{d}}> = \dfrac{1}{2\, a_1\, a_2\, a_3} \displaystyle\int_{\partial \mathcal{a}} (\underline{n} \otimes \underline{v} + \underline{v} \otimes \underline{n})\, dS$

defines the mean value of the strain rate field $\underline{\underline{d}}$ computed over the cell

• and $<\pi(\underline{\underline{d}})>$ is defined by

$$<\pi(\underline{\underline{d}})> = \frac{1}{a_1\, a_2\, a_3} \left(\int_{\mathcal{a}_m} \pi_m(\underline{\underline{d}})\, d\mathcal{a} + \int_{\mathcal{a}_f} \pi_f(\underline{\underline{d}})\, d\mathcal{a} + \int_{S_m} \pi_m(\underline{n} \; ; \; [\![\underline{v}]\!])\, dS \right.$$

$$\left. + \int_{S_f} \pi_f(\underline{n} \; ; \; [\![\underline{v}]\!])\, dS \right)$$

where

$\pi_m(\underline{\underline{d}}) = \sup_{\underline{\underline{\sigma}}} \{\underline{\underline{\sigma}} : \underline{\underline{d}} \quad , \quad F_m(\underline{\underline{\sigma}}) \leqslant 0\}$,

$\pi_f(\underline{\underline{d}})$ has the same definition with F_f ,

S denotes any jump surface for $\underline{v}$,

S_m is the portion of S lying in $(\mathcal{a}_m)$,

S_f has the same definition with $(\mathcal{a}_f)$,

$[\![\underline{v}]\!]$ is the jump of $\underline{v}$ when crossing S following its normal $\underline{n}$,

$\pi_m(\underline{n} \; ; \; [\![\underline{v}]\!]) = \sup_{\underline{\underline{\sigma}}} \{\underline{n} \cdot \underline{\underline{\sigma}} \cdot [\![\underline{v}]\!] \quad , \quad F_m(\underline{\underline{\sigma}}) \leqslant 0\}$,

$\pi_f(\underline{n}\ ;\ [\![\underline{v}]\!])$ has the same definition with F_f .

Generally speaking, the determination of the strength condition (3) would require the use of heavy numerical methods implementing either of the two above definitions. Fortunately the specific geometry of the unit cell in the case of a unidirectional composite will be exploited to derive relevant, though simple, analytical estimates for the actual macroscopic condition.

3. Lower bound estimate

Referring to the static definition given by Eq. (3) and (*4-a*) to (*4-c*), it is possible to exhibit piecewise constant stress fields satisfying all the above prescribed conditions :

$$\underline{\underline{\sigma}}(\underline{x}) = \begin{cases} \underline{\underline{\sigma}}^m & \forall \underline{x} \in \mathcal{a}_m \\ \underline{\underline{\sigma}}^m + \sigma^f\, \underline{e}_1 \otimes \underline{e}_1 & \forall \underline{x} \in \mathcal{a}_f \end{cases} \qquad (7)$$

with

$$F_m(\underline{\underline{\sigma}}^m) \leqslant 0 \quad \text{and} \quad F_f(\underline{\underline{\sigma}}^m + \sigma^f\, \underline{e}_1 \otimes \underline{e}_1) \leqslant 0 \qquad (8)$$

($\underline{e}_1$: unit vector of Ox_1 , parallel to the fiber).

This particular form of stress fields automatically complies with the periodicity condition (*4-a*) and the equilibrium equation as well since it ensures the continuity of the stress vector acting at any point upon the interface between the matrix and the fiber.

As a matter of fact, whatever $\underline{n}$ such that $\underline{n} \cdot \underline{e}_1 = 0$

$$[\![\underline{\underline{\sigma}}]\!] \cdot \underline{n} = (\sigma^f\, \underline{e}_1 \otimes \underline{e}_1) \cdot \underline{n} = \sigma^f\, \underline{e}_1 (\underline{e}_1 \cdot \underline{n}) = 0\ .$$

Denoting by η the voluminal percentage of the fibers the macroscopic stress $\underline{\underline{\Sigma}}$ defined as the volume average of such a piece-wise constant stress field writes :

$$\underline{\underline{\Sigma}} = \underline{\underline{\sigma}}^m + \eta\, \sigma^f\, \underline{e}_1 \otimes \underline{e}_1 \;. \tag{9}$$

It can be easily shown (see Appendix) that *sufficient* conditions for $\underline{\underline{\Sigma}}$ to satisfy the macroscopic criterion (3) are :

$$F_m(\underline{\underline{\sigma}}^m) \leqslant 0 \quad \text{and} \quad |\sigma^f| \leqslant K_f - K_m \;. \tag{10}$$

It follows immediately that the condition :

$$F^-(\underline{\underline{\Sigma}}) \leqslant 0$$

with :

$$F^-(\underline{\underline{\Sigma}}) = \operatorname*{Inf}_{\sigma^f} \{F_m(\underline{\underline{\Sigma}} - \eta\, \sigma^f\, \underline{e}_1 \otimes \underline{e}_1) \;;\; |\sigma^f| \leqslant K_f - K_m\} \tag{11}$$

constitutes a safe lower bound approximation to the actual macroscopic condition :

$$F^-(\underline{\underline{\Sigma}}) \leqslant 0 \quad \Rightarrow \quad F_{hom}(\underline{\underline{\Sigma}}) \leqslant 0 \;.$$

4. Upper bound estimates

4.1. A first upper bound estimate can immediately be deduced from the kinematic definition (6). Considering the homogeneous strain rate field $\underline{\underline{d}}(\underline{x}) = \underline{\underline{D}} \quad \forall \underline{x} \in \mathcal{a}$ derived from a velocity field in the form

$$\underline{v} = \underline{\underline{D}} \cdot \underline{x}$$

one gets :

$$<\pi(\underline{\underline{d}})> = (1 - \eta)\; \pi_m(\underline{\underline{D}}) + \eta\; \pi_f(\underline{\underline{D}})$$

with

$$\pi_m(\underline{\underline{D}}) = \begin{cases} K_m \left(\frac{2}{3}\, \underline{\underline{D}} : \underline{\underline{D}}\right)^{1/2} & \text{if} \quad \operatorname{tr} \underline{\underline{D}} = 0 \\ +\infty & \text{otherwise} \end{cases}$$

(analogous formulae for $\pi_f(\underline{\underline{D}})$).

Hence for any $\underline{\underline{D}}$

$$\pi_{hom}(\underline{\underline{D}}) \leqslant [(1-\eta)\ K_m + \eta\ K_f]\ (2/3\ \underline{\underline{D}} : \underline{\underline{D}})^{1/2}.$$

This proves that the strength capacities of the homogenized composite are inferior to those that would be exhibited by an isotropic von Mises material obtained by averaging the strength characteristics of the constituents.

In other words :

$$F_{hom}(\underline{\underline{\Sigma}}) \leqslant 0 \quad \Rightarrow \quad F_1^+(\underline{\underline{\Sigma}}) \leqslant 0 \tag{12-a}$$

with

$$F_1^+(\underline{\underline{\Sigma}}) = [J_2(\underline{\underline{\Sigma}})]^{1/2} - \langle K\rangle/\sqrt{3} \tag{12-b}$$

$$\langle K\rangle = (1-\eta)\ K_m + \eta\ K_f \quad .$$

4.2. Other upper bound estimates can be obtained by considering piece-wise constant velocity fields in which a discontinuity plane parallel to the fibers and running across the matrix divides the unit cell into two blocks. Due to the periodicity condition, such a plane is pres-cribed to be perpendicular to the Ox_i axis, i = 2 or 3 (Fig. 2). Denoting by $\underline{V}$ the velocity jump across this plane, one gets :

$$\underline{\underline{D}} = \langle\underline{\underline{d}}\rangle = \frac{1}{2a_i}\ (\underline{e}_i \otimes \underline{V} + \underline{V} \otimes \underline{e}_i) \qquad i = 2\ ,3$$

and

$$\langle\pi(\underline{\underline{d}})\rangle = \frac{1}{a_i}\ \pi_m(\underline{e}_i\ ;\ \underline{V})$$

with, as the matrix obeys a von Mises criterion :

$$\pi_m(\underline{e}_i\ ;\ \underline{V}) = \begin{cases} K_m\ |\underline{V}|/\sqrt{3} & \text{if} \quad \underline{V}\cdot\underline{e}_i = 0 \\ +\infty & \text{otherwise.} \end{cases}$$

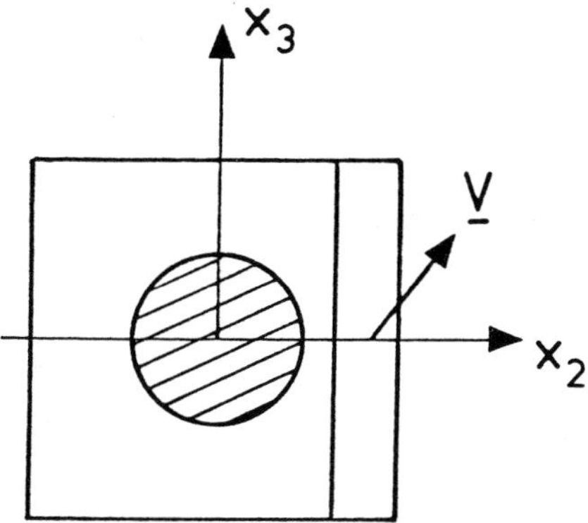

Fig. 2 : Velocity field involving a jump plane passing through the matrix.

The kinematic definition of the macroscopic criterion expressed by Eq. (5) and (6) leads to the following implication

$$F_{hom}(\underline{\underline{\Sigma}}) \leqslant 0 \Rightarrow \underline{\underline{\Sigma}} : <\underline{\underline{d}}> \leqslant <\pi(\underline{\underline{d}})>$$

and therefore

$$F_{hom}(\underline{\underline{\Sigma}}) \leqslant 0 \Rightarrow F_i^+(\underline{\underline{\Sigma}}) \leqslant 0 \qquad (13\text{-}a)$$

where for $i = 2, 3$

$$F_i^+(\underline{\underline{\Sigma}}) \leqslant 0 \Leftrightarrow (\underline{\underline{\Sigma}} \cdot \underline{e}_i) \cdot \underline{V} \leqslant \frac{K_m}{\sqrt{3}} |\underline{V}| \qquad (13\text{-}b)$$

whatever $\underline{V}$ such that $\underline{V} \cdot \underline{e}_i = 0$ (⋆) .

(⋆) $F_i^+(\underline{\underline{\Sigma}})$ *may be expressed by*

$$F_i^+(\underline{\underline{\Sigma}}) = \operatorname*{Sup}_{\underline{V}} \{(\underline{\underline{\Sigma}} \cdot \underline{e}_i) \cdot \underline{V} - \frac{K_m}{\sqrt{3}} |\underline{V}| \ ; \ \underline{V} \cdot \underline{e}_i = 0\} .$$

5. *Application to simple loading conditions*

5.1. *"Plane strain" criteria*

The concept of a "strength criterion under plane strain conditions" within the strict framework of the yield design theory, i.e. without any explicit reference to a constitutive law, has been introduced and throroughly discussed in (Salençon, 1983). It can be briefly exposed as follows in the present case.

The support function π_{hom} defined by Eq. (5) makes it possible to determine the macroscopic strength criterion through the dual formulation :

$$F_{hom}(\underline{\underline{\Sigma}}) \leqslant 0 \iff \sup_{\underline{\underline{D}}} \{\underline{\underline{\Sigma}} : \underline{\underline{D}} - \pi_{hom}(\underline{\underline{D}})\} \leqslant 0 \tag{14}$$

(cf. Frémond and Friaà, 1978).

If the latter formula is used with the considered $\underline{\underline{D}}$ being restricted to plane strain rate tensors parallel to Ox_1x_2 , i.e. satisfying the conditions :

$$D_{3i} = D_{i3} = 0 \quad , \qquad i = 1, 2, 3 \tag{15}$$

it comes out that explicitating Eq. (16)

$$\sup_{\underline{\underline{D}}} \{\underline{\underline{\Sigma}} : \underline{\underline{D}} - \pi_{hom}(\underline{\underline{D}}) \; ; \; \underline{\underline{D}} \text{ satisf. Eq. (15)}\} \leqslant 0 \tag{16}$$

will provide an upper bound estimate for $F_{hom}(\underline{\underline{\Sigma}})$ where only the components Σ_{11} , $\Sigma_{12} = \Sigma_{21}$, and Σ_{22} of $\underline{\underline{\Sigma}}$ are concerned.

As a matter of fact, Eq. (16) gives the exact restrictions imposed to those components when generated by any $\underline{\underline{\Sigma}}$ under the condition $F_{hom}(\underline{\underline{\Sigma}}) \leqslant 0$. Eq. (16) can also be seen as the projection of the strength domain defined by Eq. (3) onto the subspace of the Σ_{ij} (i , j = 1 , 2) components of $\underline{\underline{\Sigma}}$.

It can (rather abusively) be named the "plane strain strength criterion" of the composite material.

The purpose of the present section is to derive lower and upper bound estimates for the "plane strain strength criterion" from the results obtained before.

Let then Σ_I and Σ_{II} with $\Sigma_I \geqslant \Sigma_{II}$ be defined as the principal values of the two-dimensional "stress tensor" generated by the components Σ_{11} , $\Sigma_{12} = \Sigma_{21}$, Σ_{22} in plane Ox_1x_2 , and θ the inclination of the major principal axis Σ_I to the fibers direction Ox_1 . After some calculations it turns out that the previously obtained lower and upper bounds write as follows.

- *Lower bound estimate* (11)

$$F^-(\Sigma_I , \Sigma_{II} , \theta) = \Sigma_I - \Sigma_{II} - K^-(\theta) \leqslant 0 \tag{17}$$

with putting $k = \langle K \rangle / K_m \geqslant 1$

$$K^-(\theta) = K_m \begin{cases} (k-1)\,|\cos 2\theta| + \sqrt{4/3 - [(k-1)^2 \sin^2 2\theta]} \\ \qquad\qquad \text{if} \quad |\tan 2\theta| \leqslant 2/[(k-1)\sqrt{3}] \\ 2/(\sqrt{3}\sin 2\theta) \qquad \text{otherwise} . \end{cases}$$

- *Upperbound estimate* (12)

$$F_1^+(\Sigma_I , \Sigma_{II} , \theta) = \Sigma_I - \Sigma_{II} - K_1^+(\theta) \leqslant 0 \tag{18}$$

where $K_1^+(\theta) = 2\,k\,K_m/\sqrt{3}$.

- *Upperbound estimate* (13)

For $i = 2$, $\underline{V} = V\,\underline{e}_1$ and taking Eq. (14) into account (13-b) becomes

$$\Sigma_{12}\, V \leqslant \frac{K_m}{\sqrt{3}}\, |V|$$

that is

$$|\Sigma_{12}| \leqslant \frac{K_m}{\sqrt{3}}$$

or

$$F_2^+(\Sigma_I\ ,\ \Sigma_{II}\ ,\ \theta) = \Sigma_I - \Sigma_{II} - K_2^+(\theta) \leqslant 0 \tag{19}$$

where

$$K_2^+(\theta) = 2K_m / (\sqrt{3}\ \sin 2\,\theta)$$

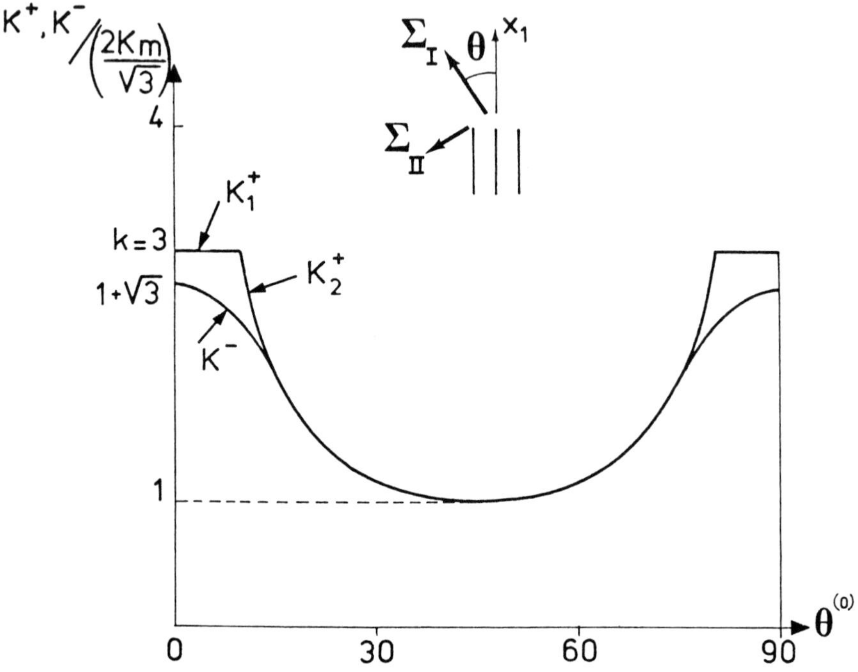

Fig. 3 : Upper and lower bound estimates for the strength of a unidirectional fiber composite under plane strain conditions.

The variations of K^- and $K^+ = \text{Min}\{K_1^+ , K_2^+\}$ as functions of the orientation θ are plotted in Fig. 3 where the numerical value of $k = 3$ has been taken by way of illustration. It appears that the relative difference between the two bounds remains lower than 20 % and even decreases to zero for θ ranging between 15° and 75°. It is worth noting in particular that the overall strength of the composite reduces to that of the matrix alone when the principal stresses are inclined at $\theta = 45°$ to the fibers direction.

5.2. Bounds on the uniaxial strength

The macroscopic solicitation writes in this case :

$$\underline{\underline{\Sigma}} = \Sigma \, \underline{e}_\theta \otimes \underline{e}_\theta$$

with $\underline{e}_\theta = \cos\theta \, \underline{e}_1 + \sin\theta \, \underline{e}_2$.

Denoting by $\Sigma^\star(\theta)$ the uniaxial tensile strength in the direction $\underline{e}_\theta$, defined by :

$$F_{hom}(\pm \Sigma^\star(\theta) \, \underline{e}_\theta \otimes \underline{e}_\theta) = 0$$

the strength condition for this particular type of solicitation can be written

$$F_{hom}(\underline{\underline{\Sigma}} = \Sigma \, \underline{e}_\theta \otimes \underline{e}_\theta) \leqslant 0 \iff |\Sigma| \leqslant \Sigma^\star(\theta) \ .$$

The following lower and upper bound estimates can be derived from the preceding results.

- *Lower bound*

From Eq. (11) one gets

$$\Sigma^-(\theta) \leqslant \Sigma^\star(\theta) \tag{20}$$

where :

$$\Sigma^-(\theta) = K_m \begin{cases} (k-1)\left(1-\frac{3}{2}\sin^2\theta\right) + \sqrt{1-3(k-1)^2\sin^2\theta\left(1-\frac{3}{4}\sin^2\theta\right)} \\ \qquad \text{if} \quad 0 \leqslant \theta \leqslant \theta^\star \\ \frac{1}{\sin\theta} \Big/ \sqrt{3\left(1-\frac{3}{4}\sin^2\theta\right)} \quad \text{if} \quad \theta^\star \leqslant \theta \leqslant 90° \end{cases}$$

where $\theta^\star$ is the value of θ for which the two above expressions give the same result.

This lower bound estimate decreases continuously from $\Sigma^- = k\,K_m$ for $\theta = 0°$ to its minimum value $\Sigma^- = K_m$ reached for $\theta \approx 57°,2$ $(\sin\theta = \sqrt{2/3})$, then increases again slightly up to $\Sigma^- = 2\,K_m/\sqrt{3}$ when $\theta = 90°$ (transverse loading) : see fig. 4.

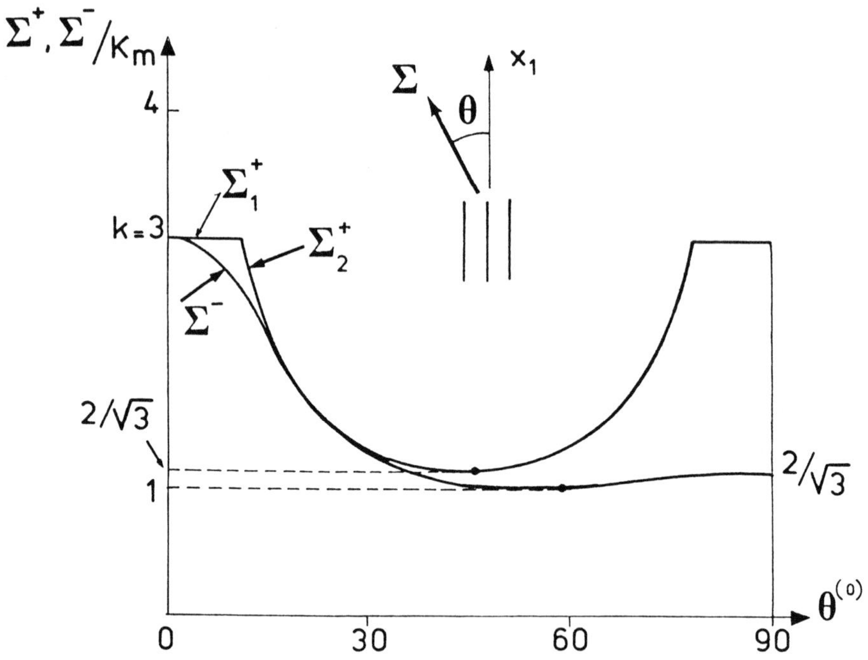

Fig. 4 : Upper and lower bounds for the uniaxial strength of a unidirectional composite.

• *Upper bounds*

A first upper bound estimate of $\Sigma^\star(\theta)$ derived from Eq. (12) is simply :

$$\Sigma^\star(\theta) \leqslant \Sigma_1^+ = k\,K_m \tag{21}$$

Making use of Eq. (13) with $i = 2$ and $\underline{V} = V\,\underline{e}_1$ yields immediately :

$$|\Sigma_{12}| \leqslant \frac{K_m}{\sqrt{3}}$$

and the upper bound $\Sigma_2^+(\theta)$:

$$\Sigma^\star(\theta) \leqslant \Sigma_2^+(\theta) = \frac{2\,K_m}{\sqrt{3}\,\sin 2\,\theta}\ . \tag{22}$$

The corresponding curves are drawn in Fig. 4 for $k = 3$. Unlike the case of plane strain conditions the gap between the two bounds widens considerably as soon as $\theta > 60°$. The actual value of the transverse strength for instance is bounded by $\Sigma^-(90°) = 2K_m/\sqrt{3}$ and $\Sigma^+(90°) = k\,K_m$.

However it can be proved (de Buhan and Taliercio, 1988 ; Taliercio, 1989) that these bounds turn out to be the exact value of the uniaxial strength of the composite for $\eta \ll 1$ (reinforcement by thin fibers) and $\eta = 1$ (matrix voluminal proportion reduced to zero) respectively. Therefore as a first approach, the uniaxial strength of the composite could be reasonably well approximated by the following interpolation formula :

$$\Sigma^\star(\theta) = (1-\eta)\,\Sigma^-(\theta) + \eta\,\Sigma^+(\theta)\ .$$

The validity of such a heuristic formula should be confirmed by a more elaborate analysis of the problem resorting for instance to numerical methods in order to solve the yield design problem defined over the unit cell and thereby to determine the macroscopic strength condition (Marigo and al., 1987 ; case of periodic porous media).

6. Conclusion

The yield design homogenization method has been successfully applied to the determination of upper and lower bounds for the macroscopic strength of a unidirectional fiber composite material. These bounds may be expressed by means of explicit analytical formulae in two circumstances : plane strain loading of the composite in a direction parallel to the fibers on the one hand and uniaxial solicitation on the other hand. The two bounds remain close to each other in the first case, thus providing a good estimate for the actual strength of the composite, whereas they diverge considerably in the second case when the composite is subjected to transverse loading. Such a drawback could be partially overcome by proposing a semi-empirical formula which linearly interpolates between the two bounds.

Appendix

Let $\underline{\underline{\Sigma}}$ be a macroscopic stress such that

$$\underline{\underline{\Sigma}} = \underline{\underline{\sigma}}^m + \eta\, \sigma^f\, \underline{e}_1 \otimes \underline{e}_1 \tag{9}$$

with

$$F_m(\underline{\underline{\sigma}}^m) = [J_2(\underline{\underline{\sigma}}^m)]^{1/2} - K_m/\sqrt{3} \leqslant 0 \tag{10-a}$$

and

$$|\sigma^f| \leqslant K_f - K_m \quad . \tag{10-b}$$

It comes out immediately :

$$\begin{aligned} & F_f(\underline{\underline{\sigma}}^m + \sigma^f\, \underline{e}_1 \otimes \underline{e}_1) \\ & = [J_2(\underline{\underline{\sigma}}^m + \sigma^f\, \underline{e}_1 \otimes \underline{e}_1)]^{1/2} - K_f/\sqrt{3} \\ & \leqslant [J_2(\underline{\underline{\sigma}}^m)]^{1/2} + [J_2(\sigma^f\, \underline{e}_1 \otimes \underline{e}_1)]^{1/2} - K_f/\sqrt{3} \end{aligned}$$

since $[J_2(\underline{\underline{\sigma}})]^{1/2}$ can be regarded as a norm for the deviatoric stress $\underline{\underline{s}}$.

Consequently on account of (10) :

$$F_f(\underline{\underline{\sigma}}^m + \sigma^f \, \underline{e}_1 \otimes \underline{e}_1) \leqslant \frac{|\sigma^f|}{\sqrt{3}} - \frac{(K_f - K_m)}{\sqrt{3}} \leqslant 0 \quad .$$

This last inequation shows that conditions (9) and (10) are sufficient to exhibit a piecewise constant stress field such as (7) in equilibrium with $\underline{\underline{\Sigma}}$ and satisfying the strength conditions defined by (8).

References

de Buhan, P., 1986. *Approche fondamentale du calcul à la rupture des ouvrages en sols renforcés*, Thesis, Université Pierre et Marie Curie, Paris.

de Buhan, P., and Salençon, J., 1987. "Yield strength of reinforced soils as anisotropic media", IUTAM symposium on *Yielding, Damage and Failure of Anisotropic Solids*, Villard-de-Lans, France. J.P. Boehler ed., to appear.

de Buhan, P., and Taliercio, A., 1988. "Critère de résistance macroscopique pour les matériaux composites à fibres", *C.R. Ac. Sc.*, 307, II, Paris, pp. 227-232.

Dvorak, G.J., and Bahei-el-Din, Y.A., 1982. "Plasticity Analysis of Fibrous Composites", *Jl. Applied Mechanics*, Vol. 49, pp. 327-335.

Frémond, M., and Friaâ , A. , 1978. Analyse limite. Comparaison des méthodes statique et cinématique, *C.R. Ac. Sc.*, Paris, t. 286, série A, pp. 107-110.

Marigo, J.J., Mialon, P., Michel, J.C., Suquet, P., 1987. "Plasticité et homogénéisation : un exemple de prévision des charges limites d'une structure hétérogène périodique", *J. Méca. Th. et Appl.*, 6, pp. 47-75.

Salençon, J., 1983. *Calcul à la rupture et analyse limite*, Presses de l'E.N.P.C., Paris.

Saurindranath, M., and Mc Laughlin, P.V., 1975. "Effects of phase geometry and volume fraction on the plane stress limit analysis of a unidirectional fiber reinforced composite", *Int. Jl. Solids Structures*, Vol. 11, pp. 777-791.

Sawicki, A., 1981. "Yield conditions for layered composites", *Int. Jl. Solids and Structures*, Vol. 17, pp. 969-979.

Suquet, P., 1985. "Elements of homogenization for inelastic solid mechanics", in *Homogenization techniques for composite media*, C.I.S.M., Udine, Italy. Springer Verlag, pp. 193-278.

Taliercio, A., 1989. *Studio del comportamento elastico e a rottura di materiali compositi a fibre e di elementi strutturali in composito*, thesis, Politecnico di Milano.

Static and Fatigue Biaxial Testing of Fiber Composites Using Thin Walled Tubular Specimens

by Tresna P.Soemardi, Dawei Lai and Claude Bathias

Abstract

A design and analysis of a multiaxial fatigue system employed for the testing of high performance continous fiber laminates are presented. A multiaxial laminate specimen based on axial-force and torsion loading of a thin walled tube has been developed which is appropriate for static and fatigue biaxial testing. The room temperature static biaxial tests and R=-1 fatigue biaxial tests at 1 Hz are also reported. In addition recent results on the biaxial failure of glass fiber/epoxy [0°] and [±45°]s tubes are presented and compared with theoretical failure envelopes.

Introduction

Acceptance and utilization of advanced fiber composites in structural application are steadily increasing, because of their combination of high performance with low density and potentially low unit cost.The composite investigated in this study is to be used by the Régie Nationale des Usines Renault-France for the automobile application of the suspension parts.The use of composites in load bearing structural parts like suspension components depends, to a large extent, on their ability to withstand multiaxial cyclic loading. It is therefore important for the designer to be well aware of the multiaxial fatigue behavior of the materials.Until now, design rules for multiaxial strength design of laminates are not wholly established.Previous investigators (see for example Owen and Griffiths (1978), Eckold et.al.(1978) ,Guess (1980), Owen and Rice (1981), Found (1985), Choo (1985) and Swanson et.al.(1988)) have reported that it is difficult to relate simple component tests, to the multiaxial strength properties of a laminated or filament-wound structure. In part this is due to the multiplicity of failure modes involving either fibers or matrixs or both, the many possible lamination sequences of interest, and certain complexities inherent in composite materials (see, Swanson, Christoforou and Colvin (1988)).

Further, there does not seem to be a well -established data base on the strength of composite laminates due to certain complexities inherent in composites such as free-edge stress concentration effect gross local buckling, local delamination and interfacial debonding, etc.(see,for example,Kim and Ebert (1978), Pluvinage(1987),Swanson,Christoforou and Colvin (1988), Soemardi, Lai and Bathias(1988,1989)). Another complexity is the large difference between the strength of the fiber and the matrix that must have fibers placed to carry the major loads. Additional complexities are that composites tend to have a sensitivity to the effect of specimen size and geometry (bar,plate,cylinder, etc.)(e.g.,see Bathias(1989)) and to stress concentration, making it somewhat more difficult to conduct well-characterized experiments (see, Swanson et.al(1988)).

A desirable approach to the problem of determining laminate multiaxial strength would seem to be that of characterizing the strengths of the individual plies under well-characterized multiaxial stress conditions in all possible failure modes. For determining laminate multiaxial strength,a number of previous investigators have used tubular specimens in order to avoid the edge-effect problem, as well as to take advantage of the ease of introducing multiaxial loading into this specimen (see, for example, Wu (1972), Grimes and Francis (1973), Guess and Gerstle (1977), Owen and Griffiths (1978), Duggen and Bailie (1980), Daniel et.al. (1980), Meshkov et.al (1982), Choo and Hull (1983), Foral and Humphrey (1985), Pluvinage(1987), Swanson, Christoforou and Colvin(1988), Soemardi, Lai and Bathias(1988,1989)). However, it is clear that the problem of end effects in tubular specimens use for composite materials must be accounted for(see, Whitney et.al(1973), Duggen and Bailie(1980), Daniel(1980), Swanson et.al(1988)).

In recent work we have developed a tubular specimen that appear to minimize these undesirable end effects, and have applied this specimen to a number of different test conditions and laminates (see, Pluvinage (1987), Soemardi, Lai and Bathias (1988,1989)). In this paper the design and fabrication of this specimen,the analysis that justifies this design are presented, and the method and the procedure of testing under static and fatigue biaxial loadings and the results of recent test programs are reported. A number of theoretical failure criteria are applied. Failure and damage stresses were compared to four criteria: three based on lamina strengths (TSAI-WU, MAXIMUM STRESS and NORRIS interaction), and one based on laminate uniaxial strength (NORRIS interaction). In fact, our research objectives were to study the damage mechanism and mechanical behavior of fiber composite tubes under static and fatigue biaxial loading. Research is continuing to study the microscopic failure mechanism in specimens by X-Ray Computed Tomography and Scanning Electron Microscope.

Specimen Design and Fabrication

Initial Tubes

The initial thin-walled glass fiber/epoxy tubes consisting of eight layers of [0°] and $[\pm 45°]_s$ laminates, which were supplied by the AEROSPATIALE-FRANCE were made of E-Glass fiber-epoxy resin 921GE5 unidirectional prepreg sheet (manufactured by CIBA-GEIGY),with 65,5% glass fiber mass content. Fig.1 shows the initial thin walled glass fiber/epoxy consisting of eight layers of [0°] and $[\pm 45°]_s$ laminates for achieving of about 1,5 mm tube wall thickness.The tubes are constructed in a nominal length of 150 mm with 54 mm gage length. The procedure for making initial thin-walled tubes was as follows. The unidirectional prepreg sheets were cut from the roll and wound around a blowable mandrel after it had been sprayed with release agent dry lubricant and the whole was placed in a rigid cylinder. And then the internal mandrel was blowed up in order to form the layers of prepreg sheets according to the external shell. The curing cycle consisted of a 135°C 2 hours and 0,4 MPa (4 bar) closed mould.The assembly was then slowly cooled to room temperature. A curing process including preheating and cooling lasted five to seven hours. Finally the tube was removed and then the wall thickness was measured at its two ends and middle circumferentially. Statistically, the real nominal wall thickness of an initial tube was 1,3±0,25 [mm] and $Rm = 1/2(R_o+R_i) = 26,75$ mm.

Cone-shaped end mounting and reinforcement

The stress concentration associated with the end-gripping region of a tubular specimen can lead to unacceptable problems, especially when used with materials such as composites that are relatively sensitive to stress concentrations. This problem has been addressed by adding material reinforcement to the end regions, and gradually tapering this reinforcement into the gage section of the specimen. The end reinforcement is then added to the specimen, consisting of fiberglass cloth overwraps with nearly the same glass content and a lower modulus epoxy. The cone-shape end mounting and reinforcement of the tube used in this study are the most desirable solution and the most feasible for the type of tube end mounting and reinforcement had been used in 1964 by Hotter,Schelling and Krauss, and followed by the other previous investigators (see, Highton and Soden (1982),Krempl and Niu (1982), Foral and Humphrey (1985) Swanson et.al (1988)). Fig.2 shows the thin-walled glass-fiber/epoxy tube with a conical or a tapered section on each end. This section is made of Stevens-Genin glass fiber/epoxy prepreg sheet. The procedure for making this conical section was as follows. A triangular prepreg tapes were cut from the prepreg and than wound around the ends of the tube. The determination of the optimum dimension of this triangular prepreg tape have asked for many trial and error models for obtaining a near perfect conical form. The assembly was then placed in two high precision moulds made by the Centre Technique Renault. Conical sections of the mould were manufactured precisely for obtaining an angle of six degrees in according to outer grip components. The moulds assembly with the reinforced tube inside was cured after it had been sprayed with release agent dry lubricant (WARLON-Spray Model-122, Fluorocarbon-release agent), with the curing cycle consisted of a 80°C 30 minutes preheat followed by 120°C and 0,6 MPa (6 bars) curing for 90 minutes. The moulds assembly was then slowly cooled and opened at 60°C (see Fig.3). This completes the specimen preparation.

Specimen Mounting

The type of the fixture used in this study (see Fig.4) compelled us to design carefully the geometry of the conical end section and its split wedge.Particularly the conical angle of this section in order to avoid sliding in torsion loading transmission. There is no problem in tension loading transmission. The 6° angle was determined based on the design calculation in the function of friction coefficient and sliding contacts between two materials, and also the ease of releasing the specimen from the fixture (see, Pluvinage (1987)). The first design outer grip was made of Aluminium with ordinary surface machined innerside. Because of the lightness of this outer grip made us easier in specimen mounting and releasing . But under maximum pressure on the specimen conical shaped end surfaces, still slippage exist under more than 600 Nm in torsion, even the tube [±45°] ultimate torsion reached 1200 Nm. We have designed a steel outer grip with a rough machined inner surface for overcoming this slippage.Of course we have to be more careful in mounting and releasing the specimen because of steel outergrip weight and inner surface roughness which can damage the specimen surface easily . An assembly drawing of the fixture with specimen is given in Fig.4. It consist of a main cover cylinder (2) which is bolted to the testing machine load cell (or the activator piston). The main cover cylinder is removed when the specimen, the attached outer grips(4) inner grip (3) are inserted. Tightening of the bolts presses the outer grips(4) and grips the specimen securely. To present crushing of the specimen an inner grip or a plug with a circle ends(3) is inserted into the specimen bore. The circle ends fit into a corresponding recess in the main cylinder(1). It is seen that torque is transmitted by friction forces on the inside surface of the inner grip and on conical outside section of the specimen . A schematic of the components, method of gripping and assembly of fixture is shown in fig.4. and photograph of biaxial-stress-test specimen in the loading frame in Fig.5.

Specimen Analysis

Contact Analysis

Contact analysis between two surfaces for determining p, the pressure on the cone-shaped surface and M, transmitted torsion were calculated from formulae,

$$p = \frac{4F}{\pi(D^2-d^2)\left[\frac{\tan\varphi}{\tan(\alpha/2)}+1\right]} \quad \text{[MPa]} \qquad (1)$$

using axial load, F(N), conical section greatest diameter, D[mm] and smallest diameter, d[mm], friction coefficient, $\mu=\tan\theta$, angle of friction, θ and conical section half -angle, $\alpha/2$,

$$M = \frac{\pi p \tan\varphi_o}{12\sin(\alpha/2)}(D^3-d^3)10^{-3} \quad \text{[MPa]} \qquad (2)$$

using adhesion coefficient, $\mu_o = \tan\varphi_o$, angle of adhesion, φ_o.

Based on equations (1) and (2) a plot of pressure on the conical surfaces, p vs axial load (F) as a function of angle (α) and also F vs the transmitted torsion (M) also as a function of angle (α)

were found.The friction coefficient (μ) between Aluminium surface and textile /epoxy was evaluated about 0,3 and the adhesion coefficient about 0,1.

Analysis on the stress distribution

In our previous work (see Pluvinage (1987)), finite-element analysis has been used to establish the specimen design, for justifying that stress distribution are uniform throughout the gage section. The results of the analysis of the current configuration will be presented here for unidirectional [0°] and angle-ply $[\pm 45°]_s$ glass fiber/epoxy laminates. The finite-element code MEF has been employed, using linear-elastic orthotropic six-node triangular elements. The wall thickness of the tube is 1,5 mm. Two computer program were used for the present investigation of the tubes. The first was Mesh pattern program (MOSAIC) capable of performing optimum mesh pattern and the second computer program (MEF) used for finite-element calculation. Both were originally developed at the Université de Technologie de Compiegne-France. Our tube specimen was symetrical about the gauge length so only one half of the tube was analysed. Axisymmetric triangular elements were available in MOSAIC for tension and torsion loading. Six-node (isoparametric) triangular elements were adopted. Finite-element calculations were carried out for axial tension and torsion loading in the elastic region. The material properties used are shown in table 1. The results are shown in Fig.6 for axial-force loading and in Fig.7 for torsion loading. The stresses for biaxial loading can be obtained by superposition of the stresses from axial-tension and axial-torsion. It can be seen that the axial and shear stresses are uniform throughout the gage section, and decrease in the vicinity of the ends. The stress is explained in normalized stress in percentage of the stress in the middle of gauge length, σ_o. In Fig.6, it can be seen that the axial stresses, σ_z are uniform in the thickness throughout about 40% of the gauge section. At the outer surface of [0°] laminate axial stresses at the end of the gauge section reach about 140% of the σ_o or 275 MPa and 60 MPa for the [±45°] laminate. At the inner surface at the same side reach 60% of the σ_o, or 118 MPa for [0°] laminate and 26 MPa for [±45°] laminate. It means that the effective gage length only 40% of the gauge section between the two conical ends. The gauge section length is 54 mm, so the effective gauge length is about 22 mm in the middle of the gauge section. This is important to note for determining the extensometer working gauge length. In Fig.7, it is seen that the shear stresses are also uniform throughout the gage section. There is a little variation , about 4% at the middle and at the end of the gauge section. It has been mentioned that there is stress variation in the tube wal thickness, however, these effects are still negligible for the present tubes. It may be noted that numerical simulation by FEA well justified specimen and fixture design. We should obtain a sufficient uniform stress distribution area in the gauge section for using an extensometer.

Method of Testing

For this paper, test objectives were to characterize the composite laminates in biaxial loading for ultimate static and fatigue strength. Axial-tensile stress, σ_z was produced by axial loading through end grips, and the shear stress $\tau_{z\theta}$ was generated by torsion loading. A total of one hundred specimens were tested at room temperature under uniaxial and biaxial static condition and under completely reversed load controlled uniaxial and biaxial loading at load frequency of 1 H_z. Proportional biaxial stressing of the specimen was achieved by adjusting the ratio of axial to torsion load with closed-loop electro hydraulic test equipment. All experiments used an INSTRON tension-torsion servohydraulic testing machine Model 1343. Electrical resistance strain gauges and an INSTRON Biaxial Extensometer (Fig.7) were used to monitor surface deformation in static tests. All the strain gauges were connected in quarter bridge formation. A data logging system was used to collect data from load(torque) transducers and strain gauges (extensometer). A micro-computer was connected to the data logger via the RS-232 interface. A real time data acquisition software, SEUC, was used for data tabulation and experimental graphics view. Only Instron biaxial extensometer was used in the fatigue test with sinusoidal load variation and zero mean load(R=-1). A fatigue test was terminated when the axial stroke went beyond 20 mm and the radial stroke went beyond 20° or catastrophic final failure was achieved. Catastrophic fracture was considered as final failure in fatigue testing. During fatigue testing, the hysteresis loop history was recorded in the data acquisition from the start until final failure, for further evaluation on stiffness degradation. In static tests, the specimens were tested by steadily increasing the load (torque) with linear variation over a period of about one minute. Firtst damage or first resin cracking of all specimens were identified by Acoustic Emission and also by visual examination aided by the light inside the specimen during fatigue testing, and the temperature on the gauge surface were also monitored.

Theoretical Consideration

Failure theories for anisotropic materials in plane stress conditions are in general empirical. The experimental values of laminate failure stresses are here compared in the tension-torsion an compression-torsion quadrants to a failure envelope based on laminate uniaxial strengths and three sets of failure envelopes calculated from lamina strengths.

Laminate failure criteria

The one chosen for comparison with experimental data was proposed by Norris and Mc Kinnon (1956):

$$\left(\frac{\sigma_z}{F_z}\right)^2 + \left(\frac{\tau_{z\theta}}{S_{z\theta}}\right)^2 = 1 \qquad (3)$$

This laminate failure criteria models the composite structure as a homogeneous, anisotropic solid, making no distinction for its layered construction and the possibility of different stress state within the individual layers. Fracture is thus interpreted in terms of average laminate uniaxial strengths (F_z in tension and $S_{z\theta}$ in shear).

Lamina failure criteria

The experimental values of ultimate strengths for individual layers or lamina within a composite structure have also been used as the bases for predicting laminate strengths. A thourough review of many failure criteria applicable to ply-by-ply analysis is presented for example by Owen and Griffiths(1978), Owen and Rice(1981) and Nahas(1986). Those considered in this paper were :

the lamina maximum stress criteria: $(\frac{\sigma_1}{X})\geq 1 \; ; \; (\frac{\sigma_2}{Y})\geq 1 \; ; \; (\frac{\tau_{12}}{S})\geq 1$ (4)

where, X,Y and S are the unidirectional strengths,

the Norris interaction: $(\frac{\sigma}{X})^2 + (\frac{\sigma_2}{Y})^2 + (\frac{\tau_{12}}{S})^2 = 1$ (5)

and Tsai-Wu criterion: $F_1\sigma_1+F_{11}\sigma_1{}^2+F_2\sigma_2+F_{22}\sigma_2{}^2+2F_{12}\sigma_1\sigma_2+F_{66}\sigma_6{}^2=1$ (6)

where, $F_1 = \frac{1}{X} - \frac{1}{X'}$, $F_2 = \frac{1}{Y} - \frac{1}{Y'}$, $F_{11} = \frac{1}{XX'}$, $F_{22} = \frac{1}{YY'}$, and F_{12} the interaction term can be treated as empirical constant (see, Tsai(1988)).

On application of lamina failure criteria

Application of a lamina failure criteria for predicting the damage and failure stresses is an analytical approach, a combined Lamination Theory (LT) and the criteria itself. Using LT, the natural co-ordinate ply stresses are found using the equation:

$$\{\sigma\}_n = |Q| \, \{\varepsilon\}_c \tag{7}$$

where $\{\sigma\}_n$ is the column matrix of natural co-ordinate stresses and $|Q|$ is the orthotropic Hooke's law matrix for the ply. The amplitude at failure of the known ratio of the three stress components can now be predicted for the individual ply using a chosen lamina failure criteria. The laminate failure load are found using (see, Snell(1978)):

$$\{N\}_f = p \, \{N\} \tag{8}$$

where $p = \frac{\sigma_{nf}}{\sigma_n}$. In this paper we used the applied stress ratio notation, $S=\frac{\text{axial stress}}{\text{shear stress}}$.

Static Test Results and Discussion

Static tests on glass fiber/epoxy [0°] tubes were performed in order to establish the lamina properties for this material. Static tests on the same material unidirectional coupon specimens were also performed in our previous work (see Pluvinage (1987)). Table 2 lists experimental and estimated properties of unidirectional [0°] glass fiber/epoxy tube using [0°] laminate. Typical stress vs strain response of [0°] tube during a tensile and torsion test are shown in Fig.8. Stress/strain response to failure were perfectly linear under axial tension load and non linear under pure torsion load. The same modulus is obtained when a [0°] tube is loaded into compression. The compressive strength is lower than the tensile strenth (65% of the UTS). The tensile strength data of [0°] tube has considerable variation. It is not surprising for non-homogeneous and low elongation composite laminate. Under torsion ,main fracture modes were resine fracture paralell to the fibers (see photo in Fig.11) and the initial micro crack was found at 50% of the ultimate shear stress. Beyond this region, the τ-γ curve is no more linear. Typical uniaxial stress/strain curves for glass fiber/epoxy [±45°]tube are also shown in Fig.8. Stress/strain responses to failure were perfectly linear under torsion load and non-linear under axial tension load. The initial slopes of the stress/strain curves in tension were used to determine the elastic modulus of [±45°] laminate in simple tension. The same modulus is obtained under axial compression and the compressive strength is significantly lower than the tensile strength (60% of the UTS). Based on these data and the lamination theory there is excellent agreement between the analytical stiffness matrix based on lamina properties and experimental results on [±45°] laminate(see Pluvinage(1987) and Soemardi et.al (1988,1989)). These composite systems follow the expected results that the stiffness matrix of composite laminates can be calculated with reasonable accuracy from laminated plate theory, stacking pattern and lamina properties as mention by previous investigators (see, Snell(1978), Guess(1980), Swanson and Trask(1980)).

The fundamental results of the biaxial static test is the failure stress diagram. These results show that there is a strong interaction between the axial-stress and the shear stress at failure of [0°] tubes (Fig.9) and [±45°] tubes (Fig.10). The stiffness of multi-directional cfrp laminates adequately determined using lamination theory, in contrast, biaxial failure stresses prediction for these laminates has hitherto been uncertain, although the recent use of various failure criteria has yielded reasonable agreement with experimental data for the biaxial strength of unidirectional composites, as also shown in Fig.9. Failure envelope in this figure correspond more to the matrix failure. The major fractures of [0°] tubes are matrix failure along fiber direction and the fracture of fibers was found only under uniaxial tension loading (see Fig.11). For the biaxial static test data of [0°] tubes in Fig.9, the failure and damage stresses in the tension-torsion quadrant are in good agreement with circular arcs based on Norris and Tsai-Wu formulas, and based on limited data in the compression-torsion quadrant the Tsai-Wu quadratic formula is seen more reasonable. It is more interesting to compare the failure data of a multidirectional laminate with the results of failure theories applied on a ply by ply basis. The four theories as applied to glass-fiber/epoxy [±45°] laminate under biaxial loading have been plotted in Fig.10 and compared with experimental biaxial damage and failure stresses. The "TSAI-WU" is an analytical approach, a combined Lamination theory (LT) and Tsai-Wu criteria. Using LT, the natural co-ordinate ply stresses are found using the equation(7). The amplitude at failure of the known ratio of the three stress components were predicted for the individual ply as shown in Fig.10 using Norris interaction,Tsai-Wu and Maximum Stress criteria. The laminate failure load are found using equation(8).The failure stresses in the tension-torsion quadrant with applied stress ratio $0 \leq S \leq 2$, and in the compression-torsion quadrant with $0 > S > -1$ (both are torsion dominant quadrant) are in excellent agreement with the Tsai-Wu criteria. Based on twelve experimental points only three points lie within the Tsai-Wu boundary with the little relative scatter variation compared with the Maximum Stress citeria. For pure tension, the Tsai-Wu is very under estimate, this may have been caused by the fiber failure which occured as the real major fracture. For pure compression, the Tsai-Wu is very over estimate, it is caused by unexpected local buckling and delamination which occured as the premature damage. The Maximum Stress and Norris laminate fit the data in all stress ratio equally well and more reasonable for predicting failure under pure tension and compression. In contrast, the Norris lamina failure criteria is very underestimate in shear dominant quadrant with the both failure and damage biaxial stresses. Nevertheless, the Norris lamina is seen applicable for predicting the uniaxial failure and damage of the laminate under pure tension and compression ($S=\infty$ and $S=-\infty$). An alternative, Norris laminate criteria which consider laminate as a whole fit also the data equally well with the maximum stress. The disadvantage of the Norris laminate, however, is that each different laminate of interest would have to be examined separately. With the ply by ply approach, there is the hope that failure rules can be identified that will hold for other laminates. Further, the laminate failure criteria are empirical and provide no insight into the lamina modes of failure, the sequence in which they occur, or single (or combination of) lamina failure modes that lead to total fracture of the laminate. For predicting the biaxial damage in all applied stress ratio, the Tsai-wu is seen most reliable and is in excellent agreement with the biaxial damage data.Typical fracture specimens under static loading are photographed in Fig 11. The fracture surfaces of the [±45°] tubes differed more significantly than the [0°] tubes, depending on the type of test. Fibers failure normal to the longitudinal axis of the [±45°] tube was found under uniaxial tension and under pure torsion. A gage wall striction was observed beyond 75% of the UTS, and remained permanently after failure. Laminate failure path with angle of 45° (angle of principal stresses under torsion load) was found under pure torsion and under biaxial loading. Under axial compression, local delamination of the outer layers appear to initiate laminate failure normal to the longitudinal axis.

Fatigue Test Results and Discussion

Fatigue tests on [0°] tubes were initially conducted under uniaxial and biaxial completely reversed load (R=-1) at 1 Hz in order to obtain the laminate fatigue behavior. Based on these results and the prediction on the lamina transverse direction fatigue behavior, we can establish the lamina fatigue life surface with a certain likely failure mode:

$$\{\sigma_{1a}, \sigma_{2a}, \tau_{12a}\} = f\ \{N\} \qquad (9)$$

Fig.12 and Fig.13 show the results of the uniaxial fatigue tests on the composite for [0°] tubes and [±45°] tubes under Completely Reversed Tension-Compression (CRTC) and under Completely Reversed Torsion-Torsion (CRTT). The alternating stress amplitude, σ_a is plotted as a function of the logarithm of the Number of cycle to failure,N (catastrophic rupture was considered as failure). In both Fig.12 and Fig.13 is seen that glass-fiber composite exhibit significant degradation of strength with number of cycles. The likely fracture modes of [0°] tubes under CRTC and under CRTT are matrix failure along fiber direction (see photograph in Fig 20), in contrast fiber failure are the major fractures for [±45°] tubes. For both specimens the fractures occur within the gage section, it indicate that specimen design is adequate for biaxial fatigue testing purpose. The typical damage process was found during axial fatigue testing on the [±45°] tube. As reported in our previous works under monotonic tension load the [±45°] tube exhibited a non-linear response with a high elongation and a time-dependent deformation. Similar phenomenon have been reported by Krempl and Niu (1982). Other particular, typical striction occured under high tension stress. Under CRTC at short lifes ($N \leq 10^5$ cycles) the tube wall exhibited also a visible striction during tension phase. Due to this significant striction effects carry over to the higher stress condition on the inner surface of the tube wall. This phenomenon and the completely reversed loading caused the initial damage in the form of inner surface ply failure normal to the axis of the tube. The crack propagation process was observed started from the inner surface circumferentially, and the reduction of load bearing effective area carry over to the high local elongation prior to failure. During axial fatigue testing a hysterisis loop developed which continously changes. Under load control, the axial strain and diametral strain can change with a gradual progression. This hysterisis indicated the "local damage" process explained above (see corresponding fracture mode in Fig.20). This typical local damage process of [±45°] tube in axial fatigue testing, was accelerated also by the rise in temperature prior to fracture. It showed that local irreversible damage cause a frictional heating. At failure the temperature has exceeded 80°C with the axial stiffness degradation about 10% nearly before total fracture. Based on this ramification, the significant intacts of stiffness degradation and rise in temperature will be a potential indicator for determining the starting damage in axial fatigue testing of composite [±45°] tube. Fig.14 shows the theoritical life surface of the lamina in terms of alternating axial stress amplitude (σ_{xa}), alternating shear stress amplitude (τ_{xya}) and Number of cycles (N). It is plotted based on the Norris interaction by equation(5),where,X(=$\sigma_{xa}{}^{u}$) and S(=$\tau_{xya}{}^{u}$) are the lamina life strength under CRTC loading and under CRTT loading, were predicted based on the S-N curves for R=-1 uniaxial and torsional fatigue showed in Fig.12 and Fig.13. It is seen that this simple formula fits good the biaxial experimental results particularly for the long fatigue life (10^5,10^6,...). It is important to note that this life surface represent the matrix failure as likely fracture mode for the [0°] tubes in uniaxial and biaxial fatigue testing. Fig 15 shows the theoritical life surface of [±45°] laminate based on the same formula in equation(5) with $\sigma_{xa}{}^{u}$ and $\tau_{xya}{}^{u}$ applied on a laminate basis. This theoritical life surface fits good also the results. It is more interesting to compare with the theoritical life surface applied on a ply by ply basis for well contributing the engineering design need on composite materials.

The literature on multiaxial fatigue behavior of composite is not very abundant, particularly for the fatigue included compression loading as also reported by Krempl and Niu(1982). Some references have reported the application of fatigue failure theories on a lamina basis(see Francis et.al(1977,1977a,1979),Owen and Griffiths(1978),Owen and Rice(1981),Found(1985)).

It was not practicable to cover all the condition of the fatigue program for exploring the theoritical damage and failure of [±45°] laminate under biaxial fatigue loading. Other consideration, at high alternating stress amplitude, the role of compression in damage process is dominant. In this case, the results will not be reasonable with existing failure theories. A reasonable starting for justifying the validity of existing failure theories in the compression included fatigue (negative R) is to conduct fatigue experiment more in long life level (N= 10^5,10^6,...). In this life level the compression is in balance with the tension in the damage process. Fig.16 shows the comparison between theoritical interactions based on lamina basis and the 10^5 cycle fatigue damage and failure stresses. The experimental 10^5 cycle failure stresses represent the fiber failure as a likely fatigue fracture mode of [±45°] tubes (see Fig.20). At 10^5 cycle life, the critical matrix damage was found at about 8,5x10^4 cycles indicated by significant intact degradation in stiffness and rise in temperature which accelerate the damage process to failure. In this case, the matrix failure life and total failure (fiber failure mode) life is not much difference. With the above explained compression and heating effects, Fig 16 shows the results which indicate failure stresses at 10^5 cycles are not in in agreement with the theoritical interaction particularly in the axial stress dominant quadrant.The failure theory for the compression included fatigue loading is in good agreement only in the low axial stress quadrant. Nevertheless, based on limited test data the damage stresses at 10^5 cycles are in good agreement with the two lamina criterion Tsai-Wu and Maximum Stress and the Norris laminate criteria. Fig 17 shows the maximum stress theory fits fairly good the fatigue results for the rupture and the onset of damage in [±45°] tubes at 10^6 cycles. Fig.19 demonstrates the illustration for the compression and heating effects on the life stresses of [±45°] tube. Fig.18

shows the N scatter for the life stresses in Fig 16 and Fig 17, and table 3 shows the principal fatigue strength of lamina for failure theory prediction at 10^5 and 10^6 cycles.

Summary and Conclusions

A detailed design has been presented for a biaxial test tubular specimen, subjected to combinations of torsional and axial loads, that appears to furnish valid data on the response and failure properties of composites. We have addressed the problem of end effects by reinforcing the ends with a cone-shaped end reinforcement blends into the gage section of the specimen.A method have been presented for designing, analysing and fabricating end reinforcement and grips for composite tubular specimen, and by this means it was possible to ensure that failure occured within the gage length. Fixtures are described appropriate for static and fatigue testing of thin-walled composite tubes of 55 mm outer diameter and for use in an Instron servohydraulic tension-torsion system. The methods described could be readily applied to a variety of tubes, reinforcement and loading conditions. Experiments include the results of one hundred tests on glass-fiber/epoxy [0°] and [±45°] tubes together with a number results for flat specimens have been performed. The room temperature static axial, shear and biaxial properties of tubes of eight layers of [0°] and [±45°]s glass-fiber/epoxy were determined. The agreement between theoritical interaction an experimental data was in most cases quite good. For the [±45°]tubes, the Tsai-Wu quadratic formula agree with the measured failure data particularly in the shear dominant quadrant ($-1<S<2$), and is in excellent agreement with the biaxial damage data. For biaxial fatigue loading conditions using R=-1 (completely reversed loading) and frequency of 1 Hz, the simple lamina failure theory which is the maximum stress criteria fits the data in most cases reasonable at long life (10^6 cycles). Since the maximum stress theory is more conservative and simpler, its use is recommended for engineering design. At the short life ($N \leq 10^5$cycles), the compression and heating effects is important in fatigue damage process of glass fiber/epoxy [±45°] tube in the axial load dominant quadrant . In this case the theoritical interactions were in poor agreement with the fatigue failure data.

Acknowledgement

This work was supported by Régie Nationale des Usines Renault (RNUR) and the Agence Français pour la Maitrise de l'Energie (AFME).

References

Ashkenazi,E.K., 1959, "On the Problem of Strength Orthotropy of Construction Materials",*Soviet Physics-Tech Physics,* Vol.4, No.9,pp.333-338.

Bathias,C.,1989, " The Fatigue of High Performance Composite Materials", *Advances in Fatigue Science and Technology,* C. Moura Branco and L. Guerra Rosa, Eds., pp.659-676.

Choo,V.K.S. and Hull,D.,1983,"Influence of Radial Compressive Stress Owing to Pressure on the Failure Modes of Composite Tube Specimens", *Journal of Composite Materials*,pp.344-356.

Choo,V.K.S.,1985, "Effect of Loading Path on the Failure of Fiber Reinforced Composite Tubes", *Journal of Composite Materials,* Vol.19, pp.525-532.

Daniel,I.M. et.al.,1980 "Analysis of Tubular Spêcimen for Biaxial Testing of Composite Laminates", *Proc.3rd ICCM*, Paris-1, 840-855.

Duggen,M.F. and Bailie,J.A., 1980, "A New Test Specimen Geometry for Achieving Uniform Biaxial Stress Distribution in Laminated Composite Cylinders", *Proc.3rd ICCM*, Paris-1, pp.900-913.

Eckold, G.C., Leadbetter,D., Soden,P.D. and Grigss,P.R., 1978, "Lamination Theory in the Prediction of Failure Envelopes for Filament Wound Materials Subjected to Biaxial Loading", *Composites*, pp.243.

Foral,R.F. and Humphrey,W.D.,1985, "Biaxial Stress Testing of Intraply Hybrid Composites", *Journal Comp.Tech. and Res.*.7(1), pp.19-25.

Found,M.S.,*1985*, "A Review of the Multiaxial Fatigue Testing of Fiber Reinforced Plastics", *Multiaxial Fatigue*,ASTM STP 853, K.J. Miller and M.W. Brown, Eds., *American Society for Testing and Materials*, Philadelphia, pp. 381-395.

Francis,P.H.,et.al., 1977, "Biaxial Fatigue Loading of Notched Composites", *NASA-CR*-145198.

Francis,P.H.,Walrath,D.E., Sims,D.F. and Weed,D.N.,1977,"Biaxial Fatigue Loading of Notched Composites",*J of Composite Materials*, pp.488-501.

Francis,P.H.,Walrath,D.E., and Weed,D.N.,1979, "First Ply Fatigue of G/E Laminates under biaxial Loadings", *Fibre Science and Technology*, 12,pp.97-110.

Fujczak,R.R.,1978,"Torsional Fatigue Behavior of Graphite-Epoxy Cylinders", *Proceeding of International Conference on Composite Materials*, Toronto, American Inst. of Mining, Metallurgical and Petroleum Engineers Corp.,pp.635-647.

Guess,T.R. and Gerstle, F.P., 1977, "Deformation and Fracture of Resin Matrix Composites in Combined Stress States", *Journal of Composite Materials*,11, pp. 146-163.

Guess,T.R., 1980, "Biaxial Testing of Composite Cylinders Experimental-Theoritical Comparison", *Composites*, pp.139-148.

Hashin,Z., and Rotem,Z., 1973," A Fatigue Failure Criterion for Fiber Reinforced Materials", *j of Composite Materials*, Vol 7, pp.448-464.

Highton,J. and Soden,P.D., 1982, "End Reinforcement and Grips for Anisotropic Tubes", *J. of Strain Analysis*, Vol.17, No.1,pp.31-43.

Hitchon,J.W. and Philips,D.C., 1978, "The Effect of Specimen Size on the Strength of cfrp", *Composites*, pp. 119-124.

Hotter,U., Scheling.H., Krauss,H., 1964,"An Experimental Study to Determine Failure Enveloppe of Composite Material with Tubular Specimens under Combined Loads and Comparison between Several Classical Criteria", *Agard CP 163*,pp.3.1-3.11.

Kim,H.C. and Ebert,I.J., "Axial Fatigue Failure Sequence and Mechanism in Unidirectional Fiber Glass Composites", *J.of Composites Materials*,Vol.12.

Krempl,E. and Niu,Tim,1982, "Graphite/Epoxy $[\pm 45°]_S$ Tubes Their Static Axial and Shear Properties and Their Fatigue Behavior under Completely Reversed Load Controlled Loading", J. of *Composite Materials*, Vol.16,p.172.

Meshkov,E.V.et.al., 1982, "Effect of Structural Parameters of Fiber-Glass on its Strength under a Complex Stressed State", *Mechanics of Composite Materials*,18(3),pp.299-302.

Nahas,M.N.,1986."Survey of Failure and Post Failure Theories of Laminated Fiber-Reinforced Composites" *J of Comp.Tech & Res.* vol.8.No.4.pp.138-153.

Norris,C.B.,,1950, "Strength of Orthotropic Materials Subjected to Combined Stresses", *Report 1816*, Forest Product Lab.Madison,NIS.

Norris,C.B. and Mc Kinnon,P.F.1956. "Compression,Tension and Shear Tests on Yellow Poplar Plywood Panels of Sizes that do not Buckle with tests Made at Variou Angle to the Face Grain",*Report No.1328*, US Forest Product Lab.

Owen, M.J. and Griffiths, J.R., 1978, "Evaluation of Biaxial Stress Failure Surfaces for a Glass Fabric Reinforced Polyester Resin under Static and Fatigue Loading", *J. of Material Science*, 13, No.7, pp.1521-1537.

Owen,M.J. and Rice,D.J.,1984, "Biaxial Strength Behavior of Glass Fabric-Reinforced Polyester Resins", *Composites*, pp.13-25.

Pluvinage,P.,1987,"Etude du Comportement en Fatigue Biaxiale de Cylindres Creux en Composite Verre/Epoxy", *DEA Thesis* ,Université de Technologie de Compiegne,France.

Snell,M.B., 1978, "Strength and Elastic Response of Symmetric Angle-Ply CFRP", *Composites*, pp.167-176

Soemardi,T.P.,Lai,D., and Bathias,C.,1988,"Etude du Comportement et de l'Endommagement en Fatigue Biaxiale de Cylindres Creux en Composite Verre/Epoxy, *Rapport Intermediare ITMA-RNUR*, pp.67-78.

Soemardi,T.P., Lai, D., and Bathias, C., 1989, " Etude du Comportement Mecanique de Tube en Materiaux Composites Soumis à des Sollicitations Axiales at Biaxèe", *Rapport Final ITMA-RNUR*, pp.25-26.

Swanson, S.R., Messick,M., and Tian,Z., 1987, " Failure of Carbon/Epoxy Laminate under Combined Stress", *J. of Composite Materials*,Vol.21,pp.619-630.

Swanson,S.R., and Trask,B.C., 1989, "Strength of Quasi-Isotropic Laminates under Off-Axis Loading", *Comp.Sci. and Tech.*,34,19-34.

Swanson,S.R., Christoforou,A.P., and Colvin, Jr., G.E., 1988, "Biaxial Testing of Fiber Composites Using Tubular Specimens", *Experimental Mechanics*, pp.238-243.

Tennyson,R.C.,MacDonald,D., and Nanyard,A.P., 1978, "Evaluation of the Tensor Polynomial Failure Criterion for Composite Materials", *J.of Composite Materials*,12,pp.63-75.
Tsai,S.W., and WU,E.M., 1971, "A General Theory of Strength of Anisotropic Materials", *J. of Composite Materials*, 5, pp.58-80.
Tsai,S.W.,1988, "Composites Design 1988", *Think Composites*,Dayton.OH.
Whitney,J.M.,Grimes,G.C., and Francis,P.H.,1973, "Effect of End Attachment on the Strength of Fiber-Reinforced Composite Cylinders", Experimental Mechanics, 13(5),pp.185-192.
Wu,E.H.,1972,"Optimal Experimental Measurement of Anisotropic Failure Tensors", *J. of Composite Materials*, pp.472-489.

Tables and Figures

Table 1. Elastic properties used in specimen analysis

Property	Value
E_{11}	40 GPa
E_{22}	10 GPa
G_{12}	4 GPa
ν_{12}	0,30

Table 2. Summary of glass fiber/epoxy lamina properties.

Property	Value	Property	Value	
Volume fraction	0,65			
Elastic constants		Strengths	Total failure	Resin cracking
Ex	43 GPa	X	550 MPa	225 MPa
Ey	11 GPa	X'	-350 MPa	-200 MPa
Gxy	4,3 GPa	Y	40 MPa	20 MPa
nx	0,28	Y'	118 MPa	60 MPa
ny	0,07	S	45 MPa	20 MPa

Table 3. Fatigue principal strengths of lamina for failure theory prediction at N cycles.

N	Resin cracking strengths [MPa]			Rupture strengths [MPa]		
cycles	Xa	Ya	Sa	Xa	Ya	Sa
10^5	150	10	10	199	12.5	19
10^6	145	9	9	195	11.6	17

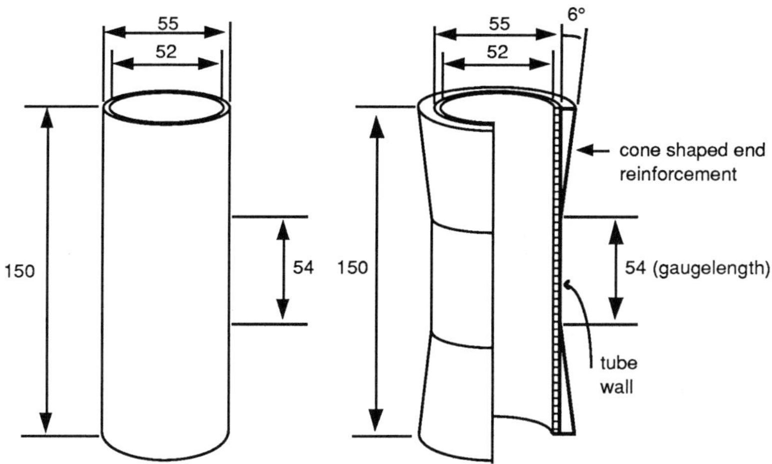

Fig.1 Initial Tube

Fig.2 Thin walled tube Specimen

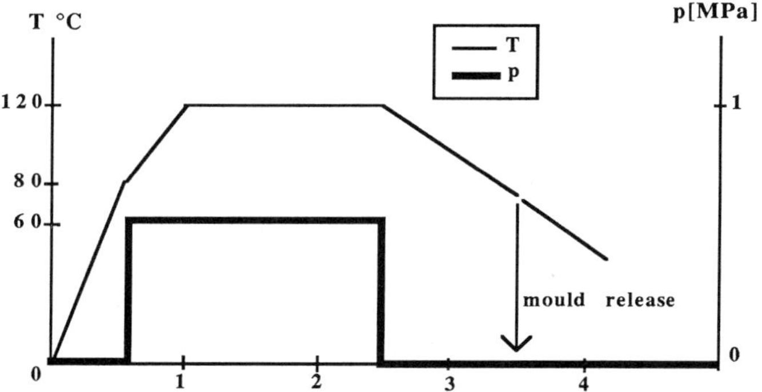

Fig.3 End reinforcement curing process procedure

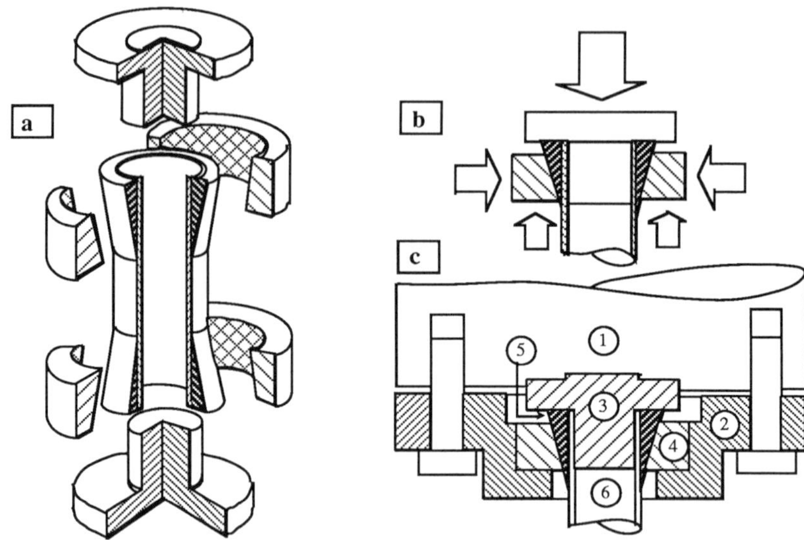

Fig.4.(a) Components, (b) Method of gripping, (c) Assembly of fixture (1)testing machine load cell (2) main cover cylinder (3) Inner grip (4) outer grip (5) conical end reinforcement (6) tube wall

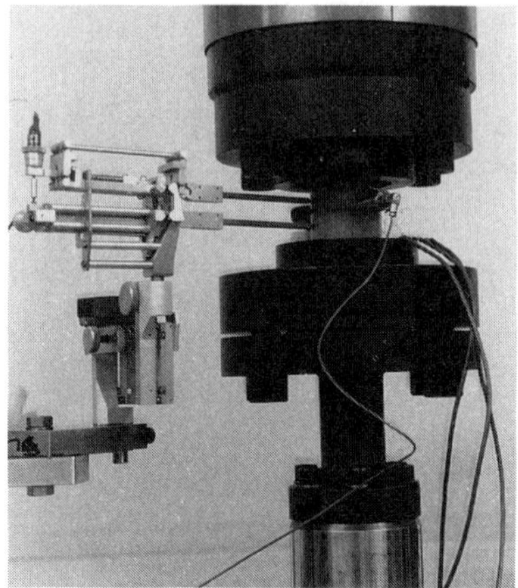

Fig.5. Biaxial-stress specimen in test loading frame

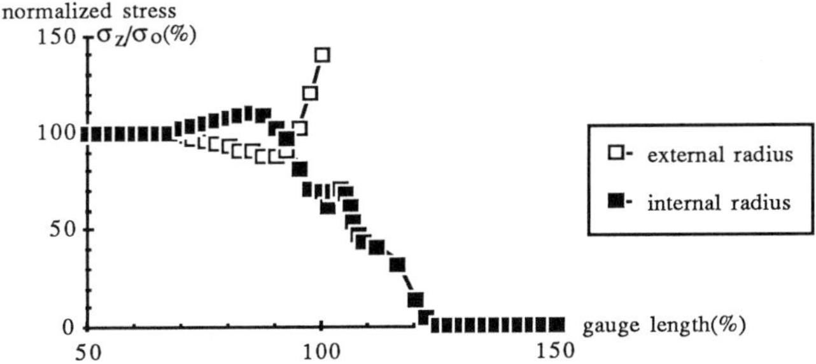

Fig.6 Finite-Element Solution for stress distribution in a composite tubular test specimen under axial tension (imposed deformation 0,1 mm)

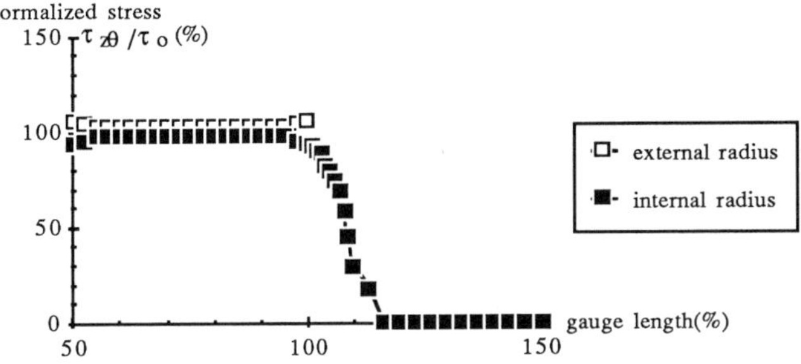

Fig.7 Finite-Element Solution for stress distribution in a composite tubular test specimen under torsion (imposed torsion 17 Nm).

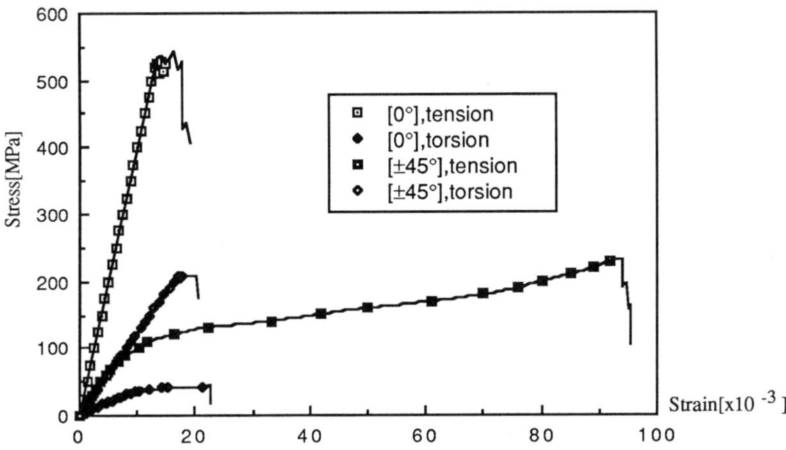

Fig.8 Typical stress/strain responses for [0°] and [±45°] tubes during tensile and torsion tests.

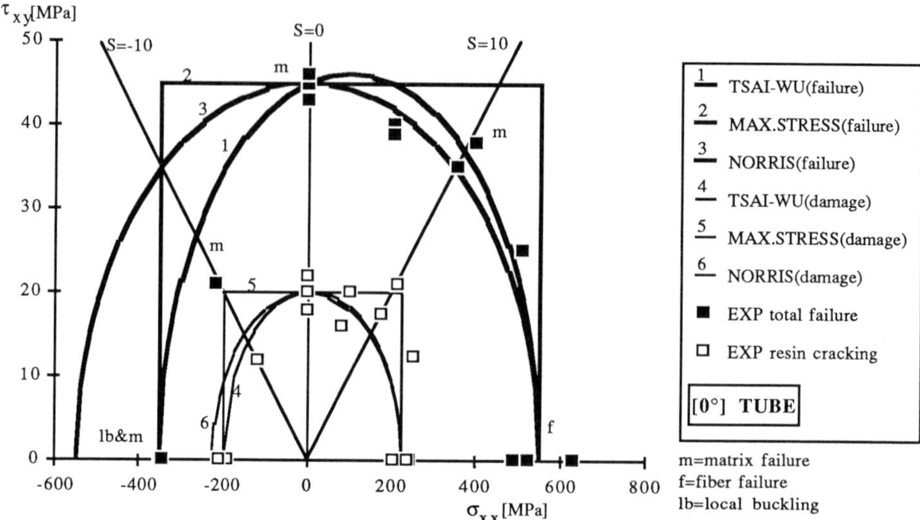

Fig.9 Experimental biaxial failure stresses of [0°] tube and the theoritical interactions.

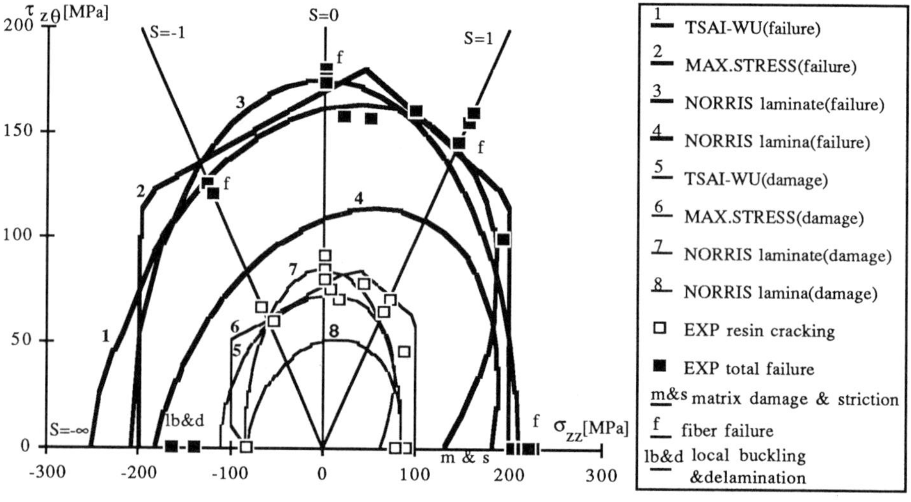

Fig.10 Experimental biaxial failure stresses of [±45°] tube and the comparison with some theoritical criteria based on lamina properties.

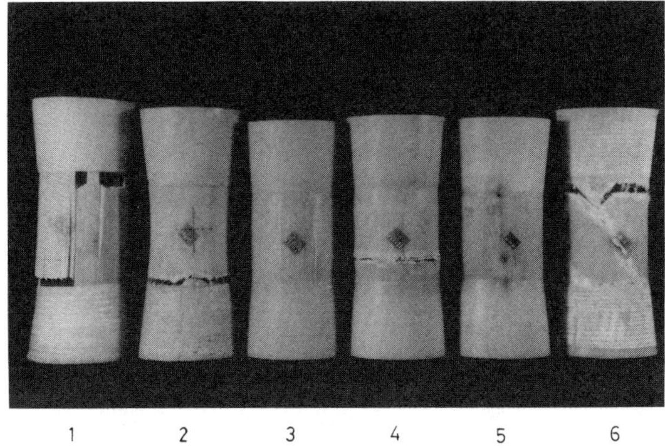

Fig 11. Photograph of typical failed specimens under static loading.
(1,3,5) [0°] tubes under monotonic axial tension, compression and torsion
(2,4,6) [±45°] tubes under monotonic axial tension,compression and torsion

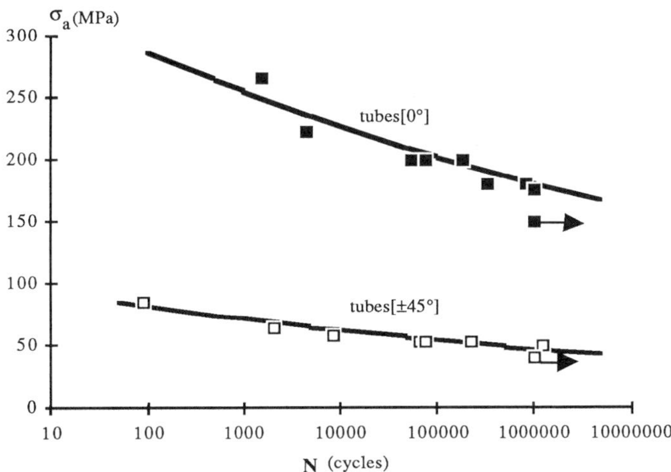

Fig.12 Typical axial fatigue results for [0°] tubes and [±45°] tubes, R=-1, 1 Hz.

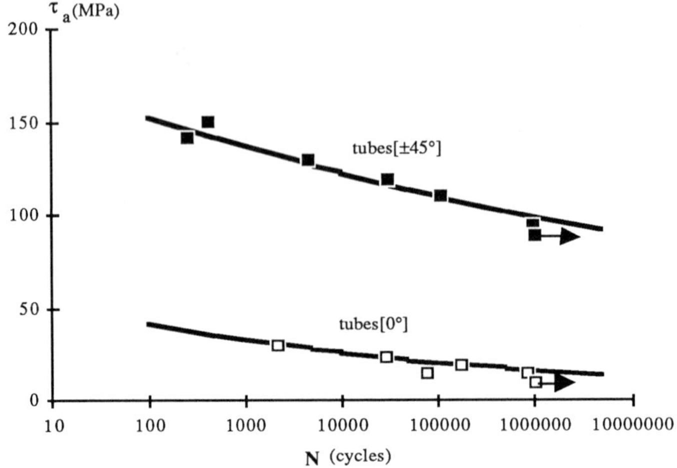

Fig.13 Typical torsional fatigue results for [0°] tubes and [±45°] tubes, R=-1, 1 Hz.

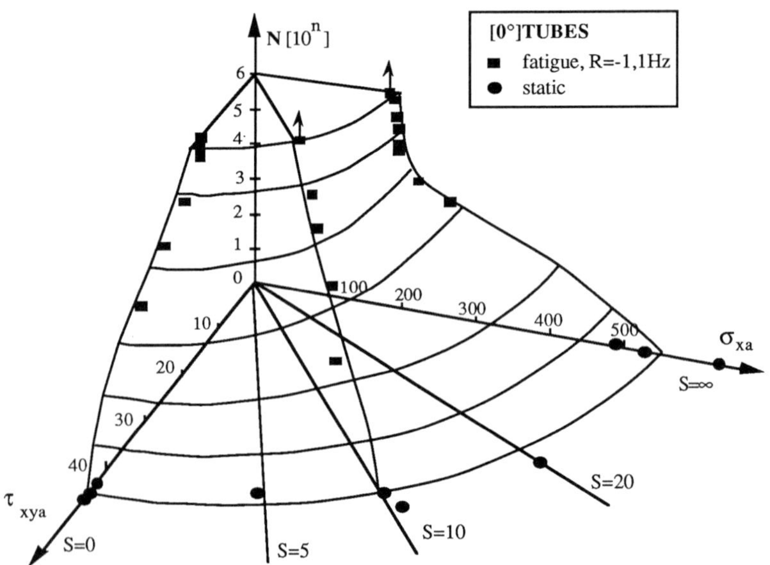

Fig.14 Fatigue results for rupture of [0°]tubes presented in life surface diagram.

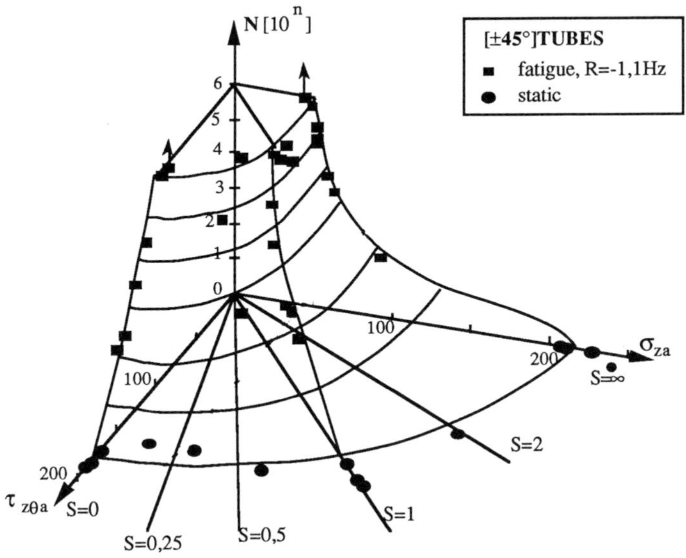

Fig.15 Fatigue results for rupture of [±45°] tubes presented in life surface diagram.

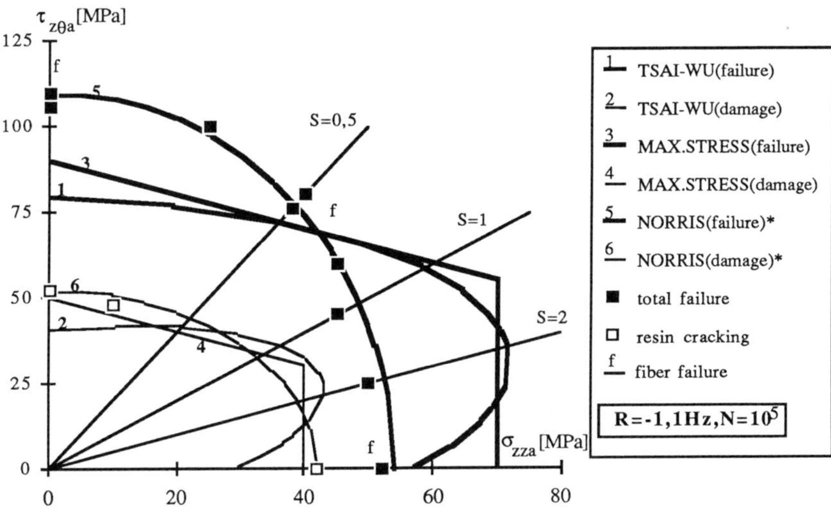

Fig.16 Fatigue results for the rupture and for the onset of damage in [±45°] tubes at 10^5 cycles compared with the theoritical interaction on lamina basis and laminate(*) basis.

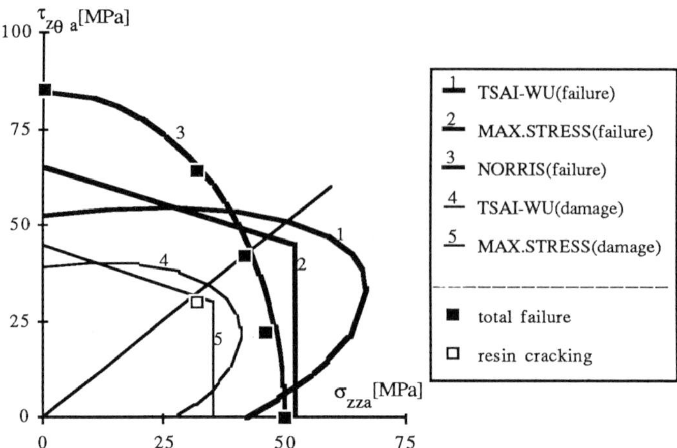

Fig.17 Fatigue results for the rupture and for the onset of damage in [±45°] tubes at 10^6 cycles compared with the theoritical interaction on a lamina basis.

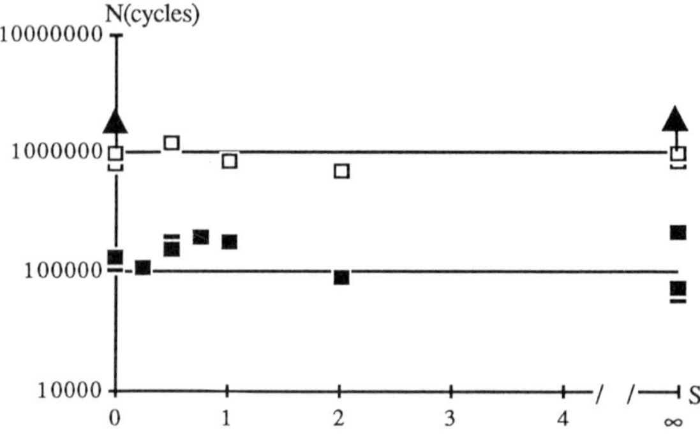

Fig.18 Scatter of N for the biaxial stresses plotted in Fig.16 and Fig.17 .

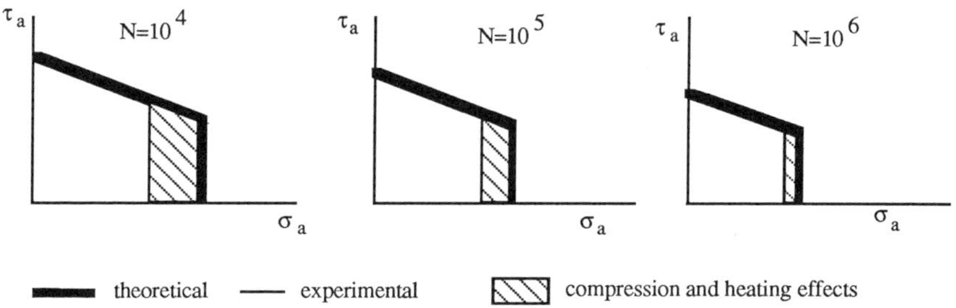

Fig.19 Ilustration for the completely reversed loading(R=-1) and heating effecs in[±45°] tubes on the agreement with the theoritical interaction on a lamina basis.

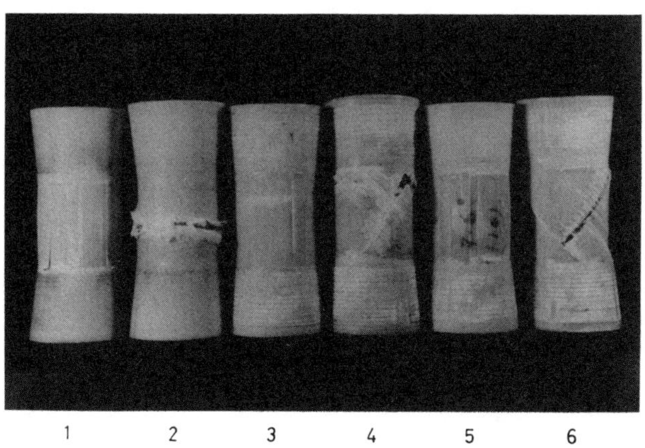

1 2 3 4 5 6

Fig 20. Photograph of typical failed specimens under fatigue loading.
(1,3,5) [0°]tubes under tension-compression, torsion-torsion and tension/compression-torsion
(2,4,6)[±45°]tubes under tension-compression, torsion-torsion and tension/compression-torsion

Fracture

Mesomodeling of Damage for Laminate Composites: Application to Delamination

P. LADEVEZE - O. ALLIX - L. DAUDEVILLE

Laboratoire de Mécanique et Technologie
ENS Cachan/CNRS/Université Paris 6
61 avenue du Président Wilson 94235 CACHAN CEDEX

For Laminate Composite Structures such as T300-914 composites a material mechanical modeling is proposed.When included in a structural analysis code, it allows to simulate the failure of a structure and more generally to estimate its damage state compared to one or several ultimate states.

Damage refers to the more or less gradual development of microvoids and microcracks. Brittle and progressive damage mechanisms are both present.

For laminate composite structures, the first idea is to reduce the behavior of any composite laminate to the modeling of two constituents : the elementary single layer and the interface. The latter is a zero thickness entity which depends on the angle between two adjacent layers. Another point is that the damage state is constant within the thickness for the single layer. Thus, we have chosen a particular scale, named mesoscale, for damage modeling. It is an intermediate and preferential scale between the macro and micro scales where damage phenomena can be described in a simple way.

The modelings include anisotropic unilateral damage and anelastic behavior following a general damage mechanics approach for composites given in Ladeveze 1983 - 1986 - 1989. Then, new variables, named damage variables and related to the change of elastic constants are added to the classical ones in order to describe the state of the material. This type of damage indicator has been introduced by Kachanov 1958 and Rabatnov 1968 and has been developed since 10 years by many authors.

The modeling of the elementary single layer, aside the rupture of the fibers, takes into account the microcracking of the matrix and the deterioration of the fiber-matrix bonding, which are considered as separate damage mechanisms (see Le Dantec 1989). Some more or less qualitative information is thus transfered at the single layer level by means of a homogenization technique. The observed compression behavior in the fiber direction which is different from the tension one is also included. For static loadings, the classical laminate theory enables us to identify the material constants and functions by means of three tensile tests $[+45,-45]_{2s}$, $[0,90]_{2s}$, $[62.5,-62.5]_{2s}$ and a four-point bending test on $[0,90]_{8s}$. The paper completes previous works Ladeveze 1986, Gilletta and al. 1986 and gives a final modeling. Our aim is to insist on the possibilities of such a simple modeling : several examples are detailed which allow a comparison with other approaches. For two materials T300-914 and IM6-914, our modeling has been checked on various stacking sequences with various tensile directions. This comparison which is satisfying is reported here.

The interface is a two dimensional entity which ensures displacement and traction transfers from one layer to another. Its influence is located near edges or defects where a three-dimensional stress state may occur and lead to delamination. A simple damage modeling with anelastic behavior is given.

A first application on a mode I delamination problem is presented. The numerical results allow to appreciate the connection between this approach and the classical one which uses linear Fracture Mechanics.

To end, the approach is used to compute until rupture a composite structure. The rupture phenomenon happens after two phases. In a first step, it is the initiation damage stage. From the critical point (or from a point just beside), the strain and also the damage variables become more and more localized ; a macrocrack appears and grows until becoming unstable. If the first stage is for any material well described with a damage approach, the full calculation of the rupture leads to severe difficulties. For our approach of laminate composites, such difficulties vanish. First computation results obtained by Allix 1989 on the delamination analysis of an initially circular hole are given.

1 - Mesomodeling of a Composite Laminate - General Ideas

The composite may be schematized by :

- a single layer which is homogeneous in the thickness
- an interface which is a surface entity connecting two adjacent layers and which depends on the relative directions of their fibers.

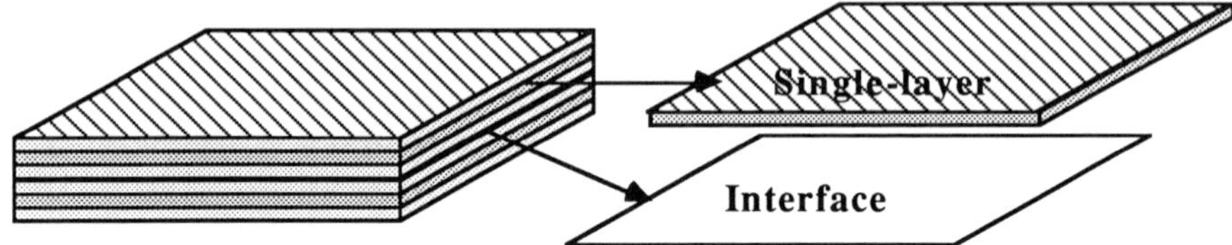

Figure 1 : Laminate modeling

These entities being modelized and identified, the mechanical behavior reconstitution of any laminate is then a relatively easy task.

We consider here single layers with only one reinforced direction. Let us note that the single layer is also analyzed at the level of its constituents : fibers, matrix, interfaces. Some more or less qualitative information is thus transfered at the single layer level.

2 - Modeling of the Single Layer

2.1 Tension and compression behaviors in the fiber direction

These behaviors are different. To get the compression one, we use a four-point bending test described in Vittecoq 1990. The main result is that the behavior is purely elastic but non-linear. The compression modulus can be written :

$$E_1^c = E_1^0 \left[1 - \frac{\alpha}{E_1^0} \langle -\sigma_{11} \rangle_+\right]$$

where α is a material constant and $\langle\ \rangle_+$ the positive part.

2.2 Damage kinematics

The identification and the modeling of the single layer is made with the assumption of in-plane stresses. In what follows subscript 1, 2 and 3 designate respectively the fiber direction, the transverse direction inside the layer, and the normal direction.

Aside brittle fractures in the fiber direction, the matrix and the fiber-matrix interfaces inside the layer are deteriorated in a very particular manner ; the microcracks are parallel to the fiber direction. A homogenization calculation shows that the only moduli which are modified are the transverse modulus E_2 and the shear modulus G_{12}. The other independent elastic characteristics E_1 and ν_{12} remain constant. These properties are confirmed by experimental observations.

The undamaged material strain energy, is written in the following form, reached by splitting up the energy into "tension" energy and "compression" energy :

$$E_D = \frac{1}{2}\left[\frac{\langle\sigma_{11}\rangle_+^2}{E_1^0} + \frac{\varphi(\langle-\sigma_{11}\rangle_+)}{E_1^0} - \left(\frac{\nu_{12}^2}{E_1^0} + \frac{\nu_{21}^2}{E_2^0}\right)\sigma_{11}\,\sigma_{22} + \frac{\langle\sigma_{22}\rangle_+^2}{E_2^0} + \frac{\langle-\sigma_{22}\rangle_+^2}{E_2^0} + \frac{\sigma_{12}^2}{G_{12}^0}\right]$$

Where φ is a material function defined such that : $\dfrac{\partial^2 E_D}{\partial^2\sigma_{11}} = \dfrac{1}{E_1^c}$ (see 2.1)

The transverse rigidity in compression being supposed equal to E_2^0, one obtains the following energy for the damaged material :

$$E_D = \frac{1}{2}\left[\frac{\langle\sigma_{11}\rangle_+^2}{E_1^0} + \frac{\varphi(\langle-\sigma_{11}\rangle_+)}{E_1^0} - \left(\frac{\nu_{12}^2}{E_1^0} + \frac{\nu_{21}^2}{E_2^0}\right)\sigma_{11}\,\sigma_{22} + \frac{\langle\sigma_{22}\rangle_+^2}{(1-d')E_2^0} + \frac{\langle-\sigma_{22}\rangle_+^2}{E_2^0} + \frac{\sigma_{12}^2}{(1-d)G_{12}^0}\right]$$

d, d' are two scalar internal variables which are constant within the thickness. They define the damage of the single layer. The conjugate variables associated with the mechanical dissipation are :

$$Y_d = -\rho\frac{\partial \langle\langle\psi\rangle\rangle}{\partial d}\Big|_{\tilde{\sigma}:cst} = \frac{\partial}{\partial d}\langle\langle E_D\rangle\rangle\Big|_{\sigma:cst} = \frac{1}{2}\frac{\langle\langle \sigma_{12}^2\rangle\rangle}{G_{12}^0(1-d)^2}$$

$$Y_{d'} = -\rho\frac{\partial \langle\langle\psi\rangle\rangle}{\partial d'}\Big|_{\tilde{\sigma}:cst} = \frac{\partial}{\partial d'}\langle\langle E_D\rangle\rangle\Big|_{\sigma:cst} = \frac{1}{2}\frac{\langle\langle \langle\sigma_{22}\rangle_+^2\rangle\rangle}{E_{22}^0(1-d')^2}$$

ψ is the free energy and << >> denotes the mean value within the thickness. $\tilde{\sigma}$ is the effective stress tensor ; we will see its expression later.

2.2 Damage evolution law

For static loadings, we have :

$$d\big|_t = A_d\left(Y_d\big|_\tau, Y_{d'}\big|_\tau \quad \tau \le t\right) ; \qquad d'\big|_t = A_{d'}\left(Y_d\big|_\tau, Y_{d'}\big|_\tau \quad \tau \le t\right)$$

where the operators A_d, $A_{d'}$ are material characteristics.

Let us define :

$$\underline{Y}_d = \sup|_{\tau\leq t} Y_d \qquad \text{and} \qquad \underline{Y}_{d'} = \sup|_{\tau\leq t} Y_{d'}$$

A simple law satisfied for carbon-expoy materials is defined :

$$d = \frac{1}{Y_c^{1/2}} < (\underline{Y}_d + b\,\underline{Y}_{d'})^{1/2} - Y_0^{1/2} >_+ \quad \text{if } d < 1 \qquad ; d = 1 \text{ otherwise}$$

$$d' = b'd \qquad \text{if } d' < 1 \qquad \text{and} \qquad Yd' < Yc' \qquad ; d' = 1 \text{ otherwise}$$

Y_0, Y_c, $Y_{c'}$, b are damage characteristic constants of the material. In practice, damages equal to 1 are rarely encountered : rupture appears far below that value and is associated with instability. Hence $\underline{Yd}$ and $\underline{Y}_{d'}$ appear as the quantities which govern the damage increasing and then the transverse fracture of the single layer.

Two different damage mechanisms are introduced (see Le Dantec 1989). The first one is a progressive damage mode related to the micro-cracking of the matrix and the deterioration of the fiber-matrix interfaces. The second mode is a brittle one ; it concerns the deterioration of the fiber-matrix interfaces submitted to a transverse tension stress σ_{22}.

An important variant of the previous damage evolution law is :

$$\dot{d} = k < \frac{(Y_d + bY_{d'})^{1/2} - \underset{\sim}{Y}^{1/2}(d)}{Y_c^{1/2}} >_+^n \qquad \text{if } d < 1 \text{ ; } d = 1 \text{ otherwise}$$

$$\text{with} \quad \underset{\sim}{Y}^{1/2}(d) = Y_0^{1/2} + Y_c^{1/2}\, d$$

$$\dot{d}' = b\dot{d} + k' < \frac{Y_{d'}^{1/2} - Y_{c'}^{1/2}}{Y_{c'}^{1/2}} >_+^{n'} \qquad \text{if } d' < 1 \text{ ; } d' = 1 \text{ otherwise}$$

where n, n', k, k' are material parameters. These modelings introduce delay effects ; they differ from the previous one only if the damage rates are very high.

Remark : In order to simplify the general case where out-plane stresses are present, Young's modulus E_3, the shear modulus G_{13} and G_{23} remain constant and thus damage effects of out-plane stresses are included in the interface behavior only.

2.3 Damage-plasticity (or viscoplasticity) coupling

The microcracking of matrix and the deterioration of fiber-matrix interfaces lead to sliding with friction and then to anelastic strains. A way to modelize these phenomena is to use plasticity or viscoplasticity mechanical modeling.

The idea which seems to work quite well is to build the modeling upon quantities which are called "effective" :

- effective stress $\tilde{\sigma}$
- effective anelastic strain rate $\dot{\tilde{\varepsilon}}_p$

which verify : $\quad \mathrm{Tr}\,[\,\tilde{\sigma}\,\dot{\tilde{\varepsilon}}_p\,] = \mathrm{Tr}\,[\,\sigma\,\dot{\varepsilon}_p\,]$

We define for this family of composites :

$$\tilde{\sigma}_{12} = \frac{\sigma_{12}}{(1 - d')} \quad ; \quad \tilde{\sigma}_{22} = \frac{\sigma_{22}}{(1 - d')} \quad ; \quad \tilde{\sigma}_{11} = \sigma_{11}$$

$$\dot{\tilde{\varepsilon}}^p_{12} = \dot{\varepsilon}^p_{12}(1 - d) \quad ; \quad \dot{\tilde{\varepsilon}}^p_{22} = \dot{\varepsilon}^p_{22}\,(1 - d') \quad ; \quad \tilde{\varepsilon}^p_{11} = \varepsilon^p_{11}$$

From a homogenization calculation, and assuming that only the matrix has an anelastic behavior, we have $\varepsilon^p_{11} = 0$. The elastic domain is defined by :

$$f = (\tilde{\sigma}^2_{12} + a^2\,\tilde{\sigma}^2_{12})^{1/2} - R - R_0$$

The hardening is assumed to be isotropic, which means that the threshold R is a function of the cumulated strain p ; p $\rightarrow$ R(p) is a characteristic function of the material. The yield conditions are written as follows :

$$\dot{\tilde{\varepsilon}}^p_{12} = \frac{1}{2}\,\dot{p}\,\frac{\tilde{\sigma}_{12}}{R + R_0} \quad ; \quad \dot{\tilde{\varepsilon}}^p_{22} = a^2\,\dot{p}\,\frac{\tilde{\sigma}_{22}}{R + R_0} \quad ; \quad (\dot{p} \geq 0)$$

An example of such a hardening curve is given, for T300-914 material, in figure 2.

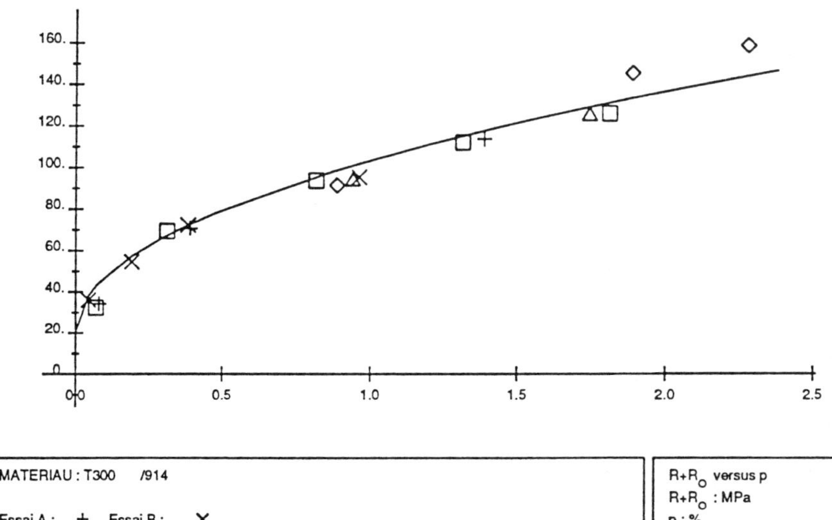

Figure 2 : Hardening curve for T300-914

2.4 Identification and checking

The modeling described above, aside the initial elastic constants, depends on :

- the rupture tension and compression strains in the direction of the fibers ε_T, ε_c and the compression coefficient α
- the hardening curve $p \rightarrow R_0 + R(p)$ and the constant a
- the b, Y_0, Y_c, $Y_{c'}$ constants which define the damage evolution laws.

The model has been identified for several laminates, in particular the T300-914 and IM6-914. α and ε_c are determined through a bending test. For the damage modeling the main test is a tensile test on a $[+45,-45]_{2s}$ specimen. A complementary test, for example on a $[+67.5,-67.5]_{2s}$ specimen is used to identify the constants a^2 and b which precise the coupling between the shear stress and the transverse stress.

This modeling has been checked out on several experimental tests :
$[0,90]_{2s}$, $[0]_8$, $[45]_8$, $[\pm 45]_{2s}$, $[\pm 67.5]_{2s}$, $[67.5,22.5]_{2s}$, $[-12,78]_{2s}$, $[\pm 22.5]_{2s}$, $[\pm 10]_{2s}$

For the last two tests, the delamination seems to be important, then the rupture experimental values are lower than those given by our modeling. This fact is not surprising because delamination is connected with the interlaminar interface deterioration. Such a damage mode is taken into account thanks to the interface modeling (see 4).

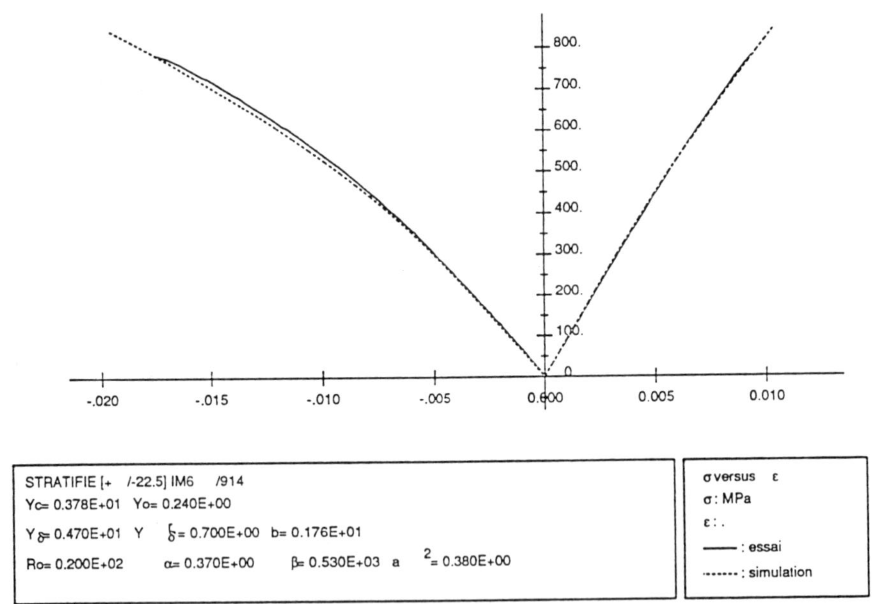

Figure 3 : Verification on a $[\pm 22.5]_{2s}$ test for IM6-914 material

3 - Analysis of some simple laminate structures

3.1 Tension test on $[90,0]_{2s}$ specimen

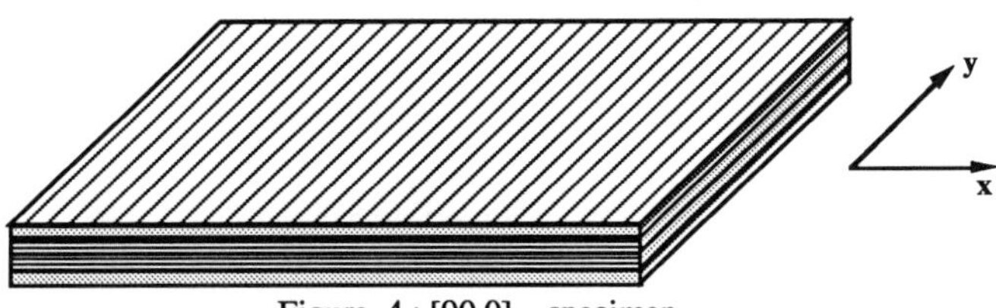

Figure 4 : $[90,0]_{2s}$ specimen

We note :

- ε^* : longitudinal strain (x direction)
- σ^* : longitudinal mean stress (x direction)
- σ^0, ε^0 : stress - strain related to the ply with fibers in the x direction
- σ^{90}, ε^{90} : stress - strain related to the ply with fibers in the y direction

Subscripts 1 and 2 denote respectively the fiber and transverse directions of each ply.

The damage modeling gives for the transverse stresses :

$$\sigma_{22}^{0} \;\#\; E_2 \nu_{12} \varepsilon^* \geq 0 \quad ; \quad \sigma_{22}^{90} \;\#\; E_2 \varepsilon^* \geq 0$$

$$\text{and} \quad Y_d^0 = 0 \quad ; \quad Y_{d'}^0 \;\#\; \frac{1}{2} E_2 \varepsilon^{*2} \nu_{12}^2$$

$$Y_d^{90} = 0 \quad ; \quad Y_{d'}^{90} \;\#\; \frac{1}{2} E_2 \varepsilon^{*2}$$

Aside fiber rupture, two damage mechanisms are present. For both 0° ply and 90° ply, a degradation of matrix and fiber-matrix interface appears, which can be the main phenomena for fatigue loading. For static loading the most important degradation is related to the 90° ply ; for certain values of the damage material constant $Y_{c'}$ this mode is the first rupture mode.

3.2 Tension test on $[+45,-45]_{2s}$ specimen

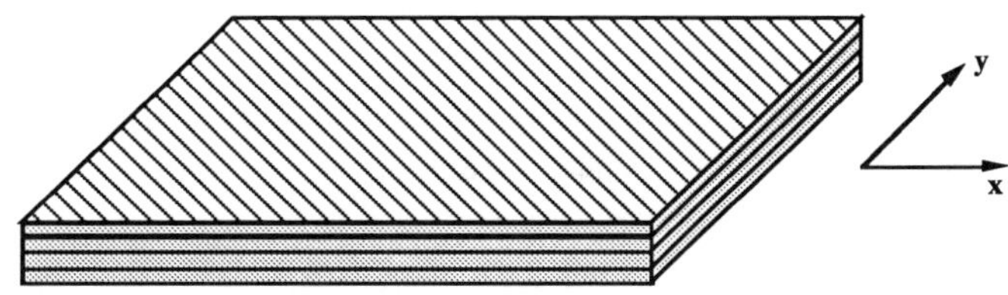

Figure 5 : $[+45, -45]_{2s}$ specimen

ε^*, σ^* denote respectively the longitudinal strain and the longitudinal mean stress. The stresses for a ply are :

$$\sigma_{11} = \sigma_{12} = \sigma_{22} = \frac{\sigma^*}{2}$$

The anelastic strains verify :

$$\dot{\tilde{\varepsilon}}^p_{22} = a^2 \dot{\tilde{\varepsilon}}^p_{12} \frac{1 - d}{1 - d'}$$

where the constant a^2 is equal to 0.38 for T300-914 and IM6-914. Then :

$$\dot{\varepsilon}^p_{22} = a^2 \left(\frac{1 - d}{1 - d'}\right)^2 \varepsilon^p_{12}$$

The damage variable d reaches large values around 0.6 - 0.7. The maximum value is associated to an instability phenomenon. Around this critical point, one has :

$$\varepsilon^p_{22} << \varepsilon^p_{12} \quad \text{and} \quad \varepsilon_{22}, \varepsilon_{11} << \varepsilon_{12} \# \varepsilon^*$$

Moreover :

$$R(p) + R_0 = \sqrt{1 + a^2}\, \frac{\sigma^*}{2} \quad ; \quad p = \sqrt{2}\, \tilde{\varepsilon}^p_{12} \quad ; \quad \dot{\tilde{\varepsilon}}^p_{12} = (1\text{-}d)\, \dot{\varepsilon}^p_{12}$$

$$\sigma^* \# 2\, G^0_{12} (1 - d) (\varepsilon^* - \varepsilon^p_{12})$$

$$d = \left\langle \frac{\sigma^*}{2} \left(\frac{b}{2E^0_2 (1\text{-}bd)^2} + \frac{1}{G^0_{12} (1\text{-}d)}\right)^{1/2} - Y_0^{1/2} \right\rangle_+ / Y_c^{1/2} \quad ; \quad d' = bd$$

4 - Interface modeling

4.1 Interface definition

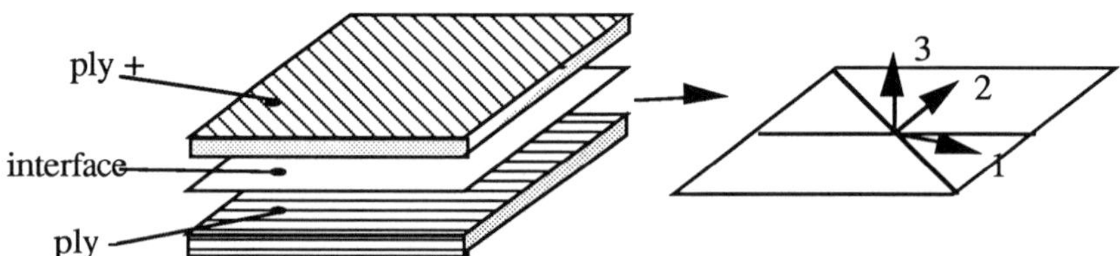

Figure 6 : Interface modeling

The interface is a surface entity (of zero thickness) which ensures stress and displacement transfers from one ply to another. In elasticity the interface is schematized by means of a model which has been used in order to modelize the fiber-matrix interface.

It depends on the relative orientation of the upper and lower plies. The displacement discontinuities are denoted by :

$$[U] = U^+ - U^- = [U]_1\, \vec{N}_1 + [U]_2\, \vec{N}_2 + [W]\, \vec{N}_3$$

The $(\vec{N}_1, \vec{N}_2)$ axes are associated with the bisectrix of the fiber directions. The undamaged energy of the interface is :

$$E_D = \frac{1}{2}[k^0 [W]^2 + k_1^0[U_1]^2 + k_2^0[U_2]^2] = \frac{1}{2}[\frac{\sigma_{33}^2}{k^0} + \frac{\sigma_{32}^2}{k_2^0} + \frac{\sigma_{31}^2}{k_1^0}]$$

k^0, k_1^0, k_2^0 are initial elastic characteristics.

4.2 Kinematics and evolution of the damages

The deterioration of the interface can be described by :

$$E_D = \frac{1}{2}[\frac{<-\sigma_{33}>_+^2}{k^0} + \frac{<-\sigma_{33}>_+^2}{k^0(1-d)} + \frac{\sigma_{32}^2}{k_2^0(1-d_2)} + \frac{\sigma_{31}^2}{k_1^0(1-d_1)}]$$

The conjugate variables associated with the dissipation are :

$$Y_d = \frac{1}{2}\frac{<\sigma_{33}>_+^2}{k^0(1-d)^2} \quad ; \quad Y_{d1} = \frac{\sigma_{33}^2}{k_1^0\,(1-d_1)^2} \quad ; \quad Y_{d2} = \frac{\sigma_{32}^2}{k_2^0\,(1-d_2)^2}$$

A simple modeling is to consider that the damage evolution is governed by :

$$\underline{Y} = \sup|_{\tau\leq t}(Y_d + \gamma_1\, Y_{d1} + \gamma_2\, Y_{d2})$$

where γ_1, γ_2 are coupling constants. In term of delamination modes, the first term is associated with the first opening mode, the other two with the second and third modes. The damage evolution law is defined through a material function w, such that :

$$d = w(\underline{Y}) \quad \text{if } d < 1 \quad ; \quad d = 1 \text{ otherwise}$$
$$d_1 = \gamma_1\, w(\underline{Y}) \quad \text{if } d_1 < 1 \quad ; \quad d_1 = 1 \text{ otherwise}$$
$$d_2 = \gamma_2\, w(\underline{Y}) \quad \text{if } d_2 < 1 \quad ; \quad d_2 = 1 \text{ otherwise}$$

A variant with delay effect can also be introduced. An anelastic response can be taken into account by introducing an anelastic part of the displacement discontinuities : $[U_1]^p$; $[U_2]^p$

Let us consider the following effective quantities :

$$[\dot{\tilde{U}}_1]^p = [\dot{U}_1]^p\,(1-d_1) \quad ; \quad [\dot{\tilde{U}}_2]^p = [\dot{U}_2]^p\,(1-d_2)$$

$$\tilde{\sigma}_1 = \frac{\sigma_{13}}{1-d_1} \quad ; \quad \tilde{\sigma}_2 = \frac{\sigma_{23}}{1-d_2} \quad ; \quad \tilde{\sigma}_3 = \frac{\sigma_{33}}{1-d}$$

We build a plasticity like model upon effective quantities :

$$f(\tilde{\sigma}_1, \tilde{\sigma}_2, \tilde{\sigma}_3, R) = \sqrt{a_1^2\,\tilde{\sigma}_1^{\,2} + a_2^2\,\tilde{\sigma}_2} - \tilde{\sigma}_3 - R(p) \leq 0$$

where a_1, a_2 are two material constants and R the "hardening" curve. Then, we have :

$$[\tilde{\dot{U}}_1]^p = \dot{p}\frac{a_1^2\,\tilde{\sigma}_1}{\sqrt{a_1^2\,\tilde{\sigma}_1^2 + a_2^2\,\tilde{\sigma}_2}} \quad ; \qquad [\tilde{\dot{U}}_2]^p = \dot{p}\frac{a_2^2\,\tilde{\sigma}_2}{\sqrt{a_1^2\,\tilde{\sigma}_1^2 + a_2^2\,\tilde{\sigma}_2}} \quad \text{and} \quad \dot{p} \geq 0$$

5 - A First Analysis of delamination

Let us consider a DCB specimen constituted with two elastic layers loaded with a pure mode I solicitation.

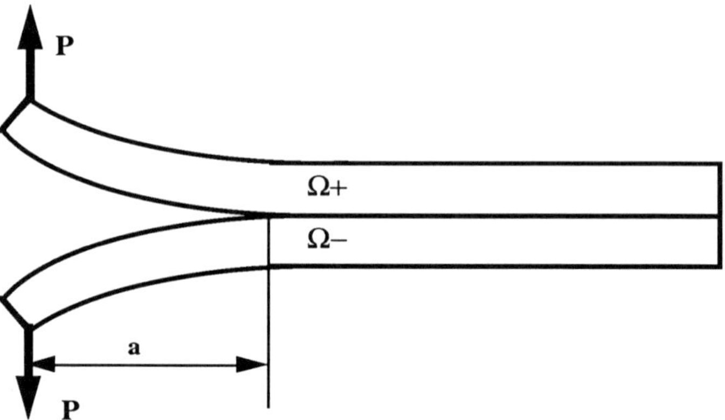

Figure 7 : DCB specimen

The layers are connected with an interface Γ, whose elastic damageable behavior is defined by :

$$w(d) = \frac{Y}{Y_c}$$

where Y_c is a material constant. The delaminated area length is a. Figure 8 gives the computed energy release rate as a function of a ; for a >> h it remains constant. Then, Fracture Mechanics appears as a simplified tool for the delamination study in the case of an established front of elastic layers. Moreover the previous results enable us to identify behaviors of different materials from experimental results. Such an approach is made with the assumption of elastic layers, thus it can be a first approach only.

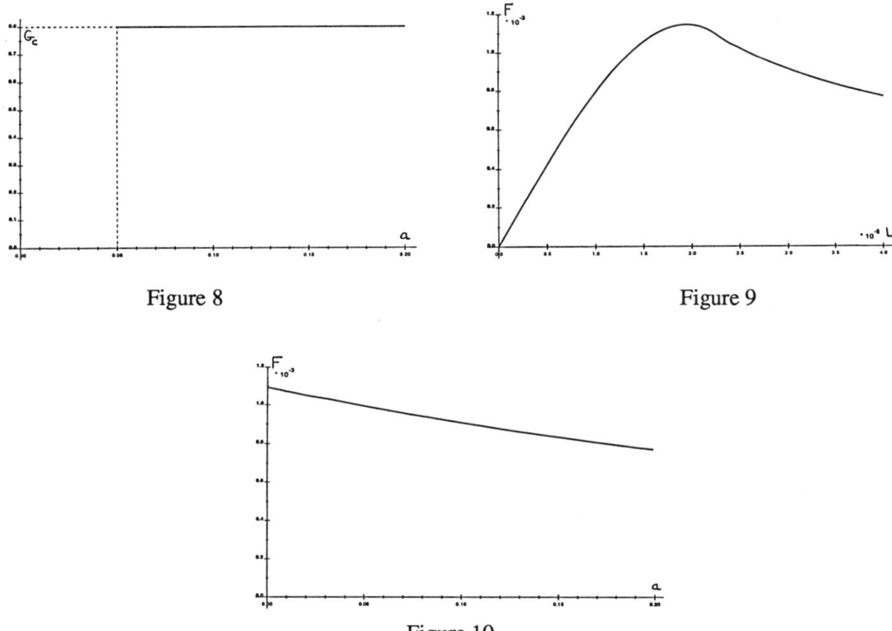

Figure 8

Figure 9

Figure 10

Figure 8 : Values of the computed energy release rate with damage modeling (large a)

Figure 9 : Value of the load with respect to an increasing displacement (initiation case)

Figure 10 : Value of the load with respect to the crack length a

Starting from a = 0, the load reaches a maximum value which can be associated to the delamination initiation.

6 - Rupture Computation

The rupture phenomenon happens after two phases. In a first step, the micro-voids and micro-cracks growth is nearly uniform : it is the initiation stage. From the critical point (or from a point just beside) the strain and also the damages become more and more localized ; a macrocrack appears and grows until becoming instable. If the first stage is well described thanks to this damage approach, the full simulation of the rupture leads to severe difficulties. The goal is to build up a true Rupture Theory for composites laminates.

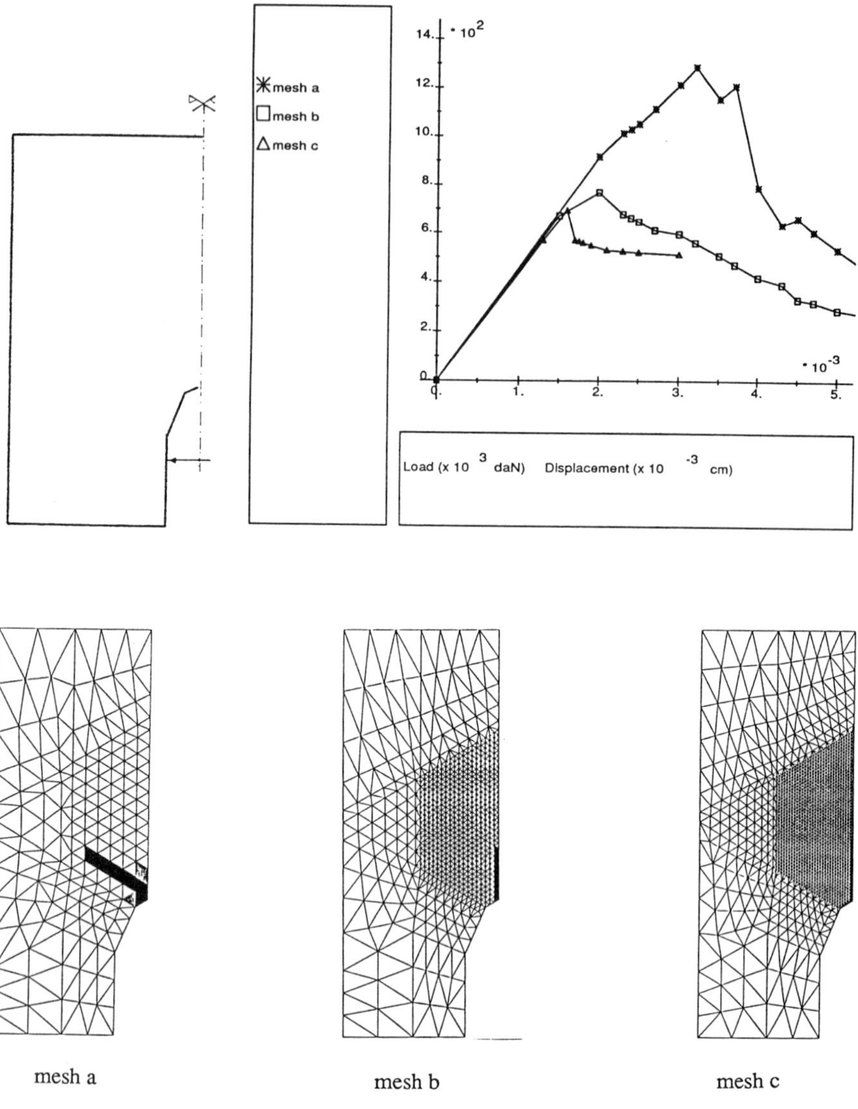

Figure n° 11 : Mesh sensivity observed in the computation of a damaged structure

Figure 11 explains these difficulties in the case of a concrete structure. It has been obtained by Saouridis 1988. The mesh dependence is very important during the second stage of the rupture process (after the critical point). The load-displacement curves are very different for the three meshes. It is also the case for the three totally damage areas which simulate the crack. This surprising result exists also for other instabilities such as localized necking.

Thus a problem is : how to follow a post critical solution independently of the element size ? An answer is given by the localization limiter concept introduced in Saouridis 1988, Bazant and Pijaudier-Cabot 1987, Belytschko and Lasry 1988. It is a regularization procedure ; the additional terms are built up from a second gradient approach or from a non local approach. Delay terms can also be introduced. To go further in the understanding of these difficulties, let us consider an exact solution.

Pure tension analysis of a composite plate :

The fibers are supposed to be perpendicular to the loading direction (see figure 12). From the critical point, numerous different solutions can occur. It is a non-denombrable family parametered by the length of the localization zone.

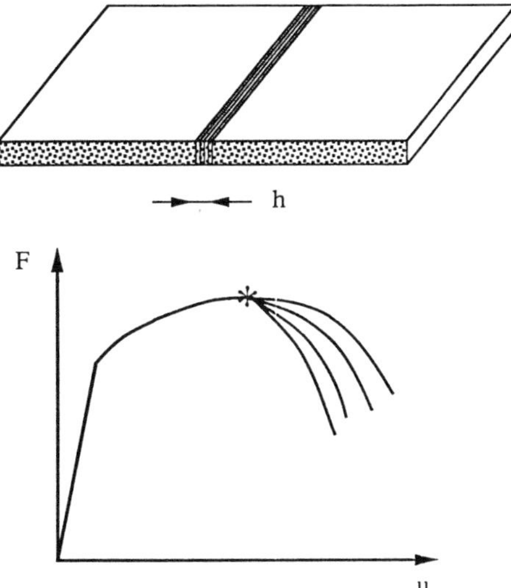

Figure 12 : Tension test for a $[90]_S$ specimen

This example shows that the theoretical post-critical solution are much more numerous than in the reality. Something homogeneous to a length is missing. For laminate composites, these difficulties partially vanish :

- for the interface modeling
- for the single layer modeling if one prescribes constant damages through the thickness. To avoid them completely one can use damage modeling with delay effects.

Let us consider the computation of a laminate structure with an initially circular hole. This delamination problem tales into account all the different damage mechanisms described previously. It is a non-linear three-dimensional problem which has been achieved in Allix 1988, who introduced a semi-analytical computational method. The whole problem is about 100000 degrees of freedom. Several results are given further for a $[0,90]_S$ laminate loaded in pure mode I. A normal displacement is prescribed on the edge of the hole (radius r_0) under the following form $U(r_0, t) = \lambda(t)\, U_0$ where $\lambda(t)=t/T$. Figure 13 shows the strain energy divided by λ, with respect to λ. Through a global instability condition, this curve allows to predict the delamination initiation and the Gauss point at which it happens. Figure 13 also shows the evolution of the peeling stress at this Gauss point.

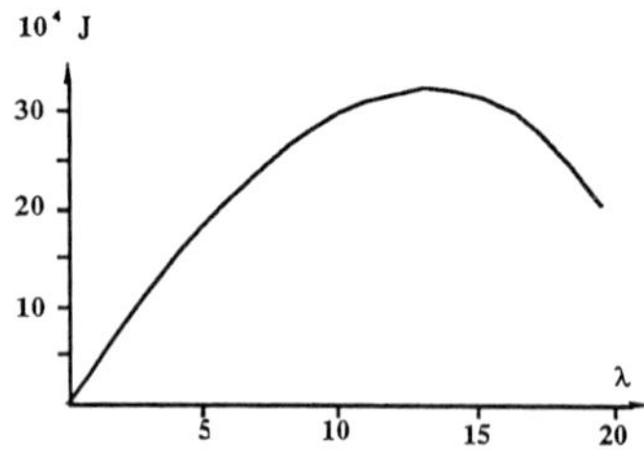

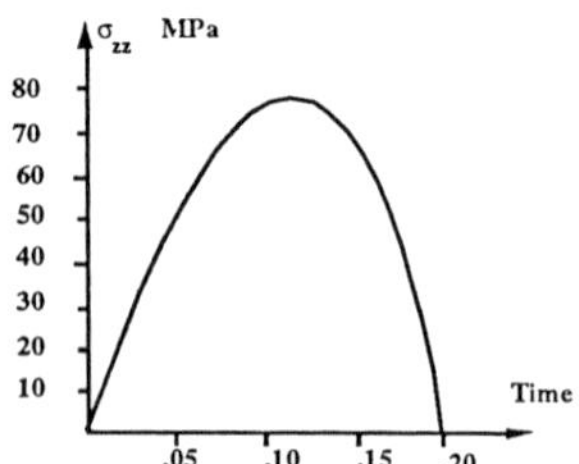

Figure 13 : Delamination analysis for a $[0,90]_S$ laminate with a hole

7 - Conclusion

This Damage Mechanics Approach seems to be a powerful tool for the prediction of delamination and more generally for the prediction of complex structures deterioration. The use of damage meso-modeling avoids the main computational difficulties. Further research is of course necessary to solve completely the computational and theoretical problems in order to achieve a true Rupture Theory.

REFERENCES

[1] ALLEN D.H., HARRIS C.E., GROVES S.E., 1987
A Thermomechanical constitutive theory for elastic composites with distributed damage
Int. J. of Solids and Structures, Vol. 23-9, p 1301 - 1338

[2] ALLIX O., 1989
Délaminage par la mécanique de l'Endommagement
Calcul des structures et Intelligence Artificielle, Vol. 1, Pluralis, Ed. J.M. Fouet, P. Ladevèze, R. Ohayon

[3] ALLIX O., LADEVEZE P., GILLETTA D., OHAYON R., 1989
A Damage Prediction Method for Composite Structures
Int. Journal for Num. Meth. in Engineering, Vol. 27, p 271-283.

[4] BAZANT Z.P., PIJAUDIER-CABOT G., 1987
Non local damage : Continuum Model and Localization Instability
Report n°87-2/ Northwestern University, Evanston, also in J. of Appl. Mech., ASM

[5] BELYTSCHKO T., LASRY D., 1988
Localization limiters and numerical strategies for strain-softening materials
in Cracking and Damage, Edited by J. Mazars and Z.P. Bazant, Elsevier, p 349-362.

[6] DUMONT J.P., LADEVEZE P., POSS M. and REMOND Y., 1987
Damage mechanics for 3D composites
Composite Structure 8, p 119-141

[7] HASHIN Z., 1980
Failure criteria for unidirectional fiber composites
J. of Appl. Mechanics, Vol. 47.

[8] GILLETTA D., GIRARD H., LADEVEZE P., 1986
Composites 2D à fibres à haute résistance : modélisation mécanique de la couche élémentaire
JNC 5, Pluralis, Paris, p 685-697.

[9] KACHANOV L.M., 1958
Time of the rupture process under creep conditions
Izv Akad Nauk S.S.R. otd Tech Nauk, 8, pp 26-31

[10] LADEVEZE P., 1983
Sur une théorie de l'endommagement anisotrope
Rapport Interne n°34, Laboratoire de Mécanique et Technologie, Cachan

[11] LADEVEZE P., 1986
Sur la Mécanique de l'Endommagement des Composites
JN5, Pluralis, Paris, p 667-683

[12] LADEVEZE P., 1989
About a Damage Mechanics Approach

Proceeding Int. Conf. on Mechanics and Mechanisms of Damage in Composites and Multimaterials - MECAMAT - St Etienne

[13] LE DANTEC E., 1989
Contribution à la Modélisation du Comportement Mécanique des Composites Stratifiés.
Thèse de l'Université Paris 6 - Cachan

[14] LEMAITRE J., 1984
How to use Damage Mechanics
Nuclear Engineering and Design, 80, p 233-245

[15] NEEDLEMAN A., 1987
Material rate dependence and mesh sensitivity in localization problems
Comp. Meth. Appl. Mech. Engng., 67, p 68-85.

[16] ODORICO J., CUNY J. VANCON F., BENEDIC B., SOULEZELLE T., 1984
Etude et comparaison des différentes méthodes de visualisation des endommagements successifs des matériaux composites carbone-epoxy
J.NC. 4, Editor Pluralis, Paris, p 191-208.

[17] RABOTNOV Y.N., 1968
Creep rupture
Proc. XII, Int. Cong. Appl., Mech., Stanford-Springer.

[18] REIFFNIDER K., 1980
Stiffness reduction mechanism in composite materials
ASTM-STP 775 - novembre

[19] SAOURIDIS C., 1988
Identification et numérisation objectifves des comportements adoucissants : une approche multiéchelle de l'endommagement du béton
Thèse de l'Université Paris 6

[20] TALREJA R., 1985
Transverse cracking and stiffness reduction in composite laminates
Journal of composite materials, Vol. 19, July

[21] VITTECOQ E., 1990
Comparison between compression and tension behaviors of composite laminates, thesis, to appear.

[22] WANG S.S. and CHOI I., 1982
Boundary layer effects in composite laminates, I Free edge stress singularities, and II Free edge stress solutions and basic characteristics
Journal Appl. Mech., 49, I, 541-548, and II 549-560.

[23] WANG S.S., 1983
Fracture Mechanics for delamination problems in composite laminates
Journal of Composite materials, Vol. 17 (3), 210-213.

Constitutive Relations of Hybrid Fiber Reinforced Plastics of GFRP/CFRP and GFRP/AFRP under Combined Stress State

Kozo Ikegami and Matsuo Yoshida

Research Laboratory of Precision Machinery and Electronics
Tokyo Institute of Technology
Nagatsuta, Midoriku, Yokohama, Japan

Abstract

The stress-strain relations of hybrid reinforced plastics are experimentally determined by subjecting filament wound tubes with various material constitutions to axial load, torsion and internal pressure. The testing materials are hybrid reinforced plastics of two combination from glass fiber, carbon fiber, and aramid fiber reinforced plastics. The stress-strain curves of constituent reinforced plastics are formulated by using the third order equations of stress components. The stress-strain relations of hybrid reinforced plastics are derived from the relations of constituent reinforced plastics on the assumption of perfect bonding between two layers of the constituent reinforced plastics. The stress-strain curves calculated by obtained relations represent the characteristic behavior of experimental curves under combined stress states as well as uniaxial stress state.

1 Introduction

Fiber reinforced plastics have the advantage of changing mechanical properties by the material constitutions. Rational material design is possible by use of the variable properties. Hybrid fiber reinforced plastics give wide possibility of material design for tailoring structural components by combining different fiber reinforced plastics. The constitutive relations are necessary for the material design process.

There are several papers on the deformation of hybrid reinforced plastics. The elastic constants are derived by Sokolov et al.(1979), Ishikawa et al.(1982) and Maksimov et al.(1982). The tensile behavior is examined by Aveston et al.(1976) and Amagi et al.(1985). Wagner et al.(1982) and Nagai et al.(1984) investigate the bending properties. Furue et al.(1986) conduct creep experiments. But there are few papers on the stress-strain relations of hybrid reinforced plastics under combined stress states which are necessary for stress analysis of hybrid reinforced components. In this paper, the stress-strain relations of hybrid reinforced plastics experimentally examined under combined stress state as well as uniaxial stress state and the relations are formulated from the macroscopic point of view.

2 Experimental Procedure

The specimens used in the experiments are thin-walled tubes which are fabricated by the filament winding method. The impregnated continuous fibers are passed through a vacuum chamber to reduce voids. The fibers are glass(E-glass), carbon(Torayca T300), and aramid(Kevlar T968). The matrix is epoxy resin(Epikote 828). The hardener system consists of a BF_3-monoethylamine complex.

The specimen has two layers of different fiber reinforcement in the radial direction. The inner layer is carbon fiber reinforced plastics(CFRP) and the outer layer is glass fiber reinforced plastics(GFRP) or aramid fiber reinforced plastics(AFRP). The two layers are moulded to the same thickness. The inner layer of the specimen is conditioned for 16 hours at room temperature and then the outer layer is wound on the surface of the inner layer. The specimen is cured at 125 centigrade for 2.5 hours and then at 150 centigrade for 2 hours.

The dimensions of the tubular specimen are given in Fig. 1. The length of the parallel part is about 50mm. The inner diameter(D_1) is 30.1mm and the outer diameter is about 1.3mm for the combination of CFRP and AFRP, and 1.6mm for the combination of CFRP and GFRP. The gripped part of the specimen is reinforced with glass fiber. The ply angles of fiber are 90 degrees to the axial direction(x) of the specimen in the inner layer, and 90, 61 and 50 degrees to the circumferential

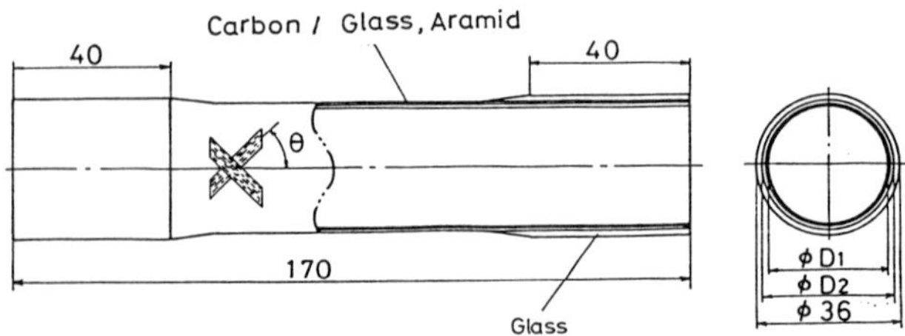

Figure 1. Configuration and dimensions of the specimen

direction(y) of the specimen in the outer layer. The fiber volume fraction of the specimens is about 50% in both layers.

The tubular specimens are subjected to axial load, torsion and internal pressure by the combined stress testing machine. The deformation and failure behavior of the specimens is examined under various stress states.

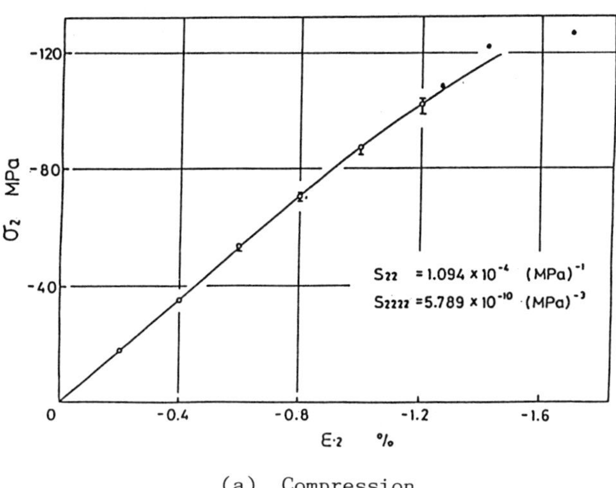

(a) Compression

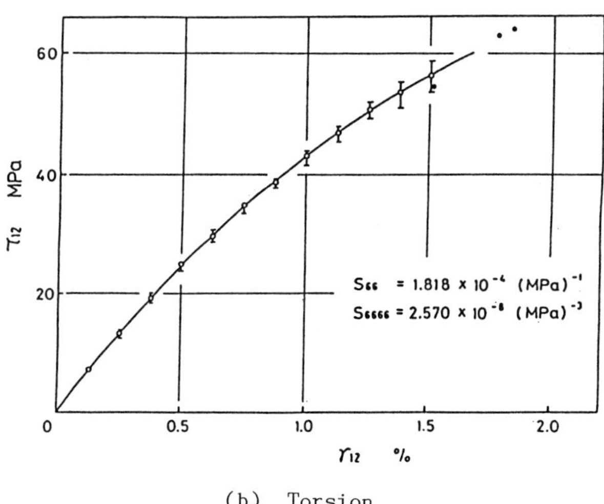

(b) Torsion

Figure 2. Uniaxial stress-strain curves in transverse directions of CFRP

3 Constitutive Relations of Unidirectionally Reinforced Plastics

Fiber reinforced plastics have the non-linearity in their stress-strain relations depending on material constitutions and fiber orientations. Figures 2 to 4 show several examples of non-linear behavior for compression and shear in the direction transverse to fibers of GFRP, CFRP and AFRP. The experiments are conducted by subjecting circumferentially reinforced tubes to internal pressure, compressive

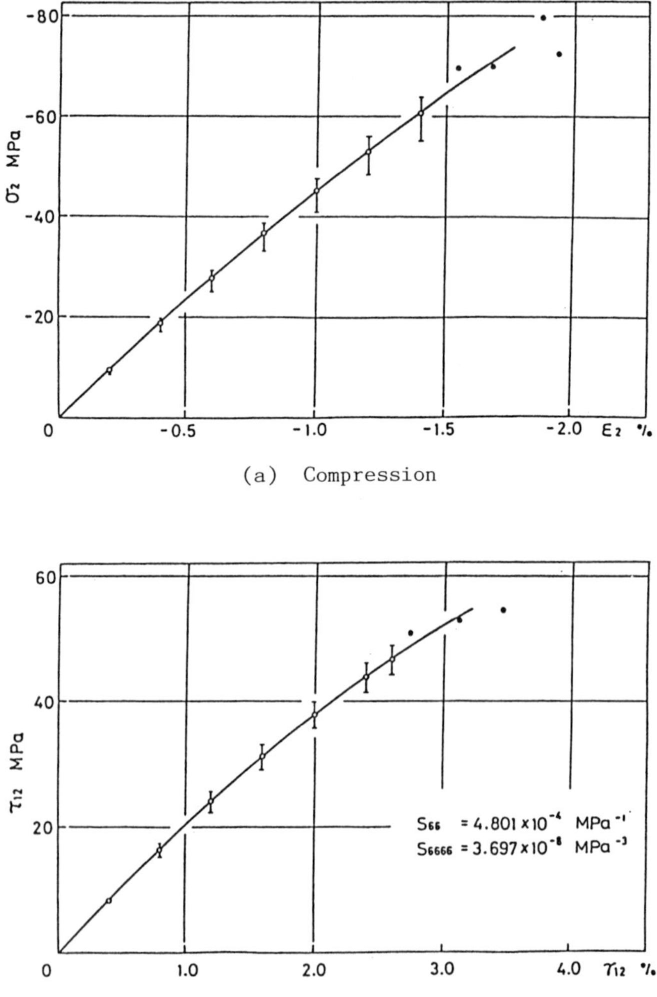

(a) Compression

(b) Torsion

Figure 3. Uniaxial stress-strain curves in transverse directions of AFRP

load and torsional load. The stress-strain curves in fibers direction and transverse direction to fibers are experimentally determined by using several specimens for one loading condition. There are two methods of representing the non-linear stress-strain relations. One is a method of dividing the non-linear stress strain curves to piece-linearwise and the other is a method of using polynomial equations between stress and strain components. For the convenience of ease applications, the non-linear stress-strain curves with respect to the principal axis of anisotropy, denoted as the axes 1 and 2, are represented by the third order polynomial equations of stress components as follows.

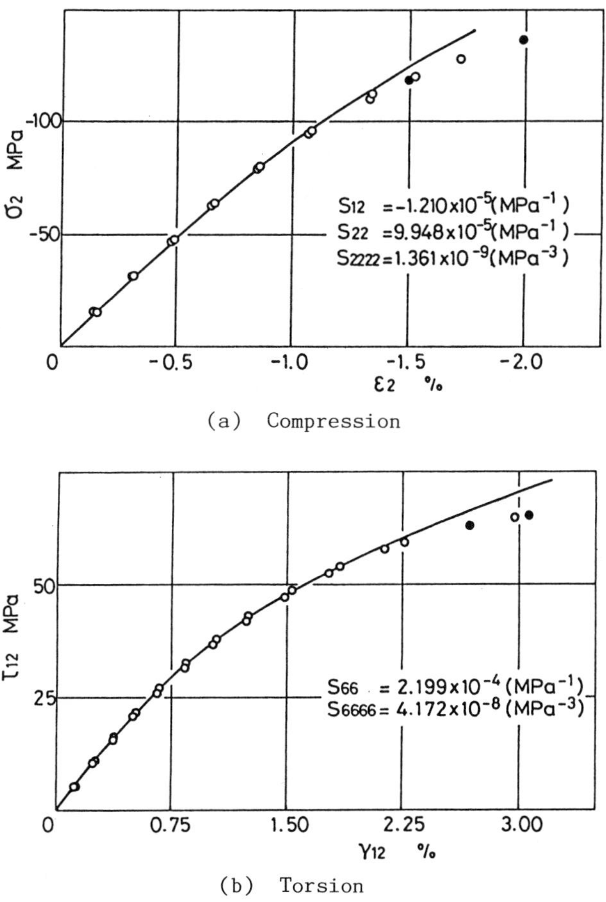

(a) Compression

(b) Torsion

Figure 4. Uniaxial stress-strain curves in transverse directions of GFRP

$$\begin{bmatrix} \varepsilon_1 \\ \varepsilon_2 \\ \gamma_{12} \end{bmatrix} = \begin{bmatrix} S_{11} & S_{12} & 0 \\ S_{12} & S_{22} & 0 \\ 0 & 0 & S_{66} \end{bmatrix} \begin{bmatrix} \sigma_1 \\ \sigma_2 \\ \tau_{12} \end{bmatrix} + \begin{bmatrix} 0 & 0 & 0 \\ 0 & S_{2222} & 0 \\ 0 & 0 & S_{6666} \end{bmatrix} \begin{bmatrix} \sigma_1^3 \\ \sigma_2^3 \\ \tau_{12}^3 \end{bmatrix} \quad (1)$$

The equation contains the non-linear terms of the stress components in compressive deformation as well as shear deformation. The stress-strain relations of Eq. (1) fit to the experimental stress-strain relations. Table 1 gives the values of the coefficient in Eq. (1) which are determined by fitting the experimental curves. The curves in Figs. 2 to 4 are the curves calculated by use of the values in Table 1.

The stress-strain curves of tension and compression are symmetric each other with respect to the origin of stress and strain plane for CFRP, AFRP and GFRP. The same values for S_{22} and S_{2222} in Eq.(1) are adopted for both tensile and compressive loading. The value of S_{11} for GFRP is larger than those for CFRP and AFRP. The values of S_{22} and S_{66} for GFRP are similar to those for CFRP. The magnitude of the compliance S depends on the elastic moduli of the fibers as well as the anisotropic properties of the fibers.

The non-linearity of the stress-strain curves is compared by use of the ratios of S_{2222}/S_{22} and S_{6666}/S_{66}. The degree of the non-linearity of the stress-strain curves in the direction transverse to the fibers is more remarkable in shear loading than in tensile or compressive loading. The ratio of the non-linearity in shear loading is approximately evaluated as 2 : 1 : 3 for CFRP, AFRP and GFRP.

The inversed equation of the stress component ε_2 is obtained from Eq.(1) as follows.

Table 1. Coefficients of the stress-strain relations

	CFRP	AFRP	GFRP	DIMENSION
S_{11}	8.197×10^{-6}	1.498×10^{-5}	2.784×10^{-5}	MPa^{-1}
S_{12}	-3.258×10^{-6}	-5.068×10^{-6}	-1.210×10^{-5}	MPa^{-1}
S_{22}	1.094×10^{-4}	2.121×10^{-4}	9.948×10^{-5}	MPa^{-1}
S_{66}	1.818×10^{-4}	4.801×10^{-4}	2.199×10^{-4}	MPa^{-1}
S_{2222}	5.739×10^{-10}	5.430×10^{-9}	1.361×10^{-9}	MPa^{-3}
S_{6666}	2.570×10^{-8}	3.697×10^{-8}	4.172×10^{-8}	MPa^{-3}

$$\varepsilon_2 = (S_{12}/S_{11})\varepsilon_1 + [(S_{11}S_{22}-S_{12}^2)/S_{11}]\sigma_2 + S_{2222}\sigma_2^3 \quad (2)$$

From Eq.(2) the stress component σ_2 is represented by the strain components as follows,

$$\sigma_2 = [a + f(e)]e \quad (3)$$

where $a = S_{11}/(S_{11}S_{22}-S_{12}^2)$, $e = \varepsilon_2 - (S_{12}/S_{11})\,\varepsilon_1$, and $f(e)$ is the root of the following cubic equation.

$$y^3 + 3ay^2 + [1/(S_{2222}e^2a) + 3a^2)y + a^3 = 0 \quad (4)$$

The relation inversed the shear strain components to the stress components is similarly obtained as follows,

$$\tau_{12} = [1/S_{66} + g(\gamma_{12})]\gamma_{12} \quad (5)$$

where $g(\gamma_{12})$ is the root of the following cubic equation.

$$y^3 + 3/S_{66}y^2 + [3/S_{66}^2 + S_{66}/(S_{6666}\gamma_{12})] + 1/S_{66}^3 = 0 \quad (6)$$

The inversed equation of Eq.(1) is obtained by using Eqs.(3) and (5) as follows.

$$\begin{bmatrix} \sigma_1 \\ \sigma_2 \\ \tau_{12} \end{bmatrix} = \begin{bmatrix} Q_{11} & Q_{12} & 0 \\ Q_{12} & Q_{22} & 0 \\ 0 & 0 & Q_{66} \end{bmatrix} \begin{bmatrix} \varepsilon_1 \\ \varepsilon_2 \\ \gamma_{12} \end{bmatrix}$$
$$+ \begin{bmatrix} (S_{12}/S_{11})_2 f(e) & -S_{12}/S_{11} f(e) & 0 \\ -(S_{12}/S_{11})f(e) & f(e) & 0 \\ 0 & 0 & g(\gamma_{12}) \end{bmatrix} \begin{bmatrix} \varepsilon_1 \\ \varepsilon_2 \\ \gamma_{12} \end{bmatrix} \quad (7)$$

The strain-stress relation with respect to the coordinate x and y which axes are rotated by the angle θ from the principal axis of anisotropy is obtained by transforming the stress and strain components into the rotated coordinate as follows.

$$\begin{bmatrix} \varepsilon_x \\ \varepsilon_y \\ \gamma_{xy} \end{bmatrix} = \begin{bmatrix} \bar{S}_{11} & \bar{S}_{12} & \bar{S}_{16} \\ \bar{S}_{12} & \bar{S}_{22} & \bar{S}_{26} \\ \bar{S}_{16} & \bar{S}_{26} & \bar{S}_{66} \end{bmatrix} \begin{bmatrix} \sigma_x \\ \sigma_y \\ \tau_{xy} \end{bmatrix}$$

$$+ S_{2222}\sigma_2^2 \begin{bmatrix} \sin^4\theta & \sin^2 2\theta/4 & -\sin2\theta \sin^2\theta \\ & \cos^4\theta & \sin2\theta \cos^2\theta \\ \text{symmetric} & & \sin^2 2\theta \end{bmatrix} \begin{bmatrix} \sigma_x \\ \sigma_y \\ \tau_{xy} \end{bmatrix}$$

$$+ S_{6666}\tau_{12}^2 \begin{bmatrix} (1-\cos4\theta)/8 & -(1-\cos4\theta)/8 & -(\sin4\theta)/4 \\ & (1-\cos4\theta)/8 & (\sin4\theta)/4 \\ \text{symmetric} & & (1-\cos4\theta)/2 \end{bmatrix} \begin{bmatrix} \sigma_x \\ \sigma_y \\ \tau_{xy} \end{bmatrix} \quad (8)$$

The stress-strain relation is similarly obtained as follows.

$$\begin{bmatrix} \sigma_x \\ \sigma_y \\ \tau_{xy} \end{bmatrix} = \begin{bmatrix} \bar{Q}_{11} & \bar{Q}_{12} & \bar{Q}_{16} \\ \bar{Q}_{12} & \bar{Q}_{22} & \bar{Q}_{26} \\ \bar{Q}_{16} & \bar{Q}_{26} & \bar{Q}_{66} \end{bmatrix} \begin{bmatrix} \varepsilon_x \\ \varepsilon_y \\ \gamma_{xy} \end{bmatrix}$$

$$+ f^*(\varepsilon_x, \varepsilon_y, \gamma_{xy}, \theta) \begin{bmatrix} Q_{11}' & Q_{12}' & Q_{16}' \\ Q_{12}' & Q_{22}' & Q_{26}' \\ Q_{16}' & Q_{26}' & Q_{66}' \end{bmatrix} \begin{bmatrix} \varepsilon_x \\ \varepsilon_y \\ \gamma_{xy} \end{bmatrix}$$

$$+ g^*(\varepsilon_x, \varepsilon_y, \gamma_{xy}, \theta) \begin{bmatrix} (1-\cos4\theta)/2 & -(1-\cos4\theta)/2 & -(\sin4\theta)/2 \\ & (1-\cos4\theta)/2 & (\sin4\theta)/2 \\ \text{symmetric} & & (1-\cos4\theta)/2 \end{bmatrix} \begin{bmatrix} \varepsilon_x \\ \varepsilon_y \\ \gamma_{xy} \end{bmatrix} \quad (9)$$

where f^* and g^* are the roots of the cubic equations (2) and (4), respectively, which strain components are transformed into the coordinate x and y. The values of $\overline{S}_{ij}$ of Eq.(8) and $\overline{Q}_{ij}$ of Eq.(9) are determined by the transform rule of elastic constants. The values of Q_{ij}' are given by the following relations.

$$\begin{aligned} &Q_{11}' = A^2, \quad Q_{12}' = AB, \quad Q_{22}' = B^2, \\ &Q_{16}' = RAC, \quad Q_{26}' = RBC, \quad Q_{66}' = R^2C, \end{aligned} \tag{10}$$

where $A = \sin^2\theta - (S_{12}/S_{11})\cos^2\theta$, $B = \cos^2\theta - (S_{12}/S_{11})\sin^2\theta$, $C = \sin 2\theta$ $R = -1/2[(S_{12}/S_{11}) + 1]$.

4 Constitutive Relations of Symmetrically Laminated Plastics

A plate which is laminated by two uniaxially reinforced laminas of the same thickness are laminated with the ply angle α is considered. The constitutive relations are derived on the assumption of perfect bonding between two laminas by use of Eq.(9) as follows.

$$\begin{bmatrix} \sigma_x \\ \sigma_y \\ \tau_{xy} \end{bmatrix} = \begin{bmatrix} \overline{Q}_{11} & \overline{Q}_{12} & 0 \\ \overline{Q}_{12} & \overline{Q}_{22} & 0 \\ 0 & 0 & \overline{Q}_{66} \end{bmatrix} \begin{bmatrix} \varepsilon_x \\ \varepsilon_y \\ \gamma_{xy} \end{bmatrix} + \begin{bmatrix} f_1 Q_{11}' & f_1 Q_{12}' & f_2 Q_{16}' \\ & f_1 Q_{22}' & f_2 Q_{26}' \\ \text{symmetric} & & f_1 Q_{66}' \end{bmatrix} \begin{bmatrix} \varepsilon_x \\ \varepsilon_y \\ \gamma_{xy} \end{bmatrix}$$
$$+ \begin{bmatrix} g_1(1-\cos 4\alpha)/2 & -g_1(1-\cos 4\alpha)/2 & -g_2(\sin 4\alpha)/2 \\ & g_1(1-\cos 4\alpha)/2 & g_2(\sin 4\alpha)/2 \\ \text{symmetric} & & g_1(1+\cos 4\alpha)/2 \end{bmatrix} \begin{bmatrix} \varepsilon_x \\ \varepsilon_y \\ \gamma_{xy} \end{bmatrix} \tag{11}$$

where $f_1,\ f_2 = 1/2[f^*(\varepsilon_x,\ \varepsilon_y,\ \gamma_{xy},\ +\alpha\ ) \pm f^*(\varepsilon_x,\ \varepsilon_y,\ \gamma_{xy},\ -\alpha)$ and

$$g_1,\ g_2 = 1/2[g^*(\varepsilon_x,\ \varepsilon_y,\ \gamma_{xy},\ +\alpha\ ) \pm g^*(\varepsilon_x,\ \varepsilon_y,\ \gamma_{xy},\ -\alpha). \tag{12}$$

The strain-stress relations can be obtained by inversing Eq. (11) for combined normal stress state and for the shear stress state. But the inversion of Eq. (11) is difficult for general plane stress states. Numerical calculations are necessary to determine the strain state for a given plane stress state.

5 Stress-strain relations for hybrid angle ply laminates

A hybrid reinforced plate laminated symmetrically by two different reinforced plastics is considered. A reinforced plastics laminate with angle α_1 is laminated on the both sides of another reinforced plastics laminas with ply angle α_2. The total thickness of the laminates is t. The skin laminate is t_1 and core laminate is t_2 in thickness, respectively. The skin laminate is divided into both sides of core laminate in equal thickness. On the assumption of perfect bonding among laminates, applied stresses to the laminates are transferred to the constituent laminate according to the thickness, and the strain components in the constituent are equal to the components of overall laminates. Then the stress-strain relations for the hybrid laminates are obtained by use of Eq. (11) as follows,

$$\begin{bmatrix} \sigma_x \\ \sigma_y \\ \tau_{xy} \end{bmatrix} = V_1 [Q_1^*(\alpha_1)] \begin{bmatrix} \varepsilon_x \\ \varepsilon_y \\ \gamma_{xy} \end{bmatrix} + V_2 [Q_2^*(\alpha_2)] \begin{bmatrix} \varepsilon_x \\ \varepsilon_y \\ \gamma_{xy} \end{bmatrix} \tag{13}$$

where $V_1 = t_1/t$ and $V_2 = t_2/t$, and $Q_i^*(\alpha_i)$ is the sum of three matrixes of Eq. (11). It is difficult to inverse Eq. (13) with respect to the strain components. The strain-stress relations are numerically calculated by use of Eq. (13). The stress-strain relations for hybrid reinforced plastics of different laminates of n kinds are similarly obtained as follows,

$$\begin{bmatrix} \sigma_x \\ \sigma_y \\ \tau_{xy} \end{bmatrix} = \sum_{i=1}^{n} V_i [Q_i^*(\alpha_i)] \begin{bmatrix} \varepsilon_x \\ \varepsilon_y \\ \gamma_{xy} \end{bmatrix} \tag{14}$$

where V_i and Q_i^* are the thickness ratio and the sum of matrixes of stress-strain relations for the constituent laminates.

The experimental stress-strain curves of CFRP/GFRP and CFRP/AFRP for tensile and shear loading are compared with the curves calculated by Eq. (13) in Figs. 5 to 8. The suffixes x and y indicate the axial and circumferential directions of tubular specimens, respectively. The fiber directions of CFRP, GFRP and AFRP layers are indicated by the

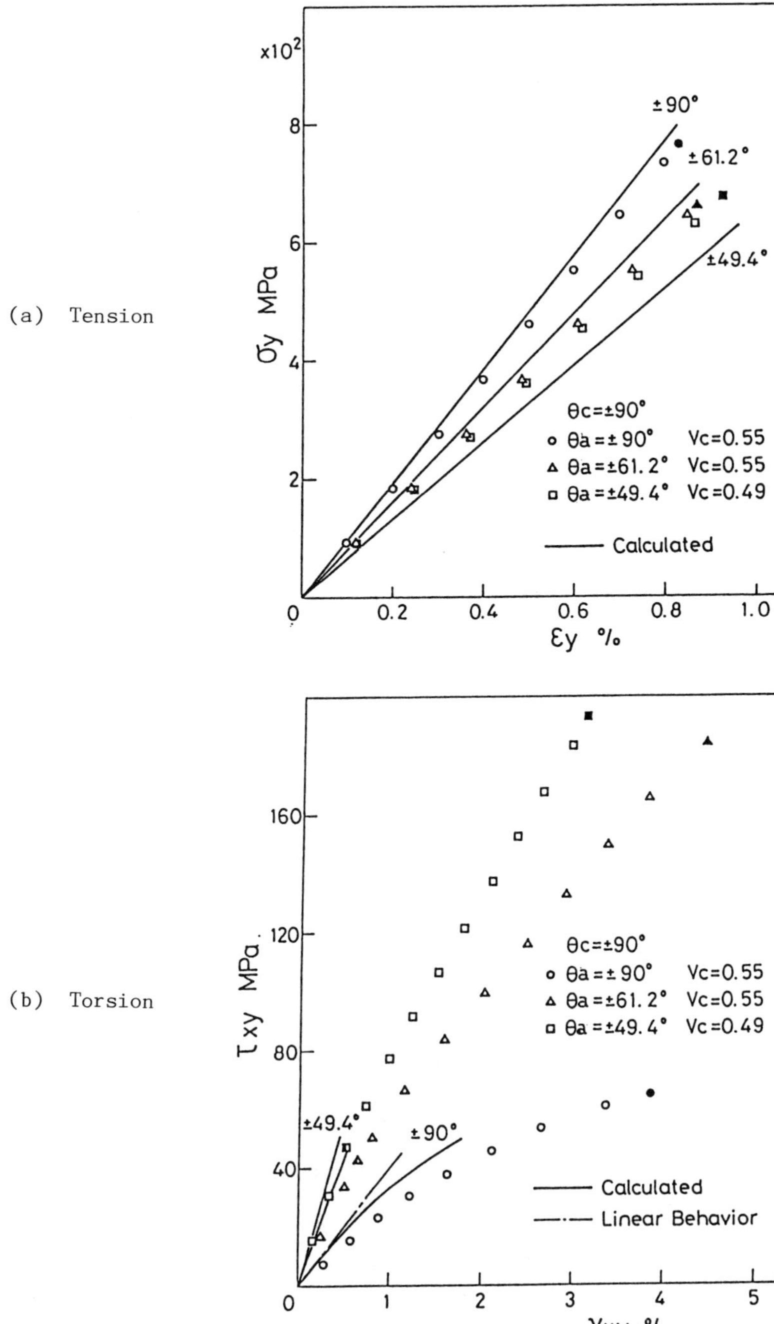

Figure 5. Uniaxial stress-strain curves of CFRP/AFRP

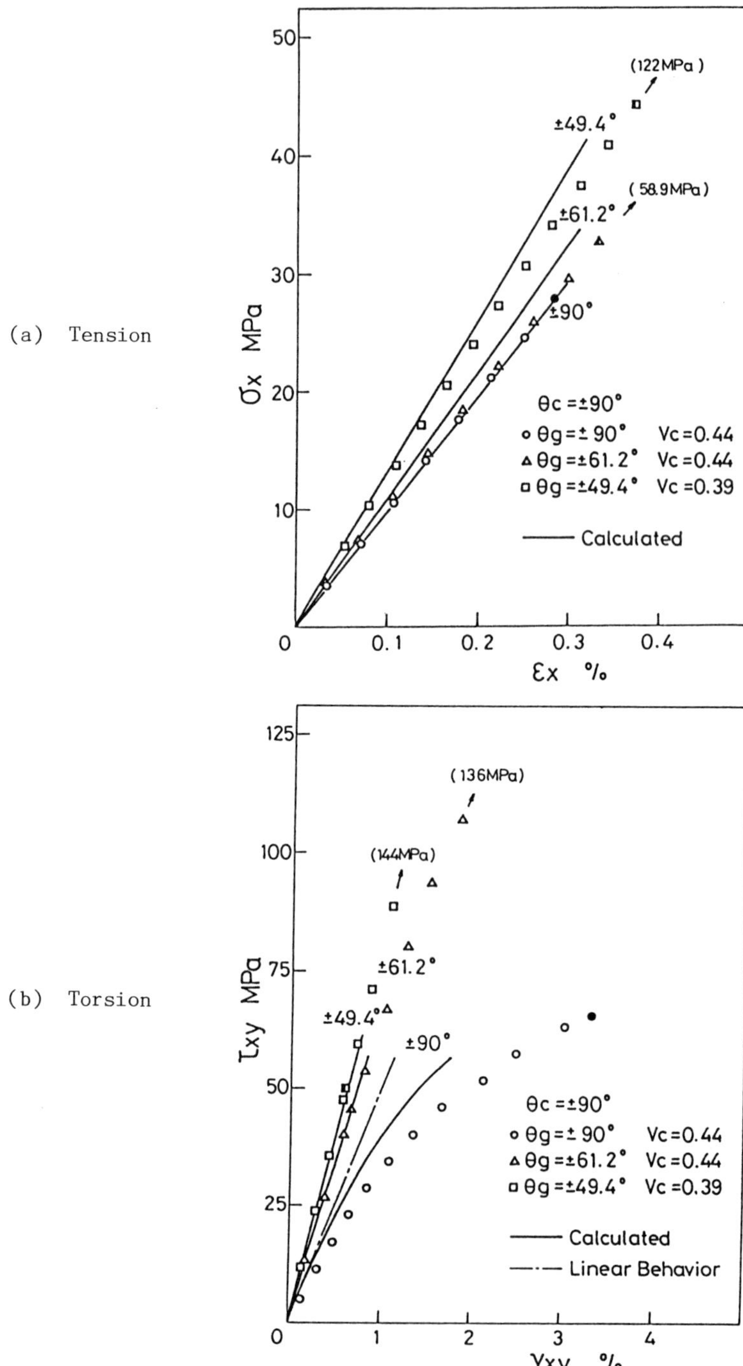

Figure 6. Uniaxial stress-strain curves of CFRP/GFRP

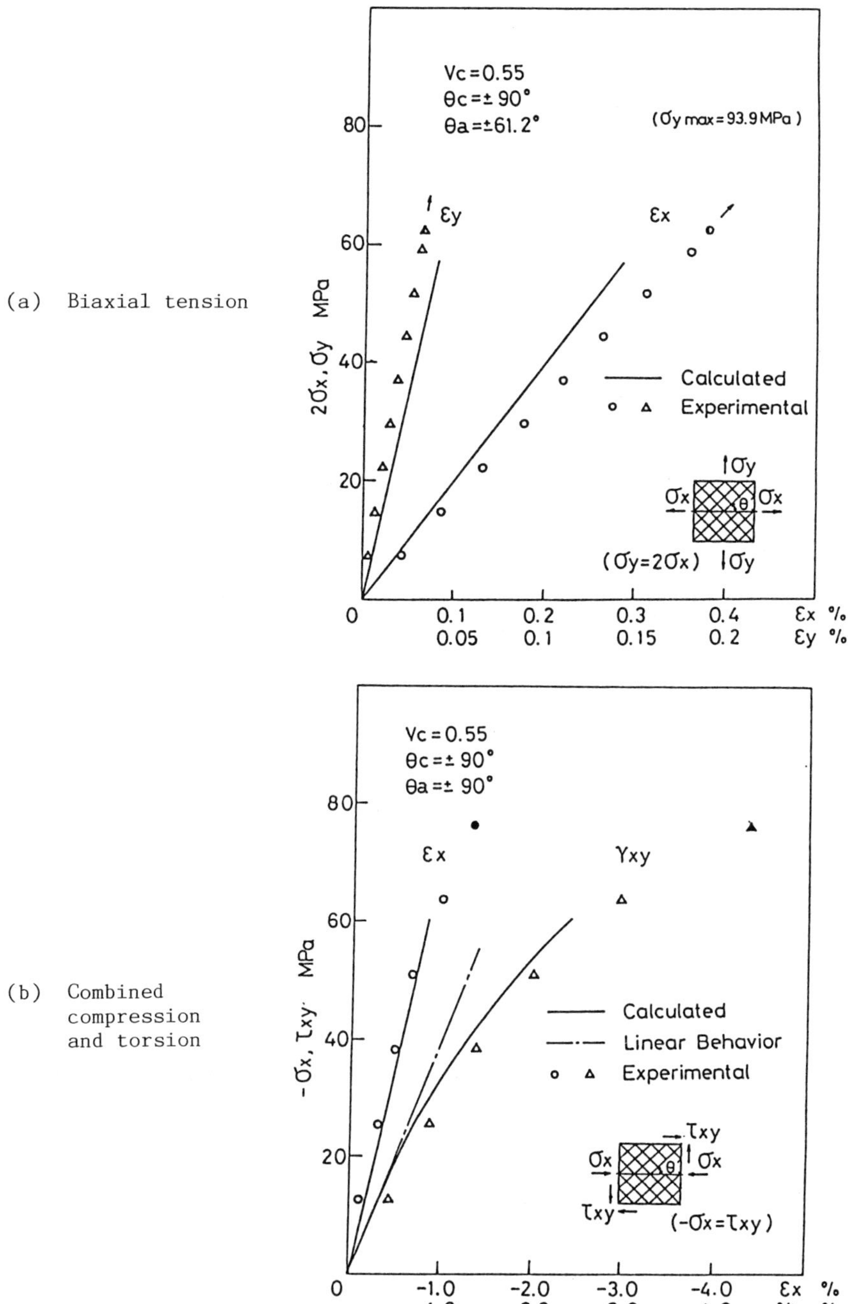

Figure 7. Combined stress-strain curves of CFRP/AFRP

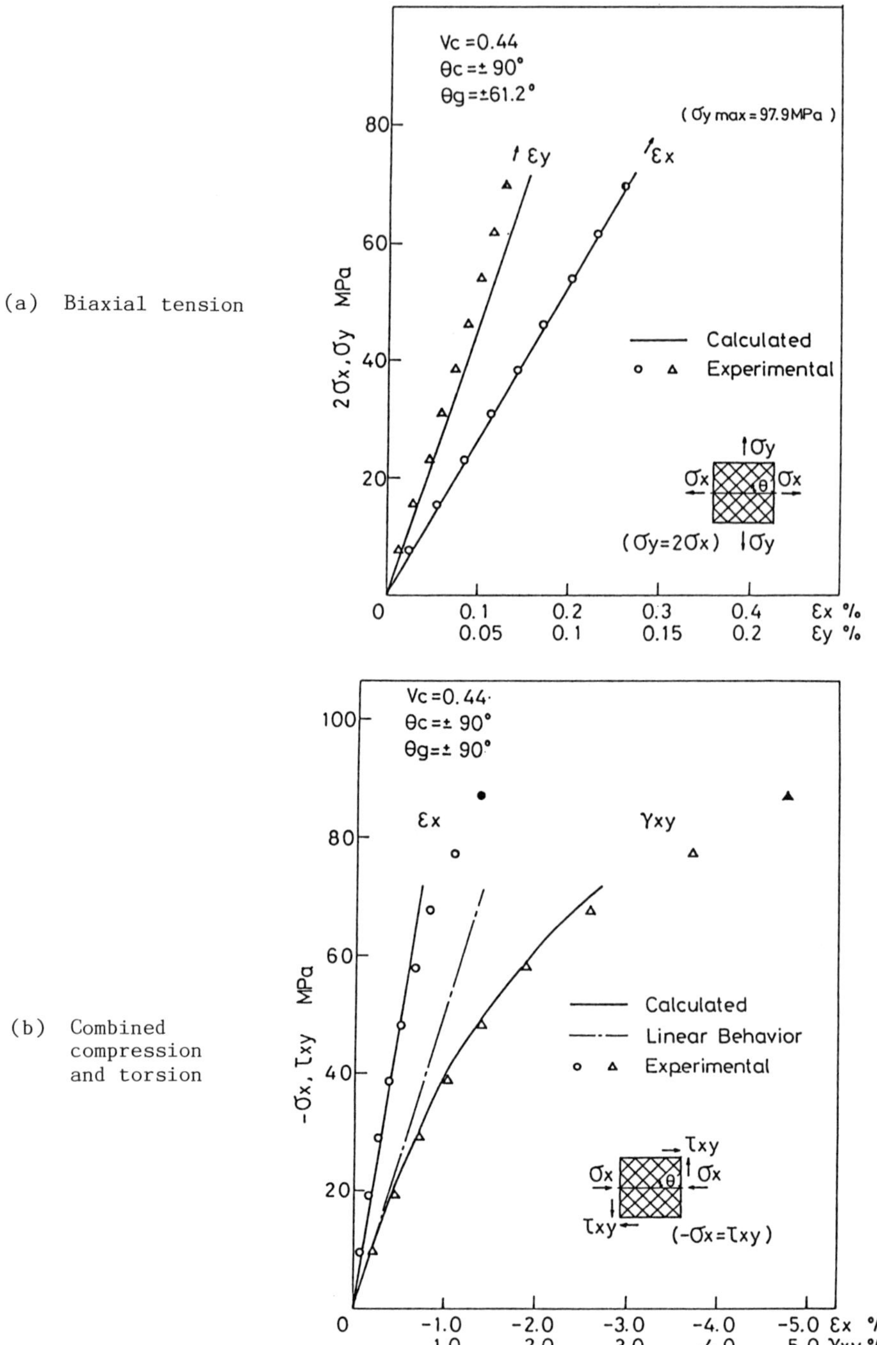

Figure 8 Combined stress–strain curves of CFRP/GFRP

suffixes c, g and a. The angle 90 degrees corresponds to circumferential direction of tubular specimens. The thickness ratio of CFRP layer to specimen thickness is indicated by the notation V_c. The solid and chain lines are the curves calculated with and without nonlinear behavior, respectively. In Figs. 5(a) and 6(a) for tensile loading, there is almost no difference between the curves with and without nonlinear behavior, but the difference is found in Figs. 5(b) and 6(b) for shear loading and the curves including nonlinear behavior give better tendencies for the experimental relations.

Figures 7 and 8 are the comparison between experimental and calculated stress-strain curves of CFRP/GFRP and CFRP/AFRP for combined loading. The figures are illustrated by the manner similar to Figs. 5 and 6. The stress-strain curves for biaxial loading show almost linear relations for both CFRP/GFRP and CFRP/AFRP. The calculated curves fit for the experimental relations. The non-linear behavior is found in the stress-strain curves for combined axial and shear loading. There is a good agreement between the experimental and calculated curves for CFRP/GFRP, but the calculated curves of CFRP/AFRP is slightly larger than the experimental relations. The disagreement between the experimental and calculated curves causes inaccuracy of fiber volume fraction, internal stress in hybrid layers and imperfect bonding between reinforced plastics. The effects on stress-strain relations are difficult to be quantitatively estimated at present.

6 Conclusions

The stress-strain relations of hybrid reinforced plastics between CFRP/GFRP and CFRP/AFRP are experimentally investigated by subjecting thin-walled tubular specimens to combined axial load, torsion and internal pressure. The relations are formulated by using the equations between total stress and total strain.

1. The linear stress-strain relations for tensile and compressive loading are found in the fiber directions of uniaxially reinforced plastics.
2. The non-linear behavior of the stress-strain curves of uniaxially reinforced plastics is observed in the transverse loading to the fiber directions. The non-linearity is remarkable in the relations for torsional loading.
3. The hybrid reinforced plastics show also non-linear stress-strain curves when the loading is applied to the direction transverse to fibers.

4. The stress-strain curves of uniaxially reinforced plastics are formulated by using the third order equations of stress components to represent the non-linear behavior.
5. The constitutive relations of hybrid reinforced plastics, which are derived from the stress-strain relations of constituent reinforced plastics, show the characteristic behavior of the experimental stress-strain curves.

References

Amagi, S., and Miyano, Y., 1985, "Tensile Properties of Carbon/Aramid Unidirectional Reinforced FRP," Composite Materials, Vol. 11, pp. 62-67.

Aveston, J., and Sillwood, J. W., 1976, "Synergistic Fiber Strengthening in Hybrid Composites," Journal of Material Science, Vol. 11, pp. 1877-1883.

Furue, H., and Noguchi, Y., 1986, "Creep Properties of CF/GF Hybrid Laminates," Journal of Japan Society of Materials Sciences, Vol. 35, pp. 545-549.

Ishikawa, T., and Chou, T., 1982, "Elastic Behavior of Woven Hybrid Composites," Journal of Composite Materials, Vol. 16, pp.2-20.

Maksimov, R. D., Plume, E. Z., and Ponomarev, V. M., 1983, "Elasticity Characteristics of Unidirectionally Reinforced Hybrid Composites," Mekhanika Kompozitnykh Materialov, Vol.19, pp. 13-19.

Nagai, M., Miyairi, H., and Sato, S., 1984, "Flexural Properties of CF/GF Hybrid Composites with Different Laminate Constitutions," Journal of Society of Material Sciences, Japan, Vol. 33, pp. 61-66.

Sokolov, E. A., and Maksimov, R. D., 1979, "Predicting the Elastic Characteristics of a Hybrid Textolite," Mekhanika Kompozitnykh Materialov, Vol.15, pp. 705-711.

Wagner, H. D., Roman, I., and Maron, G., 1982, "Hybrid Effects in the Bending Stiffness of Graphite/Glass-reinforced Composite," Journal of Material Sciences, pp. 1359-1363.

Predictions of the Critical Strain for Matrix Cracking of Ceramic Matrix Composites

Wen-Shyong Kuo and Tsu-Wei Chou

Center for Composite Materials and Department of Mechanical engineering

University of Delaware, Newark DE 19716

ABSTRACT

The critical strain for matrix cracking of ceramic matrix composites has been studied; emphasis is placed on the effects of the fiber/matrix debonding length and interfacial debonding energy. Based on a modified shear-lag model, stresses in the fiber and matrix have been found for both the bonded and debonded regions. An energy balance approach, based upon the stress field, is then adopted to evaluate the critical strain for matrix cracking. From the general equation for the critical strain, close form solutions have been deduced for two limiting cases: complete debonding and perfect bonding. Numerical solutions are given for the cases of nonzero debonding energy and partial fiber debonding. The results show that the interfacial debonding energy, which has been ignored by most of the investigators, is an important factor in determining both the critical strain and the debonding length.

INTRODUCTION

Fiber-reinforced ceramic matrix composites have recently drawn considerable attention for their potential of high temperature applications. The fibers dispersed in the matrix tend to prevent catastrophic failures and enhance the toughness of the composite by providing various energy dissipation mechanisms; one of the dominant mechanisms is matrix cracking, which occurs at the strain lower than that for fiber failure (Chou, McCullough and Pipe, 1988). Fiber/matrix interfacial debonding and fiber sliding often accompany with matrix cracking; both have strong influence on not only the stress-strain relation of the composite (Kuo and Chou, 1990) but also the critical strain for matrix cracking.

Assuming a constant interfacial stress, negligible debonding energy and complete fiber debonding throughout the interface, Aveston, Cooper and Kelly (ACK)(1971) gave a close form solution for the critical strain. This solution has widely been accepted by many investigators (see Donald and McMillan, 1976; Prewo, 1988; Kerans, Hay and Pagano, 1989; Sutcu, 1989). In their subsequent work, Aveston and Kelly (1973) also found a close-form solution for the prediction of matrix cracking in unidirectional composites with perfect bonding. Budiansky, Hutchinson and Evans (1986) discussed both unbonded and perfectly bonded cases of fiber/matrix interface for evaluating the critical stress of matrix cracking. Although initial stresses were taken into account, they neglected the contribution of the normal stress in the matrix, which may be significant for composites with high modulus matrices. Marshall, Cox and Evans (MCE) (1985) considered the problem from the fracture mechanics point of view. A non-dimensional relation between the critical stress for matrix cracking and the crack length was obtained. More recently McCartney modified the MCE model by introducing a distributed traction on the crack surface to represent the bridging fibers; a solution same as the ACK equation has been given by McCartney (1987) for long matrix cracking; the fiber/matrix interface energy was not considered in this investigation.

This paper adopts the composite stress fields found by Kuo and Chou (1990) and an energy balance method to study the effect of fiber debonding length and fiber/matrix interface energy on the critical strain for matrix cracking. The composite of silicon-carbide fiber reinforced lithium-alumino-silicate is adopted as a model system.

STRESS DISTRIBUTIONS

To find the stress fields in a composite with matrix cracking, the fiber and matrix are modelled as concentric cylinders with radius r_o and R, respectively, and the composite is assumed to be composed of series of unit cells (Fig.1). The crack spacing and debonding length are denoted by L and L_d, respectively.

The stress fields in the fiber and matrix are evaluated through the application of a modified shear-lag model which takes into account the contribution of the normal stress in the matrix (Kuo and Chou, 1990). The results are as follows.

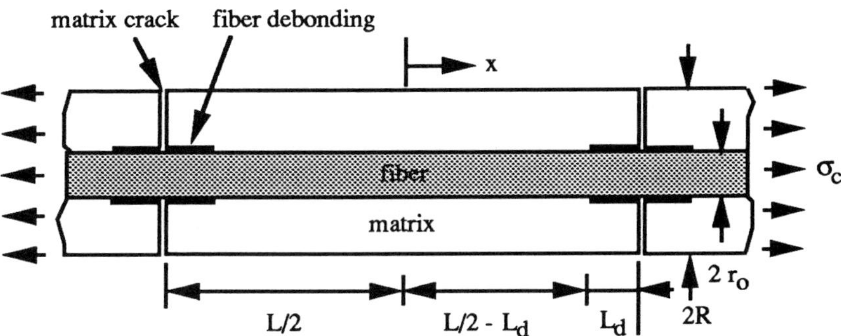

Fig.1 Unit cell for analysis

(1) Bonded region : $|x| \leq L/2 - L_d$

$$\sigma_f = \frac{\cosh \beta x}{\cosh \beta(L/2-L_d)} \left[\frac{E_m V_m}{E_c V_f} \sigma_c - 2 \frac{L_d}{r_o} \tau_s \right] + \frac{E_f}{E_c} \sigma_c$$

$$\sigma_m = \frac{\cosh \beta x}{\cosh \beta(L/2-L_d)} \left[2 \frac{V_f}{V_m} \frac{L_d}{r_o} \tau_s - \frac{E_m}{E_c} \sigma_c \right] + \frac{E_m}{E_c} \sigma_c \qquad (1)$$

(2) Debonded region : $L/2 - L_d \leq |x| \leq L/2$

$$\sigma_f = \frac{\sigma_c}{V_f} - \frac{2}{r_o} \tau_s \, (L/2 - |x|)$$

$$\sigma_m = \frac{V_f}{V_m} \frac{2}{r_o} \tau_s \, (L/2 - |x|) \qquad (2)$$

where E indicates Young's modulus and V denotes volume fraction. The subscripts f,m and c stand for fiber, matrix and composite, respectively; τ_s is the interfacial shear strength in the debonded regions; β is a constant defined as (see Kuo and Chou, 1990)

$$\beta = \left(\frac{2 E_c}{E_f (1+\nu_m) \left(\ln \frac{1}{V_f} + V_f - 1 \right)} \right)^{1/2} \frac{1}{r_o} \qquad (3)$$

where ν_m is the Poisson's ratio of the matrix and $E_c = E_f V_f + E_m V_m$.

ENERGY BALANCE APPROACH

The energies involved in creating a matrix crack (Aveston, Cooper and Kelly, 1971) are the matrix fracture surface energy (U_c), the fiber/matrix debonding energy(U_{db}), the work done against frictional force in fiber sliding

(U_s), the strain energy increment in the fiber(ΔU_f), the work done by external load(ΔW) and the decrease in matrix strain energy (ΔU_m). The relationship among the energy terms is

$$U_c + U_{db} + U_s + \Delta U_f = \Delta U_m + \Delta W \tag{4}$$

All the terms in Eq.(4) are positive. The energy terms for the unit cell with crack spacing L and debonding length L_d are:

$$U_c = \gamma_m \; 2\pi (R^2 - r_o^2) \tag{5}$$

$$U_{db} = \gamma_{db} \; 4\pi r_o L_d \tag{6}$$

$$U_s = \tau_s \; 4\pi r_o \int_{L/2-L_d}^{L/2} (u_f - u_m)\, dx \tag{7}$$

$$\Delta U_f = \pi r_o^2 \frac{1}{2E_f}\left[\int_{-L/2}^{L/2} \sigma_f^2\, dx - \left(\frac{E_f}{E_c}\sigma_c\right)^2 L\right] \tag{8}$$

$$\Delta U_m = \pi (R^2 - r_o^2) \frac{1}{2E_m}\left[\left(\frac{E_m}{E_c}\sigma_c\right)^2 L - \int_{-L/2}^{L/2} \sigma_m^2\, dx\right] \tag{9}$$

$$\Delta W = \pi R^2 \sigma_c \int_{-L/2}^{L/2} \left(\varepsilon_f - \frac{E_m}{E_c}\sigma_c\right) dx \tag{10}$$

where γ_m is the matrix surface energy per unit area; γ_{db} is the fiber/matrix interface energy per unit area. Substituting the stress fields (Eqs.(1) and (2)) into Eqs.(7-10), and using Eq.(4), the following general equation for the critical strain of matrix cracking can be obtained,

$$f_1(L,L_d)\, \varepsilon_{mu}^2 + f_2(L,L_d)\, \varepsilon_{mu} = f_3(L,L_d) \tag{11}$$

where the coefficients f_1, f_2 and f_3 are functions of L and L_d. Their explicit expressions are:

$$f_1(L,L_d) = \frac{V_m}{E_m V_f} [E_m^2 L_d - \frac{1}{4} (L-2L_d)C_7^2 + \frac{2}{\beta} E_m C_7 \sinh \beta(\frac{L}{2}-L_d)$$
$$- \frac{1}{4\beta} \sinh \beta(L-2L_d) C_7^2] + 2 \frac{E_c}{E_f V_f} [\frac{1}{\beta} C_1 \sinh \beta(\frac{L}{2}-L_d) - E_f L_d$$
$$+ C_3 L_d] - \frac{1}{E_f} [\frac{1}{4} (L-2L_d + \frac{1}{\beta} \sinh \beta(L-2L_d)) C_1^2$$
$$+ \frac{2}{\beta} C_1 E_f \sinh \beta(\frac{L}{2}-L_d) + C_3^2 L_d - E_f^2 L_d]$$

$$f_2(L,L_d) = \frac{V_m}{E_m V_f} [\frac{1}{2} (L-2L_d)C_6 C_7 - \frac{2}{\beta} E_m C_6 \sinh \beta(\frac{L}{2}-L_d)$$
$$+ \frac{1}{2\beta} C_6 C_7 \sinh \beta(L-2L_d)] + 2 \frac{E_c}{E_f V_f} [- \frac{1}{\beta} C_2 \sinh \beta(\frac{L}{2}-L_d)$$
$$- C_4 L_d + \frac{1}{2} C_5 L_d (L-L_d)] - \frac{1}{E_f} [- \frac{1}{2} (L-2L_d$$
$$+ \frac{1}{\beta} \sinh \beta(L-2L_d)) C_1 C_2 - \frac{2}{\beta} C_2 E_f \sinh \beta(\frac{L}{2}-L_d) - 2 C_3 C_4 L_d$$
$$+ C_3 C_5 L_d (L-L_d)] - \tau_s \frac{2}{r_o} L_d^2 \frac{E_c}{E_f V_f}$$

$$f_3(L,L_d) = 2 \gamma_m \frac{V_m}{V_f} + 4 \gamma_{db} \frac{L_d}{r_o} - \frac{8}{3} \frac{E_c}{E_f E_m V_m} (\frac{L_d}{r_o})^2 L_d \tau_s^2$$
$$+ \frac{1}{E_f} [\frac{1}{4} (L-2L_d + \frac{1}{\beta} \sinh \beta(L-2L_d)) C_2^2 + L_d C_4^2 - L_d (L-L_d) C_4 C_5$$
$$+ \frac{1}{3} C_5^2 L_d (\frac{3}{4} L^2 - \frac{3}{2} L L_d + L_d^2)] + \frac{V_m}{E_m V_f} [\frac{1}{4} (L-2L_d) C_6^2$$
$$+ \frac{1}{4\beta} C_6^2 \sinh \beta(L-2L_d) + L_d C_8^2 - C_8 C_9 L_d (L-L_d)$$
$$+ \frac{1}{3} C_9^2 L_d (\frac{3}{4} L^2 - \frac{3}{2} L L_d + L_d^2)]$$

where

$$[C_1, C_2, C_6, C_7] = \frac{1}{\cosh \beta(\frac{L}{2}-L_d)} [\frac{E_m V_m}{V_f}, \frac{2}{r_o} L_d \tau_s, \frac{2}{r_o} L_d \tau_s \frac{V_f}{V_m}, E_m]$$

$$[C_4, C_5, C_8, C_9] = \frac{2}{r_o} \tau_s [\frac{L}{2}, 1, \frac{L}{2} \frac{V_f}{V_m}, \frac{V_f}{V_m}]$$

$$C_3 = \frac{E_c}{V_f}$$

The elastic properties of the composite in Eq.(11) are treated as constants; the lengths L and L_d are independent variables except for the limiting cases discussed below.

LIMITING CASES

Case 1: Complete Debonding

Assume that the fiber is completely debonded and there is no energy dissipation at debonding, i.e. L_d=X, L=2X and U_{db}=0, where X is the maximum possible debonding length. By assuming that the maximum matrix stress after matrix cracking is equal to the matrix stress in the uncracked composite, X can be found as

$$X=\frac{E_m V_m r_o}{2\tau_s V_f}\,\varepsilon_{mu} \tag{12}$$

The solution for Eq.(11) for this case is

$$\varepsilon_{mu} = \left(\frac{12\tau_s\,\gamma_m\,E_f V_f^2}{E_c E_m^2\,r_o V_m}\right)^{1/3} \tag{13}$$

Equation (13) is an approximation for the critical strain of the composite with relatively weak interfacial strength and low interface energy and it is the same as the solution given by Aveston, Cooper and Kelly (1971).

Case 2: Perfect Bonding

Assume no fiber/matrix interfacial debonding and infinite crack spacing, i.e., L_d=0 and L=∞. The solution of the Eq.(11) is

$$\varepsilon_{mu} = \left(\frac{4}{3}\frac{\beta\gamma_m E_f V_f}{E_m E_c}\right)^{1/2} \tag{14}$$

The above equation best describes the critical strain for the composite with relatively strong interfacial strength.

NUMERICAL RESULTS FOR INTERMEDIATE DEBONDING LENGTHS

The material properties for the composite of silicon-carbide fibers in a lithium-alumino-silicate glass matrix (see Budiansky, Hutchinson and Evans, 1986) are adopted for the numerical calculations.

$E_f = 200$ GPa $\quad\quad \nu_m = 0.25$

$E_m = 85$ GPa $\quad\quad r_o = 8.0{*}10^{-6}$ m

$\gamma_m = 44.1 {*}10^{-9}$ GPa·m $\quad\quad V_f = 0.4$

$\tau_s = 2{*}10^{-3}$ GPa

The maximum possible debonding length, X, is evaluated from Eqs.(12) and (13). The results represented by the ratio of X to r_o are shown in Fig.2 as a function of τ_s and γ_{db}. It can be seen that X/r_o increases with a decrease in τ_s or an increase in γ_{db}. It should be noted that X and ε_{mu} are solved simultaneously from Eqs.(11) and (12) by using an iterative method. Eq.(13), for complete interfacial debonding and $\gamma_{db}=0$, is a special case of the present solution.

Once X is known, Eq.(11) can be solved numerically for intermediate debonding lengths, i.e., $0 \le L_d \le X$, $2L_d \le L \le \infty$. It is found that once L_d is fixed, the critical strain is generally insensitive to the crack spacing L. By using

$\tau_s = 2*10^{-3}$ GPa and L=2X, the solutions for the critical strain, ε_{mu} , are shown in Fig.3 for different values of L_d/X ratio and γ_{db}.

The lowest value of ε_{mu} , for each γ_{db} , is of basic importance. The ACK solution (Eq.(13)) corresponds to the case of $L_d/X=1$ and $\gamma_{db}=0$. An important observation of Fig.3 is that the location and magnitude of the minimum value of ε_{mu} is sensitive to the value of γ_{db}. This is because the debonded interfacial area is much larger than that of matrix cracking surface. Figure 3 also indicates that the location of the minimum value of ε_{mu} tends to $L_d/X=0$ as γ_{db} increases; thus, the degree of fiber debonding is reduced as γ_{db} becomes larger.

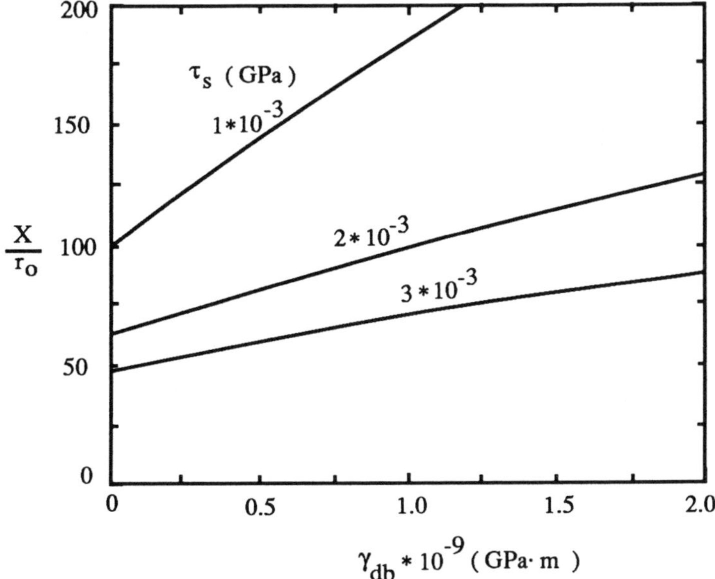

Fig.2 The maximum debonding length X versus γ_{db} and τ_s.

CONCLUSIONS

(1) The fiber/matrix interfacial shear strength $\tau_{s,}$ the interface debonding energy γ_{db} and the matrix surface energy γ_m are the key factors in determining the critical strain for matrix cracking.

(2) The ratio of L_d/X for the minimum value of ε_{mu} depends on the combination of $\tau_{s,}$ γ_{db} and γ_m; it tend to $L_d/X=0$ as τ_s and γ_{db} increase.

(3) It is improper to neglect the energy dissipated due to fiber/matrix interfacial debonding in the case that the interfacial energy is significant and the composite undergoes extensive fiber debonding.

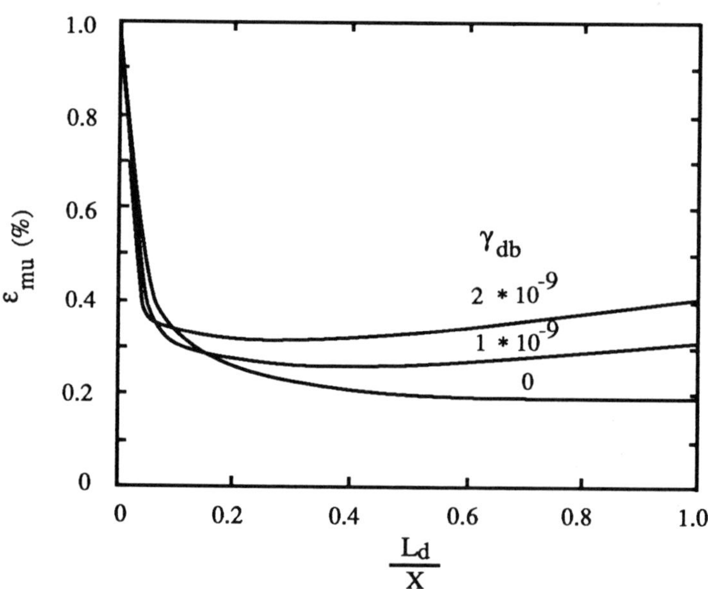

Fig.3 The critical strain for matrix cracking versus γ_{db} and L_d.

REFERENCES

Aveston, J., Cooper, G. A. and Kelly, A., 1971, "Single and Multiple Fracture," Conference Proceedings, National Physical Laboratory, IPC Science and Technology Press Ltd., pp. 15-26.

Aveston, J. and Kelly, A., 1973,"Theory of Multiple Fracture of Fibrous Composites," J. of Material Science, Vol.8, pp.352-362.

Budiansky, B. , Hutchinson, W.J. and Evans, A.G., 1986,"Matrix Fracture in Fiber-Reinforced Ceramics," J. Mech. Phys. Solids, Vol. 34, pp. 167-189.

Chou, T.W. ,McCullough, R. L. and Pipe R. B. , 1988, "Composites," Scientific American, 255, pp. 192-203.

Donald, I.W. and McMillan, P.W. ,1976,"Review: Ceramic Matrix Composites," Journal of Materials Science, Vol. 11, pp. 949-972.

Kerans, R.J., Hay, R.S. and Pagano, N.J.,1989,"The Role of the Fiber-Matrix Interface in Ceramic Composites," Ceramic Bulletin, Vol. 68, pp. 429-442.

Kuo, W.S. and Chou,T.W., 1990, "Modeling of the Thermo-mechanical Behavior of Glass-Matrix Composites," Proceeding of TMS International Conference on Advanced Metal and Ceramic Composites, Feb. 19-22, Anaheim, California.

Kuo, W.S. and Chou, T.W. "Analysis of the Thermo-mechanical Behavior of Ceramic Matrix Composites," Submitted for publication.

Marshall, D. B., Cox, B. N. and Evans, A.G.,1985, "The Mechanics of Matrix Cracking in Brittle-Matrix Fiber Composites," Acta Metall. Vol. 33, pp. 2013-2021.

McCartney, L.N. 1987,"Mechanics of Matrix Cracking in Brittle-Matrix Fibre-Reinforced Composites," Proc. R. Soc. Lond. A409, pp. 329-350.

Prewo, K.M., 1988, "Carbon Fiber Reinforced Glass Matrix Composite Tension and Flexure Properties," Journal of Materials Science, Vol.23 , pp. 2745-2752.

Sutcu, M., 1989,"Weibull Statistic Applied to Fiber Failure in Ceramic Composites and Work of Fracture," Acta Metall., Vol.37, pp. 651-661.

Acknowledgements

The authors wish to thank the support of NASA-Lewis Research Center for Tsu-Wei Chou (NAS3-25971, Dr. John P. Gyekenyesi, Program Manager) and the Center for Composite Materials of University of Delaware for Wen-Shyong Kuo.

Inelastic Behavior VI

Shear Characterisation and Inelastic Torsion of Fibre-Reinforced Materials

T. G. ROGERS

Department of Theoretical Mechanics,
University of Nottingham,
University Park,
Nottingham, NG7 2RD, U.K.

ABSTRACT

We present a relatively simple stress analysis of the torsion of thin flat specimens of composite materials continuously reinforced by fibres. Centred and non-centred torsion are considered and analyses are given for both viscoelastic and elastic-plastic material responses.

1. INTRODUCTION

A dominant feature of the mechanical response of any fibre-reinforced composite material is its anisotropy, and this is particularly true when the reinforcement consists of continuous fibres. Even when both matrix and fibres are themselves isotropic, the macroscopic response of the composite will be transversely isotropic about the local fibre-direction $\underset{\sim}{a}$ if there is just one family of fibres distributed throughout the matrix, and will be monoclinic or, more typically, orthotropic if there are two such families.

Furthermore, these composites are usually manufactured so that a relatively weak matrix with otherwise desirable properties may be stiffened in one or two particular directions by introducing relatively stiff and strong reinforcing fibres in those directions. The response of such composites is then not only anisotropic but highly anisotropic, with the extensional modulus in a fibre-direction being much greater than any other extensional or shear modulus. An important, and generally unappreciated, consequence of this high anisotropy is to make the accurate evaluation of that particular extensional modulus much less significant and that of the shear moduli much more important. This is because in general such a composite will respond to any applied tractions by taking, if possible, the 'soft' options of shearing along and/or across the fibres rather than by extending in the much stiffer fibre-directions.

To a theoretical stress analyst, it can seem to be a simple matter to measure such shear moduli; in fact, very little direct and accurate measurement of shear moduli appears to have been reported in the literature. Most the available data have been determined indirectly from tensile tests, by measuring the various Poisson ratios (see Christensen 1979, for example) in axial tests or by using 'off-axis' tests (Pagano and Halpin 1968). Unfortunately, in the former method the shear moduli so deduced will be subject to large relative experimental error, whilst the resulting S-shape of the test specimens in the 'off-axis' test implies a deformation that is far different from the homogeneous deformation assumed for the relevant simple stress analysis. Furthermore, in the context of any composite that exhibits a substantial amount of creep or viscous effects, a tensile test is impractical anyway.

For isotropic materials such difficulties can be overcome by means of simple torsion tests, which are usually oscillatory (Groves 1989) for viscoelastic response or for fluid-like behaviour (as for thermoplastics at elevated temperatures), but

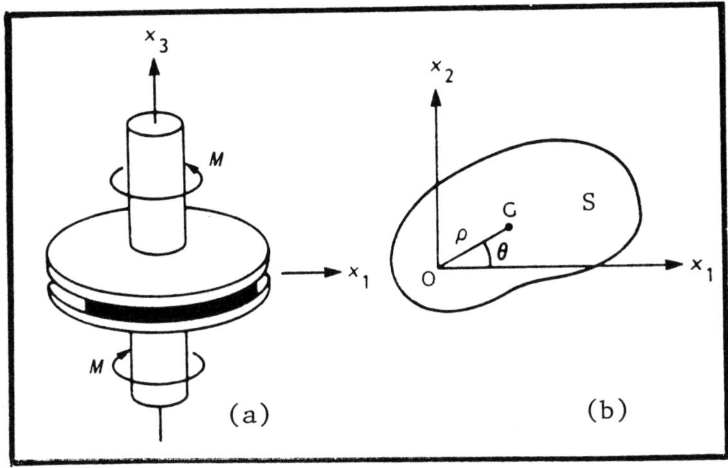

Fig. 1. Schematic diagram of torsion test. (a) Physical set–up; (b) plan view of specimen

need not be so. A typical torsion testing set–up is shown schematically in Figure 1. A thin flat specimen (shown in black) is constrained between the two parallel faces of the machine–platens, and is subjected to either steady state oscillatory or quasi–static torsion by applying a measured torque M about the rotational axis of the platens. The resulting twist or angular displacement is then measured, and the shear modulus (or viscosity, for a fluid) deduced in an accurate and direct manner. However, for anisotropic materials, there are at least two important and distinct modes of shear, which can be associated respectively with simple shearing along and transverse to the fibres. Hence the conventional torsion test would give insufficient data, providing information only about a specific combination of the two relevant shear moduli or viscosities.

In this paper we describe modifications of such torsion tests which, with the relevant analyses, can be used to determine the moduli relevant to these two modes of shear deformation. These moduli, denoted by μ_L and μ_T, are both constant for a linearly elastic composite, and are time-dependent functions for viscoelastic response. Elastic-plastic behaviour is characterised by not only μ_L and μ_T but also k_L and k_T, the yield stresses in simple shear along and transverse to the fibre-direction. The modifications involve either varying the shape of the specimen but keeping its centre of mass on the axis of torque (centred torsion), or altering the specimen's position and orientation (off-centred torsion).

For ease of presentation, we first review the relevant constitutive equations (Section 2) and then describe in Section 3 the elementary kinematics for the torsion tests to be considered. Due to the usual elastic-viscoelastic correspondence principle (Christensen 1979) and since plastic deformation is usually preceded by elastic behaviour, it is convenient to then present, in Section 4, the various analyses for elastic composites. This is followed in Section 5 by a straightforward generalisation to take into account linearly viscoelastic effects; viscous responses are also considered, albeit very briefly. The final section treats the effect of rate-independent plastic flow induced by continued torsional loading after initial yield, for both Mises and Tresca type functions, and concludes with a brief description of the relevant analysis when an elastic-plastic specimen is subjected to cyclic loading. In all sections, the analysis assumes that the rate of loading is such that inertial effects may be neglected and for convenience we restrict attention to composites reinforced by a single family of fibres.

2. CONSTITUTIVE EQUATIONS

We treat the composite as an anisotropic continuum in which the fibre-reinforcement introduces a preferred direction $\underset{\sim}{a}$ (the 'fibre-direction') at every point $\underset{\sim}{x}$ in the body. This direction, which will be associated with a large extensional modulus (compared with the shear moduli for example), is usually constant in the undeformed configuration, but is not necessarily so; in general it can change with position $\underset{\sim}{x}$.

Most of the constitutive equations which are given in this section have been derived and treated in much greater detail elsewhere (e.g. Spencer 1972, 1983, Rogers 1987). We use standard notation, referring all vector and tensor components to a rectangular system of Cartesian axes with position coordinates x_i $(i=1,2,3)$; u_i and v_i denote the components of displacement and velocity respectively, and the infinitesimal strain and rate-of-strain components are given by

$$e_{ij} = \tfrac{1}{2}\left[\frac{\partial u_i}{\partial x_j} + \frac{\partial u_j}{\partial x_i}\right], \quad d_{ij} = \tfrac{1}{2}\left[\frac{\partial v_i}{\partial v_j} + \frac{\partial v_j}{\partial x_i}\right] . \qquad (2.1)$$

The stress components are denoted by σ_{ij} and the components of fibre-direction are a_i; δ_{ij} represents the Kroneker delta and the usual repeated suffix summation convention is used whenever necessary. For convenience, we also introduce

$$\underset{\sim}{A} = \underset{\sim}{a}\,\underset{\sim}{a} \equiv \underset{\sim}{a} \otimes \underset{\sim}{a} \quad \text{with } A_{ij} = a_i a_j . \qquad (2.2)$$

(a) Linearly elastic composites

In this case a convenient intrinsic form of the constitutive equation is

$$\underset{\sim}{\sigma} = (\lambda \operatorname{tr} \underset{\sim}{e} + \alpha \operatorname{tr} \underset{\sim}{A}\underset{\sim}{e})\underset{\sim}{I} + 2\mu_T \underset{\sim}{e} + 2(\mu_L - \mu_T)(\underset{\sim}{A}\underset{\sim}{e} + \underset{\sim}{e}\underset{\sim}{A}) + \\ + (\alpha \operatorname{tr} \underset{\sim}{e} + \beta \operatorname{tr} \underset{\sim}{A}\underset{\sim}{e})\underset{\sim}{A} . \qquad (2.3)$$

Here $\operatorname{tr} \underset{\sim}{e}$ and $\operatorname{tr} \underset{\sim}{A}\underset{\sim}{e}$ are the traces of $\underset{\sim}{e}$ and $\underset{\sim}{A}\underset{\sim}{e}$, with

$$\operatorname{tr} \underset{\sim}{e} = e_{ii}, \quad \operatorname{tr} \underset{\sim}{A}\underset{\sim}{e} = A_{ij}\, e_{ij} = a_i\, a_j\, e_{ij} .$$

The component form of (2.3) is

$$\sigma_{ij} = (\lambda e_{kk} + \alpha a_k a_\ell e_{k\ell})\delta_{ij} + 2\mu_T e_{ij} + \\ + 2(\mu_L - \mu_T)(a_i a_k a_{kj} + a_j a_k e_{ki}) + \\ + (\alpha e_{kk} + \beta a_k a_\ell e_{k\ell}) a_i a_j . \qquad (2.4)$$

When the fibre-direction coincides with one of the coordinate axes, this form reduces to the usual form of transverse isotropy.

The intrinsic form, though apparently more complicated, is convenient in that it is independent of choice of axes, and highlights the role of the fibre-direction. The particular form (2.3) also distinguishes the shear moduli μ_L and μ_T from the other elastic constants α, β and λ, which can be related to the Poisson ratios and extensional moduli of the material. If the material is effectively incompressible, then the imposition of an arbitrary hydrostatic pressure field produces no change in volume, so that tr $\underset{\sim}{e} = 0$ and (2.3) reduces to

$$\underset{\sim}{\sigma} = -p\underset{\sim}{I} + \beta\underset{\sim}{A} \text{ tr } \underset{\sim}{A}\underset{\sim}{e} + 2\mu_T\underset{\sim}{e} + 2(\mu_L-\mu_T)(\underset{\sim}{A}\underset{\sim}{e}+\underset{\sim}{e}\underset{\sim}{A}). \quad (2.5)$$

If the composite is not incompressible, but is sufficiently stiff in the fibre-direction that a superimposed arbitrary tension in that direction produces negligible extension ("fibre-inextensible"), then

$$a_i a_j e_{ij} = \text{tr } \underset{\sim}{A}\underset{\sim}{e} = 0$$

and (2.3) reduces to

$$\underset{\sim}{\sigma} = T\underset{\sim}{A} + \lambda\underset{\sim}{I} \text{ tr } \underset{\sim}{e} + 2\mu_T \underset{\sim}{e} + 2(\mu_L-\mu_T)(\underset{\sim}{A}\underset{\sim}{e}+\underset{\sim}{e}\underset{\sim}{A}). \quad (2.6)$$

If both approximations (incompressibility and fibre-inextensibility) are assumed applicable, then the constitutive equation reduces to

$$\underset{\sim}{\sigma} = -p\underset{\sim}{I} + T\underset{\sim}{A} + 2\mu_T\underset{\sim}{e} + 2(\mu_L-\mu_T)(\underset{\sim}{A}\underset{\sim}{e}+\underset{\sim}{e}\underset{\sim}{A}). \quad (2.7)$$

This very special case has been termed an 'ideal fibre-reinforced material', and many useful solutions of stress analysis problems have been obtained by adopting it. We also note that, in all the special cases listed above, the terms involving the shear moduli are unaltered.

(b) Viscoelastic and viscous responses

The constitutive equations for isothermal linear viscoelastic materials are a simple generalisation (Rogers and Pipkin 1963) of those of linear elasticity, with algebraic products of moduli and strains being replaced by convolutions of the relaxation functions and the strain-rates. Thus a term such as $2\mu_T \underset{\sim}{e}$ in (2.3) - (2.7) is simply replaced by

$$2\mu_T * \underset{\sim}{d} = 2\int_{-\infty}^{t} \mu_T(t-\tau)\underset{\sim}{d}(\underset{\sim}{x},\tau)\, d\tau = 2\int_{\tau=-\infty}^{\tau=t} \mu_T(t-\tau)\, d\underset{\sim}{e}(\underset{\sim}{x},\tau) \tag{2.8}$$

where $\mu_T(t)$ now represents the time-dependent relaxation modulus associated with simple shearing transverse to the fibre-direction. Elastic behaviour is then obviously a special case for which the relaxation functions are constant.

Such equations are appropriate for small strain behaviour of viscoelastic solids, in which case the "correspondence principle" of viscoelasticity usually holds so that elastic solutions can be readily extended to give the solutions to the equivalent problem in viscoelasticity. The reaction terms involving p and T are unaffected, apart from being dependent on the time t as well as on $\underset{\sim}{x}$.

The special case of an anisotropic viscous fluid is given by letting the relaxation functions be impulse functions, so that, for example,

$$\mu_T(t) = \eta_T\delta(t) , \qquad \mu_L(t) = \eta_L\delta(t) \tag{2.9}$$

where $\delta(t)$ is the Dirac delta function and η_T, η_L are the appropriate viscosities. Thus an incompressible transversely isotropic fluid may be characterised by (cf.(2.5))

$$\underset{\sim}{\sigma} = -p\underset{\sim}{I} + 2\eta_T\underset{\sim}{d} + 2(\eta_L-\eta_T)(\underset{\sim}{A}\underset{\sim}{d}+\underset{\sim}{d}\underset{\sim}{A}) + \eta_E\underset{\sim}{A}\ \mathrm{tr}\ \underset{\sim}{A}\underset{\sim}{d}. \tag{2.10}$$

This equation is appropriate for a fibre-reinforced linear viscous fluid, with the extensional viscosity measure η_E much larger than the shear viscosities η_T and η_L. Such a response could be expected for a composite comprising a Newtonian fluid matrix reinforced by aligned chopped strand fibres. If, however, the reinforcement were to consist of continuous 'long' filaments then an appropriate model is the elastic fibre–reinforced viscous fluid in which the extensional behaviour in the fibre–direction is linearly elastic but the shear behaviour is viscous. The constitutive equation for the linearly viscous case is then, from (2.5),

$$\sigma = -p\mathbf{I} + 2\eta_T \mathbf{d} + 2(\eta_L - \eta_T)(\mathbf{A}\mathbf{d} + \mathbf{d}\mathbf{A}) + \beta \mathbf{A} \, \mathrm{tr} \, \mathbf{A}\mathbf{e} . \quad (2.11)$$

We note that although both of these fluids (described by (2.10) and (2.11)) have different extensional responses in the fibre–direction, in each case the idealised limit of inextensibility in that direction yields the same constitutive equation:

$$\sigma = -p\mathbf{I} + T\mathbf{A} + 2\eta_T \mathbf{d} + 2(\eta_L - \eta_T)(\mathbf{A}\mathbf{d} + \mathbf{d}\mathbf{A}) , \quad (2.12)$$

termed the 'ideal fibre–reinforced linearly viscous fluid'. We also note that despite being linear, all three equations (2.10)–(2.12) describe non–Newtonian responses, being highly anisotropic.

For completeness we note that the generalisation of equation (2.12) to include nonlinear effects is the ideal fibre–reinforced fluid, analogous to the 'purely viscous' non–Newtonian isotropic fluid. It can be shown – by comparison with plasticity theory (Spencer 1983, Rogers 1987), for instance – that the most general relevant constitutive equation for such an incompressible, fibre–inextensible fluid can be written in the form

$$\sigma = -p\mathbf{I} + T\mathbf{A} + \mathbf{s}(\mathbf{d}, \mathbf{a}) , \quad (2.13)$$

where the extra–stress $\mathbf{s}$ is given by

$$\mathbf{s} = \varphi_1 \mathbf{d} + \varphi_2(\mathbf{A}\mathbf{d} + \mathbf{d}\mathbf{A}) + \varphi_3 \mathbf{d}\mathbf{A}\mathbf{d} + \varphi_4(\mathbf{A}\mathbf{d}^2 + \mathbf{d}^2\mathbf{A}) . \quad (2.14)$$

Here $\varphi_1, \ldots, \varphi_4$ are functions of the strain–rate invariants

$$D_1 = \tfrac{1}{2}\mathrm{tr}\, \underset{\sim}{d}^2 - \mathrm{tr}\, \underset{\sim}{A}\underset{\sim}{d}^2 , \quad D_2 = \mathrm{tr}\, \underset{\sim}{A}\underset{\sim}{d}^2 , \quad D_3 = \mathrm{tr}\, \underset{\sim}{d}^3 . \tag{2.15}$$

A further simple generalisation to allow linear <u>elastic</u> extensibility in the fibre-direction gives

$$\underset{\sim}{\sigma} = -p\underset{\sim}{I} + \underset{\sim}{s}(\underset{\sim}{d},\underset{\sim}{a}) + E\underset{\sim}{A}\, \mathrm{tr}\, \underset{\sim}{A}\underset{\sim}{e} . \tag{2.16}$$

It is elementary to confirm that the linear case (2.12) is recovered by making φ_1 and φ_2 constant with φ_3 and φ_4 both zero.

(c) Plastic and elastic-plastic behaviour

Most permanent plastic deformation associated with composites is found in those with a metal matrix. Accordingly we consider a rate-independent plasticity theory which, as in metal plasticity, postulates the existence of a yield criterion and an associated flow rule. We assume that a scalar yield function $f(\underset{\sim}{\sigma})$ exists, such that

$$f(\underset{\sim}{\sigma}) \leqslant 1 . \tag{2.17}$$

If $f < 1$, or if $f = 1$ and the material time derivative $Df/Dt < 0$, then the deformation is elastic; if $f = 1$ and $Df/Dt = 0$, the material is being plastically deformed. The associated flow rule determines the plastic strain-rate $\underset{\sim}{d}^p$ $(= \underset{\sim}{d}-\underset{\sim}{d}^e)$ as

$$d^p_{ij} = \begin{cases} 0 & \text{elastic state} \\ \dot{\lambda}\, \partial f/\partial\sigma_{ij} & \text{plastic state} \end{cases} \tag{2.18}$$

where $\dot{\lambda}$ is a positive scalar multiplier, possibly dependent on the history of the irreversible plastic deformation. The elastic strain-rate $\underset{\sim}{d}^e$ is related to the stress-rate through the relevant elasticity equations. As is usual in metal plasticity, we also assume plastic incompressibility, so that

$$\mathrm{tr}\, \underset{\sim}{d}^p = 0 .$$

Here, as usual, the parameter t (and "velocity" $\underset{\sim}{v}$) is not now related to 'real' time, but represents a 'plastic' time which determines the sequence of events as the deformation develops.

For the locally transversely isotropic materials considered in this paper, the most general smooth quadratic form of f is

$$f_M(\sigma) = \frac{J_1}{k_T^2} + \frac{J_2}{k_L^2} + \frac{J_4^2}{Y^2} . \tag{2.19}$$

Here the stress invariants J_1, J_2 and J_4 are defined, for convenience (Spencer 1972, Rogers 1987), by

$$J_1 = \tfrac{1}{2}\,\mathrm{tr}\,\underset{\sim}{s}^2 - \mathrm{tr}\,\underset{\sim}{A}\underset{\sim}{s}^2, \quad J_2 = \mathrm{tr}\,\underset{\sim}{A}\underset{\sim}{s}^2, \quad J_4 = \tfrac{3}{2}\,\mathrm{tr}\,\underset{\sim}{A}\underset{\sim}{\sigma}', \tag{2.20}$$

where $\underset{\sim}{s}$ and the deviatoric stress tensor $\underset{\sim}{\sigma}'$ are given by

$$\begin{aligned} \underset{\sim}{s} &= \underset{\sim}{\sigma} - \tfrac{1}{2}(\mathrm{tr}\,\underset{\sim}{\sigma} - \mathrm{tr}\,\underset{\sim}{A}\underset{\sim}{\sigma})\underset{\sim}{I} + \tfrac{1}{2}(\mathrm{tr}\,\underset{\sim}{\sigma} - 3\,\mathrm{tr}\,\underset{\sim}{A}\underset{\sim}{\sigma})\underset{\sim}{A} \\ \underset{\sim}{\sigma}' &= \underset{\sim}{\sigma} - \tfrac{1}{3}(\mathrm{tr}\,\underset{\sim}{\sigma})\underset{\sim}{I} . \end{aligned} \tag{2.21}$$

Choosing such a set as (2.20) emphasizes, in any analysis, any difference of properties along and transverse to the fibres. Thus in (2.19) k_L and k_T are the plasticity analogues of the elastic moduli μ_L and μ_T, and correspond to the yield stress in shear along and transverse to the fibre-direction $\underset{\sim}{a}$; Y denotes the yield stress for simple tension in the fibre-direction and is typically much larger that k_L and k_T. This yield function reduces to the von Mises form for isotropic plasticity.

A piecewise smooth quadratic form that corresponds to a generalisation of Tresca's yield function is

$$f_T(\sigma) = \begin{cases} J_1/k_T^2 , & |J_2| \leq k_L^2 , \quad |J_4| \leq Y \\ J_2/k_L^2 , & |J_1| \leq k_T^2 , \quad |J_4| \leq Y \\ J_4^2/Y^2 , & |J_1| \leq k_T^2 , \quad |J_2| \leq k_L^2 . \end{cases} \tag{2.22}$$

In both cases, $Y \to \infty$ corresponds to the material being insensitive to a superposed tension in the fibre-direction. We note that, through (2.18), this is equivalent to $\mathrm{tr}\,\underset{\sim}{A}\underset{\sim}{d}^P = 0$, i.e. fibre-inextensibility.

These forms are just two simple examples of possible yield functions. More general forms, including some that include the Bauschinger effect and dependence on hydrostatic stress, have been proposed and discussed elsewhere (e.g. Rogers 1987). However, comparison with experimental data (Dvorak et al. 1988; also refer Spencer 1972) on initial yield (i.e. failure data) has shown that (2.19) and (2.22) can be excellent models for real fibre-reinforced composites.

3. TORSION TESTING

The typical torsion test has already been described in the Introduction. The stress analysis is very simple, and is based on the observations that the plate specimens are thin (in the sense that the ratio of thickness to typical length is small), that the fibre reinforcement is unidirectional and lies in the plane of the plate, and that the plate surfaces are constrained by adhesion to the parallel faces of the rigid circular platens of the testing machine. The deformation caused by the relative motion of the two platens is then essentially pure torsion with zero warping, apart from 'edge' effects localised to the edge of the specimen. If no 'squeezing' takes place, so that the displacement normal to the plane of the platens is zero, then the displacement field for small deformations is

$$u_1 = -\theta x_2 x_3, \quad u_2 = \theta x_1 x_3, \quad u_3 = 0 . \tag{3.1}$$

Here the coordinate axes are such that the x_1–x_2 plane coincides with the mid-plane of the specimen with the x_1-axis parallel to the fibres, and the x_3-axis is coincident with the axis of rotation; thus $\underset{\sim}{a} = (1,0,0)$. The angle θ is the angle of twist per unit thickness of specimen. If θ varies with time then the velocity field is given by

$$v_1 = -\dot{\theta} x_2 x_3 , \quad v_2 = \dot{\theta} x_1 x_3 , \quad v_3 = 0 . \tag{3.2}$$

The strain and strain-rate components are then

$$e_{13} = -\tfrac{1}{2}\theta x_2 , \qquad e_{23} = \tfrac{1}{2}\theta x_1 ,$$
$$d_{13} = -\tfrac{1}{2}\dot{\theta} x_2 , \qquad d_{23} = \tfrac{1}{2}\dot{\theta} x_1 , \tag{3.3}$$

with the remaining components all zero. Straightforward substitution of (3.3) into the constitutive equations in the previous section gives the stress components, which are all independent of x_3. The components relevant to torsion are σ_{13} and σ_{23} which satisfy, through equilibrium,

$$\partial\sigma_{13}/\partial x_1 + \partial\sigma_{23}/\partial x_2 = 0 . \tag{3.4}$$

The applied torque M about the x_3-axis is

$$M = \iint_S (x_1\sigma_{23} - x_2\sigma_{13}) \, dA , \tag{3.5}$$

where S denotes the lateral surface of the plate, and the stress components take their surface values.

The distinction between centred torsion and non-centred torsion is that in the former the mass-centre G of the plate specimen lies at the origin O; for off-centred torsion it is required that O and G should not coincide. We further note that the 'plate' need not consist of just one piece but can be the sum of several pieces; thus the region S need not be simply connected.

4. ELASTIC TORSION

For an elastic composite, substituting (3.3) into any or all of (2.3)-(2.7) gives

$$\sigma_{13} = -\mu_L\theta x_2 , \qquad \sigma_{23} = \mu_T\theta x_1 , \tag{4.1}$$

so that, from (3.5)

$$M = \left\{\iint_S (x_2^2\mu_L + x_1^2\mu_T) \, dA\right\}\theta . \tag{4.2}$$

This is more conveniently expressed as

$$M = (\mu_L I_1 + \mu_T I_2)\theta , \tag{4.3}$$

where I_1 and I_2 are the second moments of area about the x_1 and x_2-axes respectively.

The moment-twist relation (4.3) shows that when $I_1 = I_2$, as for the conventional circular or square test specimens, then only data for $\mu_L+\mu_T$ can be obtained. However, it also makes clear that if the moments of inertia are varied, then both μ_L and μ_T may be determined. In practice, I_1 and I_2 can be altered in one of two ways – either the shape of the specimen can be changed or its positioning can.

The most sensible shape to use is that of a rectangle in which each edge is either parallel or perpendicular to the fibre-direction. Such a shape is easiest to manufacture and leads to particularly simple forms of the analysis.

For centred torsion of a plate of length 2L in the direction of the fibres and of width 2H, equation (4.3) gives

$$M = \tfrac{4}{3}LH^3(\mu_L + \frac{L^2}{H^2}\mu_T)\theta = \tfrac{4}{3}L^3H(\mu_T + \frac{H^2}{L^2}\mu_L)\theta \ . \qquad (4.4)$$

Hence by determining M/θ for various aspect ratios L/H, a plot of $3M/(4L^3H\theta)$ against H^2/L^2 should then give a straight line with slope μ_L and value μ_T as the intercept as $H/L \to 0$. The same data could also be used to plot $3M/(4LH^3\theta)$ against L^2/H^2 which should also produce a straight line, but this time with slope μ_T and intercept μ_L as $L/H \to 0$.

Off-centred torsion avoids the problems associated with preparing a number of specimens of different aspect ratios; it also has the advantage of allowing the use of the same specimen for a number of tests. If ρ denotes the distance of G from O, with OG making an angle α with the fibre-direction (refer Fig. 1(b)), then we obtain (Kaprielian and Rogers 1989)

$$M/\theta = \mu_L\bar{I}_1 + \mu_T\bar{I}_2 + (\mu_L\sin^2\alpha + \mu_T\cos^2\alpha)\rho^2S \qquad (4.5)$$

or equivalently

$$M/\theta = \mu_L\bar{I}_1 + \mu_T\bar{I}_2 + \tfrac{1}{2}(\mu_L+\mu_T)\rho^2 S - \tfrac{1}{2}(\mu_L-\mu_T)\rho^2 S \cos 2\alpha. \qquad (4.6)$$

Here $\bar{I}_1$ and $\bar{I}_2$ are the second moments of area about the relevant axes throughout G, not O, and S is the surface area of the specimen. Obviously $\rho = 0$ corresponds to centred torsion.

If a series of tests were carried out using the same shape and off-centre position of G and simply varying the orientation α of the specimen, then (4.6) shows that a plot of M/θ against $\cos 2\alpha$ would again give a straight line. This time the slope would be proportional to $\mu_L-\mu_T$ and the intercept for $\alpha = \frac{1}{4}\pi$ (on the M/θ axis) would give another independent relation between μ_L and μ_T, from which the individual values of the shear moduli could be calculated. Alternatively (or additionally) the orientation α could be fixed and the distance ρ varied. Equation (4.5) shows that in that case a straight line fit should be obtained from a plot of M/θ against $\rho^2 S$, with its slope and intercept (at $\rho=0$) providing two simultaneous equations for μ_L and μ_T. The simplest relations are provided by the conventionally shaped specimens, such as square and circular plates, with $\bar{I}_1 = \bar{I}_2 = \bar{I}$, say. Then (4.5), for example, with $\alpha = 0$ (i.e. with OG in the fibre-direction) gives a straight line plot of M/θ against $\rho^2 S$, with slope μ_T and an intercept that is proportional to $\mu_L+\mu_T$.

The unbalanced nature of off-centred torsion can sometimes lead to experimental difficulties associated with the machine platens ceasing to be parallel, and hence producing deformations for which the simple analysis based on (3.1) is not valid. These problems can be rectified by introducing one or more balancing 'sleepers' of the same material which remain in the <u>same</u> positions and orientations throughout the series of tests. It is an easy matter (Kaprielian and Rogers 1989) to reinterpret the resulting data.

If an off–centred specimen is balanced by an identical specimen with the same orientation, then the resulting configuration is in fact equivalent to a single centred specimen. Varying the distance ρ of each of the two centres of area then provides yet another way of changing the 'shape' but this time without having to change the shape of each piece.

Finally we note that the stress field (4.1) will not in general satisfy the actual edge conditions of zero traction on the edge surfaces. In fact it is a simple matter to show that the conditions are satisfied only in the very special case of an elliptically–shaped specimen whose semi–axes are in the ratio of $(\mu_L/\mu_T)^{\frac{1}{2}}$, a quantity that is known only after the torsion test has been carried out! However, the resultant traction per unit length of edge is only of order h, the plate thickness, and acts in a direction normal to the lateral surfaces. Accordingly it is anticipated that the assumption of pure torsion leads to an edge effect which is negligible. Although this assertion has not been proved rigorously, it is very similar to the assumption made for the conventional simple shear test in which the specimen is sheared between two parallel plates. The same comments also apply to the viscoelastic and plastic–elastic cases treated in the next two sections.

5. VISCOELASTIC ANALYSIS

The correspondence principle of linear viscoelasticity, or the elastic–viscoelastic analogue, immediately leads to equation (4.3) being replaced for viscoelastic composites by

$$M(t) = I_1\ \mu_L * \dot{\theta} + I_2\ \mu_T * \dot{\theta}$$

$$= I_1 \int_{-\infty}^{t} \mu_L(t-\tau)\dot{\theta}(\tau)\ d\tau + I_2 \int_{-\infty}^{t} \mu_T(t-\tau)\dot{\theta}(\tau)\ d\tau . \tag{5.1}$$

For a static torsion test, in which the test specimen is subjected to an initial twist θ_0 which is then held constant, (5.1) shows that the torque relaxes according to

$$M(t) = \{I_1\mu_L(t) + I_2\mu_T(t)\}\theta_0 . \tag{5.2}$$

Comparison with (4.13) immediately shows that the subsequent analysis and results (4.4)–(4.6) for elastic torsion still hold, except that now μ_L and μ_T are time-dependent functions. Accordingly the data obtained from static tests can be analyzed in the same ways as those described in Section 4, though now separate M/θ_0 plots need to be made for each time point required.

A more standard procedure for viscoelastic measurement is the forced vibration test, in which the specimen is subjected to a sinusoidally varying torque of frequency Ω and amplitude $M_0(\Omega)$, and the resulting twist $\theta_0(\Omega)$ and phase lag $\delta(\Omega)$ are measured. Then by substituting

$$\theta(t) = \theta_0(\Omega)\exp\ i\Omega t\ , \quad M(t) = M_0(\Omega)\exp\ i(\Omega t+\delta) \tag{5.3}$$

into (5.1), we obtain

$$M_0\ \exp\ i\delta = \{I_1\mu_L^*(\Omega) + I_2\mu_T^*(\Omega)\}\theta_0\ , \tag{5.4}$$

where μ_L^* and μ_T^* are the so-called complex shear moduli. These are related to the storage moduli μ_L', μ_T' and loss moduli μ_L'', μ_T'' through

$$\mu_L^* = \mu_L' + i\mu_L''\ , \qquad \mu_T^* = \mu_T'' + i\mu_T'' .$$

Hence from (5.4)

$$\begin{aligned} M_0\ \cos\ \delta &= (I_1\mu_L' + I_2\mu_T')\theta_0\ , \\ M_0\ \sin\ \delta &= (I_1\mu_L'' + I_2\mu_T'')\theta_0 \end{aligned} \tag{5.5}$$

and comparison of (5.5) with the form of (4.3) again shows how the methods described in the previous sections may be used to analyse the data. Thus M is replaced by $M_0 \cos \delta$ for determining μ_L' and μ_T' and by $M_0 \sin \delta$ for μ_L'' and μ_T'', and the relevant plots are made for each required frequency Ω.

The special case of a fibre-reinforced linearly viscous fluid is analyzed by substituting (2.9) into (5.1), so that in this case

$$M = (I_1 \eta_L + I_2 \eta_T)\dot{\theta} \cdot \qquad (5.6)$$

Then in the forced vibration test the torque and twist will be exactly out of phase ($\delta = \frac{1}{2}\pi$) and

$$M_0/\theta_0 = (I_1 \eta_L + I_2 \eta_T)\Omega \ . \qquad (5.7)$$

The data can clearly be analyzed as before, for both centred and non-centred torsion tests.

The case of a nonlinear fibre-reinforced viscous fluid is much more complicated. Substituting (3.3) into (2.14) and (2.15) shows that

$$\sigma_{13} = -\tfrac{1}{2}(\varphi_1+\varphi_2)x_2\dot{\theta} \ , \quad \sigma_{23} = \tfrac{1}{2}\varphi_1 x_1 \dot{\theta}$$

where φ_1 and φ_2 can be functions of the three strain-rate invariants

$$D_1 = x_1^2\dot{\theta}^2 \ , \quad D_2 = x_2^2\dot{\theta}^2 \ , \quad D_3 = 0 \ .$$

The torque is therefore independent of the functions φ_3 and φ_4, and hence the torsion test data will provide data about just φ_1 and φ_2. If φ_1 and φ_2 are constants, then the analysis will again go through as above, even though the constitutive relation will still contain the nonlinear φ_3 and φ_4 terms.

Finally we note that including elastic fibre-extensibility does not affect the torsion analysis.

6. ELASTIC–PLASTIC TORSION

If the deformation is elastic–plastic, then as the applied torque M increases from zero the specimen first deforms elastically until at some point the yield function attains its critical value (here unity). Further increase in M then results in further increase in θ and the plastic region gradually spreads. Hence during loading the role of 'plastic' time can be taken by θ, so that $\dot{\theta} = 1$.

We consider only the Mises and Tresca–type behaviours characterised by $f = f_M$ and $f = f_T$ respectively. It can be verified *a posteriori* that in such cases, and in the absence of 'squeezing', the only non–zero stress components are σ_{13} and σ_{23} so that (2.20) gives

$$J_1 = \sigma_{23}^2 , \quad J_2 = \sigma_{13}^2 , \quad J_4 = 0 . \tag{6.1}$$

So for the Mises criterion $f_M \leqslant 1$, yielding first occurs when $\theta = \theta_c$, say, at the point (or points) on the boundary of the specimen for which the elastic stresses satisfy $f_M = 1$. Equations (4.1), (2.19) and (6.1) show that these points satisfy

$$\frac{x_1^2}{a_c^2} + \frac{x_2^2}{b_c^2} = 1 , \quad a_c = \left[\frac{k_T}{\mu_T}\right]\frac{1}{\theta} , \quad b_c = \left[\frac{k_L}{\mu_L}\right]\frac{1}{\theta} . \tag{6.2}$$

As θ increases with M, so do a_c and b_c decrease, so that the elastic–plastic boundary is a contracting ellipse Γ_c described by (6.2) with all points of the specimen inside Γ_c still deforming elastically and all points outside Γ_c being plastic. In the plastic region, denoted by S^p, the stress field must satisfy the yield condition $f_M = 1$:

$$\frac{\sigma_{13}^2}{k_L^2} + \frac{\sigma_{23}^2}{k_T^2} = 1 , \tag{6.3}$$

together with the flow rule (2.18) which, with $\underset{\sim}{d} = \underset{\sim}{d}^e + \underset{\sim}{d}^p$, gives

$$2\,\frac{\dot\lambda}{k_L^2}\,\sigma_{13} + \frac{1}{\mu_L}\,\dot\sigma_{13} = -x_2 \tag{6.4}$$

and

$$2\,\frac{\dot\lambda}{k_T^2}\,\sigma_{23} + \frac{1}{\mu_T}\,\dot\sigma_{23} = x_1 \ . \tag{6.5}$$

Eliminating $\dot\lambda$ and integrating shows that these equations (6.3)– (6.5) have a solution

$$\sigma_{13} = -\frac{k_L^2 x_2}{(x_1^2 k_T^2 + x_2^2 k_L^2)^{\frac{1}{2}}} \ , \ \sigma_{23} = \frac{k_T^2 x_1}{(x_1^2 k_T^2 + x_2^2 k_L^2)^{\frac{1}{2}}} \ , \tag{6.6}$$

provided we assume non–hardening behaviour (i.e. k_L and k_T both constant). This solution can be deduced most easily by investigating whether such a (plastic) time–independent solution is possible. Unfortunately, the solution should also satisfy uniqueness of traction across the elastic–plastic interface Γ_C, and this can be done by (6.6) only in the very special case of

$$k_L^2/k_T^2 = \mu_L/\mu_T \ . \tag{6.7}$$

In general, when (6.7) is not satisfied, the analysis is much more complicated and (6.6) is the long–time solution. The solution can be written in terms of a stress parameter $\varphi = \varphi(\theta)$ where

$$\sigma_{13} = -k_L \sin\varphi \ , \quad \sigma_{23} = k_T \cos\varphi,$$

with

$$\left[\frac{k_L^2}{\mu_L}\cos^2\varphi + \frac{k_L^2}{\mu_T}\sin^2\varphi\right]\dot\varphi = k_L x_2 \cos\varphi - k_T x_1 \sin\varphi, \tag{6.8}$$

subject to $\varphi = \varphi_C$ when $\theta = \theta_C$, given by

$$\tan\varphi_C = \frac{k_L \mu_T x_2}{k_T \mu_L x_1} \ , \quad \theta_C = \left[\frac{\mu_T^2 x_1^2}{k_T^2} + \frac{\mu_L x_2^2}{k_L^2}\right]^{-\frac{1}{2}} \ , \tag{6.9}$$

In principle, (6.8) and (6.9) determine φ, and hence σ_{13} and

σ_{23}, in terms of x_1, x_2 and 'time' θ. Then since the stresses in the elastic region S^e are still given by (4.1), the total torque at any loading stage of the deformation is determined by

$$M = \left\{\iint_{S^e} (x_1^2\mu_T + x_2^2\mu_L)\ dA\right\}\theta + \iint_{S^p} (k_T x_1 \cos\varphi + k_L x_2 \sin\varphi)\ dA. \tag{6.10}$$

However, even for practical shapes such as rectangles, it is clear that (6.10) does not offer a convenient analytical expression for M as a function of aspect ratio, as in the elastic and viscoelastic analyses, nor even for M in terms of θ.

The Tresca-type material $f_T \leq 1$ provides a much simpler analysis, leading to convenient analytical expressions. In this case, the elastic-plastic boundary Γ_c is a rectangle with the elastic region confined to

$$|x_1| \leq a_c\ , \qquad |x_2| \leq b_c\ . \tag{6.11}$$

It is now very straightforward to compute the torque, provided care is taken to distinguish between the eight distinct parts (Rogers 1987) of the plastic region which correspond to each of the sides and vertices of the Tresca rectangle defined by

$$|\sigma_{13}| = k_L\ , \qquad |\sigma_{23}| \leq k_T \tag{6.12}$$

and

$$|\sigma_{23}| = k_T\ , \qquad |\sigma_{13}| \leq k_L\ . \tag{6.13}$$

When the specimen is a centred rectangular plate, it is an elementary, if somewhat tedious, exercise to compute the torque. In fact, the results for off-centred specimens are more useful and no more complicated if the sides of the rectangle are parallel to the fibre-direction. Thus if ρ is again the distance of the centre from the origin with $\alpha = 0$, and assuming for illustration that

$$(\rho+L)\mu_T k_L > H\mu_L k_T\ , \tag{6.14}$$

an inequality that ensures that the elastic plastic boundary is at $x_1 = a_C < \rho+L$, then S^e is the region $\rho-L \leqslant x_1 \leqslant a_C$ and S^p is $a_C \leqslant x_1 \leqslant \rho+L$. The torque is easily evaluated, with M given by (4.5) until θ increases to

$$\theta_C = k_T/(\rho+L)\mu_L$$

when the edge $x_1 = \rho+L$ yields; then for $\theta > \theta_C$ we obtain

$$M = \iint_{S^e} (\mu_T x_1^2 + \mu_L x_2^2)\theta \; dA + \iint_{S^p} (k_T x_1 + \mu_L \theta x_2^2) \; dA$$

$$= \tfrac{2}{3}H\{2LH^2\mu_L - (\rho-L)^3\mu_T\}\theta + 4HL^2k_T - \tfrac{1}{3}Hk_T^3/\mu_T^2\theta^2. \quad (6.15)$$

This expression holds until b_C decreases to H, after which it is easy to see that the plastic regime will consist of five different parts. The subsequent form for M is still easy to obtain, though more cumbersome than (6.15), and includes terms involving k_L. Hence after determining k_T from (6.15) – as the solution to a cubic, for example, since it does not involve the unknown k_L – it might be feasible to increase θ beyond θ_C and use the data to further determine k_L. However, it is clearly preferable to simply rotate the specimen through $\frac{1}{2}\pi$ so that the fibres are parallel to the x_2-axis; then the analysis is as above, but with μ_L and μ_T interchanged and k_T replaced by k_L.

Finally, we note that the analysis for cyclic loading of a Tresca-type plastic-elastic composite has been reported elsewhere (Kaprielian and Rogers 1987). The analysis for the unloading stage is exceedingly cumbersome, but it does show that reverse yielding takes place and that in fact the new plastic-elastic boundary reaches the same position as it was at the end of the loading stage ($\theta = \theta_m$, say). The subsequent re-loading part of the cycle shows that yielding again occurs before $\theta = \theta_m$ and that the stress field at $\theta = \theta_m$ is exactly as at the end of the first loading sequence. Subsequent cycling just results in repetition of this stress cycle.

REFERENCES

Christensen, R.M., 1979, *Mechanics of Composite Materials,* John Wiley and Sons, New York.

Dvorak, G.J., Bahei-el-Din, Y.A., Macharet, Y., and Liu, C.H., 1988, "An experimental study of elastic-plastic behaviour of a fibrous boron-aluminium composite", *J. Mech. Phys. Solids*, Vol.36, pp.655-688.

Groves, D., 1989, "A characterisation of shear flow in continuous fibre thermoplastic laminates", *Composites*, Vol.19, pp.28-33.

Kaprielian, P.V., and Rogers, T.G., 1987, "Cyclic elastic-plastic torsion of fibre-reinforced materials", JRS 1/87., Dept. Theoretical Mechanics, University of Nottingham.

Kaprielian, P.V., and Rogers, T.G., 1989, "Determination of the shear moduli of fibre-reinforced materials by centred and off-centred torsion", *SAMPE*, Vol.34, pp.1442-1449.

Pagano, N.J., and Halpin, J.C., 1968, "Influence of end constraints in the testing of anisotropic bodies", *J. Composite Materials*, Vol.2, pp.18-31.

Rogers, T.G., 1987, "Yield criteria, flow rules and hardening in anisotropic plasticity", *Proc. IUTAM/ICM Symp. on Yielding, Damage and Failure of Anisotropic Solids*, Villard-de-Lans, France.

Rogers, T.G., and Pipkin, A.C., 1963, "Asymmetric relaxation and compliance matrices in linear viscoelasticity", *ZAMP*, Vol.14, pp.334-343.

Spencer, A.J.M., 1972, *Deformation of Fibre-reinforced Materials*, Clarendon Press, Oxford.

Spencer, A.J.M., 1983, "Yield conditions and hardening rules for fibre-reinforced materials with plastic response", *Failure Criteria of Structured Media*, J.P. Boehler ed., A.A. Balkema Publishers, Rotterdam, The Netherlands.

Spencer, A.J.M., et al, 1984, *Continuum Theory of the Mechanics of Fibre-reinforced Composites*, CISM 282, Springer-Verlag, Wien - New York.

Inelastic Deformation and Fatigue Damage of Composite under Multiaxial Loading

S. MURAKAMI, Y. KANAGAWA, T. ISHIDA
Department of Mechanical Engineering
Nagoya University
Furo-cho, Chikusa-ku
Nagoya 464, JAPAN
and
E. TSUSHIMA
Corporate Research & Development Laboratory
Tonen Corporation
1-3-1 Nishi-Tsurugaoka
Ohi-cho, Iruma-gun
Saitama 354, JAPAN

ABSTRACT

Inelastic deformation and damage process of Graphite/Epoxy $[\pm 45°]_4$ laminate tubes under axial and combined axial and torsional loadings are discussed. The inelastic deformation and rupture properties under combined loading are discussed first. Significant viscoplastic deformation is induced by axial loading due to fiber rotation and matrix dominant deformation. The compressive strength of the laminate tubes is about 15% smaller than the tensile strength. Direction of twist has no noticeable influence on the torsional strength of the specimens. In the case of cyclic axial loadings, significant hysteresis loops and salient viscoplastic deformation are induced due to matrix dominant deformation. The negative stress ratios have marked effects on hysteresis curves, stiffness reduction and on fatigue life. These phenomena are attributable to the matrix dominant deformation of laminas and the accelerated viscoplastic deformation of polymer matrix under reversed loading. However, the fatigue strength is governed mainly by the maximum (or minimum) stress, and is influenced by the stress ratios. The fatigue strength under cyclic compression is 10 ~ 20% smaller than that of cyclic tension. The isochronous stress loci for fatigue life are also discussed.

1. INTRODUCTION

Long fiber reinforced plastic laminates are characterized by complicated microscopic anisotropy governed by the orientation of fibers and plies and by salient inhomogeneous microstructures. Particularly when the composite laminates are subjected to uniaxial loads in the direction almost coincident with that of their fibers under monotonic loading or tension–tension cyclic loading, they have been observed to show excellent static and fatigue performance. However, the composite laminates in practical applications are usually subject to the multiaxial and cyclic loading in various stress level; *i.e.*, they may be subject to loads acting in the directions deviated from those of fibers, or subjected to cyclic loads with negative stress ratios $R=\sigma_{min}/\sigma_{max}$.

The papers which elucidated the inelastic deformation and fatigue properties of composites under multiaxial loading are rather limited [1–5]. Krempl and his coworkers [3,4], in particular, performed a series of static and fatigue tests on tubular specimens of Graphite/Epoxy and Kevlar/Epoxy composites under uniaxial and multiaxial loadings. In the case of Graphite/Epoxy $[\pm 45°]_S$ tubes under cyclic loading, they observed marked viscoelastic and hysteresis behaviors as well as significant degradation of fatigue performance, especially for negative stress ratios.

The present paper is concerned with the inelastic deformation and the fatigue damage process of $[\pm 45°]_4$ Graphite/Epoxy laminate tubes under axial and combined axial and torsional loading. The tests were performed for several combinations of multiaxial stress ratios τ/σ and cyclic stress ratios $R=\sigma_{min}/\sigma_{max}$ for different stress levels, and their effects on the evolution of hysteresis behavior, specific energy dissipation, stiffness reduction and fatigue behavior were discussed in detail.

2. EXPERIMENTAL PROCEDURE

2.1 Properties of Test Materials

In order to facilitate the static and fatigue tests under multiaxial state of stress, the experiments throughout the present paper will be performed for tubular specimens shown in Fig. 1, where inside and outside diameters and gauge length were 15mm, 17mm and 40mm, respectively. The stacking sequence of the specimens is $[\pm45°]_4$, fabricated from prepreg Toray P3052(T300/2500). The volume fraction of fiber is 59 percent in the prepreg. The tubular specimens facilitate also the compression tests excluding edge effects and buckling. Furthermore, the fiber orientation of [±45°] is most sensitive in inelastic deformation to external axial loading.

Preliminary tests were performed also for the specimens of $[0,90°]_4$ and $[0°]_4$ to elucidate the mechanical response of the lamina in the fiber direction and in that perpendicular to the fibers. Fig.2 and Table 1 show the results of the tests, and represent

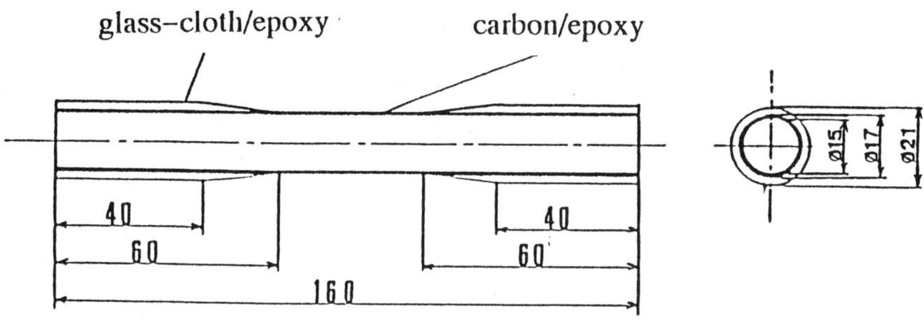

Prepreg : TORAY P3052(T300/2500)

Fig. 1. Graphite/Epoxy Laminate Tubular Specimen.

Table 1. Mechanical Properties of Graphite/Epoxy Laminate.

	E	σ_{TB}	ε_{TB}
$[0°]_4$	180GPa	1715MPa	0.93%
$[0,90°]_4$	77GPa	819MPa	1.05%
$[\pm45°]_4$	15GPa	198MPa	6.13%

typical elastic responses of the fiber. Young's modulus of [0,90°] laminate is about half of [0°] laminate and is slightly smaller than the value estimated by the law of mixture. These results suggest that contributions of the matrix to the strength of [0°] specimens may be small, and that these may have degrading effects of cross ply lamination on the mechanical properties of the laminates, due to stress concentration between plies with different fiber orientation.

2.2 Test Conditions

Combined axial–torsional loads were applied to the tubular specimens of Fig. 1. Fatigue tests were performed under stress controlled cyclic loading at the stress rate of 10MPa/sec. Loading conditions were characterized by means of two parameters, *i.e.*, multiaxial stress ratios τ/σ and the cyclic stress ratios $R=\sigma_{min}/\sigma_{max}$. Strains were measured by use of plastic strain gauges in the case of monotonic multiaxial loading, and of a clip–on extensometer in the case of cyclic tension–compression loading.

Stress–strain hysteresis loops were observed continuously in the course of combined tension–compression cyclic loading. The evolution of stiffness and the strain energy dissipation were calculated from these hysteresis loops.

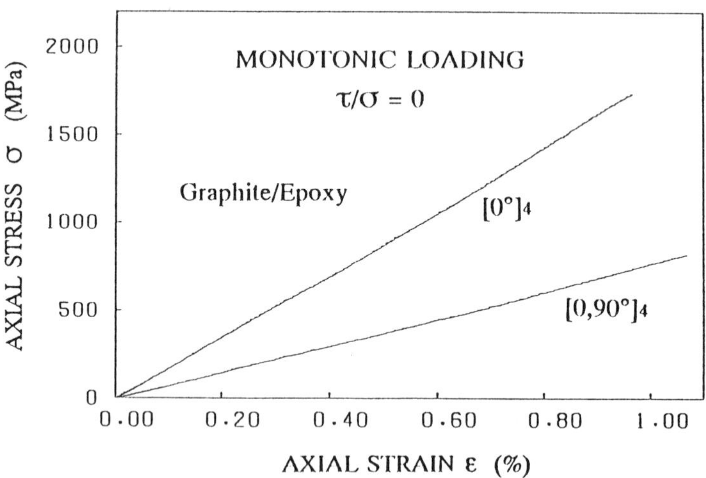

Fig. 2. Stress–Strain Curves of Graphite/Epoxy $[0°]_4$ and $[0,90°]_4$.

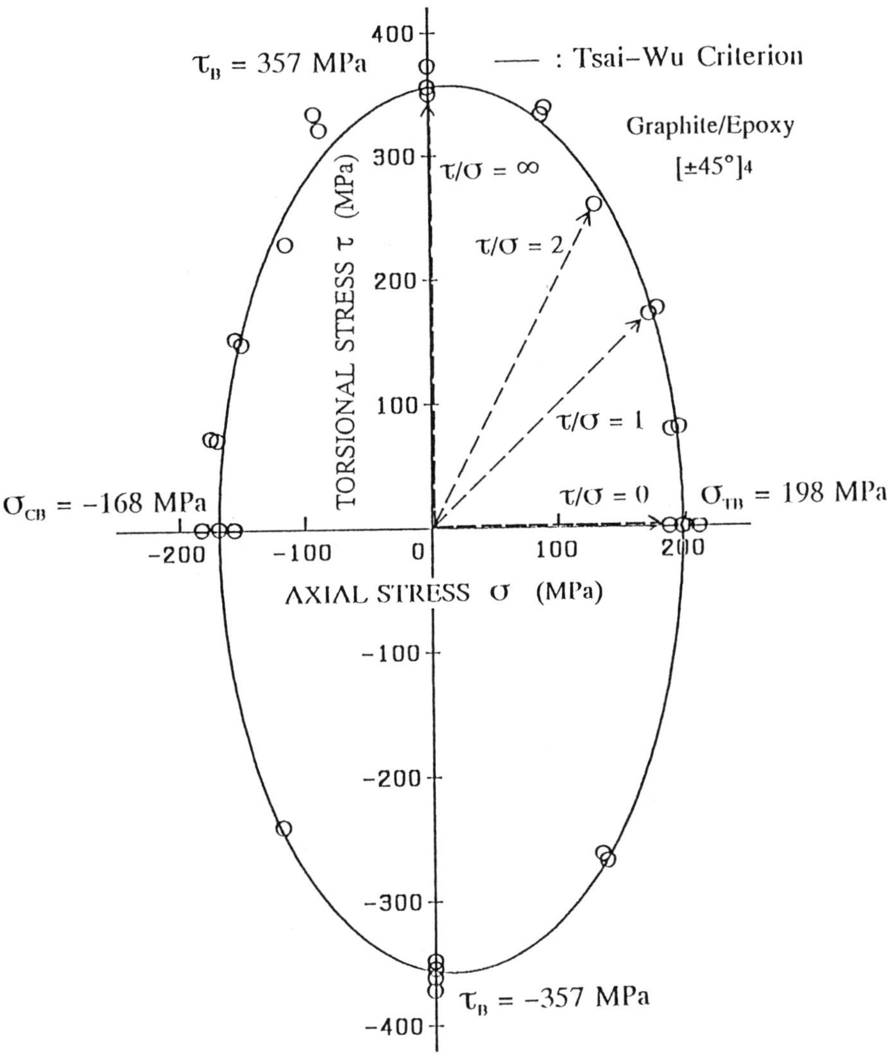

Fig. 3. Fracture Locus under Multiaxial Loading and Proportional Stress Paths.

3. INELASTIC BEHAVIOR UNDER MULTIAXIAL MONOTONIC LOADING

Monotonic loading tests on tubular specimens of $[\pm 45°]_4$ were performed to elucidate plastic behavior and final rupture under combined tension and torsion. Fig. 3 shows the resulting fracture locus. The ultimate strength in torsion τ_B was independent of the directions of twist. This is in contrast to results of Krempl[3] *et al* for Graphite/Epoxy tubes of $[\pm 45°]_2$, where the twist which would induce compression on the fibers of outer layer show larger torsional strength than the reversed torsion. This may be accounted for partly by the fact that the present tubes have 8 layers and hence the role of the outer layer may be smaller than the specimens employed by Krempl et al. The compressive strength σ_{CB}, on the other hand, is about 15% smaller than the tensile strength σ_{TB}, and Tsai–Wu criterion is applied. In view of ply angles of the specimens, the reduced strength of the compression tests may be due to the matrix dominant local buckling of the specimens.

The inelastic deformation of the tubular specimens under monotonic combined tension and torsion was observed for the stress paths entered in Fig. 3. The strain was measured by use of plastic strain gauges. The corresponding strain paths in the $\varepsilon-\gamma$ space is indicated in Fig. 4. The strain paths for the proportional loading $\tau/\sigma=2$ and $\tau/\sigma=1$ are no more proportional, as observed in this figure. The corresponding stress–strain relations for each components, $\tau-\gamma$ and $\sigma-\varepsilon$ are shown in Fig. 5 (a) and (b).

In the case of purely axial tension $\tau/\sigma=0$ [Fig. 5(a)], noticeable plastic deformation starts at the stress about 25 percent of the tensile strength. This is attributable mainly to the tilting of fibers and the matrix dominant deformation. Though the incipient part of the curves of $\tau/\sigma=1$ and $\tau/\sigma=2$ almost coincide with each other, it will be difficult to conclude any definite results concerning plastic deformation from these figures.

For pure torsion $\tau/\sigma=\infty$ in Fig. 5 (b), on the other hand, since the fiber direction coincides with the principal stress direction, $\tau-\gamma$ curve is almost linear. In view of the tensile fracture strain

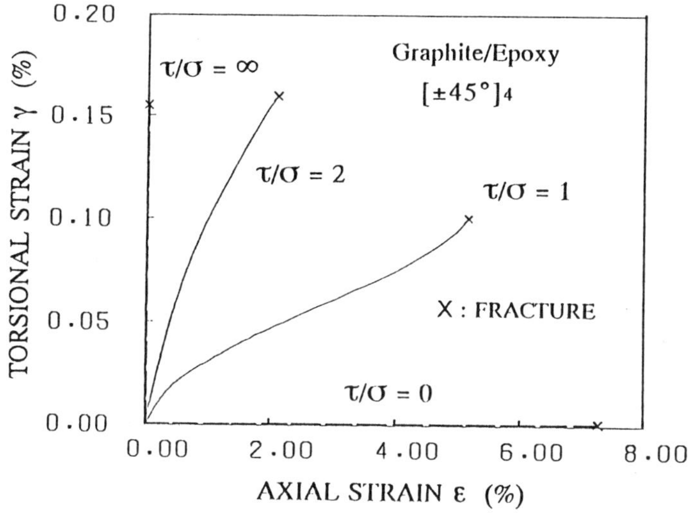

Fig. 4. Strain Paths under Proportional Loading.

ε_B=0.93% of the 0° specimens, the fracture strain γ_B=0.16% might be rather small. This may be due to the local buckling of compressed fibers and the resulting torsional buckling of tubular specimen. In the case of stress paths $\tau/\sigma=1$ and $\tau/\sigma=2$, on the other hand, noticeable plastic deformation is observed at the stress about 30% of the torsional strength. The reason of plastic shear strain in these cases may be two hold; *i.e.*, the deviation of principal stress direction from fiber direction and tilting of fibers in laminas due to tension. This is because the rotation of fibers due to axial strain makes the deviation of fiber direction from the principal stress direction. Addition of slight amount of axial stress to the specimens induces considerable plastic strain. In other words, tension in a direction slightly different from the fiber direction induces significant plastic deformations in the matrix.

4. CYCLIC PLASTICITY AND FATIGUE DAMAGE UNDER CYCLIC TENSION–COMPRESSION LOADING

It has been observed that composites laminates of [±45°] stacking show salient hysteresis behavior under cyclic axial loading for the case of R=0 and R=−1[3]. This phenomenon may be quite

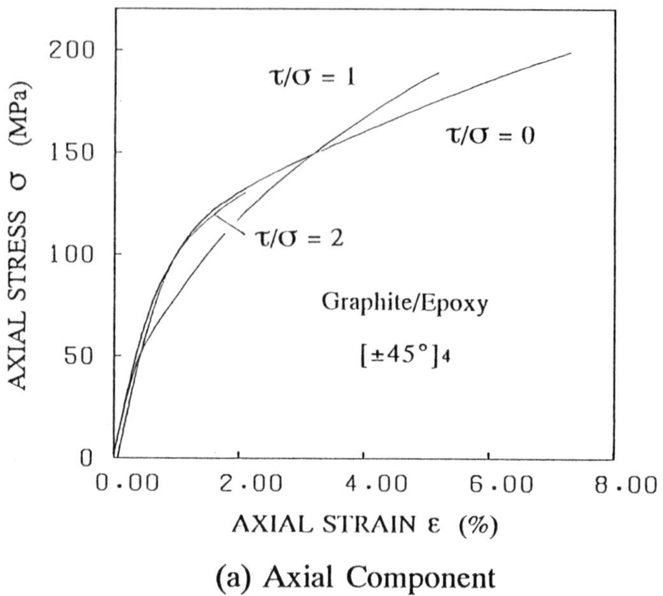

(a) Axial Component

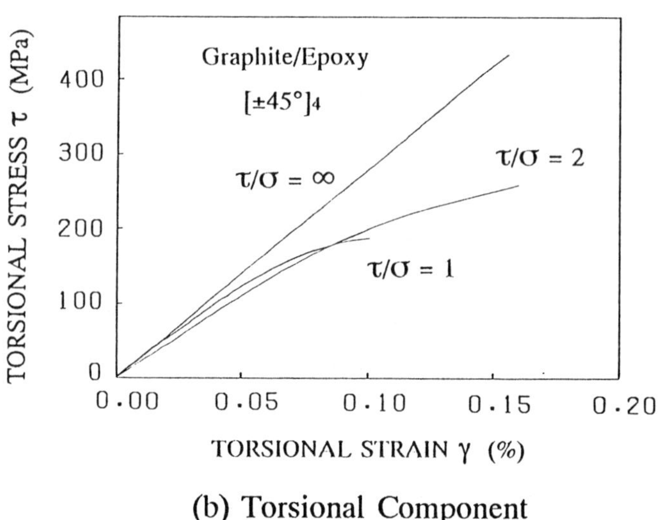

(b) Torsional Component

Fig. 5. Corresponding Stress–Strain Relations.

important because it has close relation to the microstructural change of the composites, which in turn accounts for plastic deformation and the damage of the material.

Thus, in order to elucidate the process of cyclic plasticity and fatigue damage of composites, three series of cyclic tension–compression tests were performed; *i.e.*, four different cases of constant stress range $\Delta\sigma$ combined with four different R, a case of constant maximum stress with four R and that of constant minimum stress with four R.

4.1 Effect of Stress Amplitude and Stress Ratio on Hysteresis Loops

Firstly, let us discuss the effects of stress ratio on the hysteresis loops for constant stress amplitude $\Delta\sigma/2$=67.5MPa (*i.e.*, $\Delta\sigma/\sigma_{TB}$=0.675). The typical hysteresis loops under cyclic loading are shown in Fig. 6 (a), (b) for stress ratios R=0 and R=−0.5. In the case of R=0 [Fig. 6(a)], after the first cycle of large hysteresis, the decrease of the width of the loop and the concurrent cyclic creep are observed. Since the deformation of this stacking is induced mainly in matrix laminas, the decrease of the width may be accounted for by the strain–hardening of the matrix. After N=4000 cycles, softening of the composites starts, and leads to the final fracture. The bulging of the hysteresis loop observed immediately after stress reversals implies the time–dependent viscoplastic deformation. Therefore, the tangent moduli at the unloading branches immediately after the unloading do not give a proper indication of the change of stiffness. Hence, the tangent moduli will be determined in Section 4.3 at mean stress on the loading branch of the hysteresis curves. The average fatigue life for these tests is Nf=9000.

Fig. 6(b) shows the hysteresis loops for the case of R=−0.5, but with the identical stress amplitude. Though the maximum stress is 2/3 of that of Fig.6(a) and the incipient hysteresis is less significant, the strain range increases rapidly with stress cycles. The phenomenon leads to hysteresis loops of larger size and gives the practically identical fatigue life of Nf=9000.

The hysteresis loops at fatigue limits for various stress amplitudes and stress ratios are given in Fig. 7(a), (b) and (c). It

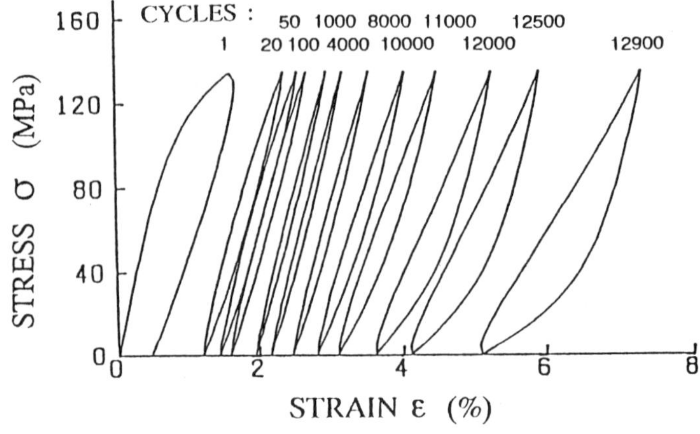

(a) Cyclic Tension
$\Delta\sigma/2$ = 67.5MPa ($\Delta\sigma/\sigma_{TB}$=0.675)
R = 0, Nf = 12900 (Nf = 9000)

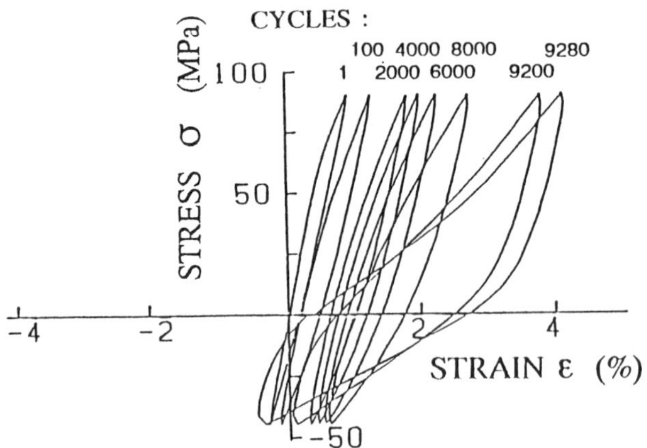

(b) Cyclic Tension–Compression
$\Delta\sigma/2$ = 67.5MPa ($\Delta\sigma/\sigma_{TB}$=0.675)
R = −0.5, Nf = 9280 (Nf = 9000)

Fig. 6. Hysteresis Loops under Cyclic Loading.

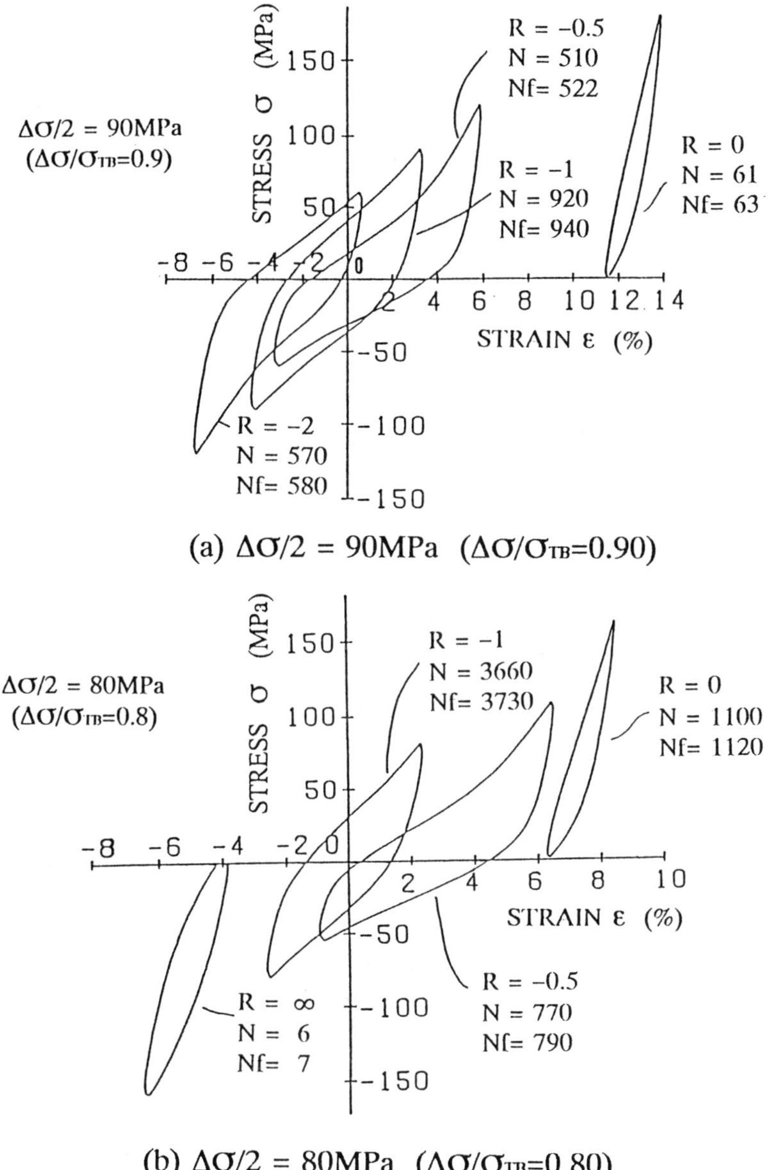

(a) $\Delta\sigma/2$ = 90MPa ($\Delta\sigma/\sigma_{TB}$=0.90)

(b) $\Delta\sigma/2$ = 80MPa ($\Delta\sigma/\sigma_{TB}$=0.80)

Fig. 7. Hysteresis Loops at Fatigue Limits.

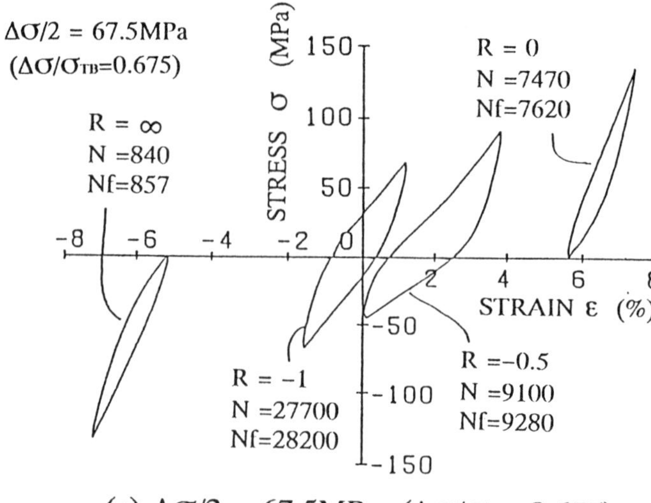

(c) $\Delta\sigma/2$ = 67.5MPa ($\Delta\sigma/\sigma_{TB}$=0.675)

Fig. 7. Hysteresis Loops at Fatigue Limits.

will be observed that the hysteresis loops of cyclic tension–compression tests(*i.e.*, R<0) show quite different loops from that of cyclic tension or of cyclic compression (*i.e.*, R≥0). Namely, in contrast to the case of cyclic tension, the cases of cyclic tension–compression indicate marked plastic(viscoplastic) deformation. For smaller stress amplitude, the effects of stress ratio are less significant, as will be observed in Fig. 7(c). The deformation in composites in these cases is matrix dependent, and is governed largely by the viscoplastic deformation of the matrix. It has been observed that viscoplastic deformation in polymer is remarkably accelerated by the reversal stressing[7,8]. Thus, this may partly account for the salient effects of negative R.

Micrographical observations on the sections of specimens after final fracture were performed by use of the surface microscope. However up to now, any significant differences in the internal defects of R≥0 and R<0 could not be found; a number of matrix cracks, but little interlaminar cracks and fiber breakage were observed in either cases.

4.2 Evolution of Specific Strain Energy Dissipation

The strain energy dissipation per one cycle might be an important measure of damage growth. However, as shown in

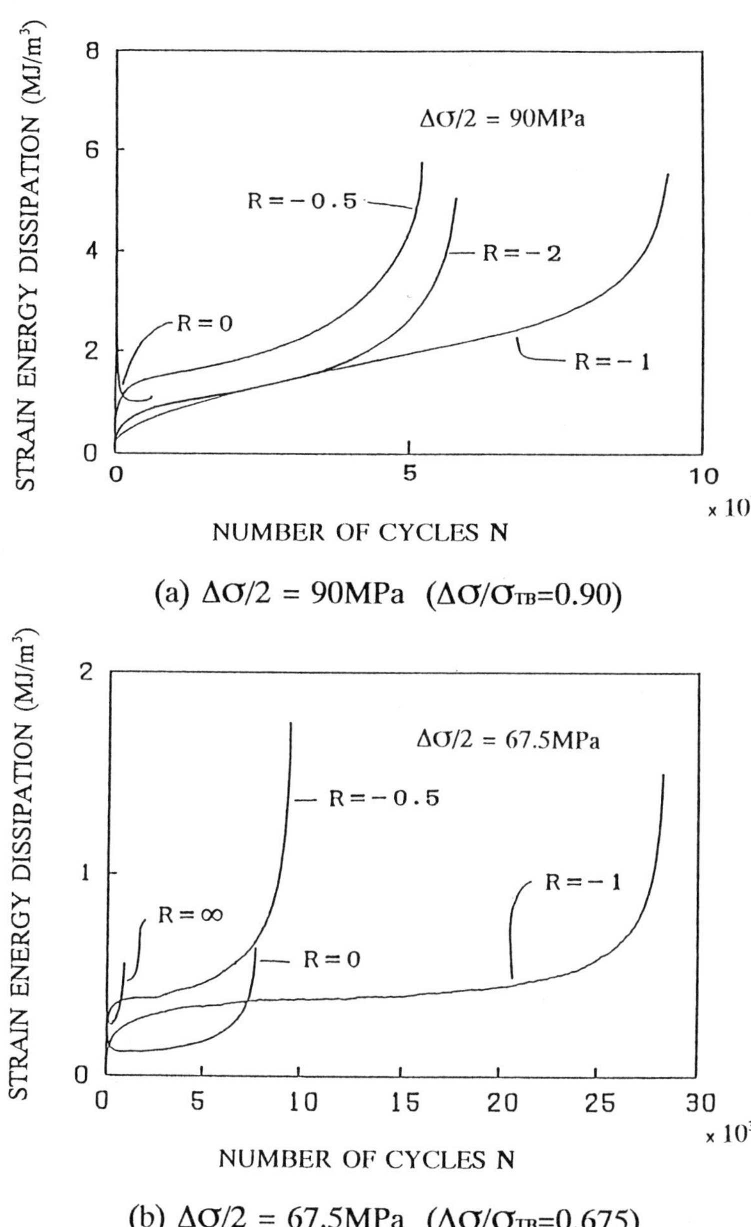

(a) $\Delta\sigma/2$ = 90MPa ($\Delta\sigma/\sigma_{TB}$=0.90)

(b) $\Delta\sigma/2$ = 67.5MPa ($\Delta\sigma/\sigma_{TB}$=0.675)

Fig 8. Evolution of Specific Strain Energy Dissipation.

Fig. 8(a) and (b), different trends are observed between the cases of R≥0 and R<0. This feature is attributable again to the interpretation of Section 4.2.

4.3 Evolution of Tangent Modulus at Mean Stress

Stiffness change has been often employed as an indication of damage in composites. Therefore the tangent moduli at mean stress were calculated from the hysteresis loops measured by use of clip-on gauges for each tests. The curves of Fig. 9 represent the evolution of tangent moduli on the loading branch of hysteresis curves for the stress ratios of r=0, −0.5, −1 and ∞. The tangent moduli, especially those for negative R, decrease significantly from the beginning. This is because these moduli have been detected at the mean stress of the hysteresis curves where the tangents decrease quickly as observed in Figs. 7 and 8.

Similar tests for R=0, −0.5 and −1 with constant maximum stress σ_{max}=120MPa and those for R=−1, −2 and ∞ with constant minimum stress σ_{min}=−120MPa were also performed. These two sets of tests revealed similar results to each other; *i.e.*, the cases of

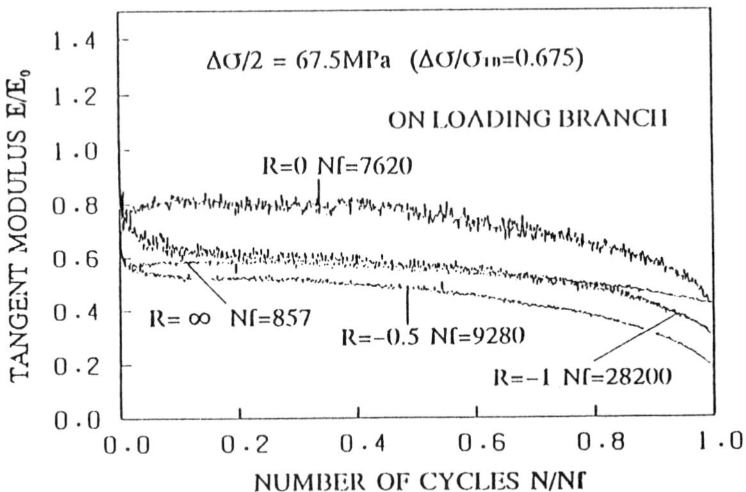

Fig. 9. Evolution of Tangent Modulus at Mean Stress.

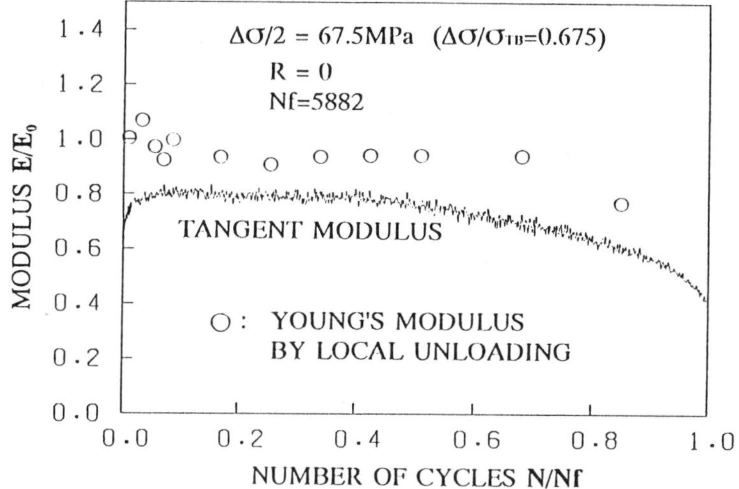

(a) $\Delta\sigma/2$ = 67.5MPa ($\Delta\sigma/\sigma_{TB}$=0.675), R=0

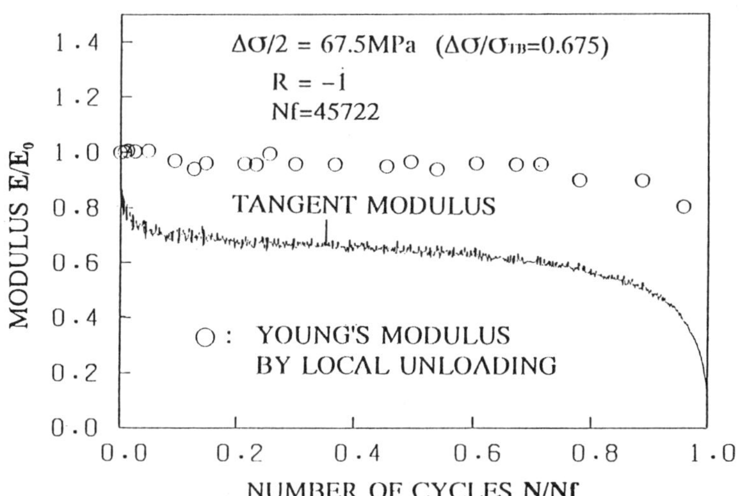

(b) $\Delta\sigma/2$ = 67.5MPa ($\Delta\sigma/\sigma_{TB}$=0.675), R=−1

Fig. 10. Reductiòn of Young's Modulus.

R=0 (σ_{max}=120MPa) and R=∞ (σ_{min}=−120MPa) show gradual decrease of tangent moduli from the beginning, while the cases of R=−0.5 and −1 (σ_{max}=120MPa) and R=−1 and −2 (σ_{min}=−120MPa) show incipient drop of about 50% and subsequent gradual decrease in the tangent moduli. However the fatigue lives of the cases of σ_{min}=−120MPa are much smaller than the case of σ_{max}=120MPa, because of the larger possibility of local buckling due to predominant compressive stresses.

Since the hysteresis behavior includes the inelastic properties of matrix, the change of tangent modulus do not necessarily a proper indication of damage[3]. Thus, also the Young's moduli were detected in the cases of R=0 and −1 by performing local unloading at certain stages of cycling. These results are entered by small circles in Fig. 10(a) and (b). Since the effect of viscoplastic behavior of matrix has been excluded from these results, the results of these circles will provide more rational measure of damage of composites.

4.4 Effect of Stress Amplitude and Maximum Stress on Fatigue Limit

The relation of stress amplitude versus number of cycles is shown in Fig. 11(a). It will be observed that the cases of negative stress ratios (R=−0.5 and −1) show steeper fatigue curves, which were observed also for Kevlar/Epoxy composites[4].

Fig. 11(b), on the other hand, represents the relation between maximum (or minimum) stresses and the fatigue life. This figure shows that the fatigue life is correlated well by the maximum (or minimum) stress, and is influenced by the stress ratio R. It should be noticed that cyclic loading with stress reversal (*i.e.*, $R<0$) is more deleterious than compressive stress itself.

5. FATIGUE UNDER MULTIAXIAL STATE OF STRESS

Finally we performed fatigue tests under the proportional and combined axial and torsional loading. Fig. 12 shows the isochronous stress loci for the fatigue lives Nf=10^2, 10^4 and 10^5 for stress ratio of R=0. The stresses corresponding to the specified value of Nf were determined by the interpolation of the

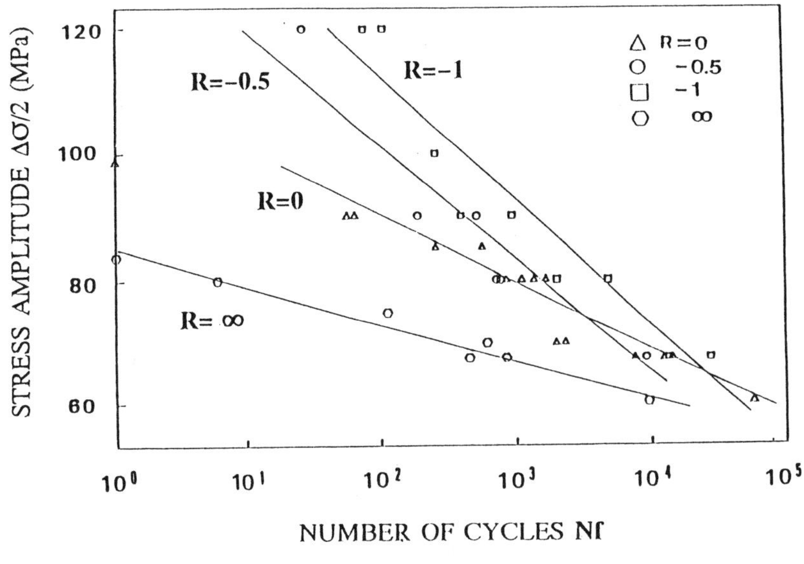

(a) Effect of Stress Amplitude

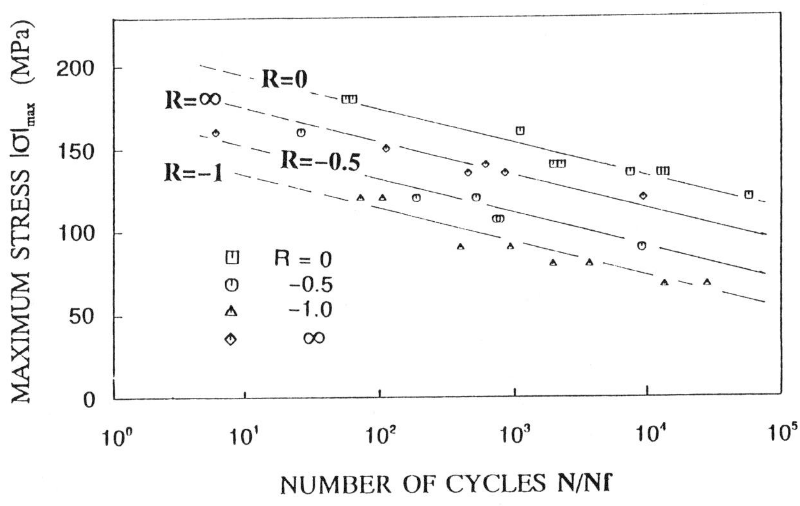

(b) Effect of Maximum (or Minimum) Stress

Fig. 11. Fatigue Limit under Axial Cyclic Loading.

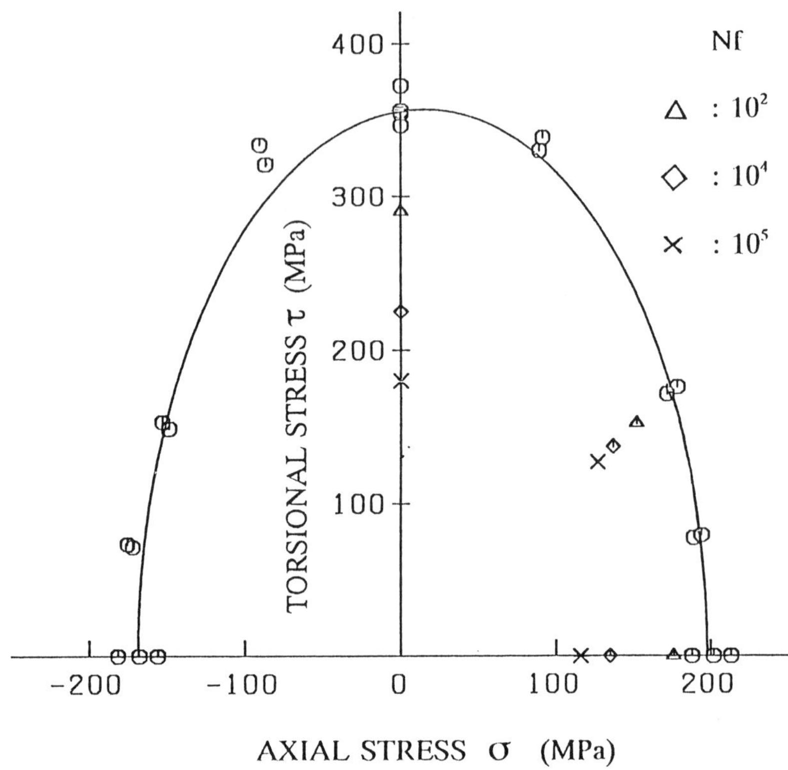

Fig. 12. Isochronous Stress Loci for Fatigue Lives.

corresponding fatigue curves.

According to the micrographical observation on the section of specimen after final fracture, a number of matrix cracks but little internal cracks and fiber breakage were detected in the specimens fatigued by axial loading, whereas the specimens fatigued by torsion showed considerable fiber breakage but scarce matrix cracks and interlaminas cracks.

6. CONCLUSIONS

The results of monotonic and cyclic loading tests on Graphite/Epoxy $[\pm 45°]_4$ laminate tubes under uniaxial and combined axial and torsional loads are summarized as follows:

1. Fracture locus was described by Tsai–Wu criterion, and

compressive strength of the laminate tubes was about 15 percent smaller than the tensile strength. The direction of twist has no noticeable influence on the torsional strength.

2. Except for the fiber dominant deformation of the torsional tests $\tau/\sigma=\infty$, salient plastic (viscoplastic) deformation was observed for the stress paths of $\tau/\sigma=2$, 1 and 0, both in axial and torsional strains. Axial stress slightly deviated from the fiber direction induces large plastic deformation in the matrix; it is attributable to the tilting of fibers and matrix dominant deformation in laminas.

3. Cyclic axial loading induces considerable hysteresis loops. This is significant especially for the cases of negative stress ratios R, probably because of the acceleration in viscoplastic deformation due to the stress reversal.

4. Since tangent moduli and specific strain energy dissipation of hysteresis loops are related to the viscoplastic deformation of the matrix, they do not provide an indication of material damage. Elastic modulus determined by performing local unloading during stress cycles may be a better measure for this purpose.

5. The fatigue life of $[\pm 45°]_4$ laminate composites was correlated well by the maximum (or minimum) stress, and was influenced by the stress ratio R. Stress reversal (negative R) in cyclic loading was more deleterious in fatigue strength than compressive stress itself.

REFERENCES

(1) R. R. Fujczak, "Torsional Fatigue Behavior of Graphite-Epoxy Cylinders", Proceedings of 1978 International Conference on Composite Materials(Toronto), American Inst. of Mining, Metallurgical and Petroleum Engineers Corp., p. 635-647

(2) M.J. Owen and J.R. Griffiths, "Evaluation of Biaxial Stress Fatigue Surface for a Glass Fabric Reinforced Polyester Resin under Static and Fatigue Loading", *J. Material Science, 13,* p. 1521-1537 (1978)

(3) E. Krempl and Tyan-Min Niu, "Graphite/Epoxy $[\pm 45°]_S$ Tubes. Their Static Axial and Shear Properties and Their Fatigue Behavior under Completely Reversed Load Controlled Loading", *J. Composite Materials, **16**,* p. 172-178 (1982)

(4) E. Krempl, D. M. Elzey, B. Z. Hong, T. Ayar and R. G. Loewy, "Uniaxial and Biaxial Fatigue Properties of Thin-Walled Composite Tubes", *J. American Helicopter Society, **33**,* p.3-10 (1988)

(5) K. Ikegami and M. Yoshida, "Constitutive Relations of Hybrid Fiber Reinforced Plastics of GFRP/CFRP and GFRP/AFRP under Combined Stress

State", Proc. IUTAM Symposium on Inelastic Deformation of Composite Materials (ed. by G. J. Dvorok), Springer–Verlag (in press)
(6) R. E. Rowlands, "Strength (Failure) Theories and Their Experimental Correlation", Failure Mechanics of Composites (ed. by G. C. Sih and A. M. Skudra), North–Holland, 1985, p. 96
(7) H. Tobushi, S. Narumi and K. Osawa, "Biaxial Creep Deformation of Softened Celluloid under Variable Stresses", *Bull. JSME, 29–248*, p. 348–354 (1986)
(8) H. Tobushi, Y. Ohashi and K. Osawa, "Creep Deformation of Softened Celluloid under Nonproportional Variable Stresses", *Bull. JSME, 29–257*, p. 3665–3671 (1986)

A Hybrid Model for Nonlinear Characterization of Composite Materials

Ryszard Pyrz

Institute of Mechanical Engineering
University of Aalborg
Pontoppidanstræde 101
9220 Aalborg, Denmark

ABSTRACT

Experimental and analytical techniques are employed in the present study to investigate the influence of microstructure rearrangements on tensile properties of woven glass fiber reinforced composite. One-dimensional constitutive model is proposed which contains a variable reflecting the influence of microdamages on macroscopic behaviour. This variable is estimated by the use of stereological methods and its relation to acoustic emission signature is enlighted.

INTRODUCTION

Among the problems of material modelling probably the most important is the evaluation of that point in the deformational process and time when failure will occur. The failure process is the consequence of two main mechanisms, namely, the generation of defects and their growth and coalescence into macroscopic failure state. Therefore it would appear that failure criterion should be in the form of a process equation rather than as a limiting condition. This is particulary important for composite materials where the generic factors in the component phases and on the boundary between them nucleate micro-damages at the early stages of the deformation. Hence, development of realistic mathematical models for describing the deformation and failure behaviour of composites must account for deformation-induced damages.

Despite a large variety of micro-damages modes influenced by the composite build up, the constituent properties as well as by interactions among the constituents, significant progress has been made in understanding the microfailure mechanisms that lead to damage nucleation. The most common formalism used in the prediction of particular damage mode is that of fracture mechanics and/or micromechanical modelling. Although adding a great deal to our understanding of various micro-damage mechanisms, the complexity of such modelling has made its application to practical problems of predicting composite performance rather limited. Moreover, concepts of fracture mechanics as applied at micro-level in highly inhomogeneous composite materials are subjected to some doubts (Reifsnider, 1988). In order to be useful in practical applications the micromechanical models must predict the performance of the composite on macro-level. This is by no means a simple task if one comtemplates to use f.exp. self-consistent scheme (Laws et al., 1983; Dvorak et al., 1985; Attiogbe, et al., 1987). Since the number of micro-damages is extremely large and their locations are mostly random it is inevitable to treat the densely cracked material as a continuum. The damage mechanics has emerged as a viable framework for continuum theories that introduce damage variables reflecting the influence of density, distribution, shape and orientation of microdefects on the macroscopic response (Talreja, 1985, 1989; Kamimura, 1985, Peng et al., 1986, Shen et al., 1986, Weitsman, 1987, 1988; Allen et al., 1987, 1988). The damage variable, though macroscopic quantity, reflects the influence of microstructure in an appropriately smoothed sense, retaining the salient aspects of the phenomenon and ignoring the less important details. This can be achieved by defining the damage variable over a certain representative volume, which is large enough to give details of microstructure and small enough for allowing the treatment of damage variable as a continuous quantity. Usually the damage variable is defined as a tensor quantity which creates the problems in experimental determination of damage variable components. Then traditionally the general theory is limited to verification in simple state of stress where reduced number of damage variable components are active.

It appears also inevitable to include a time effects in the damage description since a viscoelastic deformation plays a significant role in predicting the nonlinear and long-term behaviours of composites (Schapery, 1980, 1981, 1982, 1987), and more research should be addressed to physical observations of damages, not only at the end-of-life, but also at different fractions of lifetime by non-destructive measurements. Research of this type may suggest how to relate tensor-valued damages by scalar parameters. In view of a strong temptation to minimize the numerical effort a phenomenological continuum theory which reflects in averaged sense the microstructure and microstructural kinetic has a better chance to be accepted as a design tool than a purely micromechanical theory, particularly if one considers to implement it in a finite element code. Further, as was remarked by Kestin et al., 1988, the description of damages in simplest cases must be first throughly understood both from conceptual and experimental points of view before proceeding to multimode damages in complex state of stress.

In the present contribution a simple unidimensional model is proposed for nonlinear modelling of uniaxial deformation in woven composite. The constitutive equation is equipped with variable which reflects the microdamage pattern and its evolution. The microdamage variable is determined from microphotografic monitoring with utilization of stereological principles. The relation of the variable to some acoustic emission parameters is discussed, as well.

EXPERIMENTAL PROCEDURE

The material used in the present study was woven glass fabric-polyimide composite with plane weave fabric. Specimens were mounted on the Inströn load frame and tested at a constant strain rate both in ascending and descending part of the loading path, while load and strain were recorded as parametric inputs to the acoustic emission aquisition system, LOCAN-AT6. Acoustic emission and microphotographs were employed for detection and monitoring of damage development. Microphotographs were mainly used for the determination of the microstructure parameter entering the constitutive equation. In order to limit the modes of damages only one layer of reinforced fabric with thickness 0.16 mm was used, the warp and weft yarns being oriented at 45° to the loading direction. Since the glass content in the warp yarns was about 20% larger than in the weft yarns, the predominant mode of damage in the form of matrix cracking along the warp direction was observed, Fig. 1.

Fig. 1. Damage pattern at strain $8 \cdot 10^3$ μs and $32 \cdot 10^3$ μs, respectively.

The acoustic emission was detected with the wide-band transducers supplemented with quard censors in order to eliminate extraordinary signals coming from gripping area.

THE THEORETICAL MODEL

Most polymer-based composite materials exhibit viscoelastic phenomena under certain loading and environmental conditions. Moreover, if such material suffer simultaneously from the development of microdefects, the constitutive model must unify adverse concepts of fading memory characteristic for viscoelastic flow and permanent memory of damages. This statement is valid for the construction of constitutive models within the "functional" theory framework as opposite to the internal variable theory (Pyrz, 1989). Under the absence of intensive healing the material "remember" the damaging influence indefinitely, so that its response to an identical deforming influence applied at some later time may be different from its original response. Thus the functional form of the constitutive relation must change with deformation or in other words it must be equipped with repository for damage history, (Krempl, 1975). This repository should be estimated from the knowledge of the part played by the microstructure of the material on its macroscopic behaviour. Let us assume that the stress response of a material is formally described by a nonlinear operator $\mathfrak{J}$

$$\sigma(t) = \mathfrak{J}(t,\ D)\ \epsilon(t) \tag{1}$$

where D represents the influence of microstructure on the current response $\sigma(t)$. It is further assumed that at the beginning of the time interval $[0,t]$ the material is in its natural state i.e. $\sigma(0) = \epsilon(0) = 0$. We divide the time interval considered into elementary parts $[\tau_{n-1}, \tau_n]$ and denote an arbitrary intermediate time instant from this interval by τ_n*. The stress contribution to the material state at instant t caused by the change of strain component $\Delta\epsilon_n$ in the time interval $[\tau_{n-1}, \tau_n]$ is obtained by writing

$$\Delta_n\sigma(\tau) = R_n^*[t,\ D(t),\ \tau_n^*,\ D(\tau_n^*)]\Delta\epsilon_n \tag{2}$$

where

$$\Delta\epsilon_n = \epsilon(\tau_n) - \epsilon(\tau_{n-1})$$

and R^* is the influence function at an arbitrary instant τ_n^* dependent on the microstructure parameter D and the choice of the instant t, in which the deformation is considered. The contributions of the type (2) when added, give the approximative Stieltjes sum:

$$\sigma_N(t) = \sum_{n=1}^{N} \Delta_n \sigma(\tau) = \sum_{n=1}^{N} R_n^* \Delta\epsilon_n. \qquad (3)$$

Further, if there exists the limit of all sums of the above type independently of the mode of dividing into partial intervals and of the choice of intermediate points τ_n^*, when the lengths of intervals tend to zero with increasing N, then we can write

$$\sigma(t) = \lim_{\substack{N\to\infty \\ \max[\tau_{n-1},\tau_n]\to 0}} \sum_{n=1}^{N} R_n^* \Delta\epsilon_n = \int_0^t R[t, \tau, D(t), D(\tau)]\, d\epsilon(\tau). \qquad (4)$$

In particular, when the strain ϵ is differentiable and has an integrable derivative the integral (4) reduces to the Riemann integral

$$\sigma(t) = \int_0^t R[t, \tau, D(t), D(\tau)] \frac{d\epsilon(\tau)}{d\tau}\, d\tau. \qquad (5)$$

If it was possible to constraint D to be zero during deformation, the material would not sustain any microstructural changes and would behave as a viscoelastic material the stress being given by the relation

$$\sigma(t) = \int_0^t R[t, \tau, 0, 0] \frac{d\epsilon(\tau)}{dt}\, d\tau. \qquad (6)$$

For a given strain history the stresses in (5) and (6) will generally be different, the difference being due to the growth of the microstructurre parameter D. Conversely, for composites with rather brittle matrix the time and rate effects may be neglected giving rise to the equation

$$\sigma(t) = \int_0^t R[D(t), D(\tau)] \frac{d\epsilon(\tau)}{d\tau}\, d\tau. \qquad (7)$$

It is appearent from equation (7) that the stress is explicitely lineary related to the strain with nonlinearity totally resided in the microstructure parameter D. Thus the influence function R may be looked upon as an operational elasticity modulus possessing softening character due to damages. In what follows, we will identify the predominant cracks with geometrical quantities i.e. consider them as curves distributed over the plane of the specimen. Let us take a planar domain S of area A which contains a set of curves B showing some preferred orientation determined by an angle α, Fig. 2.

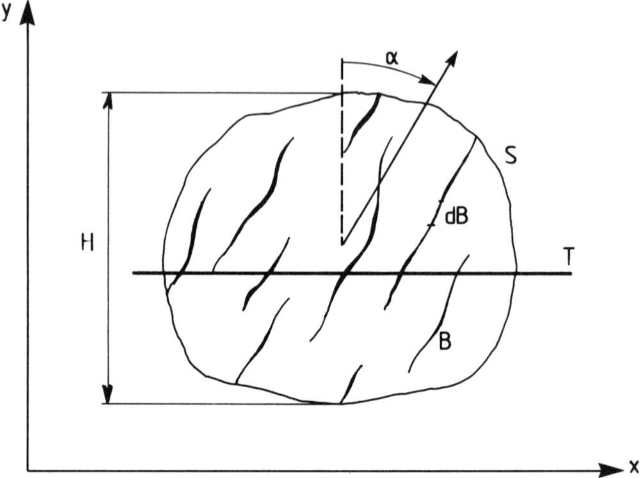

Fig. 2. Estimation of length density of planar specimen.

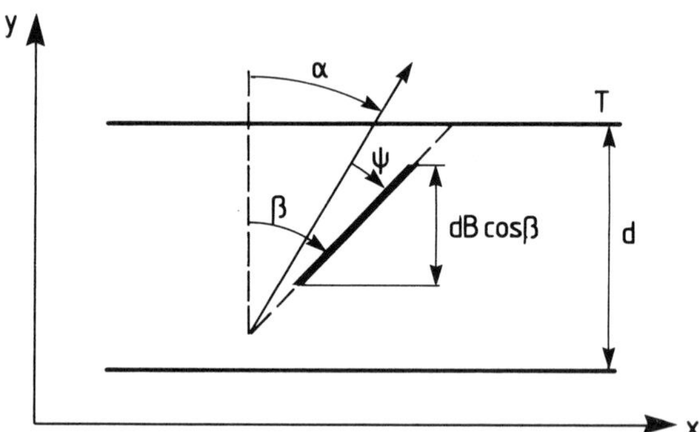

Fig. 3. Buffon problem for oriented segments.

An infinitesimal element of a curved line has the length dB and its orientation is measured by an angle β to the y-axis of the coordinate system. The grid of parallel testing lines of spacing d intersects the field S, Fig. 3. Infinitesimal elements dB for any curved line B deviate from preferred orientation α by an angle ψ, i.e. $\beta = \alpha + \psi$. Designating by $f(\beta)$ the probability density function of β we can obtain the probability of having an intersection between the test line T and dB as an integral over conditional probability at fixed β and the distribution function $f(\beta)$

$$\mathrm{pr}(dB,T) = \frac{1}{2} \int_0^{2\pi} \mathrm{pr}(d\beta, T|\beta) f(\beta) d\beta. \tag{8}$$

The probability that $dB(\beta)$ intersects T is

$$pr(dB,T|\beta) = \frac{dB\cos\beta}{d} , \tag{9}$$

thus Eq. (8) becomes

$$pr(dB,T) = \frac{dB}{2d} \int_0^{2\pi} |\cos\beta| \; f(\beta)d\beta. \tag{10}$$

This is the solution to the Buffon's needle problem for non-random directions (Weibel, 1980). By definition number of intersections between dB and T is one if they intersect and zero otherwise so that the expected number of intersections is equal to probability (10). The average number E of intersections I between lines B and the testing line T is found by integration of Eq. (10) which yields

$$E(I) = \frac{\angle(B)}{2H} \int_0^{2\pi} |\cos\beta| \; f(\beta)d\beta \tag{11}$$

where $\angle(B)$ is the total length of curves in S and H is the caliper diameter of S in the direction normal to T, representing the space within which T can be randomly positioned. On the other hand the average length of the test line contained in S is

$$E\{\angle(T)\} = \frac{A}{H} , \tag{12}$$

so that rearranging Eqs. (11) and (12) we obtain the length density as

$$\frac{\angle(B)}{A} = \frac{2\;E(I)}{E\{\angle(T)\}} \; \{ \int_0^{2\pi} |\cos\beta| \; f(\beta)d\beta\}^{-1}. \tag{13}$$

The preceding development assumed that T was a random testing line. More practical procedure for calculations is to take a grid of parallel lines of spacing d and count the number of intersections per unit length of the test lines.

In order to facilitate the numerical calculations of the integral in Eq. (13) we select the grid of testing lines being perpendicular to the preferred direction, $\alpha = 0$. It is immediately seen from Fig. 3 that $\beta = \psi$ in this case and the Eq. (13) yields

$$D \equiv \frac{\angle(B)}{A} = 2N_T \; \{ \int_0^{2\pi} |\cos\psi| \; f(\psi)d\psi\}^{-1}, \tag{14}$$

where N_T is the number of intersections per unit length of testing lines and we identify the length density with microstructure parameter D. A probability density function $f(\psi)$ must reflect a strong directionality of curves and a good choice for $f(\psi)$ is the von Mises distribution, (Mardia, 1972)

$$f(\psi) = \frac{1}{2\pi I_o(k)} \exp[k\cos\psi], \tag{15}$$

where I_o is a modified Bessel function and k is the concentration parameter which determines the spread of the distribution; for $k = 0$, $f(\psi) = 1/\pi$ and all directions for the position of dB are equally probable, but if $k \to \infty$ all lines become parallel to the preferred orientation.

RESULTS AND DISCUSSION

During all tensile experiments performed with constant strain rate $5 \cdot 10^{-5}$ [1/s] the pattern of microdamages was monitored by taking microphotographs at predetermined time intervals. Further it was accepted that the area of microphotographs is representative for length density calculations under the assumption that the length density is uniformly distributed over the specimen area at fixed strain level. Performing the calculations according to the Eq. (14) at each selected strain level results in the accumulated distribution for microstructure parameter D as indicated in Fig. 4. The concentration parameter k of the von Mises distribution was selected in a manner giving a negligible value of distribution for angles ψ larger than 10°, as observed experimentally. The calculation procedure and appropriate tables are contained in Batschelet, 1981.

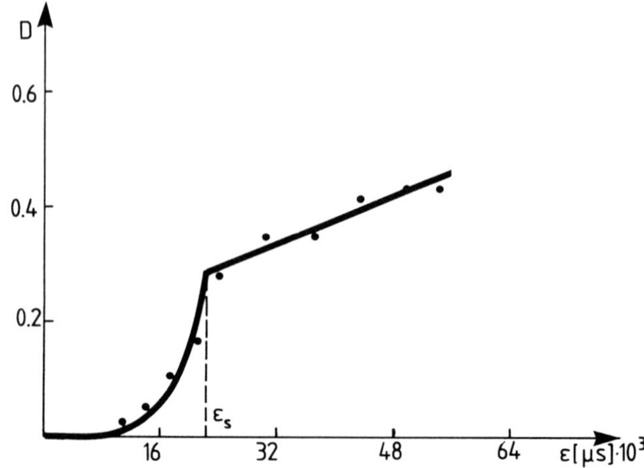

Fig. 4. Cumalative distribution of microstructure parameter D.

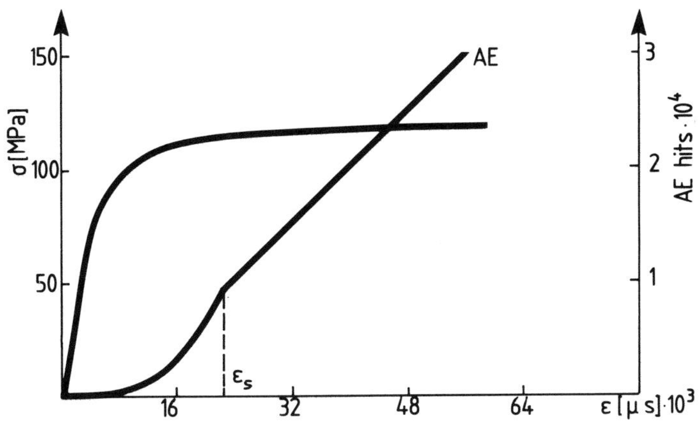

Fig. 5. Stress-strain diagram and AE-hits.

The development of the damage parameter is divided into two stages. In a primary stage the nucleation of individual microcracks of average length corresponding to the dimension of the plane weave cell is observed and terminates at strain ϵ_S. During the second stage microcracks tend to join along the preferred direction. At the same time a number of small surface microcracks appear in the direction perpendicular to preferred direction. Figure 5 shows typical stress-strain diagram together with acoustic emission hits recorded under monotonic loading. The interesting mirroring of damage curve by the acoustic emission hits is clearly seen. A quantitative correlation between these curves faces significant difficulties due to errors introduced in the estimation of the parameter D on the one hand and due to extraordinary AE signals coming from secondary microcracks not included in the determination of parameter D. Nevertheless, AE supports the finding of the limiting strain ϵ_S with good accuracy. It is interesting to note that also AE counts and amplitude distribution separate the damage development into two distinctive stages. Development of the damage parameter D was fitted with the following equation

$$D = \begin{cases} a\, ||\epsilon(t)||_{\infty}^{m} & , \quad \epsilon \le \epsilon_S \\ b\, ||\epsilon(t)-\epsilon_S||_{\infty} & , \quad \epsilon > \epsilon_S \end{cases} \tag{16}$$

where a, b, m are constants and $||\ \ ||_{\infty}$ designates the Lebesque norm with the property that it takes a present value of the argument when the argument is monotonically increasing and it is constant, keeping its previous maximum value otherwise (Fitzgerald et al., 1973; Schapery, 1984; Pyrz, 1989). The kernel function in Eq. (5) is of the form:

$$R(t) = \frac{Et^n}{\Gamma(1+n)} \exp\{-rD(t)\} \quad , \tag{17}$$

where E, n and r are constants and Γ is the gamma function. The values of constants used in calculations are listed in Table 1.

Table 1

E(GPa)	k	a	b	m	n	r
16.5	17	$9.6 \cdot 10^7$	$5.58 \cdot 10^{-3}$	5.15	-0.27	0.25

The data in Fig. 6 are fist-cycle results. The response predicted by the model and represented by dashed lines overestimates the experimental data. The deviation is particularly pronounced for well developed deformation. It is believed that at this stage of deformation a significant interaction between primary and secondary microcracks takes place. AE parameters obey the Kaiser effect for unloading - reloading cycle in the first stage of damages development but fail to do so in the second stage, where the material starts to emit AE signals at strains about 10% lower than previously attained maximum. This effect is believed to be caused by the secondary cracks which are not included in the determination of the damage parameter D.

However, it is apparent that the model qualitatively resembles the stress-strain curve. Further improvement is possible, if the errors involved in calculations of the damage parameter will be minimized. The concentration parameter k should be found by a maximum likelihood estimates and then goodness fit should be determined based on the micropho-

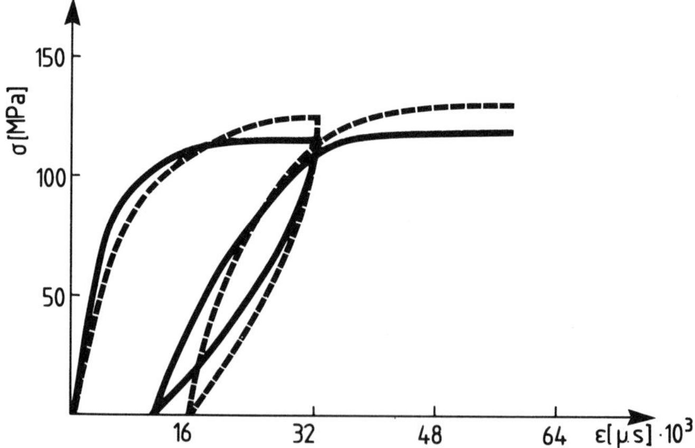

Fig. 6. Stress-strain response to a sawtooth loading.

tographs taken at different specimen sides. Measures of the damage parameter were obtained by a tedious manual treatment of microphotographs. A pattern recognition technique would greatly facilitate the accuracy of the analysis. Further effort should also be directed towards a quantitative identification of the damage parameter by the AE monitoring, the area which needs extensive research using new approaches.

REFERENCES

Allen, D.H., Harris, C.E., Groves, S.E., 1987, "A Thermomechanical Constitutive Theory for Elastic Composites With Distributed Damage - I. Theoretical Development", Int. J. Solids Structures, vol. 23, pp. 1301-1318.

Allen, D.H., Harris, C.E., Groves, S.E., 1987, "A Thermomechanical Constitutive Theory for Elastic Composites With Distributed Damage-II. Application to Matrix Cracking in Laminated Composites", Int. J. Solids Structures, vol. 23, pp. 1319-1338.

Allen, D.H., Harris, C.E. Groves, S.E., 1988, "Characteristics of Stiffness Loss in Cross Ply Laminates With Curved Matrix Cracks", J. Composite Materials, vol. 22, pp. 71-80.

Attiogbe, E.K., Darwin, D., 1987, "Self-Consistent Model for Transversely Isotropic Cracked Solid", ASCE Journal of Engineering Mechanics, vol. 113, pp. 984-999.

Batschelet, E., 1981, "Circular Statistics in Biology", Academic Press, New York.

Dvorak, G.J., Laws, N., Hejazi, M., 1985, "Analysis of Progressive Matrix Cracking in Composite Laminates, I. Thermoelastic Properties of a Ply with Cracks", Journal of Composite Materials, vol. 19, pp. 216-234.

Fitzgerald, J.E., Vakili, J., 1973, "Nonlinear Characterization of Sand-Asfalt Concrete by Means of Permanent Memory Norms", Experimental Mechanics, vol. 13, pp. 504-510.

Kamimura, K., 1985, "Continuum Damage Approach to Mechanical Behaviour of Damaged Laminate and a Modelling of Damage Parameter", Mechanical Characterisation of Load Bearing Fibre Composite Laminates, A.H. Cardon, et al., eds., Elsevier, New York, pp. 115-126.

Kestin, J., Herrmann, G., 1988, "The Fundamental Elements of an Exact Thermodynamic Theory of Damage in Elastic Solids", contributed paper presented at XVIIth IUTAM Congress, Grenoble.

Krempl, E., 1975, "On the Interaction of Rate and History Dependence in Structural Metals", Acta Mechanica, vol. 23, pp. 53-90.

Laws, N., Dvorak, G.J., Hejazi, M., 1983, "Stiffness Changes in Unidirectional Composites Caused by Crack System", Mechanics of Materials, vol. 2, pp. 123-137.

Mardia, K.V., 1972, "Statistics of Directional Data", Academic Press, New York.

Peng, L, Yang, F., Xiao, Z., Zhu, H., 1986, "Anisotropic Damage in Orthotropic Fiber Reinforced Composite Plate", Proceedings of Int. Symposium on Composite Materials and Structures, T.T. Loo, C.T. Sun, eds., Beijing, pp. 530-536.

Pyrz, R., 1989, "A Hereditary Integral Equation - A Potential Use for Mechanical Modelling and Damage in Composite Materials", Special Report No. 2, Institute of Mechanical Engineering, Aalborg.

Reifschnider, K.L., 1988, "Life Prediction Methods for Notched Composite Laminates", Proceedings 4th US-Japan Conference on Composite Materials, Technomic, pp. 265-275.

Schapery, R.A., 1980, "On Constitutive Equations for Viscoelastic Composite Materials With Damage", Workshop on A Continuum Mechanics Approach to Damage and Life Prediction, D.C., Stouffer et.al., eds., pp. 119-133.

Schapery, R.A., 1981, "On Viscoelastic Deformation and Failure Behaviour of Composite Materials With Distributed Flaws", Advances in Aerospace Structures and Materials, S.S. Wang, W.J. Renton, eds., ASME AD-01, pp. 5-20.

Schapery, R.A., 1982, "Models for Damage Growth and Fracture in Nonlinear Viscoelastic Particulate Composites", Proceeding 9th U.S. National Congress of Applied Mechanics, ASME, pp. 237-245.

Schapery, R.A., 1984, "Correspondence Principles and a Generalized J Integral for Large Deformation and Fracture Analysis of Viscoelastic Media", Int. Journal of Fracture, vol. 25, pp. 195-223.

Schen, W., Yang, F., Shen, Z., 1986, "An Anisotropic Damage Model for Anisotropic Materials", Proceedings of Int. Symposium on Composite Materials and Structures", T.T. Loo, C.T. Sun, eds., Beijing, pp. 549-555.

Talreja, R., 1985, "A Continuum Mechanics Characterisation of Damage in Composite Materials", Proc. Roy. Soc. Lon., Ser. A, Vol. 399, pp. 195-216.

Talreja, R., 1989, "Damage Development in Composites: Mechanisms and Modelling", to appear in Journal of Strain Analysis.

Tonda, R.D., Schapery, R.A., 1987, "A Method for Studying Composites With Changing Damage by Correcting for the Effects of Matrix Viscoelasticity", Damage Mechanics in Composites, A.S.D. Wang, G.K. Hanitos, eds., ASME, pp. 45-51.

Weibel, E.R., 1980, "Stereological Methods - Theoretical Formulation", vol. 2, Academic Press, New York.

Weitsman, Y., 1987, "Coupled Damage and Moisture-Transport in Fiber-Reinforced Polymer Composites", Int. Journal of Solids and Structures, vol. 23, pp. 1003-1025.

Weistman, Y., 1988, "Damage Coupled With Heat Conduction in Uniaxially Reinforced Composites", ASME Journal of Applied Mechanics, vol. 55, pp. 641-647.

Inelastic Behavior VII

Admissible Deformations in Diaphragm Forming of Continuous Fibre Reinforced Thermoplastics

P. J. Mallon[1]
C. M. O'Bradaigh[2]
M. R. Monaghan[1]
R. B. Pipes[2]

1.Department of Mechanical Engineering, University College Galway, Ireland.
2. Center for Composite Materials, University of Delaware, Newark, Delaware. 19716.

Abstract

An experimental programme on diaphragm forming of thermoplastic composite laminates has successfully demonstrated the critical deformation mechanisms which occur during the forming stage of the process. Both single and complex curvature parts were formed in an experimental autoclave which had sophisticated process control facilities for the independent control of both pressurisation and forming rates. When a 90° bend was formed into a female mould the mechanism of interply slip was clearly demonstrated. These 90° bends were formed from both unidirectional and cross ply layups of APC-2 and in all cases the degree of post forming spring forward was between 1.5° and 2.5°. The factors which influence laminate buckling during the forming of complex curvature components were shown to be ; the rate of forming, the ratio of layup area to the forming area and the stiffness of the diaphragms. The higher forming pressures required for the stiffer diaphragms indirectly caused severe squeeze flow and surface ply erosion. The latter effect was alleviated by the use of a slip ply between the laminate and the diaphragm in single curvature parts.

Introduction

The availability of high performance thermoplastic materials as matrices for continuous-fiber reinforced composites has aroused much interest within the composites industry over the past few years. Thermoplastics offer improvements over thermosets in impact strength, environmental resistance, shelf-life and the potential for substantially reduced fabrication times. The current barrier to widespread use of continuous fiber reinforced thermoplastics is mainly a lack of fully-developed manufacturing technology. The processes which are currently being developed for these new composites are thermoforming(1), tape laying(2) and filament winding(3). The most promising and exciting variation of the thermoforming processes is diaphragm forming(4) and this process has been developed on the principles of vacuum forming of thermoplastic sheet. In diaphragm forming the laminate is placed between two thin deformable diaphragms which, when clamped around the edges, maintain biaxial tensions on the laminate during deformation, consequently restricting laminate wrinkling and buckling.

The results presented here detail many observations of the various mechanisms which are present when a laminate, above the melt temperature of the matrix, is deformed into both simple and complex curvature moulds. The autoclave used in this experimental programme allowed control of both pressurisation and deformation rates and it was this capability which has allowed the unique effects of the process variables on part quality to be established. The quality aspects considered were degree of fibre buckling, thickness variation, surface finish and conformity of part to the shape of the mould. The continuous fibre reinforced thermoplastics used in these experiments was carbon fibre reinforced polyetheretherketone (APC-2), supplied by Imperial Chemical Industries.

Deformation Mechanisms

Theoretical approaches to model the sheet forming of continuous fibre thermoplastic composites during processing must simulate the important forming mechanisms. The dominant characteristic of such a material is the high stiffness of the carbon fibres in comparison to that of the viscous matrix. Flow processes that occur at the polymer melt temperature will be highly anisotropic, due to the continuous reinforcements. Experimental studies by the authors (5,6) and by Barnes and Cogswell (7) have shown that, in practical forming processes such as diaphragm forming, the composite layup is essentially inextensible in the fibre direction, with deformations being accommodated by shearing and transverse elongational mechanisms.

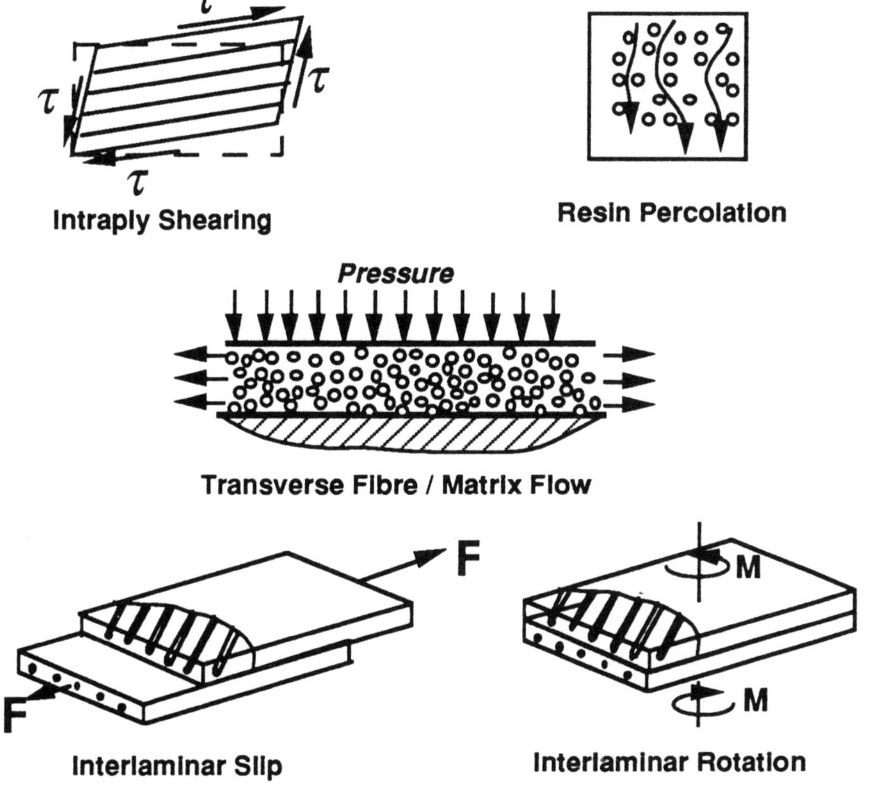

Figure 1. Deformation mechanisms in diaphragm forming

Fig.1 illustrates the primary deformation mechanisms that occur during forming of complex-curvature components. The first mechanism, intraply shearing, is needed when a shearing strain occurs in the plane of the lamina, thus allowing part conformity to complex curvature geometries. Fabrics, by contrast, are usually interlocked at fibre crossover points, limiting the shear strain to the locking angle of the unit cell. Theoretically, there is no limit to the amount of intraply shear deformation in unidirectional laminae which can be accommodated by this mechanism. Rheological studies of APC-2 by Cogswell (8), indicate an initial yield stress for the intraply shearing mode, followed by an approximately Newtonian response.

The second mechanism, resin percolation, is necessary for consolidation, but is not usually as important in the processing of fully impregnated thermoplastic composites as it is with thermoset composites. The relatively high thermoplastic matrix viscosity reduces significant polymer flow transverse to, and along the length of the fibres, although small pools of resin were observed by the authors (6) at the end of plies which were diaphragm formed over a 90° bend. Transverse squeeze flow of matrix and fibres in a direction perpendicular to the fibres, is a very important forming mechanism, dictating the final thickness distribution of the formed part, as fibres and matrix flow transversely in response to pressure gradients.

The mechanisms which are required for the forming of multi-ply laminates are both interlaminar slip and rotation. Interply (or interlaminar) shearing occurs when a laminate is deformed in a single-curvature manner. As the fibres are essentially inextensible, the discrete layers are forced to slip past one another. Cogswell (8) notes that this shearing action occurs in a thin 'resin-rich layer' (thickness - 10 microns) which migrates to each lamina surface during consolidation. An apparent yield stress is also determined for this deformation mode, followed by an approximate Newtonian response. Furthermore, Muzzy (9) carried out a three-point flexural loading on a single ply at processing temperature and discovered an initial elastic response which could correspond to this initial yield stress. In their analysis of interply slip, Tam andGutowski (10) noted the importance of resin-rich layers at intervals through the thickness in decreasing the forces needed to deform a stacked elastic/viscous model. Clearly, this mechanism is essential in practical forming applications, and has been observed by the authors microscopically in a formed 90° bend (6). The final mechanism, interlaminar rotation, is perhaps of most importance in carrying out a sheet forming analysis. Most complex-curvature parts require a change of initial fibre orientation between adjacent plies.

While the analysis of the forming process has recently attracted attention, there is still a substantial amount of work to be done before it will be possible to accurately predict the performance of a formed complex curvature component. The authors have recently made a significant advance in the development of a constitutive approach to predict the stresses and deformations throughout the process stage (11).

Experimental

Diaphragm Forming Process

The polymeric diaphragm forming process is illustrated in Fig.2 where the double diaphragms plus a laminate are being formed into a mould within an autoclave. The pressure system consists of two sides, the vessel pressure and the cavity pressure. Forming is effected by applying a greater pressure in the vessel than in the cavity, as shown in the illustration. Forming may be carried out after consolidation, by applying equal pressures on the two sides initially and then decreasing the cavity pressure, or it may be followed by consolidation, by simply applying pressure only on the vessel side. Vacuum, exerted between the diaphragms, is necessary in order to extract all air that may be present, but also to remove any gases that may evolve during the melting and consolidation of the matrix.

A typical processing cycle for production of a polymeric diaphragm formed APC-2 component is shown in Fig.3, with an overall cycle time of

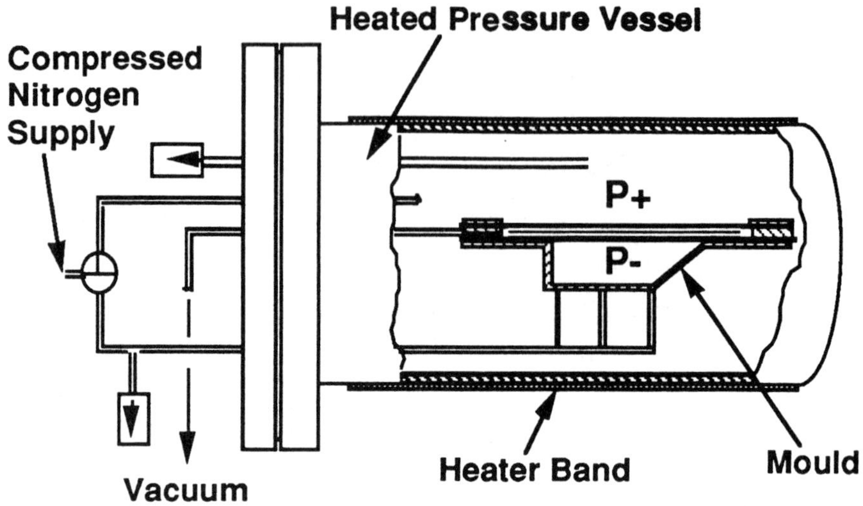

Figure 2. Schematic of Diaphragm Forming Process in an Autoclave

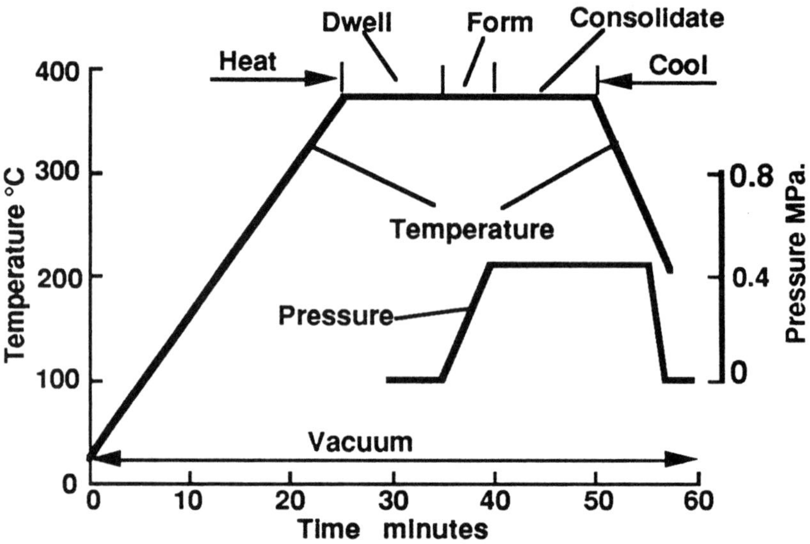

Figure 3. Typical Diaphragm Forming Process Cycle.

about 60 minutes, provided that the autoclave has been preheated to processing temperature.

Description of Diaphragm Forming Apparatus

Two separate autoclaves were used during the course of this work, one based in the University of Delaware and the other in University College Galway (UCG). The Delaware autoclave is described in detail in reference (4) while the autoclave based at UCG is described in reference (12) and shown in Fig.4. This is a large experimental autoclave with internal dimensions of 0.6 metres diameter by 1.0 metre in length. It can operate at pressures up to 2.0MPa and temperatures up to 400°C and is designed primarily for sheet forming of thermoplastic composites. It is fully computer controlled and features advanced process monitoring and control of temperature, pressure and displacement. The pressure control system uses two pressure transducers and a simple on-off normally closed solenoid valve, operated from the computer. This control system allows programmed pressure profiles to be reproduced in the autoclave.

Figure 4. Diaphragm Forming Autoclave

The deformation of the forming part is measured using a high temperature Linear Variable Differential Transformer (LVDT), model LIN256, manufactured by RDP Electronics Ltd. It is capable of the measurement of very small displacements and is used to record the central deflection of the forming part. The central moving core rests on top of the top diaphragm and moves downward with the diaphragm during forming. During the forming rate control experiments the displacement of the LVDT core was programmed from the computer and the desired displacement rate effected by controlled pressure increase.

Results and Discussion

Interply Slip

A $(0°/90°)_{4S}$ stack of prepreg, 200mm long by 100mm wide, was tacked together and both ends ground until all 0° fibres were of equal length. This prepreg stack was then diaphragm formed over a 90° bend. The part was

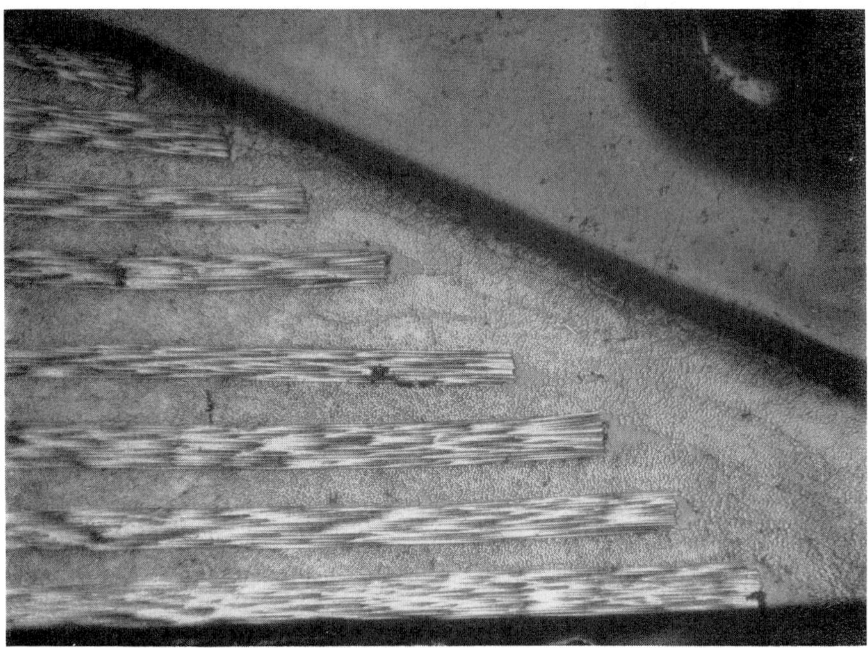

Figure 5. Interlaminar Slip in a 90° Bend

subsequently cut along the 0° fibre direction and small sections taken from each end for microscopic examination. These specimen ends were ground and polished and a photomicrograph of one is presented in Fig.5. The interlaminar slip which has occurred between the axial plies is quite clear and has been measured to be approximately equal to that which would be expected from the geometric considerations of the part. This demonstrates that the interlaminar shear forces were insufficient to cause fiber deformation at the viscosities which occur in the region of the melt temperature. Careful examination of the photomicrograph also illustrates another fundamental mode of deformation, resin percolation along the 0° fibers, causing small pools of resin at the ply ends.

Spring Forward

This is an effect which is experienced when a composite laminate above the melt temperature of the matrix is formed into a mould and subsequently cooled in the deformed shape, results in the final enclosed angle of the part being generally less than that of the mould. This effect is known as spring forward and occurs when cooling a curved laminate where there is a significant difference in coefficients of thermal expansion of the fibres and matrix.

In the experimental programme both male and female moulds were used to form 90° bends from APC-2 materials using unidirectional and cross-ply layups. In all cases the included angle of the formed part was between 1.5° and 2.5° less than the included angle of the mould. These results agree closely with the theoretical prediction of O'Neill, Rogers and Spencer (13).

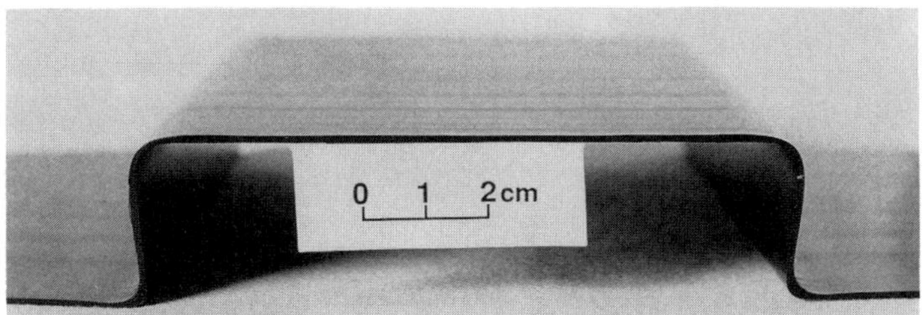

Figure 6. Channel Part Showing Spring Forward

The spring forward effect has also been observed in a selection of parts which include a channel as shown in Fig.6 and a cylindrical dish as shown in Fig.7. In addition to the channel a sine wave part demonstrates that the combination of two bends, where one in the reverse of the other, serves to cancel the overall effect of spring forward. In the case of the sine wave part the total spring forward of the straight end sections was approximately 2°. In a complex curvature part such as the cylindrical dish, spring forward is evidenced by the concave nature of the top surface which was formed onto a flat mould surface. This highlights the significance of spring forward on mould design, where in the case of the cylindrical dish, the top surface of the mould should have been convex if a flat top surface was required on the part. A complete analytical prediction of spring forward for complex curvature parts is required to enable this effect to be catered for in mould design. If the fibre orientation in the finished part was available at all locations then the task of predicting spring forward should not be difficult. However, the prediction of the fibre orientation in a complex curvature part, which has been formed from a flat layup, presents a much greater challenge and was referred to earlier under deformation mechanisms.

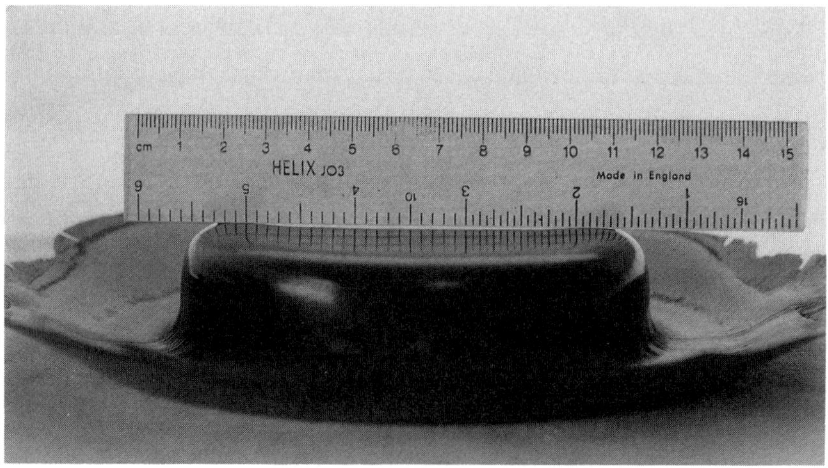

Figure 7. Cylindrical Dish Part Showing Spring Forward.

Laminate Buckling

When a 228mm diameter $(0°/90°)_{4S}$ layup was formed into each of the hemispherical moulds, shown in Fig. 8, buckling was observed at 45° to the fibre direction of the parts formed in the A and B moulds. This was most severe on the A parts as shown in Fig. 9. When a similar circular lay-up 152mm. diameter was formed no such problems existed. It therefore appears that the buckling is caused by the excess material which remains outside the formed area, and is related to the ratio of the lay-up area to the formed area, A1/A2. A 228 mm. circular lay-up formed into female mould A has a ratio, A1/A2 = 6.0, whereas the 152 mm. lay-up formed into the same mould has a ratio of only 2.67. Parts C and D, also formed from a 228 mm. lay-up, show no buckling with ratios of 2.57 and 2.0 respectively. It can be concluded that for a circular 16-ply cross-ply lay-up, formed into a female hemispherical mould, shear-buckling will be dependent on the ratio of the layup area to the formed area.

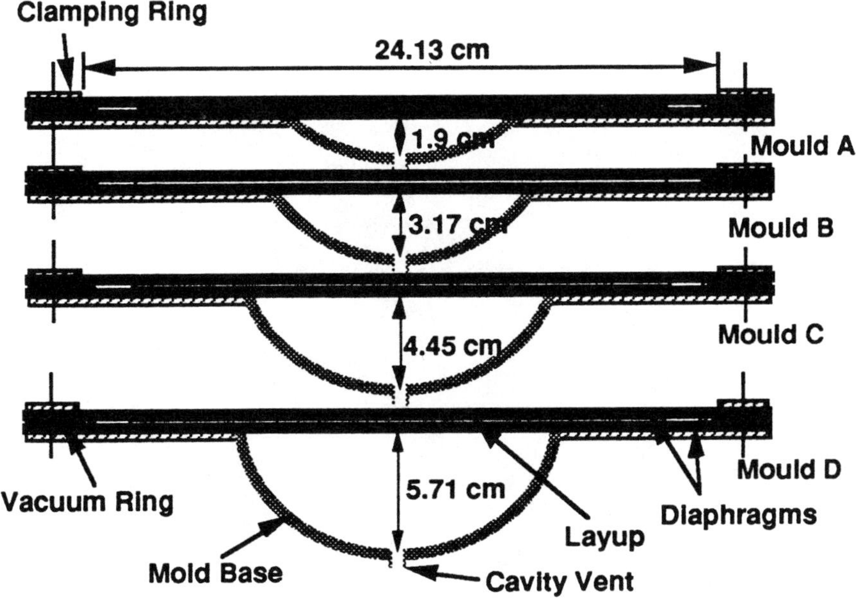

Figure 8. Set of Hemispherical Moulds

The female hemispherical mould C was used in a series of tests where the central deflection rate was varied from 3 mm/min. to 25 mm/min. The desired rate for each test was programmed into the computer and the forming rate controlled as described earlier. Upilex-R diaphragms were used to form 8 ply quasi-isotropic layups of APC-2 at the following forming rates; 3, 6, 9, 12 and 25 mm/min. The observations of degree of buckling versus forming rate is shown in Fig.10 and it can be seen that the part formed at 25 mm/min. suffered from severe out-of-plane buckling while the part formed at 3 mm/min. had no out-of-plane buckling and very little surface wrinkling. From an inspection of the other parts it was observed that the out-of-plane buckling was absent at the forming rate of 6 mm/min. while at 12 mm/min. out-of-plane buckling was present as shown in Fig.11a. When the layup was changed to a $(0°/90°)_{2S}$ layup, parts formed at 12 mm/min. did not buckle as shown in Fig.11b.These results demonstrate the need to consider the layup as well as part geometry in establishing the optimum forming rate.

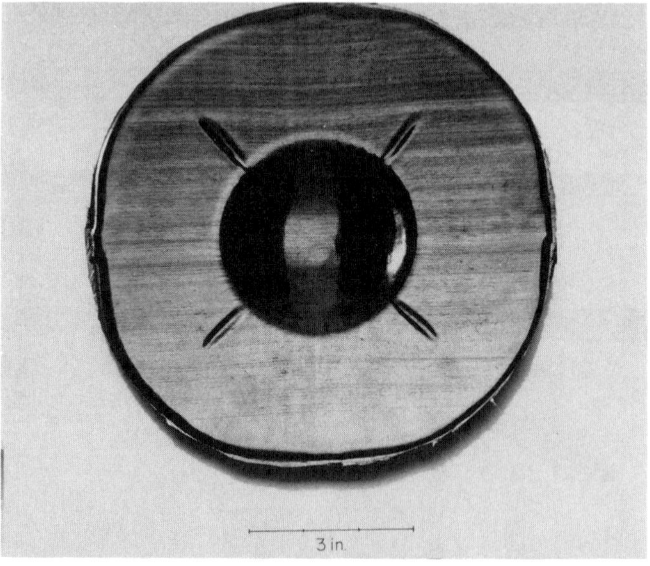

Figure 9. $(0°/90°)_{4S}$ Hemispherical 'A' Part Showing Buckling.

(a) Formed at 3 mm/min. (b) Formed at 25mm/min.

Figure 10. $(0°/+45°/-45°/90°)_S$ Hemispherical 'C' Parts.

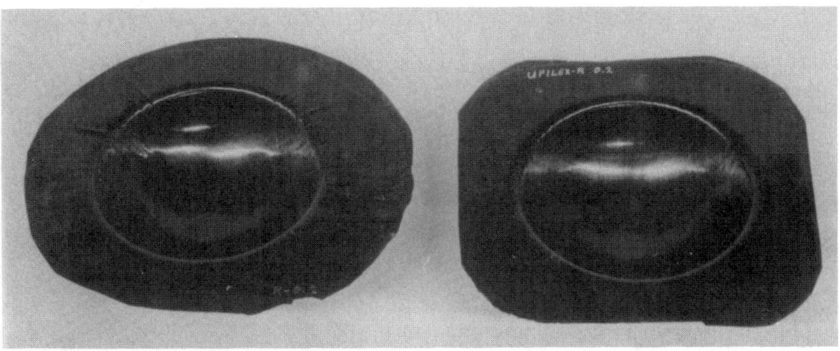

(a) $(0°/+45°/-45°/90°)_S$ (b) $(0°/90°)_{2S}$

Figure 11. Hemispherical 'C' Parts Formed at 12 mm/min.

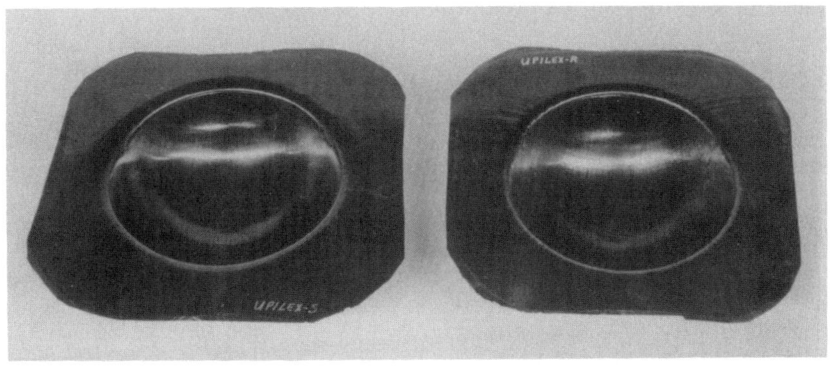

(a) Formed using Upilex-S (b) Formed using Upilex-R

Figure 12. $(0°/90°)_{2S}$ Hemispherical 'C', Formed at 12 mm/min.

It has been shown that the use of stiffer diaphragms can improve the surface quality of diaphragm formed parts(14). Fig.12 shows the surface quality of two hemispherical parts formed at 12 mm/min. deformation rates but with Upilex-R and Upilex-S diaphragm materials. The surface of the part formed using the stiffer diaphragms, Upilex-S, has no buckling or surface wrinkles, while the part formed using Upilex-R had a glossy surface with some wrinkles present. This wrinkling appears to be a direct effect of the lack of tension on the laminate during forming.

Squeeze Flow

The squeeze flow which occurs during diaphragm forming is a result of normal pressure causing transverse fibre movement (14) and the level of normal pressure generated is related to the diaphragm stiffness. A cylindrical dish mould shown in Fig.13 was used to investigate this effect using Upilex-R, Upilex-S and Supral diaphragms materials with a $(0°/90°)_{4S}$ APC-2 laminate. Fig.14 shows sections through the cylindrical dish parts and it can be seen clearly that the degree of squeeze flow, as evidenced by increased thickness at the lower corner, is much more severe from the Upilex-S and Supral diaphragms as opposed to the Upilex-R diaphragms.

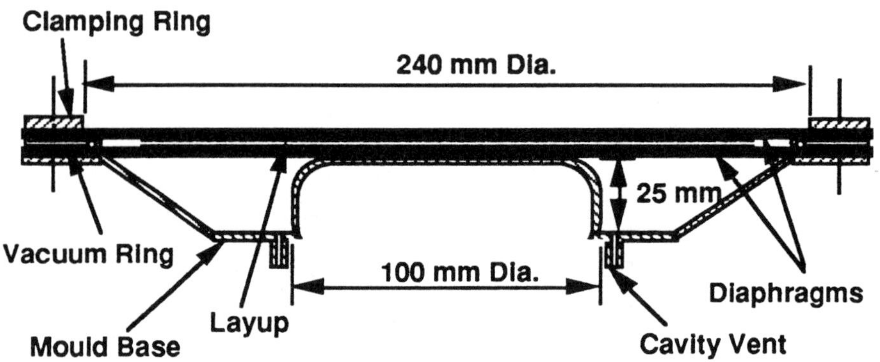

Figure 13. Cylindrical Dish Mould

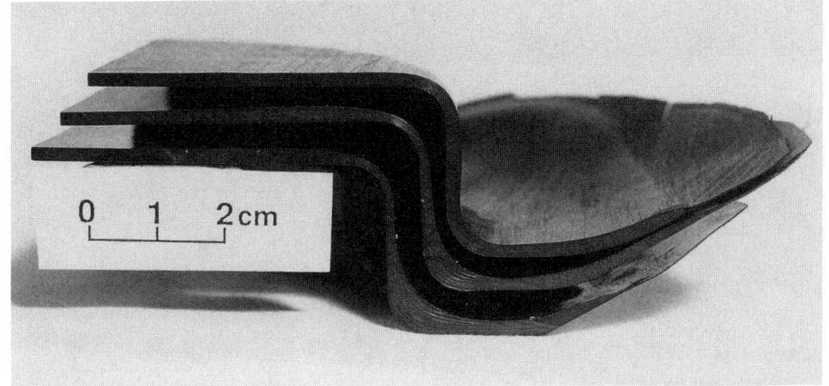

Top: Upilex-R **Centre: Supral** **Bottom: Upilex-S**

Figure 14. Cylindrical Dish Parts - Different Diaphragms.

To diaphragm form this cylindrical dish part using Upilex-R diaphragms required an autoclave pressure less than 0.1 MPa, while a pressure of 1.7 MPa was required to form the parts using Upilex-S and Supral diaphragms. These higher pressures obviously exaggerate the degree of squeeze flow present.

Diaphragm / Surface Ply Interaction

When the diaphragms and laminate deform into a mould the diaphragms undergo biaxial elongation. In the direction of the inextensible surface fibres the diaphragm extension is restrained by the contact with the laminate and consequently higher strain occurs in the diaphragm between the laminate end and the clamping ring. The diaphragm stretching in the direction perpendicular to the surface fibres is more uniform and is accommodated by transverse fibre movement which causes severe erosion of the surface ply. This effect is clearly shown in Fig.15 where a model trailer was formed into a female mould. Fig.16 shows two 90° bends where one was formed without any normal pressure and the other formed with a normal pressure of 0.14 MPa. applied throughout the forming. It can be seen that the errosion of the surface ply was more severe when additional normal pressure was applied. This effect demonstrates that the normal pressure influences the interaction between the diaphragms and the surface ply.

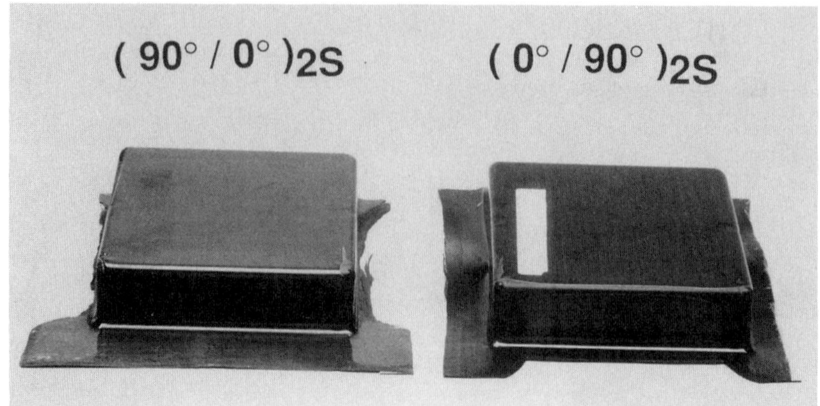

Figure 15. Model Trailer Showing Surface Ply Erosion

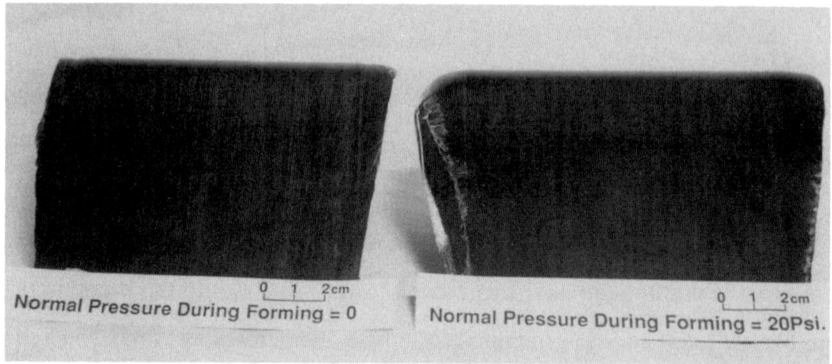

Figure 16. Effect of Normal Pressure on Surface Ply Erosion.

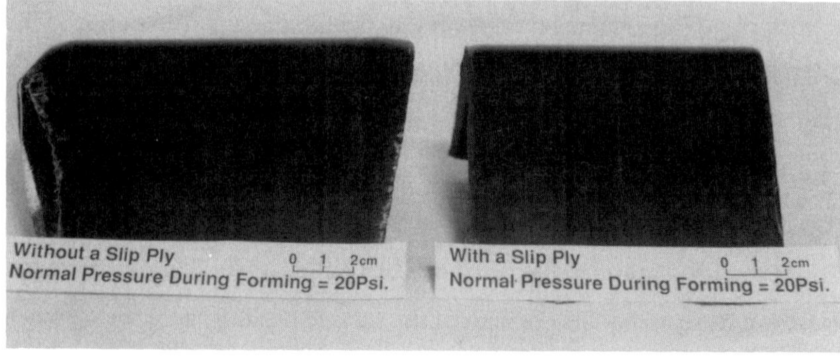

Figure 17. Surface Ply Erosion Alleviated by the use of Slip Plies.

When slip plies were introduced between the diaphragms and the laminate the transverse spreading was eliminated as shown in Fig.17. The slip plies were 0.005" thick Upilex-R film and cut such that they were just slightly larger than the layup and were placed above and below the layup. The function of these slip plies was to allow slipping to take place between the diaphragms and the slip plies and hence partially isolate the laminate from surface tension.

Conclusions

The diaphragm forming of both single and complex curvature components clearly demonstrated the important deformation mechanisms which occur during the forming phase. The tension imparted to the laminate by the deforming diaphragm enabled the primary mechanism of interply slip in single curvature parts to take place fully.

The post forming spring forward was measured in both single and complex curvature components and found to in the range 1.5° to 2.5°.

It has been demonstrated that laminate buckling which occurs during the forming stage is influenced by; the rate of forming, the ratio of layup area to the formed area and the stiffness of the diaphragms.

The squeeze flow effect in parts formed using Upilex-R diaphragms was evidenced by a thickness variation in the parts of approximately ± 10%. When stiffer diaphragms were used the part thickness variation increased to approximately ±100 % due to increased forming pressures.

The effect of surface ply erosion was greatly increased by increased normal pressure. This effect was alleviated by the use of slip plies.

References

1. Okine, R. K., 1989, "Analysis of forming parts from advanced thermoplastic composite sheet material," SAMPE Journal, 25, No.3.

2. Beyler, E. P., Phillips, W. and Guceri, S. I., 1988, "Experimental investigation of laser assisted thermoplastic tape laying consolidation," J. Thermoplastic Comp. Mater., Vol. 1, pp. 107-121.

3. Cattanach, J. B. and Cogswell, F. N., 1986, "Processing with aromatic polymer composites," Developments in Reinforced Plastics-5 (Applied Science Publishers).

4. Mallon, P. J. and O'Bradaigh, C. M., 1988, "Development of a pilot autoclave for polymeric diaphragm forming of continuous fibre-reinforced thermoplastics," Composites, Vol. 19, No. 1, pp. 37-47.

5. Mallon, P. J., O'Bradaigh, C. M. and Pipes, R.B., 1989, "Polymeric diaphragm forming of complex curvature thermoplastic composite parts," Composites, Vol.20, No.1, pp. 48-56.

6. O'Bradaigh, C. M. and Mallon, P. J., 1989, "Effect of forming temperature on the properties of polymeric diaphragm formed components," Composites Science and Technology, Vol 35, No.3, pp. 235-255.

7. Barnes, J. A. and Cogswell, F. N., 1989. "Transverse flow processes in continuous fibre reinforced thermoplastic composites," Composites, Vol. 20, No. 1, pp. 38-42.

8. Cogswell, F. N., 1987, "The processing science of thermoplastic structural composites," Int. Polymer Processing, Vol. 1, No.4, pp. 157-165.

9. Muzzy, J. D., Wu, X. and Colton, J. S., 1989, Proc. S.P.E. Conf., Antec '89.

10. Tam, A. S. and Gutowski, T. G., 1989, J. Composite Materials, Vol. 23, 587.

11. O' Bradaigh, C. M., 1990, " Ph.D. Thesis (in preparation), University of Delaware, Newark, Delaware 19716.

12. Monaghan, M. R., Mallon, P. J., 1990, "Development of a computer controlled autoclave for forming thermoplastic composites," Composites Manufacturing, Vol.1, No.1, pp. 8-14

13. O'Neill, J. M., Rogers, T. G. and Spencer, A. J. M., 1989, "Thermally induced distortions in the moulding of laminated channel sections," Mathematical Engineering in Industry, Vol. 2, No. 1, pp.65-72.

14. Monaghan, M. R., O'Bradaigh, C. M., Mallon, P. J. and Pipes, R. B., 1990, "The effect of diaphragm stiffness on the quality of diaphragm formed thermoplastic composite components,"Proc.35 th. Int. SAMPE Symposium, Anaheim, California.

Viscoelastic Creep Post Buckling Behavior of AS4/J1 Thermoplastic-Matrix Composite Laminates

Y. Nakajo , Ashikaga Institute of Technology, Japan

1. INTRODUCTION

Above-critical behavior of AS4/J1 thermoplastic-matrix composite laminates at elevated temperature is analyzed taking time-dependency into consideration.

In the previous study, fundamental creep buckling behaviors of the composite panels are investigated as time-dependent bifurcation buckling adopting the correspondence principle. In this study, the anisotropic time-temperature dependent viscoelastic constitutive equations of the composite in the previous study is used again to give mathematical expressions for compliances and Poisson's ratio for the analyses.

The advanced structural applications of high-performance thermoplastic-matrix composites such as AS4/J1 require accurate analyses and clear understanding of the above-critical behavior as well as critical behavior as structural elements. An elastic solution of post buckling of simply-supported composite laminate subject to biaxial compression is obtained in terms of end-shortening versus loads by solving the Karman type equation assuming the displacements in forms of products of trigonometric functions. The viscoelastic response to a step input of the load is obtained immediately from the time-temperature dependent constitutive equations of the composite along with the elastic post-buckling solution. Then viscoelastic solutions adopting the Duhamel integral based on the Boltzmann superposition principle is obtained for a general time-dependent load input.. The axial loads can be any functions of time because the integral is carried out by numerical method. In the present study, the axial loads are assumed to be logarithmically linear function of time as an example. The results clarify the load carrying capability of the composite panels at elevated temperature in buckled conditions.

2. METHODS OF APPROACH

2.1 ELASTIC POST BUCKLING SOLUTION

Firstly, we assume that the composite laminates are symmetric and the number of layer is large enough to cancel the coupling between bending and twisting. The basic equations are;

$$\frac{\partial^2 M_x}{\partial x^2} + 2\frac{\partial^2 M_{xy}}{\partial x \partial y} + \frac{\partial^2 M_y}{\partial y^2} + N_x \kappa_x + N_{xy}\kappa_{xy} + N_y \kappa_y = 0 \qquad (2\text{-}1)$$

$$\frac{\partial N_x}{\partial x} + \frac{\partial N_{xy}}{\partial y} = 0$$

$$\frac{\partial N_y}{\partial y} + \frac{\partial N_{xy}}{\partial x} = 0 \qquad (2\text{-}2)$$

$$N_x = A_{xx}\varepsilon_x + A_{xy}\varepsilon_y\,,\ N_y = A_{xy}\varepsilon_x + A_{yy}\varepsilon_y\,,\ N_{xy} = A_{ss}\varepsilon_{xy} \qquad (2\text{-}3)$$

$$M_x = D_{xx}\kappa_x + D_{xy}\kappa_y\,,\ M_y = D_{xy}\kappa_x + D_{yy}\kappa_y\,,\ M_{xy} = D_{ss}\kappa_{xy} \qquad (2\text{-}4)$$

$$\varepsilon_x = \frac{\partial u}{\partial x} + \frac{1}{2}\left(\frac{\partial w}{\partial x}\right)^2,\ \varepsilon_y = \frac{\partial v}{\partial y} + \frac{1}{2}\left(\frac{\partial w}{\partial y}\right)^2$$

$$\varepsilon_{xy} = \frac{\partial u}{\partial y} + \frac{\partial v}{\partial x} + \left(\frac{\partial w}{\partial x}\right)\left(\frac{\partial w}{\partial y}\right) \qquad (2\text{-}5)$$

$$\kappa_x = -\frac{\partial^2 w}{\partial x^2}\,,\ \kappa_y = -\frac{\partial^2 w}{\partial y^2}\,,\ \kappa_{xy} = -2\frac{\partial^2 w}{\partial x \partial y} \qquad (2\text{-}6)$$

where

N_x, N_y, N_{xy} : Compressive and Shear Loads

M_x, M_y, M_{xy} : Bending and Twisting Moments

u, v, w : Displacements in x, y, z directions respectively

ε_x, ε_y, ε_{xy} : Strains

κ_x, κ_y, κ_{xy} : Curvatures,

and

$$A_{xx} = h\,[E'_{11}\cos^4\phi + 2E'_{11}\nu_{21}\sin^2\phi\cos^2\phi + E'_{22}\sin^4\phi + 4G_{66}\sin^2\phi\cos^2\phi]$$

$$A_{xy} = h\,[(E'_{11}+E'_{22})\sin^2\phi\cos^2\phi + E'_{11}\nu_{21}(\sin^4\phi+\cos^4\phi) - 4G_{66}\sin^2\phi\cos^2\phi]$$

$$A_{yy} = h\,[E'_{11}\sin^4\phi + 2E'_{11}\nu_{21}\sin^2\phi\cos^2\phi + E'_{22}\cos^4\phi + 4G_{66}\sin^2\phi\cos^2\phi] \qquad (2\text{-}7)$$

$$A_{ss} = h\,[(E'_{11}+E'_{22}-2E'_{11}\nu_{21})\sin^2\phi\cos^2\phi + G_{66}(\cos^2\phi-\sin^2\phi)^2]$$

$$D_{pq} = \frac{h^2}{12}A_{pq} \quad (p,q = x,y,s)$$

where

$$E'_{11,22} = \frac{E_{11,22}}{1-\nu_{12}\nu_{21}}.$$

By assuming the deflection in the next form and substituting the eqs. (2-3),(2-5), and (2-8) into the basic equation (2-1), and solving for the displacements u and v, the next solutions are obtained.

$$w_{as} = C_{mn}\sin\lambda_m x\sin\lambda_n y \qquad (2\text{-}8)$$

where

$\lambda_m = \dfrac{m\pi}{a}$, $\lambda_n = \dfrac{n\pi}{b}$ (a,b : length of plate in x,y directions respectively)

$$u = C_1 x - \frac{C_{mn}^2}{16\lambda_m}\left[\lambda_m^2(1-\cos 2\lambda_n y) - \frac{A_{xy}}{A_{xx}}\lambda_n^2\right]\sin 2\lambda_m x$$

$$v = C_2 y - \frac{C_{mn}^2}{16\lambda_n}\left[\lambda_n^2(1-\cos 2\lambda_m x) - \frac{A_{xy}}{A_{yy}}\lambda_m^2\right]\sin 2\lambda_n y \qquad (2\text{-}9)$$

Boundary conditions for the forces

$$\int_0^b N_x dy = bN_x^0 \quad \text{at } x=0,\ x=a$$

$$\int_0^a N_y dx = aN_y^0 \quad \text{at } y=0,\ y=b \qquad (2\text{-}10)$$

give next relations.

$$C_1 = \frac{(A_{xy} N_y^0 - A_{yy} N_x^0)}{A_{xx}A_{yy} - A_{xy}^2} - \frac{1}{3}\lambda_m^2 C_{mn}^2$$

$$C_2 = \frac{(A_{xy} N_x^0 - A_{xx} N_y^0)}{A_{xx}A_{yy} - A_{xy}^2} - \frac{1}{3}\lambda_n^2 C_{mn}^2 \qquad (2\text{-}11)$$

A potential energy of the system is expressed as

$$\Pi = \frac{1}{2}\int_0^a\int_0^b (N_x\varepsilon_x + N_y\varepsilon_y + N_{xy}\varepsilon_{xy} + M_x\kappa_x + M_y\kappa_y + M_{xy}\kappa_{xy})dxdy$$

$$- \left[N_x^0\int_0^a \left(\frac{\partial u}{\partial x}\right) dx + N_y^0\int_0^b \left(\frac{\partial v}{\partial y}\right) dy \right] \qquad (2\text{-}12)$$

Rewriting the loads and strains in terms of displacements and substituting the assumed deflections and the displacements in the form of eq.(2-8) and adopting the Principle of minimum potential energy, the coefficient C_{mn} is determined as follows.

$$C_{mn}^2 = \frac{8\lambda_m^2(N_x^0 - N_{mn})}{\left[\frac{C}{2A_{xx}A_{yy}} (A_{yy}\lambda_n^4 + A_{xx}\lambda_m^4) + \frac{25}{9} (A_{yy}\lambda_n^4 + A_{xx}\lambda_m^4 + 2A_{xy} \lambda_n^2\lambda_m^2)\right]}$$

where

$$N_{mn} = \lambda_n^2\left(D_{xx}\frac{\lambda_m^2}{\lambda_n^2} + 2D_{xy} + 4D_{ss} + D_{yy}\frac{\lambda_n^2}{\lambda_m^2} \right) \qquad (2\text{-}13)$$

2.2 VISCOELASTIC PROPERTIES OF THE COMPOSITE

The viscoelastic properties of the composite laminates can be expressed with those of unidirectional composite i.e., J_{11}, J_{22}, J_{66}, and ν_{12}. These properties are determined through a series of experiments in a separate study. From the experimental data, time-temperature dependent constitutive equations of the composite are obtained in the form of exponential series or generalized Voigt model as follows.

$$J_{ij}(T,t) = C_{ij}^{(0)}(T) + \sum_{k=1}^{n} C_{ij}^{(k)}(T)\exp(\mu_k(T)t) \tag{2-14}$$

$$\nu_{ij}(T,t) = D_{ij}^{(0)}(T) + \sum_{k=1}^{n} D_{ij}^{(k)}(T)\exp(\gamma_k(T)t) \tag{2-15}$$

where C_{ij}, D_{ij}, μ_k and γ_k are coefficients dependent on the temperature and fiber orientations.

In the above equations, J_{ij}'s are compliances and ν_{ij}'s are Poisson's ratios of the composite. To express the experimental data of the anisotropic composite properties in the above forms, Sequential Unconstrained Minimization Technique (SUMT) is used. Curve fitting for a longer time range requires the introduction of several new features in the solution procedure, including (1) use of sufficient number of terms, (2) introduction of additional constraint conditions, and (3) selection of proper initial values. The SUMT method is employed to solve the problem efficiently with these three requirements. The procedure has demonstrated to minimize the discrepancies between experimental data and the exponential series solution under numerous constraint conditions. Details of the SUMT can be found in the reference[1]. The results of the curve fitting are shown in Figs.(2-1) to (2-3) and Tables (2-2) to (2-4) with the basic properties of the composite at room temperature in Table (2-1).

2.3 VISCOELASTIC CREEP POST BUCKLING SOLUTIONS

The elastic solution in the form of end shortening versus applied load is modified to viscoelastic response to a

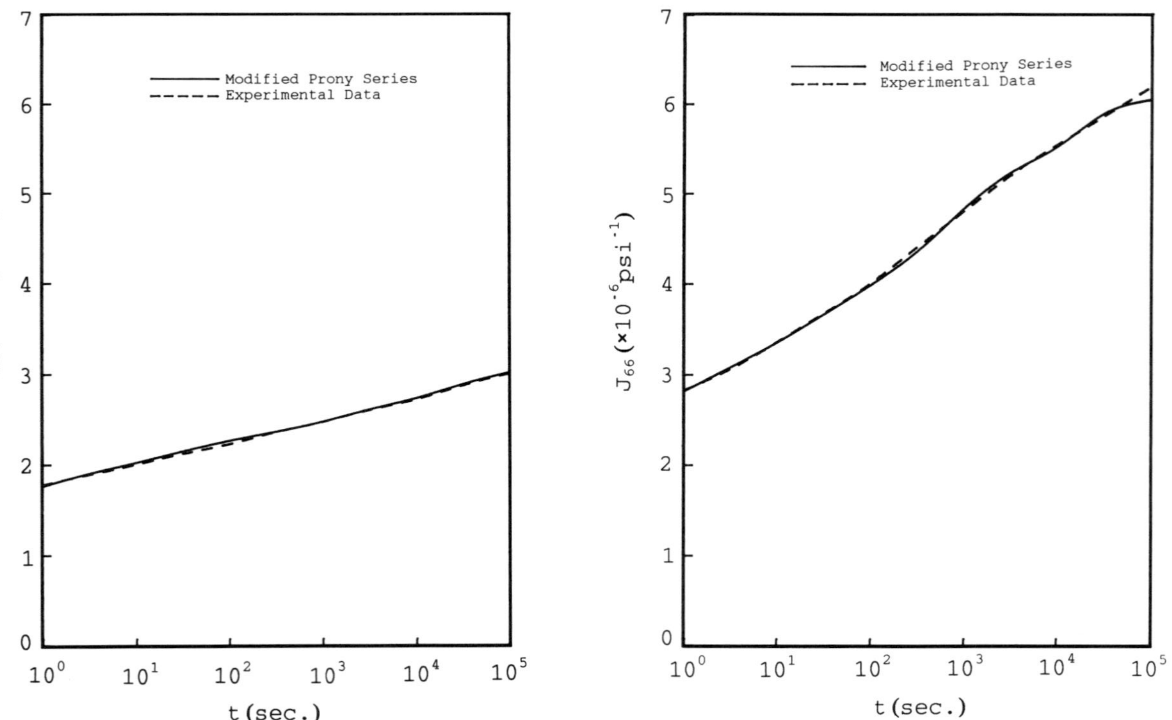

Fig. 2-1. High-Temperature Creep Experimental Data and the Modified Prony Series Solution for AS4/J1 Thermoplastic Composite at T=145 C.

Fig. 2-2. High-Temperature Creep Experimental Data and the Modified Prony Series Solution for AS4/J1 Thermoplastic Composite at T=145 C.

Table 2-1 Properties of Fiber and Matrix.

	AS4 Fiber	J1 Polymer
Tensile Strength (GPa)	3.79	69.0
Tensile Modulus (GPa)	234	2.21
Tensile Failure Strain (%)	1.5	5
Density (g/cm^3)	1.80	1.04

Table 2-2 Various Terms in Modified Prony Series for Composite Creep Compliance J_{22} at T=145°C.

k	$C_{22}^{(k)}$ (Mpa^{-1})	μ_k
0	4.40336 E-4	
1	-2.10554 E-5	-2.51442
2	-1.66286 E-5	-1.27771
3	-1.33932 E-5	-7.36571 E-1
4	-9.76488 E-6	-3.16115 E-1
5	-8.68564 E-6	-1.56968 E-1
6	-9.74617 E-6	-8.36424 E-2
7	-1.23113 E-5	-4.53550 E-2
8	-1.48929 E-5	-2.20002 E-2
9	-1.65546 E-5	-7.45779 E-3
10	-1.72521 E-5	-6.62765 E-4
11	-1.74391 E-5	-1.06532 E-3
12	-1.77117 E-5	-1.60279 E-4
13	-1.78031 E-5	-2.86918 E-5
14	-1.78031 E-5	-2.86918 E-5
15	-1.78176 E-5	-2.86918 E-5

Table 2-3 Various Terms in Modified Prony Series for Composite Creep Compliance J_{66} at T=145°C.

k	$C_{66}^{(k)}$ (Mpa^{-1})	μ_k
0	8.78526 E-4	
1	-2.90565 E-5	-2.50119
2	-2.40482 E-5	-1.27583
3	-2.10859 E-5	-7.35502 E-1
4	-1.94604 E-5	-3.15107 E-1
5	-2.18587 E-5	-1.55352 E-1
6	-2.75514 E-5	-8.06181 E-2
7	-3.52814 E-5	-3.99268 E-2
8	-4.19311 E-5	-1.45174 E-2
9	-4.48006 E-5	-3.26728 E-3
10	-4.52936 E-5	-1.37041 E-5
11	-4.53792 E-5	-5.62197 E-5
12	-4.53690 E-5	-5.62197 E-5
13	-4.53386 E-5	-5.62197 E-5
14	-4.53096 E-5	-7.79696 E-4
15	-4.53023 E-5	-1.15559 E-3

Table 2-4 Various Terms in Modified Prony Series for Poisson's ratio ν_{12} at T=145°C.

k	$D^{(k)}$	γ_k
0	4.9051 E-1	
1	-1.0151 E-2	-2.49794
2	-6.7556 E-3	-1.27491
3	-5.3963 E-3	-7.34993 E-1
4	-3.6905 E-3	-3.14815 E-1
5	-2.0699 E-3	-1.55492 E-1
6	-1.4857 E-3	-8.28316 E-2
7	-4.7518 E-3	-4.71662 E-2
8	-8.3283 E-3	-2.80405 E-2
9	-1.0999 E-2	-1.58641 E-2
10	-1.2835 E-2	-5.16678 E-3
11	-1.5104 E-2	-4.04642 E-5
12	-1.5977 E-2	-3.41099 E-4
13	-1.6218 E-2	-4.04642 E-5

* E_{11} is 17.5 Msi and time-independent.

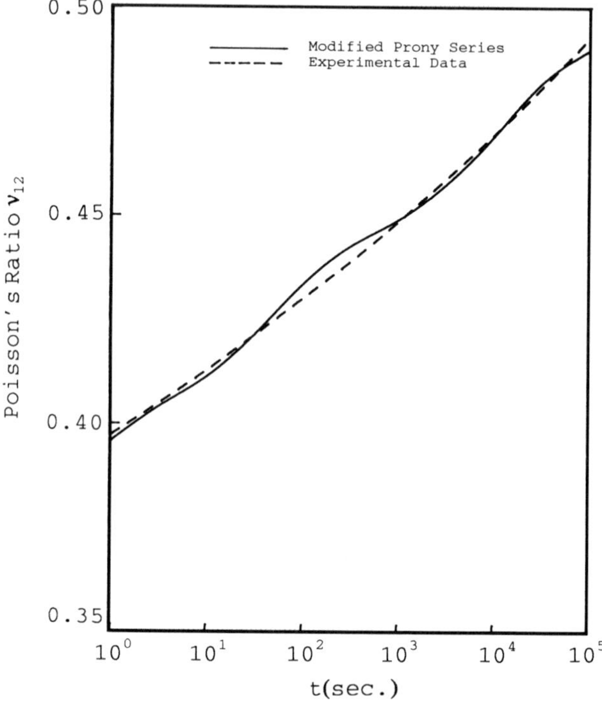

Fig. 2-3. High-Temperature Creep Experimental Data and the Modified Prony Series Solution for AS4/J1 Thermoplastic Composite at T=145 C.

step input of the load by directly substituting the viscoelastic properties of the composite in eqs.(2-14) and (2-15) To extend the solution to a much generalized form corresponding to time-dependent load input, Duhamel integral is adopted. If a response to a step input is written in next form,

$$e = \phi(\sigma, t) \tag{2-16}$$

then, a generalized solution for any input is expressed in the next form using the convolution integral.

$$e(t) = \int_0^t \frac{\partial \phi}{\partial \sigma}(\sigma, t-\tau) \frac{d\sigma(\tau)}{d\tau} d\tau \tag{2-17}$$

The above theory is based on Boltzmann's superposition principle and valid for any linear materials.

3 RESULTS AND DISCUSSIONS

Figure (3-1) shows solutions in the form of end-shortenings of the unidirectional composite panel versus biaxially applied load with time and biaxiality as parameters. The curves in this figure represent responses to step load inputs. Therefore, a horizontal section at given load in the figure shows growth of end-shortening with time when the load is suddenly applied at t=0. The cusps in the figure represent buckling points.

Figure (3-2) shows solutions for ±45 degree angle ply similarly to Fig.(3-1). In this case the composite panel shows surpassing load bearing capability after the buckling provided the end-shortenings are representative of the performance although the degradation of it is much significant affected by the shear modulus which is also time-dependent (Fig.(2-3)).

Solutions for the unidirectional composite panel subject to transverse loading only are shown in Fig.(3-3). The load

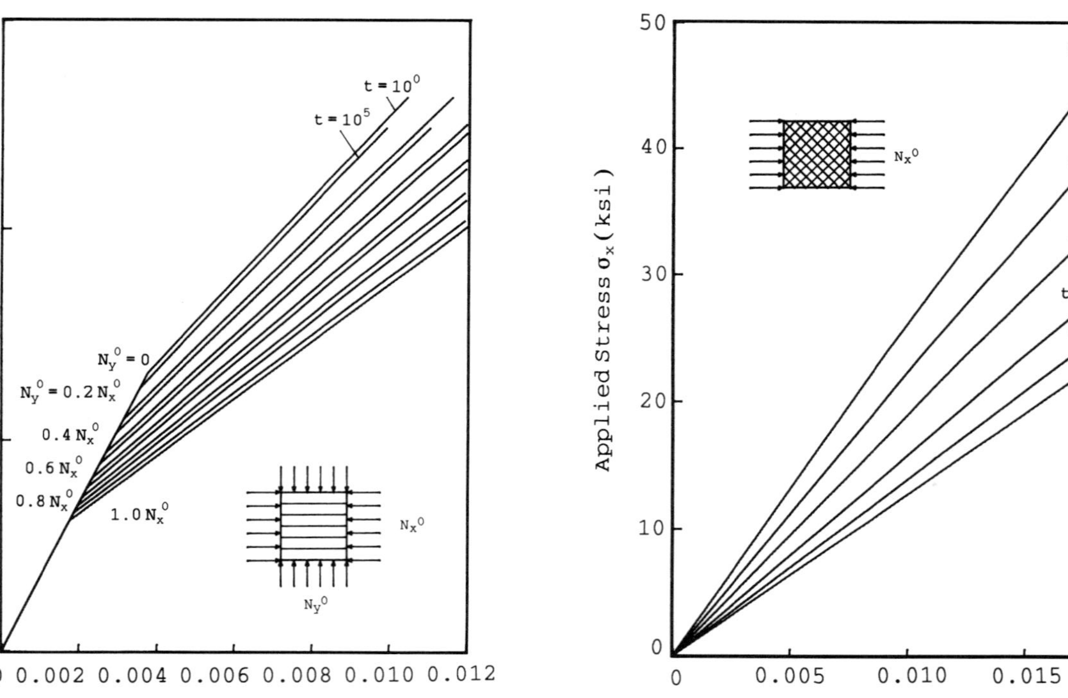

Fig.3-1. Creep Post Buckling Failure of AS4/J1 Thermoplastic-Matrix Composite Panel under Biaxial Compression of Various Stress Biaxiality Ratios at T = 145C.

Fig.3-2. Creep Post Buckling Failure of AS4/J1 Thermoplastic-Matrix Composite Angle Ply under Axial Compression at T = 145C.

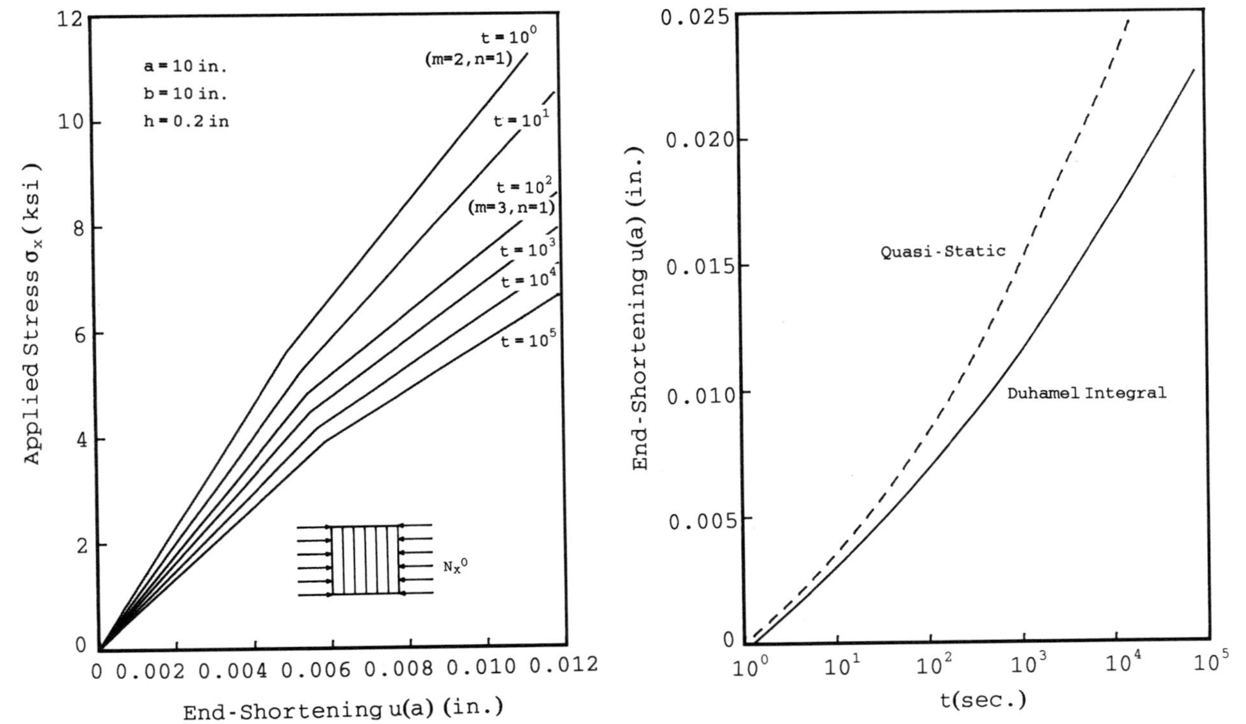

Fig.3-3. Creep Post Buckling Failure of AS4/J1 Thermoplastic-Matrix Composite panel under Axial Compression at T = 145C.

Fig.3-4. End shortening of AS4/J1 Thermoplastic-Matrix Composite Angle Ply at T = 145C subject to Logalithmically Linear Compression $N_x^0(t) = n_x^0 \log(t)$ $(n_x^0 = N_{mn}/3)$.

bearing capability is not high because the panel is compressed in a weak direction. Moreover, the growth of end-shortenings drastically change with the change of buckling mode. Thus, the mode change can happen when the transverse compressive load is large compared to the longitudinal one.

Figure (3-4) is an example solution for general load input which is a function of time. In this case, a logarithmically linear load input which reaches the elastic buckling load at t = 10^3 sec. ($N_x^0 = (N_{mn}/3)\log(t)$) is employed to demonstrate the Duhamel integral. The dotted line represents the quasi-static solution obtained by substituting the material properties and the load at the instant directly into the elastic solution. The solid line in the figure represents the solution obtained by the convolution integral, which gives rather small displacement than the quasi-static solution because of contribution of the material up to the moment. The curve shows a moderate change of slope at t = 10^3 sec.

4. CONCLUSION

Based on the analytical methods developed for studying the creep postbuckling of fiber composites and the results obtained for the AS4/J1 thermoplastic-matrix composite panels subject to biaxial compression, the following conclusions are made in this study.

1. In the case of angle ply, the degradation of load bearing capability with time is significant affected by the time-dependent shear modulus.
2. In the case of transverse compression, or in the case where large transverse component appears, the mode change during the deformation can greatly degrade the performance of the composite.
3. Duhamel integral is performed numerically utilizing the creep postbuckling step response of the composite panel and it reveals obvious discrepancy with the quasi-static solution.

5. REFERENCES

Azikov, N. S., and Vasil'ev, V. V., 1986, "Stability and Above-Critical Behavior of Compressed Composite Panels," *Izv. AN SSSR. Mekhanika Tverdogo Tela*, Vol. 21, No. 5, pp. 152-158.

Booker, J., R., et al., 1974, "Stability of Viscoelastic Structural Members," *Civil Engng. Trans.*, p.45.

Fiacco, A., V., and McCormick, G., P., 1968, Nonlinear Programming Sequential Unconstrained Minimization Techniques, John Wiley & Sons. Inc., New York, NY.

Flugge, W., 1967, *Viscoelasticity*, Blaisdell Pub. Co., Waltham, MA, P.92.

Hashin, Z., 1979, "Analysis of Properties of Fiber Composites with Anisotropic Constituents,", *J. of Applied Mech.*, Vol. 46, p.543.

Lekhnitskii, S., G., 1968, *Anisotropic Plates*, Gorden and Breach Science Publishers, New York, NY, p.382, p.437.

Massonnet, Ch., 1974, "Buckling Behavior of Imperfect Elastic and Linearly Viscoelastic Structures," *Int. J. of Solid and Structures*, Vol. 10, p.755.

Miyase, A., Chen, A., W.-L., Geil, P., H., and Wang, S., S., 1986, "Anelastic Deformation and Fracture of Thermoplastic-Matrix Fiber Composite at Elevated Temperature., Part II Unidirectional Composite Laminate," American Chemical Society 20th. Great Lake Regional Meeting, Milwaukee, June.

Nakajo, Y., and Wang, S., S., 1988, "Elevated-Temperature Creep Buckling of Thermoplastic-Matrix Fiber Composites Under Biaxial Loading," Proc. of 4th. Japan-US Conference on Composite Materials, Technomic Publishing Company, Inc., New Holland, PA, p.696.

Powell, P., C., 1969, "Principle for Using Design Data," *Thermoplastics*, R. M. Ogorkiewicz (editor), CRC Press, Cleveland, OH, p.211

Wang, S., S., Klang, E., C., Miyase, A., and Dasgupta, A., 1987, "Viscoelastic Shear Behavior of the Thermoplastic Matrix Composite at Elevated Temperatures," ASME, Winter Annual Meeting, Boston, December.

Williams, 1973, J., G., *Stress Analysis of Polymers*, Longman Group Limited, London, U.K. p.148.

Asymmetrical Growth of Edge Delaminations in CFRP Tensile Specimens

Peter A. Klumpp and Eckart Schnack

Institute of Solid Mechanics, Karlsruhe University (FR Germany)

Quasistatic tension tests were carried out on special delamination specimens made from carbon fiber / epoxy using a PTFE foil as a start delamination. The development of edge delaminations can easily be observed by two modern optical techniques (shearography and reflection grating technique) which are briefly described.
The observation of pointsymmetrical growth patterns indicates a micromechanical fracture mechanism of angleply/crossply ($\pm\theta$ / 90°) interfaces at the specimen edges. It is active only for one of the two delamination tips which is arrested for certain small strain intervals. The result suggests that edge delamination growth cannot be described sufficiently on a macromechanical level.

1. Introduction

A substantial part of fracture mechanical research in the field of carbon fiber/epoxy laminates has been dealing with the ply separations known as delaminations (Pagano and Pipes 1973, Reifsnider et al. 1977, O'Brien 1984). If this damage mode appears in practice, e.g. after buckling deformation, a catastrophic loss of bending stiffness may result on the respective machine part.

The effect of edge delamination in a tensile specimen can be demonstrated in a quite spectacular way, but such experiments are generally not representative for the conditions under which technical CFRP structures are used and fatigued. This refers to the geometrical form of the structures, to the stacking sequencies and to the loading conditions.

On the other hand, delamination plays an important role in the typical damage process which combines different fracture modes (Reifsnider et al. 1983). The analysis of the real fracture process is not only complicated by the statistic features of the single modes, but also especially by their interactions. A satisfactory description will have to comprise both micro- and macromechanical aspects of the inhomogeneous fiber/matrix composite; for example, it should be able to describe the interaction of a proceeding delamination front with a matrix crack that has developed independently in one of the adjacent plies. In view of the complexity of this problem, the analysis of simplified model systems for the different fracture modes becomes important. In the case of edge delamination, they allow fundamental insights in nucleation and development of the ply separation.

In order to keep the interaction with the preceding fracture modes in the specimen as small as possible, the specimen structure has to be chosen in a way that prefers the transformation of the external load into the driving force for the selected fracture mode. Edge delaminations are favoured in stacking sequencies that develop high interlaminar stress concentrations at the edges of a tensile specimen (Pipes and Pagano 1970, Pinchas and Pian 1979, Rohwer 1982, Herakovich et al. 1985) ; the simplest case is a pure mode-1 fracture in the midplane of a symmetrical specimen due to a high tensile stress concentration ('peeling stress') under tensile load of the specimen. Matrix cracking as a concurrent mode can be restricted to high longitudinal strains using stacking sequencies with small orientation angles (e.g. $|\theta| \leq 30°$).

2. Specimen design

A stacking sequency which is optimum for delamination experiments has already been discussed in detail (Pagano and Pipes 1973). The construction principle is the combination of laminae with great and with small Poisson numbers ν known as 'Poisson mismatch' in a symmetrical balanced stacking sequency. These laminates are orthotropic in the specimen axis system, i.e. they show neither torsion nor shearing under longitudinal stress.

To give an impression of the values of ν realizable with a typical carbon fiber/epoxy system, fig. 1 shows the function $\nu(\theta) = -\varepsilon_x / \varepsilon_z$ for

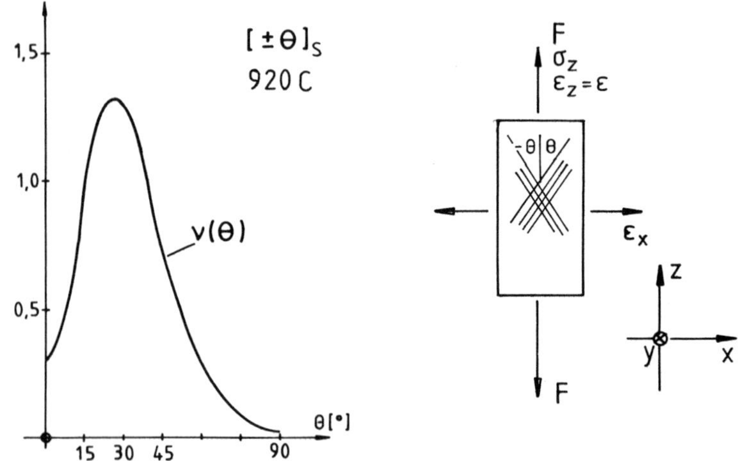

Fig. 1
Poisson number for the laminates of the stacking sequency $[\pm\theta]_s$

the laminates of the group $[\pm\theta]_s$. We have calculated the curve using classical laminate theory (Jones 1975) for the case of pure longitudinal stress in z-direction. The elastic moduli of the unidirectional laminae stem from our measurements on the CFRP system 920 C-TS-5-34 manufactured by CIBA-GEIGY (E_1 = 117 GPa, E_2 = 7.6 GPa, $G_{12} \cong 5$ GPa, ν_{12} = 0.34). The specimens used in the experiments were produced from this prepreg which assumed a thickness d_L = 0.143 mm of the single lamina after curing at 125° C.

It is obvious from fig. 1 that the Poisson mismatch will be maximum in a laminate consisting of cross plies (θ = 90°) and of angle plies with $\pm\ \theta$ = 25°...30°. The ratio of the Poisson numbers falls to a minimum for $\nu(90°)/\nu(27°) \cong 2 : 130$. If the cross plies are placed at the midplane and the angle plies on the outsides, the laminate shows a positive stress concentration near the edges in the midplane under tension; the cleaved halves will curve outward after separation (Pagano and Pipes 1973). It can be shown in the frame of classical lamination theory that this curvature is proportional to the longitudinal strain if the curvature is not restrained by the mechanical boundary conditions; the proportionality constant can be used for estimates of the relative delamination tendency of different stacking sequencies.

An optimum thickness ratio of both components in the stacking sequency has been estimated for θ = 25° (Harris and Orringer 1979) ; for production technical reasons we have chosen a nearly optimum 10-ply laminate of the stacking sequency $[(\pm30)_2\ 90]_s$ with a longitudinal modulus of E_z = 50.3 GPa for our experiments. It must be noticed that the matrix cracks will appear on the first damage stage (Reifsnider et al. 1983) at low strain levels in the 90°-plies of this laminate; they may influence the following delamination process.

Subjected to quasistatic tension, the chosen laminate shows spontaneous complete edge delamination at a longitudinal strain of $\varepsilon_c \cong$ 0.85%. ...0.90%.; the index c stands for complete delamination. Our estimate for the strain interval between nucleation and complete development is $\Delta\varepsilon \leq$ 0.05% based on the observation of the 'delamination nuclei' which were predicted with the energy release rate concept (Wang and Crossman 1980, see also chapter 5 below).

The process can be expected to proceed more continuously for dynamical loading (O'Brien 1982). In order to keep the loading apparatus simple

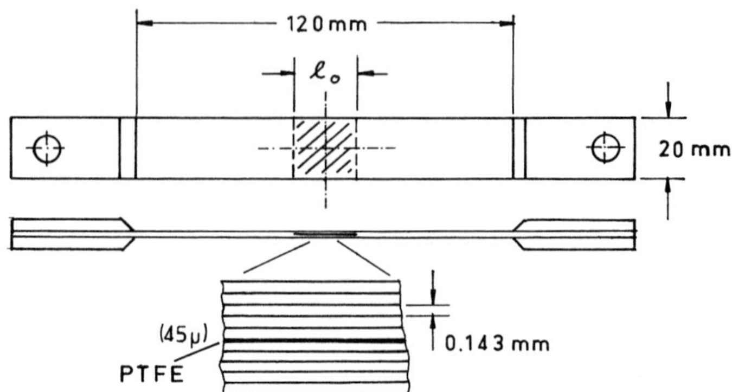

Fig. 2
Specimen design for the delamination experiments

and to achieve short testing times we have used a PTFE ('Teflon') foil as start delamination in the midplane of the laminate (Wang and Slomiana 1982, Wang et al. 1985). These authors noticed that the small strain interval between nucleation and complete development of edge delaminations is both shifted to lower strain levels and enlarged by a symmetrical start delamination on both specimen edges. Moreover, only one delamination develops on each edge. We have used foils of 45 μm thickness and lenghts ℓ_0 = 5 mm, 10 mm or 20 mm in specimens of width b = 20 mm (see fig. 2).

The start delaminations begin to grow visibly at $\varepsilon \cong 0.3\%...0.4\%$ and reach the regions where their further growth is restricted by the specimen end tabs at $\varepsilon_c \cong 0.60\%...0.75\%$.; the mean value for ε_c increases for decreasing ℓ_0. The crack tips leave the influence region of the PTFE insert above $\varepsilon \cong 0.45\%$, hence a strain interval of $\Delta\varepsilon \cong 0.15\%...0.3\%$ length becomes accessible in which a relatively undisturbed growth of edge delaminations can be observed under quasistatic tension.

3. Interferometric defect detection

The surface deformation on the damaged tensile specimens can be measured in form of the out-of-plane displacement field w(x,z) or its derivatives. Several coherent and incoherent optical techniques are available for displacement, inclination and curvature measurements; holographic interferometry (Vest 1979), speckle interferometry (Jones and Wykes 1983) and shearography (Hung and Taylor 1974, Hung and Liang 1979) are coherent optical methods using the interference of laser wavefronts, while moiré techniques (Sciamarella 1982) use the superposition of material line structures with pitches down to the order of a light wavelength (Post 1983). The reflection grating method (Ritter and Hahn 1983, Andresen et al. 1985) is related to the classical moiré techniques and works with diffuse incoherent white light.

The common result of these interferometric techniques is a fringe pattern with isolines of the measured variable. The effort for the quantitative evaluation of this picture can be quite high. Useful methods are for example the extraction of moiré lines by suppression of the carrier frequency (De Haas and Loof 1966) and image digitalization with subsequent digital processing (Kreis 1980).

A direct mesurement of $w(x,z)$ has the disadvantage that global deformations of the specimen like torsion or bending will appear superposed to the local deformation field of the damaged region. They can be caused by imperfect realization of the clamping conditions or an asymmetric growth of the damage state. This effect reduces measurement precision for shape and position of the damage zone, i.e. the position of the delamination front. It can be reduced if the slowly variable global terms are differentiated out in interferograms of the inclination or curvature fields of the object surface. Nevertheless, great care is necessary in construction and realization of the tension loading apparatus for optical NDT purposes in order to keep the global deformations as small as possible. We have found a solution by clamping one specimen end directly to the machine frame, while the other one is pulled in a clamp guided by means of a rugged precision translator stage that is mounted to the frame (fig. 6 below shows this machine).

We have concentrated on two inclination measurement techniques measuring $\partial w/\partial x$ in our delamination experiments. They avoid the disadvantage of a displacement measurement mentioned above. A second differentiation (i.e. a curvature measurement) seems not to be necessary according to our experimental results; it would be possible with a multiaperture interferometer (Sharma et al. 1984, Sharma et al. 1986).

The reflection grating technique indicates object surface inclinations in real time in form of distortions of the virtual image of a material grating with parallel black and white lines. This image is formed by reflection at the object surface. The surface must of course be specularly reflecting in this case; this can be achieved on CFRP specimens by application of a thin epoxy film which is the replica of a flat glass surface (Schütze 1985). The results of this procedure depend critically on the preparation of the glass surface with a suitable parting compound like PTFE spray.

The shearographic measurement technique (Hung and Liang 1979) uses an involved storage and reconstruction process for information encoded in the stochastic interference patterns appearing in diffusely reflected laser light (speckle patterns, see Dainty 1984). An efficient white light reconstruction process has been described recently (Klumpp 1989, Klumpp and Schnack 1989). Up to now, holographic film with a resolution of more than 1000 lines/mm served as storage medium.

Like the other coherent optical techniques, shearography is very sensitive to in-plane object motions which ruin the contrast of the interferogram and must be restricted to less than some 10 µm. This disadvantage against the reflection grating technique necessitates a testing apparatus of higher precision, but on the other hand a coat of diffuse white paint on the specimen surface is sufficient instead of an epoxy film with smooth surface.

The resolution of both techniques is different in practical applications. Fig. 3 shows the reflection grating setup, fig. 4 its realization.

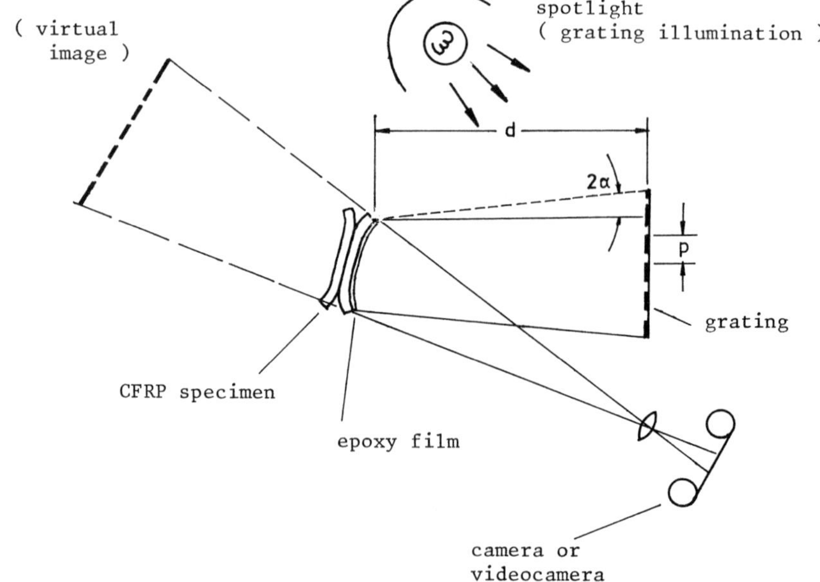

Fig. 3
Setup for the reflection grating method

Fig. 4
Realization of the setup shown in fig. 3

If the grating constant (distance dark line / dark line) is denoted by p and the distance object surface-grating by d, a fringe shift of N = 1 periods indicates an inclination angle α_1 of the object surface defined by

$$\tan 2\alpha_1 = p / d \tag{1}$$

where $|\alpha| << 1$ allows the approximation

$$\alpha_1 \cong p / (2d) \tag{2}$$

With the values p = 0.65 mm and d = 165 mm used in the experiments the angle α_1 corresponding to a fringe order shift of N = 1 results with about 2 mrad (≅ 0.1°).

The corresponding formula for shearography can be derived from the relation between interference phase and displacement vector valid for holographic interferometry (Hung and Taylor 1974). Fig. 5 shows the shearographic camera, fig. 6 again its realization together with a suitable tension testing apparatus. We can derive the following equation from the basic relation between interference phase and measured variable $\partial w/\partial x$ (Hung and Taylor 1974) under assumption of pure out-of-plane displacement of a plane object:

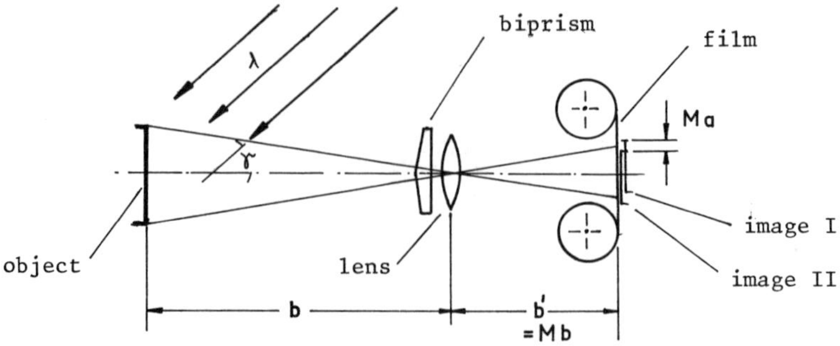

coherent light illumination
λ
biprism
film
Ma
γ
object
lens
image I
image II
b
b'
=Mb

Fig. 5
Shearographic camera setup

Fig. 6
Realization of fig. 5, with high precision tension testing apparatus on the left (10 kN)

$$\alpha_1 = \lambda \,/\, (\, a(\, 1 + \cos\gamma\,)\,) \tag{3}$$

where a denotes the image translation (= 'shearing') measured in object coordinates, λ the wavelength of the monochromatic laser light and γ the incident angle of the object illumination. For typical values of γ ($|\gamma| \le 30°$) the above formula can be approximated well by

$$\alpha_1 \cong \lambda \,/\, (2a) \tag{4}$$

For our experimental parameters of a = 3 mm and λ = 0.633 μm, α_1 becomes about 0.1 mrad (≅ 0.005°). The inclination resolution was hence higher for shearography by a factor of 20.

Eqs. (2) and (4) for the resolvable inclination of the two techniques do not indicate physical limits for α_1. The resolution of the reflection grating technique can be decreased arbitrarily ($p \rightarrow \infty$), but an upper limit is set by the threedimensional distortion of the virtual image. It can only be registered as a sharp real image with a registration system (eye, camera) of sufficient depth of focus, i.e. with a system of small optical resolution. Therefore, the registration system for the interferogram sets a lower limit for the ratio p/d in eqn. (2). The resolution in shearography can be varied with the image shearing a. If a is chosen too large, the first order approximation used for the derivation of the measured variable $\partial w/\partial x$ (Hung and Taylor 1974) is no longer valid and the interferogram cannot be interpreted as an inclination map. On the other hand, the shearing Ma in the image plane where M denotes the image scale of the shearographic camera must be kept greater than the diameter of the speckles (Dainty 1984) generated in this plane.

These resolution restrictions suggest the use of both techniques according to the magnitude of the measured inclination fields. Shearographic measurements were therefore performed at the inclination difference resulting from a tiny strain increment of $\Delta\varepsilon$ = 0.006%...0.02% starting at the actual ε value in the tension test. The reflection grating technique uses the flat initial state of the specimen as reference and measures hence the absolute inclination of damaged regions at longitudinal strains of ε = 0.3%...1%; the resulting image is visible in real time, but its quantitative evaluation is difficult.

The double exposure shearograms must be chemically developed before reconstruction, but the resulting interferograms can be evaluated directly in terms of lines of constant inclination change. A reflection grating technique using double exposure which shows lines of equal inclination as moiré fringes is also possible (Ritter and Hahn 1983). The problem with this technique is that the deformed grating lines in the two virtual images must be well-focused in order to yield an evaluable moiré superposition; furthermore, the measuring range is limited by the restriction of the moiré line density to a fraction of the grating line density 1/p (see fig. 1).

4. Results

Fig. 7 shows a series of reconstructed shearograms from a quasistatic delamination test, fig. 8 the same experiment monitored with the reflection grating technique. In both cases, the growth of the two mirror-symmetrical initial defects could be observed clearly. Deviations from the initial symmetry to the x and z-axis (see fig. 2) were first explained by statistical influences of the material microstructure on the fracture process. The diagram in fig. 9 shows the delamination lengths $\ell_D(\varepsilon)$ measured on shearograms for three specimens, i.e. six delaminations. All curves start at $\ell_D(\varepsilon{=}0) = \ell_0 = 20$ mm defined by the dimensions of the PTFE insert; a precision of ± 0.5 mm was assumed for the position measurement of the delamination tips on the specimen edges. Delaminations with $\ell_D \geq 60$ mm were defined as total delaminations for which a common value of $\ell_D = 60$ mm was assumed. The interval of free delamination growth is about $0.45\% < \varepsilon < 0.60\%$. The diagram shows clearly statistical deviations between the individual delamination processes.

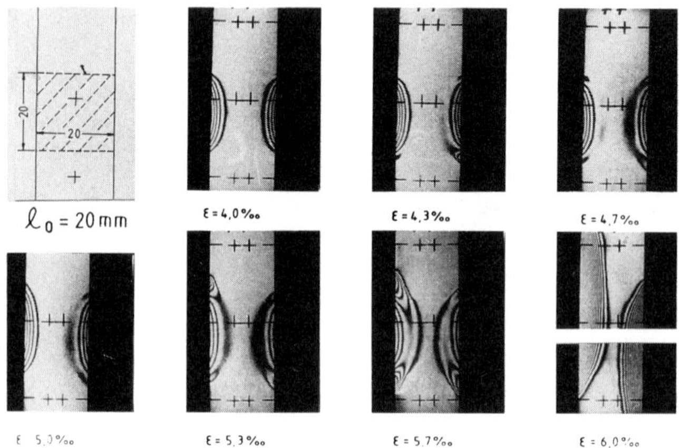

Fig. 7
Delamination growth under quasistatic tension
(shearographic observation)

Apart from these deviations, the scheme shown in fig. 10 gives a qualitative characterization of our experimental observations averaged over 10 tension tests. The point symmetrical growth pattern was already observed just before complete delamination in figs. 7 and 8; fig. 11 shows an example for the typical drop shape at an advanced delamination state ($\varepsilon = 0.57\%$). The different ℓ_D values of the two delaminations (49 mm and 33 mm, respectively) illustrate the statistical deviations plotted in fig. 9.

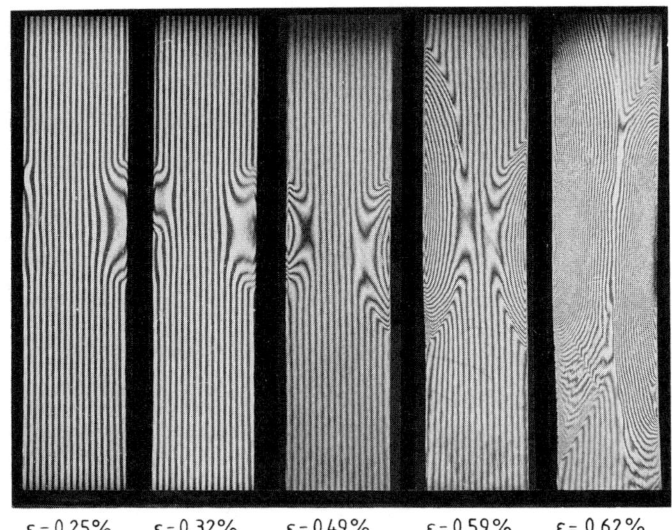

Fig. 8
Delamination growth under quasistatic tension
(reflection grating observation)

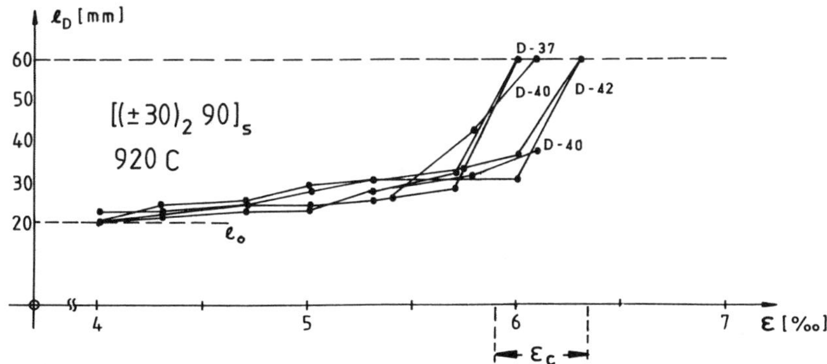

Fig. 9
Delamination length as function of longitudinal strain for three specimens

A striking asymmetry is visible on the photo of the edge of a specimen after complete delamination (fig. 12); the edge was painted white before the tension test in order to improve contrast. Near the start delamination visible at the center of fig. 12, the laminate has split into two 5-ply parts $[(\pm 30)_2\ 90]$ and $[90\ (\mp 30)_2]$. Outside this start region, the delamination develops from a pure mode-1 crack at the (90°/90°)-interface into a mixed-mode crack at the two (-30°/90°)-interfaces which are obviously preferred by energetical reasons (O'Brien 1984). The crack appears as a straigth line on the right hand side of fig. 12. We have also observed stepwise straight crack courses in this position on other specimens (see fig. 13). Contrarily, the crack wiggles between both interfaces on the left hand side of fig. 12. The growth in this direction was found to be retarded relative to the other direction (see fig. 10).

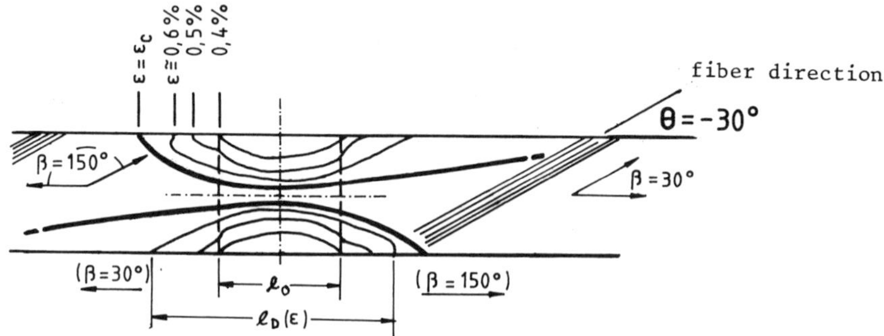

Fig. 10
Scheme for the average delamination growth

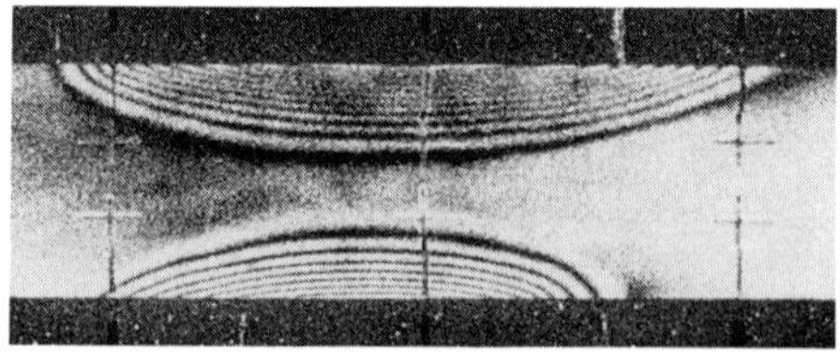

Fig. 11
Delaminations of drop shape at ε = 0.57% (shearogram)

Fig. 12
Specimen edge showing the course of a complete delamination

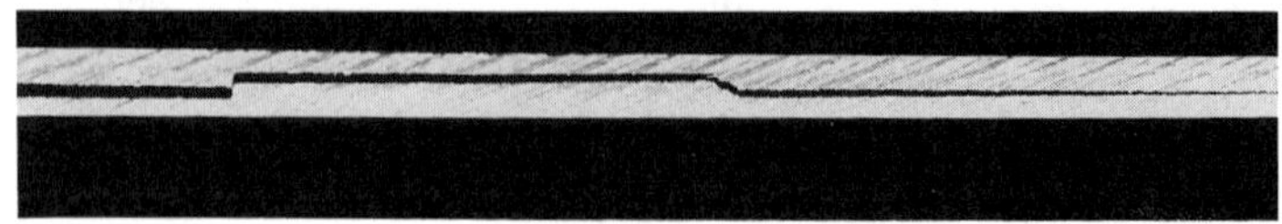

Fig. 13
Stepwise straight delamination course observed at specimen edge

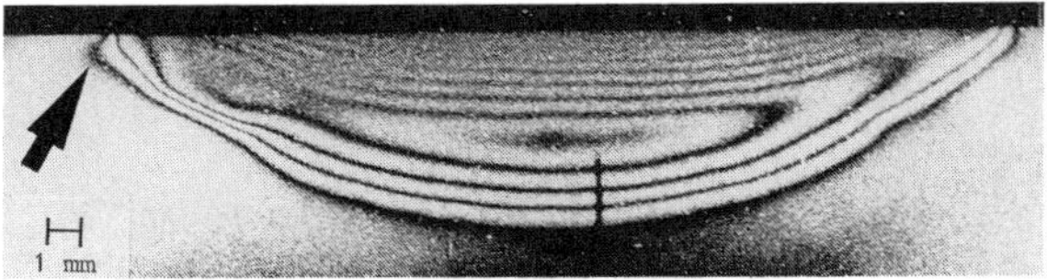

Fig. 14
Shearogram showing delamination front retardation at the specimen edge

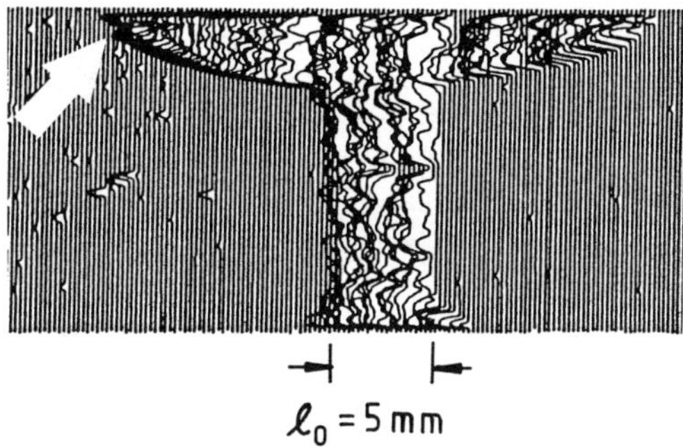

Fig. 15
C-Scan of edge delamination

We can conclude that the delamination growth in the (-30°/90°)-interfaces shows a significant asymmetry between the two growth directions along the specimen edge. It appears retarded in the direction where the fibers hit the edge under an obtuse angle (β = 150°, see fig. 10), while it develops faster in the direction where β = 30° is an acute angle.

This suggests a micromechanical interaction between crack tip and angle ply fibers at the specimen edge. Hints for a model of the mechanism came from shearographic and ultrasonic inspection of the delamination tips with β = 150°. Fig. 14 shows a shearogram of a delamination where the fringe at the left crack tip curves backward (→). This indicates that the crack has been arrested at the edge while it was able to proceed further in the region one or two millimeters below. Fig. 15 shows the C-scan of another delaminated specimen with two edge delaminations of quite different ℓ_D (courtesy Dr. H. Eggers, DLR Braunschweig, West Germany). Though the resolution is not optimum, an indentation (→) is visible in the crack front on the left hand side about 1 mm away from the specimen edge. An x-ray test with contrast enhancement of the same specimen gave no indication of this irregularity.

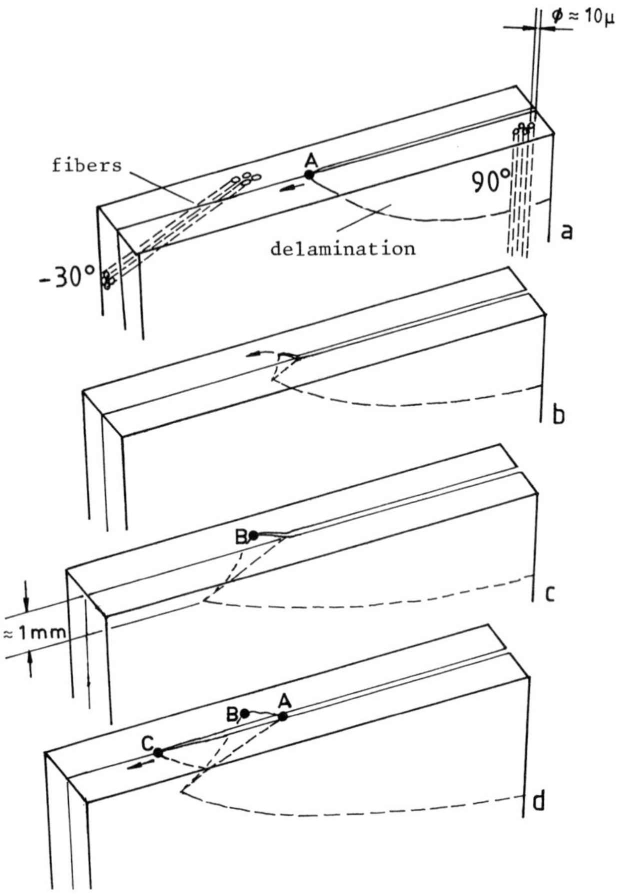

Fig. 16
Model for the crack tip arrest mechanism at a (-30°/90°)-interface

Our model for the arrest mechanism is shown in fig. 16. Obviously, the crack can penetrate into the -30°-layer from the (-30°/90°)-interface (fig. 16a) without breaking fibers if it runs in the direction with β = 150°, while it cannot for the opposite direction. After a small crack has developed along the fiber direction (figs. 16b and 16c), the tip is arrested. The stresses responsible the crack development decrease with the distance of the crack tip from the (-30°/90°)-interface (Rohwer 1982), and the crack in its form can only return to this interface by means of fiber breakage. Hence the crack development continues at the point A where the tip moved into the angle ply (fig. 16d). The stresses at this point were relieved by the crack nearby; the crack will continue only after some strain increment. This stress reduction may also be the reason for the tendency of the new tip to move to the other (-30°/90°)-interface where the same trap process is repeated (wiggled course in fig. 12). The modeled mechanism will retard the delamination development in one direction, while it is not active in the other one.

It should be added that a change between both (-30°/90°)-interfaces is easily possible for the delamination at matrix cracks of the two 90°-plies. This explains the observation of fig. 13. The effect is present in both growth directions along the laminate edge, but it cannot be observed within the wiggled course in the direction with retarded growth.

5. Experiments with other stacking sequencies

The results described above raised the question whether asymmetrical growth is also observed in the laminate $[(\pm\theta)_2\ 90]_s$ for $\theta \neq 30°$. We have tested specimens with $\theta = 15°$ and $\theta = 45°$ using the reflection grating method and found clear evidence for asymmetrical growth in both cases. This result was expected from the postulated arrest mechanism; because of strong statistical deviations of the individual fracture processes, we cannot give any estimate of the relative strength of the effect for the three cases $\theta = 15°, 30°, 45°$.

We have further tried to extend the tests to stacking sequencies that are more relevant for technical applications than $[(\pm\theta)_2\ 90]_s$. The group of symmetrical quasiisotropic laminates has often been used for fracture mechanical research (Whitcomb and Raju 1985, Masters and Reifsnider 1982, Reifsnider et al. 1977). The delamination tendency under tension of 8-ply quasiisotropic laminates depends on the exact stacking sequency. Out of the 12 possible nonequivalent sequencies, $[\pm45\ 0\ 90]_s$ shows the strongest delamination tendency and was hence chosen for tension tests with and without PTFE insert.

Fig. 17 shows the delamination of a specimen without start delamination. As in the case of the laminate $[(\pm30)_2\ 90]_s$, spontaneous complete delamination is observed. At $\varepsilon = 0.83\%$, no delamination is visible; the edge delamination that has opened after the strain was increased by $\Delta\varepsilon = 0.02\%$ to $\varepsilon_c = 0.85\%$ has grown further into the specimen at $\varepsilon = 0.89\%$. Small delamination nuclei of about 0.5 mm width (theoretical prediction by Wang and Crossman 1980) can now be seen at the right specimen edge (→). They indicate that a delamination of the right edge is imminent for further strain increments. The same stacking sequency prepared with a

PTFE insert shows continuous delamination growth (fig. 18, ℓ_0 = 10 mm). Asymmetries were not observed because of the mirror symmetry of the (0°/90°)-interfaces (β = 0° and β = 180° are equivalent).

This result suggests that the experiments described in chapter 4 can be extended to more general stacking sequencies; the start delamination realized by a PTFE insert is a useful tool to initiate a relatively continuous delamination growth under quasistatic tension.

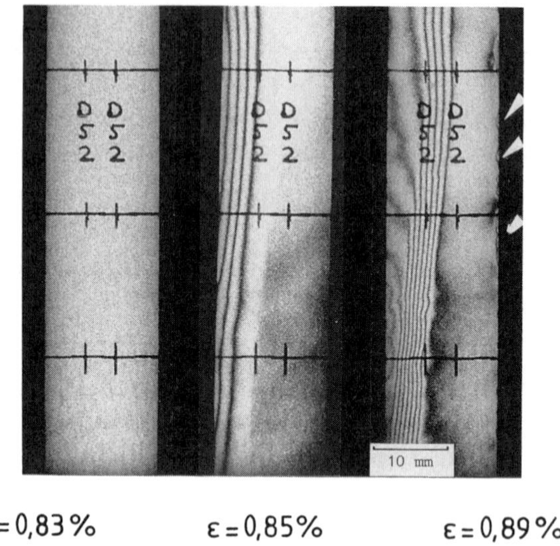

Fig. 17
Spontaneous delamination of the laminate [±45 0 90] without PTFE insert

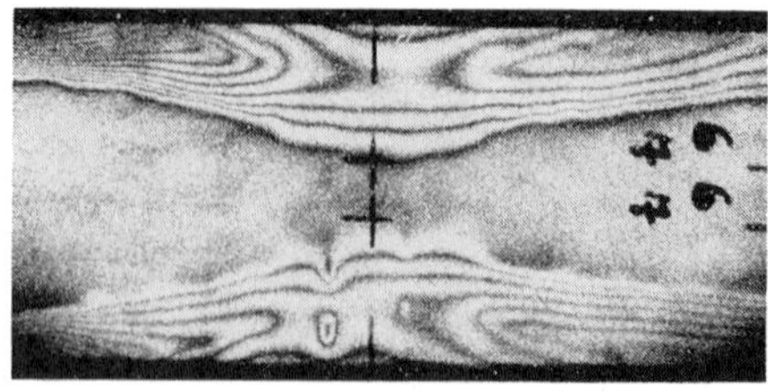

Fig. 18
Continuous delamination of the laminate [±45 0 90] with PTFE insert
($\varepsilon = \varepsilon_c$ = 0.89%)

6. Conclusion and Acknowledgements

Quasistatic tension experiments on CFRP laminates of the stacking sequency $[(\pm\theta)_2\ 90]_s$ with a PTFE insert acting as a start delamination should maintain the initial mirror symmetry in the growth pattern, if the fracture process of the anisotropic material is considered on macromechanical level. Our experimental observation of point symmetrical growth patterns in (θ/90°)-interfaces for θ = ±15°...±45° indicates a micromechanical interaction of the delamination tips with the fiber ends at the laminate edge. A model for this effect that retards the growth in one of the possible two directions at the edge has been postulated. Though this effect will be less prominent in technical applications of CFRP structures, its observation suggests that a purely macromechanical description of the real delamination process (e.g., by the energy release rate concept, Wang and Crossman 1980) will remain insufficient. Two modern optical techniques have been used for the observation of the delamination growth. Both have proven to be reliable and fast NDT tools. They can be used directly with the tension loading system, if the guidance of the specimens is sufficiently precise; shearography requires more effort in this respect than the reflection grating method.

The authors are grateful to the students M. Heizelmann, A. Majorek and H. Tomanek for their contributions to this work. The same applies to the German Research Community (DFG) for the support of important preparatory work by the contract Schn 245/2-1/2 KERB-FASER.

7. References

Andresen, K., Ritter, R., and Schütze, R., 1985, "Application of grating methods for testing of material and quality control including digital image processing," *SPIE Proceedings*, Vol. 599, pp. 251-258.

Dainty, J.C., ed., 1984, *Laser Speckle and Related Phenomena*, Springer Verlag, Berlin, Germany.

De Haas, H.M., and Loof, H.W., 1966, "An optical method to facilitate the interpretation of moiré pictures," *VDI Berichte*, Vol. 120, pp. 75-80.

Harris, A., and Orringer, O., 1979, "Investigation of angle-ply delamination specimen for interlaminar strength test," *Journ. Comp. Mat.*, Vol. 12, pp. 285-299.

Herakovich, C.T., Post, D., Buczek, M.B., and Czarnek, R., 1985, "Free edge strain concentrations in real composite laminates: experimental-theoretical correlation," *Journ. Appl. Mech.*, Vol. 52, pp. 787-793.

Hung, Y.Y., and Taylor, C.E., 1974, "Measurement of slopes of structural deflections by speckle-shearing interferometry," *Experimental Mechanics*, Vol. 14, pp. 281-285.

Hung, Y.Y, and Liang, C.Y., 1979, "Image-shearing camera for direct measurements of surface strains," *Applied Optics*, Vol. 18, pp. 1046-1051.

Jones, R., and Wykes, C., 1983, *Holographic and Speckle Inteferometry*, Cambridge Univ. Press, Cambridge, UK.
Jones, R.M., 1975, *Mechanics of Composite Materials*, Scripta Book Comp., Washington, DC.
Klumpp, P.A., 1989, "Simple spatial filtering for shearograms," *Optics & Laser Technology*, Vol. 21, pp. 105-111.
Klumpp, P.A., and Schnack, E., 1989, "White light reconstruction setup for shearograms," *Applied Optics*, Vol. 28, pp. 3270-3271.
Kreis, T.M., and Kreitlow, H., 1980, "Quantitative evaluation of holographic interference patterns under image processing aspects," *SPIE Proceedings*, Vol. 210, pp. 196-202.
Masters, J.E., and Reifsnider, K.L., 1982, "An investigation of cumulative damage development in quasi-isotropic graphite/epoxy laminates", *ASTM-STP*, Vol. 775, K.L. Reifsnider, ed., pp. 40-62.
O'Brien, T.K., 1984, "Mixed-mode strain energy release rate effects on edge delamination of composites," *ASTM-STP*, Vol. 836, pp. 125-142.
O'Brien, T.K., 1980, "Characterization of delamination onset and growth in a composite laminate," *ASTM-STP*, Vol. 775, pp. 140-167.
Pagano, N.J., and Pipes, R.B., 1973, "Some observations on the interlaminar strength of composite laminates," *Int. Journ. of Mechanical Sciences*, Vol. 15, pp. 679-688.
Pinchas, B.Y., and Pian, T.H.H., 1981, "Calculation of interlaminar stress concentration in composite laminates," *Journ. Comp. Mater.*, Vol. 15, pp. 225-239.
Pipes, R.B., and Pagano, N.J., 1970, "Interlaminar stresses in composite laminates under uniform axial extension", *Journ. Comp. Mater.*, Vol. 4, pp. 538-548.
Post, D., 1983, "Moiré interferometry at VPI and SU", *Experimental Mechanics*, Vol. 23, pp. 203-210.
Reifsnider, K.L., Henneke, E.G., and Stinchcomb, W.W., 1977, "Delamination in quasiisotropic graphite-epoxy laminates," *ASTM-STP*, Vol. 617, pp. 93-105.
Reifsnider, K.L., Henneke, E.G., Stinchcomb, W.W., and Duke, J.C., 1983, "Damage mechanics and NDE of composite laminates," *Proceedings of IUTAM Symposium on mechanics of composite materials*, Blacksburg, VA, Z. Hashin and C.T. Herakovich, eds., pp. 399-420.
Ritter, R., and Hahn, R., 1983, "Contribution to the analysis of the reflection grating method", *Optics and Lasers in Engineering*, Vol. 4, pp. 13-24.
Rohwer, K., 1982, "Stresses and deformations in laminated test specimens of carbon fiber reinforced composites," *DFVLR-Forschungsbericht*, No. 82-15, DFVLR Braunschweig / FRG, 82 pages.
Schütze, R., and Goetting, H.C., 1985, "On-line measurement of onset and growth of edge-delaminations in CFRP-laminates by an optical grating reflection method," *Zeitschrift für Werkstofftechnik*, Vol. 16, pp. 306-310.
Sciamarella, C.A., 1982, "The moiré method - a review," *Experimental Mechanics*, Vol. 22, pp. 418-433.
Sharma, D.K., Sirohi, R.S., and Kothiyal, M.P., 1984, "Simultaneous measurement of slope and curvature with a three-aperture speckle shearing interferometer," *Applied Optics*, Vol. 23, pp. 1542-1546.
Sharma, D.K., Krishna Mohan, N., and Sirohi, R.S., 1986, "A holographic speckle shearing technique for the measurement of out-of-plane displacement, slope and curvature," *Opt. Communications*, Vol. 57, pp. 230-235.
Vest, C.M., 1979, *Holographic Interferometry*, Wiley and Sons, NY.
Wang, A.S.D., and Crossman, F.W., 1980, "Initiation and growth of transverse cracks and edge delaminations in composite laminates", *Journ. Comp. Mat. Supplement*, Vol. 14, Part 1: pp. 71-87, Part 2: pp. 88-106.
Wang, A.S.D., and Slomiana, M., 1982, "Fracture mechanics of delamination - initiation and growth", *NADC-TR*-79056-60, Naval Air Development

Center, Warminster, PA, 169 pages.
Wang, A.S.D., Slomiana, M. and Bucinell, R.B., 1985, "Delamination crack growth in composite laminates," *ASTM-STP*, Vol. 876, W.S. Johnson, ed., pp. 135-167.
Whitcomb, J.D., and Raju, I.S., 1985, "Analysis of interlaminar stresses in thick composite laminates with and without edge delamination," *ASTM-STP*, Vol. 876, W.S. Johnson, ed., pp. 69-94.

Matrix Mean-Field and Local-Field Approaches in the Analysis of Metal Matrix Composites

Jacob Aboudi[1]

Marek-Jerzy Pindera[2]

Abstract

A micromechanical investigation of the inelastic response of metal matrix composites analyzed by two different methodologies is presented. The first method is based on the mean stress field in the entire ductile matrix phase, while the second one is based on the local stress field. The present study is a continuation of a previous investigation in which a micromechanics model based on a periodic array of fibers was employed to generate yield surfaces of metal matrix composites using local and mean matrix stresses. In this paper, we extend the aforementioned analysis to the prediction of the inelastic stress-strain response of metal matrix composites subjected to different loading histories. Results for the overall elastoplastic response of the investigated metal matrix composites indicate that the mean-field approach may lead to significant deviations of the effective composite behavior as compared either to finite element results or measured data. The predictions of the effective composite response generated by the two approaches are compared with experimental and numerical data on unidirectional boron/aluminum and graphite/aluminum.

Introduction

In a previous investigation, Pindera and Aboudi (1988) discussed the use of average matrix stress in determining initial yield surfaces of metal matrix composites. Specifically, the micromechanics model proposed by Aboudi (1986) was employed to generate initial yield surfaces of unidirectional and multidirectional (cross-ply) boron/aluminum laminates under a variety of loading conditions using two different approaches. In the first approach, overall yielding of

---

[1]Professor and Dean, Faculty of Engineering, Tel-Aviv University, Ramat-Aviv 69978, Israel

[2]Assistant Professor, SEAS, University of Virginia, Charlottesville, VA 22903, USA

the composite was assumed to take place when the yield condition in the matrix phase was fulfilled locally in one of the matrix subcells of the representative volume element (RVE) used to model the composite. The von Mises criterion was the chosen yield condition. The yield surfaces generated in this fashion correlated very well with finite element predictions obtained by Dvorak and co-workers (1973, 1974) for most loading directions. In the second approach, initial yielding of the composite was assumed to occur when the average stress in the entire matrix phase was fulfilled. Comparison of the initial yield surfaces generated in this fashion with the corresponding predictions based on local matrix yielding and finite element analysis illustrated that significant differences may occur in the predicted surfaces for certain loading directions. In general, the initial yield surfaces obtained on the basis of the average stress in the entire matrix phase predicted higher overall yield stresses than those obtained on the basis of local matrix yielding. For certain other directions however, the use of the average matrix stress in calculating overall yield surfaces appeared to be a good approximation.

In the present paper, we extend the aforementioned micromechanical investigation to the prediction of inelastic response of metal matrix composites using the two different approaches to calculate inelastic strains in the matrix phase. The investigation is motivated by the recent development of approximate three-dimensional micromechanical models for the initial yielding and inelastic response of metal matrix composites based on the assumption that the entire matrix phase is uniformly strained (Wakashima, et al. (1979), Pindera and Herakovich (1982), Dvorak and Bahei-El-Din (1982)). In such models, the effect of local stresses is neglected, and yielding and subsequent inelastic response is thus governed in the first approximation by the average stress in the matrix. The present investigation was undertaken to quantify the effect of this assumption on the elastoplastic response of two types of metal matrix composites under selected loading paths using a micromechanics model whose predictive capability has been demonstrated in previous investigations (see Aboudi (1989) for a recent review).

It is shown that the use of the mean-field stress in the matrix phase to compute the effective behavior of metal matrix composites may produce results that significantly deviate from the response based on a more rigorous basis, and may lead to erroneous conclusions.

Micromechanics Model

The micromechanics model is based on the analysis of a RVE of a doubly periodic array that consists of a fiber surrounded by three matrix subcells, Figure 1. The fibers are parallel to the 1 - direction and the plane of isotropy is the 2 - 3 plane. Approximate analysis of the deformation and stress fields in the RVE yields closed form expressions for the effective response in terms of the constituent properties and geometry of the phases (Aboudi (1987)). Specifically, a first order (linear) displacement expansion is employed in each of the subcells which is sufficient for the extraction of the effective or average stress-strain equations for the composite. Continuity of tractions and displacements across the boundaries of the individual subcells of the RVE is imposed in an average sense in the course of generating the effective stress-strain curves. Due to the geometric arrangement of the fibers, the isotropic behavior in the transverse (2 - 3) plane is enforced through an averaging process. The availability of closed form expressions ensures ease of programming and manageable computation times for modeling both material and structural response of composites. Incorporation of different constitutive models for the response of the individual phases is also readily accomplished for the same reason.

The micromechanical analysis of stress and deformation fields in the RVE yields the following form of effective, elastoplastic stress-strain-temperature equations for a unidirectional, transversely isotropic fiber-reinforced composite :

$$\underset{\sim}{\bar{\sigma}} = \underset{\sim}{E}\,(\underset{\sim}{\bar{\varepsilon}} - \underset{\sim}{\bar{\varepsilon}}^{PL}) - \underset{\sim}{U}\,\Delta T \tag{1}$$

which are referred to a Cartesian coordinate system ($x_1, x_2, x_3,$) where x_1 is aligned with the fiber direction.

In the above,

$$\underset{\sim}{\bar{\sigma}} = [\, \bar{\sigma}_{11}, \bar{\sigma}_{22}, \bar{\sigma}_{33}, \bar{\sigma}_{12}, \bar{\sigma}_{13}, \bar{\sigma}_{23} \,] \tag{2}$$

is the average stress ;

$$\underset{\sim}{\bar{\varepsilon}} = [\, \bar{\varepsilon}_{11}, \bar{\varepsilon}_{22}, \bar{\varepsilon}_{33}, 2\bar{\varepsilon}_{12}, 2\bar{\varepsilon}_{13}, 2\bar{\varepsilon}_{23} \,] \tag{3}$$

is the average total strain, with the plastic strain $\underset{\sim}{\bar{\varepsilon}}^{PL}$ having the same form as $\underset{\sim}{\bar{\varepsilon}}$, and ;

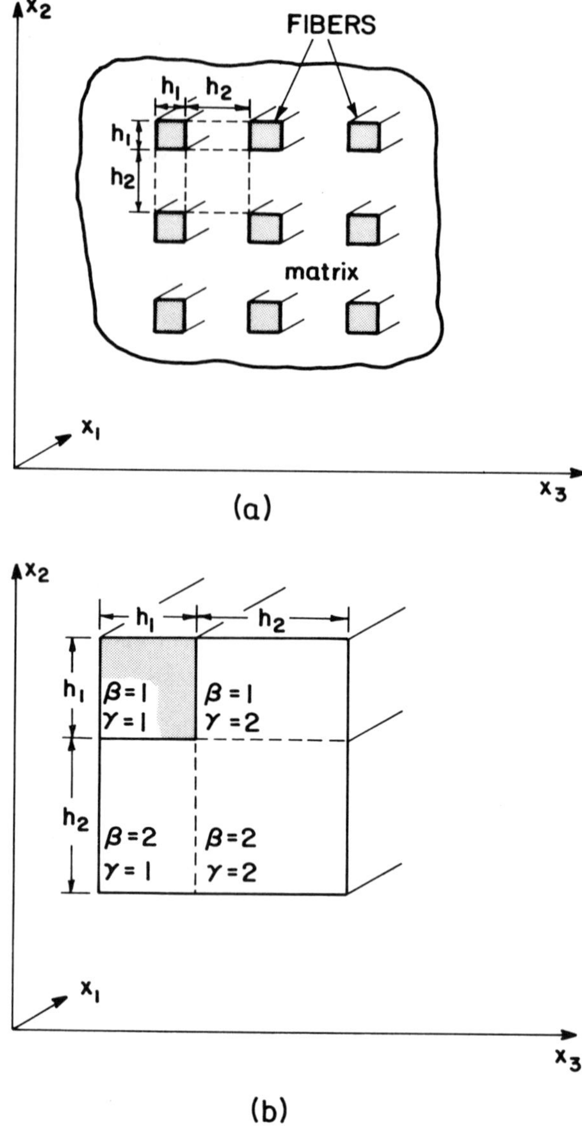

Figure 1. Doubly periodic array of fibers (a), and the associated representative volume element (b).

$$\underset{\sim}{E} = \begin{bmatrix} e_{11} & e_{12} & e_{12} & 0 & 0 & 0 \\ & e_{22} & e_{23} & 0 & 0 & 0 \\ & & e_{22} & 0 & 0 & 0 \\ & & & e_{44} & 0 & 0 \\ & sym & & & e_{44} & 0 \\ & & & & & e_{66} \end{bmatrix} \tag{4}$$

where e_{ij} are the five effective elastic constants of the composite (with $e_{66} = 1/2\,(e_{22} - e_{23})$) which are given explicitly in terms of the elastic stiffness elements C_{ij}^m and C_{ij}^f of the matrix and fiber phases and their volume fractions, $c_m, c_f = 1 - c_m$ by Aboudi (1987). That is,

$$\underset{\sim}{E} = \underset{\sim}{E}\,(C_{ij}^m, C_{ij}^f, c_f) \tag{5}$$

It can easily be verified that the predicted effective moduli for many unidirectional composites are in excellent agreement with those computed using elasticity analysis (Pickett (1968), Chen and Chang (1970) and Behrens (1971)). Furthermore, the effective transverse shear modulus of a unidirectional glass/polyester composite predicted using the present micromechanics approach coincides with the corresponding modulus computed on the basis of the three-phase model (Christensen (1979)). This is a significant result since the prediction of the transverse shear modulus or, equivalently, the transverse Young's modulus of unidirectional composites provides a critical check on the validity of a micromechanics model in the elastic region.

Similarly, the effective plastic strain $\underset{\sim}{\bar{\varepsilon}}^{PL}$ can also be expressed as :

$$\underset{\sim}{\bar{\varepsilon}}^{PL} = \underset{\sim}{\bar{\varepsilon}}^{PL}\,(C_{ij}^m, C_{ij}^f, c_f, L_{ij}^{(\beta\gamma)}) \tag{6}$$

where $L_{ij}^{(\beta\gamma)}$ are the plastic strain components in the individual matrix subcells $(\beta\gamma)$, Figure 1b. The effect of thermal loading on the composite stresses is represented by the quantity $\underset{\sim}{U}\,\Delta T$, where ΔT is the applied temperature change and $\underset{\sim}{U}$ depends on the properties of the phases in the following fashion :

$$\underset{\sim}{U} = \underset{\sim}{U}\,(C_{ij}^m, C_{ij}^f, \alpha_m, \alpha_f, c_f) \tag{7}$$

where α_m and α_f are the coefficients of thermal expansion of the matrix and fibers, respectively. The explicit expressions for $\underset{\sim}{E}$, $\underset{\sim}{\bar{\varepsilon}}^{PL}$ and $\underset{\sim}{U}$ in terms of the properties of the phases are not given here because they are too lengthy and have already been provided in previous papers (cf. Aboudi (1987, 1989)).

In the present investigation, the unified theory of plasticity proposed by Bodner and Partom (1975) was employed to model the constitutive response of the matrix phase. Let $S_{ij}^{(\beta\gamma)}$ represent the average stresses in the subcell $(\beta\gamma)$. The local rate of plastic strain in a given matrix subcell is expressed as :

$$\dot{L}_{ij}^{(\beta\gamma)} = \Lambda_{(\beta\gamma)} \, \hat{S}_{ij}^{(\beta\gamma)} \quad , \quad \beta + \gamma \neq 2 \tag{8}$$

where $\hat{S}_{ij}^{(\beta\gamma)}$ are the components of stress deviators and $\Lambda_{(\beta\gamma)}$ is the flow rule function of the matrix phase. The explicit form of the flow rule function for the matrix is given by :

$$\Lambda_{(\beta\gamma)} = D_0 \exp \{-\hat{n} \, [Z_{(\beta\gamma)}^2 / (3J_2^{(\beta\gamma)})]^n\} \, / \, \sqrt{J_2^{(\beta\gamma)}} \tag{9}$$

where $\hat{n} = 0.5\,(n+1)/n$ and $J_2^{(\beta\gamma)} = \frac{1}{2} \hat{S}_{ij}^{(\beta\gamma)} \hat{S}_{ij}^{(\beta\gamma)}$ is the second invariant of the local matrix subcell stresses. D_0 and n are inelastic parameters, and $Z_{(\beta\gamma)}$ is a state variable given by :

$$Z_{(\beta\gamma)} = Z_1 + (Z_0 - Z_1) \exp [-m \, W_p^{(\beta\gamma)} / Z_0] \tag{10}$$

where $W_p^{(\beta\gamma)}$ is the plastic work per unit volume.

The five parameters D_0, Z_0, Z_1, n and m in Equations (9) and (10) have the following physical meaning. The parameter Z_0 is related to the "yield stress" of the material in simple tension and Z_1 is proportional to the ultimate stress. The material parameter m determines the rate of work-hardening and the rate sensitivity of the material is controlled by the material constant n. It should be noted however, that the results of numerical experiments indicate that, with $n = 10$, the response is essentially rate independent for strain rates less than $10/sec$. Finally, D_0 is the limiting strain rate and usually can be arbitrarily chosen as $D_0^{-1} = 10^{-4}\ sec$.

Equation (8) is the flow rule for the plastic strains in the individual subcells of the matrix phase. Alternatively, a flow rule based on the average stress in the matrix phase can be formulated as follows. Let us define the average stresses in the matrix constituent in the form :

$$\bar{\sigma}_{ij}^{(m)} = [\, v_{12} \, S_{ij}^{(12)} + v_{21} \, S_{ij}^{(21)} + v_{22} \, S_{ij}^{(22)} \,] \, / \, (\, v_{12} + v_{21} + v_{22} \,) \tag{11}$$

where v_{12}, v_{21} and v_{22} are the volumes of the matrix subcells (i.e. $v_{ij} = h_i h_j$). The flow rule based on the average matrix stress that is the counterpart of Equation (8) thus becomes :

$$\dot{L}_{ij}^{(\beta\gamma)} = \Lambda_{(m)} \hat{\sigma}_{ij}^{(m)} \quad , \quad \beta + \gamma \neq 2 \tag{12}$$

where $\hat{\sigma}_{ij}^{(m)}$ are the stress deviators of $\bar{\sigma}_{ij}^{(m)}$, and $\Lambda_{(m)}$ is the flow function of the matrix based on the average stresses $\bar{\sigma}_{ij}^{(m)}$. Thus $\Lambda_{(m)}$ is given by Equation (9) but with $J_2^{(m)} = \frac{1}{2} \hat{\sigma}_{ij}^{(m)} \hat{\sigma}_{ij}^{(m)}$, and ;

$$Z_{(m)} = Z_1 + (Z_0 - Z_1) \exp\,[-m\, W_p^{(m)} / Z_0] \tag{13}$$

We note that Equation (12) implies that $L_{ij}^{(12)} = L_{ij}^{(21)} = L_{ij}^{(22)}$, contrary to the previous formulation given by Equation (8) in which the plastic strains in the matrix subcells are independent of each other.

Analytical Results

Stress-strain curves of two types of unidirectional metal matrix composites have been generated using the local-field and mean-field approaches outlined in the preceding section for loading by different combinations of externally applied stresses. In particular, the response to loading by longitudinal and transverse normal stresses, longitudinal shear stress and axisymmetric normal stresses applied separately in the principal material coordinate system was investigated. The response to combined stresses was also investigated by considering the stress-strain behavior of off-axis specimens. The generated results are subsequently compared with numerical calculations based on the finite element method and experimental data obtained by various investigators.

Figures 2 and 3 present the stress-strain response of unidirectional graphite/aluminum (T-50/2024-T4) composite (0.3 fiber volume fraction) under externally applied transverse normal ($\bar{\sigma}_{22}$) and longitudinal shear ($\bar{\sigma}_{12}$) stresses, respectively. In Figure 2, the response based on the mean-field and local-field approaches obtained from the micromechanics model is presented, while Figure 3 also includes the finite element solution of Hashin and Humphreys (1981). The material parameters of the fiber and matrix phases employed to generate the curves are given in Table 1. In the case of the transverse response, Figure 2, it is evident that the mean-field approach significantly overestimates the initiation of yielding and, in turn, produces plastic strains that deviate significantly from those obtained using the local-field approach. In the case of the longitudinal shear response, on the other hand, there is little difference between the predictions based on the two approaches in the considered range of strains. It is interesting to note that the results obtained with the more rigorous finite element

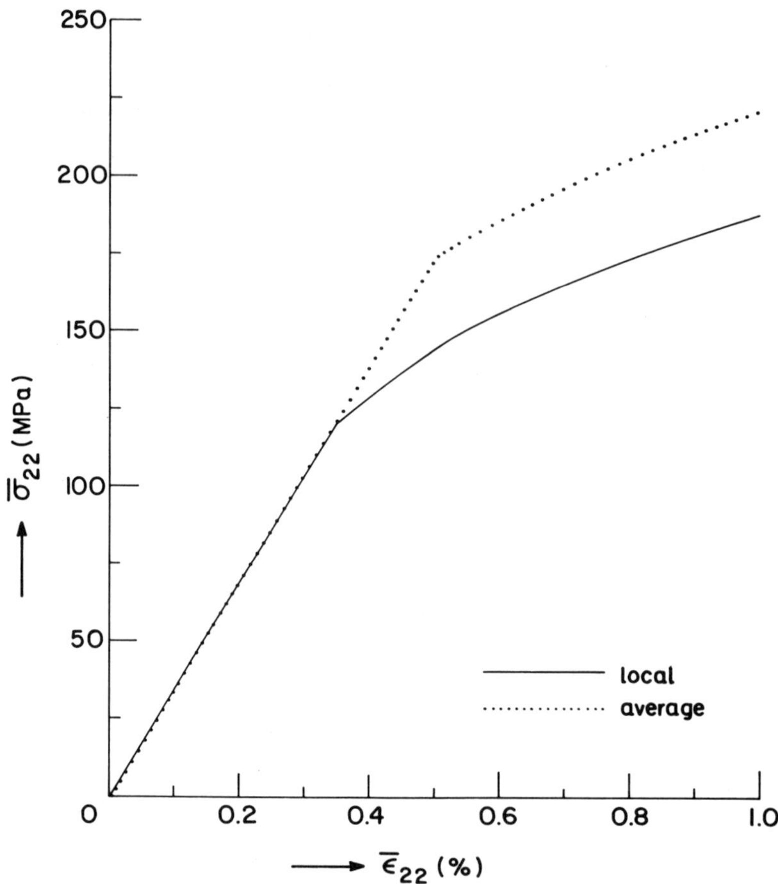

Figure 2. Transverse response of a T50/2024-T4 graphite/aluminum.

Material	E_A (GPa)	E_T (GPa)	G_A (GPa)	ν_A	ν_T
T-50 graphite	388.2	7.6	14.9	0.41	0.45

Material	E (GPa)	ν	D_0^{-1} (sec)	Z_0 (MPa)	Z_1 (MPa)	m	n
2024-T4 aluminum	72.49	0.33	10^{-4}	340	435	300	10

Table 1. Material parameters of graphite fibers and aluminum matrix.

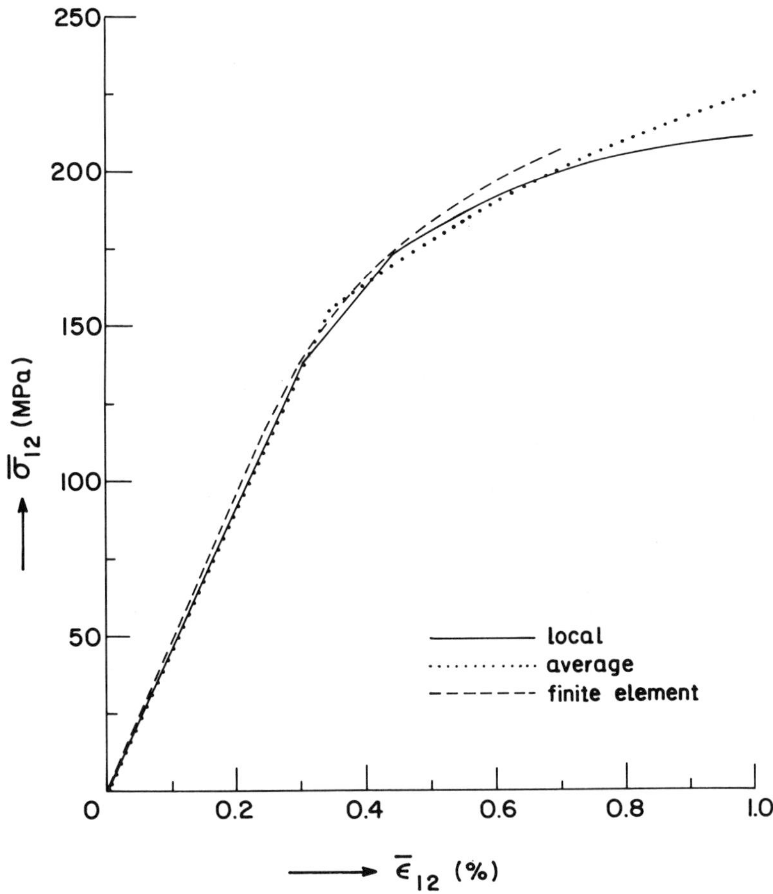

Figure 3. Longitudinal shear response of a T50/2024-T4 graphite/aluminum.

technique are not significantly different from the results generated with the present micromechanics model.

The results for the transverse normal and longitudinal shear response of unidirectional boron/aluminum (0.5 fiber volume fraction) are presented in Figures 4 and 5, respectively, using the material parameters given in Table 2 and noting that the aluminum matrix was assumed to be elastic perfectly-plastic. In this case, the results obtained with the micromechanics model using the local- and mean-field approaches are compared with the finite element solution obtained by Foye (1973) for both loading situations. We note that as the finite

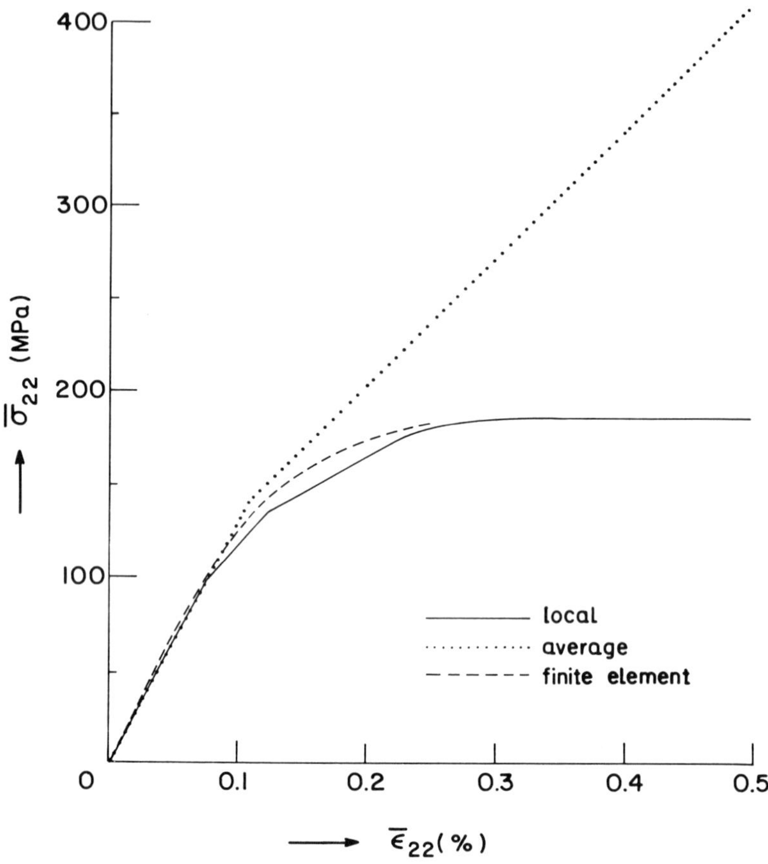

Figure 4. Transverse response of a boron/aluminum composite.

Material	E (GPa)	ν
Boron fibers	413.7	0.20

Material	E (GPa)	ν	D_0^{-1} (sec)	Z_0 (MPa)	Z_1 (MPa)	m	n
Aluminum	55.16	0.30	10^{-4}	103.42	103.42	-	10

Table 2. Material parameters of boron fibers and aluminum matrix.

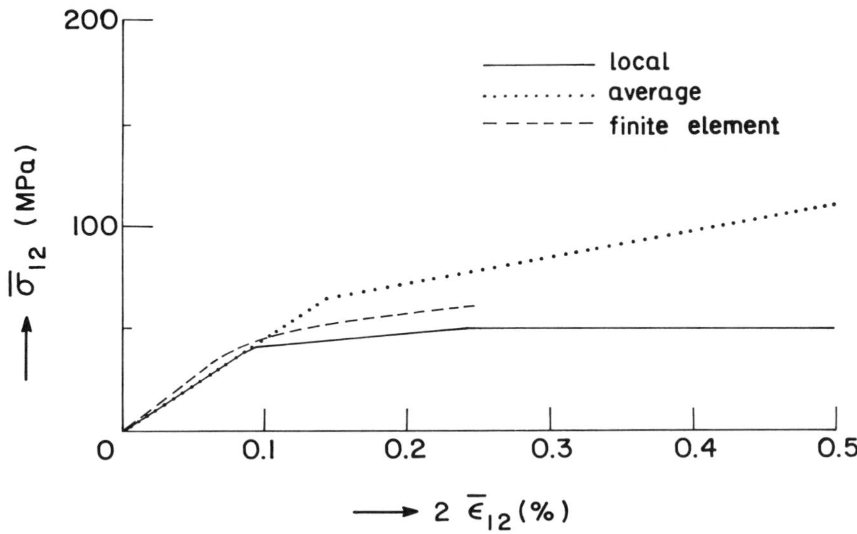

Figure 5. Longitudinal shear response of a boron/aluminum composite.

element solution has been generated for a square array of fibers, the averaging procedure outlined by Aboudi (1987) to obtain the effective response of a transversely isotropic continuum has not been applied in this case. The correlation between the transverse response obtained with the finite element analysis and the micromechanics model based on the local-field approach, Figure 4, is very good. The prediction of the micromechanics model based on the mean-field approach presented in the same figure is drastically different from the latter. The deviation in this case is significantly more pronounced than in the transverse response of graphite/aluminum illustrated in Figure 2. Similar observations can be made regarding the longitudinal shear response, Figure 5, which can be contrasted with the longitudinal shear response of graphite/aluminum given in Figure 3. It appears that the extent of deviation between the results obtained on the basis of local-field and mean-field quantities depends on the mismatch in the elastic properties of the constituent phases and possibly on the work hardening of the aluminum phase.

The predictions of the micromechanics model have also been compared with the experimentally generated results obtained by Pindera, et al. (1989) for unidirectional boron/aluminum composite (0.46 fiber volume fraction) whose

Material	E (GPa)	ν	α (/ °C)
Boron fibers	400	0.20	$5.0 \cdot 10^{-6}$

Material	E (GPa)	ν	α (/ °C)	D_0^{-1} (sec)	Z_0 (MPa)	Z_1 (MPa)	m	n
Aluminum	72.5	0.33	$23.0 \cdot 10^{-6}$	10^{-4}	100	190	70	10

Table 3. Material parameters of boron fibers and aluminum matrix.

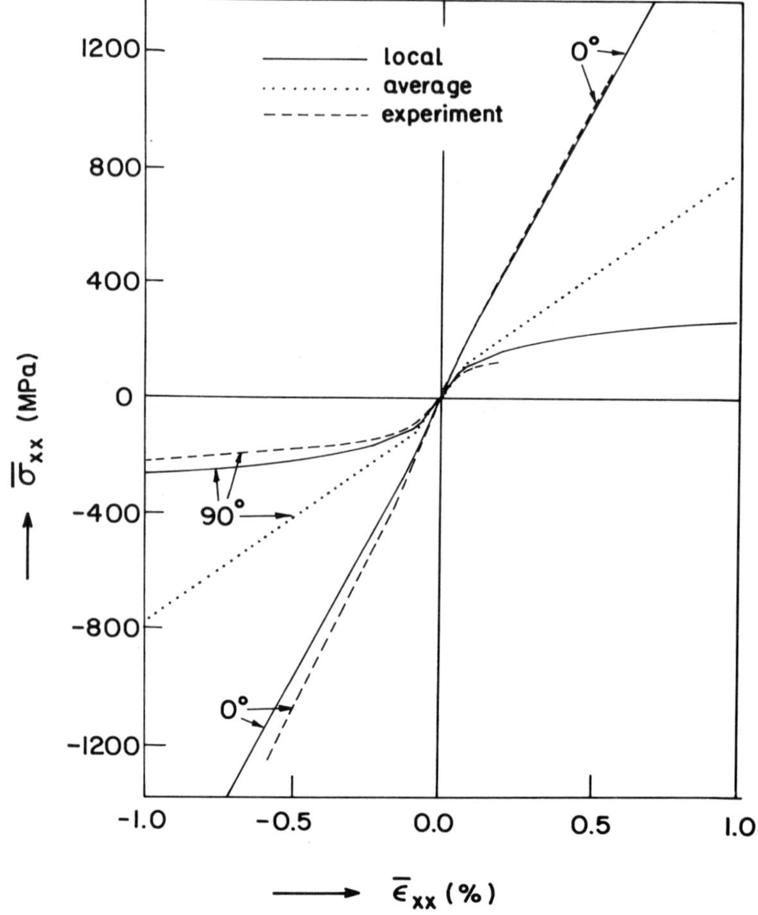

Figure 6. Longitudinal and transverse response of a boron/aluminum composite.

constituent properties are given in Table 3. The longitudinal and transverse response in tension and compression obtained from the 0° and 90° specimens is illustrated in Figure 6. The figure gives the comparison between the analytical predictions based on the two approaches and the measured data. It can clearly be observed that the mean-field prediction significantly underestimates the extent of plastic flow in the case of the transverse loading. The longitudinal response, on the other hand, is predicted with sufficient accuracy by the two approaches. This is not surprising since it has been shown by Mulhern, et al. (1967) that the longitudinal response of an elastoplastic composite can accurately be predicted by assuming that the entire matrix phase yields uniformly (cf. Hill (1964)). It should be noted however, that a true measure of a micromechanics model's predictive capability is given by the accuracy of the prediction for the matrix dominated response such as transverse or shear behavior. Thus it can be concluded that the mean-field approach is not useful in this particular case.

The off-axis response of 10°, 15° and 45° boron/aluminum composite discussed above is given in Figure 7, where comparison between the measured data and the two micromechanics approaches is shown. The dramatic differences between the predictions based on the local-field and mean-field calculations are clear. Whereas the local-field predictions correlate very well with the experimentally observed behavior, the mean-field data exhibits unacceptably large deviations. It should be noted that the state of stress in the principal material directions of the employed off-axis specimens includes significant transverse normal and axial shear stress components. These stresses result in a matrix dominated response of the composite (either in shear and/or transverse tension).

The last example deals with the response of unidirectional boron/aluminum composite subjected to axisymmetric loading by the longitudinal stress $\bar{\sigma}_{11}$ and all-around transverse stress $\bar{\sigma}_{22} = \bar{\sigma}_{33}$. The material parameters for this composite, used in the calculations, have been provided in Table 3. Initial yield surfaces for this composite were generated for the following two cases : $\Delta T = 0$ and $\Delta T = 60°C$. The resulting predictions based on the local-field and mean-field matrix stresses are given in Figure 8. The axes in the figure have been normalized with respect to the yield stress of the matrix in simple tension, $Y = 84$ MPa. It can be seen that the prediction based on the mean-field approach results in an open yield surface in the direction along which no yielding takes place. This direction is denoted by arrows in the figure and is determined by requiring that :

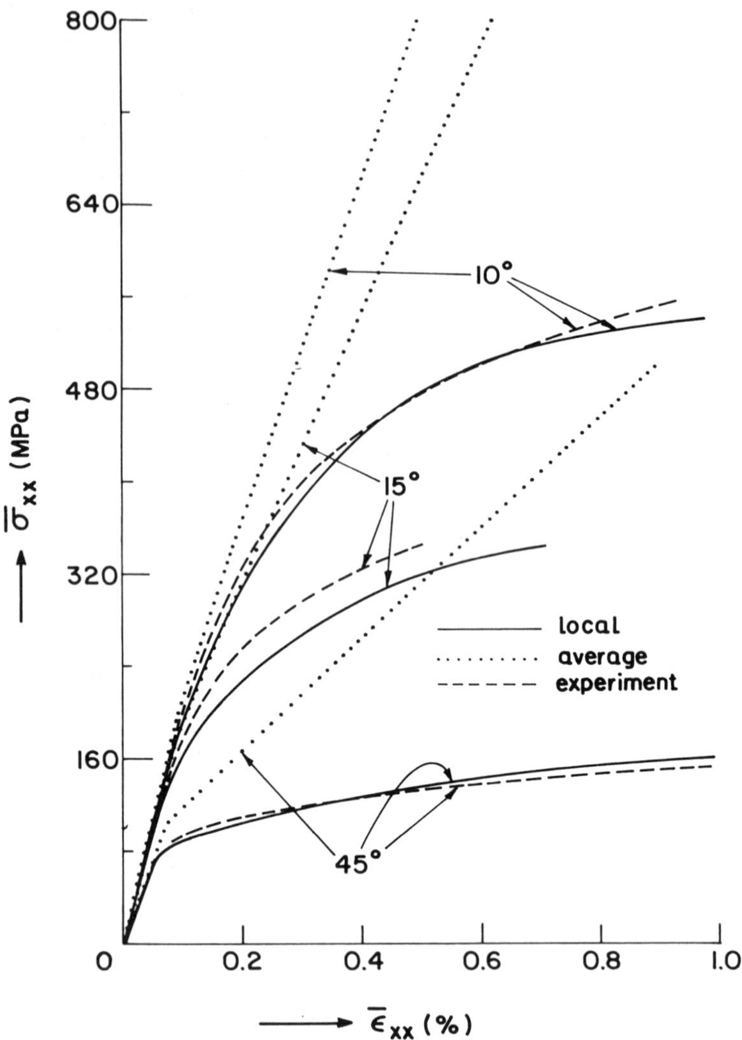

Figure 7. Off-axis response of a boron/aluminum composite.

$$\overline{\sigma}_{11}^{(m)} = \overline{\sigma}_{22}^{(m)} = \overline{\sigma}_{33}^{(m)} \tag{14}$$

The above relation ensures that the matrix is under a hydrostatic state of stress which produces no yielding in the case of the von Mises criterion employed to generate Figure 8. The explicit expression for this direction in the case of no temperature change was given by Pindera and Aboudi (1988). Thermal loading

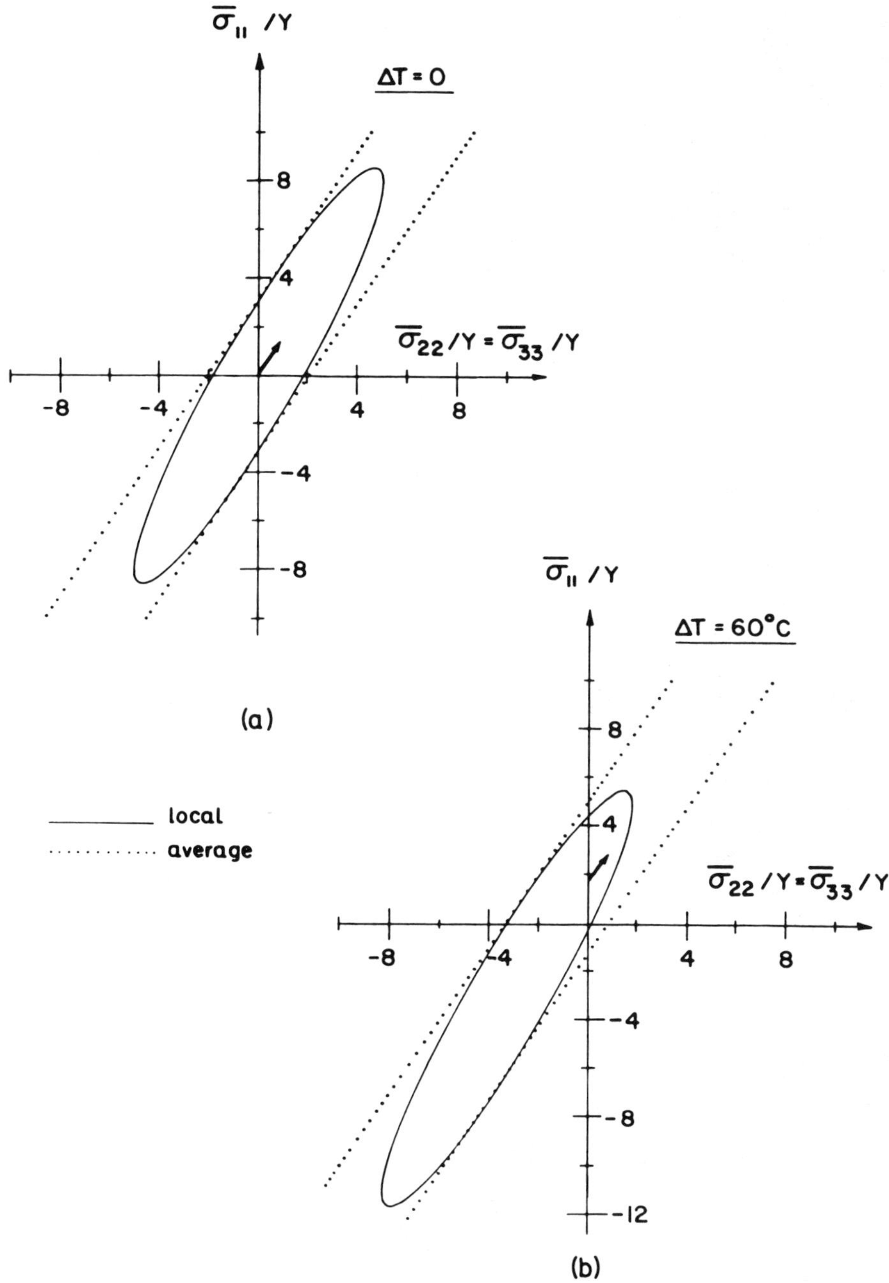

Figure 8. Initial yield surfaces of a boron/aluminum under axisymmetric loading.

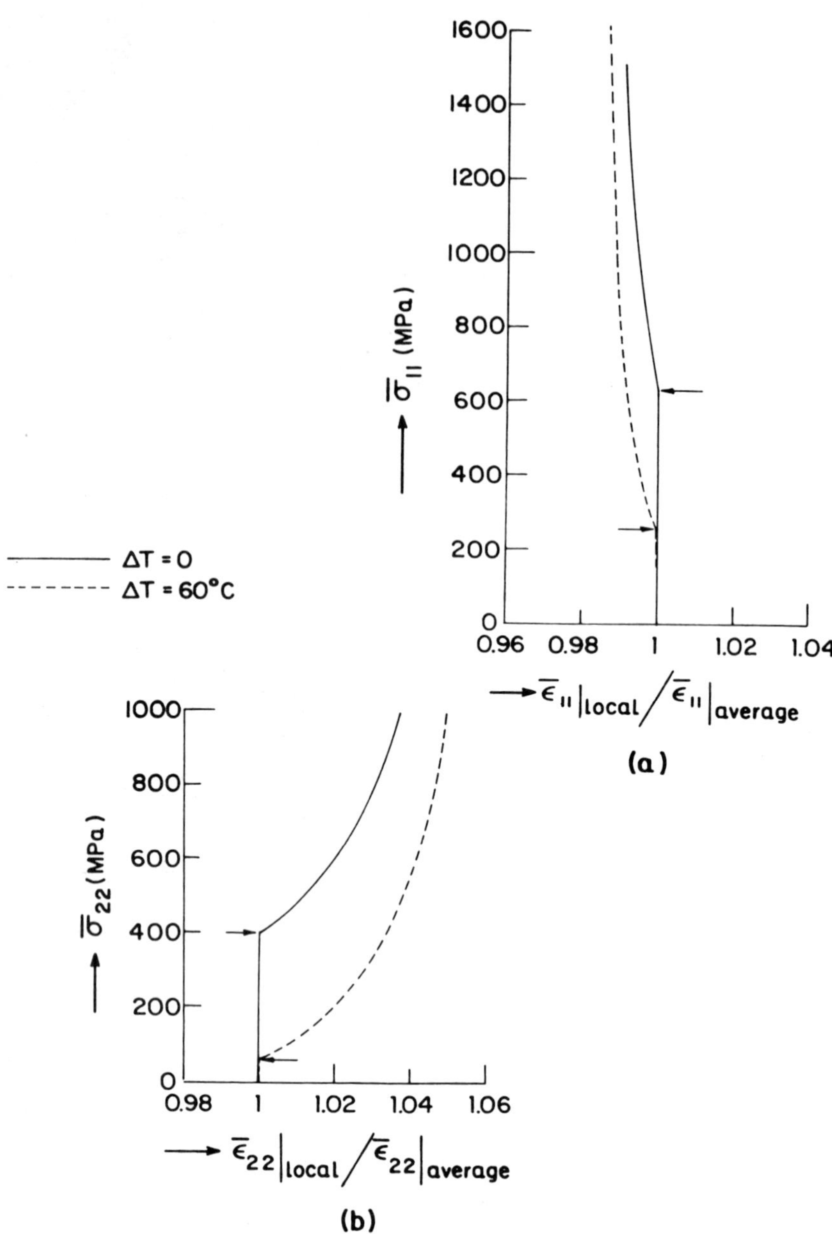

Figure 9. Response of a boron/aluminum under axisymmetric loading.

translates the initial yield surface but does not change the direction along which yielding does not occur if the mean-field approach is used. It can be shown that for the considered boron/aluminum composite, the following relation can be established in conjunction with Equation (14) :

$$\bar{\sigma}_{11} = 2.597 \cdot 10^6 \, \Delta T + 1.513 \, \bar{\sigma}_{22} \tag{15}$$

which provides the necessary condition for axisymmetric and thermal loading such that no yielding takes place.

Another pronounced difference between the local-field and the mean-field approach is illustrated by the temperature change at which the composite yields in the absence of any external mechanical loading. For the considered composite, the temperature changes calculated on the basis of local-field and mean-field matrix stresses that result in yielding of the composite are $\Delta T = 63.7°$ and $100.9°C$, respectively. The difference in the temperature changes predicted by the two methods may have pronounced effect on the subsequent elastoplastic response of the composite. For the case of axisymmetric loading given by Equation (15), the elastoplastic response of the composite is illustrated in Figure 9 for $\Delta T = 0°$ and $60°C$. The figure gives the normal stresses as a function of normal strains calculated using the local-field approach, normalized with respect to the corresponding normal strains calculated using the mean-field approach. The arrows in the figure indicate the initiation of yielding and plastic flow which are not present in the response obtained using the mean-field matrix stresses for this loading situation. As is clearly seen, thermal loading plays a significant role in the differences between the responses obtained by the two methods. As expected, this difference is considerably greater for the transverse response component than the longitudinal component.

Conclusions

The results of the present investigation indicate that the use of the average stresses in the matrix to generate yield surfaces and subsequent elastoplastic response of metal matrix composites may lead to significant deviation from the behavior predicted on the basis of local matrix stresses. The extent of deviation depends on the direction of loading and the mismatch in the material parameters of the constituent phases. This was established by comparison with numerical and measured data. The effect of residual stresses produced by temperature changes in the case an axisymmetric loading magnified the extent of the deviation between the elastoplastic responses obtained with the two methods.

Acknowledgements

The first author is grateful to Professor C. T. Herakovich for the visiting appointment in the Civil Engineering Department at the University of Virginia, which made this investigation possible.

References

Aboudi, J., 1986, "Elastoplasticity Theory for Composite Materials. *Solid Mechanics Archives*, Vol.11, pp. 141 - 183.

Aboudi, J., 1987, "Closed Form Constitutive Equations for Metal Matrix Composites," *Int. J. Engng. Sci.*, Vol. 25, pp. 1229 - 1240.

Aboudi, J., 1989, "Micromechanical Analysis of Composites by the Method of Cells," *Applied Mechanics Reviews*, Vol.42, pp. 193 - 221.

Behrens, E., 1971, "Elastic Constants of Fiber-reinforced Composites with Transversely Isotropic Constituents," *Journal of Applied Mechanics*, Vol. 38, pp. 1062 - 1065.

Bodner, S. R. and Partom, Y., 1975, "Constitutive Equations for Elastic-Viscoplastic Strain Hardening Material," *Journal of Applied Mechanics*, Vol. 42, pp. 385 - 389.

Chen, C. H. and Cheng, S., 1970, "Mechanical Properties of Anisotropic Fiber-reinforced Composites," *Journal of Applied Mechanics*, Vol. 37, pp. 186 - 189.

Christensen, R. M., 1979, *Mechanics of Composite Materials*, Wiley-Interscience, New York.

Dvorak, G. J., Rao, M. S. M.,and Tarn, J. Q., 1973, "Yielding in Unidirectional Composites under External Loads and Temperature Changes," *Journal of Composite Materials*, Vol. 7, pp. 194 - 216.

Dvorak, G. J. and Bahei-El-Din, Y. A., 1982, "Plasticity Analysis of Fibrous Composites," *Journal of Applied Mechanics*, Vol. 49, pp. 327 - 335.

Foye, R. L., 1973, "Theoretical Post-Yielding Behavior of Composite Laminates. 1. Inelastic Micromechanics," *Journal of Composite Materials*, Vol. 7, pp. 178 - 193.

Hashin, Z. and Humphreys, E. A. (1980). "Elevated Temperature Behavior of Metal Matrix Composites," *MSC TPR 1130/1502*.

Hill, R., 1964, "Theory of Mechanical Properties of Fibre-Strengthened Materials. Part 2 - Inelastic Behaviour," *Journal of the Mechanics and Physics of Solids*, Vol.12, pp. 213 - 218.

Mulhern, J. F., Rogers, T. G., Spencer, A. J. M., 1967, "Cyclic Extension of an Elastic Fibre with an Elastic-Plastic Coating," *Journal of Institute of Mathematics and Its Applications*, Vol. 3, pp. 22 - 40.

Pickett, G., 1968, "Elastic Moduli of Fiber Reinforced Plastic Composites," *Fundamental Aspects of fiber Reinforced Plastic Composites*, R. T. Schwartz and H. S. Schwartz, eds.,

Interscience.

Pindera, M-J. and Herakovich, C. T., 1982, "Influence of Stress Interaction on the Behavior of Off-Axis Unidirectional Composites," *Fracture of Composite Materials*, G. C. Sih and V. P. Tamuzs, eds., Martinus Nijhoff, pp. 185 - 196.

Pindera, M-J. and Aboudi, J., 1988, "Micromechanical Analysis of Yielding of Metal Matrix Composites," *Int. Journal of Plasticity*, Vol. 4, pp. 195 - 214.

Pindera, M-J., Herakovich, C. T., Becker, W. and Aboudi, J., 1989, "Nonlinear Response of Unidirectional Boron/Aluminum," *Journal of Composite Materials*, Vol. 24, pp. 2 - 21.

Rao, M. S. M., 1974, "Plasticity Theory of Fiber-reinforced Metal Matrix Composites," Ph. D. dissertation, Duke University.

Wakashima, K., Suzuki, Y. and Umekawa, S., 1979, "A Micromechanical Prediction of Initial Yield Surfaces of Unidirectional Composites," *Journal of Composite Materials*, Vol.13, pp. 288 - 302.